Lines

Slope of line through (x_1, y_1) and (x_2, y_2):

$$m = \frac{y_2 - y_1}{x_2 - x_1}$$

Point-slope equation of line through (x_1, y_1) with slope m:

$$y - y_1 = m(x - x_1)$$

Slope-intercept equation of line with slope m and y-intercept b:

$$y = b + mx$$

Rules of Exponents

$$a^x a^t = a^{x+t}$$
$$\frac{a^x}{a^t} = a^{x-t}$$
$$(a^x)^t = a^{xt}$$

Definition of Natural Log

$y = \ln x \quad$ means $\quad e^y = x$
ex: $\ln 1 = 0$ since $e^0 = 1$

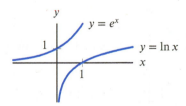

Identities

$$\ln e^x = x$$
$$e^{\ln x} = x$$

Rules of Natural Logarithms

$$\ln(AB) = \ln A + \ln B$$
$$\ln\left(\frac{A}{B}\right) = \ln A - \ln B$$
$$\ln A^p = p \ln A$$

Distance and M

Distance D between

$$D = \sqrt{(x_2 - x_1)^2 + (y_2 - y_1)^2}$$

Midpoint of (x_1, y_1) and (x_2, y_2):

$$\left(\frac{x_1 + x_2}{2}, \frac{y_1 + y_2}{2}\right)$$

Quadratic Formula

If $ax^2 + bx + c = 0$, then

$$x = \frac{-b \pm \sqrt{b^2 - 4ac}}{2a}$$

Factoring Special Polynomials

$$x^2 - y^2 = (x + y)(x - y)$$
$$x^3 + y^3 = (x + y)(x^2 - xy + y^2)$$
$$x^3 - y^3 = (x - y)(x^2 + xy + y^2)$$

Circles

Center (h, k) and radius r:

$$(x - h)^2 + (y - k)^2 = r^2$$

Ellipse

$$\frac{x^2}{a^2} + \frac{y^2}{b^2} = 1$$

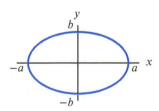

Hyperbola

$$\frac{x^2}{a^2} - \frac{y^2}{b^2} = 1$$

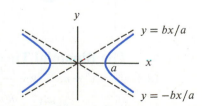

Geometric Formulas

Conversion Between Radians and Degrees: π radians $= 180°$

Triangle

$A = \frac{1}{2}bh$

$= \frac{1}{2}ab\sin\theta$

Circle

$A = \pi r^2$

$C = 2\pi r$

Sector of Circle

$A = \frac{1}{2}r^2\theta$ (θ in radians)

$s = r\theta$ (θ in radians)

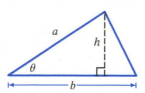

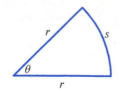

Sphere

$V = \frac{4}{3}\pi r^3$ $A = 4\pi r^2$

Cylinder

$V = \pi r^2 h$

Cone

$V = \frac{1}{3}\pi r^2 h$

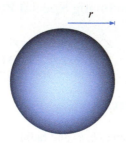

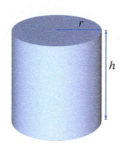

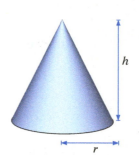

Trigonometric Functions

$\sin\theta = \dfrac{y}{r}$

$\cos\theta = \dfrac{x}{r}$

$\tan\theta = \dfrac{y}{x}$

$\tan\theta = \dfrac{\sin\theta}{\cos\theta}$

$\cos^2\theta + \sin^2\theta = 1$

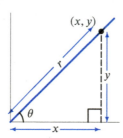

$\sin(A\pm B) = \sin A\cos B\pm\cos A\sin B$

$\cos(A\pm B) = \cos A\cos B\mp\sin A\sin B$

$\sin(2A) = 2\sin A\cos A$

$\cos(2A) = 2\cos^2 A - 1 = 1 - 2\sin^2 A$

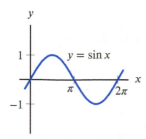

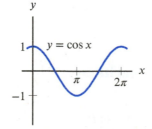

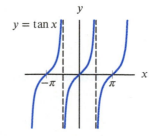

The Binomial Theorem

$(x+y)^n = x^n + nx^{n-1}y + \dfrac{n(n-1)}{1\cdot 2}x^{n-2}y^2 + \dfrac{n(n-1)(n-2)}{1\cdot 2\cdot 3}x^{n-3}y^3 + \cdots + nxy^{n-1} + y^n$

$(x-y)^n = x^n - nx^{n-1}y + \dfrac{n(n-1)}{1\cdot 2}x^{n-2}y^2 - \dfrac{n(n-1)(n-2)}{1\cdot 2\cdot 3}x^{n-3}y^3 + \cdots \pm nxy^{n-1} \mp y^n$

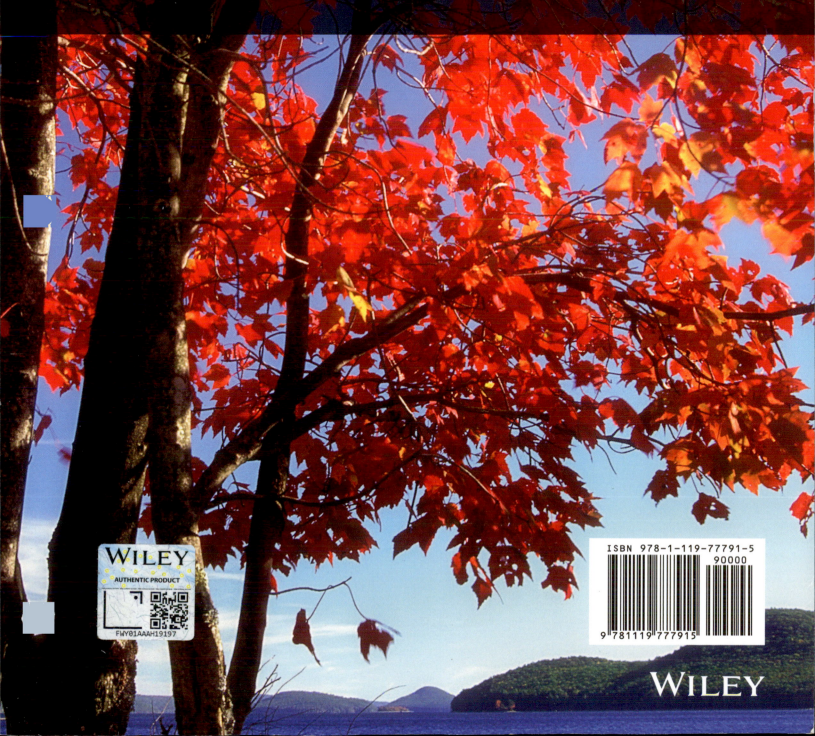

Wiley Loose-Leaf Print Edition

MULTIVARIABLE

CALCULUS

EIGHTH EDITION

McCALLUM · HUGHES-HALLETT · GLEASON et al.

ISBN 978-1-119-77791-5

90000

9 781119 777915

WILEY

MULTIVARIABLE CALCULUS

Eighth Edition

We dedicate this book to Andrew M. Gleason.

*His brilliance and the extraordinary kindness and
dignity with which he treated others made an
enormous difference to us, and to many, many people.
Andy brought out the best in everyone.*

*Deb Hughes Hallett
for the Calculus Consortium*

MULTIVARIABLE CALCULUS

Eighth Edition

Produced by the Calculus Consortium and initially funded by a National Science Foundation Grant.

William G. McCallum
University of Arizona

Deborah Hughes-Hallett
University of Arizona

Daniel Flath

Brad G. Osgood
Stanford University

Andrew M. Gleason
Harvard University

Cody L. Patterson
Texas State University

Selin Kalaycıoğlu
New York University

Douglas Quinney
University of Keele

Brigitte Lahme
Sonoma State University

Ayşe Şahin
Wright State University

Patti Frazer Lock
St. Lawrence University

Adam H. Spiegler
University of Colorado Denver

Guadalupe I. Lozano
University of Arizona

Jeff Tecosky-Feldman
Haverford College

Jerry Morris
Sonoma State University

Thomas W. Tucker
Colgate University

David Mumford
Brown University

Aaron D. Wootton
University of Portland

with the assistance of
Adrian Iovita
University of Washington

Coordinated by
Elliot J. Marks

SR. DIRECTOR	Michelle Renda
SR. ACQUISITIONS EDITOR	John LaVacca
SR. MANAGING EDITOR	Mary Donovan
EXECUTIVE MANAGING EDITOR	Valerie Zaborski
EDITORIAL ASSISTANT	Hannah Larkin
SR. COURSE CONTENT DEVELOPER	Melissa Whelan
SR. MANAGER, COURSE DEVELOPMENT AND PRODUCTION	Svetlana Barskaya
SR. COURSE PRODUCTION OPERATIONS SPECIALIST	Laura Abrams
SR. PHOTO EDITOR	Avinash Singh
SR. CREATIVE PRODUCT DESIGNER	Tom Nery
COVER DESIGNER	Tom Nery
COVER AND CHAPTER OPENING PHOTO	©Patrick Zephyr/Patrick Zephyr Nature Photography

Problems from Calculus: The Analysis of Functions, by Peter D. Taylor (Toronto: Wall & Emerson, Inc., 1992). Reprinted with permission of the publisher.

This book was set in STIX by the Consortium using TEX, Mathematica, and the package ASTEX, which was written by Alex Kasman.

Founded in 1807, John Wiley & Sons, Inc. has been a valued source of knowledge and understanding for more than 200 years, helping people around the world meet their needs and fulfill their aspirations. Our company is built on a foundation of principles that include responsibility to the communities we serve and where we live and work. In 2008, we launched a Corporate Citizenship Initiative, a global effort to address the environmental, social, economic, and ethical challenges we face in our business. Among the issues we are addressing are carbon impact, paper specifications and procurement, ethical conduct within our business and among our vendors, and community and charitable support. For more information, please visit our website: www.wiley.com/go/citizenship.

This material is based upon work supported by the National Science Foundation under Grant No. DUE-9352905. Opinions expressed are those of the authors and not necessarily those of the Foundation.

ISBN-13 978-1-119-77789-2
Library of Congress Catalog Number: LCCN 2020041522 (print) | LCCN 2020041523 (ebook)

Printed in the United States of America

SKY10022076_102920

PREFACE

Calculus is one of the greatest achievements of the human intellect. Inspired by problems in astronomy, Newton and Leibniz developed the ideas of calculus 300 years ago. Since then, each century has demonstrated the power of calculus to illuminate questions in mathematics, the physical sciences, engineering, and the social and biological sciences.

Calculus has been so successful both because its central theme—change—is pivotal to an analysis of the natural world and because of its extraordinary power to reduce complicated problems to simple procedures. Therein lies the danger in teaching calculus: it is possible to teach the subject as nothing but procedures—thereby losing sight of both the mathematics and of its practical value. This edition of *Calculus* continues our effort to promote courses in which understanding and computation reinforce each other. It reflects the input of users at research universities, four-year colleges, community colleges, and secondary schools, as well as of professionals in partner disciplines such as engineering and the natural and social sciences.

Flexibility in a New Era of Teaching and Learning

The world has changed and the education system has changed with it. With little or no training, instructors and students have adjusted to distance learning. As instructors ourselves, we saw how challenging this adjustment has been, especially for our students. These experiences taught us first-hand the importance of being able to adapt our classes to a variety of formats, from regular in-person classes, to online courses or some hybrid version in-between. The basis of the Eighth Edition is to provide a text and companion resources that are flexible enough to support an active and engaging experience in each of these formats.

Active Learning: Good Problems in Different Formats

Active participation in solving well-crafted problems promotes student learning. Since its inception, the hallmark of our text has been its innovative and engaging problems. These problems probe student understanding in ways often taken for granted. Praised for their creativity and variety, these problems have had influence far beyond the users of our textbook.

The Eighth Edition continues this tradition by providing an array of new problems, with many drawing on data from timely real-world applications. The Eighth Edition also expands on this tradition by adapting existing and new problems into an online format that retains the original pedagogical goals of the problem. Under our approach, which we call the "Rule of Four," ideas are presented graphically, numerically, symbolically, and verbally, thereby encouraging students to deepen their understanding.

Problems types in this text include:

- **Strengthen Your Understanding** problems at the end of every section. These problems ask students to reflect on what they have learned by deciding "What is wrong?" with a statement and to "Give an example" of an idea. Many of these problems have been adapted into WileyPLUS.

- **ConcepTests** promote active learning in the classroom. These can be used with polling software, and have been shown to dramatically improve student learning. Available at www.WileyPLUS.com. All ConcepTests have been adapted into WileyPLUS problems so can serve as a tool to measure student understanding in a virtual classroom.

- **Class Worksheets** allow instructors to engage students in individual or group class-work. Samples are available in the Instructor's Manual, and at www.WileyPLUS.com.

- **Data and Models** Many examples and problems throughout the text involve data-driven models.

Mathematical Thinking Supported by Theory and Modeling

The first stage in the development of mathematical thinking is the acquisition of a clear intuitive picture of the central ideas. In the next stage, the student learns to reason with the intuitive ideas in plain English. After this foundation has been laid, there is a choice of direction. All students benefit from both theory and modeling, but the balance may differ for different groups. Some students, such as mathematics majors, may prefer more theory, while others may prefer more modeling. For instructors wishing to emphasize the connection between calculus and other fields, the text includes:

- A variety of problems from the **physical sciences** and **engineering**.
- Examples from the **biological sciences** and **economics**.
- Models from the **health sciences**.
- Problems on **sustainability** and **climate change**.

Enhanced Online Content

This Eigth Edition provides opportunities for students to experience the concepts of calculus in ways that are not possible in a traditional textbook. The E-Text of *Calculus*, powered by VitalSource, and WileyPLUS provide a wealth of resources such as interactive demonstrations of concepts, embedded videos that illustrate problem-solving techniques, and built-in assessments that allow students to check their understanding as they read.

Specific resources include:

- Worked example **videos** by Donna Krawczyk at the University of Arizona, which provide students the opportunity to see and hear hundreds of the book's examples being explained in detail.

- Homework management tools, which enable the instructor to assign questions easily and grade them automatically, using a rich set of options and controls.

- Pre-designed homework assignments. Use them as-is or customize them to fit the needs of your classroom.

- Set up for Success questions, in which students are prompted for responses as they step through a problem solution and receive targeted feedback based on those responses.

- Algebra & Trigonometry Refresher material provides students with an opportunity to brush up on material necessary to master Calculus.

- Embedded **Interactive Explorations**, applets that present and explore key ideas graphically and dynamically—especially useful for display of three-dimensional graphs.

- Material that reviews and extends the major ideas of each chapter: Extra problems for many section, Review Exercises and Problems for each chapter, CAS Challenge Problems, and Projects.

- Challenging problems that involve further exploration and application.

- Appendices that include ideas useful in this course such as determinants.

Flexibility and Adaptability: Varied Approaches

The Eighth Edition of *Calculus* is designed to provide flexibility for instructors who have a range of preferences regarding inclusion of topics and applications and the use of computational technology. For those who prefer the lean topic list of earlier editions, we have kept clear the main conceptual paths. For example,

- A Fundamental Tool: Vectors (Chapter 13) can be covered before Chapter 12 (Functions of Several Variables).

- Instructors can teach a course in Multivariable Calculus using Chapters 12–16, or a course in Vector Calculus using Chapters 12–14 and a selection of material from Chapters 17–21.

- Instructors who want to show how to calculate flux integrals using general parameterizations early can teach Chapter 21 (Parameters, Coordinates and Integrals) after Section 19.1.

To use calculus effectively, students need skill in both symbolic manipulation and the use of technology. The balance between the two may vary, depending on the needs of the students and the wishes of the instructor. The book is adaptable to many different combinations.

The book does not require any specific software or technology. It has been used with graphing calculators, graphing software, and computer algebra systems. Any technology with the ability to graph functions and perform numerical integration will suffice. Students are expected to use their own judgment to determine where technology is useful.

Content

This content represents our vision of how calculus can be taught. It is flexible enough to accommodate individual course needs and requirements. Topics can easily be added or deleted, or the order changed.

Chapter 12: Functions of Several Variables

This chapter introduces functions of many variables from several points of view, using surface graphs, contour diagrams, and tables. We assume throughout that functions of two or more variables are defined on regions with piecewise smooth boundaries. We conclude with a section on continuity. Chapter 13 can be taught before Chapter 12.

Chapter 13: A Fundamental Tool: Vectors

This chapter introduces vectors geometrically and algebraically and discusses the dot and cross product. Chapter 13 can be taught before Chapter 12.

Chapter 14: Differentiating Functions of Several Variables

Partial derivatives, directional derivatives, gradients, and local linearity are introduced. The chapter also discusses higher order partial derivatives, quadratic Taylor approximations, and differentiability.

Chapter 15: Optimization

The ideas of the previous chapter are applied to optimization problems, both constrained and unconstrained.

Chapter 16: Integrating Functions of Several Variables

This chapter discusses double and triple integrals in Cartesian, polar, cylindrical, and spherical coordinates.

Chapter 17: Parameterization and Vector Fields

This chapter discusses parameterized curves and motion, vector fields and flowlines.

Chapter 18: Line Integrals

This chapter introduces line integrals and shows how to calculate them using parameterizations. Conservative fields, gradient fields, the Fundamental Theorem of Calculus for Line Integrals, and Green's Theorem are discussed.

Chapter 19: Flux Integrals and Divergence

This chapter introduces flux integrals and shows how to calculate them over surface graphs, portions of cylinders, and portions of spheres. The divergence is introduced and its relationship to flux integrals discussed in the Divergence Theorem.

Chapter 20: The Curl and Stokes' Theorem

The purpose of this chapter is to give students a practical understanding of the curl and of Stokes' Theorem and to lay out the relationship between the theorems of vector calculus.

Chapter 21: Parameters, Coordinates, and Integrals

This chapter covers parameterized surfaces, the change of variable formula in a double or triple integral, and flux though a parameterized surface.

Appendices

There are online appendices on roots, accuracy, and bounds; complex numbers; Newton's method; and vectors in the plane. The appendix on vectors can be covered at any time, but may be particularly useful in the conjunction with Section **??** on parametric equations.

Supplementary Materials and Additional Resources

Supplements for the instructor can be obtained online through WileyPLUS or by contacting your Wiley representative. The following supplementary materials are available for this edition:

- **Instructor's Manual** containing teaching tips, calculator programs, overhead transparency masters, sample worksheets, and sample syllabi.
- **Computerized Test Bank**, powered by TestGen, comprised of nearly 7,000 questions, mostly algorithmically-generated, which allows for multiple versions of a single test or quiz.
- **Instructor's Solution Manual** with complete solutions to all problems.
- **Student Solution Manual** with complete solutions to half the odd-numbered problems.
- **Graphing Calculator Manual**, to help students get the most out of their graphing calculators, and to show how they can apply the numerical and graphing functions of their calculators to their study of calculus.
- **Additional Material**, elaborating specially marked points in the text and password-protected electronic versions of the instructor ancillaries, can be found at www.WileyPLUS.com.

Acknowledgements

First and foremost, we want to express our appreciation to the National Science Foundation for their faith in our ability to produce a revitalized calculus curriculum and, in particular, to our program officers, Louise Raphael, John Kenelly, John Bradley, and James Lightbourne. We also want to thank the members of our Advisory Board, Benita Albert, Lida Barrett, Simon Bernau, Robert Davis, M. Lavinia DeConge-Watson, John Dossey, Ron Douglas, Eli Fromm, William Haver, Seymour Parter, John Prados, and Stephen Rodi.

In addition, a host of other people around the country and abroad deserve our thanks for their contributions to shaping this edition. They include: Huriye Arikan, Barbara Armenta, Pau Atela, James Baglama, Ruth Baruth, Paul Blanchard, Lewis Blake, David Bressoud, Stephen Boyd, Lucille Buonocore, Matthew Michael Campbell, Jo Cannon, Ray Cannon, Phil Cheifetz, Scott Clark, Jon Clauss, Jailing Dai, Ann Darke, Ann Davidian, Tom Dick, Srdjan Divac, Tevian Dray, Steven Dunbar, Penny Dunham, David Durlach, John Eggers, Wade Ellis, Johann Engelbrecht, Brad Ernst, Sunny Fawcett, Paul Feehan, Dana Fine, Isaac Flath, Sol Friedberg, Melanie Fulton, Tom Gearhart, David Glickenstein, Chris Goff, Sheldon P. Gordon, Salim Haïdar, Elizabeth Hentges, Michael Huber, Rob Indik, Adrian Iovita, David Jackson, Sue Jensen, Alex Kasman, Matthias Kawski, Christopher Kennedy, Mike Klucznik, Donna Krawczyk, Stephane Lafortune, Andrew Lawrence, Carl Leinert, Daniel Look, Andrew Looms, Bin Lu, Alex Mallozzi, Corinne Manogue, Jay Martin, Greg Marks, Eric Mazur, Abby McCallum, Dan McGee, Ansie Meiring, Lang Moore, Jerry Morris, Hideo Nagahashi, Kartikeya Nagendra, Alan Newell, Steve Olson, John Orr, Arnie Ostebee, Wes Ostertag, Andrew Pasquale, Scott Pilzer, Wayne Raskind, Michael J.Riedinger, Maria Robinson, Laurie Rosatone, Ayse Sahin, Nataliya Sandler, Ken Santor, Anne Scanlan-Rohrer, Ellen Schmierer, Michael Sherman, Pat Shure, Ben Smith, David Smith, Ernie Solheid, Misha Stepanov, Steve Strogatz, Carl Swenson, Peter Taylor, Dinesh Thakur, Sally Thomas, Joe Thrash, Alan Tucker, Mark Turner, Doug Ulmer, Ignatios Vakalis, Bill Vélez, Joe Vignolini, Stan Wagon, Aaron Weinberg, Hannah Winkler, Debra Wood, Beth Wolf, Deane Yang, Bruce Yoshiwara, Kathy Yoshiwara, Jianying Zhang, and Paul Zorn.

Reports from the following reviewers were most helpful for the seventh edition:

Scott Adamson, Janet Beery, Tim Biehler, Lewis Blake, Mark Booth, Tambi Boyle, David Brown, Jeremy Case, Phil Clark, Patrice Conrath, Pam Crawford, Roman J. Dial, Rebecca Dibbs, Marcel B. Finan, Vauhn Foster-Grahler, Jill Guerra, Salim M. Haidar, Ryan A. Hass, Firas Hindeleh, Todd King, Mary Koshar, Dick Lane, Glenn Ledder, Oscar Levin, Tom Linton, Erich McAlister, Osvaldo Mendez, Cindy Moss, Victor Padron, Michael Prophet, Ahmad Rajabzadeh, Catherine A. Roberts, Kari Rothi, Edward J. Soares, Diana Staats, Robert Talbert, James Vicich, Wendy Weber, Mina Yavari, and Xinyun Zhu.

Reports from the following reviewers were most helpful for the eighth edition:

Amy Decelles, Beth Wolf, Erich McAlister, Glenn Ledder, Kevin Mierzwinski, Michael Prophet, Paul Kessenich, Philip Huling, Scott Adamson, Tambi Boyle, Lewis Blake, Timothy Biehler, James Vicich, Laura Stevens, Melissa Keranen, Paul Buckelew, Chris Francisco, and Tammi Marshall.

Finally, we extend our particular thanks to Jon Christensen for his creativity with our three-dimensional figures.
The Calculus Consortium

To Students: How to Learn from this Book

- This book may be different from other math textbooks that you have used, so it may be helpful to know about some of the differences in advance. This book emphasizes at every stage the *meaning* (in practical, graphical or numerical terms) of the symbols you are using. There is much less emphasis on "plug-and-chug" and using formulas, and much more emphasis on the interpretation of these formulas than you may expect. You will often be asked to explain your ideas in words or to explain an answer using graphs.

- The book contains the main ideas of multivariable calculus in plain English. Your success in using this book will depend on your reading, questioning, and thinking hard about the ideas presented. Although you may not have done this with other books, you should plan on reading the text in detail, not just the worked examples.

- There are very few examples in the text that are exactly like the homework problems. This means that you can't just look at a homework problem and search for a similar–looking "worked out" example. Success with the homework will come by grappling with the ideas of calculus.

- Many of the problems that we have included in the book are open-ended. This means that there may be more than one approach and more than one solution, depending on your analysis. Many times, solving a problem relies on common sense ideas that are not stated in the problem but which you will know from everyday life.

- Some problems in this book assume that you have access to a graphing calculator or computer; preferably one that can draw surface graphs, contour diagrams, and vector fields, and can compute multivariable integrals and line integrals numerically. There are many situations where you may not be able to find an exact solution to a problem, but you can use a calculator or computer to get a reasonable approximation.

- This book attempts to give equal weight to three methods for describing functions: graphical (a picture), numerical (a table of values) and algebraic (a formula). Sometimes you may find it easier to translate a problem given in one form into another. For example, if you have to find the maximum of a function, you might use a contour diagram to estimate its approximate position, use its formula to find equations that give the exact position, then use a numerical method to solve the equations. The best idea is to be flexible about your approach: if one way of looking at a problem doesn't work, try another.

- Students using this book have found discussing these problems in small groups very helpful. There are a great many problems which are not cut-and-dried; it can help to attack them with the other perspectives your colleagues can provide. If group work is not feasible, see if your instructor can organize a discussion session in which additional problems can be worked on.

- You are probably wondering what you'll get from the book. The answer is, if you put in a solid effort, you will get a real understanding of one of the most important accomplishments of the millennium – calculus – as well as a real sense of the power of mathematics in the age of technology.

CONTENTS

For online material, see www.WileyPLUS.com.

APPENDICES — *Online*

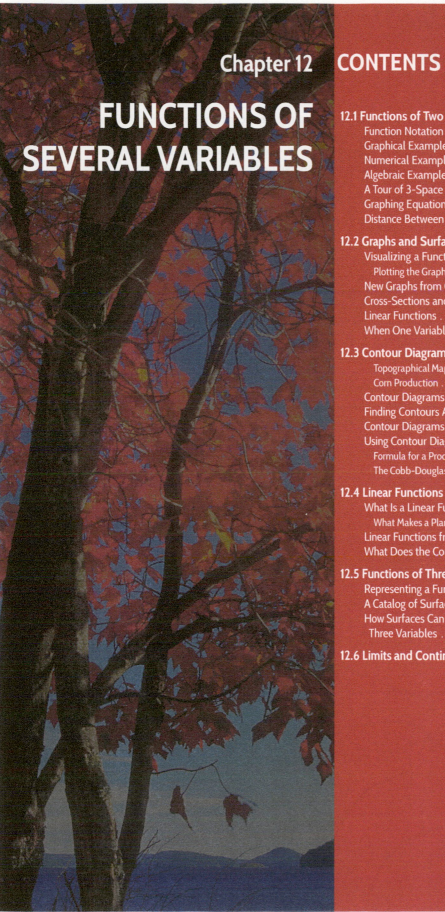

Chapter 12

FUNCTIONS OF SEVERAL VARIABLES

CONTENTS

12.1   FUNCTIONS OF TWO VARIABLES

Function Notation

Suppose you want to calculate your monthly payment on a five-year car loan; this depends on both the amount of money you borrow and the interest rate. These quantities can vary separately: the loan amount can change while the interest rate remains the same, or the interest rate can change while the loan amount remains the same. To calculate your monthly payment you need to know both. If the monthly payment is $\$m$, the loan amount is $\$L$, and the interest rate is $r\%$, then we express the fact that m is a function of L and r by writing:

$$m = f(L, r).$$

This is just like the function notation of one-variable calculus. The variable m is called the dependent variable, and the variables L and r are called the independent variables. The letter f stands for the *function* or rule that gives the value of m corresponding to given values of L and r.

A function of two variables can be represented graphically, numerically by a table of values, or algebraically by a formula. In this section, we give examples of each.

Graphical Example: A Weather Map

Figure 12.1 shows a weather map from a newspaper. What information does it convey? It displays the predicted high temperature, T, in degrees Fahrenheit (°F), throughout the US on that day. The curves on the map, called *isotherms*, separate the country into zones, according to whether T is in the 60s, 70s, 80s, 90s, or 100s. (*Iso* means same and *therm* means heat.) Notice that the isotherm separating the 80s and 90s zones connects all the points where the temperature is exactly 90°F.

Example 1   Estimate the predicted value of T in Boise, Idaho; Topeka, Kansas; and Buffalo, New York.

Solution   Boise and Buffalo are in the 70s region, and Topeka is in the 80s region. Thus, the predicted temperature in Boise and Buffalo is between 70 and 80 while the predicted temperature in Topeka is between 80 and 90. In fact, we can say more. Although both Boise and Buffalo are in the 70s, Boise is quite close to the $T = 70$ isotherm, whereas Buffalo is quite close to the $T = 80$ isotherm. So we estimate the temperature to be in the low 70s in Boise and in the high 70s in Buffalo. Topeka is about halfway between the $T = 80$ isotherm and the $T = 90$ isotherm. Thus, we guess the temperature in Topeka to be in the mid-80s. In fact, the actual high temperatures for that day were 71°F for Boise, 79°F for Buffalo, and 86°F for Topeka.

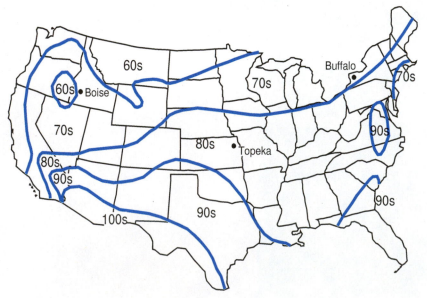

Figure 12.1: Weather map showing predicted high temperatures, T, on a summer day

The predicted high temperature, T, illustrated by the weather map is a function of (that is, depends on) two variables, often longitude and latitude, or miles east-west and miles north-south of a fixed point, say, Topeka. The weather map in Figure 12.1 is called a *contour map* or *contour diagram* of that function. Section 12.2 shows another way of visualizing functions of two variables using surfaces; Section 12.3 looks at contour maps in detail.

Numerical Example: Body Mass Index (BMI)

The body mass index (BMI) is a value that attempts to quantify a person's body fat based on their height h and weight w. In function notation, we write:

$$\text{BMI} = f(h, w).$$

Table 12.1 contains values of this function for h in inches and w in pounds. Values of w are across the top, values of h are down the left side, and corresponding values of $f(h, w)$ are in the table.[1] For example, to find the value of $f(66, 140)$, we look in the row corresponding to $h = 66$ under $w = 140$, where we find the number 22.6. Thus,

$$f(66, 140) = 22.6.$$

This means that if an individual is 66 inches tall and weighs 140 lbs, their body mass index is 22.6.

Table 12.1   *Body mass index (BMI)*

		Weight w (lbs)				
		120	140	160	180	200
	60	23.4	27.3	31.2	35.2	39.1
	63	21.3	24.8	28.3	31.9	35.4
Height h	66	19.4	22.6	25.8	29.0	32.3
(inches)	69	17.7	20.7	23.6	26.6	29.5
	72	16.3	19.0	21.7	24.4	27.1
	75	15.0	17.5	20.0	22.5	25.0

Notice how this table differs from the table of values of a one-variable function, where one row or one column is enough to list the values of the function. Here many rows and columns are needed because the function has a value for every *pair* of values of the independent variables.

Algebraic Examples: Formulas

In the weather map example there is no formula for the underlying function. That is usually the case for functions representing real-life data. On the other hand, for many models in physics, engineering, and economics, there are exact formulas.

Example 2   Give a formula for the function $M = f(B, t)$ where M is the amount of money in a bank account t years after an initial investment of B dollars, if interest is accrued at a rate of 1.2% per year compounded annually.

Solution   Annual compounding means that M increases by a factor of 1.012 every year, so

$$M = f(B, t) = B(1.012)^t.$$

Example 3   A cylinder with closed ends has radius r and height h. If its volume is V and its surface area is A, find formulas for the functions $V = f(r, h)$ and $A = g(r, h)$.

Solution   Since the area of the circular base is πr^2, we have

$$V = f(r, h) = \text{Area of base} \cdot \text{Height} = \pi r^2 h.$$

[1] http://www.cdc.gov, accessed December 30, 2019.

The surface area of the side is the circumference of the bottom, $2\pi r$, times the height h, giving $2\pi rh$. Thus,

$$A = g(r, h) = 2 \cdot \text{Area of base} + \text{Area of side} = 2\pi r^2 + 2\pi rh.$$

A Tour of 3-Space

In Section 12.2 we see how to visualize a function of two variables as a surface in space. Now we see how to locate points in three-dimensional space (3-space).

Imagine three coordinate axes meeting at the *origin*: a vertical axis, and two horizontal axes at right angles to each other. (See Figure 12.2.) Think of the xy-plane as being horizontal, while the z-axis extends vertically above and below the plane. The labels x, y, and z show which part of each axis is positive; the other side is negative. We generally use *right-handed axes* in which looking down the positive z-axis gives the usual view of the xy-plane. We specify a point in 3-space by giving its coordinates (x, y, z) with respect to these axes. Think of the coordinates as instructions telling you how to get to the point: start at the origin, go x units along the x-axis, then y units in the direction parallel to the y-axis, and finally z units in the direction parallel to the z-axis. The coordinates can be positive, zero or negative; a zero coordinate means "don't move in this direction," and a negative coordinate means "go in the negative direction parallel to this axis." For example, the origin has coordinates $(0, 0, 0)$, since we get there from the origin by doing nothing at all.

Example 4   Describe the position of the points with coordinates $(1, 2, 3)$ and $(0, 0, -1)$.

Solution   We get to the point $(1, 2, 3)$ by starting at the origin, going 1 unit along the x-axis, 2 units in the direction parallel to the y-axis, and 3 units up in the direction parallel to the z-axis. (See Figure 12.3.)

To get to $(0, 0, -1)$, we don't move at all in the x- and the y-directions, but move 1 unit in the negative z-direction. So the point is on the negative z-axis. (See Figure 12.4.) You can check that the position of the point is independent of the order of the x, y, and z displacements.

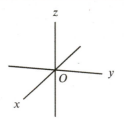

Figure 12.2: Coordinate axes in three-dimensional space

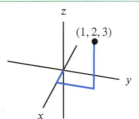

Figure 12.3: The point $(1, 2, 3)$ in 3-space

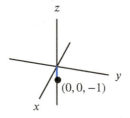

Figure 12.4: The point $(0, 0, -1)$ in 3-space

Example 5   You start at the origin, go along the y-axis a distance of 2 units in the positive direction, and then move vertically upward a distance of 1 unit. What are the coordinates of your final position?

Solution   You started at the point $(0, 0, 0)$. When you went along the y-axis, your y-coordinate increased to 2. Moving vertically increased your z-coordinate to 1; your x-coordinate did not change because you did not move in the x-direction. So your final coordinates are $(0, 2, 1)$. (See Figure 12.5.)

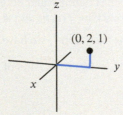

Figure 12.5: The point $(0, 2, 1)$ is reached by moving 2 along the y-axis and 1 upward

It is often helpful to picture a three-dimensional coordinate system in terms of a room. The origin is a corner at floor level where two walls meet the floor. The z-axis is the vertical intersection of the two walls; the x- and the y-axis are the intersections of each wall with the floor. Points with negative coordinates lie behind a wall in the next room or below the floor.

Graphing Equations in 3-Space

We can graph an equation involving the variables x, y, and z in 3-space; such a graph is a picture of all points (x, y, z) that satisfy the equation.

Example 6   What do the graphs of the equations $z = 0$, $z = 3$, and $z = -1$ look like?

Solution   To graph $z = 0$, we visualize the set of points whose z-coordinate is zero. If the z-coordinate is 0, then we must be at the same vertical level as the origin; that is, we are in the horizontal plane containing the origin. So the graph of $z = 0$ is the middle plane in Figure 12.6. The graph of $z = 3$ is a plane parallel to the graph of $z = 0$, but three units above it. The graph of $z = -1$ is a plane parallel to the graph of $z = 0$, but one unit below it.

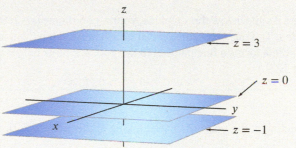

Figure 12.6: The planes $z = -1$, $z = 0$, and $z = 3$

The plane $z = 0$ contains the x- and the y-coordinate axes, and is called the xy-plane. There are two other coordinate planes. The yz-plane contains both the y- and the z-axis, and the xz-plane contains the x- and the z-axis. (See Figure 12.7.)

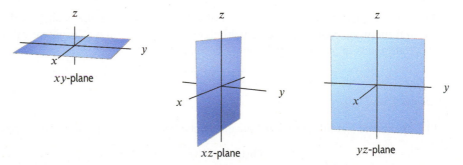

Figure 12.7: The three coordinate planes

Example 7   Which of the points $A = (1, -1, 0)$, $B = (0, 3, 4)$, $C = (2, 2, 1)$, and $D = (0, -4, 0)$ lies closest to the xz-plane? Which point lies on the y-axis?

Solution   The magnitude of the y-coordinate gives the distance to the xz-plane. The point A lies closest to that plane, because it has the smallest y-coordinate in magnitude. To get to a point on the y-axis, we move along the y-axis, but we don't move at all in the x- or the z-direction. Thus, a point on the y-axis has both its x- and z-coordinates equal to zero. The only point of the four that satisfies this is D. (See Figure 12.8.)

In general, if a point has one of its coordinates equal to zero, it lies in one of the coordinate planes. If a point has two of its coordinates equal to zero, it lies on one of the coordinate axes.

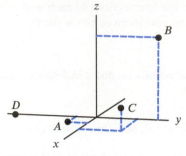

Figure 12.8: Which point lies closest to the xz-plane? Which point lies on the y-axis?

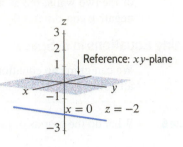

Figure 12.9: The line $x = 0$, $z = -2$

Example 8 You are 2 units below the xy-plane and in the yz-plane. What are your coordinates?

Solution Since you are 2 units below the xy-plane, your z-coordinate is -2. Since you are in the yz-plane, your x-coordinate is 0; your y-coordinate can be anything. Thus, you are at the point $(0, y, -2)$. The set of all such points forms a line parallel to the y-axis, 2 units below the xy-plane, and in the yz-plane. (See Figure 12.9.)

Example 9 You are standing at the point $(4, 5, 2)$, looking at the point $(0.5, 0, 3)$. Are you looking up or down?

Solution The point you are standing at has z-coordinate 2, whereas the point you are looking at has z-coordinate 3; hence you are looking up.

Example 10 Imagine that the yz-plane in Figure 12.7 is a page of this book. Describe the region behind the page algebraically.

Solution The positive part of the x-axis pokes out of the page; moving in the positive x-direction brings you out in front of the page. The region behind the page corresponds to negative values of x, so it is the set of all points in 3-space satisfying the inequality $x < 0$.

Distance Between Two Points

In 2-space, the formula for the distance between two points (x, y) and (a, b) is given by

$$\text{Distance} = \sqrt{(x - a)^2 + (y - b)^2}.$$

The distance between two points (x, y, z) and (a, b, c) in 3-space is represented by PG in Figure 12.10. The side PE is parallel to the x-axis, EF is parallel to the y-axis, and FG is parallel to the z-axis.

Using Pythagoras' theorem twice gives

$$(PG)^2 = (PF)^2 + (FG)^2 = (PE)^2 + (EF)^2 + (FG)^2 = (x - a)^2 + (y - b)^2 + (z - c)^2.$$

Thus, a formula for the distance between the points (x, y, z) and (a, b, c) in 3-space is

$$\boxed{\text{Distance} = \sqrt{(x - a)^2 + (y - b)^2 + (z - c)^2}.}$$

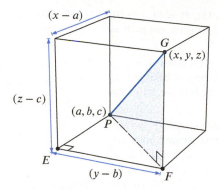

Figure 12.10: The diagonal PG gives the distance between the points (x, y, z) and (a, b, c)

Example 11 Find the distance between $(1, 2, 1)$ and $(-3, 1, 2)$.

Solution Distance $= \sqrt{(-3-1)^2 + (1-2)^2 + (2-1)^2} = \sqrt{18} = 4.243$.

Example 12 Find an expression for the distance from the origin to the point (x, y, z).

Solution The origin has coordinates $(0, 0, 0)$, so the distance from the origin to (x, y, z) is given by

$$\text{Distance} = \sqrt{(x-0)^2 + (y-0)^2 + (z-0)^2} = \sqrt{x^2 + y^2 + z^2}.$$

Example 13 Find an equation for a sphere of radius 1 with center at the origin.

Solution The sphere consists of all points (x, y, z) whose distance from the origin is 1, that is, which satisfy the equation

$$\sqrt{x^2 + y^2 + z^2} = 1.$$

This is an equation for the sphere. If we square both sides we get the equation in the form

$$x^2 + y^2 + z^2 = 1.$$

Note that this equation represents the *surface* of the sphere. The solid ball enclosed by the sphere is represented by the inequality $x^2 + y^2 + z^2 \leq 1$.

Summary for Section 12.1

- **Function notation**: $f(x, y)$ is the value of the function f with inputs x and y.
- Points in **3-space** are specified by their coordinates relative to the x, y, z-axes.
- Some **planes** in 3-space can be specified by simple equations given by one variable equal to a constant. For example, the xy-plane (where $z = 0$), the xz-plane (where $y = 0$), and the yz-plane (where $x = 0$).
- The **distance** between the points (x, y, z) and (a, b, c) in 3-space is

$$\text{Distance} = \sqrt{(x-a)^2 + (y-b)^2 + (z-c)^2}.$$

Exercises and Problems for Section 12.1   Online Resource: Additional Problems for Section 12.1
EXERCISES

1. Which of the points $P = (1, 2, 1)$ and $Q = (2, 0, 0)$ is closest to the origin?

2. Which two of the three points $P_1 = (1, 2, 3)$, $P_2 = (3, 2, 1)$ and $P_3 = (1, 1, 0)$ are closest to each other?

3. Which of the points $P_1 = (-3, 2, 15)$, $P_2 = (0, -10, 0)$, $P_3 = (-6, 5, 3)$ and $P_4 = (-4, 2, 7)$ is closest to $P = (6, 0, 4)$?

4. Which of the points $A = (1.3, -2.7, 0)$, $B = (0.9, 0, 3.2)$, $C = (2.5, 0.1, -0.3)$ is closest to the yz-plane? Which one lies on the xz-plane? Which one is farthest from the xy-plane?

5. You are at the point $(3, 1, 1)$, standing upright and facing the yz-plane. You walk 2 units forward, turn left, and walk another 2 units. What is your final position? From the point of view of an observer looking at the coordinate system in Figure 12.2 on page 696, are you in front of or behind the yz-plane? To the left or to the right of the xz-plane? Above or below the xy-plane?

6. On a set of x, y and z axes oriented as in Figure 12.5 on page 696, draw a straight line through the origin, lying in the yz-plane and such that if you move along the line with your y-coordinate increasing, your z-coordinate is increasing.

7. What is the midpoint of the line segment joining the points $(-1, 3, 9)$ and $(5, 6, -3)$?

■ In Exercises 8–11, which of (I)–(IV) lie on the graph of the equation?

I. $(2, 2, 4)$ II. $(-1, 1, 0)$
III. $(-3, -2, -1)$ IV. $(-2, -2, 4)$

8. $z = 4$ 9. $x + y + z = 0$

10. $x^2 + y^2 + z^2 = 14$ 11. $x - y = 0$

■ In Exercises 12–15 sketch graphs of the equations in 3-space.

12. $z = 4$ 13. $x = -3$

14. $y = 1$ 15. $z = 2$ and $y = 4$

16. With the z-axis vertical, a sphere has center $(2, 3, 7)$ and lowest point $(2, 3, -1)$. What is the highest point on the sphere?

17. Find an equation of the sphere with radius 5 centered at the origin.

18. Find the equation of the sphere with radius 2 and centered at $(1, 0, 0)$.

19. Find the equation of the vertical plane perpendicular to the y-axis and through the point $(2, 3, 4)$.

■ Exercises 20–22 refer to the map in Figure 12.1 on page 694.

20. Give the range of daily high temperatures for:

   (a) Pennsylvania   (b)   North Dakota
   (c) California

21. Sketch a possible graph of the predicted high temperature T on a line north-south through Topeka.

22. Sketch possible graphs of the predicted high temperature on a north-south line and an east-west line through Boise.

■ For Exercises 23–25, refer to Table 12.1 on page 695 where w is a person's weight (in lbs) and h their height (in inches).

23. Compute a table of values of BMI, with h fixed at 60 inches and w between 120 and 200 lbs at intervals of 20.

24. Medical evidence suggests that BMI values between 18.5 and 24.9 are healthy values.[2] Estimate the range of weights that are considered healthy for a woman who is 6 feet tall.

25. Estimate the BMI of a man who weighs 90 kilograms and is 1.9 meters tall.

PROBLEMS

26. The temperature adjusted for wind chill is a temperature which tells you how cold it feels, as a result of the combination of wind and temperature.[3] See Table 12.2.

Table 12.2   *Temperature adjusted for wind chill (°F) as a function of wind speed and temperature*

		Temperature (°F)							
		35	30	25	20	15	10	5	0
Wind Speed (mph)	5	31	25	19	13	7	1	-5	-11
	10	27	21	15	9	3	-4	-10	-16
	15	25	19	13	6	0	-7	-13	-19
	20	24	17	11	4	-2	-9	-15	-22
	25	23	16	9	3	-4	-11	-17	-24

(a) If the temperature is 0°F and the wind speed is 15 mph, how cold does it feel?

(b) If the temperature is 35°F, what wind speed makes it feel like 24°F?

(c) If the temperature is 25°F, what wind speed makes it feel like 12°F?

(d) If the wind is blowing at 20 mph, what temperature feels like 0°F?

■ In Problems 27–28, use Table 12.2 to make tables with the given properties.

27. The temperature adjusted for wind chill as a function of wind speed for temperatures of 20°F and 0°F.

28. The temperature adjusted for wind chill as a function of temperature for wind speeds of 5 mph and 20 mph.

[2]http://www.cdc.gov. Accessed January 10, 2016.
[3]www.nws.noaa.gov. Accessed January 10, 2016.

For Problems **29–31**, refer to Table 12.3, which contains values of beef consumption C (in pounds per week per household) as a function of household income, I (in thousands of dollars per year), and the price of beef, p (in dollars per pound). Values of p are shown across the top, values of I are down the left side, and corresponding values of beef consumption $C = f(I, p)$ are given in the table.[4]

Table 12.3 *Quantity of beef bought (pounds/household/week)*

		\multicolumn{4}{Price of beef ($/lb)}			
		3.00	3.50	4.00	4.50
Household income per year, I ($1000)	20	2.65	2.59	2.51	2.43
	40	4.14	4.05	3.94	3.88
	60	5.11	5.00	4.97	4.84
	80	5.35	5.29	5.19	5.07
	100	5.79	5.77	5.60	5.53

29. Give tables for beef consumption as a function of p, with I fixed at $I = 20$ and $I = 100$. Give tables for beef consumption as a function of I, with p fixed at $p = 3.00$ and $p = 4.00$. Comment on what you see in the tables.

30. Make a table of the proportion, P, of household income spent on beef per week as a function of price and income. (Note that P is the fraction of income spent on beef.)

31. How does beef consumption vary as a function of household income if the price of beef is held constant?

For Problems **32–35**, a person's body mass index (BMI) is a function of their weight W (in kg) and height H (in m) given by $B(W, H) = W/H^2$.

32. What is the BMI of a 1.72 m tall man weighing 72 kg?

33. A 1.58 m tall woman has a BMI of 23.2. What is her weight?

34. With a BMI less than 18.5, a person is considered underweight. What is the possible range of weights for an underweight person 1.58 m tall?

35. For weight w in lbs and height h in inches, a persons BMI is approximated using the formula $f(w, h) = 703w/h^2$. Check this approximation by converting the formula $B(W, H)$.

36. A car rental company charges $40 a day and 15 cents a mile for its cars.

   (a) Write a formula for the cost, C, of renting a car as a function, f, of the number of days, d, and the number of miles driven, m.

   (b) If $C = f(d, m)$, find $f(5, 300)$ and interpret it.

37. A cable company charges $100 for a monthly subscription to its services and $5 for each special feature movie that a subscriber chooses to watch.

   (a) Write a formula for the monthly revenue, R in dollars, earned by the cable company as a function of s, the number of monthly subscribers it serves, and m, the total number of special feature movies that its subscribers view.

   (b) If $R = f(s, m)$, find $f(1000, 5000)$ and interpret it in terms of revenue.

38. The gravitational force, F newtons, exerted on an object by the earth depends on the object's mass, m kilograms, and its distance, r meters, from the center of the earth, so $F = f(m, r)$. Interpret the following statement in terms of gravitation: $f(100, 7000000) \approx 820$.

39. A heating element is attached to the center point of a metal rod at time $t = 0$. Let $H = f(d, t)$ represent the temperature in °C of a point d cm from the center after t minutes.

   (a) Interpret the statement $f(2, 5) = 24$ in terms of temperature.

   (b) If d is held constant, is H an increasing or a decreasing function of t? Why?

   (c) If t is held constant, is H an increasing or a decreasing function of d? Why?

40. The pressure, P atmospheres, of 10 moles of nitrogen gas in a steel cylinder depends on the temperature of the gas, T Kelvin, and the volume of the cylinder, V liters, so $P = f(T, V)$. Interpret the following statement in terms of pressure: $f(300, 5) = 49.2$.

41. The monthly payment, m dollars, for a 30-year fixed rate mortgage is a function of the total amount borrowed, P dollars, and the annual interest rate, $r\%$. In other words, $m = f(P, r)$.

   (a) Interpret the following statement in the context of monthly payment: $f(300,000, 5) = 1610.46$.

   (b) If P is held constant, is m an increasing or a decreasing function of r? Why?

   (c) If r is held constant, is m an increasing or a decreasing function of P? Why?

42. Consider the acceleration due to gravity, g, at a distance h from the center of a planet of mass m.

   (a) If m is held constant, is g an increasing or decreasing function of h? Why?

   (b) If h is held constant, is g an increasing or decreasing function of m? Why?

43. A cube is located such that its top four corners have the coordinates $(-1, -2, 2)$, $(-1, 3, 2)$, $(4, -2, 2)$ and $(4, 3, 2)$. Give the coordinates of the center of the cube.

[4]Adapted from Richard G. Lipsey, *An Introduction to Positive Economics*, 3rd ed. (London: Weidenfeld and Nicolson, 1971).

44. Describe the set of points whose distance from the x-axis is 2.

45. Describe the set of points whose distance from the x-axis equals the distance from the yz-plane.

46. Find the point on the x-axis closest to the point $(3, 2, 1)$.

47. Does the line parallel to the y-axis through the point $(2, 1, 4)$ intersect the plane $y = 5$? If so, where?

48. Find a formula for the shortest distance between a point (a, b, c) and the y-axis.

49. Find the equations of planes that just touch the sphere $(x - 2)^2 + (y - 3)^2 + (z - 3)^2 = 16$ and are parallel to

   (a) The xy-plane   **(b)** The yz-plane

   (c) The xz-plane

50. Find an equation of the largest sphere contained in the cube determined by the planes $x = 2, x = 6$; $y = 5, y = 9$; and $z = -1, z = 3$.

51. A cube has edges parallel to the axes. One corner is at $A = (5, 1, 2)$ and the corner at the other end of the longest diagonal through A is $B = (12, 7, 4)$.

   (a) What are the coordinates of the other three vertices on the bottom face?

   (b) What are the coordinates of the other three vertices on the top face?

52. An equilateral triangle is standing vertically with a vertex above the xy-plane and its two other vertices at $(7, 0, 0)$ and $(9, 0, 0)$. What is its highest point?

53. **(a)** Find the midpoint of the line segment joining $A = (1, 5, 7)$ to $B = (5, 13, 19)$.

   (b) Find the point one quarter of the way along the line segment from A to B.

   (c) Find the point one quarter of the way along the line segment from B to A.

Strengthen Your Understanding

■ In Problems 54–56, explain what is wrong with the statement.

54. In 3-space, $y = 1$ is a line parallel to the x-axis.

55. The xy-plane has equation $xy = 0$.

56. The distance from $(2, 3, 4)$ to the x-axis is 2.

■ In Problems 57–58, give an example of:

57. A formula for a function $f(x, y)$ that is increasing in x and decreasing in y.

58. A point in 3-space with all its coordinates negative and farther from the xz-plane than from the plane $z = -5$.

■ Are the statements in Problems 59–72 true or false? Give reasons for your answer.

59. If $f(x, y)$ is a function of two variables defined for all x and y, then $f(10, y)$ is a function of one variable.

60. The volume V of a box of height h and square base of side length s is a function of h and s.

61. If $H = f(t, d)$ is the function giving the water temperature $H°C$ of a lake at time t hours after midnight and depth d meters, then t is a function of d and H.

62. A table for a function $f(x, y)$ cannot have any values of f appearing twice.

63. If $f(x)$ and $g(y)$ are both functions of a single variable, then the product $f(x) \cdot g(y)$ is a function of two variables.

64. The point $(1, 2, 3)$ lies above the plane $z = 2$.

65. The graph of the equation $z = 2$ is a plane parallel to the xz-plane.

66. The points $(1, 0, 1)$ and $(0, -1, 1)$ are the same distance from the origin.

67. The point $(2, -1, 3)$ lies on the graph of the sphere $(x - 2)^2 + (y + 1)^2 + (z - 3)^2 = 25$.

68. There is only one point in the yz-plane that is a distance 3 from the point $(3, 0, 0)$.

69. There is only one point in the yz-plane that is a distance 5 from the point $(3, 0, 0)$.

70. If the point $(0, b, 0)$ has distance 4 from the plane $y = 0$, then b must be 4.

71. A line parallel to the z-axis can intersect the graph of $f(x, y)$ at most once.

72. A line parallel to the y-axis can intersect the graph of $f(x, y)$ at most once.

12.2   GRAPHS AND SURFACES

The weather map on page 694 is one way of visualizing a function of two variables. In this section we see how to visualize a function of two variables in another way, using a surface in 3-space.

Visualizing a Function of Two Variables Using a Graph

For a function of one variable, $y = f(x)$, the graph of f is the set of all points (x, y) in 2-space such that $y = f(x)$. In general, these points lie on a curve in the plane. When a computer or calculator graphs f, it approximates by plotting points in the xy-plane and joining consecutive points by line segments. The more points, the better the approximation.

Now consider a function of two variables.

> The **graph** of a function of two variables, f, is the set of all points (x, y, z) such that $z = f(x, y)$. In general, the graph of a function of two variables is a surface in 3-space.

Plotting the Graph of the Function $f(x, y) = x^2 + y^2$

To sketch the graph of f we connect points as for a function of one variable. We first make a table of values of f, such as in Table 12.4.

Table 12.4   *Table of values of $f(x, y) = x^2 + y^2$*

		y						
		-3	-2	-1	0	1	2	3
	-3	18	13	10	9	10	13	18
	-2	13	8	5	4	5	8	13
	-1	10	5	2	1	2	5	10
x	0	9	4	1	0	1	4	9
	1	10	5	2	1	2	5	10
	2	13	8	5	4	5	8	13
	3	18	13	10	9	10	13	18

Now we plot points. For example, we plot $(1, 2, 5)$ because $f(1, 2) = 5$ and we plot $(0, 2, 4)$ because $f(0, 2) = 4$. Then, we connect the points corresponding to the rows and columns in the table. The result is called a *wire-frame* picture of the graph. Filling in between the wires gives a surface. That is the way a computer drew the graphs in Figures 12.11 and 12.12. As more points are plotted, we get the surface in Figure 12.13, called a *paraboloid*.

You should check to see if the sketches make sense. Notice that the graph goes through the origin since $(x, y, z) = (0, 0, 0)$ satisfies $z = x^2 + y^2$. Observe that if x is held fixed and y is allowed to vary, the graph dips down and then goes back up, just like the entries in the rows of Table 12.4. Similarly, if y is held fixed and x is allowed to vary, the graph dips down and then goes back up, just like the columns of Table 12.4.

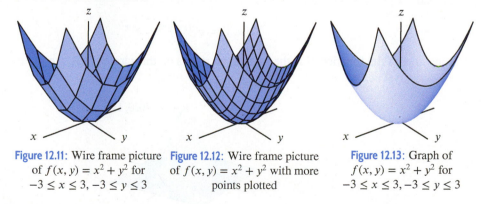

Figure 12.11: Wire frame picture of $f(x, y) = x^2 + y^2$ for $-3 \le x \le 3, -3 \le y \le 3$

Figure 12.12: Wire frame picture of $f(x, y) = x^2 + y^2$ with more points plotted

Figure 12.13: Graph of $f(x, y) = x^2 + y^2$ for $-3 \le x \le 3, -3 \le y \le 3$

New Graphs from Old

We can use the graph of a function to visualize the graphs of related functions.

Example 1   Let $f(x, y) = x^2 + y^2$. Describe in words the graphs of the following functions:
(a) $g(x, y) = x^2 + y^2 + 3$,   (b) $h(x, y) = 5 - x^2 - y^2$,   (c) $k(x, y) = x^2 + (y - 1)^2$.

Solution   We know from Figure 12.13 that the graph of f is a paraboloid, or a bowl, with its vertex at the origin. From this we can work out what the graphs of g, h, and k will look like.

(a) The function $g(x, y) = x^2 + y^2 + 3 = f(x, y) + 3$, so the graph of g is the graph of f, but raised by 3 units. See Figure 12.14.

(b) Since $-x^2 - y^2$ is the negative of $x^2 + y^2$, the graph of $-x^2 - y^2$ is a paraboloid opening downward. Thus, the graph of $h(x, y) = 5 - x^2 - y^2 = 5 - f(x, y)$ looks like a downward-opening paraboloid with vertex at $(0, 0, 5)$, as in Figure 12.15.

(c) The graph of $k(x, y) = x^2 + (y - 1)^2 = f(x, y - 1)$ is a paraboloid with vertex at $x = 0$, $y = 1$, since that is where $k(x, y) = 0$, as in Figure 12.16.

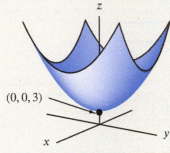

Figure 12.14: Graph of $g(x, y) = x^2 + y^2 + 3$

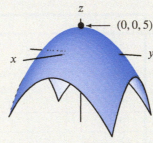

Figure 12.15: Graph of $h(x, y) = 5 - x^2 - y^2$

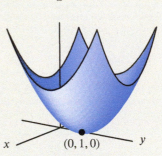

Figure 12.16: Graph of $k(x, y) = x^2 + (y - 1)^2$

Example 2   Describe the graph of $G(x, y) = e^{-(x^2+y^2)}$. What symmetry does it have?

Solution   Since the exponential function is always positive, the graph lies entirely above the xy-plane. From the graph of $x^2 + y^2$ we see that $x^2 + y^2$ is zero at the origin and gets larger as we move farther from the origin in any direction. Thus, $e^{-(x^2+y^2)}$ is 1 at the origin, and gets smaller as we move away from the origin in any direction. It can't go below the xy-plane; instead it flattens out, getting closer and closer to the plane. We say the surface is *asymptotic* to the xy-plane. (See Figure 12.17.) Now consider a point (x, y) on the circle $x^2 + y^2 = r^2$. Since

$$G(x, y) = e^{-(x^2+y^2)} = e^{-r^2},$$

the value of the function G is the same at all points on this circle. Thus, we say the graph of G has *circular symmetry*.

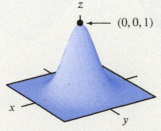

Figure 12.17: Graph of $G(x, y) = e^{-(x^2+y^2)}$

Cross-Sections and the Graph of a Function

We have seen that a good way to analyze a function of two variables is to let one variable vary while the other is kept fixed.

For a function $f(x, y)$, the function we get by holding x fixed and letting y vary is called a **cross-section** of f with x fixed. The graph of the cross-section of $f(x, y)$ with $x = c$ is the curve, or cross-section, we get by intersecting the graph of f with the plane $x = c$. We define a cross-section of f with y fixed similarly.

For example, the cross-section of $f(x, y) = x^2 + y^2$ with $x = 2$ is $f(2, y) = 4 + y^2$. The graph of this cross-section is the curve we get by intersecting the graph of f with the plane perpendicular to the x-axis at $x = 2$. (See Figure 12.18.)

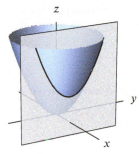

Figure 12.18: Cross-section of the surface $z = f(x, y)$ by the plane $x = 2$

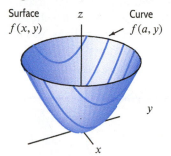

Figure 12.19: The curves $z = f(a, y)$ with a constant: cross-sections with x fixed

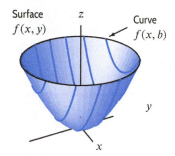

Figure 12.20: The curves $z = f(x, b)$ with b constant: cross-sections with y fixed

Figure 12.19 shows graphs of other cross-sections of f with x fixed; Figure 12.20 shows graphs of cross-sections with y fixed.

Example 3   Describe the cross-sections of the function $g(x, y) = x^2 - y^2$ with y fixed and then with x fixed. Use these cross-sections to describe the shape of the graph of g.

Solution   The cross-sections with y fixed at $y = b$ are given by

$$z = g(x, b) = x^2 - b^2.$$

Thus, each cross-section with y fixed gives a parabola opening upward, with minimum $z = -b^2$. The cross-sections with x fixed are of the form

$$z = g(a, y) = a^2 - y^2,$$

which are parabolas opening downward with a maximum of $z = a^2$. (See Figures 12.21 and 12.22.) The graph of g is shown in Figure 12.23. Notice the upward-opening parabolas in the x-direction and the downward-opening parabolas in the y-direction. We say that the surface is *saddle-shaped*.

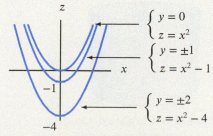

Figure 12.21: Cross-sections of $g(x, y) = x^2 - y^2$ with y fixed

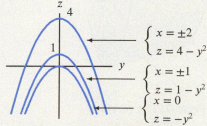

Figure 12.22: Cross-sections of $g(x, y) = x^2 - y^2$ with x fixed

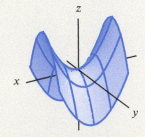

Figure 12.23: Graph of $g(x, y) = x^2 - y^2$ showing cross-sections

Linear Functions

Linear functions are central to single-variable calculus; they are equally important in multivariable calculus. You may be able to guess the shape of the graph of a linear function of two variables. (It's a plane.) Let's look at an example.

Example 4   Describe the graph of $f(x, y) = 1 + x - y$.

Solution   The plane $x = a$ is vertical and parallel to the yz-plane. Thus, the cross-section with $x = a$ is the line $z = 1 + a - y$ which slopes downward in the y-direction. Similarly, the plane $y = b$ is parallel to the xz-plane. Thus, the cross-section with $y = b$ is the line $z = 1 + x - b$ which slopes upward in the x-direction. Since all the cross-sections are lines, you might expect the graph to be a flat plane, sloping down in the y-direction and up in the x-direction. This is indeed the case. (See Figure 12.24.)

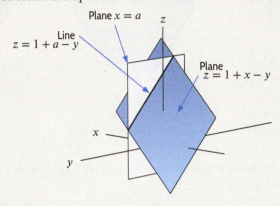

Figure 12.24: Graph of the plane $z = 1 + x - y$ showing cross-section with $x = a$

When One Variable Is Missing: Cylinders

Suppose we graph an equation like $z = x^2$ which has one variable missing. What does the surface look like? Since y is missing from the equation, the cross-sections with y fixed are all the same parabola, $z = x^2$. Letting y vary up and down the y-axis, this parabola sweeps out the trough-shaped surface shown in Figure 12.25. The cross-sections with x fixed are horizontal lines obtained by cutting the surface by a plane perpendicular to the x-axis. This surface is called a *parabolic cylinder*, because it is formed from a parabola in the same way that an ordinary cylinder is formed from a circle; it has a parabolic cross-section instead of a circular one.

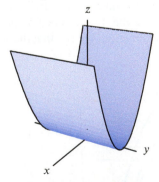

Figure 12.25: A parabolic cylinder $z = x^2$

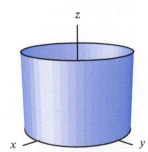

Figure 12.26: Circular cylinder $x^2 + y^2 = 1$

Example 5   Graph the equation $x^2 + y^2 = 1$ in 3-space.

Solution
: Although the equation $x^2 + y^2 = 1$ does not represent a function, the surface representing it can be graphed by the method used for $z = x^2$. The graph of $x^2 + y^2 = 1$ in the xy-plane is a circle. Since z does not appear in the equation, the intersection of the surface with any horizontal plane will be the same circle $x^2 + y^2 = 1$. Thus, the surface is the cylinder shown in Figure 12.26.

Summary for Section 12.2

- The **graph** of the function $f(x, y)$ is the set of points $(x, y, f(x, y))$ in 3-space.
- Simple **changes** to a function, such as adding or multiplying by a constant, affect the graph by shifting, stretching or flipping, just as for functions of one variable.
- A **cross-section** of a function $f(x, y)$ is the one-variable function obtained by setting x or y equal to a constant.
- A **cylinder** is the result of having one of the variables unspecified, such as $f(x, y) = x^2$.

Exercises and Problems for Section 12.2   Online Resource: Additional Problems for Section 12.2
EXERCISES

In Exercises 1–4, which of (I)–(IV) lie on the graph of the function $z = f(x, y)$?

I. $(1, 0, 1)$ II. $(\sqrt{8}, 1, 3)$
III. $(-3, 7, -3)$ IV. $(1, 1, 1/2)$

1. $f(x, y) = -3$
2. $f(x, y) = \sqrt{x^2 + y^2}$
3. $f(x, y) = 1/(x^2 + y^2)$
4. $f(x, y) = 4 - y$

5. Without a calculator or computer, match the functions with their graphs in Figure 12.27.

   (a) $z = 2 + x^2 + y^2$ (b) $z = 2 - x^2 - y^2$
   (c) $z = 2(x^2 + y^2)$ (d) $z = 2 + 2x - y$
   (e) $z = 2$

6. Without a calculator or computer, match the functions with their graphs in Figure 12.28.

   (a) $z = \dfrac{1}{x^2 + y^2}$ (b) $z = -e^{-x^2 - y^2}$

   (c) $z = x + 2y + 3$ (d) $z = -y^2$

   (e) $z = x^3 - \sin y$.

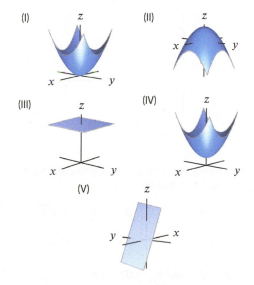

Figure 12.27

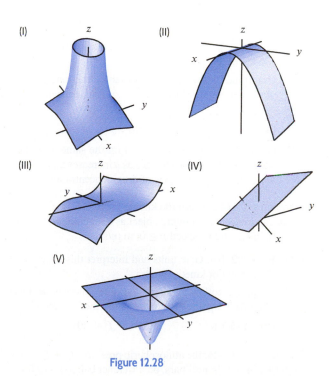

Figure 12.28

7. Figure 12.29 shows the graph of $z = f(x, y)$.

   (a) Suppose y is fixed and positive. Does z increase or decrease as x increases? Graph z against x.
   (b) Suppose x is fixed and positive. Does z increase or decrease as y increases? Graph z against y.

Figure 12.29

■ In Exercises 8–15, sketch a graph of the surface and briefly describe it in words.

8. $z = 3$

9. $x^2 + y^2 + z^2 = 9$

10. $z = x^2 + y^2 + 4$

11. $z = 5 - x^2 - y^2$

12. $z = y^2$

13. $2x + 4y + 3z = 12$

14. $x^2 + y^2 = 4$

15. $x^2 + z^2 = 4$

■ In Exercises 16–18, find the equation of the surface.

16. A cylinder of radius $\sqrt{7}$ with its axis along the y-axis.

17. A sphere of radius 3 centered at $\left(0, \sqrt{7}, 0\right)$.

18. The paraboloid obtained by moving the surface $z = x^2 + y^2$ so that its vertex is at $(1, 3, 5)$, its axis is parallel to the x-axis, and the surface opens towards negative x values.

PROBLEMS

19. Consider the function f given by $f(x, y) = y^3 + xy$. Draw graphs of cross-sections with:

    (a) x fixed at $x = -1$, $x = 0$, and $x = 1$.
    (b) y fixed at $y = -1$, $y = 0$, and $y = 1$.

■ Problems 20–22 concern the concentration, C, in mg/liter, of a drug in the blood as a function of x, the amount, in mg, of the drug given and t, the time in hours since the injection. For $0 \le x \le 4$ and $t \ge 0$, we have $C = f(x, t) = te^{-t(5-x)}$.

20. Find $f(3, 2)$. Give units and interpret in terms of drug concentration.

21. Graph the following single-variable functions and explain their significance in terms of drug concentration.

    (a) $f(4, t)$ (b) $f(x, 1)$

22. Graph $f(a, t)$ for $a = 1, 2, 3, 4$ on the same axes. Describe how the graph changes as a increases and explain what this means in terms of drug concentration.

■ Problems 23–24 concern the kinetic energy, $E = f(m, v) = \frac{1}{2}mv^2$, in joules, of a moving object as a function of its mass $m \ge 0$, in kg, and its speed $v \ge 0$, in m/sec.

23. Find $f(2, 10)$. Give units and interpret this quantity in the context of kinetic energy.

24. Graph the following single-variable functions and explain their significance in terms of kinetic energy.

    (a) $f(6, v)$ (b) $f(m, 20)$

■ In Problems 25–26, the atmospheric pressure, $P = f(y, t) = (950 + 2t)e^{-y/7}$, in millibars, on a weather balloon, is a function of its height $y \ge 0$, in km above sea level after t hours with $0 \le t \le 48$.

25. Find $f(2, 12)$. Give units and interpret this quantity in the context of atmospheric pressure.

26. Graph the following single-variable functions and explain the significance of the shape of the graph in terms of atmospheric pressure.

    (a) $f(3, t)$ (b) $f(y, 24)$

27. Without a calculator or computer, for $z = x^2 + 2xy^2$, determine which of (I)–(II) in Figure 12.30 are cross-sections with x fixed and which are cross-sections with y fixed.

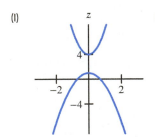 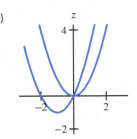

Figure 12.30

28. Without a computer or calculator, match the equations (a)–(i) with the graphs (I)–(IX).

    (a) $z = xye^{-(x^2+y^2)}$

    (b) $z = \cos\left(\sqrt{x^2 + y^2}\right)$

    (c) $z = \sin y$

    (d) $z = -\dfrac{1}{x^2 + y^2}$

    (e) $z = \cos^2 x \cos^2 y$

    (f) $z = \dfrac{\sin(x^2 + y^2)}{x^2 + y^2}$

    (g) $z = \cos(xy)$

    (h) $z = |x||y|$

    (i) $z = (2x^2 + y^2)e^{1-x^2-y^2}$

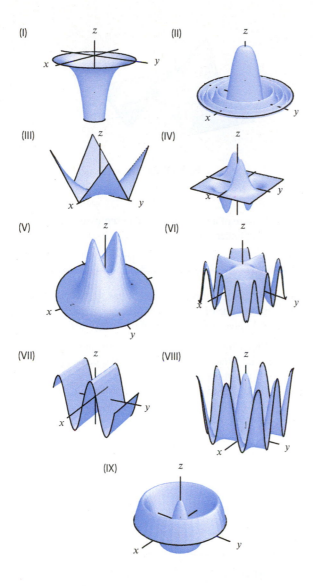

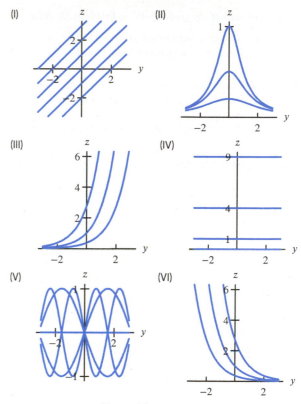

Figure 12.31

32. You like pizza and you like cola. Which of the graphs in Figure 12.32 represents your happiness as a function of how many pizzas and how much cola you have if

(a) There is no such thing as too many pizzas and too much cola?

(b) There is such a thing as too many pizzas or too much cola?

(c) There is such a thing as too much cola but no such thing as too many pizzas?

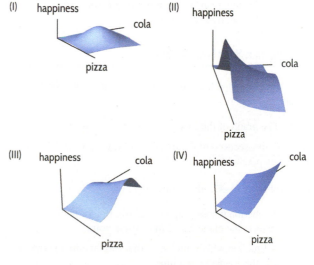

Figure 12.32

29. Decide whether the graph of each of the following equations is the shape of a bowl, a plate, or neither. Consider a plate to be any flat surface and a bowl to be anything that could hold water. Assume the positive z-axis is up.

(a) $z = x^2 + y^2$ (b) $z = 1 - x^2 - y^2$

(c) $x + y + z = 1$ (d) $z = -\sqrt{5 - x^2 - y^2}$

(e) $z = 3$

30. Sketch cross-sections for each function in Problem 29.

31. Without a calculator or computer, match the functions with their cross-sections with x fixed in Figure 12.31.

(a) $z = 1/(1 + x^2 + y^2)$ (b) $z = 1 + x + y$

(c) $z = e^{-x+y}$ (d) $z = e^{x-y}$

(e) $z = \sin(xy)$ (f) $z = x^2$.

33. For each of the graphs I–IV in Problem 32, draw:

    (a) Two cross-sections with pizza fixed

    (b) Two cross-sections with cola fixed.

■ For Problems **34–37**, give a formula for a function whose graph is described. Sketch it using a computer or calculator.

34. A bowl which opens upward and has its vertex at 5 on the z-axis.

35. A plane which has its x-, y-, and z-intercepts all positive.

36. A parabolic cylinder opening upward from along the line $y = x$ in the xy-plane.

37. A cone of circular cross-section opening downward and with its vertex at the origin.

38. Sketch cross-sections of $f(r, h) = \pi r^2 h$, first keeping h fixed, then keeping r fixed.

39. By setting one variable constant, find a plane that intersects the graph of $z = 4x^2 - y^2 + 1$ in a:

    (a) Parabola opening upward
    (b) Parabola opening downward
    (c) Pair of intersecting straight lines.

40. Sketch cross-sections of $z = y - x^2$ with x fixed and with y fixed. Use them to sketch a graph of $z = y - x^2$.

41. A wave travels along a canal. Let x be the distance along the canal, t be the time, and z be the height of the water above the equilibrium level. The graph of z as a function of x and t is in Figure 12.33.

    (a) Draw the profile of the wave for $t = -1, 0, 1, 2$. (Put the x-axis to the right and the z-axis vertical.)
    (b) Is the wave traveling in the direction of increasing or decreasing x?
    (c) Sketch a surface representing a wave traveling in the opposite direction.

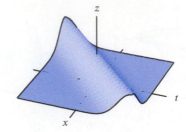

Figure 12.33

42. The pressure of a fixed amount of compressed nitrogen gas in a cylinder is given, in atmospheres, by

$$P = f(T, V) = \frac{10T}{V},$$

where T is the temperature of the gas, in Kelvin, and V is the volume of the cylinder, in liters. Figures 12.34 and 12.35 give cross-sections of the function f.

    (a) Which figure shows cross-sections of f with T fixed? What does the shape of the cross-sections tell you about the pressure?
    (b) Which figure shows cross-sections of f with V fixed? What does the shape of the cross-sections tell you about the pressure?

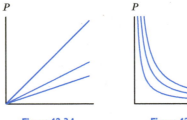

Figure 12.34　　　　　**Figure 12.35**

Strengthen Your Understanding

■ In Problems **43–44**, explain what is wrong with the statement.

43. The graph of the function $f(x, y) = x^2 + y^2$ is a circle.

44. Cross-sections of the function $f(x, y) = x^2$ with x fixed are parabolas.

■ In Problems **45–47**, give an example of:

45. A function whose graph lies above the xy-plane and intersects the plane $z = 2$ in a single point.

46. A function which intersects the xz-plane in a parabola and the yz-plane in a line.

47. A function which intersects the xy-plane in a circle.

■ Are the statements in Problems **48–61** true or false? Give reasons for your answer.

48. The function given by the formula $f(v, w) = e^v / w$ is an increasing function of v when w is a nonzero constant.

49. A function $f(x, y)$ can be an increasing function of x with y held fixed, and be a decreasing function of y with x held fixed.

50. A function $f(x, y)$ can have the property that $g(x) = f(x, 5)$ is increasing, whereas $h(x) = f(x, 10)$ is decreasing.

51. The plane $x + 2y - 3z = 1$ passes through the origin.

52. The plane $x + y + z = 3$ intersects the x-axis when $x = 3$.

53. The sphere $x^2 + y^2 + z^2 = 10$ intersects the plane $x = 10$.

54. The cross-section of the function $f(x, y) = x + y^2$ with $y = 1$ is a line.

55. The function $g(x, y) = 1 - y^2$ has identical parabolas for all cross-sections with x constant.

56. The function $g(x, y) = 1 - y^2$ has lines for all cross-sections with y constant.

57. The graphs of $f(x, y) = \sin(xy)$ and $g(x, y) = \sin(xy) + 2$ never intersect.

58. The graphs of $f(x, y) = x^2 + y^2$ and $g(x, y) = 1 - x^2 - y^2$ intersect in a circle.

59. If all the cross-sections of the graph of $f(x, y)$ with x constant are lines, then the graph of f is a plane.

60. The only point of intersection of the graphs of $f(x, y)$ and $-f(x, y)$ is the origin.

61. The point $(0, 0, 10)$ is the highest point on the graph of the function $f(x, y) = 10 - x^2 - y^2$.

62. The object in 3-space described by $x = 2$ is

(a) A point (b) A line

(c) A plane (d) Undefined.

12.3 CONTOUR DIAGRAMS

The surface which represents a function of two variables often gives a good idea of the function's general behavior—for example, whether it is increasing or decreasing as one of the variables increases. However, it is difficult to read numerical values off a surface and it can be hard to see all of the function's behavior from a surface. Thus, functions of two variables are often represented by contour diagrams like the weather map on page 694. Contour diagrams have the additional advantage that they can be extended to functions of three variables.

Topographical Maps

One of the most common examples of a contour diagram is a topographical map like that shown in Figure 12.36. It gives the elevation in the region and is a good way of getting an overall picture of the terrain: where the mountains are, where the flat areas are. Such topographical maps are frequently colored green at the lower elevations and brown, red, or white at the higher elevations.

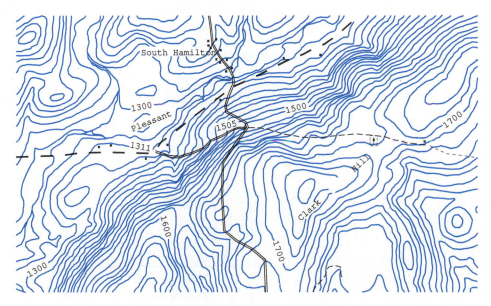

Figure 12.36: A topographical map showing the region around South Hamilton, NY

The curves on a topographical map that separate lower elevations from higher elevations are called *contour lines* because they outline the contour or shape of the land.[5] Because every point

---

[5]In fact they are usually not straight lines, but curves. They may also be in disconnected pieces.

along the same contour has the same elevation, contour lines are also called *level curves* or *level sets*. The more closely spaced the contours, the steeper the terrain; the more widely spaced the contours, the flatter the terrain (provided, of course, that the elevation between contours varies by a constant amount). Certain features have distinctive characteristics. A mountain peak is typically surrounded by contour lines like those in Figure 12.37. A pass in a range of mountains may have contours that look like Figure 12.38. A long valley has parallel contour lines indicating the rising elevations on both sides of the valley (see Figure 12.39); a long ridge of mountains has the same type of contour lines, only the elevations decrease on both sides of the ridge. Notice that the elevation numbers on the contour lines are as important as the curves themselves. We usually draw contours for equally spaced values of z.

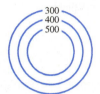

Figure 12.37: Mountain peak

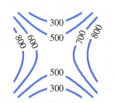

Figure 12.38: Pass between two mountains

Figure 12.39: Long valley

Figure 12.40: Impossible contour lines

Notice that two contours corresponding to different elevations cannot cross each other as shown in Figure 12.40. If they did, the point of intersection of the two curves would have two different elevations, which is impossible (assuming the terrain has no overhangs).

Corn Production

Contour maps can display information about a function of two variables without reference to a surface. Consider the effect of weather conditions on US corn production. Figure 12.41 gives corn production $C = f(R, T)$ as a function of the total rainfall, R, in inches, and average temperature, T, in degrees Fahrenheit, during the growing season.[6] At the present time, $R = 15$ inches and $T = 76°$F. Production is measured as a percentage of the present production; thus, the contour through $R = 15$, $T = 76$, has value 100, that is, $C = f(15, 76) = 100$.

Example 1   Use Figure 12.41 to estimate $f(18, 78)$ and $f(12, 76)$ and interpret in terms of corn production.

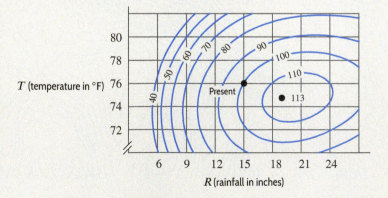

Figure 12.41: Corn production, C, as a function of rainfall and temperature

[6]Adapted from S. Beaty and R. Healy, "The Future of American Agriculture," *Scientific American* 248, No. 2, February 1983.

Solution   The point with R-coordinate 18 and T-coordinate 78 is on the contour $C = 100$, so $f(18, 78) = 100$. This means that if the annual rainfall were 18 inches and the temperature were 78°F, the country would produce about the same amount of corn as at present, although it would be wetter and warmer than it is now.

The point with R-coordinate 12 and T-coordinate 76 is about halfway between the $C = 80$ and the $C = 90$ contours, so $f(12, 76) \approx 85$. This means that if the rainfall fell to 12 inches and the temperature stayed at 76°, then corn production would drop to about 85% of what it is now.

Example 2   Use Figure 12.41 to describe in words the cross-sections with T and R constant through the point representing present conditions. Give a common-sense explanation of your answer.

Solution   To see what happens to corn production if the temperature stays fixed at 76°F but the rainfall changes, look along the horizontal line $T = 76$. Starting from the present and moving left along the line $T = 76$, the values on the contours decrease. In other words, if there is a drought, corn production decreases. Conversely, as rainfall increases, that is, as we move from the present to the right along the line $T = 76$, corn production increases, reaching a maximum of more than 110% when $R = 21$, and then decreases (too much rainfall floods the fields).

If, instead, rainfall remains at the present value and temperature increases, we move up the vertical line $R = 15$. Under these circumstances corn production decreases; a 2°F increase causes a 10% drop in production. This makes sense since hotter temperatures lead to greater evaporation and hence drier conditions, even with rainfall constant at 15 inches. Similarly, a decrease in temperature leads to a very slight increase in production, reaching a maximum of around 102% when $T = 74$, followed by a decrease (the corn won't grow if it is too cold).

Contour Diagrams and Graphs

Contour diagrams and graphs are two different ways of representing a function of two variables. How do we go from one to the other? In the case of the topographical map, the contour diagram was created by joining all the points at the same height on the surface and dropping the curve into the xy-plane.

How do we go the other way? Suppose we wanted to plot the surface representing the corn production function $C = f(R, T)$ given by the contour diagram in Figure 12.41. Along each contour the function has a constant value; if we take each contour and lift it above the plane to a height equal to this value, we get the surface in Figure 12.42.

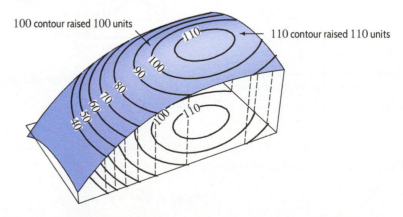

100 contour raised 100 units

110 contour raised 110 units

Figure 12.42: Getting the graph of the corn yield function from the contour diagram

Notice that the raised contours are the curves we get by slicing the surface horizontally. In general, we have the following result:

> Contour lines, or level curves, are obtained from a surface by slicing it with horizontal planes. A contour diagram is a collection of level curves labeled with function values.

Finding Contours Algebraically

Algebraic equations for the contours of a function f are easy to find if we have a formula for $f(x, y)$. Suppose the surface has equation

$$z = f(x, y).$$

A contour is obtained by slicing the surface with a horizontal plane with equation $z = c$. Thus, the equation for the contour at height c is given by:

$$f(x, y) = c.$$

Example 3   Find equations for the contours of $f(x, y) = x^2 + y^2$ and draw a contour diagram for f. Relate the contour diagram to the graph of f.

Solution   The contour at height c is given by

$$f(x, y) = x^2 + y^2 = c.$$

This is a contour only for $c \geq 0$, For $c > 0$ it is a circle of radius $\sqrt{c}$. For $c = 0$, it is a single point (the origin). Thus, the contours at an elevation of $c = 1, 2, 3, 4, \ldots$ are all circles centered at the origin of radius $1, \sqrt{2}, \sqrt{3}, 2, \ldots$. The contour diagram is shown in Figure 12.43. The bowl–shaped graph of f is shown in Figure 12.44. Notice that the graph of f gets steeper as we move further away from the origin. This is reflected in the fact that the contours become more closely packed as we move further from the origin; for example, the contours for $c = 6$ and $c = 8$ are closer together than the contours for $c = 2$ and $c = 4$.

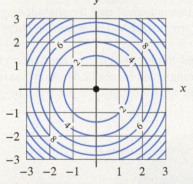

Figure 12.43: Contour diagram for $f(x, y) = x^2 + y^2$ (even values of c only)

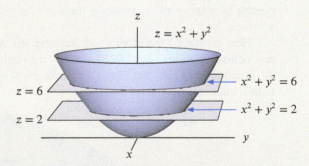

Figure 12.44: The graph of $f(x, y) = x^2 + y^2$

Example 4   Draw a contour diagram for $f(x, y) = \sqrt{x^2 + y^2}$ and relate it to the graph of f.

Solution   The contour at level c is given by

$$f(x, y) = \sqrt{x^2 + y^2} = c.$$

For $c > 0$ this is a circle, just as in the previous example, but here the radius is c instead of $\sqrt{c}$. For $c = 0$, it is the origin. Thus, if the level c increases by 1, the radius of the contour increases by 1. This means the contours are equally spaced concentric circles (see Figure 12.45) which do not become more closely packed further from the origin. Thus, the graph of f has the same constant slope as we move away from the origin (see Figure 12.46), making it a cone rather than a bowl.

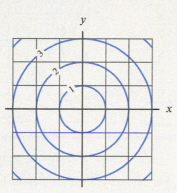

Figure 12.45: A contour diagram for
$f(x, y) = \sqrt{x^2 + y^2}$

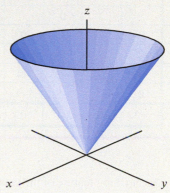

Figure 12.46: The graph of
$f(x, y) = \sqrt{x^2 + y^2}$

In both of the previous examples the level curves are concentric circles because the surfaces have circular symmetry. Any function of two variables which depends only on the quantity $(x^2 + y^2)$ has such symmetry: for example, $G(x, y) = e^{-(x^2+y^2)}$ or $H(x, y) = \sin\left(\sqrt{x^2 + y^2}\right)$.

Example 5   Draw a contour diagram for $f(x, y) = 2x + 3y + 1$.

Solution   The contour at level c has equation $2x + 3y + 1 = c$. Rewriting this as $y = -(2/3)x + (c - 1)/3$, we see that the contours are parallel lines with slope $-2/3$. The y-intercept for the contour at level c is $(c - 1)/3$; each time c increases by 3, the y-intercept moves up by 1. The contour diagram is shown in Figure 12.47.

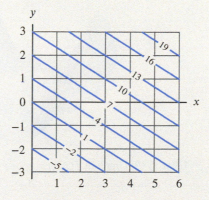

Figure 12.47: A contour diagram for $f(x, y) = 2x + 3y + 1$

Contour Diagrams and Tables

Sometimes we can get an idea of what the contour diagram of a function looks like from its table.

Example 6   Relate the values of $f(x, y) = x^2 - y^2$ in Table 12.5 to its contour diagram in Figure 12.48.

Table 12.5   *Table of values of $f(x, y) = x^2 - y^2$*

	-3	-2	-1	0	1	2	3
3	0	-5	-8	-9	-8	-5	0
2	5	0	-3	-4	-3	0	5
1	8	3	0	-1	0	3	8
y 0	9	4	1	0	1	4	9
-1	8	3	0	-1	0	3	8
-2	5	0	-3	-4	-3	0	5
-3	0	-5	-8	-9	-8	-5	0

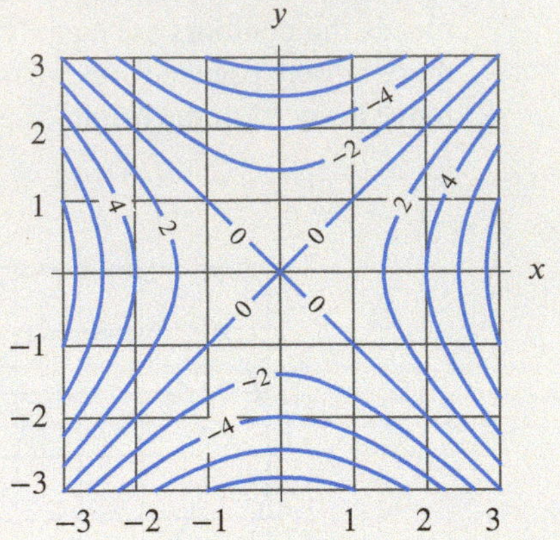

x

Figure 12.48: Contour map of $f(x, y) = x^2 - y^2$

Solution   One striking feature of the values in Table 12.5 is the zeros along the diagonals. This occurs because $x^2 - y^2 = 0$ along the lines $y = x$ and $y = -x$. So the $z = 0$ contour consists of these two lines. In the triangular region of the table that lies to the right of both diagonals, the entries are positive. To the left of both diagonals, the entries are also positive. Thus, in the contour diagram, the positive contours lie in the triangular regions to the right and left of the lines $y = x$ and $y = -x$. Further, the table shows that the numbers on the left are the same as the numbers on the right; thus, each contour has two pieces, one on the left and one on the right. See Figure 12.48. As we move away from the origin along the x-axis, we cross contours corresponding to successively larger values. On the saddle-shaped graph of $f(x, y) = x^2 - y^2$ shown in Figure 12.49, this corresponds to climbing out of the saddle along one of the ridges. Similarly, the negative contours occur in pairs in the top and bottom triangular regions; the values get more and more negative as we go out along the y-axis. This corresponds to descending from the saddle along the valleys that are submerged below the xy-plane in Figure 12.49. Notice that we could also get the contour diagram by graphing the family of hyperbolas $x^2 - y^2 = 0, \pm 2, \pm 4, \ldots$.

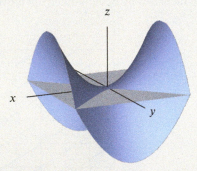

Figure 12.49: Graph of $f(x, y) = x^2 - y^2$ showing plane $z = 0$

Using Contour Diagrams: The Cobb-Douglas Production Function

Suppose you decide to expand your small printing business. Should you start a night shift and hire more workers? Should you buy more expensive but faster computers which will enable the current staff to keep up with the work? Or should you do some combination of the two?

Obviously, the way such a decision is made in practice involves many other considerations—such as whether you could get a suitably trained night shift, or whether there are any faster computers available. Nevertheless, you might model the quantity, P, of work produced by your business as a function of two variables: your total number, N, of workers, and the total value, V, of your equipment. What might the contour diagram of the production function look like?

Example 7 Explain why the contour diagram in Figure 12.50 does not model the behavior expected of the production function, whereas the contour diagram in Figure 12.51 does.

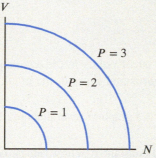

Figure 12.50: Incorrect contours for printing production

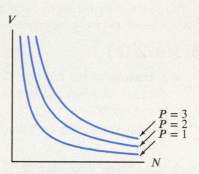

Figure 12.51: Correct contours for printing production

Solution Look at Figure 12.50. Notice that the contour $P = 1$ intersects the N- and the V-axes, suggesting that it is possible to produce work with no workers or with no equipment; this is unreasonable. However, no contours in Figure 12.51 intersect either the N- or the V-axis.

In Figure 12.51, fixing V and letting N increase corresponds to moving to the right, crossing contours less and less frequently. Production increases more and more slowly because hiring additional workers does little to boost production if the machines are already used to capacity.

Similarly, if we fix N and let V increase, Figure 12.51 shows production increasing, but at a decreasing rate. Buying machines without enough people to use them does not increase production much. Thus Figure 12.51 fits the expected behavior of the production function best.

Formula for a Production Function

Production functions are often approximated by formulas of the form

$$P = f(N, V) = cN^\alpha V^\beta$$

where P is the quantity produced and $c, \alpha,$ and β are positive constants, $0 < \alpha < 1$ and $0 < \beta < 1$.

Example 8 Show that the contours of the function $P = cN^\alpha V^\beta$ have approximately the shape of the contours in Figure 12.51.

Solution The contours are the curves where P is equal to a constant value, say P_0, that is, where

$$cN^\alpha V^\beta = P_0.$$

Solving for V, we get

$$V = \left(\frac{P_0}{c}\right)^{1/\beta} N^{-\alpha/\beta}.$$

Thus, V is a power function of N with a negative exponent, so its graph has the general shape shown in Figure 12.51.

The Cobb-Douglas Production Model

In 1928, Cobb and Douglas used a similar function to model the production of the entire US economy in the first quarter of this century. Using government estimates of P, the total yearly production between 1899 and 1922, of K, the total capital investment over the same period, and of L, the total labor force, they found that P was well approximated by the *Cobb-Douglas production function*

$$P = 1.01L^{0.75}K^{0.25}.$$

This function turned out to model the US economy surprisingly well, both for the period on which it was based and for some time afterward.[7]

Summary for Section 12.3

- A **contour** of the function $f(x, y)$ is the set of points in the xy-plane satisfying $f(x, y) =$ constant. Contours can be thought of as horizontal slices of the graph of a function at a particular height.
- A **contour diagram** for a function $f(x, y)$ is a graph of several contours for a selection of constants.
- In a contour diagram with equally-spaced constant function values, contours that are **closer together** represent more rapid change of the function.
- To find a contour **algebraically**, set the formula for $f(x, y)$ equal to a constant.
- Sometimes contours can be seen **numerically** in a table of values by seeing where the same values occur in the table.
- A **Cobb-Douglas production function** has the form

$$f(N, V) = cN^{\alpha}V^{\beta}.$$

Exercises and Problems for Section 12.3

EXERCISES

■ In Exercises 1–4, sketch a possible contour diagram for each surface, marked with reasonable z-values. (Note: There are many possible answers.)

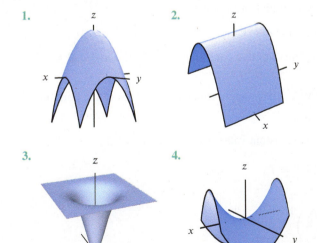

1.

2.

3.

4.

■ In Exercises 5–8, use the contour diagram of $f(x, y)$ given in Figure 12.52.

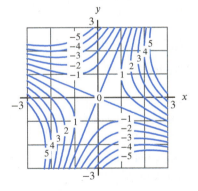

Figure 12.52

5. Find $f(1, 3)$.

6. Find $f(-2, 2)$.

7. Find a point where $f(x, y) = 0$.

8. Find a point where $f(x, y) = 2$.

9. Let $f(x, y) = 3x^2y + 7x + 20$. Find an equation for the contour that goes through the point $(5, 10)$.

10. **(a)** For $z = f(x, y) = xy$, sketch and label the level curves $z = \pm 1$, $z = \pm 2$.
    (b) Sketch and label cross-sections of f with $x = \pm 1$, $x = \pm 2$.
    (c) The surface $z = xy$ is cut by a vertical plane containing the line $y = x$. Sketch the cross-section.

[7]C. Cobb and P. Douglas, "A Theory of Production", *American Economic Review* 18 (1928: Supplement), pp. 139–165.

11. Match the surfaces (a)–(e) in Figure 12.53 with the contour diagrams (I)–(V) in Figure 12.54.

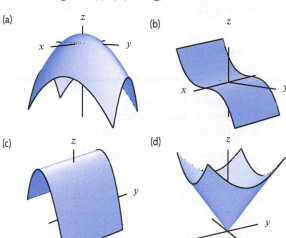

(a)

(b)

(c)

(d)

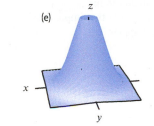

(e)

Figure 12.53

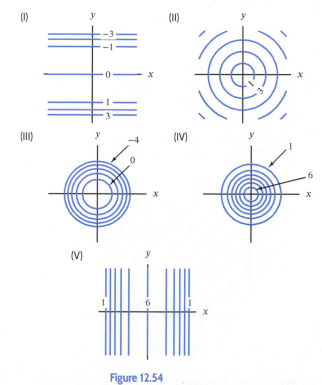

(I)

(II)

(III)

(IV)

(V)

Figure 12.54

12. Figure 12.55 shows the contour diagram of $z = f(x, y)$. Which of the points (I)–(VI) lie on the graph of $z = f(x, y)$?

I. $(1, 0, 2)$ II. $(1, 1, 1)$
III. $(0, -1, -2)$ IV. $(-1, 0, -2)$
V. $(0, 1, 1)$ VI. $(-1, -1, 0)$

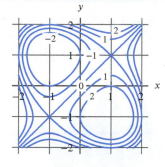

Figure 12.55

13. Match Tables 12.6–12.9 with contour diagrams (I)–(IV) in Figure 12.56.

Table 12.6

$y \backslash x$	-1	0	1
-1	2	1	2
0	1	0	1
1	2	1	2

Table 12.7

$y \backslash x$	-1	0	1
-1	0	1	0
0	1	2	1
1	0	1	0

Table 12.8

$y \backslash x$	-1	0	1
-1	2	0	2
0	2	0	2
1	2	0	2

Table 12.9

$y \backslash x$	-1	0	1
-1	2	2	2
0	0	0	0
1	2	2	2

(I)

(II)

(III)

(IV)

Figure 12.56

■ In Exercises 14–22, sketch a contour diagram for the function with at least four labeled contours. Describe in words the contours and how they are spaced.

14. $f(x, y) = x + y$

15. $f(x, y) = 3x + 3y$

16. $f(x, y) = x^2 + y^2$

17. $f(x, y) = -x^2 - y^2 + 1$

18. $f(x, y) = xy$

19. $f(x, y) = y - x^2$

20. $f(x, y) = x^2 + 2y^2$

21. $f(x, y) = \sqrt{x^2 + 2y^2}$

22. $f(x, y) = \cos \sqrt{x^2 + y^2}$

PROBLEMS

23. Figure 12.57 shows a graph of $f(x, y) = (\sin x)(\cos y)$ for $-2\pi \le x \le 2\pi$, $-2\pi \le y \le 2\pi$. Use the surface $z = 1/2$ to sketch the contour $f(x, y) = 1/2$.

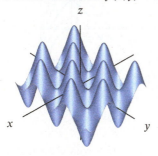

Figure 12.57

24. Total sales, Q, of a product are a function of its price and the amount spent on advertising. Figure 12.58 shows a contour diagram for total sales. Which axis corresponds to the price of the product and which to the amount spent on advertising? Explain.

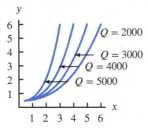

Figure 12.58

25. Each contour diagram (a)–(c) in Figure 12.59 shows satisfaction with quantities of two items X and Y combined. Match (a)–(c) with the items in (I)–(III).

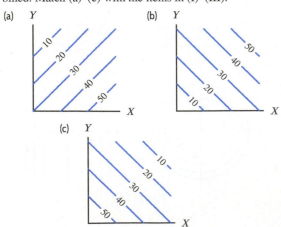

Figure 12.59

(I) X: Income; Y: Leisure time

(II) X: Income; Y: Hours worked

(III) X: Hours worked; Y: Time spent commuting

26. Figure 12.60 shows a contour plot of job satisfaction as a function of the hourly wage and the safety of the workplace (higher values mean safer). Match the jobs at points P, Q, and R with the three descriptions.

(a) The job is so unsafe that higher pay alone would not increase my satisfaction very much.

(b) I could trade a little less safety for a little more pay. It would not matter to me.

(c) The job pays so little that improving safety would not make me happier.

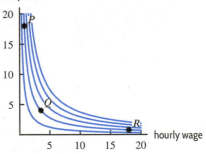

Figure 12.60

27. Figure 12.61 shows a contour diagram of Dan's happiness with snacks of different numbers of cherries and grapes.

(a) What is the slope of the contours?

(b) What does the slope tell you?

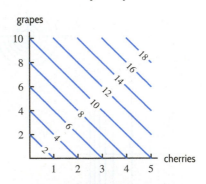

Figure 12.61

28. Figure 12.62 shows contours of $f(x, y) = 100e^x - 50y^2$. Find the values of f on the contours. They are equally spaced multiples of 10.

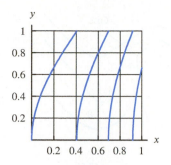

Figure 12.62

29. Figure 12.63 shows contours for a person's body mass index, BMI $= f(w, h) = 703w/h^2$, where w is weight in pounds and h is height in inches. Find the BMI contour values bounding the *underweight* and *normal* regions.

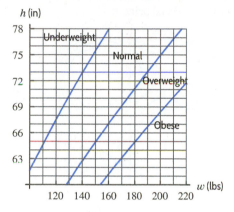

Figure 12.63

30. The wind chill tells you how cold it feels as a function of the air temperature and wind speed. Figure 12.64 is a contour diagram of wind chill (°F).

(a) If the wind speed is 15 mph, what temperature feels like −20°F?

(b) Estimate the wind chill if the temperature is 0°F and the wind speed is 10 mph.

(c) Humans are at extreme risk when the wind chill is below −50°F. If the temperature is −20°F, estimate the wind speed at which extreme risk begins.

(d) If the wind speed is 15 mph and the temperature drops by 20°F, approximately how much colder do you feel?

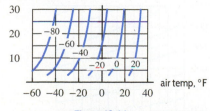

Figure 12.64

31. Match the functions (a)–(f) with the level curves (I)–(VI):

(a) $f(x, y) = x^2 - y^2 - 2x + 4y - 3$

(b) $g(x, y) = x^2 + y^2 - 2x - 4y + 15$

(c) $h(x, y) = -x^2 - y^2 + 2x + 4y - 8$

(d) $j(x, y) = -x^2 + y^2 + 2x - 4y + 3$

(e) $k(x, y) = \sqrt{(x - 1)^2 + (y - 2)^2}$

(f) $l(x, y) = -\sqrt{(x - 1)^2 + (y - 2)^2}$

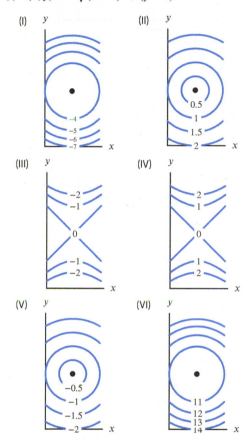

32. Figure 12.65 shows contour diagrams of $f(x, y)$ and $g(x, y)$. Sketch the smooth curve with equation $f(x, y) = g(x, y)$.

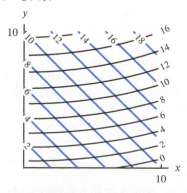

Figure 12.65: Black: $f(x, y)$. Blue: $g(x, y)$.

33. Figure 12.66 shows the level curves of the temperature H in a room near a recently opened window. Label the three level curves with reasonable values of H if the house is in the following locations.

　(a) Minnesota in winter (where winters are harsh).
　(b) San Francisco in winter (where winters are mild).
　(c) Houston in summer (where summers are hot).
　(d) Oregon in summer (where summers are mild).

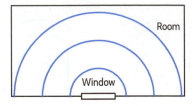

Figure 12.66

34. You are in a room 30 feet long with a heater at one end. In the morning the room is 65°F. You turn on the heater, which quickly warms up to 85°F. Let $H(x,t)$ be the temperature x feet from the heater, t minutes after the heater is turned on. Figure 12.67 shows the contour diagram for H. How warm is it 10 feet from the heater 5 minutes after it was turned on? 10 minutes after it was turned on?

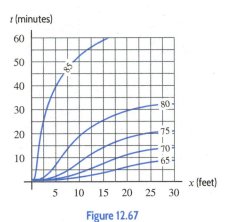

Figure 12.67

35. Using the contour diagram in Figure 12.67, sketch the graphs of the one-variable functions $H(x,5)$ and $H(x,20)$. Interpret the two graphs in practical terms, and explain the difference between them.

36. Figure 12.68 shows a contour map of a hill with two paths, A and B.

　(a) On which path, A or B, will you have to climb more steeply?
　(b) On which path, A or B, will you probably have a wider view of the horizon? (Assume trees do not block your view.)
　(c) Alongside which path is there more likely to be a stream?

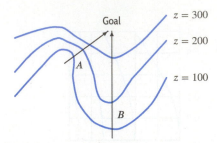

Figure 12.68

■ In Problems **37–40**, for the two given points:
　(a) Find the distance h from the first to the second point.
　(b) Use Figure 12.69, the contour diagram of $f(x,y)$, to find Δf, the difference between the values of f from the first to the second point.
　(c) Find $\Delta f/h$, the average rate of change of $f(x,y)$ from the first to the second point.

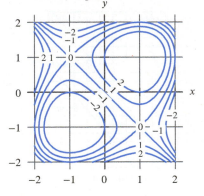

Figure 12.69

37. $(0,0)$ and $(2,0)$　　　　　**38.** $(0,-1)$ and $(1,0)$
39. $(-1,1)$ and $(1,2)$　　　　**40.** $(0,-2)$ and $(0,2)$

41. Figure 12.70 is a contour diagram of the monthly payment on a 5-year car loan as a function of the interest rate and the amount you borrow. The interest rate is 13% and you borrow \$6000 for a used car.

　(a) What is your monthly payment?
　(b) If interest rates drop to 11%, how much more can you borrow without increasing your monthly payment?
　(c) Make a table of how much you can borrow without increasing your monthly payment, as a function of the interest rate.

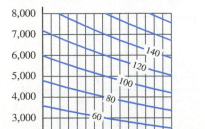

Figure 12.70

42. Hiking on a level trail going due east, you decide to leave the trail and climb toward the mountain on your left. The farther you go along the trail before turning off, the gentler the climb. Sketch a possible topographical map showing the elevation contours.

43. The total productivity $f(n, T)$ of an advertising agency (in ads per day) depends on the number n of workers and the temperature T of the office in degrees Fahrenheit. More workers create more ads, but the farther the temperature from 75°F, the slower they work. Draw a possible contour diagram for the function $f(n, T)$.

44. Match the functions (a)–(d) with the shapes of their level curves (I)–(IV). Sketch each contour diagram.

(a) $f(x, y) = x^2$ (b) $f(x, y) = x^2 + 2y^2$

(c) $f(x, y) = y - x^2$ (d) $f(x, y) = x^2 - y^2$

I.   Lines   II.   Parabolas

III.   Hyperbolas   IV.   Ellipses

45. Match the functions (a)–(d) with the shapes of their typical level curves (I)–(IV).

(a) $f(x, y) = \dfrac{y}{x^2 + 1}$ (b) $f(x, y) = \dfrac{1}{x^2 + 2y^2}$

(c) $f(x, y) = \dfrac{x^2 + 1}{y^2 + 1}$ (d) $f(x, y) = \dfrac{x}{x^2 + y^2 + 1}$

I.   Circles   II.   Parabolas

III.   Hyperbolas   IV.   Ellipses

46. Figure 12.71 shows the density of the fox population P (in foxes per square kilometer) for southern England.[8] Draw two different cross-sections along a north-south line and two different cross-sections along an east-west line of the population density P.

47. A manufacturer sells two goods, one at a price of \$3000 a unit and the other at a price of \$12,000 a unit. A quantity q_1 of the first good and q_2 of the second good are sold at a total cost of \$4000 to the manufacturer.

(a) Express the manufacturer's profit, π, as a function of q_1 and q_2.

(b) Sketch curves of constant profit in the $q_1 q_2$-plane for $\pi = 10,000$, $\pi = 20,000$, and $\pi = 30,000$ and the break-even curve $\pi = 0$.

48. A shopper buys x units of item A and y units of item B, obtaining satisfaction $s(x, y)$ from the purchase. (Satisfaction is called *utility* by economists.) The contours $s(x, y) = xy = c$ are called *indifference curves* because they show pairs of purchases that give the shopper the same satisfaction.

(a) A shopper buys 8 units of A and 2 units of B. What is the equation of the indifference curve showing the other purchases that give the shopper the same satisfaction? Sketch this curve.

(b) After buying 4 units of item A, how many units of B must the shopper buy to obtain the same satisfaction as obtained from buying 8 units of A and 2 units of B?

(c) The shopper reduces the purchase of item A by k, a fixed number of units, while increasing the purchase of B to maintain satisfaction. In which of the following cases is the increase in B largest?

- Initial purchase of A is 6 units
- Initial purchase of A is 8 units

49. Match each Cobb-Douglas production function (a)–(c) with a graph in Figure 12.72 and a statement (D)–(G).

(a) $F(L, K) = L^{0.25} K^{0.25}$

(b) $F(L, K) = L^{0.5} K^{0.5}$

(c) $F(L, K) = L^{0.75} K^{0.75}$

(D) Tripling each input triples output.

(E) Quadrupling each input doubles output.

(G) Doubling each input almost triples output.

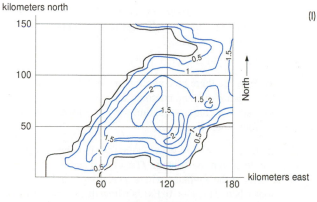

Figure 12.71

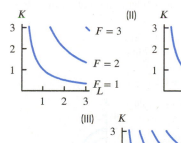
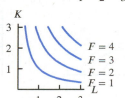

Figure 12.72

---

[8]From J. D. Murray et al, "On the Spatial Spread of Rabies Among Foxes", *Proc. R. Soc. Lond. B*, 229: 111–150, 1986.

50. A Cobb-Douglas production function has the form

$$P = cL^\alpha K^\beta \quad \text{with } \alpha, \beta > 0.$$

What happens to production if labor and capital are both scaled up? For example, does production double if both labor and capital are doubled? Economists talk about

- *increasing returns to scale* if doubling L and K more than doubles P,
- *constant returns to scale* if doubling L and K exactly doubles P,
- *decreasing returns to scale* if doubling L and K less than doubles P.

What conditions on α and β lead to increasing, constant, or decreasing returns to scale?

51. (a) Match $f(x, y) = x^{0.2}y^{0.8}$ and $g(x, y) = x^{0.8}y^{0.2}$ with the level curves in Figures (I) and (II). All scales on the axes are the same.

(b) Figure (III) shows the level curves of $h(x, y) = x^\alpha y^{1-\alpha}$ for $0 < \alpha < 1$. Find the range of possible values for α. Again, the scales are the same on both axes.

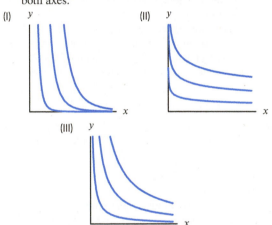

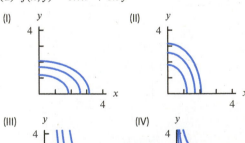

52. Match the functions (a)–(d) with the contour diagrams in Figures I–IV.

(a) $f(x, y) = 0.7 \ln x + 0.3 \ln y$
(b) $g(x, y) = 0.3 \ln x + 0.7 \ln y$
(c) $h(x, y) = 0.3x^2 + 0.7y^2$
(d) $j(x, y) = 0.7x^2 + 0.3y^2$

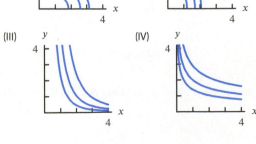

53. Figure 12.73 is the contour diagram of $f(x, y)$. Sketch the contour diagram of each of the following functions.

(a) $3f(x, y)$ **(b)** $f(x, y) - 10$
(c) $f(x - 2, y - 2)$ **(d)** $f(-x, y)$

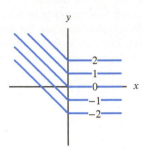

Figure 12.73

54. Figure 12.74 shows part of the contour diagram of $f(x, y)$. Complete the diagram for $x < 0$ if

(a) $f(-x, y) = f(x, y)$ **(b)** $f(-x, y) = -f(x, y)$

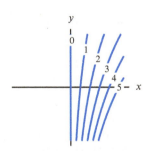

Figure 12.74

55. The contour at level 0 of $f(x, y) = (x+2y)^2 - (3x-4y)^2$ consists of two intersecting lines in the xy-plane. Find equations for the lines.

56. Let $z = f(x, y) = x^2/(x^2 + y^2)$.

(a) Why are there no contours for $z < 0$?
(b) Why are there no contours for $z > 1$?
(c) Sketch a contour diagram for $f(x, y)$ with at least four labeled contours.

57. Let $f(x, y) = x^2 - y^2 = (x - y)(x + y)$. Use the factored form to sketch the contour $f(x, y) = 0$ and to find the regions in the xy-plane where $f(x, y) > 0$ and the regions where $f(x, y) < 0$. Explain how this sketch shows that the graph of $f(x, y)$ is saddle-shaped at the origin.

58. Use Problem 57 to find a formula for a "monkey saddle" surface $z = g(x, y)$ which has three regions with $g(x, y) > 0$ and three with $g(x, y) < 0$.

59. The power P produced by a windmill is proportional to the square of the diameter d of the windmill and to the cube of the speed v of the wind.[9]

(a) Write a formula for P as a function of d and v.
(b) A windmill generates 100 kW of power at a certain

wind speed. If a second windmill is built having twice the diameter of the original, what fraction of the original wind speed is needed by the second windmill to produce 100 kW?

(c) Sketch a contour diagram for P.

Strengthen Your Understanding

In Problems 60–61, explain what is wrong with the statement.

60. A contour diagram for $z = f(x, y)$ is a surface in xyz-space.

61. The functions $f(x, y) = \sqrt{x^2 + y^2}$ and $g(x, y) = x^2 + y^2$ have the same contour diagram.

In Problems 62–63, give an example of:

62. A function $f(x, y)$ whose $z = 10$ contour consists of two or more parallel lines.

63. A function whose contours are all parabolas.

Decide if the statements in Problems 64–68 must be true, might be true, or could not be true. The function $z = f(x, y)$ is defined everywhere.

64. The level curves corresponding to $z = 1$ and $z = -1$ cross at the origin.

65. The level curve $z = 1$ consists of the circle $x^2 + y^2 = 2$ and the circle $x^2 + y^2 = 3$, but no other points.

66. The level curve $z = 1$ consists of two lines which intersect at the origin.

67. If $z = e^{-(x^2+y^2)}$, there is a level curve for every value of z.

68. If $z = e^{-(x^2+y^2)}$, there is a level curve through every point (x, y).

Are the statements in Problems 69–76 true or false? Give reasons for your answer.

69. Two isotherms representing distinct temperatures on a weather map cannot intersect.

70. A weather map can have two isotherms representing the same temperature that do not intersect.

71. The contours of the function $f(x, y) = y^2 + (x-2)^2$ are either circles or a single point.

72. If the contours of $g(x, y)$ are concentric circles, then the graph of g is a cone.

73. If the contours for $f(x, y)$ get closer together in a certain direction, then f is increasing in that direction.

74. If all of the contours of $f(x, y)$ are parallel lines, then the graph of f is a plane.

75. If the $f = 10$ contour of the function $f(x, y)$ is identical to the $g = 10$ contour of the function $g(x, y)$, then $f(x, y) = g(x, y)$ for all (x, y).

76. The $f = 5$ contour of the function $f(x, y)$ is identical to the $g = 0$ contour of the function $g(x, y) = f(x, y) - 5$.

12.4 LINEAR FUNCTIONS

What Is a Linear Function of Two Variables?

Linear functions played a central role in one-variable calculus because many one-variable functions have graphs that look like a line when we zoom in. In two-variable calculus, a *linear function* is one whose graph is a plane. In Chapter 14, we see that many two-variable functions have graphs which look like planes when we zoom in.

What Makes a Plane Flat?

What makes the graph of the function $z = f(x, y)$ a plane? Linear functions of *one* variable have straight line graphs because they have constant slope. On a plane, the situation is a bit more complicated. If we walk around on a tilted plane, the slope is not always the same: it depends on the direction in which we walk. However, at every point on the plane, the slope is the same as long as we choose the same direction. If we walk parallel to the x-axis, we always find ourselves walking up or down with the same slope;[10] the same is true if we walk parallel to the y-axis. In other words, the slope ratios $\Delta z/\Delta x$ (with y fixed) and $\Delta z/\Delta y$ (with x fixed) are each constant.

---

[9]From www.ecolo.org/documents/documents_in_english/WindmillFormula.htm, accessed January 1, 2020.
[10]To be precise, walking in a vertical plane parallel to the x-axis while rising or falling with the plane you are on.

Example 1   A plane cuts the z-axis at $z = 5$ and has slope 2 in the x-direction and slope -1 in the y-direction. What is the equation of the plane?

Solution   Finding the equation of the plane means constructing a formula for the z-coordinate of the point on the plane directly above the point (x, y) in the xy-plane. To get to that point, start from the point above the origin, where $z = 5$. Then walk x units in the x-direction. Since the slope in the x-direction is 2, the height increases by $2x$. Then walk y units in the y-direction; since the slope in the y-direction is -1, the height decreases by y units. Since the height has changed by $2x - y$ units, the z-coordinate is $5 + 2x - y$. Thus, the equation for the plane is

$$z = 5 + 2x - y.$$

For any linear function, if we know its value at a point (x_0, y_0), its slope in the x-direction, and its slope in the y-direction, then we can write the equation of the function. This is just like the equation of a line in the one-variable case, except that there are two slopes instead of one.

> If a **plane** has slope m in the x-direction, has slope n in the y-direction, and passes through the point (x_0, y_0, z_0), then its equation is
>
> $$z = z_0 + m(x - x_0) + n(y - y_0).$$
>
> This plane is the graph of the **linear function**
>
> $$f(x, y) = z_0 + m(x - x_0) + n(y - y_0).$$
>
> If we write $c = z_0 - mx_0 - ny_0$, then we can write $f(x, y)$ in the equivalent form
>
> $$f(x, y) = c + mx + ny.$$

Just as in 2-space a line is determined by two points, so in 3-space a plane is determined by three points, provided they do not lie on a line.

Example 2   Find the equation of the plane passing through the points $(1, 0, 1)$, $(1, -1, 3)$, and $(3, 0, -1)$.

Solution   The first two points have the same x-coordinate, so we use them to find the slope of the plane in the y-direction. As the y-coordinate changes from 0 to -1, the z-coordinate changes from 1 to 3, so the slope in the y-direction is $n = \Delta z / \Delta y = (3 - 1)/(-1 - 0) = -2$. The first and third points have the same y-coordinate, so we use them to find the slope in the x-direction; it is $m = \Delta z / \Delta x = (-1 - 1)/(3 - 1) = -1$. Because the plane passes through $(1, 0, 1)$, its equation is

$$z = 1 - (x - 1) - 2(y - 0) \quad \text{or} \quad z = 2 - x - 2y.$$

You should check that this equation is also satisfied by the points $(1, -1, 3)$ and $(3, 0, -1)$.

Example 2 was made easier by the fact that two of the points had the same x-coordinate and two had the same y-coordinate. An alternative method, which works for any three points, is to substitute the x, y, and z-values of each of the three points into the equation $z = c + mx + ny$. The resulting three equations in c, m, n are then solved simultaneously.

Linear Functions from a Numerical Point of View

To avoid flying planes with empty seats, airlines sell some tickets at full price and some at a discount. Table 12.10 shows an airline's revenue in dollars from tickets sold on a particular route, as a function of the number of full-price tickets sold, f, and the number of discount tickets sold, d.

In every column, the revenue jumps by \$40,000 for each extra 200 discount tickets. Thus, each column is a linear function of the number of discount tickets sold. In addition, every column has the same slope, $40,000/200 = 200$ dollars/ticket. This is the price of a discount ticket. Similarly, each row is a linear function and all the rows have the same slope, 450, which is the price in dollars of a full-fare ticket. Thus, R is a linear function of f and d, given by:

$$R = 450f + 200d.$$

We have the following general result:

A **linear function** can be recognized from its table by the following features:
- Each row and each column is linear.
- All the rows have the same slope.
- All the columns have the same slope (although the slope of the rows and the slope of the columns are generally different).

Example 3   The table contains values of a linear function. Fill in the blank and give a formula for the function.

$x\backslash y$	1.5	2.0
2	0.5	1.5
3	−0.5	?

Solution   In the first column the function decreases by 1 (from 0.5 to −0.5) as x goes from 2 to 3. Since the function is linear, it must decrease by the same amount in the second column. So the missing entry must be $1.5 - 1 = 0.5$. The slope of the function in the x-direction is -1. The slope in the y-direction is 2, since in each row the function increases by 1 when y increases by 0.5. From the table we get $f(2, 1.5) = 0.5$. Therefore, the formula is

$$f(x, y) = 0.5 - (x - 2) + 2(y - 1.5) = -0.5 - x + 2y.$$

Table 12.10   *Revenue from ticket sales (dollars)*

		Full-price tickets (f)			
		100	200	300	400
	200	85,000	130,000	175,000	220,000
	400	125,000	170,000	215,000	260,000
Discount tickets (d)	600	165,000	210,000	255,000	300,000
	800	205,000	250,000	295,000	340,000
	1000	245,000	290,000	335,000	380,000

What Does the Contour Diagram of a Linear Function Look Like?

The formula for the airline revenue function in Table 12.10 is $R = 450f + 200d$, where f is the number of full fares and d is the number of discount fares sold.

Notice that the contours of this function in Figure 12.75 are parallel straight lines. What is the practical significance of the slope of these contour lines? Consider the contour $R = 100,000$; that means we are looking at combinations of ticket sales that yield \$100,000 in revenue. If we move down and to the right on the contour, the f-coordinate increases and the d-coordinate decreases, so we sell more full fares and fewer discount fares. This is because to receive a fixed revenue of $100,000, we must sell more full fares if we sell fewer discount fares. The exact trade-off depends on the slope of the contour; the diagram shows that each contour has a slope of about -2. This means that for a fixed revenue, we must sell two discount fares to replace one full fare. This can also be seen by comparing prices. Each full fare brings in $450; to earn the same amount in discount fares we need to sell $450/200 = 2.25 \approx 2$ fares. Since the price ratio is independent of how many of each type of fare we sell, this slope remains constant over the whole contour map; thus, the contours are all parallel straight lines.

Notice also that the contours are evenly spaced. Thus, no matter which contour we are on, a fixed increase in one of the variables causes the same increase in the value of the function. In terms of revenue, no matter how many fares we have sold, an extra fare, whether full or discount, brings the same revenue as before.

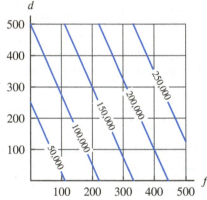

Figure 12.75: Revenue as a function of full and discount fares, $R = 450f + 200d$

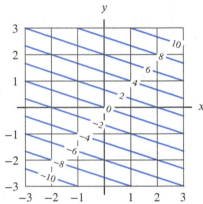

Figure 12.76: Contour map of linear function $f(x, y)$

Example 4   Find the equation of the linear function whose contour diagram is in Figure 12.76.

Solution   Suppose we start at the origin on the $z = 0$ contour. Moving 2 units in the y-direction takes us to the $z = 6$ contour, so the slope in the y-direction is $\Delta z/\Delta y = 6/2 = 3$. Similarly, a move of 2 units in the x-direction from the origin takes us to the $z = 2$ contour, so the slope in the x-direction is $\Delta z/\Delta x = 2/2 = 1$. Since $f(0, 0) = 0$, we have $f(x, y) = x + 3y$.

Summary for Section 12.4

- A **linear function** with slope m in the x-direction, slope y in the y-direction, and z-intercept c has formula
$$f(x, y) = c + mx + ny.$$
- A linear function can be recognized from a **table** by having a constant x-slope with y held constant, and a (possibly different) constant y-slope with x held constant.
- **Contours** of linear functions are parallel lines, evenly spaced.

Exercises and Problems for Section 12.4   Online Resource: Additional Problems for Section 12.4
EXERCISES

■ Exercises 1–2 each contain a partial table of values for a linear function. Fill in the blanks.

■ In Exercises 3–6, could the tables of values represent a linear function?

1.

$x\backslash y$	0.0	1.0
0.0		1.0
2.0	3.0	5.0

2.

$x\backslash y$	−1.0	0.0	1.0
2.0	4.0		
3.0		3.0	5.0

3.

		y	
	0	1	2
0	0	1	4
x 1	1	0	1
2	4	1	0

4.

		y	
	0	1	2
0	10	13	16
x 1	6	9	12
2	2	5	8

5.

	y		
	0	1	2
0	0	5	10
x 1	2	7	12
2	4	9	14

6.

	y		
	0	1	2
0	5	7	9
x 1	6	9	12
2	7	11	15

7. Find the equation of the linear function $z = c + mx + ny$ whose graph contains the points $(0, 0, 0)$, $(0, 2, -1)$, and $(-3, 0, -4)$.

8. Find the linear function whose graph is the plane through the points $(4, 0, 0)$, $(0, 3, 0)$ and $(0, 0, 2)$.

9. Find an equation for the plane containing the line in the xy-plane where $y = 1$, and the line in the xz-plane where $z = 2$.

10. Find the equation of the linear function $z = c + mx + ny$ whose graph intersects the xz-plane in the line $z = 3x + 4$ and intersects the yz-plane in the line $z = y + 4$.

11. Suppose that z is a linear function of x and y with slope 2 in the x-direction and slope 3 in the y-direction.

   (a) A change of 0.5 in x and -0.2 in y produces what change in z?

   (b) If $z = 2$ when $x = 5$ and $y = 7$, what is the value of z when $x = 4.9$ and $y = 7.2$?

12. (a) Find a formula for the linear function whose graph is a plane passing through point $(4, 3, -2)$ with slope 5 in the x-direction and slope -3 in the y-direction.

(b) Sketch the contour diagram for this function.

◼ In Exercises **13–14**, could the contour diagram represent a linear function?

13.

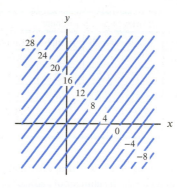

14.

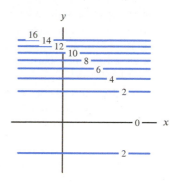

PROBLEMS

15. An internet video streaming company offers a basic and premium monthly streaming subscription package. Figure 12.77 shows the revenue (in dollars per month) of the company as a function of the number, c, of basic subscribers and the number, d, of premium subscribers it has. What is the price of a basic subscription? What is the price of a premium subscription?

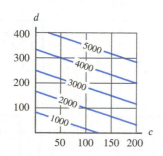

Figure 12.77

16. The charge, C, in dollars, for access to a company's 4G LTE network is a function of m, the number of months

of use, and t, the total number of gigabytes used:

$$C = f(m, t) = 99 + 30m + 10t.$$

   (a) Is f a linear function?

   (b) Give units for the coefficients of m and t, and interpret them as charges.

   (c) Interpret the intercept 99 as a charge.

   (d) Find $f(3, 8)$ and interpret your answer.

17. A manufacturer makes two products out of two raw materials. Let q_1, q_2 be the quantities sold of the two products, p_1, p_2 their prices, and m_1, m_2 the quantities purchased of the two raw materials. Which of the following functions do you expect to be linear, and why? In each case, assume that all variables except the ones mentioned are held fixed.

   (a) Expenditure on raw materials as a function of m_1 and m_2.

   (b) Revenue as a function of q_1 and q_2.

   (c) Revenue as a function of p_1 and q_1.

■Problems **18–20** concern Table 12.11, which gives the number of calories burned per minute for someone roller-blading, as a function of the person's weight and speed.[11]

Table 12.11

Calories burned per minute				
Weight	8 mph	9 mph	10 mph	11 mph
120 lbs	4.2	5.8	7.4	8.9
140 lbs	5.1	6.7	8.3	9.9
160 lbs	6.1	7.7	9.2	10.8
180 lbs	7.0	8.6	10.2	11.7
200 lbs	7.9	9.5	11.1	12.6

18. Does the data in Table 12.11 look approximately linear? Give a formula for B, the number of calories burned per minute in terms of the weight, w, and the speed, s. Does the formula make sense for all weights or speeds?

19. Who burns more total calories to go 10 miles: A 120-lb person going 10 mph or a 180-lb person going 8 mph? Which of these two people burns more calories per pound for the 10-mile trip?

20. Use Problem 18 to give a formula for P, the number of calories burned per pound, in terms of w and s, for a person weighing w lbs roller-blading 10 miles at s mph.

■For Problems **21–22**, find a possible equation for a linear function with the given contour diagram.

21.

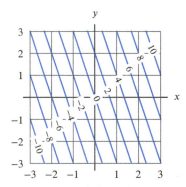

22.

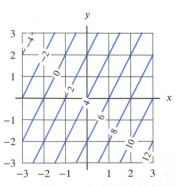

■In Problems **23–24**, could the contour diagram represent a linear function? If so, find an equation for that function.

23.

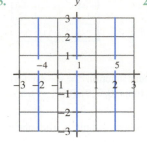

24.

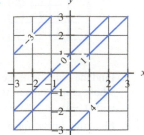

■For Problems **25–26**, find an equation for the linear function with the given values.

25.

$x\backslash y$	−1	0	1	2
0	1.5	1	0.5	0
1	3.5	3	2.5	2
2	5.5	5	4.5	4
3	7.5	7	6.5	6

26.

$x\backslash y$	10	20	30	40
100	3	6	9	12
200	2	5	8	11
300	1	4	7	10
400	0	3	6	9

■In Problems **27–34**, could the table of values represent a linear function? If so, find a possible formula for the function. If not, give a reason why not.

27.

		y	
	1	2	3
x 1	1	5	9
2	2	6	10
3	3	7	11

28.

		y	
	0	1	2
x 0	1	2	1
1	2	3	2
2	3	4	3

29.

		y	
	-2	0	2
x -2	2	2	2
0	5	5	5
2	8	8	8

30.

		y	
	2	4	6
x 1	0	3	6
3	1	4	7
5	4	7	10

31.

		y	
	0	1	2
x 0	-5	-7	-9
2	-2	-4	-6
4	1	-1	-3

32.

		y	
	1	2	3
x 1	1	2	3
2	4	5	6
4	7	8	9

[11]From the August 28, 1994, issue of *Parade Magazine*.

33.

	0	2	5
0	3	5	8
1	5	7	10
3	9	11	14

y (across top), *x* (down side)

34.

	0	2	5
0	0	4	10
1	1	5	11
2	4	6	14

y (across top), *x* (down side)

In Problems 35–38, use the contours of the linear function $z = f(x, y)$ in Figure 12.78 to create possible contour labels for the linear function $z = g(x, y)$ satisfying the given condition.

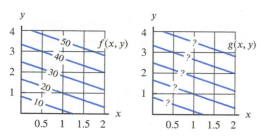

Figure 12.78

35. The graph of g is parallel but different from the graph of f.

36. The graph of g is parallel to the graph of f and passes through the point $(2, 2, 0)$.

37. The graph of g has the same contour as f for the value $z = 30$ but is different from the graph of f.

38. The graph of g has the same contour as f for the value $z = 40$, and a negative slope in the x-direction.

In Problems 39–42, graph the linear function by plotting the x, y, and z-intercepts and joining them by a triangle as in Figure 12.79. This shows the part of the plane in the octant where $x \geq 0$, $y \geq 0$, $z \geq 0$. If the intercepts are not all positive, the same method works if the x, y, and z-axes are drawn from a different perspective.

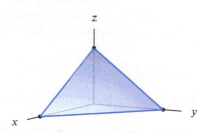

Figure 12.79

39. $z = 2 - 2x + y$

40. $z = 2 - x - 2y$

41. $z = 4 + x - 2y$

42. $z = 6 - 2x - 3y$

43. Figure 12.80 is the contour diagram of a linear function $f(x, y) = mx + 4y + c$. What is the value of m?

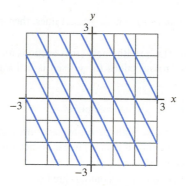

Figure 12.80

44. For the contour diagrams (I)–(IV) on $-2 \leq x \leq 2$, $-2 \leq y \leq 2$, pick the corresponding function.

$f(x, y) = 2x + 3y + 10$ $k(x, y) = -2x + 3y + 12$

$g(x, y) = 2x + 3y + 60$ $m(x, y) = -2x + 3y + 60$

$h(x, y) = 2x - 3y + 12$ $n(x, y) = -2x - 3y + 14$

$j(x, y) = 2x - 3y + 60$ $p(x, y) = -2x - 3y + 60$

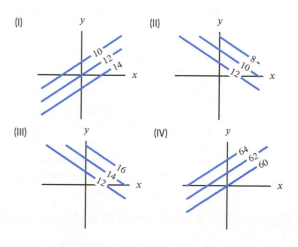

45. A linear function has the formula $f(x, y) = a + 10x - 5y$, but you don't know the value of a. Give a numerical value for the following, if possible.

(a) $f(50, 62)$

(b) $f(51, 60)$

(c) $f(51, 60) - f(50, 62)$

Strengthen Your Understanding

In Problems 46–47, explain what is wrong with the statement.

46. If the contours of f are all parallel lines, then f is linear.

47. A function $f(x, y)$ with linear cross-sections for x fixed and linear cross-sections for y fixed is a linear function.

In Problems 48–49, give an example of:

48. A table of values, with three rows and three columns, for a nonlinear function that is linear in each row and in each column.

49. A linear function whose contours are lines with slope 2.

Are the statements in Problems 50–62 true or false? Give reasons for your answer.

50. The planes $z = 3 + 2x + 4y$ and $z = 5 + 2x + 4y$ intersect.

51. The function represented in Table 12.12 is linear.

Table 12.12

$u \backslash v$	1.1	1.2	1.3	1.4
3.2	11.06	12.06	13.06	14.06
3.4	11.75	12.82	13.89	14.96
3.6	12.44	13.58	14.72	15.86
3.8	13.13	14.34	15.55	16.76
4.0	13.82	15.10	16.38	17.66

52. Contours of $f(x, y) = 3x + 2y$ are lines with slope 3.

53. If f is a non-constant linear function, then the contours of f are parallel lines.

54. If $f(0, 0) = 1, f(0, 1) = 4, f(0, 3) = 5$, then f cannot be linear.

55. The graph of a linear function is always a plane.

56. The cross-section $x = c$ of a linear function $f(x, y)$ is always a line.

57. There is no linear function $f(x, y)$ with a graph parallel to the xy-plane.

58. There is no linear function $f(x, y)$ with a graph parallel to the xz-plane.

59. A linear function $f(x, y) = 2x + 3y - 5$, has exactly one point (a, b) satisfying $f(a, b) = 0$.

60. In a table of values of a linear function, the columns have the same slope as the rows.

61. There is exactly one linear function $f(x, y)$ whose $f = 0$ contour is $y = 2x + 1$.

62. If the contours of $f(x, y) = c + mx + ny$ are vertical lines, then $n = 0$.

12.5 FUNCTIONS OF THREE VARIABLES

In applications of calculus, functions of any number of variables can arise. The density of matter in the universe is a function of three variables, since it takes three numbers to specify a point in space. Models of the US economy often use functions of ten or more variables. We need to be able to apply calculus to functions of arbitrarily many variables.

One difficulty with functions of more than two variables is that it is hard to visualize them. The graph of a function of one variable is a curve in 2-space, the graph of a function of two variables is a surface in 3-space, so the graph of a function of three variables would be a solid in 4-space. Since we can't easily visualize 4-space, we won't use the graphs of functions of three variables. On the other hand, it is possible to draw contour diagrams for functions of three variables, only now the contours are surfaces in 3-space.

Representing a Function of Three Variables Using a Family of Level Surfaces

A function of two variables, $f(x, y)$, can be represented by a family of level curves of the form $f(x, y) = c$ for various values of the constant, c.

> A **level surface**, or **level set** of a function of three variables, $f(x, y, z)$, is a surface of the form $f(x, y, z) = c$, where c is a constant. The function f can be represented by the family of level surfaces obtained by allowing c to vary.

The value of the function, f, is constant on each level surface.

Example 1   The temperature, in °C, at a point (x, y, z) is given by $T = f(x, y, z) = x^2 + y^2 + z^2$. What do the level surfaces of the function f look like and what do they mean in terms of temperature?

Solution   The level surface corresponding to $T = 100$ is the set of all points where the temperature is 100°C. That is, where $f(x, y, z) = 100$, so

$$x^2 + y^2 + z^2 = 100.$$

This is the equation of a sphere of radius 10, with center at the origin. Similarly, the level surface corresponding to $T = 200$ is the sphere with radius $\sqrt{200}$. The other level surfaces are concentric spheres. The temperature is constant on each sphere. We may view the temperature distribution as a set of nested spheres, like concentric layers of an onion, each one labeled with a different temperature, starting from low temperatures in the middle and getting hotter as we go out from the center. (See Figure 12.81.) The level surfaces become more closely spaced as we move farther from the origin because the temperature increases more rapidly the farther we get from the origin.

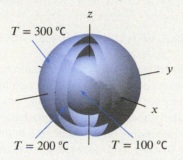

Figure 12.81: Level surfaces of $T = f(x, y, z) = x^2 + y^2 + z^2$, each one having a constant temperature

Example 2   What do the level surfaces of $f(x, y, z) = x^2 + y^2$ and $g(x, y, z) = z - y$ look like?

Solution   The level surface of f corresponding to the constant c is the surface consisting of all points satisfying the equation

$$x^2 + y^2 = c.$$

Since there is no z-coordinate in the equation, z can take any value. For $c > 0$, this is a circular cylinder of radius $\sqrt{c}$ around the z-axis. The level surfaces are concentric cylinders; on the narrow ones near the z-axis, f has small values; on the wider ones, f has larger values. See Figure 12.82.

The level surface of g corresponding to the constant c is the plane

$$z - y = c.$$

Since there is no x variable in the equation, these planes are parallel to the x-axis and cut the yz-plane in the line $z - y = c$. See Figure 12.83.

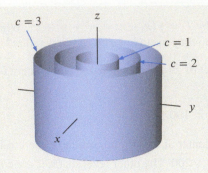

Figure 12.82: Level surfaces of $f(x, y, z) = x^2 + y^2$

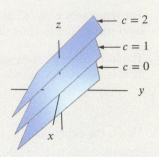

Figure 12.83: Level surfaces of $g(x, y, z) = z - y$

We say $g(x, y, z) = z - y$ in Example 2 is a *linear function* of the three variables x, y, z, whereas $f(x, y) = x^2 + y^2$ and $f(x, y, z) = x^2 + y^2 + z^2$ are *quadratic functions* of three variables.

Example 3   What do the level surfaces of $f(x, y, z) = x^2 + y^2 - z^2$ look like?

Solution   In Section 12.3, we saw that the two-variable quadratic function $g(x, y) = x^2 - y^2$ has a saddle-shaped graph and three types of contours. The contour equation $x^2 - y^2 = c$ gives a hyperbola opening right-left when $c > 0$, a hyperbola opening up-down when $c < 0$, and a pair of intersecting lines when $c = 0$. Similarly, the three-variable quadratic function $f(x, y, z) = x^2 + y^2 - z^2$ has three types of level surfaces depending on the value of c in the equation $x^2 + y^2 - z^2 = c$.

Suppose that $c > 0$, say $c = 1$. Rewrite the equation as $x^2 + y^2 = z^2 + 1$ and think of what happens as we cut the surface perpendicular to the z-axis by holding z fixed. The result is a circle, $x^2 + y^2 = $ constant, of radius at least 1 (since the constant $z^2 + 1 \geq 1$). The circles get larger as z gets larger. If we take the $x = 0$ cross-section instead, we get the hyperbola $y^2 - z^2 = 1$. The result is shown in Figure 12.87, with $a = b = c = 1$.

Suppose instead that $c < 0$, say $c = -1$. Then the horizontal cross-sections of $x^2 + y^2 = z^2 - 1$ are again circles except that the radii shrink to 0 at $z = \pm 1$ and between $z = -1$ and $z = 1$ there are no cross-sections at all. The result is shown in Figure 12.88 with $a = b = c = 1$.

When $c = 0$, we get the equation $x^2 + y^2 = z^2$. Again the horizontal cross-sections are circles, this time with the radius shrinking down to exactly 0 when $z = 0$. The resulting surface, shown in Figure 12.89 with $a = b = c = 1$, is the cone $z = \sqrt{x^2 + y^2}$ studied in Section 12.3, together with the lower cone $z = -\sqrt{x^2 + y^2}$.

A Catalog of Surfaces

For later reference, here is a small catalog of the surfaces we have encountered.

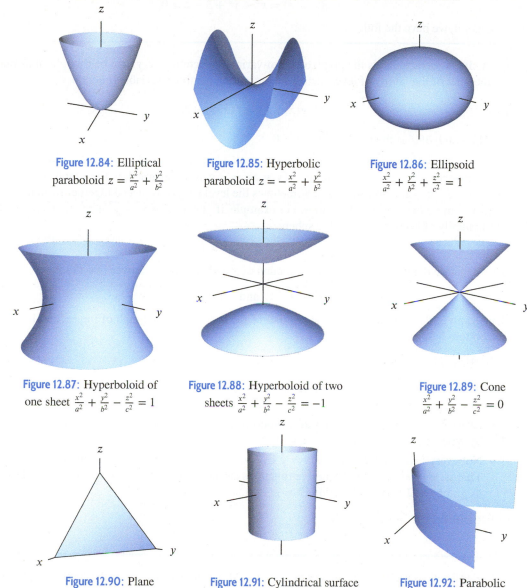

Figure 12.84: Elliptical paraboloid $z = \frac{x^2}{a^2} + \frac{y^2}{b^2}$

Figure 12.85: Hyperbolic paraboloid $z = -\frac{x^2}{a^2} + \frac{y^2}{b^2}$

Figure 12.86: Ellipsoid $\frac{x^2}{a^2} + \frac{y^2}{b^2} + \frac{z^2}{c^2} = 1$

Figure 12.87: Hyperboloid of one sheet $\frac{x^2}{a^2} + \frac{y^2}{b^2} - \frac{z^2}{c^2} = 1$

Figure 12.88: Hyperboloid of two sheets $\frac{x^2}{a^2} + \frac{y^2}{b^2} - \frac{z^2}{c^2} = -1$

Figure 12.89: Cone $\frac{x^2}{a^2} + \frac{y^2}{b^2} - \frac{z^2}{c^2} = 0$

Figure 12.90: Plane $ax + by + cz = d$

Figure 12.91: Cylindrical surface $x^2 + y^2 = a^2$

Figure 12.92: Parabolic cylinder $y = ax^2$

(These are viewed as equations in three variables x, y, and z.)

How Surfaces Can Represent Functions of Two Variables and Functions of Three Variables

You may have noticed that we have used surfaces to represent functions in two different ways. First, we used a *single* surface to represent a two-variable function $f(x, y)$. Second, we used a *family* of level surfaces to represent a three-variable function $g(x, y, z)$. These level surfaces have equation $g(x, y, z) = c$.

What is the relation between these two uses of surfaces? For example, consider the function

$$f(x, y) = x^2 + y^2 + 3.$$

Define

$$g(x, y, z) = x^2 + y^2 + 3 - z$$

The points on the graph of f satisfy $z = x^2 + y^2 + 3$, so they also satisfy $x^2 + y^2 + 3 - z = 0$. Thus the graph of f is the same as the level surface

$$g(x, y, z) = x^2 + y^2 + 3 - z = 0.$$

In general, we have the following result:

> A single surface that is the graph of a two-variable function $f(x, y)$ can be thought of as one member of the family of level surfaces representing the three-variable function
>
> $$g(x, y, z) = f(x, y) - z.$$
>
> The graph of f is the level surface $g = 0$.

Conversely, a single level surface $g(x, y, z) = c$ can be regarded as the graph of a function $f(x, y)$ if it is possible to solve for z. Sometimes the level surface is pieced together from the graphs of two or more two-variable functions. For example, if $g(x, y, z) = x^2 + y^2 + z^2$, then one member of the family of level surfaces is the sphere

$$x^2 + y^2 + z^2 = 1.$$

This equation defines z implicitly as a function of x and y. Solving it gives two functions:

$$z = \sqrt{1 - x^2 - y^2} \quad \text{and} \quad z = -\sqrt{1 - x^2 - y^2}.$$

The graph of the first function is the top half of the sphere and the graph of the second function is the bottom half.

Summary for Section 12.5

- A **level surface** of a 3-variable function $f(x, y, z)$ is the set of points in 3-space that satisfy $f(x, y, z) = \text{constant}$.
- Every surface that is the **graph of a function** $z = g(x, y)$ **can be rewritten as a level surface** by writing $f(x, y, z) = g(x, y) - z = 0$.
- Not every level surface $f(x, y, z) = c$ can be rewritten as the graph of a function $z = g(x, y)$. That is, level surfaces of 3-variable functions can describe more surfaces than can be described as graphs of 2-variable functions $z = g(x, y)$.

Exercises and Problems for Section 12.5

EXERCISES

1. Match the following functions with the level surfaces in Figure 12.93.

   (a) $f(x, y, z) = y^2 + z^2$ (b) $h(x, y, z) = x^2 + z^2$.

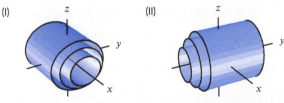

Figure 12.93

2. Match the functions with the level surfaces in Figure 12.94.

   (a) $f(x, y, z) = x^2 + y^2 + z^2$
   (b) $g(x, y, z) = x^2 + z^2$.

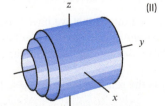

Figure 12.94

3. Write the level surface $x + 2y + 3z = 5$ as the graph of a function $f(x, y)$.

4. Find a formula for a function $f(x, y, z)$ whose level surface $f = 4$ is a sphere of radius 2, centered at the origin.

5. Write the level surface $x^2 + y + \sqrt{z} = 1$ as the graph of a function $f(x, y)$.

6. Find a formula for a function $f(x, y, z)$ whose level surfaces are spheres centered at the point (a, b, c).

7. Which of the graphs in the catalog of surfaces on page 735 is the graph of a function of x and y?

In Exercises 8–11, use the catalog on page 735 to identify the surface.

8. $x^2 + y^2 - z = 0$

9. $-x^2 - y^2 + z^2 = 1$

10. $x + y = 1$

11. $x^2 + y^2/4 + z^2 = 1$

In Exercises 12–15, decide if the given level surface can be expressed as the graph of a function, $f(x, y)$.

12. $z - x^2 - 3y^2 = 0$

13. $2x + 3y - 5z - 10 = 0$

14. $x^2 + y^2 + z^2 - 1 = 0$

15. $z^2 = x^2 + 3y^2$

16. Match the functions (a)–(d) with the descriptions of their level surfaces in I–IV.

(a) $f(x, y, z) = \sqrt{9 - x^2 - y^2}$

(b) $f(x, y, z) = \sqrt{x^2 + y^2 + z^2}$

(c) $f(x, y, z) = \dfrac{1}{x^2 + y^2 + z^2}$

(d) $f(x, y, z) = 5 + y^2 + z^2$

I. Cylinders that get larger as the function value increases

II. Cylinders that get smaller as the function value increases

III. Spheres that get larger as the function value increases

IV. Spheres that get smaller as the function value increases

PROBLEMS

In Problems 17–19, represent the surface whose equation is given as the graph of a two-variable function, $f(x, y)$, and as the level surface of a three-variable function, $g(x, y, z) = c$. There are many possible answers.

17. The plane $4x - y - 2z = 6$

18. The top half of the sphere $x^2 + y^2 + z^2 - 10 = 0$

19. The bottom half of the ellipsoid $x^2 + y^2 + z^2/2 = 1$

20. The balance, B, in dollars, in a bank account depends on the amount deposited, A dollars, the annual interest rate, $r\%$, and the time, t, in months since the deposit, so $B = f(A, r, t)$.

(a) Is f an increasing or decreasing function of A? Of r? Of t?

(b) Interpret the statement $f(1250, 1, 25) \approx 1276$. Give units.

21. A person's basal metabolic rate (BMR) is the minimal number of daily calories needed to keep their body functioning at rest. The BMR (in kcal/day) of a man of mass m (in kg), height h (in cm) and age a (in years) can be approximated by[12]

$$P = f(m, h, a) = 14m + 5h - 7a + 66$$

and for women by

$$P = g(m, h, a) = 10m + 2h - 5a + 655.$$

(a) What is the BMR of a 28-year-old man 180 cm tall weighing 59 kg?

(b) What is the BMR of a 43-year-old woman 162 cm tall weighing 52 kg?

[12] www.wikipedia.org, accessed December 30, 2019.

(c) Describe the level surface $P = 2000$ for a woman and explain what the points on this level surface represent.

(d) If a 40-year-old man 175 cm tall weighing 77 kg restricts himself to a diet with a daily caloric intake of 1600 kcal, should he expect to lose weight?

22. The monthly payments, P dollars, on a mortgage in which A dollars were borrowed at an annual interest rate of $r\%$ for t years is given by $P = f(A, r, t)$. Is f an increasing or decreasing function of A? Of r? Of t?

23. The balance in a bank account, B dollars, is given by $B = f(P, r, t) = P(1 + 0.01r)^t$, where P dollars is the principal amount invested, $r\%$ is the annual interest rate, and t years is the time since the investment was made.

(a) Find a formula for the level surface of f containing the point $(P, r, t) = (1000, 5, 20)$, and explain the significance of this surface in terms of balance.

(b) Find another point on the level surface in part (a), and explain the significance of this point in terms of balance.

24. The pressure of gas in a storage container, in atmospheres, is given by

$$P = f(n, T, V) = \frac{82nT}{V},$$

where n is the amount of gas, in kilomoles, T is the temperature of the gas, in Kelvin, and V is the volume of the storage container, in liters.

(a) Find a formula for the level surface of f containing the point $(n, T, V) = (1, 270, 20)$, and explain the significance of this surface in terms of pressure.

(b) Find another point on the level surface in part (a), and explain the significance of this point in terms of pressure.

25. The mass, in grams, of a rod in the shape of a right circular cylinder, is given by $m = f(r, h, \rho) = \pi r^2 h \rho$, where the rod has a radius of r cm, a height of h cm, and a uniform density of ρ gm/cm^3.

    (a) Find a formula for the level surface of f containing the point $(r, h, \rho) = (2, 10, 3)$, and explain the significance of this surface in terms of mass.

    (b) Find another point on the level surface in part (a), and explain the significance of this point in terms of mass.

26. Find a function $f(x, y, z)$ whose level surface $f = 1$ is the graph of the function $g(x, y) = x + 2y$.

27. Find two functions $f(x, y)$ and $g(x, y)$ so that the graphs of both together form the ellipsoid $x^2 + y^2/4 + z^2/9 = 1$.

28. Find a formula for a function $g(x, y, z)$ whose level surfaces are planes parallel to the plane $z = 2x + 3y - 5$.

29. Which of the following functions have planes as level surfaces?

$$f(x, y, z) = e^{x+z} \qquad r(x, y, z) = x^3$$
$$g(x, y, z) = e^x + z \qquad m(x, y, z) = \ln(x + z)$$

30. The surface S is the graph of $f(x, y) = \sqrt{1 - x^2 - y^2}$.

    (a) Explain why S is the upper hemisphere of radius 1, with equator in the xy-plane, centered at the origin.

    (b) Find a level surface $g(x, y, z) = c$ representing S.

31. The surface S is the graph of $f(x, y) = \sqrt{1 - y^2}$.

    (a) Explain why S is the upper half of a circular cylinder of radius 1, centered along the x-axis.

    (b) Find a level surface $g(x, y, z) = c$ representing S.

32. A cone C, with height 1 and radius 1, has its base in the xz-plane and its vertex on the positive y-axis. Find a function $g(x, y, z)$ such that C is part of the level surface $g(x, y, z) = 0$. [Hint: The graph of $f(x, y) = \sqrt{x^2 + y^2}$ is a cone which opens up and has vertex at the origin.]

33. Describe the level surface $f(x, y, z) = x^2/4 + z^2 = 1$ in words.

34. Describe the level surface $g(x, y, z) = x^2 + y^2/4 + z^2 = 1$ in words. [Hint: Look at cross-sections with constant x, y, and z values.]

35. Describe in words the level surfaces of the function $g(x, y, z) = x + y + z$.

36. Describe in words the level surfaces of $f(x, y, z) = \sin(x + y + z)$.

37. Describe the surface $x^2 + y^2 = (2 + \sin z)^2$. In general, if $f(z) \geq 0$ for all z, describe the surface $x^2 + y^2 = (f(z))^2$.

38. What do the level surfaces of $f(x, y, z) = x^2 - y^2 + z^2$ look like? [Hint: Use cross-sections with y constant instead of cross-sections with z constant.]

39. Describe in words the level surfaces of $g(x, y, z) = e^{-(x^2+y^2+z^2)}$.

40. Describe in words the level surfaces of $f(x, y, z) = z/x$.

41. Show that the level surfaces of $g(x, y, z) = ax + by + cz$ where $c \neq 0$ are parallel planes.

42. Sketch and label level surfaces of $h(x, y, z) = e^{z-y}$ for $h = 1, e, e^2$.

43. Sketch and label level surfaces of $f(x, y, z) = 4 - x^2 - y^2 - z^2$ for $f = 0, 1, 2$.

44. Sketch and label level surfaces of $g(x, y, z) = 1 - x^2 - y^2$ for $g = 0, -1, -2$.

45. What is the relationship between the level surfaces of $g(x, y, z) = f(x, y) - z$ and the graph of $z = f(x, y)$?

46. Describe the level surfaces of $g(x, y, z) = y - f(x)$.

Strengthen Your Understanding

■ In Problems 47–49, explain what is wrong with the statement.

47. The graph of a function $f(x, y, z)$ is a surface in 3-space.

48. The level surfaces of $f(x, y, z) = x^2 - y^2$ are all saddle-shaped.

49. The level surfaces of $f(x, y, z) = x^2 + y^2$ are paraboloids.

■ In Problems 50–53, give an example of:

50. A function $f(x, y, z)$ whose level surfaces are equally spaced planes perpendicular to the yz-plane.

51. A function $f(x, y, z)$ whose level sets are concentric cylinders centered on the y-axis.

52. A nonlinear function $f(x, y, z)$ whose level sets are parallel planes.

53. A function $f(x, y, z)$ whose level sets are paraboloids.

■ Are the statements in Problems 54–64 true or false? Give reasons for your answer.

54. The graph of the function $f(x, y) = x^2 + y^2$ is the same as the level surface $g(x, y, z) = x^2 + y^2 - z = 0$.

55. The graph of $f(x, y) = \sqrt{1 - x^2 - y^2}$ is the same as the level surface $g(x, y, z) = x^2 + y^2 + z^2 = 1$.

56. Any surface which is the graph of a two-variable function $f(x, y)$ can also be represented as the level surface of a three-variable function $g(x, y, z)$.

57. Any surface which is the level surface of a three-variable function $g(x, y, z)$ can also be represented as the graph of a two-variable function $f(x, y)$.

58. The level surfaces of the function $g(x, y, z) = x + 2y + z$ are parallel planes.

59. The level surfaces of $g(x, y, z) = x^2 + y + z^2$ are cylinders with axis along the y-axis.

60. A level surface of a function $g(x, y, z)$ cannot be a single point.

61. If $g(x, y, z) = ax + by + cz + d$, where a, b, c, d are nonzero constants, then the level surfaces of g are planes.

62. If the level surfaces of g are planes, then $g(x, y, z) = ax + by + cz + d$, where a, b, c, d are constants.

63. If the level surfaces $g(x, y, z) = k_1$ and $g(x, y, z) = k_2$ are the same surface, then $k_1 = k_2$.

64. If $x^2 + y^2 + z^2 = 1$ is the level surface $g(x, y, z) = 1$, then $x^2 + y^2 + z^2 = 4$ is the level surface $g(x, y, z) = 4$.

12.6 LIMITS AND CONTINUITY

The sheer face of Half Dome, in Yosemite National Park in California, was caused by glacial activity during the Ice Age. (See Figure 12.95.) As we scale the rock from the west, the height of the terrain rises abruptly by nearly 5000 feet from the valley floor, 2000 feet of it vertical.

If we consider the function h giving the height of the terrain above sea level in terms of longitude and latitude, then h has a *discontinuity* along the path at the base of the cliff of Half Dome. Looking at the contour map of the region in Figure 12.96, we see that in most places a small change in position results in a small change in height, except near the cliff. There, no matter how small a step we take, we get a large change in height. (You can see how crowded the contours get near the cliff; some end abruptly along the discontinuity.)

This geological feature illustrates the ideas of continuity and discontinuity. Roughly speaking, a function is said to be *continuous* at a point if its values at places near the point are close to the value at the point. If this is not the case, the function is said to be *discontinuous*.

The property of continuity is one that, practically speaking, we usually assume of the functions we are studying. Informally, we expect (except under special circumstances) that values of a function do not change drastically when making small changes to the input variables. Whenever we model a one-variable function by an unbroken curve, we are making this assumption. Even when functions come to us as tables of data, we usually make the assumption that the missing function values between data points are close to the measured ones.

In this section we study limits and continuity a bit more formally in the context of functions of several variables. For simplicity we study these concepts for functions of two variables, but our discussion can be adapted to functions of three or more variables.

One can show that sums, products, and compositions of continuous functions are continuous,

Figure 12.95: Half Dome in Yosemite National Park

Figure 12.96: A contour map of Half Dome

while the quotient of two continuous functions is continuous everywhere the denominator function is nonzero. Thus, each of the functions

$$\cos(x^2 y), \qquad \ln(x^2 + y^2), \qquad \frac{e^{x+y}}{x + y}, \qquad \ln(\sin(x^2 + y^2))$$

is continuous at all points (x, y) where it is defined. As for functions of one variable, the graph of a continuous function over an unbroken domain is unbroken—that is, the surface has no holes or rips in it.

Example 1   From Figures 12.97–12.100, which of the following functions appear to be continuous at $(0, 0)$?

(a) $f(x, y) = \begin{cases} \dfrac{x^2 y}{x^2 + y^2}, & (x, y) \neq (0, 0), \\ 0, & (x, y) = (0, 0). \end{cases}$

(b) $g(x, y) = \begin{cases} \dfrac{x^2}{x^2 + y^2}, & (x, y) \neq (0, 0), \\ 0, & (x, y) = (0, 0). \end{cases}$

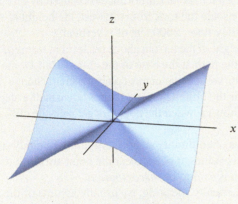

Figure 12.97: Graph of $z = x^2 y/(x^2 + y^2)$

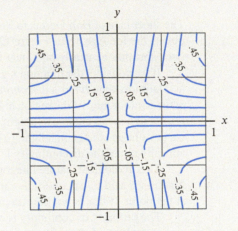

Figure 12.98: Contour diagram of $z = x^2 y/(x^2 + y^2)$

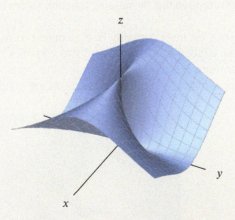

Figure 12.99: Graph of $z = x^2/(x^2 + y^2)$

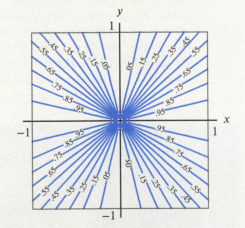

Figure 12.100: Contour diagram of $z = x^2/(x^2 + y^2)$

Solution   (a) The graph and contour diagram of f in Figures 12.97 and 12.98 suggest that f is close to 0 when (x, y) is close to $(0, 0)$. That is, the figures suggest that f is continuous at the point $(0, 0)$; the graph appears to have no rips or holes there.

However, the figures cannot tell us for sure whether f is continuous. To be certain we must investigate the limit analytically, as is done in Example 2(a) on page 741.

(b) The graph of g and its contours near $(0,0)$ in Figure 12.99 and 12.100 suggest that g behaves differently from f: The contours of g seem to "crash" at the origin and the graph rises rapidly from 0 to 1 near $(0,0)$. Small changes in (x, y) near $(0,0)$ can yield large changes in g, so we expect that g is not continuous at the point $(0,0)$. Again, a more precise analysis is given in Example 2(b).

The previous example suggests that continuity *at* a point depends on a function's behavior *near* the point. To study behavior near a point more carefully we need the idea of a limit of a function of two variables. Suppose that $f(x, y)$ is a function defined on a set in 2-space, not necessarily containing the point (a, b), but containing points (x, y) arbitrarily close to (a, b); suppose that L is a number.

> The function f has a **limit** L at the point (a, b), written
>
> $$\lim_{(x,y)\to(a,b)} f(x, y) = L,$$
>
> if $f(x, y)$ is as close to L as we please whenever the distance from the point (x, y) to the point (a, b) is sufficiently small, but not zero.

We define continuity for functions of two variables in the same way as for functions of one variable:

> A function f is **continuous at the point** (a, b) if
>
> $$\lim_{(x,y)\to(a,b)} f(x, y) = f(a, b).$$
>
> A function is **continuous on a region** R in the xy-plane if it is continuous at each point in R.

Thus, if f is continuous at the point (a, b), then f must be defined at (a, b) and the limit, $\lim_{(x,y)\to(a,b)} f(x, y)$, must exist and be equal to the value $f(a, b)$. If a function is defined at a point (a, b) but is not continuous there, then we say that f is *discontinuous* at (a, b).

We now apply the definition of continuity to the functions in Example 1, showing that f is continuous at $(0, 0)$ and that g is discontinuous at $(0, 0)$.

Example 2 Let f and g be the functions in Example 1. Use the definition of the limit to show that:

(a) $\lim\limits_{(x,y)\to(0,0)} f(x, y) = 0$ (b) $\lim\limits_{(x,y)\to(0,0)} g(x, y)$ does not exist.

Solution To investigate these limits of f and g, we consider values of these functions near, but not at, the origin, where they are given by the formulas

$$f(x, y) = \frac{x^2 y}{x^2 + y^2} \qquad g(x, y) = \frac{x^2}{x^2 + y^2}.$$

(a) The graph and contour diagram of f both suggest that $\lim_{(x,y)\to(0,0)} f(x, y) = 0$. To use the definition of the limit, we estimate $|f(x, y) - L|$ with $L = 0$:

$$|f(x, y) - L| = \left| \frac{x^2 y}{x^2 + y^2} - 0 \right| = \left| \frac{x^2}{x^2 + y^2} \right| |y| \leq |y| \leq \sqrt{x^2 + y^2}.$$

Now $\sqrt{x^2 + y^2}$ is the distance from (x, y) to $(0, 0)$. Thus, to make $|f(x, y) - 0| < 0.001$, for example, we need only require that (x, y) be within 0.001 of $(0, 0)$. More generally, for any positive number u, no matter how small, we are sure that $|f(x, y) - 0| < u$ whenever (x, y) is no farther than u from $(0, 0)$. This is what we mean by saying that the difference $|f(x, y) - 0|$ can be made as small as we wish by choosing the distance to be sufficiently small. Thus, we conclude that

$$\lim_{(x,y)\to(0,0)} f(x, y) = \lim_{(x,y)\to(0,0)} \frac{x^2 y}{x^2 + y^2} = 0.$$

Notice that since this limit equals $f(0, 0)$, the function f is continuous at $(0, 0)$.

(b) Although the formula defining the function g looks similar to that of f, we saw in Example 1 that g's behavior near the origin is quite different. If we consider points $(x, 0)$ lying along the x-axis near $(0, 0)$, then the values $g(x, 0)$ are equal to 1, while if we consider points $(0, y)$ lying along the y-axis near $(0, 0)$, then the values $g(0, y)$ are equal to 0. Thus, within any distance (no matter how small) from the origin, there are points where $g = 0$ and points where $g = 1$. Therefore the limit $\lim_{(x,y)\to(0,0)} g(x, y)$ does not exist, and thus g is not continuous at $(0, 0)$.

While the notions of limit and continuity look formally the same for one- and two-variable functions, they are somewhat more subtle in the multivariable case. The reason for this is that on the line (1-space), we can approach a point from just two directions (left or right) but in 2-space there are an infinite number of ways to approach a given point.

Summary for Section 12.6

- Informally, a function is *continuous* at a point if values of the function at nearby points in all directions approach the value of the function at the point.

- The function f has a **limit** L at the point (a, b), written

$$\lim_{(x,y)\to(a,b)} f(x, y) = L,$$

if $f(x, y)$ is as close to L as we please whenever the distance from the point (x, y) to the point (a, b) is sufficiently small, but not zero.

- A function f is **continuous at the point** (a, b) if

$$\lim_{(x,y)\to(a,b)} f(x, y) = f(a, b).$$

- A function is **continuous on a region** R in the xy-plane if it is continuous at each point in R.

Exercises and Problems for Section 12.6

EXERCISES

In Exercises 1–6, is the function continuous at all points in the given region?

1. $\dfrac{1}{x^2 + y^2}$ on the square $-1 \le x \le 1, -1 \le y \le 1$

2. $\dfrac{1}{x^2 + y^2}$ on the square $1 \le x \le 2, 1 \le y \le 2$

3. $\dfrac{y}{x^2 + 2}$ on the disk $x^2 + y^2 \le 1$

4. $\dfrac{e^{\sin x}}{\cos y}$ on the rectangle $-\frac{\pi}{2} \le x \le \frac{\pi}{2}, 0 \le y \le \frac{\pi}{4}$

5. $\tan(xy)$ on the square $-2 \le x \le 2, -2 \le y \le 2$

6. $\sqrt{2x - y}$ on the disk $x^2 + y^2 \le 4$

In Exercises 7–11, find the limit as $(x, y) \to (0, 0)$ of $f(x, y)$. Assume that polynomials, exponentials, logarithmic, and trigonometric functions are continuous.

7. $f(x, y) = e^{-x-y}$

8. $f(x, y) = x^2 + y^2$

9. $f(x, y) = \dfrac{x}{x^2 + 1}$

10. $f(x, y) = \dfrac{x + y}{(\sin y) + 2}$

11. $f(x, y) = \dfrac{\sin(x^2 + y^2)}{x^2 + y^2}$ [Hint: $\lim\limits_{t\to 0} \dfrac{\sin t}{t} = 1$.]

In Exercises **12–15**, use the contour diagram for $f(x, y)$ in Figure 12.101 to suggest an estimate for the limit, or explain why it may not exist.

12. $\displaystyle\lim_{(x,y)\to(2,1)} f(x, y)$ 13. $\displaystyle\lim_{(x,y)\to(-1,2)} f(x, y)$

14. $\displaystyle\lim_{(x,y)\to(-2,0)} f(x, y)$ 15. $\displaystyle\lim_{(x,y)\to(0,0)} f(x, y)$

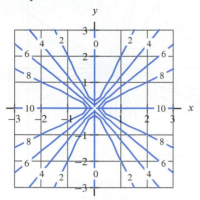

Figure 12.101

PROBLEMS

In Problems **16–17**, show that the function $f(x, y)$ does not have a limit as $(x, y) \to (0, 0)$. [Hint: Use the line $y = mx$.]

16. $f(x, y) = \dfrac{x + y}{x - y}, \qquad x \neq y$

17. $f(x, y) = \dfrac{x^2 - y^2}{x^2 + y^2}$

18. By approaching the origin along the positive x-axis and the positive y-axis, show that the following limit does not exist:
$$\lim_{(x,y)\to(0,0)} \frac{2x - y^2}{2x + y^2}.$$

19. Show that $f(x, y)$ has no limit as $(x, y) \to (0, 0)$ if
$$f(x, y) = \frac{xy}{|xy|}, \qquad x \neq 0 \text{ and } y \neq 0.$$

20. Show that the function f does not have a limit at $(0, 0)$ by examining the limits of f as $(x, y) \to (0, 0)$ along the curve $y = kx^2$ for different values of k:
$$f(x, y) = \frac{x^2}{x^2 + y}, \qquad x^2 + y \neq 0.$$

21. Let $f(x, y) = \begin{cases} \dfrac{|x|}{x} y & \text{for } x \neq 0 \\ 0 & \text{for } x = 0. \end{cases}$

Is $f(x, y)$ continuous

   (a) On the x-axis?   (b) On the y-axis?
   (c) At $(0, 0)$?

In Problems **22–23**, determine whether there is a value for the constant c making the function continuous everywhere. If so, find it. If not, explain why not.

22. $f(x, y) = \begin{cases} c + y, & x \leq 3, \\ 5 - x, & x > 3. \end{cases}$

23. $f(x, y) = \begin{cases} c + y, & x \leq 3, \\ 5 - y, & x > 3. \end{cases}$

24. Is the following function continuous at $(0, 0)$?
$$f(x, y) = \begin{cases} x^2 + y^2 & \text{if } (x, y) \neq (0, 0) \\ 2 & \text{if } (x, y) = (0, 0) \end{cases}$$

25. What value of c makes the following function continuous at $(0, 0)$?
$$f(x, y) = \begin{cases} x^2 + y^2 + 1 & \text{if } (x, y) \neq (0, 0) \\ c & \text{if } (x, y) = (0, 0) \end{cases}$$

26. (a) Use a computer to draw the graph and the contour diagram of the following function:
$$f(x, y) = \begin{cases} \dfrac{xy(x^2 - y^2)}{x^2 + y^2}, & (x, y) \neq (0, 0), \\ 0, & (x, y) = (0, 0). \end{cases}$$

   (b) Do your answers to part (a) suggest that f is continuous at $(0, 0)$? Explain your answer.

27. The function f, whose graph and contour diagram are in Figures 12.102 and 12.103, is given by

$$f(x, y) = \begin{cases} \dfrac{xy}{x^2 + y^2}, & (x, y) \neq (0, 0), \\ 0, & (x, y) = (0, 0). \end{cases}$$

(a) Show that $f(0, y)$ and $f(x, 0)$ are each continuous functions of one variable.

(b) Show that rays emanating from the origin are contained in contours of f.

(c) Is f continuous at $(0, 0)$?

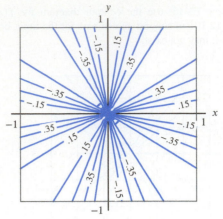

Figure 12.103: Contour diagram of
$z = xy/(x^2 + y^2)$

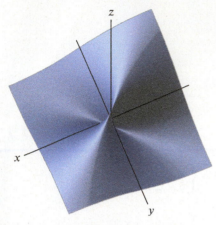

Figure 12.102: Graph of $z = xy/(x^2 + y^2)$

Strengthen Your Understanding

In Problems **28–29**, explain what is wrong with the statement.

28. If a function $f(x, y)$ has a limit as (x, y) approaches (a, b), then it is continuous at (a, b).

29. If both f and g are continuous at (a, b), then so are $f + g$, fg and f/g.

In Problems **30–31**, give an example of:

30. A function $f(x, y)$ which is continuous everywhere except at $(0, 0)$ and $(1, 2)$.

31. A function $f(x, y)$ that approaches 1 as (x, y) approaches $(0, 0)$ along the x-axis and approaches 2 as (x, y) approaches $(0, 0)$ along the y-axis.

In Problems **32–34**, construct a function $f(x, y)$ with the given property.

32. Not continuous along the line $x = 2$; continuous everywhere else.

33. Not continuous at the point $(2, 0)$; continuous everywhere else.

34. Not continuous along the curve $x^2 + y^2 = 1$; continuous everywhere else.

Online Resource: Review Problems and Projects

Are the statements in Problems **35–40** true or false? Give reasons for your answer.

35. If the limit of $f(x, y)$ is 1 as (x, y) approaches $(0, 0)$ along the x-axis, and the limit of $f(x, y)$ is 1 as (x, y) approaches $(0, 0)$ along the y-axis, then

$$\lim_{(x,y)\to(0,0)} f(x, y) \text{ exists.}$$

36. If $f(1, 0) = 2$, then $\displaystyle\lim_{(x,y)\to(1,0)} f(x, y) = 2$.

37. If $f(x, y)$ is continuous and $f(1, 0) = 2$, then

$$\lim_{(x,y)\to(1,0)} f(x, y) = 2.$$

38. If $\displaystyle\lim_{(x,y)\to(0,0)} f(x, y) = 3$, then the limit of $f(x, y)$ is 3 as (x, y) approaches $(0, 0)$ along the x-axis.

39. If $f(x, y)$ is continuous at (a, b), then its limit exists at (a, b).

40. If $\displaystyle\lim_{(x,y)\to(a,b)} f(x, y)$ exists then $f(x, y)$ is continuous at (a, b).

Chapter 13

A FUNDAMENTAL TOOL: VECTORS

CONTENTS

13.1   DISPLACEMENT VECTORS

Suppose you are a pilot planning a flight from Dallas to Pittsburgh. There are two things you must know: the distance to be traveled (so you have enough fuel to make it) and in what direction to go (so you don't miss Pittsburgh). Both these quantities together specify the displacement or *displacement vector* between the two cities.

> The **displacement vector** from one point to another is an arrow with its tail at the first point and its tip at the second. The **magnitude** (or length) of the displacement vector is the distance between the points and is represented by the length of the arrow. The **direction** of the displacement vector is the direction of the arrow.

Figure 13.1 shows a map with the displacement vectors from Dallas to Pittsburgh, from Albuquerque to Oshkosh, and from Los Angeles to Buffalo, SD. These displacement vectors have the same length and the same direction. We say that the displacement vectors between the corresponding cities are the same, even though they do not coincide. In other words,

> Displacement vectors which point in the same direction and have the same magnitude are considered to be the same, even if they do not coincide.

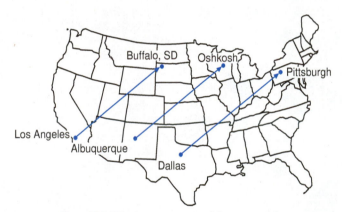

Figure 13.1: Displacement vectors between cities

Notation and Terminology

The displacement vector is our first example of a vector. Vectors have both magnitude and direction; in comparison, a quantity specified only by a number, but no direction, is called a *scalar*.[1] For instance, the time taken by the flight from Dallas to Pittsburgh is a scalar quantity. Displacement is a vector since it requires both distance and direction to specify it.

In this book, vectors are written with an arrow over them, $\vec{v}$, to distinguish them from scalars. Other books use a bold **v** to denote a vector. We use the notation $\overrightarrow{PQ}$ to denote the displacement vector from a point P to a point Q. The magnitude, or length, of a vector $\vec{v}$ is written $\|\vec{v}\|$.

Addition and Subtraction of Displacement Vectors

Suppose NASA commands a robot on Mars to move 75 meters in one direction and then 50 meters in another direction. (See Figure 13.2.) Where does the robot end up? Suppose the displacements are represented by the vectors $\vec{v}$ and $\vec{w}$, respectively. Then the sum $\vec{v} + \vec{w}$ gives the final position.

---

[1] So named by W. R. Hamilton because they are merely numbers on the *scale* from $-\infty$ to ∞.

The **sum**, $\vec{v} + \vec{w}$, of two vectors $\vec{v}$ and $\vec{w}$ is the combined displacement resulting from first applying $\vec{v}$ and then $\vec{w}$. (See Figure 13.3.) The sum $\vec{w} + \vec{v}$ gives the same displacement.

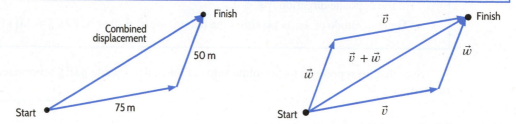

Figure 13.2: Sum of displacements of robots on Mars

Figure 13.3: The sum $\vec{v} + \vec{w} = \vec{w} + \vec{v}$

Suppose two different robots start from the same location. One moves along a displacement vector $\vec{v}$ and the second along a displacement vector $\vec{w}$. What is the displacement vector, $\vec{x}$, from the first robot to the second? (See Figure 13.4.) Since $\vec{v} + \vec{x} = \vec{w}$, we define $\vec{x}$ to be the difference $\vec{x} = \vec{w} - \vec{v}$. In other words, $\vec{w} - \vec{v}$ gets you from the first robot to the second.

The **difference**, $\vec{w} - \vec{v}$, is the displacement vector that, when added to $\vec{v}$, gives $\vec{w}$. That is, $\vec{w} = \vec{v} + (\vec{w} - \vec{v})$. (See Figure 13.4.)

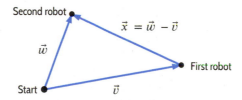

Figure 13.4: The difference $\vec{w} - \vec{v}$

If the robot ends up where it started, then its total displacement vector is the *zero vector*, $\vec{0}$. The zero vector has no direction.

The **zero vector**, $\vec{0}$, is a displacement vector with zero length.

Scalar Multiplication of Displacement Vectors

If $\vec{v}$ represents a displacement vector, the vector $2\vec{v}$ represents a displacement of twice the magnitude in the same direction as $\vec{v}$. Similarly, $-2\vec{v}$ represents a displacement of twice the magnitude in the opposite direction. (See Figure 13.5.)

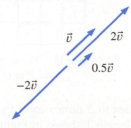

Figure 13.5: Scalar multiples of the vector $\vec{v}$

> If λ is a scalar and $\vec{v}$ is a displacement vector, the **scalar multiple of $\vec{v}$ by** λ, written $\lambda\vec{v}$, is the displacement vector with the following properties:
> - The displacement vector $\lambda\vec{v}$ is parallel to $\vec{v}$, pointing in the same direction if $\lambda > 0$ and in the opposite direction if $\lambda < 0$.
> - The magnitude of $\lambda\vec{v}$ is $|\lambda|$ times the magnitude of $\vec{v}$, that is, $\|\lambda\vec{v}\| = |\lambda|\,\|\vec{v}\|$.

Note that $|\lambda|$ represents the absolute value of the scalar λ while $\|\lambda\vec{v}\|$ represents the magnitude of the vector $\lambda\vec{v}$.

Example 1   Explain why $\vec{w} - \vec{v} = \vec{w} + (-1)\vec{v}$.

Solution   The vector $(-1)\vec{v}$ has the same magnitude as $\vec{v}$, but points in the opposite direction. Figure 13.6 shows that the combined displacement $\vec{w} + (-1)\vec{v}$ is the same as the displacement $\vec{w} - \vec{v}$.

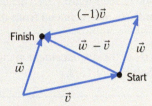

Figure 13.6: Explanation for
why $\vec{w} - \vec{v} = \vec{w} + (-1)\vec{v}$

Parallel Vectors

Two vectors $\vec{v}$ and $\vec{w}$ are *parallel* if one is a scalar multiple of the other, that is, if $\vec{w} = \lambda\vec{v}$, for some scalar λ.

Components of Displacement Vectors: The Vectors $\vec{i}$, $\vec{j}$, and $\vec{k}$

Suppose that you live in a city with equally spaced streets running east-west and north-south and that you want to tell someone how to get from one place to another. You'd be likely to tell them how many blocks east-west and how many blocks north-south to go. For example, to get from P to Q in Figure 13.7, we go 4 blocks east and 1 block south. If $\vec{i}$ and $\vec{j}$ are as shown in Figure 13.7, then the displacement vector from P to Q is $4\vec{i} - \vec{j}$.

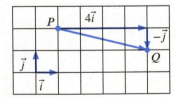

Figure 13.7: The displacement vector
from P to Q is $4\vec{i} - \vec{j}$

We extend the same idea to 3 dimensions. First we choose a Cartesian system of coordinate axes. The three vectors of length 1 shown in Figure 13.8 are the vector $\vec{i}$, which points along the positive x-axis, the vector $\vec{j}$, along the positive y-axis, and the vector $\vec{k}$, along the positive z-axis.

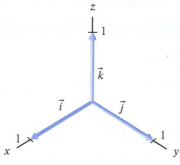

Figure 13.8: The vectors $\vec{i}$, $\vec{j}$ and $\vec{k}$ in 3-space

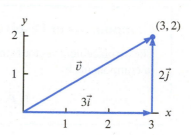

Figure 13.9: We resolve $\vec{v}$ into components by writing $\vec{v} = 3\vec{i} + 2\vec{j}$

Writing Displacement Vectors Using $\vec{i}, \vec{j}, \vec{k}$

Any displacement in 3-space or the plane can be expressed as a combination of displacements in the coordinate directions. For example, Figure 13.9 shows that the displacement vector $\vec{v}$ from the origin to the point $(3, 2)$ can be written as a sum of displacement vectors along the x- and y-axes:

$$\vec{v} = 3\vec{i} + 2\vec{j}.$$

This is called *resolving $\vec{v}$ into components*. In general:

> We **resolve** $\vec{v}$ into components by writing $\vec{v}$ in the form
>
> $$\vec{v} = v_1\vec{i} + v_2\vec{j} + v_3\vec{k},$$
>
> where v_1, v_2, v_3 are scalars. We call $v_1\vec{i}$, $v_2\vec{j}$, and $v_3\vec{k}$ the **components** of $\vec{v}$.

An Alternative Notation for Vectors

Many people write a vector in three dimensions as a string of three numbers, that is, as

$$\vec{v} = (v_1, v_2, v_3) \quad \text{instead of} \quad \vec{v} = v_1\vec{i} + v_2\vec{j} + v_3\vec{k}.$$

Since the first notation can be confused with a point and the second cannot, we usually use the second form.

Example 2   Resolve the displacement vector, $\vec{v}$, from the point $P_1 = (2, 4, 10)$ to the point $P_2 = (3, 7, 6)$ into components.

Solution   To get from P_1 to P_2, we move 1 unit in the positive x-direction, 3 units in the positive y-direction, and 4 units in the negative z-direction. Hence $\vec{v} = \vec{i} + 3\vec{j} - 4\vec{k}$.

Example 3   Decide whether the vector $\vec{v} = 2\vec{i} + 3\vec{j} + 5\vec{k}$ is parallel to each of the following vectors:

$$\vec{w} = 4\vec{i} + 6\vec{j} + 10\vec{k}, \quad \vec{a} = -\vec{i} - 1.5\vec{j} - 2.5\vec{k}, \quad \vec{b} = 4\vec{i} + 6\vec{j} + 9\vec{k}.$$

Solution   Since $\vec{w} = 2\vec{v}$ and $\vec{a} = -0.5\vec{v}$, the vectors $\vec{v}, \vec{w}$, and $\vec{a}$ are parallel. However, $\vec{b}$ is not a multiple of $\vec{v}$ (since, for example, $4/2 \neq 9/5$), so $\vec{v}$ and $\vec{b}$ are not parallel.

In general, Figure 13.10 shows us how to express the displacement vector between two points in components:

Components of Displacement Vectors

The displacement vector from the point $P_1 = (x_1, y_1, z_1)$ to the point $P_2 = (x_2, y_2, z_2)$ is given in components by

$$\overrightarrow{P_1 P_2} = (x_2 - x_1)\vec{i} + (y_2 - y_1)\vec{j} + (z_2 - z_1)\vec{k}.$$

Position Vectors: Displacement of a Point from the Origin

A displacement vector whose tail is at the origin is called a *position vector*. Thus, any point (x_0, y_0, z_0) in space has associated with it the position vector $\vec{r}_0 = x_0\vec{i} + y_0\vec{j} + z_0\vec{k}$. (See Figure 13.11.) In general, a position vector gives the displacement of a point from the origin.

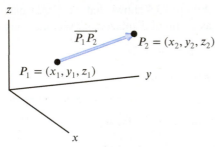

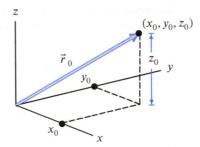

Figure 13.10: The displacement vector
$\overrightarrow{P_1 P_2} = (x_2 - x_1)\vec{i} + (y_2 - y_1)\vec{j} + (z_2 - z_1)\vec{k}$

Figure 13.11: The position vector
$\vec{r}_0 = x_0\vec{i} + y_0\vec{j} + z_0\vec{k}$

The Components of the Zero Vector

The zero displacement vector has magnitude equal to zero and is written $\vec{0}$. So $\vec{0} = 0\vec{i} + 0\vec{j} + 0\vec{k}$.

The Magnitude of a Vector in Components

For a vector, $\vec{v} = v_1\vec{i} + v_2\vec{j}$, the Pythagorean theorem is used to find its magnitude, $\|\vec{v}\|$. (See Figure 13.12.) The angle θ gives the direction of $\vec{v}$.

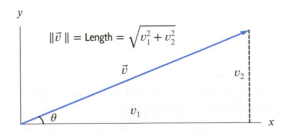

Figure 13.12: Magnitude, $\|\vec{v}\|$, of a 2-dimensional vector, $\vec{v}$

In three dimensions, for a vector $\vec{v} = v_1\vec{i} + v_2\vec{j} + v_3\vec{k}$, we have

---

Magnitude of $\vec{v} = \|\vec{v}\| = $ Length of the arrow $= \sqrt{v_1^2 + v_2^2 + v_3^2}.$

For instance, if $\vec{v} = 3\vec{i} - 4\vec{j} + 5\vec{k}$, then $\|\vec{v}\| = \sqrt{3^2 + (-4)^2 + 5^2} = \sqrt{50}$.

Addition and Scalar Multiplication of Vectors in Components

Suppose the vectors $\vec{v}$ and $\vec{w}$ are given in components:

$$\vec{v} = v_1\vec{i} + v_2\vec{j} + v_3\vec{k} \quad \text{and} \quad \vec{w} = w_1\vec{i} + w_2\vec{j} + w_3\vec{k}.$$

Then

$$\vec{v} + \vec{w} = (v_1 + w_1)\vec{i} + (v_2 + w_2)\vec{j} + (v_3 + w_3)\vec{k},$$

and

$$\lambda\vec{v} = \lambda v_1\vec{i} + \lambda v_2\vec{j} + \lambda v_3\vec{k}.$$

Figures 13.13 and 13.14 illustrate these properties in two dimensions. Finally, $\vec{v} - \vec{w} = \vec{v} + (-1)\vec{w}$, so we can write $\vec{v} - \vec{w} = (v_1 - w_1)\vec{i} + (v_2 - w_2)\vec{j} + (v_3 - w_3)\vec{k}$.

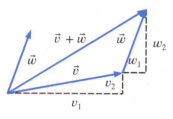

Figure 13.13: Sum $\vec{v} + \vec{w}$ in components

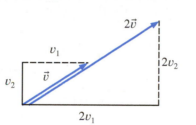

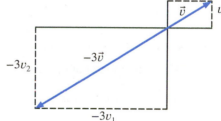

Figure 13.14: Scalar multiples of vectors showing $\vec{v}, 2\vec{v}$, and $-3\vec{v}$

How to Resolve a Vector into Components

You may wonder how we find the components of a 2-dimensional vector, given its length and direction. Suppose the vector $\vec{v}$ has length v and makes an angle of θ with the x-axis, measured counterclockwise, as in Figure 13.15. If $\vec{v} = v_1\vec{i} + v_2\vec{j}$, Figure 13.15 shows that

$$v_1 = v\cos\theta \quad \text{and} \quad v_2 = v\sin\theta.$$

Thus, we resolve $\vec{v}$ into components by writing

$$\vec{v} = (v\cos\theta)\vec{i} + (v\sin\theta)\vec{j}.$$

Vectors in 3-space are resolved using direction cosines; see Problem 66 (available online).

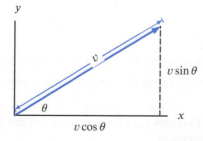

Figure 13.15: Resolving a vector: $\vec{v} = (v\cos\theta)\vec{i} + (v\sin\theta)\vec{j}$

Example 4   Resolve $\vec{v}$ into components if $\|\vec{v}\| = 2$ and $\theta = \pi/6$.

Solution   We have $\vec{v} = 2\cos(\pi/6)\vec{i} + 2\sin(\pi/6)\vec{j} = 2\left(\sqrt{3}/2\right)\vec{i} + 2(1/2)\vec{j} = \sqrt{3}\vec{i} + \vec{j}$.

Unit Vectors

A *unit vector* is a vector whose magnitude is 1. The vectors $\vec{i}$, $\vec{j}$, and $\vec{k}$ are unit vectors in the directions of the coordinate axes. It is often helpful to find a unit vector in the same direction as a given vector $\vec{v}$. Suppose that $\|\vec{v}\| = 10$; a unit vector in the same direction as $\vec{v}$ is $\vec{v}/10$. In general, a unit vector in the direction of any nonzero vector $\vec{v}$ is

$$\vec{u} = \frac{\vec{v}}{\|\vec{v}\|}.$$

Example 5   Find a unit vector, $\vec{u}$, in the direction of the vector $\vec{v} = \vec{i} + 3\vec{j}$.

Solution   If $\vec{v} = \vec{i} + 3\vec{j}$, then $\|\vec{v}\| = \sqrt{1^2 + 3^2} = \sqrt{10}$. Thus, a unit vector in the same direction is given by

$$\vec{u} = \frac{\vec{v}}{\sqrt{10}} = \frac{1}{\sqrt{10}}(\vec{i} + 3\vec{j}) = \frac{1}{\sqrt{10}}\vec{i} + \frac{3}{\sqrt{10}}\vec{j} \approx 0.32\vec{i} + 0.95\vec{j}.$$

Example 6   Find a unit vector at the point (x, y, z) that points directly outward away from the origin.

Solution   The vector from the origin to (x, y, z) is the position vector

$$\vec{r} = x\vec{i} + y\vec{j} + z\vec{k}.$$

Thus, if we put its tail at (x, y, z) it will point away from the origin. Its magnitude is

$$\|\vec{r}\| = \sqrt{x^2 + y^2 + z^2},$$

so a unit vector pointing in the same direction is

$$\frac{\vec{r}}{\|\vec{r}\|} = \frac{x\vec{i} + y\vec{j} + z\vec{k}}{\sqrt{x^2 + y^2 + z^2}} = \frac{x}{\sqrt{x^2 + y^2 + z^2}}\vec{i} + \frac{y}{\sqrt{x^2 + y^2 + z^2}}\vec{j} + \frac{z}{\sqrt{x^2 + y^2 + z^2}}\vec{k}.$$

Summary for Section 13.1

- A **vector** is a quantity that has both magnitude (or length) and direction.
- A **unit vector** is any vector with length 1. If $\vec{v}$ is nonzero, then $\frac{1}{\|\vec{v}\|}\vec{v}$ is a unit vector in the same direction as $\vec{v}$.
- The **component vectors** $\vec{i}$, $\vec{j}$, $\vec{k}$ are unit vectors and point in the direction of the positive x-, y-, and z-axes, respectively.
- The **vector sum** $\vec{v} + \vec{w}$ is the displacement vector given by following a displacement of $\vec{v}$ with a displacement of $\vec{w}$.
- The **scalar multiple** $\lambda\vec{v}$ is the vector with the same direction as $\vec{v}$ (if $\lambda > 0$) or the opposite direction (if $\lambda < 0$) and length scaled by a factor of $|\lambda|$.
- Two vectors are **parallel** if one is a scalar multiple of the other.
- Vectors in 3-space can be represented in **components** as $\vec{v} = a\vec{i} + b\vec{j} + c\vec{k}$.

- **Vector addition and scalar multiplication** can be computed componentwise: If

$$\vec{v} = v_1\vec{i} + v_2\vec{j} + v_3\vec{k} \quad \text{and} \quad \vec{w} = w_1\vec{i} + w_2\vec{j} + w_3\vec{k},$$

then

$$\vec{v} + \vec{w} = (v_1 + w_1)\vec{i} + (v_2 + w_2)\vec{j} + (v_3 + w_3)\vec{k}$$

and

$$\lambda\vec{v} = \lambda v_1\vec{i} + \lambda v_2\vec{j} + \lambda v_3\vec{k}.$$

- The **zero vector** has magnitude equal to zero and is written $\vec{0}$. In components, $\vec{0} = 0\vec{i} + 0\vec{j} + 0\vec{k}$.
- The **length** of $\vec{v} = v_1\vec{i} + v_2\vec{j} + v_3\vec{k}$ is given by

$$\|\vec{v}\| = \sqrt{v_1^2 + v_2^2 + v_3^2}.$$

- The **displacement vector** from the point $P_1 = (x_1, y_1, z_1)$ to the point $P_2 = (x_2, y_2, z_2)$ is given in components by

$$\overrightarrow{P_1P_2} = (x_2 - x_1)\vec{i} + (y_2 - y_1)\vec{j} + (z_2 - z_1)\vec{k}.$$

- A displacement vector whose tail is at the origin is called a **position vector**. A point (x_0, y_0, z_0) in 3-space has position vector $\vec{r}_0 = x_0\vec{i} + y_0\vec{j} + z_0\vec{k}$.

Exercises and Problems for Section 13.1   Online Resource: Additional Problems for Section 13.1
EXERCISES

In Exercises **1–6**, resolve the vectors into components.

1.

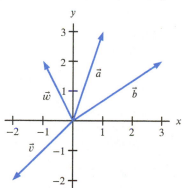

2.

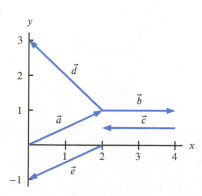

3. A vector starting at the point $Q = (4, 6)$ and ending at the point $P = (1, 2)$.

4. A vector starting at the point $P = (1, 2)$ and ending at the point $Q = (4, 6)$.

5.   **6.**

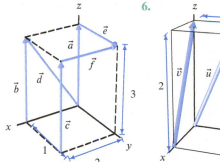

For Exercises **7–14**, perform the indicated computation.

7. $(4\vec{i} + 2\vec{j}) - (3\vec{i} - \vec{j})$

8. $(\vec{i} + 2\vec{j}) + (-3)(2\vec{i} + \vec{j})$

9. $-4(\vec{i} - 2\vec{j}) - 0.5(\vec{i} - \vec{k})$

10. $2(0.45\vec{i} - 0.9\vec{j} - 0.01\vec{k}) - 0.5(1.2\vec{i} - 0.1\vec{k})$

11. $(3\vec{i} - 4\vec{j} + 2\vec{k}) - (6\vec{i} + 8\vec{j} - \vec{k})$

12. $(4\vec{i} - 3\vec{j} + 7\vec{k}) - 2(5\vec{i} + \vec{j} - 2\vec{k})$

13. $(0.6\vec{i} + 0.2\vec{j} - \vec{k}) + (0.3\vec{i} + 0.3\vec{k})$

14. $\frac{1}{2}(2\vec{i} - \vec{j} + 3\vec{k}) + 3(\vec{i} - \frac{1}{6}\vec{j} + \frac{1}{2}\vec{k})$

In Exercises **15–19**, find the length of the vectors.

15. $\vec{v} = \vec{i} - \vec{j} + 2\vec{k}$ **16.** $\vec{z} = \vec{i} - 3\vec{j} - \vec{k}$

17. $\vec{v} = \vec{i} - \vec{j} + 3\vec{k}$

18. $\vec{v} = 7.2\vec{i} - 1.5\vec{j} + 2.1\vec{k}$

19. $\vec{v} = 1.2\vec{i} - 3.6\vec{j} + 4.1\vec{k}$

■For Exercises **20–25**, perform the indicated operations on the following vectors:

$$\vec{a} = 2\vec{j} + \vec{k}, \quad \vec{b} = -3\vec{i} + 5\vec{j} + 4\vec{k}, \quad \vec{c} = \vec{i} + 6\vec{j},$$

$$\vec{x} = -2\vec{i} + 9\vec{j}, \quad \vec{y} = 4\vec{i} - 7\vec{j}, \quad \vec{z} = \vec{i} - 3\vec{j} - \vec{k}.$$

20. $4\vec{z}$ **21.** $5\vec{a} + 2\vec{b}$ **22.** $\vec{a} + \vec{z}$

23. $2\vec{c} + \vec{x}$ **24.** $2\vec{a} + 7\vec{b} - 5\vec{z}$ **25.** $\|\vec{y} - \vec{x}\|$

26. (a) Draw the position vector for $\vec{v} = 5\vec{i} - 7\vec{j}$.
  (b) What is $\|\vec{v}\|$?
  (c) Find the angle between $\vec{v}$ and the positive x-axis.

27. Find the unit vector in the direction of $0.06\vec{i} - 0.08\vec{k}$.

28. Find the unit vector in the opposite direction to $\vec{i} - \vec{j} + \vec{k}$.

29. Find a unit vector in the opposite direction to $2\vec{i} - \vec{j} - \sqrt{11}\vec{k}$.

30. Find a vector with length 2 that points in the same direction as $\vec{i} - \vec{j} + 2\vec{k}$.

PROBLEMS

31. Find the value(s) of a making $\vec{v} = 5a\vec{i} - 3\vec{j}$ parallel to $\vec{w} = a^2\vec{i} + 6\vec{j}$.

32. (a) For $a = 1, 2$, and 3, draw position vectors for
  (i) $\vec{v} = a^2\vec{i} + 6\vec{j}$ (ii) $\vec{w} = 5\vec{i} - a^2\vec{j}$
  (b) Explain why there is no value of a that makes $\vec{v}$ and $\vec{w}$ parallel.

33. (a) Find a unit vector from the point $P = (1, 2)$ and toward the point $Q = (4, 6)$.
  (b) Find a vector of length 10 pointing in the same direction.

34. If north is the direction of the positive y-axis and east is the direction of the positive x-axis, give the unit vector pointing northwest.

35. Resolve the following vectors into components:

  (a) The vector in 2-space of length 2 pointing up and to the right at an angle of $\pi/4$ with the x-axis.
  (b) The vector in 3-space of length 1 lying in the xz-plane pointing upward at an angle of $\pi/6$ with the positive x-axis.

36. (a) From Figure 13.16, read off the coordinates of the five points, A, B, C, D, E, and thus resolve into components the following two vectors: $\vec{u} = (2.5)\overrightarrow{AB} + (-0.8)\overrightarrow{CD}$, $\vec{v} = (2.5)\overrightarrow{BA} - (-0.8)\overrightarrow{CD}$.
  (b) What is the relation between $\vec{u}$ and $\vec{v}$? Why was this to be expected?

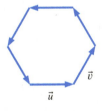

Figure 13.16

37. Find the components of a vector $\vec{p}$ that has the same direction as $\overrightarrow{EA}$ in Figure 13.16 and whose length equals two units.

38. For each of the four statements below, answer the following questions: Does the statement make sense? If yes, is it true for all possible choices of $\vec{a}$ and $\vec{b}$? If no, why not?

  (a) $\vec{a} + \vec{b} = \vec{b} + \vec{a}$
  (b) $\vec{a} + \|\vec{b}\| = \|\vec{a} + \vec{b}\|$
  (c) $\|\vec{b} + \vec{a}\| = \|\vec{a} + \vec{b}\|$
  (d) $\|\vec{a} + \vec{b}\| = \|\vec{a}\| + \|\vec{b}\|$.

39. For each condition, find unit vectors $\vec{a}$ and $\vec{b}$ or explain why no such vectors exist.

  (a) $\|\vec{a} + \vec{b}\| = 0$ **(b)** $\|\vec{a} + \vec{b}\| = 1$
  (c) $\|\vec{a} + \vec{b}\| = 2$ **(d)** $\|\vec{a} + \vec{b}\| = 3$

40. Two adjacent sides of a regular hexagon are given as the vectors $\vec{u}$ and $\vec{v}$ in Figure 13.17. Label the remaining sides in terms of $\vec{u}$ and $\vec{v}$.

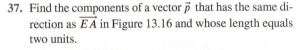

Figure 13.17

41. For what values of t are the following pairs of vectors parallel?

  (a) $2\vec{i} + (t^2 + \frac{2}{3}t + 1)\vec{j} + t\vec{k}$, $6\vec{i} + 8\vec{j} + 3\vec{k}$
  (b) $t\vec{i} + \vec{j} + (t - 1)\vec{k}$, $2\vec{i} - 4\vec{j} + \vec{k}$
  (c) $2t\vec{i} + t\vec{j} + t\vec{k}$, $6\vec{i} + 3\vec{j} + 3\vec{k}$.

42. Show that the unit vector $\vec{v} = x\vec{i} + y\vec{j}$ is not parallel to $\vec{w} = y\vec{i} - x\vec{j}$ for any choice of x and y.

43. Find all unit vectors $\vec{v} = x\vec{i} + y\vec{j}$ parallel to $\vec{w} = y\vec{i} + x\vec{j}$.

44. Find all vectors $\vec{v}$ in 2 dimensions having $\|\vec{v}\| = 5$ such that the $\vec{i}$-component of $\vec{v}$ is $3\vec{i}$.

45. **(a)** Find the point on the x-axis closest to the point (a, b, c).
    (b) Find a unit vector that points from the point you found in part (a) toward (a, b, c).

46. Figure 13.18 shows a molecule with four atoms at O, A, B and C. Check that every atom in the molecule is 2 units away from every other atom.

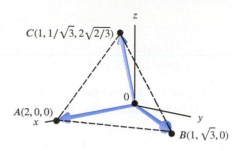

Figure 13.18

Strengthen Your Understanding

■ In Problems 47–50, explain what is wrong with the statement.

47. If $\|\vec{u}\| = 1$ and $\|\vec{v}\| > 0$, then $\|\vec{u} + \vec{v}\| \geq 1$.

48. The vector $c\vec{u}$ has the same direction as $\vec{u}$.

49. $\|\vec{v} - \vec{u}\|$ is the length of the shorter of the two diagonals of the parallelogram determined by $\vec{u}$ and $\vec{v}$.

50. Given three vectors $\vec{u}$, $\vec{v}$, and $\vec{w}$, if $\vec{u} + \vec{w} = \vec{u}$ then it is possible for $\vec{v} + \vec{w} \neq \vec{v}$.

■ In Problems 51–53, give an example of:

51. A vector $\vec{v}$ of length 2 with a positive $\vec{k}$-component and lying on a plane parallel to the yz-plane.

52. Two unit vectors $\vec{u}$ and $\vec{v}$ for which $\vec{v} - \vec{u}$ is also a unit vector.

53. Two vectors $\vec{u}$ and $\vec{v}$ that have difference vector $\vec{w} = 2\vec{i} + 3\vec{j}$.

■ Are the statements in Problems 54–63 true or false? Give reasons for your answer.

54. There is exactly one unit vector parallel to a given nonzero vector $\vec{v}$.

55. The vector $\dfrac{1}{\sqrt{3}}\vec{i} + \dfrac{-1}{\sqrt{3}}\vec{j} + \dfrac{2}{\sqrt{3}}\vec{k}$ is a unit vector.

56. The length of the vector $2\vec{v}$ is twice the length of the vector $\vec{v}$.

57. If $\vec{v}$ and $\vec{w}$ are any two vectors, then $\|\vec{v} + \vec{w}\| = \|\vec{v}\| + \|\vec{w}\|$.

58. If $\vec{v}$ and $\vec{w}$ are any two vectors, then $\|\vec{v} - \vec{w}\| = \|\vec{v}\| - \|\vec{w}\|$.

59. The vectors $2\vec{i} - \vec{j} + \vec{k}$ and $\vec{i} - 2\vec{j} + \vec{k}$ are parallel.

60. The vector $\vec{u} + \vec{v}$ is always larger in magnitude than both $\vec{u}$ and $\vec{v}$.

61. For any scalar c and vector $\vec{v}$ we have $\|c\vec{v}\| = c\|\vec{v}\|$.

62. The displacement vector from $(1, 1, 1)$ to $(1, 2, 3)$ is $-\vec{j} - 2\vec{k}$.

63. The displacement vector from (a, b) to (c, d) is the same as the displacement vector from (c, d) to (a, b).

13.2 VECTORS IN GENERAL

Besides displacement, there are many quantities that have both magnitude and direction and are added and multiplied by scalars in the same way as displacements. Any such quantity is called a *vector* and is represented by an arrow in the same manner we represent displacements. The length of the arrow is the *magnitude* of the vector, and the direction of the arrow is the direction of the vector.

Velocity Versus Speed

The speed of a moving body tells us how fast it is moving, say 80 km/hr. The speed is just a number; it is therefore a scalar. The velocity, on the other hand, tells us both how fast the body is moving and the direction of motion; it is a vector. For instance, if a car is heading northeast at 80 km/hr, then its velocity is a vector of length 80 pointing northeast.

> The **velocity vector** of a moving object is a vector whose magnitude is the speed of the object and whose direction is the direction of its motion.

The velocity vector is the displacement vector if the object moves at constant velocity for one unit of time.

Example 1   A car is traveling north at a speed of 100 km/hr, while a plane above is flying horizontally southwest at a speed of 500 km/hr. Draw the velocity vectors of the car and the plane.

Solution   Figure 13.19 shows the velocity vectors. The plane's velocity vector is five times as long as the car's, because its speed is five times as great.

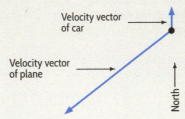

Figure 13.19: Velocity vector of the car is 100 km/hr north and of the plane is 500 km/hr southwest

The next example illustrates that the velocity vectors for two motions add to give the velocity vector for the combined motion, just as displacements do.

Example 2   A riverboat is moving with velocity $\vec{v}$ and a speed of 8 km/hr relative to the water. In addition, the river has a current $\vec{c}$ and a speed of 1 km/hr. (See Figure 13.20.) What is the physical significance of the vector $\vec{v} + \vec{c}$?

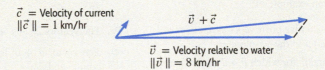

Figure 13.20: Boat's velocity relative to the river bed is the sum $\vec{v} + \vec{c}$

Solution   The vector $\vec{v}$ shows how the boat is moving relative to the water, while $\vec{c}$ shows how the water is moving relative to the riverbed. During an hour, imagine that the boat first moves 8 km relative to the water, which remains still; this displacement is represented by $\vec{v}$. Then imagine the water moving 1 km while the boat remains stationary relative to the water; this displacement is represented by $\vec{c}$. The combined displacement is represented by $\vec{v} + \vec{c}$. Thus, the vector $\vec{v} + \vec{c}$ is the velocity of the boat relative to the riverbed.

Note that the effective speed of the boat is not necessarily 9 km/hr unless the boat is moving in the direction of the current. Although we add the velocity vectors, we do not necessarily add their lengths.

Scalar multiplication also makes sense for velocity vectors. For example, if $\vec{v}$ is a velocity vector, then $-2\vec{v}$ represents a velocity of twice the magnitude in the opposite direction.

Example 3   A ball is moving with velocity $\vec{v}$ when it hits a wall at a right angle and bounces straight back, with its speed reduced by 20%. Express its new velocity in terms of the old one.

Solution   The new velocity is $-0.8\vec{v}$, where the negative sign expresses the fact that the new velocity is in the direction opposite to the old.

We can represent velocity vectors in components in the same way we did on page 751.

Example 4   Represent the velocity vectors of the car and the plane in Example 1 using components. Take north to be the positive y-axis, east to be the positive x-axis, and upward to be the positive z-axis.

Solution   The car is traveling north at 100 km/hr, so the y-component of its velocity is $100\vec{j}$ and the x-component is $0\vec{i}$. Since it is traveling horizontally, the z-component is $0\vec{k}$. So we have

$$\text{Velocity of car} = 0\vec{i} + 100\vec{j} + 0\vec{k} = 100\vec{j}.$$

The plane's velocity vector also has $\vec{k}$ component equal to zero. Since it is traveling southwest, its $\vec{i}$ and $\vec{j}$ components have negative coefficients (north and east are positive). Since the plane is traveling at 500 km/hr, in one hour it is displaced $500/\sqrt{2} \approx 354$ km to the west and 354 km to the south. (See Figure 13.21.) Thus,

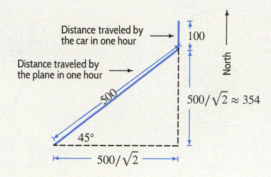

Figure 13.21: Distance traveled by the plane and car in one hour

$$\text{Velocity of plane} = -(500\cos 45°)\vec{i} - (500\sin 45°)\vec{j} \approx -354\vec{i} - 354\vec{j}.$$

Of course, if the car were climbing a hill or if the plane were descending for a landing, then the $\vec{k}$ component would not be zero.

Acceleration

Another example of a vector quantity is acceleration. Acceleration, like velocity, is specified by both a magnitude and a direction — for example, the acceleration due to gravity is 9.81 m/sec² vertically downward.

Force

Force is another example of a vector quantity. Suppose you push on an open door. The result depends both on how hard you push and in what direction. Thus, to specify a force we must give its magnitude (or strength) and the direction in which it is acting. For example, the gravitational force exerted on an object by the earth is a vector pointing from the object toward the center of the earth; its magnitude is the strength of the gravitational force.

Example 5   The earth travels around the sun in an ellipse. The gravitational force on the earth and the velocity of the earth are governed by the following laws:

Newton's Law of Gravitation: The gravitational attraction, $\vec{F}$, of a mass m_1 on a mass m_2 at a distance r has magnitude $||\vec{F}|| = Gm_1m_2/r^2$, where G is a constant, and is directed from m_2 toward m_1.
Kepler's Second Law: The line joining a planet to the sun sweeps out equal areas in equal times.

(a) Sketch vectors representing the gravitational force of the sun on the earth at two different positions in the earth's orbit.
(b) Sketch the velocity vector of the earth at two points in its orbit.

Solution   (a) Figure 13.22 shows the earth orbiting the sun. Note that the gravitational force vector always points toward the sun and is larger when the earth is closer to the sun because of the r^2 term in the denominator. (In fact, the real orbit looks much more like a circle than we have shown here.)

(b) The velocity vector points in the direction of motion of the earth. Thus, the velocity vector is tangent to the ellipse. See Figure 13.23. Furthermore, the velocity vector is longer at points of the orbit where the planet is moving quickly, because the magnitude of the velocity vector is the speed. Kepler's Second Law enables us to determine when the earth is moving quickly and when it is moving slowly. Over a fixed period of time, say one month, the line joining the earth to the sun sweeps out a sector having a certain area. Figure 13.23 shows two sectors swept out in two different one-month time-intervals. Kepler's law says that the areas of the two sectors are the same. Thus, the earth must move farther in a month when it is close to the sun than when it is far from the sun. Therefore, the earth moves faster when it is closer to the sun and slower when it is farther away.

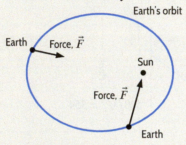

Figure 13.22: Gravitational force, $\vec{F}$, exerted by the sun on the earth: Greater magnitude closer to sun

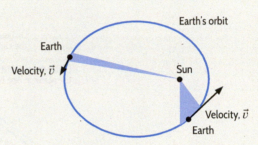

Figure 13.23: The velocity vector, $\vec{v}$, of the earth: Greater magnitude closer to the sun

Properties of Addition and Scalar Multiplication

In general, vectors add, subtract, and are multiplied by scalars in the same way as displacement vectors. Thus, for any vectors $\vec{u}$, $\vec{v}$, and $\vec{w}$ and any scalars α and β, we have the following properties:

Commutativity
1. $\vec{v} + \vec{w} = \vec{w} + \vec{v}$

Associativity
2. $(\vec{u} + \vec{v}) + \vec{w} = \vec{u} + (\vec{v} + \vec{w})$
3. $\alpha(\beta\vec{v}) = (\alpha\beta)\vec{v}$

Distributivity
4. $(\alpha + \beta)\vec{v} = \alpha\vec{v} + \beta\vec{v}$
5. $\alpha(\vec{v} + \vec{w}) = \alpha\vec{v} + \alpha\vec{w}$

Identity
6. $1\vec{v} = \vec{v}$
7. $0\vec{v} = \vec{0}$
8. $\vec{v} + \vec{0} = \vec{v}$
9. $\vec{w} + (-1)\vec{v} = \vec{w} - \vec{v}$

Problems 28–35 at the end of this section ask for a justification of these results in terms of displacement vectors.

Using Components

Example 6   A plane, heading due east at an airspeed of 600 km/hr, experiences a wind of 50 km/hr blowing toward the northeast. Find the plane's direction and ground speed.

Solution   We choose a coordinate system with the x-axis pointing east and the y-axis pointing north. See Figure 13.24.

The airspeed tells us the speed of the plane relative to still air. Thus, the plane is moving due east with velocity $\vec{v} = 600\vec{i}$ relative to still air. In addition, the air is moving with a velocity $\vec{w}$. Writing $\vec{w}$ in components, we have

$$\vec{w} = (50\cos 45°)\vec{i} + (50\sin 45°)\vec{j} = 35.4\vec{i} + 35.4\vec{j}.$$

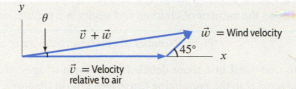

Figure 13.24: Plane's velocity relative to the ground is the sum $\vec{v} + \vec{w}$

The vector $\vec{v} + \vec{w}$ represents the displacement of the plane in one hour relative to the ground. Therefore, $\vec{v} + \vec{w}$ is the velocity of the plane relative to the ground. In components, we have

$$\vec{v} + \vec{w} = 600\vec{i} + \left(35.4\vec{i} + 35.4\vec{j}\right) = 635.4\vec{i} + 35.4\vec{j}.$$

The direction of the plane's motion relative to the ground is given by the angle θ in Figure 13.24, where

$$\tan\theta = \frac{35.4}{635.4},$$

so

$$\theta = \arctan\left(\frac{35.4}{635.4}\right) = 3.2°.$$

The ground speed is the speed of the plane relative to the ground, so

$$\text{Ground speed} = ||\vec{v} + \vec{w}|| = \sqrt{635.4^2 + 35.4^2} = 636.4 \text{ km/hr.}$$

Thus, the speed of the plane relative to the ground has been increased slightly by the wind. (This is as we would expect, as the wind has a positive component in the direction in which the plane is traveling.) The angle θ shows how far the plane is blown off course by the wind.

Vectors in n Dimensions

Using the alternative notation $\vec{v} = (v_1, v_2, v_3)$ for a vector in 3-space, we can define a vector in n dimensions as a string of n numbers. Thus, a vector in n dimensions can be written as

$$\vec{c} = (c_1, c_2, \ldots, c_n).$$

Addition and scalar multiplication are defined by the formulas

$$\vec{v} + \vec{w} = (v_1, v_2, \ldots, v_n) + (w_1, w_2, \ldots, w_n) = (v_1 + w_1, v_2 + w_2, \ldots, v_n + w_n)$$

and

$$\lambda\vec{v} = \lambda(v_1, v_2, \ldots, v_n) = (\lambda v_1, \lambda v_2, \ldots, \lambda v_n).$$

Why Do We Want Vectors in n Dimensions?

Vectors in two and three dimensions can be used to model displacement, velocities, or forces. But what about vectors in n dimensions? There is another interpretation of 3-dimensional vectors (or 3-vectors) that is useful: they can be thought of as listing three different quantities—for example, the displacements parallel to the x-, y-, and z-axes. Similarly, the n-vector

$$\vec{c} = (c_1, c_2, \ldots, c_n)$$

can be thought of as a way of keeping n different quantities organized. For example, a *population* vector $\vec{N}$ shows the number of children and adults in a population:

$$\vec{N} = (\text{Number of children, Number of adults}),$$

or, if we are interested in a more detailed breakdown of ages, we might give the number in each ten-year age bracket in the population (up to age 110) in the form

$$\vec{N} = (N_1, N_2, N_3, N_4, \ldots, N_{10}, N_{11}),$$

where N_1 is the population aged 0–9, and N_2 is the population aged 10–19, and so on.

A *consumption* vector

$$\vec{q} = (q_1, q_2, \ldots, q_n)$$

shows the quantities q_1, q_2, ..., q_n consumed of each of n different goods. A *price* vector

$$\vec{p} = (p_1, p_2, \ldots, p_n)$$

contains the prices of n different items.

In 1907, Hermann Minkowski used vectors with four components when he introduced *space-time coordinates*, whereby each event is assigned a vector position $\vec{v}$ with four coordinates, three for its position in space and one for time:

$$\vec{v} = (x, y, z, t).$$

Example 7   Suppose the vector $\vec{I}$ represents the number of copies, in thousands, made by each of four copy centers in the month of December and $\vec{J}$ represents the number of copies made at the same four copy centers during the previous eleven months (the "year-to-date"). If $\vec{I} = (25, 211, 818, 642)$, and $\vec{J} = (331, 3227, 1377, 2570)$, compute $\vec{I} + \vec{J}$. What does this sum represent?

Solution   The sum is

$$\vec{I} + \vec{J} = (25 + 331, 211 + 3227, 818 + 1377, 642 + 2570) = (356, 3438, 2195, 3212).$$

Each term in $\vec{I} + \vec{J}$ represents the sum of the number of copies made in December plus those in the previous eleven months, that is, the total number of copies made during the entire year at that particular copy center.

Example 8   The price vector $\vec{p} = (p_1, p_2, p_3)$ represents the prices in dollars of three goods. Write a vector that gives the prices of the same goods in cents.

Solution   The prices in cents are $100p_1$, $100p_2$, and $100p_3$ respectively, so the new price vector is

$$(100p_1, 100p_2, 100p_3) = 100\vec{p}.$$

Summary for Section 13.2

- **Velocity, acceleration** and **force** are all examples of vector quantities.
- Vector addition and scalar multiplication satisfy many algebraic properties, such as **commutativity** and **associativity**.
- **Vectors in higher dimensions** can be used to represent collections of related quantities. For example, a price vector $\vec{p} = (p_1, p_2, \ldots, p_n)$ can represent the prices of n different commodities.

Exercises and Problems for Section 13.2 Online Resource: Additional Problems for Section 13.2

EXERCISES

In Exercises 1–5, say whether the given quantity is a vector or a scalar.

1. The population of the US.

2. The distance from Seattle to St. Louis.

3. The temperature at a point on the earth's surface.

4. The magnetic field at a point on the earth's surface.

5. The populations of each of the 50 states.

6. Give the components of the velocity vector for wind blowing at 10 km/hr toward the southeast. (Assume north is in the positive y-direction.)

7. Give the components of the velocity vector of a boat that is moving at 40 km/hr in a direction 20° south of west. (Assume north is in the positive y-direction.)

8. A car is traveling at a speed of 50 km/hr. The positive y-axis is north and the positive x-axis is east. Resolve the car's velocity vector (in 2-space) into components if the car is traveling in each of the following directions:

   (a) East
   (b) South
   (c) Southeast
   (d) Northwest.

9. Which is traveling faster, a car whose velocity vector is $21\vec{i} + 35\vec{j}$ or a car whose velocity vector is $40\vec{i}$, assuming that the units are the same for both directions?

10. What angle does a force of $\vec{F} = 15\vec{i} + 18\vec{j}$ make with the x-axis?

PROBLEMS

11. The velocity of the current in a river is $\vec{c} = 0.6\vec{i} + 0.8\vec{j}$ km/hr. A boat moves relative to the water with velocity $\vec{v} = 8\vec{i}$ km/hr.

   (a) What is the speed of the boat relative to the riverbed?
   (b) What angle does the velocity of the boat relative to the riverbed make with the vector $\vec{v}$? What does this angle tell us in practical terms?

12. The current in Problem 11 is twice as fast and in the opposite direction. What is the speed of the boat with respect to the riverbed?

13. A boat is heading due east at 25 km/hr (relative to the water). The current is moving toward the southwest at 10 km/hr.

   (a) Give the vector representing the actual movement of the boat.
   (b) How fast is the boat going, relative to the ground?
   (c) By what angle does the current push the boat off of its due east course?

14. A truck is traveling due north at 30 km/hr approaching a crossroad. On a perpendicular road a police car is traveling west toward the intersection at 40 km/hr. Both vehicles will reach the crossroad in exactly one hour. Find the vector currently representing the displacement from the police car to the truck.

15. An airplane heads northeast at an airspeed of 700 km/hr, but there is a wind blowing from the west at 60 km/hr. In what direction does the plane end up flying? What is its speed relative to the ground?

16. Two forces, represented by the vectors $\vec{F}_1 = 8\vec{i} - 6\vec{j}$ and $\vec{F}_2 = 3\vec{i} + 2\vec{j}$, are acting on an object. Give a vector representing the force that must be applied to the object if it is to remain stationary.

17. An airplane is flying at an airspeed of 500 km/hr in a wind blowing at 60 km/hr toward the southeast. In what direction should the plane head to end up going due east? What is the airplane's speed relative to the ground?

18. The current in a river is pushing a boat in direction 25° north of east with a speed of 12 km/hr. The wind is pushing the same boat in a direction 80° south of east with a speed of 7 km/hr. Find the velocity vector of the boat's engine (relative to the water) if the boat actually moves due east at a speed of 40 km/hr relative to the ground.

19. A large ship is being towed by two tugs. The larger tug exerts a force which is 25% greater than the smaller tug and at an angle of 30 degrees north of east. Which direction must the smaller tug pull to ensure that the ship travels due east?

20. An object P is pulled by a force $\vec{F}_1$ of magnitude 15 lb at an angle of 20 degrees north of east. Give the components of a force $\vec{F}_2$ of magnitude 20 lb to ensure that P moves due east.

21. An object is to be moved vertically upward by a crane. As the crane cannot get directly above the object, three ropes are attached to guide the object. One rope is pulled parallel to the ground with a force of 100 newtons in a direction 30° north of east. The second rope is pulled parallel to the ground with a force of 70 newtons in a direction 80° south of east. If the crane is attached to the third rope and can pull with a total force of 3000 newtons, find the force vector for the crane. What is the resulting (total) force on the object? (Assume vector $\vec{i}$ points east, vector $\vec{j}$ points north, and vector $\vec{k}$ points vertically up.)

22. The earth is at the origin, the moon is at the point $(384, 0)$, and a spaceship is at $(280, 90)$, where distance is in thousands of kilometers.

    (a) What is the displacement vector of the moon relative to the earth? Of the spaceship relative to the earth? Of the spaceship relative to the moon?
    (b) How far is the spaceship from the earth? From the moon?
    (c) The gravitational force on the spaceship from the earth is 461 newtons and from the moon is 26 newtons. What is the resulting force?

23. A particle moving with speed v hits a barrier at an angle of $60°$ and bounces off at an angle of $60°$ in the opposite direction with speed reduced by 20 percent. See Figure 13.25. Find the velocity vector of the object after impact.

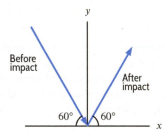

Figure 13.25

24. There are five students in a class. Their scores on the midterm (out of 100) are given by the vector $\vec{v} = (73, 80, 91, 65, 84)$. Their scores on the final (out of 100) are given by $\vec{w} = (82, 79, 88, 70, 92)$. If the final counts twice as much as the midterm, find a vector giving the total scores (as a percentage) of the students.

25. The price vector of beans, rice, and tofu is $(1.6, 1.28, 2.60)$ in dollars per pound. Express it in dollars per ounce.

26. An object is moving counterclockwise at a constant speed around the circle $x^2 + y^2 = 1$, where x and y are measured in meters. It completes one revolution every minute.

    (a) What is its speed?

    (b) What is its velocity vector 30 seconds after it passes the point $(1, 0)$? Does your answer change if the object is moving clockwise? Explain.

27. An object is attached by a string to a fixed point and rotates 30 times per minute in a horizontal plane. Show that the speed of the object is constant but the velocity is not. What does this imply about the acceleration?

■ In Problems 28–35, use the geometric definition of addition and scalar multiplication to explain each of the properties.

28. $\vec{w} + \vec{v} = \vec{v} + \vec{w}$
29. $(\alpha + \beta)\vec{v} = \alpha\vec{v} + \beta\vec{v}$
30. $\alpha(\vec{v} + \vec{w}) = \alpha\vec{v} + \alpha\vec{w}$
31. $\alpha(\beta\vec{v}) = (\alpha\beta)\vec{v}$
32. $\vec{v} + \vec{0} = \vec{v}$
33. $1\vec{v} = \vec{v}$
34. $\vec{v} + (-1)\vec{w} = \vec{v} - \vec{w}$
35. $(\vec{u} + \vec{v}) + \vec{w} = \vec{u} + (\vec{v} + \vec{w})$

36. In the game of laser tag, you shoot a harmless laser gun and try to hit a target worn at the waist by other players. Suppose you are standing at the origin of a three-dimensional coordinate system and that the xy-plane is the floor. Suppose that waist-high is 3 feet above floor level and that eye level is 5 feet above the floor. Three of your friends are your opponents. One is standing so that his target is 30 feet along the x-axis, another lying down so that his target is at the point $x = 20$, $y = 15$, and the third lying in ambush so that his target is at a point 8 feet above the point $x = 12$, $y = 30$.

    (a) If you aim with your gun at eye level, find the vector from your gun to each of the three targets.
    (b) If you shoot from waist height, with your gun one foot to the right of the center of your body as you face along the x-axis, find the vector from your gun to each of the three targets.

37. A car drives northeast downhill on a $5°$ incline at a constant speed of 60 miles per hour. The positive x-axis points east, the y-axis north, and the z-axis up. Resolve the car's velocity into components.

Strengthen Your Understanding

■ In Problems 38–39, explain what is wrong with the statement.

38. Two vectors in 3-space that have equal $\vec{k}$-components and the same magnitude must be the same vector.

39. A vector $\vec{v}$ in the plane whose $\vec{i}$-component is 0.5 has smaller magnitude than the vector $\vec{w} = 2\vec{i}$.

■ In Problems 40–41, give an example of:

40. A nonzero vector $\vec{F}$ on the plane that when combined with the force vector $\vec{G} = \vec{i} + \vec{j}$ results in a combined force vector $\vec{R}$ with a positive $\vec{i}$-component and a negative $\vec{j}$-component.

41. Nonzero vectors $\vec{u}$ and $\vec{v}$ such that $\|\vec{u} + \vec{v}\| = \|\vec{u}\| + \|\vec{v}\|$.

■ In Problems 42–47, is the quantity a vector? Give a reason for your answer.

42. Velocity
43. Speed
44. Force
45. Area
46. Acceleration
47. Volume

13.3   THE DOT PRODUCT

We have seen how to add vectors; can we multiply two vectors together? In the next two sections we will see two different ways of doing so: the *scalar product* (or *dot product*), which produces a scalar, and the *vector product* (or *cross product*), which produces a vector.

Definition of the Dot Product

The dot product links geometry and algebra. We already know how to calculate the length of a vector from its components; the dot product gives us a way of computing the angle between two vectors. For any two vectors $\vec{v} = v_1\vec{i} + v_2\vec{j} + v_3\vec{k}$ and $\vec{w} = w_1\vec{i} + w_2\vec{j} + w_3\vec{k}$, shown in Figure 13.26, we define a scalar as follows:

> The following two definitions of the **dot product**, or **scalar product**, $\vec{v} \cdot \vec{w}$, are equivalent:
> - **Geometric definition**
> $$\vec{v} \cdot \vec{w} = \|\vec{v}\|\|\vec{w}\| \cos \theta \qquad \text{where } \theta \text{ is the angle between } \vec{v} \text{ and } \vec{w} \text{ and } 0 \le \theta \le \pi.$$
> - **Algebraic definition**
> $$\vec{v} \cdot \vec{w} = v_1 w_1 + v_2 w_2 + v_3 w_3.$$
> Notice that the dot product of two vectors is a *number*, not a vector.

Why don't we give just one definition of $\vec{v} \cdot \vec{w}$? The reason is that both definitions are equally important; the geometric definition gives us a picture of what the dot product means and the algebraic definition gives us a way of calculating it.

How do we know the two definitions are equivalent—that is, they really do define the same thing? First, we observe that the two definitions give the same result in a particular example. Then we show why they are equivalent in general.

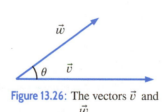

Figure 13.26: The vectors $\vec{v}$ and $\vec{w}$

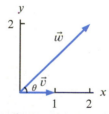

Figure 13.27: Calculating the dot product of the vectors $v = \vec{i}$ and $\vec{w} = 2\vec{i} + 2\vec{j}$ geometrically and algebraically gives the same result

Example 1   Suppose $\vec{v} = \vec{i}$ and $\vec{w} = 2\vec{i} + 2\vec{j}$. Compute $\vec{v} \cdot \vec{w}$ both geometrically and algebraically.

Solution   To use the geometric definition, see Figure 13.27. The angle between the vectors is $\pi/4$, or $45°$, and the lengths of the vectors are given by

$$\|\vec{v}\| = 1 \quad \text{and} \quad \|\vec{w}\| = 2\sqrt{2}.$$

Thus,

$$\vec{v} \cdot \vec{w} = \|\vec{v}\|\|\vec{w}\| \cos \theta = 1 \cdot 2\sqrt{2} \cos \left(\frac{\pi}{4}\right) = 2.$$

Using the algebraic definition, we get the same result:

$$\vec{v} \cdot \vec{w} = 1 \cdot 2 + 0 \cdot 2 = 2.$$

Why the Two Definitions of the Dot Product Give the Same Result

In the previous example, the two definitions give the same value for the dot product. To show that the geometric and algebraic definitions of the dot product always give the same result, we must show that, for any vectors $\vec{v} = v_1\vec{i} + v_2\vec{j} + v_3\vec{k}$ and $\vec{w} = w_1\vec{i} + w_2\vec{j} + w_3\vec{k}$ with an angle θ between them:

$$\|\vec{v}\,\|\|\vec{w}\,\|\cos\theta = v_1w_1 + v_2w_2 + v_3w_3.$$

One method follows; a method that does not use trigonometry is given in Problem 109 (available online).

Using the Law of Cosines. Suppose that $0 < \theta < \pi$, so that the vectors $\vec{v}$ and $\vec{w}$ form a triangle. (See Figure 13.28.) By the Law of Cosines, we have

$$\|\vec{v} - \vec{w}\,\|^2 = \|\vec{v}\,\|^2 + \|\vec{w}\,\|^2 - 2\|\vec{v}\,\|\|\vec{w}\,\|\cos\theta.$$

This result is also true for $\theta = 0$ and $\theta = \pi$. We calculate the lengths using components:

$$\|\vec{v}\,\|^2 = v_1^2 + v_2^2 + v_3^2$$
$$\|\vec{w}\,\|^2 = w_1^2 + w_2^2 + w_3^2$$
$$\|\vec{v} - \vec{w}\,\|^2 = (v_1 - w_1)^2 + (v_2 - w_2)^2 + (v_3 - w_3)^2$$
$$= v_1^2 - 2v_1w_1 + w_1^2 + v_2^2 - 2v_2w_2 + w_2^2 + v_3^2 - 2v_3w_3 + w_3^2.$$

Substituting into the Law of Cosines and canceling, we see that

$$-2v_1w_1 - 2v_2w_2 - 2v_3w_3 = -2\|\vec{v}\,\|\|\vec{w}\,\|\cos\theta.$$

Therefore we have the result we wanted, namely that:

$$v_1w_1 + v_2w_2 + v_3w_3 = \|\vec{v}\,\|\|\vec{w}\,\|\cos\theta.$$

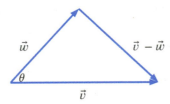

Figure 13.28: Triangle used in the justification of $\|\vec{v}\,\|\|\vec{w}\,\|\cos\theta = v_1w_1 + v_2w_2 + v_3w_3$

Properties of the Dot Product

The following properties of the dot product can be justified using the algebraic definition; see Problem 105. For a geometric interpretation of Property 3, see Problem 107 (both available online).

Properties of the Dot Product. For any vectors $\vec{u}$, $\vec{v}$, and $\vec{w}$ and any scalar λ,

1. $\vec{v} \cdot \vec{w} = \vec{w} \cdot \vec{v}$
2. $\vec{v} \cdot (\lambda\vec{w}) = \lambda(\vec{v} \cdot \vec{w}) = (\lambda\vec{v}) \cdot \vec{w}$
3. $(\vec{v} + \vec{w}) \cdot \vec{u} = \vec{v} \cdot \vec{u} + \vec{w} \cdot \vec{u}$

Perpendicularity, Magnitude, and Dot Products

Two vectors are perpendicular if the angle between them is $\pi/2$ or $90°$. Since $\cos(\pi/2) = 0$, if $\vec{v}$ and $\vec{w}$ are perpendicular, then $\vec{v} \cdot \vec{w} = 0$. Conversely, provided that $\vec{v} \cdot \vec{w} = 0$, then $\cos\theta = 0$, so $\theta = \pi/2$ and the vectors are perpendicular. Thus, we have the following result:

Two nonzero vectors $\vec{v}$ and $\vec{w}$ are **perpendicular**, or **orthogonal**, if and only if

$$\vec{v} \cdot \vec{w} = 0.$$

For example: $\vec{i} \cdot \vec{j} = 0, \vec{j} \cdot \vec{k} = 0, \vec{i} \cdot \vec{k} = 0.$

If we take the dot product of a vector with itself, then $\theta = 0$ and $\cos\theta = 1$. For any vector $\vec{v}$:

Magnitude and dot product are related as follows:

$$\vec{v} \cdot \vec{v} = \|\vec{v}\|^2.$$

For example: $\vec{i} \cdot \vec{i} = 1, \vec{j} \cdot \vec{j} = 1, \vec{k} \cdot \vec{k} = 1.$

Using the Dot Product

Depending on the situation, one definition of the dot product may be more convenient to use than the other. In Example 2, the geometric definition is the only one that can be used because we are not given components. In Example 3, the algebraic definition is used.

Example 2   Suppose the vector $\vec{b}$ is fixed and has length 2; the vector $\vec{a}$ is free to rotate and has length 3. What are the maximum and minimum values of the dot product $\vec{a} \cdot \vec{b}$ as the vector $\vec{a}$ rotates through all possible positions in the plane? What positions of $\vec{a}$ and $\vec{b}$ lead to these values?

Solution   The geometric definition gives $\vec{a} \cdot \vec{b} = \|\vec{a}\|\|\vec{b}\|\cos\theta = 3 \cdot 2\cos\theta = 6\cos\theta.$ Thus, the maximum value of $\vec{a} \cdot \vec{b}$ is 6, and it occurs when $\cos\theta = 1$ so $\theta = 0$, that is, when $\vec{a}$ and $\vec{b}$ point in the same direction. The minimum value of $\vec{a} \cdot \vec{b}$ is -6, and it occurs when $\cos\theta = -1$ so $\theta = \pi$, that is, when $\vec{a}$ and $\vec{b}$ point in opposite directions. (See Figure 13.29.)

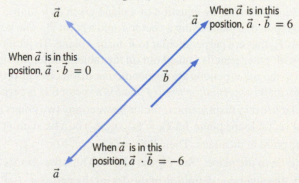

Figure 13.29: Maximum and minimum values of $\vec{a} \cdot \vec{b}$ obtained from a fixed vector $\vec{b}$ of length 2 and rotating vector $\vec{a}$ of length 3

Example 3   Which pairs from the following list of 3-dimensional vectors are perpendicular to one another?

$$\vec{u} = \vec{i} + \sqrt{3}\,\vec{k}, \quad \vec{v} = \vec{i} + \sqrt{3}\,\vec{j}, \quad \vec{w} = \sqrt{3}\,\vec{i} + \vec{j} - \vec{k}.$$

Solution   The geometric definition tells us that two vectors are perpendicular if and only if their dot product is zero. Since the vectors are given in components, we calculate dot products using the algebraic definition:

$$\vec{v} \cdot \vec{u} = (\vec{i} + \sqrt{3}\,\vec{j} + 0\vec{k}\,) \cdot (\vec{i} + 0\vec{j} + \sqrt{3}\,\vec{k}\,) = 1 \cdot 1 + \sqrt{3} \cdot 0 + 0 \cdot \sqrt{3} = 1,$$

$$\vec{v} \cdot \vec{w} = (\vec{i} + \sqrt{3}\,\vec{j} + 0\vec{k}\,) \cdot (\sqrt{3}\,\vec{i} + \vec{j} - \vec{k}\,) = 1 \cdot \sqrt{3} + \sqrt{3} \cdot 1 + 0(-1) = 2\sqrt{3},$$

$$\vec{w} \cdot \vec{u} = (\sqrt{3}\,\vec{i} + \vec{j} - \vec{k}\,) \cdot (\vec{i} + 0\vec{j} + \sqrt{3}\,\vec{k}\,) = \sqrt{3} \cdot 1 + 1 \cdot 0 + (-1) \cdot \sqrt{3} = 0.$$

So the only two vectors that are perpendicular are $\vec{w}$ and $\vec{u}$.

Example 4   Compute the angle between the vectors $\vec{v}$ and $\vec{w}$ from Example 3.

Solution   We know that $\vec{v} \cdot \vec{w} = \|\vec{v}\|\|\vec{w}\| \cos\theta$, so $\cos\theta = \dfrac{\vec{v} \cdot \vec{w}}{\|\vec{v}\|\|\vec{w}\|}$. From Example 3, we know that $\vec{v} \cdot \vec{w} = 2\sqrt{3}$. This gives:

$$\cos\theta = \frac{2\sqrt{3}}{\|\vec{v}\|\|\vec{w}\|} = \frac{2\sqrt{3}}{\sqrt{1^2 + \left(\sqrt{3}\right)^2 + 0^2}\sqrt{\left(\sqrt{3}\right)^2 + 1^2 + (-1)^2}} = \frac{\sqrt{3}}{\sqrt{5}}$$

so $\theta = \arccos\left(\dfrac{\sqrt{3}}{\sqrt{5}}\right) = 39.2315°.$

Normal Vectors and the Equation of a Plane

In Section 12.4 we wrote the equation of a plane given its x-slope, y-slope and z-intercept. Now we write the equation of a plane using a vector $\vec{n}$ and a point P_0. The key idea is that all the displacement vectors from P_0 that are perpendicular to $\vec{n}$ form a plane. To picture this, imagine a pencil balanced on a table, with other pencils fanned out on the table in different directions. The upright pencil is $\vec{n}$, its base is P_0, the other pencils are perpendicular displacement vectors, and the table is the plane.

More formally, a *normal vector* to a plane is a vector that is perpendicular to the plane, that is, it is perpendicular to every displacement vector between any two points in the plane. Let $\vec{n} = a\vec{i} + b\vec{j} + c\vec{k}$ be a normal vector to the plane, let $P_0 = (x_0, y_0, z_0)$ be a fixed point in the plane, and let $P = (x, y, z)$ be any other point in the plane. Then $\overrightarrow{P_0P} = (x - x_0)\vec{i} + (y - y_0)\vec{j} + (z - z_0)\vec{k}$ is a vector whose head and tail both lie in the plane. (See Figure 13.30.) Thus, the vectors $\vec{n}$ and $\overrightarrow{P_0P}$ are perpendicular, so $\vec{n} \cdot \overrightarrow{P_0P} = 0$. The algebraic definition of the dot product gives $\vec{n} \cdot \overrightarrow{P_0P} = a(x - x_0) + b(y - y_0) + c(z - z_0)$, so we obtain the following result:

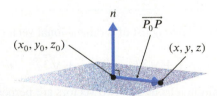

Figure 13.30: Plane with normal $\vec{n}$ and containing a fixed point (x_0, y_0, z_0)

> The **equation of the plane** with normal vector $\vec{n} = a\vec{i} + b\vec{j} + c\vec{k}$ and containing the point $P_0 = (x_0, y_0, z_0)$ is
>
> $$a(x - x_0) + b(y - y_0) + c(z - z_0) = 0.$$
>
> Letting $d = ax_0 + by_0 + cz_0$ (a constant), we can write the equation of the plane in the form
>
> $$ax + by + cz = d.$$

Example 5   (a) Find the equation of the plane perpendicular to $\vec{n} = -\vec{i} + 3\vec{j} + 2\vec{k}$ and passing through the point $(1, 0, 4)$.
(b) Find a vector parallel to the plane.

Solution   (a) The equation of the plane is

$$-(x - 1) + 3(y - 0) + 2(z - 4) = 0,$$

which can be written as

$$-x + 3y + 2z = 7.$$

(b) Any vector $\vec{v}$ that is perpendicular to n is also parallel to the plane, so we look for any vector satisfying $\vec{v} \cdot \vec{n} = 0$; for example, $\vec{v} = 3\vec{i} + \vec{j}$. There are many other possible vectors.

Example 6   Find a normal vector to the plane with equation   (a) $x - y + 2z = 5$ (b) $z = 0.5x + 1.2y$.

Solution   (a) Since the coefficients of $\vec{i}$, $\vec{j}$, and $\vec{k}$ in a normal vector are the coefficients of x, y, and z in the equation of the plane, a normal vector is $\vec{n} = \vec{i} - \vec{j} + 2\vec{k}$.
(b) Before we can find a normal vector, we rewrite the equation of the plane in the form

$$0.5x + 1.2y - z = 0.$$

Thus, a normal vector is $\vec{n} = 0.5\vec{i} + 1.2\vec{j} - \vec{k}$.

The Dot Product in n Dimensions

The algebraic definition of the dot product can be extended to vectors in higher dimensions.

> If $\vec{u} = (u_1, \ldots, u_n)$ and $\vec{v} = (v_1, \ldots, v_n)$ then the dot product of $\vec{u}$ and $\vec{v}$ is the **scalar**
>
> $$\vec{u} \cdot \vec{v} = u_1 v_1 + \cdots + u_n v_n.$$

Example 7   A video store sells videos, tapes, CDs, and computer games. We define the quantity vector $\vec{q} = (q_1, q_2, q_3, q_4)$, where q_1, q_2, q_3, q_4 denote the quantities sold of each of the items, and the price vector $\vec{p} = (p_1, p_2, p_3, p_4)$, where p_1, p_2, p_3, p_4 denote the price per unit of each item. What does the dot product $\vec{p} \cdot \vec{q}$ represent?

Solution   The dot product is $\vec{p} \cdot \vec{q} = p_1 q_1 + p_2 q_2 + p_3 q_3 + p_4 q_4$. The quantity $p_1 q_1$ represents the revenue received by the store for the videos, $p_2 q_2$ represents the revenue for the tapes, and so on. The dot product represents the total revenue received by the store for the sale of these four items.

Resolving a Vector into Components: Projections

In Section 13.1, we resolved a vector into components parallel to the axes. Now we see how to resolve a vector, $\vec{v}$, into components, called $\vec{v}_{\text{parallel}}$ and $\vec{v}_{\text{perp}}$, which are parallel and perpendicular, respectively, to a given nonzero vector, $\vec{u}$. (See Figure 13.31.)

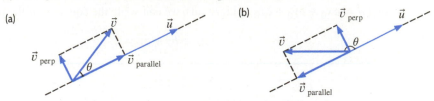

Figure 13.31: Resolving $\vec{v}$ into components parallel and perpendicular to $\vec{u}$
(a) $0 < \theta < \pi/2$ (b) $\pi/2 < \theta < \pi$

The projection of $\vec{v}$ on $\vec{u}$, written $\vec{v}_{\text{parallel}}$, measures (in some sense) how much the vector $\vec{v}$ is aligned with the vector $\vec{u}$. The length of $\vec{v}_{\text{parallel}}$ is the length of the shadow cast by $\vec{v}$ on a line in the direction of $\vec{u}$.

To compute $\vec{v}_{\text{parallel}}$, we assume $\vec{u}$ is a unit vector. (If not, create one by dividing by its length.) Then Figure 13.31(a) shows that, if $0 \leq \theta \leq \pi/2$:

$$\|\vec{v}_{\text{parallel}}\| = \|\vec{v}\| \cos \theta = \vec{v} \cdot \vec{u} \qquad (\text{since } \|\vec{u}\| = 1).$$

Now $\vec{v}_{\text{parallel}}$ is a scalar multiple of $\vec{u}$, and since $\vec{u}$ is a unit vector,

$$\vec{v}_{\text{parallel}} = (\|\vec{v}\| \cos \theta)\vec{u} = (\vec{v} \cdot \vec{u})\vec{u}.$$

A similar argument shows that if $\pi/2 < \theta \leq \pi$, as in Figure 13.31(b), this formula for $\vec{v}_{\text{parallel}}$ still holds. The vector $\vec{v}_{\text{perp}}$ is specified by

$$\vec{v}_{\text{perp}} = \vec{v} - \vec{v}_{\text{parallel}}.$$

Thus, we have the following results:

Projection of $\vec{v}$ on the Line in the Direction of the Unit Vector $\vec{u}$

If $\vec{v}_{\text{parallel}}$ and $\vec{v}_{\text{perp}}$ are components of $\vec{v}$ that are parallel and perpendicular, respectively, to $\vec{u}$, then

$$\text{Projection of } \vec{v} \text{ onto } \vec{u} = \vec{v}_{\text{parallel}} = (\vec{v} \cdot \vec{u})\vec{u} \qquad \text{provided } \|\vec{u}\| = 1$$

and $\vec{v} = \vec{v}_{\text{parallel}} + \vec{v}_{\text{perp}}$ so $\vec{v}_{\text{perp}} = \vec{v} - \vec{v}_{\text{parallel}}.$

Example 8   Figure 13.32 shows the force the wind exerts on the sail of a sailboat. Find the component of the force in the direction in which the sailboat is traveling.

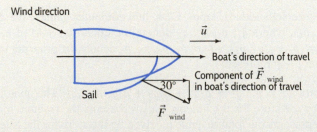

Figure 13.32: Wind moving a sailboat

Solution

Let $\vec{u}$ be a unit vector in the direction of travel. The force of the wind on the sail makes an angle of $30°$ with $\vec{u}$. Thus, the component of this force in the direction of $\vec{u}$ is

$$\vec{F}_{\text{parallel}} = (\vec{F} \cdot \vec{u})\vec{u} = \|\vec{F}\|(\cos 30°)\vec{u} = 0.87\|\vec{F}\|\vec{u}.$$

Thus, the boat is being pushed forward with about 87% of the total force due to the wind. (In fact, the interaction of wind and sail is much more complex than this model suggests.)

A Physical Interpretation of the Dot Product: Work

In physics, the word "work" has a different meaning from its everyday meaning. In physics, when a force of magnitude F acts on an object through a distance d, we say the *work*, W, done by the force is

$$W = Fd,$$

provided the force and the displacement are in the same direction. For example, if a 1 kg body falls 10 meters under the force of gravity, which is 9.8 newtons, then the work done by gravity is

$$W = (9.8 \text{ newtons}) \cdot (10 \text{ meters}) = 98 \text{ joules}.$$

What if the force and the displacement are not in the same direction? Suppose a force $\vec{F}$ acts on an object as it moves along a displacement vector $\vec{d}$. Let θ be the angle between $\vec{F}$ and $\vec{d}$. First, we assume $0 \leq \theta \leq \pi/2$. Figure 13.33 shows how we can resolve $\vec{F}$ into components that are parallel and perpendicular to $\vec{d}$:

$$\vec{F} = \vec{F}_{\text{parallel}} + \vec{F}_{\text{perp}}.$$

Then the work done by $\vec{F}$ is defined to be

$$W = \|\vec{F}_{\text{parallel}}\| \|\vec{d}\|.$$

We see from Figure 13.33 that $\vec{F}_{\text{parallel}}$ has magnitude $\|\vec{F}\| \cos \theta$. So the work is given by the dot product:

$$W = (\|\vec{F}\| \cos \theta)\|\vec{d}\| = \|\vec{F}\|\|\vec{d}\| \cos \theta = \vec{F} \cdot \vec{d}.$$

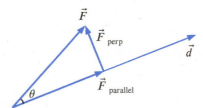

Figure 13.33: Resolving the force $\vec{F}$ into two forces, one parallel to $\vec{d}$, one perpendicular to $\vec{d}$

The formula $W = \vec{F} \cdot \vec{d}$ holds when $\pi/2 < \theta \leq \pi$ also. In that case, the work done by the force is negative and the object is moving against the force. Thus, we have the following definition:

> The **work**, W, done by a force $\vec{F}$ acting on an object through a displacement $\vec{d}$ is given by
>
> $$W = \vec{F} \cdot \vec{d}.$$

Example 9

How much work does the wind do on the sailboat from Example 8 if the boat moves 20 m and the wind's force is 120 newtons?

Solution   From Example 8, we know that the force of the wind $\vec{F}$ makes a 30° angle with the boat's displacement $\vec{d}$. Since $\|\vec{F}\| = 120$ and $\|\vec{d}\| = 20$, the work done by the wind on the boat is

$$W = \vec{F} \cdot \vec{d} = \|\vec{F}\|\|\vec{d}\| \cos 30° = 2078.461 \text{ joules}.$$

Notice that if the vectors $\vec{F}$ and $\vec{d}$ are parallel and in the same direction, with magnitudes F and d, then $\cos\theta = \cos 0 = 1$, so $W = \|\vec{F}\|\|\vec{d}\| = Fd$, which is the original definition. When the vectors are perpendicular, $\cos\theta = \cos(\pi/2) = 0$, so $W = 0$ and no work is done in the technical definition of the word. For example, if you carry a heavy box across the room at the same horizontal height, no work is done by gravity because the force of gravity is vertical but the motion is horizontal.

Summary for Section 13.3

- For any two vectors $\vec{v} = v_1\vec{i} + v_2\vec{j} + v_3\vec{k}$ and $\vec{w} = w_1\vec{i} + w_2\vec{j} + w_3\vec{k}$, we have two equivalent definitions of the **dot product**, $\vec{v} \cdot \vec{w}$:
  - **Geometric definition**:
    $\vec{v} \cdot \vec{w} = \|\vec{v}\|\|\vec{w}\| \cos\theta$ where θ is the angle between $\vec{v}$ and $\vec{w}$ and $0 \leq \theta \leq \pi$.
  - **Algebraic definition**:
    $\vec{v} \cdot \vec{w} = v_1w_1 + v_2w_2 + v_3w_3$.
- For any vectors $\vec{u}$, $\vec{v}$, and $\vec{w}$ and any scalar λ, the dot product has **properties**:
  1. $\vec{v} \cdot \vec{w} = \vec{w} \cdot \vec{v}$
  2. $\vec{v} \cdot (\lambda\vec{w}) = \lambda(\vec{v} \cdot \vec{w}) = (\lambda\vec{v}) \cdot \vec{w}$
  3. $(\vec{v} + \vec{w}) \cdot \vec{u} = \vec{v} \cdot \vec{u} + \vec{w} \cdot \vec{u}$
- Two nonzero vectors $\vec{v}$ and $\vec{w}$ are **perpendicular**, or **orthogonal**, if and only if

$$\vec{v} \cdot \vec{w} = 0.$$

- The **length** of a vector can be computed using the dot product:

$$\vec{v} \cdot \vec{v} = \|\vec{v}\|^2.$$

- The **equation of the plane** with normal vector $\vec{n} = a\vec{i} + b\vec{j} + c\vec{k}$ and containing the point $P_0 = (x_0, y_0, z_0)$ is

$$a(x - x_0) + b(y - y_0) + c(z - z_0) = 0.$$

or we can combine constants and write

$$ax + by + cz = d.$$

- Given vectors $\vec{u}$ and $\vec{v}$, the vectors $\vec{v}_{\text{parallel}}$ and $\vec{v}_{\text{perp}}$ are the components of $\vec{v}$ that are parallel and perpendicular, respectively, to $\vec{u}$. We have $\vec{v} = \vec{v}_{\text{parallel}} + \vec{v}_{\text{perp}}$.
- The vector $\vec{v}_{\text{parallel}}$ is also called **the projection** of $\vec{v}$ onto the vector $\vec{u}$.
- The **projection** of $\vec{v}$ onto a **unit vector** $\vec{u}$ can be calculated using

$$\text{Projection of } \vec{v} \text{ onto } \vec{u} = \vec{v}_{\text{parallel}} = (\vec{v} \cdot \vec{u})\vec{u}.$$

- We can find the **component** of $\vec{v}$ **perpendicular** to $\vec{u}$ by

$$\vec{v}_{\text{perp}} = \vec{v} - \vec{v}_{\text{parallel}}.$$

Exercises and Problems for Section 13.3   Online Resource: Additional Problems for Section 13.3
EXERCISES

In Exercises **1–4**, evaluate the dot product.

1. $(3\vec{i} + 2\vec{j} - 5\vec{k}) \cdot (\vec{i} - 2\vec{j} - 3\vec{k})$

2. $(\vec{i} + \vec{j} + \vec{k}) \cdot (4\vec{i} + 5\vec{j} + 6\vec{k})$

3. $(3\vec{i} - 2\vec{j} - 4\vec{k}) \cdot (3\vec{i} - 2\vec{j} - 4\vec{k})$

4. $(2i + 5\vec{k}) \cdot 10\vec{j}$

In Exercises **5–6**, evaluate $\vec{u} \cdot \vec{w}$.

5. $\|\vec{u}\| = 3$, $\|\vec{w}\| = 5$; the angle between $\vec{u}$ and $\vec{w}$ is $45°$.

6. $\|\vec{u}\| = 10$, $\|\vec{w}\| = 20$; the angle between $\vec{u}$ and $\vec{w}$ is $120°$.

For Exercises **7–15**, perform the following operations on the given 3-dimensional vectors.

$$\vec{a} = 2\vec{j} + \vec{k} \qquad \vec{b} = -3\vec{i} + 5\vec{j} + 4\vec{k} \qquad \vec{c} = \vec{i} + 6\vec{j}$$
$$\vec{y} = 4\vec{i} - 7\vec{j} \qquad \vec{z} = \vec{i} - 3\vec{j} - \vec{k}$$

7. $\vec{a} \cdot \vec{y}$

8. $\vec{c} \cdot \vec{y}$

9. $\vec{a} \cdot \vec{b}$

10. $\vec{a} \cdot \vec{z}$

11. $\vec{c} \cdot \vec{a} + \vec{a} \cdot \vec{y}$

12. $\vec{a} \cdot (\vec{c} + \vec{y})$

13. $(\vec{a} \cdot \vec{b})\vec{a}$

14. $(\vec{a} \cdot \vec{y})(\vec{c} \cdot \vec{z})$

15. $((\vec{c} \cdot \vec{c})\vec{a}) \cdot \vec{a}$

In Exercises **16–20**, find a normal vector to the plane.

16. $2x + y - z = 5$

17. $2(x - z) = 3(x + y)$

18. $1.5x + 3.2y + z = 0$

19. $z = 3x + 4y - 7$

20. $\pi(x - 1) = (1 - \pi)(y - z) + \pi$

In Exercises **21–27**, find an equation of a plane that satisfies the given conditions.

21. Through $(1, 5, 2)$ perpendicular to $3\vec{i} - \vec{j} + 4\vec{k}$.

22. Through $(2, -1, 3)$ perpendicular to $5\vec{i} + 4\vec{j} - \vec{k}$.

23. Through $(1, 3, 5)$ and normal to $\vec{i} - \vec{j} + \vec{k}$.

24. Perpendicular to $5\vec{i} + \vec{j} - 2\vec{k}$ and passing through $(0, 1, -1)$.

25. Parallel to $2x + 4y - 3z = 1$ and through $(1, 0, -1)$.

26. Through $(-2, 3, 2)$ and parallel to $3x + y + z = 4$.

27. Perpendicular to $\vec{v} = 2\vec{i} - 3\vec{j} + 5\vec{k}$ and through $(4, 5, -2)$.

In Exercises **28–32**, compute the angle between the vectors.

28. $\vec{i} + \vec{j} + \vec{k}$ and $\vec{i} - \vec{j} - \vec{k}$.

29. $\vec{i} + \vec{k}$ and $\vec{j} - \vec{k}$.

30. $\vec{i} + \vec{j} - \vec{k}$ and $2\vec{i} + 3\vec{j} + \vec{k}$.

31. $\vec{i} + \vec{j}$ and $\vec{i} + 2\vec{j} - \vec{k}$.

32. $\vec{i}$ and $2\vec{i} + 3\vec{j} - \vec{k}$.

33. Match statements (a)-(c) with diagrams (I)-(III) of vectors $\vec{u}$ and $\vec{w}$ in Figure 13.34.

(a) $\vec{u} \cdot \vec{w} = 0$ (b) $\vec{u} \cdot \vec{w} > 0$ (c) $\vec{u} \cdot \vec{w} < 0$

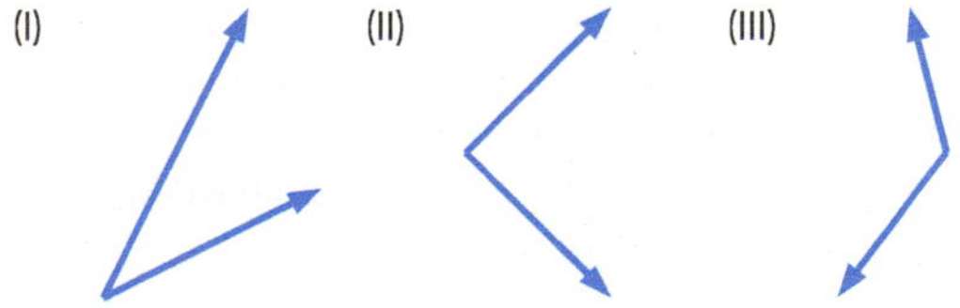

(I)   (II)   (III)

Figure 13.34

PROBLEMS

34. Are the dot products of the two-dimensional vectors in Figure 13.34 positive, negative, or zero?

(a) $\vec{a} \cdot \vec{b}$ (b) $\vec{a} \cdot \vec{c}$ (c) $\vec{b} \cdot \vec{c}$

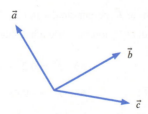

Figure 13.35

35. Give a unit vector

(a) In the same direction as $\vec{v} = 2\vec{i} + 3\vec{j}$.
(b) Perpendicular to $\vec{v}$.

36. For what value of c is the vector $3\vec{i} + 2\vec{j} + c\vec{k}$ parallel to the plane $x + 3y - z = 5$?

37. For what value of c is the vector $c\vec{i} + c\vec{j} + \vec{k}$ parallel to the plane $10 + 2x + 3y + z = 0$?

38. For what value(s) of a is the vector $a\vec{i} - a\vec{j} + \vec{k}$ parallel to the plane $ax + 2y - 15z = 0$?

39. A plane has equation $z = 5x - 2y + 7$.

(a) Find a value of λ making the vector $\lambda\vec{i} + \vec{j} + 0.5\vec{k}$ normal to the plane.
(b) Find a value of a so that the point $(a + 1, a, a - 1)$ lies on the plane.

40. A plane has equation $3x - ay + 5z = 1$.

(a) What value of a makes the vector $\vec{v} = 9\vec{i} - 4\vec{j} + 7\vec{k}$ parallel to the plane?

(b) If possible, find a value of a making the vector $\vec{v} = 9\vec{i} - 4\vec{j} + 7\vec{k}$ normal to the plane.

41. Consider the plane $5x - y + 7z = 21$.

(a) Find a point on the x-axis on this plane.
(b) Find two other points on the plane.
(c) Find a vector perpendicular to the plane.
(d) Find a vector parallel to the plane.

42. (a) Find a vector perpendicular to the plane $z = 2 + 3x - y$.

(b) Find a vector parallel to the plane.

43. (a) Find a vector perpendicular to the plane $z = 2x + 3y$.

(b) Find a vector parallel to the plane.

44. Consider the plane $x + 2y - z = 5$ and the vector $\vec{v} = 2\vec{i} - 5\vec{j} + 3\vec{k}$.

(a) Find a normal vector to the plane.
(b) What is the angle between $\vec{v}$ and the vector you found in part (a)?
(c) What is the angle between $\vec{v}$ and the plane?

45. Match the planes in (a)–(d) with one or more of the descriptions in (I)–(IV). No reasons are needed.

(a) $3x - y + z = 0$ (b) $4x + y + 2z - 5 = 0$
(c) $x + y = 5$ (d) $x = 5$

I Goes through the origin.
II Has a normal vector parallel to the xy-plane.
III Goes through the point $(0, 5, 0)$.
IV Has a normal vector whose dot products with $\vec{i}, \vec{j}, \vec{k}$ are all positive.

46. Which pairs (if any) of vectors from the following list

(a) Are perpendicular?
(b) Are parallel?
(c) Have an angle less than $\pi/2$ between them?
(d) Have an angle of more than $\pi/2$ between them?

$\vec{a} = \vec{i} - 3\vec{j} - \vec{k}$, $\vec{b} = \vec{i} + \vec{j} + 2\vec{k}$,
$\vec{c} = -2\vec{i} - \vec{j} + \vec{k}$, $\vec{d} = -\vec{i} - \vec{j} + \vec{k}$.

47. List any vectors that are parallel to each other and any vectors that are perpendicular to each other:

$\vec{v}_1 = \vec{i} - 2\vec{j}$ $\vec{v}_2 = 2\vec{i} + 4\vec{j}$
$\vec{v}_3 = 3\vec{i} + 1.5\vec{j}$ $\vec{v}_4 = -1.2\vec{i} + 2.4\vec{j}$
$\vec{v}_5 = -5\vec{i} - 2.5\vec{j}$ $\vec{v}_6 = 12\vec{i} - 12\vec{j}$
$\vec{v}_7 = 4\vec{i} + 2\vec{j}$ $\vec{v}_8 = 3\vec{i} - 6\vec{j}$
$\vec{v}_9 = 0.70\vec{i} - 0.35\vec{j}$

48. (a) Give a vector that is parallel to, but not equal to, $\vec{v} = 4\vec{i} + 3\vec{j}$.

(b) Give a vector that is perpendicular to $\vec{v}$.

49. For what values of t are $\vec{u} = t\vec{i} - \vec{j} + \vec{k}$ and $\vec{v} = t\vec{i} + t\vec{j} - 2\vec{k}$ perpendicular? Are there values of t for which $\vec{u}$ and $\vec{v}$ are parallel?

50. Let θ be the angle between $\vec{v}$ and $\vec{w}$, with $0 < \theta < \pi/2$. What is the effect on $\vec{v} \cdot \vec{w}$ of increasing each of the following quantities? Does $\vec{v} \cdot \vec{w}$ increase or decrease?

(a) $\|\vec{v}\|$ (b) θ

In Problems 51–53, for two-dimensional vectors $\vec{a}$ and $\vec{b}$, if $\|\vec{a}\| = 2$ and $\|\vec{b}\| = 4$, find $\|\vec{a} + \vec{b}\|$ for the given $\vec{a} \cdot \vec{b}$.

51. $\vec{a} \cdot \vec{b} = -8$ **52.** $\vec{a} \cdot \vec{b} = 8$ **53.** $\vec{a} \cdot \vec{b} = 0$

54. For a fixed two-dimensional vector $\vec{a}$ with $\|\vec{a}\| = 2$, determine how many vectors there are with $\|\vec{b}\| = 4$ and $\vec{a} \cdot \vec{b} = 4$.

55. Write $\vec{a} = 3\vec{i} + 2\vec{j} - 6\vec{k}$ as the sum of two vectors, one parallel and one perpendicular to $\vec{d} = 2\vec{i} - 4\vec{j} + \vec{k}$.

56. Find angle BAC if $A = (2, 2, 2)$, $B = (4, 2, 1)$, and $C = (2, 3, 1)$.

57. The points $(5, 0, 0)$, $(0, -3, 0)$, and $(0, 0, 2)$ form a triangle. Find the lengths of the sides of the triangle and each of its angles.

58. Let S be the triangle with vertices $A = (2, 2, 2)$, $B = (4, 2, 1)$, and $C = (2, 3, 1)$.

(a) Find the length of the shortest side of S.
(b) Find the cosine of the angle BAC at vertex A.

In Problems 59–61, find the work done by a force $\vec{F}$ moving an object on the line from point P to point Q. Give answers in joules and foot-pounds, using 1 joule ≈ 0.73756 ft-lb.

59. $\vec{F} = 3\vec{i} + 4\vec{j}$ newtons, $P = (3, 4)$ meters, $Q = (8, 10)$ meters

60. $\vec{F} = 4\vec{i} + 2\vec{j}$ newtons, $P = (10, 9)$ meters, $Q = (12, 2)$ meters

61. $\vec{F} = 20\vec{i} + 30\vec{j}$ pounds, $P = (9, 3)$ feet, $Q = (12, 5)$ feet

In Problems 62–67, given $\vec{v} = 3\vec{i} + 4\vec{j}$ and force vector $\vec{F}$, find:

(a) The component of $\vec{F}$ parallel to $\vec{v}$.
(b) The component of $\vec{F}$ perpendicular to $\vec{v}$.
(c) The work, W, done by force $\vec{F}$ through displacement $\vec{v}$.

62. $\vec{F} = 4\vec{i} + \vec{j}$ **63.** $\vec{F} = 0.2\vec{i} - 0.5\vec{j}$

64. $\vec{F} = 9\vec{i} + 12\vec{j}$ **65.** $\vec{F} = -0.4\vec{i} + 0.3\vec{j}$

66. $\vec{F} = -3\vec{i} - 5\vec{j}$ **67.** $\vec{F} = -6\vec{i} - 8\vec{j}$

■ In Problems 68–71, the force on an object is $\vec{F} = -20\vec{j}$. For vector $\vec{v}$, find:

(a) The component of $\vec{F}$ parallel to $\vec{v}$.

(b) The component of $\vec{F}$ perpendicular to $\vec{v}$.

(c) The work W done by force $\vec{F}$ through displacement $\vec{v}$.

68. $\vec{v} = 2\vec{i} + 3\vec{j}$ **69.** $\vec{v} = 5\vec{i} - \vec{j}$

70. $\vec{v} = 3\vec{j}$ **71.** $\vec{v} = 5\vec{i}$

72. A basketball gymnasium is 25 meters high, 80 meters wide and 200 meters long. For a half-time stunt, the cheerleaders want to run two strings, one from each of the two corners above one basket to the diagonally opposite corners of the gym floor. What is the cosine of the angle made by the strings as they cross?

73. An inner diagonal of a cube runs from one vertex through the center to the opposite vertex. For the cube with vertices $(\pm1, \pm1, \pm1)$, at what acute angle do two distinct inner diagonals intersect?

74. A 100-meter dash is run on a track in the direction of the vector $\vec{v} = 2\vec{i} + 6\vec{j}$. The wind velocity $\vec{w}$ is $5\vec{i} + \vec{j}$ km/hr. The rules say that a legal wind speed measured in the direction of the dash must not exceed 5 km/hr. Will the race results be disqualified due to an illegal wind? Justify your answer.

75. An airplane is flying toward the southeast. Which of the following wind velocity vectors increases the plane's speed the most? Which slows down the plane the most?

$$\vec{w}_1 = -4\vec{i} - \vec{j} \qquad \vec{w}_2 = \vec{i} - 2\vec{j} \qquad \vec{w}_3 = -\vec{i} + 8\vec{j}$$
$$\vec{w}_4 = 10\vec{i} + 2\vec{j} \qquad \vec{w}_5 = 5\vec{i} - 2\vec{j}$$

76. A canoe is moving with velocity $\vec{v} = 5\vec{i} + 3\vec{j}$ m/sec relative to the water. The velocity of the current in the water is $\vec{c} = \vec{i} + 2\vec{j}$ m/sec.

(a) What is the speed of the current?

(b) What is the speed of the current in the direction of the canoe's motion?

77. A planet at the point $(30, 60, 90)$ is in a circular orbit about the line through the origin in the direction of the unit vector $\vec{u} = 2/3\vec{i} + 2/3\vec{j} - 1/3\vec{k}$. For the orbit, find the

(a) Center   (b) Radius

78. Find the shortest distance between the planes $2x - 5y + z = 10$ and $z = 5y - 2x$.

79. A street vendor sells six items, with prices p_1 dollars per unit, p_2 dollars per unit, and so on. The vendor's price vector is $\vec{p} = (p_1, p_2, p_3, p_4, p_5, p_6) = (1.00, 3.50, 4.00, 2.75, 5.00, 3.00)$. The vendor sells q_1 units of the first item, q_2 units of the second item, and so on. The vendor's quantity vector is $\vec{q} = (q_1, q_2, q_3, q_4, q_5, q_6) = (43, 57, 12, 78, 20, 35)$. Find $\vec{p} \cdot \vec{q}$, give its units, and explain its significance to the vendor.

80. A course has four exams, weighted 10%, 15%, 25%, 50%, respectively. The class average on each of these exams is 75%, 91%, 84%, 87%, respectively. What do the vectors $\vec{a} = (0.75, 0.91, 0.84, 0.87)$ and $\vec{w} = (0.1, 0.15, 0.25, 0.5)$ represent, in terms of the course? Calculate the dot product $\vec{w} \cdot \vec{a}$. What does it represent, in terms of the course?

81. A consumption vector of three goods is defined by $\vec{x} = (x_1, x_2, x_3)$, where x_1, x_2 and x_3 are the quantities consumed of the three goods. A budget constraint is represented by the equation $\vec{p} \cdot \vec{x} = k$, where $\vec{p}$ is the price vector of the three goods and k is a constant. Show that the difference between two consumption vectors corresponding to points satisfying the same budget constraint is perpendicular to the price vector $\vec{p}$.

82. What does Property 2 of the dot product in the box on page 764 say geometrically?

83. Show that the vectors $(\vec{b} \cdot \vec{c})\vec{a} - (\vec{a} \cdot \vec{c})\vec{b}$ and $\vec{c}$ are perpendicular.

84. Show that if $\vec{u}$ and $\vec{v}$ are two vectors such that

$$\vec{u} \cdot \vec{w} = \vec{v} \cdot \vec{w}$$

for every vector $\vec{w}$, then

$$\vec{u} = \vec{v}.$$

85. The Law of Cosines for a triangle with side lengths a, b, and c, and with angle C opposite side c, says

$$c^2 = a^2 + b^2 - 2ab\cos C.$$

On page 764, we used the Law of Cosines to show that the two definitions of the dot product are equivalent. In this problem, use the geometric definition of the dot product and its properties in the box on page 764 to prove the Law of Cosines. [Hint: Let $\vec{u}$ and $\vec{v}$ be the displacement vectors from C to the other two vertices, and express c^2 in terms of $\vec{u}$ and $\vec{v}$.]

86. For any vectors $\vec{v}$ and $\vec{w}$, consider the following function of t:

$$q(t) = (\vec{v} + t\vec{w}) \cdot (\vec{v} + t\vec{w}).$$

(a) Explain why $q(t) \geq 0$ for all real t.

(b) Expand $q(t)$ as a quadratic polynomial in t using the properties on page 764.

(c) Using the discriminant of the quadratic, show that

$$|\vec{v} \cdot \vec{w}| \leq \|\vec{v}\| \|\vec{w}\|.$$

Strengthen Your Understanding

■ In Problems 87–89, explain what is wrong with the statement.

87. For any 3-dimensional vectors $\vec{u}, \vec{v}, \vec{w}$, we have $(\vec{u} \cdot \vec{v}) \cdot \vec{w} = \vec{u} \cdot (\vec{v} \cdot \vec{w})$.

88. If $\vec{u} = \vec{i} + \vec{j}$ and $\vec{v} = 2\vec{i} + \vec{j}$, then the component of $\vec{v}$ parallel to $\vec{u}$ is $\vec{v}_{\text{parallel}} = (\vec{v} \cdot \vec{u})\vec{u} = 3\vec{i} + 3\vec{j}$.

89. A normal vector for the plane $z = 2x + 3y$ is $2\vec{i} + 3\vec{j}$.

■ In Problems 90–91, give an example of:

90. A point (a, b) such that the displacement vector from $(1, 1)$ to (a, b) is perpendicular to $\vec{i} + 2\vec{j}$.

91. A linear function $f(x, y) = mx + ny + c$ whose graph is perpendicular to $\vec{i} + 2\vec{j} + 3\vec{k}$.

■ Are the statements in Problems 92–103 true or false? Give reasons for your answer.

92. The quantity $\vec{u} \cdot \vec{v}$ is a vector.

93. The plane $x + 2y - 3z = 5$ has normal vector $\vec{i} + 2\vec{j} - 3\vec{k}$.

94. If $\vec{u} \cdot \vec{v} < 0$ then the angle between $\vec{u}$ and $\vec{v}$ is greater than $\pi/2$.

95. An equation of the plane with normal vector $\vec{i} + \vec{j} + \vec{k}$ containing the point $(1, 2, 3)$ is $z = x + y$.

96. The triangle in 3-space with vertices $(1, 1, 0), (0, 1, 0)$ and $(0, 1, 1)$ has a right angle.

97. The dot product $\vec{v} \cdot \vec{v}$ is never negative.

98. If $\vec{u} \cdot \vec{v} = 0$ then either $\vec{u} = 0$ or $\vec{v} = 0$.

99. If $\vec{u}, \vec{v}$ and $\vec{w}$ are all nonzero, and $\vec{u} \cdot \vec{v} = \vec{u} \cdot \vec{w}$, then $\vec{v} = \vec{w}$.

100. For any vectors $\vec{u}$ and $\vec{v}$: $(\vec{u} + \vec{v}) \cdot (\vec{u} - \vec{v}) = \|\vec{u}\|^2 - \|\vec{v}\|^2$.

101. If $\|\vec{u}\| = 1$, then the vector $\vec{v} - (\vec{v} \cdot \vec{u})\vec{u}$ is perpendicular to $\vec{u}$.

102. If $\vec{u} \cdot \vec{v} = \|\vec{u}\|\|\vec{v}\|$ then $\|\vec{u} + \vec{v}\| = \|\vec{u}\| + \|\vec{v}\|$.

103. The two nonzero vectors $\vec{v} = x\vec{i} + y\vec{j}$ and $\vec{w} = y\vec{i} - x\vec{j}$ are orthogonal for any choice of x and y.

13.4   THE CROSS PRODUCT

In the previous section we combined two vectors to get a number, the dot product. In this section we see another way of combining two vectors, this time to get a vector, the *cross product*. Any two vectors in 3-space form a parallelogram. We define the cross product using this parallelogram.

The Area of a Parallelogram

Consider the parallelogram formed by the vectors $\vec{v}$ and $\vec{w}$ with an angle of θ between them. Then Figure 13.36 shows

$$\text{Area of parallelogram} = \text{Base} \cdot \text{Height} = \|\vec{v}\|\|\vec{w}\|\sin\theta.$$

How would we compute the area of the parallelogram if we were given $\vec{v}$ and $\vec{w}$ in components, $\vec{v} = v_1\vec{i} + v_2\vec{j} + v_3\vec{k}$ and $\vec{w} = w_1\vec{i} + w_2\vec{j} + w_3\vec{k}$? Project 1 (available online) shows that if $\vec{v}$ and $\vec{w}$ are in the xy-plane so that $v_3 = w_3 = 0$, then

$$\text{Area of parallelogram} = \left| v_1 w_2 - v_2 w_1 \right|.$$

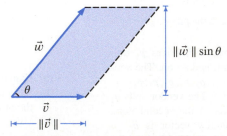

Figure 13.36: Parallelogram formed by $\vec{v}$ and $\vec{w}$ has
Area $= \|\vec{v}\|\|\vec{w}\|\sin\theta$

What if $\vec{v}$ and $\vec{w}$ do not lie in the xy-plane? The cross product will enable us to compute the area of the parallelogram formed by any two vectors.

Definition of the Cross Product

We define the cross product of the vectors $\vec{v}$ and $\vec{w}$, written $\vec{v} \times \vec{w}$, to be a vector perpendicular to both $\vec{v}$ and $\vec{w}$. The magnitude of this vector is the area of the parallelogram formed by the two vectors. The direction of $\vec{v} \times \vec{w}$ is given by the normal vector, $\vec{n}$, to the plane defined by $\vec{v}$ and $\vec{w}$. If we require that $\vec{n}$ be a unit vector, there are two choices for $\vec{n}$, pointing out of the plane in opposite directions. We pick one by the following rule (see Figure 13.37):

> **The right-hand rule:** Place $\vec{v}$ and $\vec{w}$ so that their tails coincide and curl the fingers of your right hand through the smaller of the two angles from $\vec{v}$ to $\vec{w}$; your thumb points in the direction of the normal vector, $\vec{n}$.

Like the dot product, there are two equivalent definitions of the cross product:

> The following two definitions of the **cross product** or **vector product** $\vec{v} \times \vec{w}$ are equivalent:
> - **Geometric definition**:
>   If $\vec{v}$ and $\vec{w}$ are not parallel, then
>
> $$\vec{v} \times \vec{w} = \left(\begin{array}{c}\text{Area of parallelogram} \\ \text{with edges } \vec{v} \text{ and } \vec{w}\end{array}\right) \vec{n} = (\|\vec{v}\|\|\vec{w}\|\sin\theta)\vec{n}\,,$$
>
>   where $0 \le \theta \le \pi$ is the angle between $\vec{v}$ and $\vec{w}$ and $\vec{n}$ is the unit vector perpendicular to $\vec{v}$ and $\vec{w}$ pointing in the direction given by the right-hand rule. If $\vec{v}$ and $\vec{w}$ are parallel, then $\vec{v} \times \vec{w} = \vec{0}$.
> - **Algebraic definition**:
>
> $$\vec{v} \times \vec{w} = (v_2 w_3 - v_3 w_2)\vec{i} + (v_3 w_1 - v_1 w_3)\vec{j} + (v_1 w_2 - v_2 w_1)\vec{k}$$
>
>   where $\vec{v} = v_1\vec{i} + v_2\vec{j} + v_3\vec{k}$ and $\vec{w} = w_1\vec{i} + w_2\vec{j} + w_3\vec{k}$.

Problems 78 and 81 (available online) show that the geometric and algebraic definitions of the cross product give the same result.

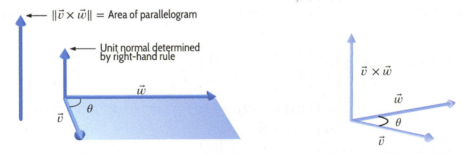

Figure 13.37: Area of parallelogram $= \|\vec{v} \times \vec{w}\|$ **Figure 13.38:** The cross product $\vec{v} \times \vec{w}$

The geometric definition shows us that the cross product is *rotation invariant*. Imagine the two vectors $\vec{v}$ and $\vec{w}$ as two metal rods welded together. Attach a third rod whose direction and length correspond to $\vec{v} \times \vec{w}$. (See Figure 13.38.) Then, no matter how we turn this set of rods, the third will still be the cross product of the first two.

The algebraic definition is more easily remembered by writing it as a 3×3 determinant. (See Appendix E.)

$$\vec{v} \times \vec{w} = \begin{vmatrix} \vec{i} & \vec{j} & \vec{k} \\ v_1 & v_2 & v_3 \\ w_1 & w_2 & w_3 \end{vmatrix} = (v_2 w_3 - v_3 w_2)\vec{i} + (v_3 w_1 - v_1 w_3)\vec{j} + (v_1 w_2 - v_2 w_1)\vec{k}.$$

Example 1   Find $\vec{i} \times \vec{j}$ and $\vec{j} \times \vec{i}$.

Solution   The vectors $\vec{i}$ and $\vec{j}$ both have magnitude 1 and the angle between them is $\pi/2$. By the right-hand rule, the vector $\vec{i} \times \vec{j}$ is in the direction of $\vec{k}$, so $\vec{n} = \vec{k}$ and we have

$$\vec{i} \times \vec{j} = \left(\|\vec{i}\| \|\vec{j}\| \sin \frac{\pi}{2} \right) \vec{k} = \vec{k}.$$

Similarly, the right-hand rule says that the direction of $\vec{j} \times \vec{i}$ is $-\vec{k}$, so

$$\vec{j} \times \vec{i} = \left(\|\vec{j}\| \|\vec{i}\| \sin \frac{\pi}{2} \right) \left(-\vec{k} \right) = -\vec{k}.$$

Similar calculations show that $\vec{j} \times \vec{k} = \vec{i}$ and $\vec{k} \times \vec{i} = \vec{j}$.

Example 2   For any vector $\vec{v}$, find $\vec{v} \times \vec{v}$.

Solution   Since $\vec{v}$ is parallel to itself, $\vec{v} \times \vec{v} = \vec{0}$.

Example 3   Find the cross product of $\vec{v} = 2\vec{i} + \vec{j} - 2\vec{k}$ and $\vec{w} = 3\vec{i} + \vec{k}$ and check that the cross product is perpendicular to both $\vec{v}$ and $\vec{w}$.

Solution   Writing $\vec{v} \times \vec{w}$ as a determinant and expanding it into three two-by-two determinants, we have

$$\vec{v} \times \vec{w} = \begin{vmatrix} \vec{i} & \vec{j} & \vec{k} \\ 2 & 1 & -2 \\ 3 & 0 & 1 \end{vmatrix} = \vec{i} \begin{vmatrix} 1 & -2 \\ 0 & 1 \end{vmatrix} - \vec{j} \begin{vmatrix} 2 & -2 \\ 3 & 1 \end{vmatrix} + \vec{k} \begin{vmatrix} 2 & 1 \\ 3 & 0 \end{vmatrix}$$

$$= \vec{i} \, (1(1) - 0(-2)) - \vec{j} \, (2(1) - 3(-2)) + \vec{k} \, (2(0) - 3(1))$$

$$= \vec{i} - 8\vec{j} - 3\vec{k}.$$

To check that $\vec{v} \times \vec{w}$ is perpendicular to $\vec{v}$, we compute the dot product:

$$\vec{v} \cdot (\vec{v} \times \vec{w}) = (2\vec{i} + \vec{j} - 2\vec{k}) \cdot (\vec{i} - 8\vec{j} - 3\vec{k}) = 2 - 8 + 6 = 0.$$

Similarly,

$$\vec{w} \cdot (\vec{v} \times \vec{w}) = (3\vec{i} + 0\vec{j} + \vec{k}) \cdot (\vec{i} - 8\vec{j} - 3\vec{k}) = 3 + 0 - 3 = 0.$$

Thus, $\vec{v} \times \vec{w}$ is perpendicular to both $\vec{v}$ and $\vec{w}$.

Properties of the Cross Product

The right-hand rule tells us that $\vec{v} \times \vec{w}$ and $\vec{w} \times \vec{v}$ point in opposite directions. The magnitudes of $\vec{v} \times \vec{w}$ and $\vec{w} \times \vec{v}$ are the same, so $\vec{w} \times \vec{v} = -(\vec{v} \times \vec{w})$. (See Figure 13.39.)

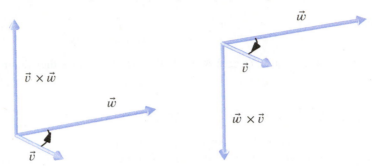

Figure 13.39: Diagram showing $\vec{v} \times \vec{w} = -(\vec{w} \times \vec{v})$

This explains the first of the following properties. The other two are derived in Problems 73, 74, and 81 (available online).

Properties of the Cross Product

For vectors $\vec{u}, \vec{v}, \vec{w}$ and scalar λ,
1. $\vec{w} \times \vec{v} = -(\vec{v} \times \vec{w})$
2. $(\lambda\vec{v}) \times \vec{w} = \lambda(\vec{v} \times \vec{w}) = \vec{v} \times (\lambda\vec{w})$
3. $\vec{u} \times (\vec{v} + \vec{w}) = \vec{u} \times \vec{v} + \vec{u} \times \vec{w}$.

The Equation of a Plane Through Three Points

As we saw on page 766, the equation of a plane is determined by a point $P_0 = (x_0, y_0, z_0)$ on the plane, and a normal vector, $\vec{n} = a\vec{i} + b\vec{j} + c\vec{k}$:

$$a(x - x_0) + b(y - y_0) + c(z - z_0) = 0.$$

However, a plane can also be determined by three points on it (provided they do not lie on the same line). In that case we can find an equation of the plane by first determining two vectors in the plane and then finding a normal vector using the cross product, as in the following example.

Example 4   Find an equation of the plane containing the points $P = (1, 3, 0)$, $Q = (3, 4, -3)$, and $R = (3, 6, 2)$.

Solution   Since the points P and Q are in the plane, the displacement vector between them, $\overrightarrow{PQ}$, is in the plane, where

$$\overrightarrow{PQ} = (3 - 1)\vec{i} + (4 - 3)\vec{j} + (-3 - 0)\vec{k} = 2\vec{i} + \vec{j} - 3\vec{k}.$$

The displacement vector $\overrightarrow{PR}$ is also in the plane, where

$$\overrightarrow{PR} = (3 - 1)\vec{i} + (6 - 3)\vec{j} + (2 - 0)\vec{k} = 2\vec{i} + 3\vec{j} + 2\vec{k}.$$

Thus, a normal vector, $\vec{n}$, to the plane is given by

$$\vec{n} = \overrightarrow{PQ} \times \overrightarrow{PR} = \begin{vmatrix} \vec{i} & \vec{j} & \vec{k} \\ 2 & 1 & -3 \\ 2 & 3 & 2 \end{vmatrix} = 11\vec{i} - 10\vec{j} + 4\vec{k}.$$

Since the point $(1, 3, 0)$ is on the plane, the equation of the plane is

$$11(x - 1) - 10(y - 3) + 4(z - 0) = 0,$$

which simplifies to

$$11x - 10y + 4z = -19.$$

You should check that P, Q, and R satisfy this equation, since they lie on the plane.

Areas and Volumes Using the Cross Product and Determinants

We can use the cross product to calculate the area of the parallelogram with sides $\vec{v}$ and $\vec{w}$. We say that $\vec{v} \times \vec{w}$ is the *area vector* of the parallelogram. The geometric definition of the cross product tells us that $\vec{v} \times \vec{w}$ is normal to the parallelogram and gives us the following result:

> **Area of a parallelogram** with edges $\vec{v} = v_1\vec{i} + v_2\vec{j} + v_3\vec{k}$ and $\vec{w} = w_1\vec{i} + w_2\vec{j} + w_3\vec{k}$ is given by
>
> $$\text{Area} = \|\vec{v} \times \vec{w}\|, \qquad \text{where} \quad \vec{v} \times \vec{w} = \begin{vmatrix} \vec{i} & \vec{j} & \vec{k} \\ v_1 & v_2 & v_3 \\ w_1 & w_2 & w_3 \end{vmatrix}.$$

Example 5   Find the area of the parallelogram with edges $\vec{v} = 2\vec{i} + \vec{j} - 3\vec{k}$ and $\vec{w} = \vec{i} + 3\vec{j} + 2\vec{k}$.

Solution   We calculate the cross product:

$$\vec{v} \times \vec{w} = \begin{vmatrix} \vec{i} & \vec{j} & \vec{k} \\ 2 & 1 & -3 \\ 1 & 3 & 2 \end{vmatrix} = (2 + 9)\vec{i} - (4 + 3)\vec{j} + (6 - 1)\vec{k} = 11\vec{i} - 7\vec{j} + 5\vec{k}.$$

The area of the parallelogram with edges $\vec{v}$ and $\vec{w}$ is the magnitude of the vector $\vec{v} \times \vec{w}$:

$$\text{Area} = \|\vec{v} \times \vec{w}\| = \sqrt{11^2 + (-7)^2 + 5^2} = \sqrt{195}.$$

Volume of a Parallelepiped

Consider the parallelepiped with sides formed by $\vec{a}$, $\vec{b}$, and $\vec{c}$. (See Figure 13.40.) Since the base is formed by the vectors $\vec{b}$ and $\vec{c}$, we have

$$\text{Area of base of parallelepiped} = \|\vec{b} \times \vec{c}\|.$$

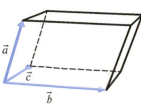

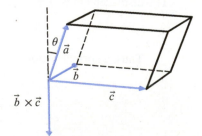

Figure 13.40: Volume of a parallelepiped

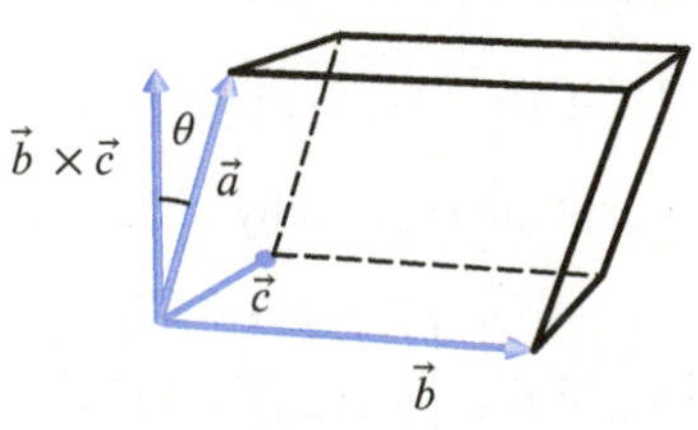

Figure 13.41: The vectors $\vec{a}$, $\vec{b}$, $\vec{c}$ are called a right-handed set

Figure 13.42: The vectors $\vec{a}$, $\vec{b}$, $\vec{c}$ are called a left-handed set

The vectors $\vec{a}$, $\vec{b}$, and $\vec{c}$ can be arranged either as in Figure 13.41 or as in Figure 13.42. In either case,

$$\text{Height of parallelepiped} = \|\vec{a}\| \cos\theta,$$

where θ is the angle shown in the figures. In Figure 13.41 the angle θ is less than $\pi/2$, so the product, $(\vec{b} \times \vec{c}) \cdot \vec{a}$, called the *triple product*, is positive. Thus, in this case

$$\text{Volume of parallelepiped} = \text{Base} \cdot \text{Height} = \|\vec{b} \times \vec{c}\| \cdot \|\vec{a}\| \cos\theta = (\vec{b} \times \vec{c}) \cdot \vec{a}.$$

In Figure 13.42, the angle, $\pi - \theta$, between $\vec{a}$ and $\vec{b} \times \vec{c}$ is more than $\pi/2$, so the product $(\vec{b} \times \vec{c}) \cdot \vec{a}$ is negative. Thus, in this case we have

$$\text{Volume} = \text{Base} \cdot \text{Height} = \|\vec{b} \times \vec{c}\| \cdot \|\vec{a}\| \cos\theta = -\|\vec{b} \times \vec{c}\| \cdot \|\vec{a}\| \cos(\pi - \theta)$$
$$= -(\vec{b} \times \vec{c}) \cdot \vec{a} = \left| (\vec{b} \times \vec{c}) \cdot \vec{a} \right|.$$

Therefore, in both cases the volume is given by $\left| (\vec{b} \times \vec{c}) \cdot \vec{a} \right|$. Using determinants, we can write

---

Volume of a parallelepiped with edges $\vec{a}$, $\vec{b}$, $\vec{c}$ is given by

$$\text{Volume} = \left| (\vec{b} \times \vec{c}) \cdot \vec{a} \right| = \text{Absolute value of the determinant} \begin{vmatrix} a_1 & a_2 & a_3 \\ b_1 & b_2 & b_3 \\ c_1 & c_2 & c_3 \end{vmatrix}.$$

Angular Velocity

Angular velocity, which describes rotation about an axis, can be represented by a vector. For example, the angular velocity of the rotating flywheel in Figure 13.43 is represented by the vector $\vec{\omega}$, whose direction is parallel to the axis of rotation in the direction given by the right-hand rule. If the fingers of the right-hand curl around the axis in the direction of the rotation, then the thumb points along the axis in the direction of $\vec{\omega}$. The magnitude $\|\vec{\omega}\|$ is the angular speed of rotation, for example in radians per unit time or revolutions per unit time.

Every point on the flywheel travels a circular orbit around the axis. Since one orbit is 2π radians,

$$\text{Time to complete one orbit} = \frac{\text{Angle traveled}}{\text{Angular speed}} = \frac{2\pi}{\|\vec{\omega}\|}.$$

In Figure 13.43, let $\vec{r}$ be the vector from the center of the orbit to the point P. In one orbit, the point P travels a distance of $2\pi\|\vec{r}\|$ around the circumference of a circle, so

$$\text{Speed} = \frac{\text{Distance}}{\text{Time}} = \frac{2\pi\|\vec{r}\|}{2\pi/\|\vec{\omega}\|} = \|\vec{\omega}\| \cdot \|\vec{r}\|.$$

The velocity vector $\vec{v}$ is tangent to the orbit, so $\vec{v}$ is perpendicular to both the axis and the radius of the orbit. The magnitude of $\vec{v}$ is the speed of P, so $\|\vec{v}\| = \|\vec{\omega}\| \cdot \|\vec{r}\|$. Since the cross product $\vec{\omega} \times \vec{r}$ has the same direction as the velocity (both $\vec{\omega} \times \vec{r}$ and $\vec{v}$ are perpendicular to $\vec{\omega}$ and to $\vec{r}$), and the same magnitude as the velocity (both magnitudes are $\|\vec{\omega}\| \cdot \|\vec{r}\|$), we have

$$\vec{v} = \vec{\omega} \times \vec{r}.$$

The formula $\vec{v} = \vec{\omega} \times \vec{r}$ holds for $\vec{r}$ a vector from any point on the axis of rotation to the point P. This is because $\vec{r}$ can be expressed as the sum of two component vectors, one parallel to the axis and the other a radius vector of the orbit. Only the radial component contributes to the cross product. See Figure 13.44.

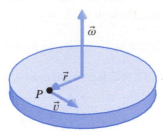

Figure 13.43: Rotating flywheel

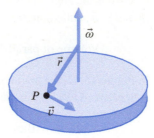

Figure 13.44: Rotating flywheel

Example 6   The world record for the fastest spin by a figure skater, 342 revolutions per minute, is held by Olivia Oliver.[2]

(a) Find Olivia's angular velocity vector, assuming she is vertical and spinning her fastest to her left.
(b) Let her skates touch the ice at the point $(0, 0, 0)$ and her left elbow be at $P = (10, 15, 110)$, where distances are in centimeters. Find the velocity of her elbow.
(c) Find the speed of her elbow in centimeters per minute.

Solution   (a) Since she is spinning around a vertical axis, we have $\vec{\omega} = c\vec{k}$ where c, the rate of rotation, is positive because she is spinning to her left. Since 1 revolution corresponds to 2π radians, we have $c = 2\pi \cdot 342 = 2149$. Hence $\vec{\omega} = 2149\vec{k}$ radians per minute.

(b) Her elbow has position vector $\vec{r} = 10\vec{i} + 15\vec{j} + 110\vec{k}$ cm. Her elbow velocity is the vector $\vec{v} = \vec{\omega} \times \vec{r} = -32{,}235\vec{i} + 21{,}490\vec{j}$ cm/min.

(c) Her elbow is moving with speed

$$\|\vec{v}\| = \sqrt{(-32{,}235)^2 + (21{,}490)^2} = 38{,}742 \text{ cm/min}.$$

This is a speed of about 6.5 meters per second.

Summary for Section 13.4

- The **direction** of $\vec{v} \times \vec{w}$ the vector orthogonal to both of them given by the **right-hand rule:** Place $\vec{v}$ and $\vec{w}$ so that their tails coincide and curl the fingers of your right hand through the smaller of the two angles from $\vec{v}$ to $\vec{w}$; your thumb points in the direction of the cross product $\vec{v} \times \vec{w}$.

- There are two equivalent definitions of the **cross product** $\vec{v} \times \vec{w}$:

   ○ **Geometric definition**:
      If $\vec{v}$ and $\vec{w}$ are not parallel, then

   $$\vec{v} \times \vec{w} = \left(\begin{array}{c} \text{Area of parallelogram} \\ \text{with edges } \vec{v} \text{ and } \vec{w} \end{array} \right) \vec{n} = (\|\vec{v}\| \|\vec{w}\| \sin \theta) \vec{n},$$

   where $0 \leq \theta \leq \pi$ is the angle between $\vec{v}$ and $\vec{w}$ and $\vec{n}$ is the unit vector perpendicular to $\vec{v}$ and $\vec{w}$ pointing in the direction given by the right-hand rule. If $\vec{v}$ and $\vec{w}$ are parallel, then $\vec{v} \times \vec{w} = \vec{0}$.

[2]From www.guinnessworldrecords.com, accessed May 12, 2016.

○ **Algebraic definition**:

$$\vec{v} \times \vec{w} = (v_2 w_3 - v_3 w_2)\vec{i} + (v_3 w_1 - v_1 w_3)\vec{j} + (v_1 w_2 - v_2 w_1)\vec{k}$$

where $\vec{v} = v_1\vec{i} + v_2\vec{j} + v_3\vec{k}$ and $\vec{w} = w_1\vec{i} + w_2\vec{j} + w_3\vec{k}$.

- **Properties of the cross product**: For vectors $\vec{u}, \vec{v}, \vec{w}$ and scalar λ
  1. $\vec{w} \times \vec{v} = -(\vec{v} \times \vec{w})$
  2. $(\lambda\vec{v}) \times \vec{w} = \lambda(\vec{v} \times \vec{w}) = \vec{v} \times (\lambda\vec{w})$
  3. $\vec{u} \times (\vec{v} + \vec{w}) = \vec{u} \times \vec{v} + \vec{u} \times \vec{w}$.
- **Area of a parallelogram** with edges $\vec{v}$ and $\vec{w}$ is given by $\|\vec{v} \times \vec{w}\|$.
- **Volume of a parallelepiped** with edges $\vec{a}, \vec{b}, \vec{c}$ is given by $\left|(\vec{b} \times \vec{c}) \cdot \vec{a}\right|$.

Exercises and Problems for Section 13.4 Online Resource: Additional Problems for Section 13.4

EXERCISES

In Exercises 1–7, use the algebraic definition to find $\vec{v} \times \vec{w}$.

1. $\vec{v} = \vec{k}, \vec{w} = \vec{j}$
2. $\vec{v} = -\vec{i}, \vec{w} = \vec{j} + \vec{k}$
3. $\vec{v} = \vec{i} + \vec{k}, \vec{w} = \vec{i} + \vec{j}$
4. $\vec{v} = \vec{i} + \vec{j} + \vec{k}, \vec{w} = \vec{i} + \vec{j} - \vec{k}$
5. $\vec{v} = 2\vec{i} - 3\vec{j} + \vec{k}, \vec{w} = \vec{i} + 2\vec{j} - \vec{k}$
6. $\vec{v} = 2\vec{i} - \vec{j} - \vec{k}, \vec{w} = -6\vec{i} + 3\vec{j} + 3\vec{k}$
7. $\vec{v} = -3\vec{i} + 5\vec{j} + 4\vec{k}, \vec{w} = \vec{i} - 3\vec{j} - \vec{k}$

In Exercises 8–9, use the geometric definition to find:

8. $2\vec{i} \times (\vec{i} + \vec{j})$

9. $(\vec{i} + \vec{j}) \times (\vec{i} - \vec{j})$

In Exercises 10–11, use the properties on page 777 to find:

10. $\left((\vec{i} + \vec{j}) \times \vec{i}\right) \times \vec{j}$

11. $(\vec{i} + \vec{j}) \times (\vec{i} \times \vec{j})$

12. For $\vec{a} = 3\vec{i} + \vec{j} - \vec{k}$ and $\vec{b} = \vec{i} - 4\vec{j} + 2\vec{k}$, find $\vec{a} \times \vec{b}$ and check that it is perpendicular to both $\vec{a}$ and $\vec{b}$.

13. If $\vec{v} = 3\vec{i} - 2\vec{j} + 4\vec{k}$ and $\vec{w} = \vec{i} + 2\vec{j} - \vec{k}$, find $\vec{v} \times \vec{w}$ and $\vec{w} \times \vec{v}$. What is the relation between the two answers?

In Exercises 14–15, find an equation for the plane through the points.

14. $(1, 0, 0), (0, 1, 0), (0, 0, 1)$.
15. $(3, 4, 2), (-2, 1, 0), (0, 2, 1)$.

In Exercises 16–19, find the volume of the parallelepiped with edges $\vec{a}, \vec{b}, \vec{c}$.

16. $\vec{a} = 3\vec{i} + 4\vec{j} + 5\vec{k}, \vec{b} = 5\vec{i} + 4\vec{j} + 3\vec{k}, \vec{c} = \vec{i} + \vec{j} + \vec{k}$.
17. $\vec{a} = -\vec{i} + \vec{j} + \vec{k}, \vec{b} = \vec{i} - \vec{j} + \vec{k}, \vec{c} = \vec{i} + \vec{j} - \vec{k}$.
18. $\vec{a} = -\vec{i} + 8\vec{j} + 7\vec{k}, \vec{b} = 2\vec{j} + 9\vec{k}, \vec{c} = 3\vec{k}$.
19. $\vec{a} = \vec{i} + \vec{j} + 2\vec{k}, \vec{b} = \vec{i} + \vec{k}, \vec{c} = \vec{j} + \vec{k}$.

In Exercises 20–23, the point is rotating around an axis through the origin with angular velocity $\vec{\omega} = 2\vec{i} + \vec{j} - 3\vec{k}$. Find its velocity vector.

20. $(1, 2, 1)$
21. $(1, 0, -1)$
22. $(2, -2, 0)$
23. $(4, 2, -6)$

PROBLEMS

24. Find a vector parallel to the line of intersection of the planes given by $2y - z = 2$ and $-2x + y = 4$.

25. Find an equation of the plane through the origin that is perpendicular to the line of intersection of the planes in Problem 24.

26. Find an equation of the plane through the point $(4, 5, 6)$ and perpendicular to the line of intersection of the planes in Problem 24.

27. Find an equation for the plane through the origin con-

taining the points $(1, 3, 0)$ and $(2, 4, 1)$.

28. Find a vector parallel to the line of intersection of the two planes $4x - 3y + 2z = 12$ and $x + 5y - z = 25$.

29. Find a vector parallel to the intersection of the planes $2x - 3y + 5z = 2$ and $4x + y - 3z = 7$.

30. Find an equation of the plane through the origin that is perpendicular to the line of intersection of the planes in Problem 29.

31. Find an equation of the plane through the point $(4, 5, 6)$ that is perpendicular to the line of intersection of the planes in Problem 29.

32. Find the equation of a plane through the origin and perpendicular to $x - y + z = 5$ and $2x + y - 2z = 7$.

33. Given the points $P = (1, 2, 3)$, $Q = (3, 5, 7)$, and $R = (2, 5, 3)$, find:

   (a) A unit vector perpendicular to a plane containing P, Q, R.
   (b) The angle between PQ and PR.
   (c) The area of the triangle PQR.
   (d) The distance from R to the line through P and Q.

34. Let $A = (-1, 3, 0)$, $B = (3, 2, 4)$, and $C = (1, -1, 5)$.

   (a) Find an equation for the plane that passes through these three points.
   (b) Find the area of the triangle determined by these three points.

35. Consider the plane $z + 2y + x = 4$.

   (a) Find a point on the x-axis on this plane.
   (b) Find a point on the y-axis on this plane.
   (c) Find a point on the z-axis on this plane.
   (d) Find the area of the region of this plane with $x \geq 0$, $y \geq 0$ and $z \geq 0$.

36. If $\vec{v}$ and $\vec{w}$ are both parallel to the xy-plane, what can you conclude about $\vec{v} \times \vec{w}$? Explain.

37. Suppose $\vec{v} \cdot \vec{w} = 5$ and $\|\vec{v} \times \vec{w}\| = 3$, and the angle between $\vec{v}$ and $\vec{w}$ is θ. Find

   (a) $\tan \theta$ (b) θ

38. If $\vec{v} \times \vec{w} = 2\vec{i} - 3\vec{j} + 5\vec{k}$, and $\vec{v} \cdot \vec{w} = 3$, find $\tan \theta$ where θ is the angle between $\vec{v}$ and $\vec{w}$.

39. Suppose $\vec{v} \cdot \vec{w} = 8$ and $\vec{v} \times \vec{w} = 12\vec{i} - 3\vec{j} + 4\vec{k}$ and that the angle between $\vec{v}$ and $\vec{w}$ is θ. Find

   (a) $\tan \theta$ (b) θ

40. If $\vec{v} \cdot (\vec{i} + \vec{j} + \vec{k}) = 6$ and $\vec{v} \times (\vec{i} + \vec{j} + \vec{k}) = \vec{0}$, find $\vec{v}$.

41. Why does a baseball curve? The baseball in Figure 13.45 has velocity $\vec{v}$ meters/sec and is spinning at ω radians per second about an axis in the direction of the unit vector $\vec{n}$. The ball experiences a force, called the Magnus force,[3] $\vec{F}_M$, that is proportional to $\omega \vec{n} \times \vec{v}$.

   (a) What is the effect on $\vec{F}_M$ of increasing ω?
   (b) The ball in Figure 13.45 is moving away from you. What is the direction of the Magnus force?

Figure 13.45: Spinning baseball

42. The London Eye Ferris wheel rotates in a counterclockwise direction when viewed from the east and completes one full rotation in 30 minutes.[4]

   (a) Let the x-axis point east, the y-axis north, and the z-axis up. Find the angular velocity, $\vec{\omega}$, of the London Eye.
   (b) The passenger capsules of the London Eye are a distance of approximately 200 feet from the center. If the center of the London Eye is at $(0, 0, 0)$, find the velocity vector of a passenger capsule at its highest point.
   (c) Find the speed of the capsule.

43. The point P in Figure 13.46 has position vector $\vec{v}$ obtained by rotating the position vector $\vec{r}$ of the point (x, y) by 90° counterclockwise about the origin.

   (a) Use the geometric definition of the cross product to explain why $\vec{v} = \vec{k} \times \vec{r}$.
   (b) Find the coordinates of P.

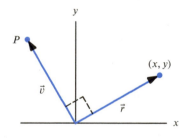

Figure 13.46

44. The points $P_1 = (0, 0, 0)$, $P_2 = (2, 4, 2)$, $P_3 = (3, 0, 0)$, and $P_4 = (5, 4, 2)$ are vertices of a parallelogram.

   (a) Find the displacement vectors along each of the four sides. Check that these are equal in pairs.
   (b) Find the area of the parallelogram.

■ In Problems 45–46, find an area vector for the parallelogram with given vertices.

45. $P = (2, 1, 1)$, $Q = (3, 3, 0)$, $R = (4, 0, 2)$, $S = (5, 2, 1)$

46. $P = (-1, -2, 0)$, $Q = (0, -1, 0)$, $R = (-2, -4, 1)$, $S = (-1, -3, 1)$

[3] Named after German physicist Heinrich Magnus, who first described it in 1853.
[4] http://en.wikipedia.org, accessed May 12, 2016.

47. A parallelogram P formed by the vectors $\vec{v} = \vec{i} - 2\vec{j} + \vec{k}$ and $\vec{w} = \vec{i} + 2\vec{j} - 2\vec{k}$ has area vector $\vec{A} = \vec{v} \times \vec{w} = 2\vec{i} + 3\vec{j} + 4\vec{k}$. Find area vectors of each of the following parallelograms and explain how they are related to $\vec{A}$.

    (a) The parallelogram obtained by projecting P onto the xy-plane.
    (b) The parallelogram obtained by projecting P onto the xz-plane.
    (c) The parallelogram obtained by projecting P onto the yz-plane.

48. Using the parallelogram in Problem 44 as a base, create a parallelopiped with side $\overrightarrow{P_1 P_5}$ where $P_5 = (1, 0, 4)$. Find the volume of this parallelepiped.

In Problems 49–51, if $0 \leq \theta \leq \pi$, what are the possible values for the angle, θ, between two nonzero vectors $\vec{v}$ and $\vec{w}$ satisfying the inequality?

49. $|\vec{v} \cdot \vec{w}| = \|\vec{v} \times \vec{w}\|$ 50. $|\vec{v} \cdot \vec{w}| < \|\vec{v} \times \vec{w}\|$

51. $|\vec{v} \cdot \vec{w}| > \|\vec{v} \times \vec{w}\|$

52. Use a parallelepiped to show that $\vec{a} \cdot (\vec{b} \times \vec{c}) = (\vec{a} \times \vec{b}) \cdot \vec{c}$ for any vectors $\vec{a}$, $\vec{b}$, and $\vec{c}$.

53. Figure 13.47 shows the tetrahedron determined by three vectors $\vec{a}, \vec{b}, \vec{c}$. The *area vector* of a face is a vector perpendicular to the face, pointing outward, whose magnitude is the area of the face. Show that the sum of the four outward-pointing area vectors of the faces equals the zero vector.

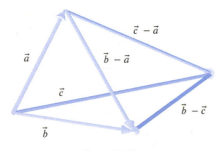

Figure 13.47

In Problems 54–56, find the vector representing the area of a surface. The magnitude of the vector equals the magnitude of the area; the direction is perpendicular to the surface. Since there are two perpendicular directions, we pick one by giving an orientation for the surface.

54. The rectangle with vertices $(0, 0, 0)$, $(0, 1, 0)$, $(2, 1, 0)$, and $(2, 0, 0)$, oriented so that it faces downward.

55. The circle of radius 2 in the yz-plane, facing in the direction of the positive x-axis.

56. The triangle ABC, oriented upward, where $A = (1, 2, 3)$, $B = (3, 1, 2)$, and $C = (2, 1, 3)$.

57. This problem relates the area of a parallelogram S lying in the plane $z = mx + ny + c$ to the area of its projection R in the xy-plane. Let S be determined by the vectors $\vec{u} = u_1 \vec{i} + u_2 \vec{j} + u_3 \vec{k}$ and $\vec{v} = v_1 \vec{i} + v_2 \vec{j} + v_3 \vec{k}$. See Figure 13.48.

    (a) Find the area of S.
    (b) Find the area of R.
    (c) Find m and n in terms of the components of $\vec{u}$ and $\vec{v}$.
    (d) Show that

$$\text{Area of } S = \sqrt{1 + m^2 + n^2} \cdot \text{Area of } R.$$

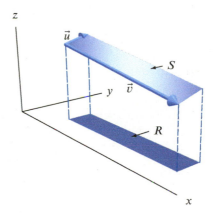

Figure 13.48

Strengthen Your Understanding

In Problems 58–59, explain what is wrong with the statement.

58. There is only one unit vector perpendicular to two non-parallel vectors in 3-space.

59. $\vec{u} \times \vec{v} = \vec{0}$ when $\vec{u}$ and $\vec{v}$ are perpendicular.

In Problems 60–61, give an example of:

60. A vector $\vec{u}$ whose cross product with $\vec{v} = \vec{i} + \vec{j}$ is parallel to $\vec{k}$.

61. A vector $\vec{v}$ such that $\|\vec{u} \times \vec{v}\| = 10$, where $\vec{u} = 3\vec{i} + 4\vec{j}$.

■ Are the statements in Problems 62–72 true or false? Give reasons for your answer.

62. $\vec{u} \times \vec{v}$ is a vector.

63. $\vec{u} \times \vec{v}$ has direction parallel to both $\vec{u}$ and $\vec{v}$.

64. $\|\vec{u} \times \vec{v}\| = \|\vec{u}\| \|\vec{v}\|$.

65. $(\vec{i} \times \vec{j}) \cdot \vec{k} = \vec{i} \cdot (\vec{j} \times \vec{k})$.

66. If $\vec{v}$ is a nonzero vector and $\vec{v} \times \vec{u} = \vec{v} \times \vec{w}$, then $\vec{u} = \vec{w}$.

67. The value of $\vec{v} \cdot (\vec{v} \times \vec{w})$ is always 0.

68. The value of $\vec{v} \times \vec{w}$ is never the same as $\vec{v} \cdot \vec{w}$.

69. The area of the triangle with two sides given by $\vec{i} + \vec{j}$ and $\vec{j} + 2\vec{k}$ is 3/2.

70. Given a nonzero vector $\vec{v}$ in 3-space, there is a nonzero vector $\vec{w}$ such that $\vec{v} \times \vec{w} = \vec{0}$.

71. It is never true that $\vec{v} \times \vec{w} = \vec{w} \times \vec{v}$.

72. Two points are circling an axis at a rate of 5 rad/sec. The point closer to the axis has the greater speed.

Online Resource: Review Problems and Projects

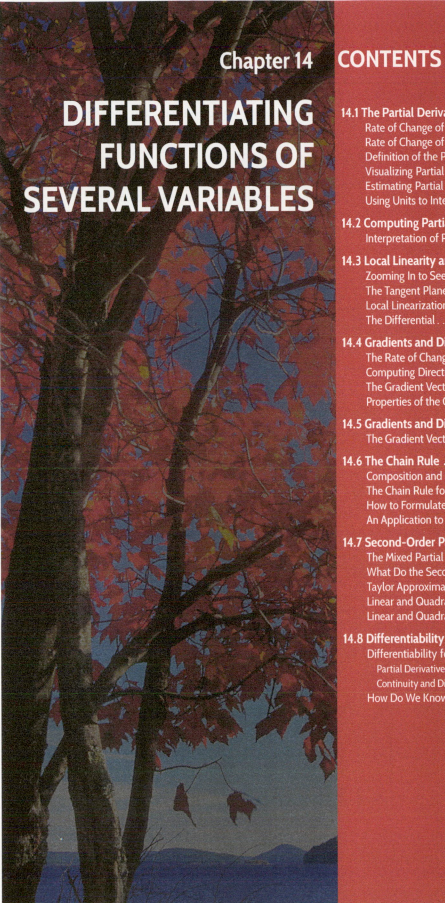

Chapter 14

DIFFERENTIATING FUNCTIONS OF SEVERAL VARIABLES

CONTENTS

14.1   THE PARTIAL DERIVATIVE

The derivative of a one-variable function measures its rate of change. In this section we see how a two-variable function has two rates of change: one as x changes (with y held constant) and one as y changes (with x held constant).

Rate of Change of Temperature in a Metal Rod: a One-Variable Problem

Imagine an unevenly heated metal rod lying along the x-axis, with its left end at the origin and x measured in meters. (See Figure 14.1.) Let $u(x)$ be the temperature (in °C) of the rod at the point x. Table 14.1 gives values of $u(x)$. We see that the temperature increases as we move along the rod, reaching its maximum at $x = 4$, after which it starts to decrease.

Figure 14.1: Unevenly heated metal rod

Table 14.1 *Temperature $u(x)$ of the rod*

x (m)	0	1	2	3	4	5
$u(x)$ (°C)	125	128	135	160	175	160

Example 1   Estimate the derivative $u'(2)$ using Table 14.1 and explain what the answer means in terms of temperature.

Solution   The derivative $u'(2)$ is defined as a limit of difference quotients:

$$u'(2) = \lim_{h \to 0} \frac{u(2+h) - u(2)}{h}.$$

Choosing $h = 1$ so that we can use the data in Table 14.1, we get

$$u'(2) \approx \frac{u(2+1) - u(2)}{1} = \frac{160 - 135}{1} = 25.$$

This means that the temperature increases at a rate of approximately 25°C per meter as we go from left to right, past $x = 2$.

Rate of Change of Temperature in a Metal Plate

Imagine an unevenly heated thin rectangular metal plate lying in the xy-plane with its lower left corner at the origin and x and y measured in meters. The temperature (in °C) at the point (x, y) is $T(x, y)$. See Figure 14.2 and Table 14.2. How does T vary near the point $(2, 1)$? We consider the horizontal line $y = 1$ containing the point $(2, 1)$. The temperature along this line is the cross section, $T(x, 1)$, of the function $T(x, y)$ with $y = 1$. Suppose we write $u(x) = T(x, 1)$.

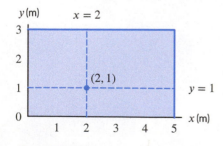

Figure 14.2: Unevenly heated metal plate

Table 14.2 *Temperature (°C) of a metal plate*

y (m)						
3	85	90	110	135	155	180
2	100	110	120	145	190	170
1	125	128	135	160	175	160
0	120	135	155	160	160	150
	0	1	2	3	4	5
			x (m)			

What is the meaning of the derivative $u'(2)$? It is the rate of change of temperature T *in the x-direction* at the point $(2, 1)$, keeping y fixed. Denote this rate of change by $T_x(2, 1)$, so that

$$T_x(2, 1) = u'(2) = \lim_{h \to 0} \frac{u(2+h) - u(2)}{h} = \lim_{h \to 0} \frac{T(2+h, 1) - T(2, 1)}{h}.$$

We call $T_x(2, 1)$ the *partial derivative of T with respect to x at the point* $(2, 1)$. Taking $h = 1$, we can read values of T from the row with $y = 1$ in Table 14.2, giving

$$T_x(2, 1) \approx \frac{T(3, 1) - T(2, 1)}{1} = \frac{160 - 135}{1} = 25°\text{C/m}.$$

The fact that $T_x(2, 1)$ is positive means that the temperature of the plate is increasing as we move past the point $(2, 1)$ in the direction of increasing x (that is, horizontally from left to right in Figure 14.2).

Example 2   Estimate the rate of change of T in the y-direction at the point $(2, 1)$.

Solution   The temperature along the line $x = 2$ is the cross-section of T with $x = 2$, that is, the function $v(y) = T(2, y)$. If we denote the rate of change of T in the y-direction at $(2, 1)$ by $T_y(2, 1)$, then

$$T_y(2, 1) = v'(1) = \lim_{h \to 0} \frac{v(1 + h) - v(1)}{h} = \lim_{h \to 0} \frac{T(2, 1 + h) - T(2, 1)}{h}.$$

We call $T_y(2, 1)$ the *partial derivative of T with respect to y at the point* $(2, 1)$. Taking $h = 1$ so that we can use the column with $x = 2$ in Table 14.2, we get

$$T_y(2, 1) \approx \frac{T(2, 1 + 1) - T(2, 1)}{1} = \frac{120 - 135}{1} = -15°\text{C/m}.$$

The fact that $T_y(2, 1)$ is negative means that at $(2, 1)$, the temperature decreases as y increases..

Definition of the Partial Derivative

We study the influence of x and y separately on the value of the function $f(x, y)$ by holding one fixed and letting the other vary. This leads to the following definitions.

Partial Derivatives of f with Respect to x and y

For all points at which the limits exist, we define the **partial derivatives at the point** $(\mathbf{a}, \mathbf{b})$ by

$$f_x(a, b) = \begin{array}{c} \text{Rate of change of } f \text{ with respect to } x \\ \text{at the point } (a, b) \end{array} = \lim_{h \to 0} \frac{f(a + h, b) - f(a, b)}{h},$$

$$f_y(a, b) = \begin{array}{c} \text{Rate of change of } f \text{ with respect to } y \\ \text{at the point } (a, b) \end{array} = \lim_{h \to 0} \frac{f(a, b + h) - f(a, b)}{h}.$$

If we let a and b vary, we have the **partial derivative functions** $f_x(x, y)$ and $f_y(x, y)$.

Just as with ordinary derivatives, there is an alternative notation:

Alternative Notation for Partial Derivatives

If $z = f(x, y)$, we can write

$$f_x(x, y) = \frac{\partial z}{\partial x} \quad \text{and} \quad f_y(x, y) = \frac{\partial z}{\partial y},$$

$$f_x(a, b) = \frac{\partial z}{\partial x}\bigg|_{(a,b)} \quad \text{and} \quad f_y(a, b) = \frac{\partial z}{\partial y}\bigg|_{(a,b)}.$$

We use the symbol ∂ to distinguish partial derivatives from ordinary derivatives. In cases where the independent variables have names different from x and y, we adjust the notation accordingly. For example, the partial derivatives of $f(u, v)$ are denoted by f_u and f_v.

Visualizing Partial Derivatives on a Graph

The ordinary derivative of a one-variable function is the slope of its graph. How do we visualize the partial derivative $f_x(a, b)$? The graph of the one-variable function $f(x, b)$ is the curve where the vertical plane $y = b$ cuts the graph of $f(x, y)$. (See Figure 14.3.) Thus, $f_x(a, b)$ is the slope of the tangent line to this curve at $x = a$.

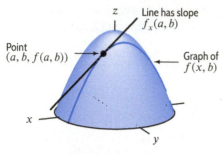

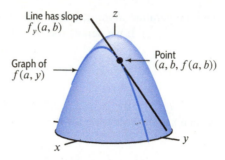

Figure 14.3: The curve $z = f(x, b)$ on the graph of f has slope $f_x(a, b)$ at $x = a$

Figure 14.4: The curve $z = f(a, y)$ on the graph of f has slope $f_y(a, b)$ at $y = b$

Similarly, the graph of the function $f(a, y)$ is the curve where the vertical plane $x = a$ cuts the graph of f, and the partial derivative $f_y(a, b)$ is the slope of this curve at $y = b$. (See Figure 14.4.)

Example 3   At each point labeled on the graph of the surface $z = f(x, y)$ in Figure 14.5, say whether each partial derivative is positive or negative.

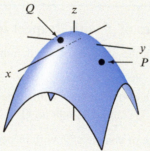

Figure 14.5: Decide the signs of f_x and f_y at P and Q

Solution   The positive x-axis points out of the page. Imagine heading off in this direction from the point marked P; we descend steeply. So the partial derivative with respect to x is negative at P, with quite a large absolute value. The same is true for the partial derivative with respect to y at P, since there is also a steep descent in the positive y-direction.

At the point marked Q, heading in the positive x-direction results in a gentle descent, whereas heading in the positive y-direction results in a gentle ascent. Thus, the partial derivative f_x at Q is negative but small (that is, near zero), and the partial derivative f_y is positive but small.

Estimating Partial Derivatives from a Contour Diagram

The graph of a function $f(x, y)$ often makes clear the sign of the partial derivatives. However, numerical estimates of these derivatives are more easily made from a contour diagram than a surface graph. If we move parallel to one of the axes on a contour diagram, the partial derivative is the rate of change of the value of the function on the contours. For example, if the values on the contours are increasing as we move in the positive direction, then the partial derivative must be positive.

Example 4   Figure 14.6 shows the contour diagram for the temperature $H(x, t)$ (in °C) in a room as a function of distance x (in meters) from a heater and time t (in minutes) after the heater has been turned on. What are the signs of $H_x(10, 20)$ and $H_t(10, 20)$? Estimate these partial derivatives and explain the answers in practical terms.

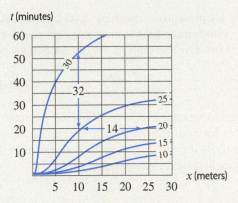

Figure 14.6: Temperature in a heated room: Heater at $x = 0$ is turned on at $t = 0$

Solution   The point $(10, 20)$ is nearly on the $H = 25$ contour. As x increases, we move toward the $H = 20$ contour, so H is decreasing and $H_x(10, 20)$ is negative. This makes sense because the $H = 30$ contour is to the left: As we move further from the heater, the temperature drops. On the other hand, as t increases, we move toward the $H = 30$ contour, so H is increasing; as t decreases H decreases. Thus, $H_t(10, 20)$ is positive. This says that as time passes, the room warms up.

To estimate the partial derivatives, use a difference quotient. Looking at the contour diagram, we see there is a point on the $H = 20$ contour about 14 units to the right of the point $(10, 20)$. Hence, H decreases by 5 when x increases by 14, so we find

$$\text{Rate of change of } H \text{ with respect to } x = H_x(10, 20) \approx \frac{-5}{14} \approx -0.36°\text{C/meter.}$$

This means that near the point 10 m from the heater, after 20 minutes the temperature drops about 0.36, or one third, of a degree, for each meter we move away from the heater.

To estimate $H_t(10, 20)$, we notice that the $H = 30$ contour is about 32 units directly above the point $(10, 20)$. So H increases by 5 when t increases by 32. Hence,

$$\text{Rate of change of } H \text{ with respect to } t = H_t(10, 20) \approx \frac{5}{32} = 0.16°\text{C/minute.}$$

This means that after 20 minutes the temperature is going up about 0.16, or 1/6, of a degree each minute at the point 10 m from the heater.

Using Units to Interpret Partial Derivatives

The meaning of a partial derivative can often be explained using units.

Example 5   Suppose that your weight w in pounds is a function $f(c, n)$ of the number c of calories you consume daily and the number n of minutes you exercise daily. Using the units for w, c and n, interpret in everyday terms the statements

$$\frac{\partial w}{\partial c}(2000, 15) = 0.02 \quad \text{and} \quad \frac{\partial w}{\partial n}(2000, 15) = -0.025.$$

Solution   The units of $\partial w/\partial c$ are pounds per calorie. The statement

$$\frac{\partial w}{\partial c}(2000, 15) = 0.02$$

means that if you are presently consuming 2000 calories daily and exercising 15 minutes daily, you will weigh 0.02 pounds more for each extra calorie you consume daily, or about 2 pounds for each extra 100 calories per day. The units of $\partial w/\partial n$ are pounds per minute. The statement

$$\frac{\partial w}{\partial n}(2000, 15) = -0.025$$

means that for the same calorie consumption and number of minutes of exercise, you will weigh 0.025 pounds less for each extra minute you exercise daily, or about 1 pound less for each extra 40 minutes per day. So if you eat an extra 100 calories each day and exercise about 80 minutes more each day, your weight should remain roughly steady.

Summary for Section 14.1

- The **partial derivatives** at the point (a, b) are:
  - $f_x(a, b) =$ Rate of change of f with respect to x at the point $(a, b) = \lim\limits_{h \to 0} \dfrac{f(a + h, b) - f(a, b)}{h}$
  - $f_y(a, b) =$ Rate of change of f with respect to y at the point $(a, b) = \lim\limits_{h \to 0} \dfrac{f(a, b + h) - f(a, b)}{h}$
- If $z = f(x, y)$, we can also write **partial derivatives** using the notation:

$$f_x(x, y) = \frac{\partial z}{\partial x} \qquad \text{and} \qquad f_y(x, y) = \frac{\partial z}{\partial y},$$

$$f_x(a, b) = \frac{\partial z}{\partial x}\bigg|_{(a,b)} \qquad \text{and} \qquad f_y(a, b) = \frac{\partial z}{\partial y}\bigg|_{(a,b)}.$$

- The **signs** of the partial derivatives indicate increase or decrease of the function in the x or y direction.
- The **units** of a partial derivative are the units of the output divided by the units of the direction variable. For example, if $z = f(x, y)$, the units of $f_x = \frac{\partial z}{\partial x}$ are the units of z over the units of x.

Exercises and Problems for Section 14.1

EXERCISES

1. Given the following table of values for $z = f(x, y)$, estimate $f_x(3, 2)$ and $f_y(3, 2)$, assuming they exist.

$x \backslash y$	0	2	5
1	1	2	4
3	−1	1	2
6	−3	0	0

2. Using difference quotients, estimate $f_x(3, 2)$ and $f_y(3, 2)$ for the function given by

$$f(x, y) = \frac{x^2}{y + 1}.$$

[Recall: A difference quotient is an expression of the form $(f(a + h, b) - f(a, b))/h$.]

3. Use difference quotients with $\Delta x = 0.1$ and $\Delta y = 0.1$ to estimate $f_x(1, 3)$ and $f_y(1, 3)$, where

$$f(x, y) = e^{-x} \sin y.$$

Then give better estimates by using $\Delta x = 0.01$ and $\Delta y = 0.01$.

4. The price P in dollars to purchase a used car is a function of its original cost, C, in dollars, and its age, A, in years.

(a) What are the units of $\partial P/\partial A$?
(b) What is the sign of $\partial P/\partial A$ and why?
(c) What are the units of $\partial P/\partial C$?
(d) What is the sign of $\partial P/\partial C$ and why?

5. Your monthly car payment in dollars is $P = f(P_0, t, r)$, where $\$P_0$ is the amount you borrowed, t is the number of months it takes to pay off the loan, and $r\%$ is the interest rate. What are the units, the financial meaning, and the signs of $\partial P/\partial t$ and $\partial P/\partial r$?

6. A drug is injected into a patient's blood vessel. The function $c = f(x, t)$ represents the concentration of the drug at a distance x mm in the direction of the blood flow measured from the point of injection and at time t seconds since the injection. What are the units of the following partial derivatives? What are their practical interpretations? What do you expect their signs to be?

   (a) $\partial c/\partial x$ (b) $\partial c/\partial t$

7. You borrow $\$A$ at an interest rate of $r\%$ (per month) and pay it off over t months by making monthly payments of $P = g(A, r, t)$ dollars. In financial terms, what do the following statements tell you?

   (a) $g(8000, 1, 24) = 376.59$

   (b) $\left.\dfrac{\partial g}{\partial A}\right|_{(8000,1,24)} = 0.047$

   (c) $\left.\dfrac{\partial g}{\partial r}\right|_{(8000,1,24)} = 44.83$

8. The sales of a product, $S = f(p, a)$, are a function of the price, p, of the product (in dollars per unit) and the amount, a, spent on advertising (in thousands of dollars).

   (a) Do you expect f_p to be positive or negative? Why?

   (b) Explain the meaning of the statement $f_a(8, 12) = 150$ in terms of sales.

9. The quantity, Q, of beef purchased at a store, in kilograms per week, is a function of the price of beef, b, and the price of chicken, c, both in dollars per kilogram.

   (a) Do you expect $\partial Q/\partial b$ to be positive or negative? Explain.

   (b) Do you expect $\partial Q/\partial c$ to be positive or negative? Explain.

   (c) Interpret the statement $\partial Q/\partial b = -213$ in terms of quantity of beef purchased.

■ In Exercises 10–15, a point A is shown on a contour diagram of a function $f(x, y)$.

(a) Evaluate $f(A)$.

(b) Is $f_x(A)$ positive, negative, or zero?

(c) Is $f_y(A)$ positive, negative, or zero?

10.   *y*

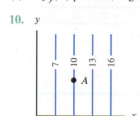

11.   *y*

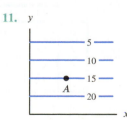

12.   *y*

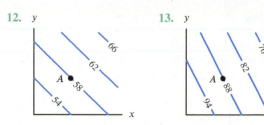

13.   *y*

14.   *y*

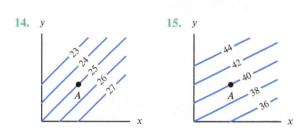

15.   *y*

■ In Exercises 16–19, determine the sign of f_x and f_y at the point using the contour diagram of f in Figure 14.7.

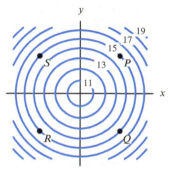

Figure 14.7

16. P 17. Q 18. R 19. S

20. Values of $f(x, y)$ are in Table 14.3. Assuming they exist, decide whether you expect the following partial derivatives to be positive or negative.

   (a) $f_x(-2, -1)$ (b) $f_y(2, 1)$

   (c) $f_x(2, 1)$ (d) $f_y(0, 3)$

Table 14.3

$x \backslash y$	-1	1	3	5
-2	7	3	2	1
0	8	5	3	2
2	10	7	5	4
4	13	10	8	7

PROBLEMS

21. Figure 14.8 is a contour diagram for $z = f(x, y)$. Is f_x positive or negative? Is f_y positive or negative? Estimate $f(2, 1)$, $f_x(2, 1)$, and $f_y(2, 1)$.

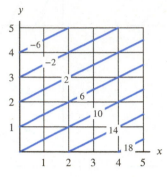

Figure 14.8

22. Approximate $f_x(3, 5)$ using the contour diagram of $f(x, y)$ in Figure 14.9.

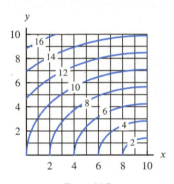

Figure 14.9

23. When riding your bike in winter, the windchill temperature is a measure of how cold you feel as a result of the induced breeze caused by your travel. If W represents windchill temperature (in °F) that you experience, then $W = f(T, v)$, where T is the actual air temperature (in °F) and v is your speed, in meters per second. Match each of the practical interpretations below with a mathematical statement that most accurately describes it below. For the remaining mathematical statement, give a practical interpretation.

  (i) "The faster you ride, the colder you'll feel."
  (ii) "The warmer the day, the warmer you'll feel."

  (a) $f_T(T, v) > 0$ **(b)** $f(0, v) \leq 0$ **(c)** $f_v(T, v) < 0$

24. People commuting to a city can choose to go either by bus or by train. The number of people who choose either method depends in part upon the price of each. Let $f(P_1, P_2)$ be the number of people who take the bus when P_1 is the price of a bus ride and P_2 is the price of a train ride. What can you say about the signs of $\partial f / \partial P_1$ and $\partial f / \partial P_2$? Explain your answers.

25. The average price of large cars getting low gas mileage ("gas guzzlers") is x and the average price of a gallon of gasoline is y. The number, q_1, of gas guzzlers bought in a year, depends on both x and y, so $q_1 = f(x, y)$. Similarly, if q_2 is the number of gallons of gas bought to fill gas guzzlers in a year, then $q_2 = g(x, y)$.

  (a) What do you expect the signs of $\partial q_1 / \partial x$ and $\partial q_2 / \partial y$ to be? Explain.
  (b) What do you expect the signs of $\partial q_1 / \partial y$ and $\partial q_2 / \partial x$ to be? Explain.

For Problems 26–28, refer to Table 12.2 on page 700 giving the temperature adjusted for wind chill, C, in °F, as a function $f(w, T)$ of the wind speed, w, in mph, and the temperature, T, in °F. The temperature adjusted for wind chill tells you how cold it feels, as a result of the combination of wind and temperature.

26. Estimate $f_w(10, 25)$. What does your answer mean in practical terms?

27. Estimate $f_T(5, 20)$. What does your answer mean in practical terms?

28. From Table 12.2 you can see that when the temperature is 20°F, the temperature adjusted for wind-chill drops by an average of about 0.8°F with every 1 mph increase in wind speed from 5 mph to 10 mph. Which partial derivative is this telling you about?

29. An experiment to measure the toxicity of formaldehyde yielded the data in Table 14.4. The values show the percent, $P = f(t, c)$, of rats surviving an exposure to formaldehyde at a concentration of c (in parts per million, ppm) after t months. Estimate $f_t(18, 6)$ and $f_c(18, 6)$. Interpret your answers in terms of formaldehyde toxicity.

Table 14.4

		Time t (months)					
		14	16	18	20	22	24
	0	100	100	100	99	97	95
Conc. c (ppm)	2	100	99	98	97	95	92
	6	96	95	93	90	86	80
	15	96	93	82	70	58	36

30. Figure 14.10 shows contours of $f(x, y)$ with values of f on the contours omitted. If $f_x(P) > 0$, find the sign:

(a) $f_y(P)$ (b) $f_y(Q)$ (c) $f_x(Q)$

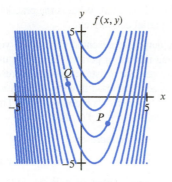

Figure 14.10

31. Figure 14.11 shows the contour diagram of $g(x, y)$. Mark the points on the contours where

(a) $g_x = 0$ (b) $g_y = 0$

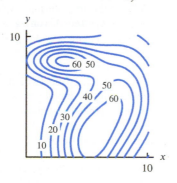

Figure 14.11

32. The surface $z = f(x, y)$ is shown in Figure 14.12. The points A and B are in the xy-plane.

(a) What is the sign of

(i) $f_x(A)$? (ii) $f_y(A)$?

(b) The point P in the xy-plane moves along a straight line from A to B. How does the sign of $f_x(P)$ change? How does the sign of $f_y(P)$ change?

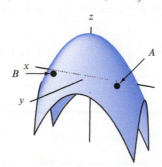

Figure 14.12

33. Figure 14.13 shows the saddle-shaped surface $z = f(x, y)$.

(a) What is the sign of $f_x(0, 5)$?
(b) What is the sign of $f_y(0, 5)$?

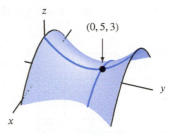

Figure 14.13

34. Figure 14.14 shows the graph of the function $f(x, y)$ on the domain $0 \le x \le 4$ and $0 \le y \le 4$. Use the graph to rank the following quantities in order from smallest to largest: $f_x(3, 2)$, $f_x(1, 2)$, $f_y(3, 2)$, $f_y(1, 2)$, 0.

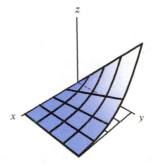

Figure 14.14

35. Figure 14.15 shows a contour diagram for the monthly payment P as a function of the interest rate, $r\%$, and the amount, L, of a 5-year loan. Estimate $\partial P/\partial r$ and $\partial P/\partial L$ at the following points. In each case, give the units and the everyday meaning of your answer.

(a) $r = 8, L = 4000$ (b) $r = 8, L = 6000$
(c) $r = 13, L = 7000$

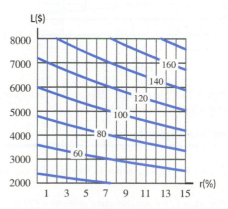

Figure 14.15

36. Figure 14.16 shows a contour diagram for the temperature T (in °C) along a wall in a heated room as a function of distance x along the wall and time t in minutes. Estimate $\partial T/\partial x$ and $\partial T/\partial t$ at the given points. Give units and interpret your answers.

(a) $x = 15, t = 20$ **(b)** $x = 5, t = 12$

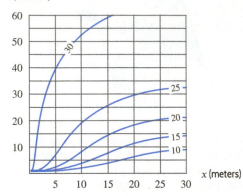

Figure 14.16

In Problems **37–39**, we use Figure 14.17 to model the heat required to clear an airport of fog by heating the air. The amount of heat, $H(T, w)$, required (in calories per cubic meter of fog) is a function of the temperature T (in degrees Celsius) and the water content w (in grams per cubic meter of fog). Note that Figure 14.17 is not a contour diagram, but shows cross-sections of H with w fixed at 0.1, 0.2, 0.3, 0.4.

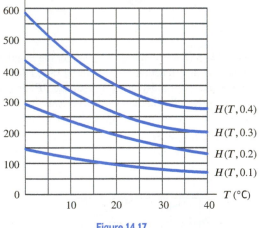

Figure 14.17

37. Use Figure 14.17 to estimate $H_T(10, 0.1)$. Interpret the partial derivative in practical terms.

38. Make a table of values for $H(T, w)$ from Figure 14.17, and use it to estimate $H_T(T, w)$ for $T = 10, 20$, and 30 and $w = 0.1, 0.2$, and 0.3.

39. Repeat Problem 38 for $H_w(T, w)$ at $T = 10, 20$, and 30 and $w = 0.1, 0.2$, and 0.3. What is the practical meaning of these partial derivatives?

40. The cardiac output, represented by c, is the volume of blood flowing through a person's heart per unit time. The systemic vascular resistance (SVR), represented by s, is the resistance to blood flowing through veins and arteries. Let p be a person's blood pressure. Then p is a function of c and s, so $p = f(c, s)$.

(a) What does $\partial p/\partial c$ represent?

Suppose now that $p = kcs$, where k is a constant.

(b) Sketch the level curves of p. What do they represent? Label your axes.
(c) For a person with a weak heart, it is desirable to have the heart pumping against less resistance, while maintaining the same blood pressure. Such a person may be given the drug nitroglycerine to decrease the SVR and the drug dopamine to increase the cardiac output. Represent this on a graph showing level curves. Put a point A on the graph representing the person's state before drugs are given and a point B for after.
(d) Right after a heart attack, a patient's cardiac output drops, thereby causing the blood pressure to drop. A common mistake made by medical residents is to get the patient's blood pressure back to normal by using drugs to increase the SVR, rather than by increasing the cardiac output. On a graph of the level curves of p, put a point D representing the patient before the heart attack, a point E representing the patient right after the heart attack, and a third point F representing the patient after the resident has given the drugs to increase the SVR.

41. In each case, give a possible contour diagram for the function $f(x, y)$ if

(a) $f_x > 0$ and $f_y > 0$ **(b)** $f_x > 0$ and $f_y < 0$
(c) $f_x < 0$ and $f_y > 0$ **(d)** $f_x < 0$ and $f_y < 0$

In Problems **42–45**, give a possible contour diagram for the function $f(x, y)$ if

42. $f_x = 0, f_y \neq 0$ **43.** $f_y = 0, f_x \neq 0$

44. $f_x = 1$ **45.** $f_y = -2$

Strengthen Your Understanding

In Problems **46–47**, explain what is wrong with the statement.

46. For $f(x, y)$, $\partial f/\partial x$ has the same units as $\partial f/\partial y$.

47. The partial derivative with respect to y is not defined for functions such as $f(x, y) = x^2 + 5$ that have a formula that does not contain y explicitly.

In Problems **48–49**, give an example of:

48. A table of values with three rows and three columns of a linear function $f(x, y)$ with $f_x < 0$ and $f_y > 0$.

49. A function $f(x, y)$ with $f_x > 0$ and $f_y < 0$ everywhere.

Are the statements in Problems 50–60 true or false? Give reasons for your answer.

50. If $f(x, y)$ is a function of two variables and $f_x(10, 20)$ is defined, then $f_x(10, 20)$ is a scalar.

51. If $f(x, y) = x^2 + y^2$, then $f_y(1, 1) < 0$.

52. If the graph of $f(x, y)$ is a hemisphere centered at the origin, then $f_x(0, 0) = f_y(0, 0) = 0$.

53. If $P = f(T, V)$ is a function expressing the pressure P (in grams/cm^3) of gas in a piston in terms of the temperature T (in degrees °C) and volume V (in cm^3), then $\partial P / \partial V$ has units of grams.

54. If $f_x(a, b) > 0$, then the values of f decrease as we move in the negative x-direction near (a, b).

55. If $g(r, s) = r^2 + s$, then for fixed s, the partial derivative g_r increases as r increases.

56. Let $P = f(m, d)$ be the purchase price (in dollars) of a used car that has m miles on its engine and originally cost d dollars when new. Then $\partial P / \partial m$ and $\partial P / \partial d$ have the same sign.

57. If $f(x, y)$ is a function with the property that $f_x(x, y)$ and $f_y(x, y)$ are both constant, then f is linear.

58. If $f(x, y)$ has $f_x(a, b) = f_y(a, b) = 0$ at the point (a, b), then f is constant everywhere.

59. If $f_x = 0$ and $f_y \neq 0$, then the contours of $f(x, y)$ are horizontal lines.

60. If the contours of $f(x, y)$ are vertical lines, then $f_y = 0$.

14.2 COMPUTING PARTIAL DERIVATIVES ALGEBRAICALLY

Since the partial derivative $f_x(x, y)$ is the ordinary derivative of the function $f(x, y)$ with y held constant and $f_y(x, y)$ is the ordinary derivative of $f(x, y)$ with x held constant, we can use all the differentiation formulas from one-variable calculus to find partial derivatives.

Example 1   Let $f(x, y) = \dfrac{x^2}{y + 1}$. Find $f_x(3, 2)$ algebraically.

Solution   We use the fact that $f_x(3, 2)$ equals the derivative of $f(x, 2)$ at $x = 3$. Since

$$f(x, 2) = \frac{x^2}{2 + 1} = \frac{x^2}{3},$$

differentiating with respect to x, we have

$$f_x(x, 2) = \frac{\partial}{\partial x}\left(\frac{x^2}{3}\right) = \frac{2x}{3}, \qquad \text{and so} \qquad f_x(3, 2) = 2.$$

Example 2   Compute the partial derivatives with respect to x and with respect to y for the following functions.
(a) $f(x, y) = y^2 e^{3x}$ (b) $z = (3xy + 2x)^5$ (c) $g(x, y) = e^{x+3y} \sin(xy)$

Solution   (a) This is the product of a function of x (namely e^{3x}) and a function of y (namely y^2). When we differentiate with respect to x, we think of the function of y as a constant, and vice versa. Thus,

$$f_x(x, y) = y^2 \frac{\partial}{\partial x}\left(e^{3x}\right) = 3y^2 e^{3x},$$

$$f_y(x, y) = e^{3x} \frac{\partial}{\partial y}(y^2) = 2y e^{3x}.$$

(b) Here we use the chain rule:

$$\frac{\partial z}{\partial x} = 5(3xy + 2x)^4 \frac{\partial}{\partial x}(3xy + 2x) = 5(3xy + 2x)^4(3y + 2),$$

$$\frac{\partial z}{\partial y} = 5(3xy + 2x)^4 \frac{\partial}{\partial y}(3xy + 2x) = 5(3xy + 2x)^4 3x = 15x(3xy + 2x)^4.$$

(c) Since each function in the product is a function of both x and y, we need to use the product rule for each partial derivative:

$$g_x(x, y) = \left(\frac{\partial}{\partial x}(e^{x+3y}) \right) \sin(xy) + e^{x+3y} \frac{\partial}{\partial x}(\sin(xy)) = e^{x+3y} \sin(xy) + e^{x+3y} y \cos(xy),$$

$$g_y(x, y) = \left(\frac{\partial}{\partial y}(e^{x+3y}) \right) \sin(xy) + e^{x+3y} \frac{\partial}{\partial y}(\sin(xy)) = 3e^{x+3y} \sin(xy) + e^{x+3y} x \cos(xy).$$

For functions of three or more variables, we find partial derivatives by the same method: Differentiate with respect to one variable, regarding the other variables as constants. For a function $f(x, y, z)$, the partial derivative $f_x(a, b, c)$ gives the rate of change of f with respect to x along the line $y = b, z = c$.

Example 3   Find all the partial derivatives of $f(x, y, z) = \dfrac{x^2 y^3}{z}$.

Solution   To find $f_x(x, y, z)$, for example, we consider y and z as fixed, giving

$$f_x(x, y, z) = \frac{2xy^3}{z}, \quad \text{and} \quad f_y(x, y, z) = \frac{3x^2 y^2}{z}, \quad \text{and} \quad f_z(x, y, z) = -\frac{x^2 y^3}{z^2}.$$

Interpretation of Partial Derivatives

Example 4   A vibrating guitar string, originally at rest along the x-axis, is shown in Figure 14.18. Let x be the distance in meters from the left end of the string. At time t seconds the point x has been displaced $y = f(x, t)$ meters vertically from its rest position, where

$$y = f(x, t) = 0.003 \sin(\pi x) \sin(2765t).$$

Evaluate $f_x(0.3, 1)$ and $f_t(0.3, 1)$ and explain what each means in practical terms.

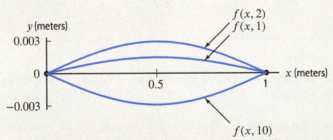

Figure 14.18: The position of a vibrating guitar string at several different times: Graph of $f(x, t)$ for $t = 1, 2, 10$.

Solution   Differentiating $f(x, t) = 0.003 \sin(\pi x) \sin(2765t)$ with respect to x, we have

$$f_x(x, t) = 0.003\pi \cos(\pi x) \sin(2765t).$$

In particular, substituting $x = 0.3$ and $t = 1$ gives

$$f_x(0.3, 1) = 0.003\pi \cos(\pi(0.3)) \sin(2765) \approx 0.002.$$

To see what $f_x(0.3, 1)$ means, think about the function $f(x, 1)$. The graph of $f(x, 1)$ in Figure 14.19 is a snapshot of the string at the time $t = 1$. Thus, the derivative $f_x(0.3, 1)$ is the slope of the string at the point $x = 0.3$ at the instant when $t = 1$.

Similarly, taking the derivative of $f(x,t) = 0.003 \sin(\pi x) \sin(2765t)$ with respect to t, we get

$$f_t(x,t) = (0.003)(2765)\sin(\pi x)\cos(2765t) = 8.3\sin(\pi x)\cos(2765t).$$

Since $f(x,t)$ is in meters and t is in seconds, the derivative $f_t(0.3, 1)$ is in m/sec. Thus, substituting $x = 0.3$ and $t = 1$,

$$f_t(0.3, 1) = 8.3\sin(\pi(0.3))\cos(2765(1)) \approx 6 \text{ m/sec.}$$

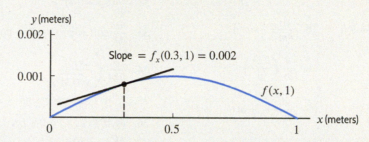

Figure 14.19: Graph of $f(x, 1)$: Snapshot of the shape of the string at $t = 1$ sec

To see what $f_t(0.3, 1)$ means, think about the function $f(0.3, t)$. The graph of $f(0.3, t)$ is a position versus time graph that tracks the up-and-down movement of the point on the string where $x = 0.3$. (See Figure 14.20.) The derivative $f_t(0.3, 1) = 6$ m/sec is the velocity of that point on the string at time $t = 1$. The fact that $f_t(0.3, 1)$ is positive indicates that the point is moving upward when $t = 1$.

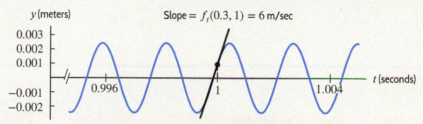

Figure 14.20: Graph of $f(0.3, t)$: Position versus time graph of the point $x = 0.3$ m from the left end of the guitar string

Summary for Section 14.2

- To calculate $f_x(x, y)$, **the partial derivative with respect to x**, algebraically, think of y as a constant, and differentiate with respect to x using the usual one-variable derivative rules.
- To calculate $f_y(x, y)$, **the partial derivative with respect to y**, algebraically, think of x as a constant, and differentiate with respect to y using the usual one-variable derivative rules.

Exercises and Problems for Section 14.2

EXERCISES

1. **(a)** If $f(x, y) = 2x^2 + xy + y^2$, approximate $f_y(3, 2)$ using $\Delta y = 0.01$.
   (b) Find the exact value of $f_y(3, 2)$.

In Exercises 2–40, find the partial derivatives. The variables are restricted to a domain on which the function is defined.

2. f_x and f_y if $f(x, y) = 5x^2y^3 + 8xy^2 - 3x^2$

3. $f_x(1, 2)$ and $f_y(1, 2)$ if $f(x, y) = x^3 + 3x^2y - 2y^2$

4. $\dfrac{\partial}{\partial y}(3x^5y^7 - 32x^4y^3 + 5xy)$

5. $\dfrac{\partial z}{\partial x}$ and $\dfrac{\partial z}{\partial y}$ if $z = (x^2 + x - y)^7$

6. f_x and f_y if $f(x, y) = A^\alpha x^{\alpha+\beta} y^{1-\alpha-\beta}$

7. f_x and f_y if $f(x, y) = \ln(x^{0.6}y^{0.4})$

8. z_x if $z = \dfrac{1}{2x^2ay} + \dfrac{3x^5abc}{y}$

9. z_x if $z = x^2 y + 2x^5 y$

10. $\dfrac{\partial}{\partial x}(a\sqrt{x})$

11. V_r if $V = \frac{1}{3}\pi r^2 h$

12. $\dfrac{\partial}{\partial T}\left(\dfrac{2\pi r}{T}\right)$

13. $\dfrac{\partial}{\partial x}(xe^{\sqrt{xy}})$

14. $\dfrac{\partial}{\partial t}e^{\sin(x+ct)}$

15. F_m if $F = mg$

16. a_v if $a = \dfrac{v^2}{r}$

17. $\dfrac{\partial A}{\partial h}$ if $A = \frac{1}{2}(a+b)h$

18. $\dfrac{\partial}{\partial m}\left(\frac{1}{2}mv^2\right)$

19. $\dfrac{\partial}{\partial B}\left(\dfrac{1}{u_0}B^2\right)$

20. $\dfrac{\partial}{\partial r}\left(\dfrac{2\pi r}{v}\right)$

21. F_v if $F = \dfrac{mv^2}{r}$

22. $\dfrac{\partial}{\partial v_0}(v_0 + at)$

23. z_x if $z = \sin(5x^3 y - 3xy^2)$

24. $\dfrac{\partial z}{\partial y}\bigg|_{(1,0.5)}$ if $z = e^{x+2y}\sin y$

25. g_x if $g(x,y) = \ln(ye^{xy})$

26. $\dfrac{\partial f}{\partial x}\bigg|_{(\pi/3,1)}$ if $f(x,y) = x\ln(y\cos x)$

27. z_x and z_y for $z = x^7 + 2^y + x^y$

28. f_x if $f(x,y) = e^{xy}(\ln y)$

29. $\dfrac{\partial F}{\partial m_2}$ if $F = \dfrac{Gm_1 m_2}{r^2}$

30. $\dfrac{\partial}{\partial x}\left(\dfrac{1}{a}e^{-x^2/a^2}\right)$

31. $\dfrac{\partial}{\partial a}\left(\dfrac{1}{a}e^{-x^2/a^2}\right)$

32. $\dfrac{\partial}{\partial t}(v_0 t + \frac{1}{2}at^2)$

33. $\dfrac{\partial}{\partial \theta}(\sin(\pi\theta\phi) + \ln(\theta^2 + \phi))$

34. $\dfrac{\partial}{\partial M}\left(\dfrac{2\pi r^{3/2}}{\sqrt{GM}}\right)$

35. f_a if $f(a,b) = e^a \sin(a+b)$

36. F_L if $F(L,K) = 3\sqrt{LK}$

37. $\dfrac{\partial V}{\partial r}$ and $\dfrac{\partial V}{\partial h}$ if $V = \frac{4}{3}\pi r^2 h$

38. u_E if $u = \frac{1}{2}\epsilon_0 E^2 + \dfrac{1}{2\mu_0}B^2$

39. $\dfrac{\partial}{\partial x}\left(\dfrac{1}{\sqrt{2\pi}\sigma}e^{-(x-\mu)^2/(2\sigma^2)}\right)$

40. $\dfrac{\partial Q}{\partial K}$ if $Q = c(a_1 K^{b_1} + a_2 L^{b_2})^\gamma$

PROBLEMS

In Problems 41–43:

(a) Find $f_x(1,1)$ and $f_y(1,1)$.

(b) Use part (a) to match $f(x,y)$ with one of the contour diagrams (I)–(III), each shown centered at $(1,1)$ with the same scale in the x and y directions.

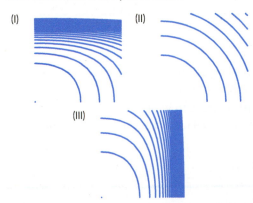

(I)   (II)

(III)

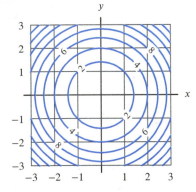

Figure 14.21

41. $f(x,y) = x^2 + y^2$

42. $f(x,y) = e^{x^2} + y^2$

43. $f(x,y) = x^2 + e^{y^2}$

44. (a) Let $f(x,y) = x^2 + y^2$. Estimate $f_x(2,1)$ and $f_y(2,1)$ using the contour diagram for f in Figure 14.21.

(b) Estimate $f_x(2,1)$ and $f_y(2,1)$ from a table of values for f with $x = 1.9, 2, 2.1$ and $y = 0.9, 1, 1.1$.

(c) Compare your estimates in parts (a) and (b) with the exact values of $f_x(2,1)$ and $f_y(2,1)$ found algebraically.

45. (a) Let $f(w,z) = e^{w\ln z}$. Use difference quotients with $h = 0.01$ to approximate $f_w(2,2)$ and $f_z(2,2)$.

(b) Now evaluate $f_w(2,2)$ and $f_z(2,2)$ exactly.

46. (a) The surface S is given, for some constant a, by

$$z = 3x^2 + 4y^2 - axy$$

Find the values of a which ensure that S is sloping upward when we move in the positive x-direction from the point $(1,2)$.

(b) With the values of a from part (a), if you move in the positive y-direction from the point $(1,2)$, does the surface slope up or down? Explain.

47. Money in a bank account earns interest at a continuous rate, r. The amount of money, $\$B$, in the account depends on the amount deposited, $\$P$, and the time, t, it has been in the bank according to the formula

$$B = Pe^{rt}.$$

Find $\partial B/\partial t$ and $\partial B/\partial P$ and interpret each in financial terms.

48. The acceleration g due to gravity, at a distance r from the center of a planet of mass m, is given by

$$g = \frac{Gm}{r^2},$$

where G is the universal gravitational constant.

(a) Find $\partial g/\partial m$ and $\partial g/\partial r$.
(b) Interpret each of the partial derivatives you found in part (a) as the slope of a graph in the plane and sketch the graph.

49. The Dubois formula relates a person's surface area, s, in m^2, to weight, w, in kg, and height, h, in cm, by

$$s = f(w, h) = 0.01 w^{0.25} h^{0.75}.$$

Find $f(65, 160)$, $f_w(65, 160)$, and $f_h(65, 160)$. Interpret your answers in terms of surface area, height, and weight.

50. The energy, E, of a body of mass m moving with speed v is given by the formula

$$E = mc^2 \left(\frac{1}{\sqrt{1 - v^2/c^2}} - 1 \right).$$

The speed, v, is nonnegative and less than the speed of light, c, which is a constant.

(a) Find $\partial E/\partial m$. What would you expect the sign of $\partial E/\partial m$ to be? Explain.
(b) Find $\partial E/\partial v$. Explain what you would expect the sign of $\partial E/\partial v$ to be and why.

51. Let $h(x, t) = 5 + \cos(0.5x - t)$ describe a wave. The value of $h(x, t)$ gives the depth of the water in cm at a distance x meters from a fixed point and at time t seconds. Evaluate $h_x(2, 5)$ and $h_t(2, 5)$ and interpret each in terms of the wave.

52. A one-meter-long bar is heated unevenly, with temperature in °C at a distance x meters from one end at time t given by

$$H(x, t) = 100 e^{-0.1t} \sin(\pi x) \qquad 0 \le x \le 1.$$

(a) Sketch a graph of H against x for $t = 0$ and $t = 1$.
(b) Calculate $H_x(0.2, t)$ and $H_x(0.8, t)$. What is the practical interpretation (in terms of temperature) of these two partial derivatives? Explain why each one has the sign it does.
(c) Calculate $H_t(x, t)$. What is its sign? What is its interpretation in terms of temperature?

53. Show that the Cobb-Douglas function

$$Q = bK^\alpha L^{1-\alpha} \quad \text{where} \quad 0 < \alpha < 1$$

satisfies the equation

$$K \frac{\partial Q}{\partial K} + L \frac{\partial Q}{\partial L} = Q.$$

■ In Problems 54–57, find all points where the partial derivatives of $f(x, y)$ are both 0.

54. $f(x, y) = x^2 + y^2$
55. $f(x, y) = xe^y$
56. $f(x, y) = e^{x^2 + 2x + y^2}$
57. $f(x, y) = x^3 + 3x^2 + y^3 - 3y$

58. Is there a function f which has the following partial derivatives? If so, what is it? Are there any others?

$$f_x(x, y) = 4x^3 y^2 - 3y^4,$$
$$f_y(x, y) = 2x^4 y - 12xy^3.$$

Strengthen Your Understanding

■ In Problems 59–60, explain what is wrong with the statement.

59. The partial derivative of $f(x, y) = x^2 y^3$ is $2xy^3 + 3y^2 x^2$.

60. For $f(x, y)$, if $\dfrac{f(0.01, 0) - f(0, 0)}{0.01} > 0$, then $f_x(0, 0) > 0$.

■ In Problems 61–63, give an example of:

61. A nonlinear function $f(x, y)$ such that $f_x(0, 0) = 2$ and $f_y(0, 0) = 3$.

62. Functions $f(x, y)$ and $g(x, y)$ such that $f_x = g_x$ but $f_y \ne g_y$.

63. A non-constant function $f(x, y)$ such that $f_x = 0$ everywhere.

■ Are the statements in Problems 64–71 true or false? Give reasons for your answer.

64. There is a function $f(x, y)$ with $f_x(x, y) = y$ and $f_y(x, y) = x$.

65. The function $z(u, v) = u \cos v$ satisfies the equation

$$\cos v \frac{\partial z}{\partial u} - \frac{\sin v}{u} \frac{\partial z}{\partial v} = 1.$$

66. If $f(x, y)$ is a function of two variables and $g(x)$ is a function of a single variable, then

$$\frac{\partial}{\partial y} (g(x) f(x, y)) = g(x) f_y(x, y).$$

67. The function $k(r, s) = rse^s$ is increasing in the s-direction at the point $(r, s) = (-1, 2)$.

68. There is a function $f(x, y)$ with $f_x(x, y) = y^2$ and $f_y(x, y) = x^2$.

69. If $f(x, y)$ has $f_y(x, y) = 0$ then f must be a constant.

70. If $f(x, y) = ye^{g(x)}$ then $f_x = f$.

71. If f is a symmetric two-variable function, that is $f(x, y) = f(y, x)$, then $f_x(x, y) = f_y(x, y)$.

72. Which of the following functions satisfy the following equation (called Euler's Equation):

$$xf_x + yf_y = f?$$

(a) x^2y^3 **(b)** $x+y+1$ **(c)** $x^2 + y^2$ **(d)** $x^{0.4}y^{0.6}$

14.3   LOCAL LINEARITY AND THE DIFFERENTIAL

In Sections 14.1 and 14.2 we studied a function of two variables by allowing one variable at a time to change. We now let both variables change at once to develop a linear approximation for functions of two variables.

Zooming In to See Local Linearity

For a function of one variable, local linearity means that as we zoom in on the graph, it looks like a straight line. As we zoom in on the graph of a two-variable function, the graph usually looks like a plane, which is the graph of a linear function of two variables. (See Figure 14.22.)

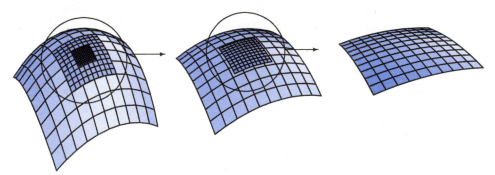

Figure 14.22: Zooming in on the graph of a function of two variables until the graph looks like a plane

Similarly, Figure 14.23 shows three successive views of the contours near a point. As we zoom in, the contours look more like equally spaced parallel lines, which are the contours of a linear function. (As we zoom in, we have to add more contours.)

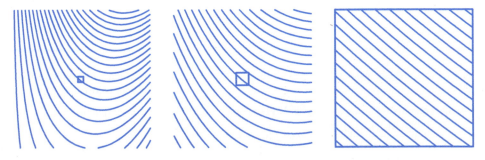

Figure 14.23: Zooming in on a contour diagram until the lines look parallel and equally spaced

This effect can also be seen numerically by zooming in with tables of values. Table 14.5 shows three tables of values for $f(x, y) = x^2 + y^3$ near $x = 2$, $y = 1$, each one a closer view than the previous one. Notice how each table looks more like the table of a linear function.

Table 14.5 *Zooming in on values of $f(x, y) = x^2 + y^3$ near $(2, 1)$ until the table looks linear*

	y		
	0	1	2
x 1	1	2	9
2	4	5	12
3	9	10	17

	y		
	0.9	1.0	1.1
x 1.9	4.34	4.61	4.94
2.0	4.73	5.00	5.33
2.1	5.14	5.41	5.74

	y		
	0.99	1.00	1.01
x 1.99	4.93	4.96	4.99
2.00	4.97	5.00	5.03
2.01	5.01	5.04	5.07

Zooming in Algebraically: Differentiability

Seeing a plane when we zoom in at a point tells us (provided the plane is not vertical) that $f(x, y)$ is closely approximated near that point by a linear function, $L(x, y)$:

$$f(x, y) \approx L(x, y).$$

The Tangent Plane

The graph of the function $z = L(x, y)$ is the tangent plane at that point. See Figure 14.24. Provided the approximation is sufficiently good, we say that $f(x, y)$ is *differentiable* at the point. Section 14.8 on page 847 defines precisely what is meant by the approximation being sufficiently good. The functions we encounter are differentiable at most points in their domain.

What is the equation of the tangent plane? At the point (a, b), the x-slope of the graph of f is the partial derivative $f_x(a, b)$ and the y-slope is $f_y(a, b)$. Thus, using the equation for a plane on page 726 of Chapter 12, we have the following result:

Tangent Plane to the Surface $z = f(x, y)$ at the Point (a, b)

Assuming f is differentiable at (a, b), the equation of the tangent plane is

$$z = f(a, b) + f_x(a, b)(x - a) + f_y(a, b)(y - b).$$

Here we are thinking of a and b as fixed, so $f(a, b)$, and $f_x(a, b)$, and $f_y(a, b)$ are constants. Thus, the right side of the equation is a linear function of x and y.

Example 1 Find the equation for the tangent plane to the surface $z = x^2 + y^2$ at the point $(3, 4)$.

Solution We have $f_x(x, y) = 2x$, so $f_x(3, 4) = 6$, and $f_y(x, y) = 2y$, so $f_y(3, 4) = 8$. Also, $f(3, 4) = 3^2 + 4^2 = 25$. Thus, the equation for the tangent plane at $(3, 4)$ is

$$z = 25 + 6(x - 3) + 8(y - 4).$$

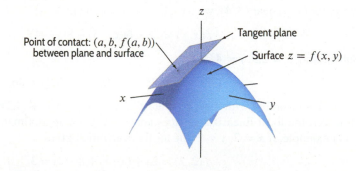

Figure 14.24: The tangent plane to the surface $z = f(x, y)$ at the point (a, b)

Local Linearization

Since the tangent plane lies close to the surface near the point at which they meet, z-values on the tangent plane are close to values of $f(x, y)$ for points near (a, b). Thus, replacing z by $f(x, y)$ in the equation of the tangent plane, we get the following approximation:

Tangent Plane Approximation to $f(x, y)$ for (x, y) Near the Point (a, b)

Provided f is differentiable at (a, b), we can approximate $f(x, y)$:

$$f(x, y) \approx f(a, b) + f_x(a, b)(x - a) + f_y(a, b)(y - b).$$

We are thinking of a and b as fixed, so the expression on the right side is linear in x and y. The right side of this approximation gives the **local linearization** of f near $x = a$, $y = b$.

Figure 14.25 shows the tangent plane approximation graphically.

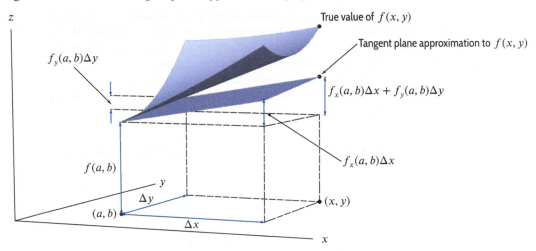

Figure 14.25: Local linearization: Approximating $f(x, y)$ by the z-value from the tangent plane

---

Example 2   Find the local linearization of $f(x, y) = x^2 + y^2$ at the point $(3, 4)$. Estimate $f(2.9, 4.2)$ and $f(2, 2)$ using the linearization and compare your answers to the true values.

Solution   Let $z = f(x, y) = x^2 + y^2$. In Example 1, we found the equation of the tangent plane at $(3, 4)$ to be

$$z = 25 + 6(x - 3) + 8(y - 4).$$

Therefore, for (x, y) near $(3, 4)$, we have the local linearization

$$f(x, y) \approx 25 + 6(x - 3) + 8(y - 4).$$

Substituting $x = 2.9$, $y = 4.2$ gives

$$f(2.9, 4.2) \approx 25 + 6(-0.1) + 8(0.2) = 26.$$

This compares favorably with the true value $f(2.9, 4.2) = (2.9)^2 + (4.2)^2 = 26.05$.

However, the local linearization does not give a good approximation at points far away from $(3, 4)$. For example, if $x = 2$, $y = 2$, the local linearization gives

$$f(2, 2) \approx 25 + 6(-1) + 8(-2) = 3,$$

whereas the true value of the function is $f(2, 2) = 2^2 + 2^2 = 8$.

Example 3   Designing safe boilers depends on knowing how steam behaves under changes in temperature and pressure. Steam tables, such as Table 14.6, are published giving values of the function $V = f(T, P)$ where V is the volume (in ft^3) of one pound of steam at a temperature T (in °F) and pressure P (in lb/in^2).

(a) Give a linear function approximating $V = f(T, P)$ for T near 500°F and P near 24 lb/in^2.

(b) Estimate the volume of a pound of steam at a temperature of 505°F and a pressure of 24.3 lb/in^2.

Table 14.6   *Volume (in cubic feet) of one pound of steam at various temperatures and pressures*

		Pressure P (lb/in^2)			
		20	22	24	26
Temperature	480	27.85	25.31	23.19	21.39
T	500	28.46	25.86	23.69	21.86
(°F)	520	29.06	26.41	24.20	22.33
	540	29.66	26.95	24.70	22.79

Solution   (a) We want the local linearization around the point $T = 500$, $P = 24$, which is

$$f(T, P) \approx f(500, 24) + f_T(500, 24)(T - 500) + f_P(500, 24)(P - 24).$$

We read the value $f(500, 24) = 23.69$ from the table.

Next we approximate $f_T(500, 24)$ by a difference quotient. From the $P = 24$ column, we compute the average rate of change between $T = 500$ and $T = 520$:

$$f_T(500, 24) \approx \frac{f(520, 24) - f(500, 24)}{520 - 500} = \frac{24.20 - 23.69}{20} = 0.0255.$$

Note that $f_T(500, 24)$ is positive, because steam expands when heated.

Next we approximate $f_P(500, 24)$ by looking at the $T = 500$ row and computing the average rate of change between $P = 24$ and $P = 26$:

$$f_P(500, 24) \approx \frac{f(500, 26) - f(500, 24)}{26 - 24} = \frac{21.86 - 23.69}{2} = -0.915.$$

Note that $f_P(500, 24)$ is negative, because increasing the pressure on steam decreases its volume. Using these approximations for the partial derivatives, we obtain the local linearization:

$$V = f(T, P) \approx 23.69 + 0.0255(T - 500) - 0.915(P - 24) \text{ ft}^3 \quad \begin{array}{l} \text{for } T \text{ near 500 °F} \\ \text{and } P \text{ near 24 lb/in}^2. \end{array}$$

(b) We are interested in the volume at $T = 505$°F and $P = 24.3$ lb/in^2. Since these values are close to $T = 500$°F and $P = 24$ lb/in^2, we use the linear relation obtained in part (a):

$$V \approx 23.69 + 0.0255(505 - 500) - 0.915(24.3 - 24) = 23.54 \text{ ft}^3.$$

Local Linearity with Three or More Variables

Local linear approximations for functions of three or more variables follow the same pattern as for functions of two variables. The local linearization of $f(x, y, z)$ at (a, b, c) is given by

$$f(x, y, z) \approx f(a, b, c) + f_x(a, b, c)(x - a) + f_y(a, b, c)(y - b) + f_z(a, b, c)(z - c).$$

The Differential

We are often interested in the change in the value of the function as we move from the point (a, b) to a nearby point (x, y). We rewrite the tangent plane approximation as

$$\underbrace{f(x, y) - f(a, b)}_{\Delta f} \approx f_x(a, b) \underbrace{(x - a)}_{\Delta x} + f_y(a, b) \underbrace{(y - b)}_{\Delta y},$$

giving us a relationship between Δf, Δx, and Δy:

$$\Delta f \approx f_x(a, b)\Delta x + f_y(a, b)\Delta y.$$

If a and b are fixed, $f_x(a, b)\Delta x + f_y(a, b)\Delta y$ is a linear function of Δx and Δy that can be used to estimate Δf for small Δx and Δy. We introduce new variables dx and dy to represent changes in x and y.

The Differential of a Function $z = f(x, y)$

The **differential**, df (or dz), at a point (a, b) is the linear function of dx and dy given by the formula

$$df = f_x(a, b)\,dx + f_y(a, b)\,dy.$$

The differential at a general point is often written $df = f_x\,dx + f_y\,dy$.

Example 4   Compute the differentials of the following functions.
(a) $f(x, y) = x^2 e^{5y}$ (b) $z = x\sin(xy)$ (c) $f(x, y) = x\cos(2x)$

Solution   (a) Since $f_x(x, y) = 2xe^{5y}$ and $f_y(x, y) = 5x^2 e^{5y}$, we have

$$df = 2xe^{5y}\,dx + 5x^2 e^{5y}\,dy.$$

(b) Since $\partial z/\partial x = \sin(xy) + xy\cos(xy)$ and $\partial z/\partial y = x^2 \cos(xy)$, we have

$$dz = (\sin(xy) + xy\cos(xy))\,dx + x^2\cos(xy)\,dy.$$

(c) Since $f_x(x, y) = \cos(2x) - 2x\sin(2x)$ and $f_y(x, y) = 0$, we have

$$df = (\cos(2x) - 2x\sin(2x))\,dx + 0\,dy = (\cos(2x) - 2x\sin(2x))\,dx.$$

Example 5   The density ρ (in g/cm^3) of carbon dioxide gas CO_2 depends upon its temperature T (in °C) and pressure P (in atmospheres). The ideal gas model for CO_2 gives what is called the state equation:

$$\rho = \frac{0.5363P}{T + 273.15}.$$

Compute the differential $d\rho$. Explain the signs of the coefficients of dT and dP.

Solution   The differential for $\rho = f(T, P)$ is

$$d\rho = f_T(T, P)\,dT + f_P(T, P)\,dP = \frac{-0.5363P}{(T + 273.15)^2}\,dT + \frac{0.5363}{T + 273.15}\,dP.$$

The coefficient of dT is negative because increasing the temperature expands the gas (if the pressure is kept constant) and therefore decreases its density. The coefficient of dP is positive because increasing the pressure compresses the gas (if the temperature is kept constant) and therefore increases its density.

Where Does the Notation for the Differential Come From?

We write the differential as a linear function of the new variables dx and dy. You may wonder why we chose these names for our variables. The reason is historical: The people who invented calculus thought of dx and dy as "infinitesimal" changes in x and y. The equation

$$df = f_x dx + f_y dy$$

was regarded as an infinitesimal version of the local linear approximation

$$\Delta f \approx f_x \Delta x + f_y \Delta y.$$

In spite of the problems with defining exactly what "infinitesimal" means, some mathematicians, scientists, and engineers think of the differential in terms of infinitesimals.

Figure 14.26 illustrates a way of thinking about differentials that combines the definition with this informal point of view. It shows the graph of f along with a view of the graph around the point $(a, b, f(a, b))$ under a microscope. Since f is locally linear at the point, the magnified view looks like the tangent plane. Under the microscope, we use a magnified coordinate system with its origin at the point $(a, b, f(a, b))$ and with coordinates dx, dy, and dz along the three axes. The graph of the differential df is the tangent plane, which has equation $dz = f_x(a, b)\,dx + f_y(a, b)\,dy$ in the magnified coordinates.

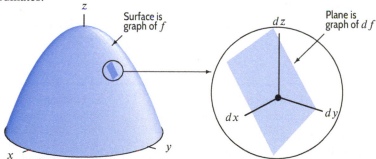

Figure 14.26: The graph of f, with a view through a microscope showing the tangent plane in the magnified coordinate system

Summary for Section 14.3

- **Local linearity** means that when zooming in near a point, a function behaves more and more like a linear function.

- A function is **differentiable** at a point if it is locally linear near the point.

- The **tangent plane** to the surface $z = f(x, y)$ at the point (a, b) is
$$z = f(a, b) + f_x(a, b)(x - a) + f_y(a, b)(y - b).$$

- The **tangent plane approximation** to $f(x, y)$ for (x, y) near the point (a, b) is
$$f(x, y) \approx f(a, b) + f_x(a, b)(x - a) + f_y(a, b)(y - b).$$

- The **differential**, df, at a point (a, b) is the linear function of dx and dy given by the formula
$$df = f_x(a, b)\,dx + f_y(a, b)\,dy.$$

The differential df computes the approximate change in f given small changes dx in x and dy in y near the point (a, b).

Exercises and Problems for Section 14.3

EXERCISES

■In Exercises 1–8, find the equation of the tangent plane at the given point.

1. $z = ye^{x/y}$ at the point $(1, 1, e)$

2. $z = \sin(xy)$ at $x = 2, y = 3\pi/4$

3. $z = \ln(x^2 + 1) + y^2$ at the point $(0, 3, 9)$

4. $z = e^y + x + x^2 + 6$ at the point $(1, 0, 9)$

5. $z = \frac{1}{2}(x^2 + 4y^2)$ at the point $(2, 1, 4)$

6. $x^2 + y^2 - z = 1$ at the point $(1, 3, 9)$

7. $x^2 y^2 + z - 40 = 0$ at $x = 2, y = 3$

8. $x^2 y + \ln(xy) + z = 6$ at the point $(4, 0.25, 2)$

■In Exercises 9–12, find the differential of the function.

9. $f(x, y) = \sin(xy)$

10. $g(u, v) = u^2 + uv$

11. $z = e^{-x} \cos y$

12. $h(x, t) = e^{-3t} \sin(x + 5t)$

■In Exercises 13–16, find the differential of the function at the point.

13. $g(x, t) = x^2 \sin(2t)$ at $(2, \pi/4)$

14. $f(x, y) = xe^{-y}$ at $(1, 0)$

15. $P(L, K) = 1.01 L^{0.25} K^{0.75}$ at $(100, 1)$

16. $F(m, r) = Gm/r^2$ at $(100, 10)$

■In Exercises 17–20, assume points P and Q are close. Estimate $\Delta f = f(Q) - f(P)$ using the differential df.

17. $df = 10\, dx - 5\, dy$, $P = (200, 400)$, $Q = (202, 405)$

18. $df = y\, dx + x\, dy$, $P = (10, 5)$, $Q = (9.8, 5.3)$

19. $df = 6\sqrt{1 + 4x + 2y}\, dx + 3\sqrt{1 + 4x + 2y}\, dy$, $P = (1, 2)$, $Q = (1.03, 2.05)$

20. $df = (2x + 2y + 5)\, dx + (2x + 3)\, dy$, $P = (0, 0)$, $Q = (0.1, -0.2)$

■In Exercises 21–24, assume points P and Q are close. Estimate $g(Q)$.

21. $P = (60, 80)$, $Q = (60.5, 82)$, $g(P) = 100$, $g_x(P) = 2$, $g_y(P) = -3$.

22. $P = (-150, 200)$, $Q = (-152, 203)$, $g(P) = 2500$, $g_x(P) = 10$, $g_y(P) = 20$.

23. $P = (5, 8)$, $Q = (4.97, 7.99)$, $g(P) = 12$, $g_x(P) = -0.1$, $g_y(P) = -0.2$.

24. $P = (30, 125)$, $Q = (25, 135)$, $g(P) = 840$, $g_x(P) = 4$, $g_y(P) = 1.5$.

PROBLEMS

25. At a distance of x feet from the beach, the price in dollars of a plot of land of area a square feet is $f(a, x)$.

    (a) What are the units of $f_a(a, x)$?
    (b) What does $f_a(1000, 300) = 3$ mean in practical terms?
    (c) What are the units of $f_x(a, x)$?
    (d) What does $f_x(1000, 300) = -2$ mean in practical terms?
    (e) Which is cheaper: 1005 square feet that are 305 feet from the beach or 998 square feet that are 295 feet from the beach? Justify your answer.

26. A student was asked to find the equation of the tangent plane to the surface $z = x^3 - y^2$ at the point $(x, y) = (2, 3)$. The student's answer was

    $$z = 3x^2(x - 2) - 2y(y - 3) - 1.$$

    (a) At a glance, how do you know this is wrong?
    (b) What mistake did the student make?
    (c) Answer the question correctly.

27. (a) Check the local linearity of $f(x, y) = e^{-x} \sin y$ near $x = 1$, $y = 2$ by making a table of values of f for $x = 0.9$, 1.0, 1.1 and $y = 1.9$, 2.0, 2.1.

    Express values of f with 4 digits after the decimal point. Then make a table of values for $x = 0.99$, 1.00, 1.01 and $y = 1.99$, 2.00. 2.01, again showing 4 digits after the decimal point. Do both tables look nearly linear? Does the second table look more linear than the first?

    (b) Give the local linearization of $f(x, y) = e^{-x} \sin y$ at $(1, 2)$, first using your tables and second using the fact that $f_x(x, y) = -e^{-x} \sin y$ and $f_y(x, y) = e^{-x} \cos y$.

28. Find the local linearization of the function $f(x, y) = x^2 y$ at the point $(3, 1)$.

29. The tangent plane to $z = f(x, y)$ at the point $(1, 2)$ is $z = 3x + 2y - 5$.

    (a) Find $f_x(1, 2)$ and $f_y(1, 2)$.
    (b) What is $f(1, 2)$?
    (c) Approximate $f(1.1, 1.9)$.

30. Find an equation for the tangent plane to $z = f(x, y)$ at $(3, -2)$ if the differential at $(3, -2)$ is $df = 5dx + dy$ and $f(3, -2) = 8$.

31. Find df at $(2, -4)$ if the tangent plane to $z = f(x, y)$ at $(2, -4)$ is $z = -3(x - 2) + 2(y + 4) + 3$.

32. Give a linear function approximating $z = f(x, y)$ near $(1, -1)$ using its contour diagram in Figure 14.27.

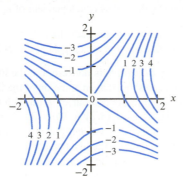

Figure 14.27

33. For the differentiable function $h(x, y)$, we are told that $h(600, 100) = 300$ and $h_x(600, 100) = 12$ and $h_y(600, 100) = -8$. Estimate $h(605, 98)$.

34. (a) Find the equation of the plane tangent to the graph of $f(x, y) = x^2 e^{xy}$ at $(1, 0)$.
    (b) Find the linear approximation of $f(x, y)$ for (x, y) near $(1, 0)$.
    (c) Find the differential of f at the point $(1, 0)$.

35. Find the differential of $f(x, y) = \sqrt{x^2 + y^3}$ at the point $(1, 2)$. Use it to estimate $f(1.04, 1.98)$.

36. (a) Find the differential of $g(u, v) = u^2 + uv$.
    (b) Use your answer to part (a) to estimate the change in g as you move from $(1, 2)$ to $(1.2, 2.1)$.

37. An unevenly heated plate has temperature $T(x, y)$ in °C at the point (x, y). If $T(2, 1) = 135$, and $T_x(2, 1) = 16$, and $T_y(2, 1) = -15$, estimate the temperature at the point $(2.04, 0.97)$.

38. A right circular cylinder has a radius of 50 cm and a height of 100 cm. Use differentials to estimate the change in volume of the cylinder if its height and radius are both increased by 1 cm.

39. Give the local linearization for the monthly car-loan payment function at each of the points investigated in Problem 35 on page 793.

40. In Example 3 on page 803 we found a linear approximation for $V = f(T, P)$ near $(500, 24)$. Now find a linear approximation near $(480, 20)$.

41. In Example 3 on page 803 we found a linear approximation for $V = f(T, p)$ near $(500, 24)$.
    (a) Test the accuracy of this approximation by comparing its predicted value with the four neighboring values in the table. What do you notice? Which predicted values are accurate? Which are not? Explain your answer.
    (b) Suggest a linear approximation for $f(T, p)$ near $(500, 24)$ that does not have the property you noticed in part (a). [Hint: Estimate the partial derivatives in a different way.]

42. In a room, the temperature is given by $T = f(x, t)$ degrees Celsius, where x is the distance from a heater (in meters) and t is the elapsed time (in minutes) since the heat has been turned on. A person standing 3 meters from the heater 5 minutes after it has been turned on observes the following: (1) The temperature is increasing by 1.2°C per minute, and (2) As the person walks away from the heater, the temperature decreases by 2°C per meter as time is held constant. Estimate how much cooler or warmer it would be 2.5 meters from the heater after 6 minutes.

43. Van der Waal's equation relates the pressure, P, and the volume, V, of a fixed quantity of a gas at constant temperature T. For a, b, n, R constants, the equation is

$$\left(P + \frac{n^2 a}{V^2}\right)(V - nb) = nRT.$$

   (a) Express P as a function of T and V.
   (b) Write a linear approximation for the change in pressure, $\Delta P = P - P_0$, resulting from a change in temperature $\Delta T = T - T_0$ and a change in pressure, $\Delta V = V - V_0$.

44. The gas equation for one mole of oxygen relates its pressure, P (in atmospheres), its temperature, T (in K), and its volume, V (in cubic decimeters, dm³):

$$T = 16.574 \frac{1}{V} - 0.52754 \frac{1}{V^2} - 0.3879P + 12.187VP.$$

   (a) Find the temperature T and differential dT if the volume is 25 dm³ and the pressure is 1 atmosphere.
   (b) Use your answer to part (a) to estimate how much the volume would have to change if the pressure increased by 0.1 atmosphere and the temperature remained constant.

45. The coefficient, β, of thermal expansion of a liquid relates the change in the volume V (in m³) of a fixed quantity of a liquid to an increase in its temperature T (in °C):

$$dV = \beta V \, dT.$$

   (a) Let ρ be the density (in kg/m³) of water as a function of temperature. (For a mass m of liquid, we have $\rho = m/V$.) Write an expression for $d\rho$ in terms of ρ and dT.
   (b) The graph in Figure 14.28 shows density of water as a function of temperature. Use it to estimate β when $T = 20$°C and when $T = 80$°C.

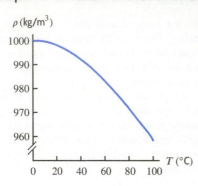

Figure 14.28

46. A fluid moves through a tube of length 1 meter and radius $r = 0.005 \pm 0.00025$ meters under a pressure $p = 10^5 \pm 1000$ pascals, at a rate $v = 0.625 \cdot 10^{-9}$ m³ per unit time. Use differentials to estimate the maximum error in the viscosity η given by

$$\eta = \frac{\pi}{8} \frac{pr^4}{v}.$$

47. The period, T, of oscillation in seconds of a pendulum clock is given by $T = 2\pi \sqrt{l/g}$, where g is the acceleration due to gravity. The length of the pendulum, l, depends on the temperature, t, according to the formula $l = l_0(1 + \alpha(t - t_0))$ where l_0 is the length of the pendulum at temperature t_0 and α is a constant which characterizes the clock. The clock is set to the correct period

at the temperature t_0. How many seconds a day does the clock gain or lose when the temperature is $t_0 + \Delta t$? Show that this gain or loss is independent of l_0.

48. Two functions that have the same local linearization at a point have contours that are tangent at this point.

(a) If $f_x(a, b)$ or $f_y(a, b)$ is nonzero, use the local linearization to show that an equation of the line tangent at (a, b) to the contour of f through (a, b) is $f_x(a, b)(x - a) + f_y(a, b)(y - b) = 0$.

(b) Find the slope of the tangent line if $f_y(a, b) \neq 0$.

(c) Find an equation for the line tangent to the contour of $f(x, y) = x^2 + xy$ at $(3, 4)$.

In Problems 49–52, the point is on the surface in 3-space.

(a) Find the differential of the equation (that is, of each side).

(b) Find dz at the point.

(c) Find an equation of the tangent plane to the surface at the point.

49. $2x^2 + 13 = y^2 + 3z^2$, $(2, 3, 2)$

50. $x^2 + y^2 + z^2 + 1 = xyz + 2x^2 + 3y^2 - 2z^2$, $(1, 1, 1)$

51. $xe^y + z^2 + 1 = \cos(x - 1) + \sqrt{z^2 + 3}$, $(1, 0, 1)$

52. $xz^2 + xy + 5 = x^2 + z^2$, $(2, -1, 1)$

Strengthen Your Understanding

In Problems 53–55, explain what is wrong with the statement.

53. An equation for the tangent plane to the surface $z = f(x, y)$ at the point $(3, 4)$ is

$$z = f(3, 4) + f_x(3, 4)x + f_y(3, 4)y.$$

54. If $f_x(0, 0) = g_x(0, 0)$ and $f_y(0, 0) = g_y(0, 0)$, then the surfaces $z = f(x, y)$ and $z = g(x, y)$ have the same tangent planes at the point $(0, 0)$.

55. The tangent plane to the surface $z = x^2y$ at the point $(1, 2)$ has equation

$$z = 2 + 2xy(x - 1) + x^2(y - 2).$$

In Problems 56–57, give an example of:

56. Two different functions with the same differential.

57. A surface in three space whose tangent plane at $(0, 0, 3)$ is the plane $z = 3$.

Are the statements in Problems 58–65 true or false? Give reasons for your answer.

58. The tangent plane approximation of $f(x, y) = ye^{x^2}$ at the point $(0, 1)$ is $f(x, y) \approx y$.

59. If f is a function with $df = 2y\,dx + \sin(xy)\,dy$, then f changes by about -0.4 between the points $(1, 2)$ and $(0.9, 2.0002)$.

60. The local linearization of $f(x, y) = x^2 + y^2$ at $(1,1)$ gives an overestimate of the value of $f(x, y)$ at the point $(1.04, 0.95)$.

61. If two functions f and g have the same differential at the point $(1, 1)$, then $f = g$.

62. If two functions f and g have the same tangent plane at a point $(1, 1)$, then $f = g$.

63. If $f(x, y)$ is a constant function, then $df = 0$.

64. If $f(x, y)$ is a linear function, then df is a linear function of dx and dy.

65. If you zoom close enough near a point (a, b) on the contour diagram of a differentiable function, the contours are *precisely* parallel and *exactly* equally spaced.

14.4   GRADIENTS AND DIRECTIONAL DERIVATIVES IN THE PLANE

The Rate of Change in an Arbitrary Direction: The Directional Derivative

The partial derivatives of a function f tell us the rate of change of f in the directions parallel to the coordinate axes. In this section we see how to compute the rate of change of f in an arbitrary direction.

Example 1   Figure 14.29 shows the temperature, in °C, at the point (x, y). Estimate the average rate of change of temperature as we walk from point A to point B.

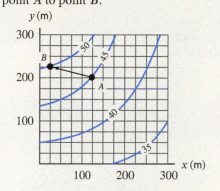

Figure 14.29: Estimating rate of change on a temperature map

Solution   At the point A we are on the $H = 45°$C contour. At B we are on the $H = 50°$C contour. The displacement vector from A to B has x component approximately $-100\vec{i}$ and y component approximately $25\vec{j}$, so its length is $\sqrt{(-100)^2 + 25^2} \approx 103$. Thus, the temperature rises by 5°C as we move 103 meters, so the average rate of change of the temperature in that direction is about $5/103 \approx 0.05°$C/m.

Suppose we want to compute the rate of change of a function $f(x, y)$ at the point $P = (a, b)$ in the direction of the unit vector $\vec{u} = u_1\vec{i} + u_2\vec{j}$. For $h > 0$, consider the point $Q = (a + hu_1, b + hu_2)$ whose displacement from P is $h\vec{u}$. (See Figure 14.30.) Since $\|\vec{u}\| = 1$, the distance from P to Q is h. Thus,

$$\begin{array}{c}\text{Average rate of change} \\ \text{in } f \text{ from } P \text{ to } Q\end{array} = \frac{\text{Change in } f}{\text{Distance from } P \text{ to } Q} = \frac{f(a + hu_1, b + hu_2) - f(a, b)}{h}.$$

Taking the limit as $h \to 0$ gives the instantaneous rate of change and the following definition:

Directional Derivative of f at (a, b) in the Direction of a Unit Vector $\vec{u}$

If $\vec{u} = u_1\vec{i} + u_2\vec{j}$ is a unit vector, we define the directional derivative, $f_{\vec{u}}$, by

$$f_{\vec{u}}(a, b) = \begin{array}{c}\text{Rate of change} \\ \text{of } f \text{ in direction} \\ \text{of } \vec{u} \text{ at } (a, b)\end{array} = \lim_{h \to 0} \frac{f(a + hu_1, b + hu_2) - f(a, b)}{h},$$

provided the limit exists. Note that the directional derivative is a scalar.

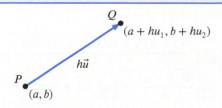

Figure 14.30: Displacement of $h\vec{u}$ from the point (a, b)

Notice that if $\vec{u} = \vec{i}$, so $u_1 = 1, u_2 = 0$, then the directional derivative is f_x, since

$$f_{\vec{i}}(a, b) = \lim_{h \to 0} \frac{f(a + h, b) - f(a, b)}{h} = f_x(a, b).$$

Similarly, if $\vec{u} = \vec{j}$ then the directional derivative $f_{\vec{j}} = f_y$.

What If We Do Not Have a Unit Vector?

We defined $f_{\vec{u}}$ for $\vec{u}$ a unit vector. If $\vec{v}$ is not a unit vector, $\vec{v} \neq \vec{0}$, we construct a unit vector $\vec{u} = \vec{v} / \|\vec{v}\|$ in the same direction as $\vec{v}$ and define the rate of change of f in the direction of $\vec{v}$ as $f_{\vec{u}}$.

Example 2   For each of the functions f, g, and h in Figure 14.31, decide whether the directional derivative at the indicated point is positive, negative, or zero, in the direction of the vector $\vec{v} = \vec{i} + 2\vec{j}$, and in the direction of the vector $\vec{w} = 2\vec{i} + \vec{j}$.

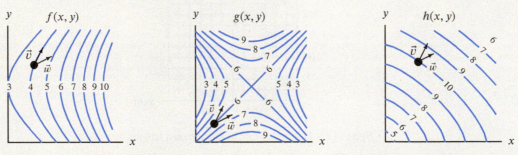

Figure 14.31: Contour diagrams of three functions with direction vectors $\vec{v} = \vec{i} + 2\vec{j}$ and $\vec{w} = 2\vec{i} + \vec{j}$ marked on each

Solution   On the contour diagram for f, the vector $\vec{v} = \vec{i} + 2\vec{j}$ appears to be tangent to the contour. Thus, in this direction, the value of the function is not changing, so the directional derivative in the direction of $\vec{v}$ is zero. The vector $\vec{w} = 2\vec{i} + \vec{j}$ points from the contour marked 4 toward the contour marked 5. Thus, the values of the function are increasing and the directional derivative in the direction of $\vec{w}$ is positive.

On the contour diagram for g, the vector $\vec{v} = \vec{i} + 2\vec{j}$ points from the contour marked 6 toward the contour marked 5, so the function is decreasing in that direction. Thus, the rate of change is negative. On the other hand, the vector $\vec{w} = 2\vec{i} + \vec{j}$ points from the contour marked 6 toward the contour marked 7, and hence the directional derivative in the direction of $\vec{w}$ is positive.

Finally, on the contour diagram for h, both vectors point from the $h = 10$ contour to the $h = 9$ contour, so both directional derivatives are negative.

Example 3   Calculate the directional derivative of $f(x, y) = x^2 + y^2$ at $(1, 0)$ in the direction of the vector $\vec{i} + \vec{j}$.

Solution   First we have to find the unit vector in the same direction as the vector $\vec{i} + \vec{j}$. Since this vector has magnitude $\sqrt{2}$, the unit vector is

$$\vec{u} = \frac{1}{\sqrt{2}}(\vec{i} + \vec{j}) = \frac{1}{\sqrt{2}}\vec{i} + \frac{1}{\sqrt{2}}\vec{j}.$$

Thus,

$$f_{\vec{u}}(1, 0) = \lim_{h \to 0} \frac{f(1 + h/\sqrt{2}, h/\sqrt{2}) - f(1, 0)}{h} = \lim_{h \to 0} \frac{(1 + h/\sqrt{2})^2 + (h/\sqrt{2})^2 - 1}{h}$$

$$= \lim_{h \to 0} \frac{\sqrt{2}h + h^2}{h} = \lim_{h \to 0}(\sqrt{2} + h) = \sqrt{2}.$$

Computing Directional Derivatives from Partial Derivatives

If f is differentiable, we will now see how to use local linearity to find a formula for the directional derivative which does not involve a limit. If $\vec{u}$ is a unit vector, the definition of $f_{\vec{u}}$ says

$$f_{\vec{u}}(a, b) = \lim_{h \to 0} \frac{f(a + hu_1, b + hu_2) - f(a, b)}{h} = \lim_{h \to 0} \frac{\Delta f}{h},$$

where $\Delta f = f(a + hu_1, b + hu_2) - f(a, b)$ is the change in f. We write Δx for the change in x, so $\Delta x = (a + hu_1) - a = hu_1$; similarly, $\Delta y = hu_2$. Using local linearity, we have

$$\Delta f \approx f_x(a, b)\Delta x + f_y(a, b)\Delta y = f_x(a, b)hu_1 + f_y(a, b)hu_2.$$

Thus, dividing by h gives

$$\frac{\Delta f}{h} \approx \frac{f_x(a, b)hu_1 + f_y(a, b)hu_2}{h} = f_x(a, b)u_1 + f_y(a, b)u_2.$$

This approximation becomes exact as $h \to 0$, so we have the following formula:

$$f_{\vec{u}}(a, b) = f_x(a, b)u_1 + f_y(a, b)u_2.$$

Example 4 Use the preceding formula to compute the directional derivative in Example 3. Check that we get the same answer as before.

Solution We calculate $f_{\vec{u}}(1, 0)$, where $f(x, y) = x^2 + y^2$ and $\vec{u} = \frac{1}{\sqrt{2}}\vec{i} + \frac{1}{\sqrt{2}}\vec{j}$.

The partial derivatives are $f_x(x, y) = 2x$ and $f_y(x, y) = 2y$. So, as before,

$$f_{\vec{u}}(1, 0) = f_x(1, 0)u_1 + f_y(1, 0)u_2 = (2)\left(\frac{1}{\sqrt{2}}\right) + (0)\left(\frac{1}{\sqrt{2}}\right) = \sqrt{2}.$$

The Gradient Vector

Notice that the expression for $f_{\vec{u}}(a, b)$ can be written as a dot product of $\vec{u}$ and a new vector:

$$f_{\vec{u}}(a, b) = f_x(a, b)u_1 + f_y(a, b)u_2 = (f_x(a, b)\vec{i} + f_y(a, b)\vec{j}) \cdot (u_1\vec{i} + u_2\vec{j}).$$

The new vector, $f_x(a, b)\vec{i} + f_y(a, b)\vec{j}$, turns out to be important. Thus, we make the following definition:

> **The Gradient Vector** of a differentiable function f at the point (a, b) is
>
> $$\text{grad } f(a, b) = f_x(a, b)\vec{i} + f_y(a, b)\vec{j}$$

The formula for the directional derivative can be written in terms of the gradient as follows:

> ## The Directional Derivative and the Gradient
>
> If f is differentiable at (a, b) and $\vec{u} = u_1\vec{i} + u_2\vec{j}$ is a unit vector, then
>
> $$f_{\vec{u}}(a, b) = f_x(a, b)u_1 + f_y(a, b)u_2 = \text{grad } f(a, b) \cdot \vec{u}.$$

The change in f corresponding to a small change $\Delta \vec{r} = \Delta x\vec{i} + \Delta y\vec{j}$ can be estimated using the gradient:

$$\Delta f \approx \text{grad } f \cdot \Delta \vec{r}.$$

Example 5   Find the gradient vector of $f(x, y) = x + e^y$ at the point $(1, 1)$.

Solution   Using the definition, we have

$$\text{grad } f = f_x \vec{i} + f_y \vec{j} = \vec{i} + e^y \vec{j},$$

so at the point $(1, 1)$

$$\text{grad } f(1, 1) = \vec{i} + e\vec{j}.$$

Alternative Notation for the Gradient

You can think of $\dfrac{\partial f}{\partial x}\vec{i} + \dfrac{\partial f}{\partial y}\vec{j}$ as the result of applying the vector operator (pronounced "del")

$$\nabla = \frac{\partial}{\partial x}\vec{i} + \frac{\partial}{\partial y}\vec{j}$$

to the function f. Thus, we get the alternative notation

$$\text{grad } f = \nabla f.$$

If $z = f(x, y)$, we can write grad z or ∇z for grad f or for ∇f.

What Does the Gradient Tell Us?

The fact that $f_{\vec{u}} = \text{grad } f \cdot \vec{u}$ enables us to see what the gradient vector represents. Suppose θ is the angle between the vectors grad$f$ and $\vec{u}$. At the point (a, b), we have

$$f_{\vec{u}} = \text{grad } f \cdot \vec{u} = \| \text{grad } f \| \underbrace{\|\vec{u}\|}_{1} \cos \theta = \| \text{grad } f \| \cos \theta.$$

Imagine that grad f is fixed and that $\vec{u}$ can rotate. (See Figure 14.32.) The maximum value of $f_{\vec{u}}$ occurs when $\cos \theta = 1$, so $\theta = 0$ and $\vec{u}$ is pointing in the direction of grad f. Then

$$\text{Maximum } f_{\vec{u}} = \| \text{grad } f \| \cos 0 = \| \text{grad } f \|.$$

The minimum value of $f_{\vec{u}}$ occurs when $\cos \theta = -1$, so $\theta = \pi$ and $\vec{u}$ is pointing in the direction opposite to grad f. Then

$$\text{Minimum } f_{\vec{u}} = \| \text{grad } f \| \cos \pi = -\| \text{grad } f \|.$$

When $\theta = \pi/2$ or $3\pi/2$, so $\cos \theta = 0$, the directional derivative is zero.

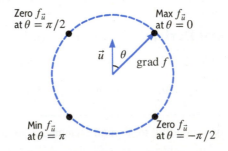

Figure 14.32: Values of the directional derivative at different angles to the gradient

Properties of the Gradient Vector

We have seen that the gradient vector points in the direction of the greatest rate of change at a point and the magnitude of the gradient vector is that rate of change.

Figure 14.33 shows that the gradient vector at a point is perpendicular to the contour through that point. If the contours represent equally spaced f-values and f is differentiable, local linearity tells us that the contours of f around a point appear straight, parallel, and equally spaced. The greatest rate of change is obtained by moving in the direction that takes us to the next contour in the shortest possible distance; that is, perpendicular to the contour. Thus, we have the following:

Geometric Properties of the Gradient Vector in the Plane

If f is a differentiable function at the point (a, b) and grad $f(a, b) \neq \vec{0}$, then:
- The direction of grad $f(a, b)$ is
  - Perpendicular[1] to the contour of f through (a, b);
  - In the direction of the maximum rate of increase of f.
- The magnitude of the gradient vector, $\| \text{grad} f \|$, is
  - The maximum rate of change of f at that point;
  - Large when the contours are close together and small when they are far apart.

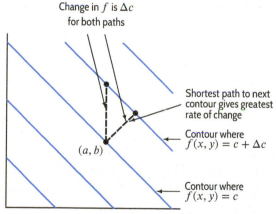

Figure 14.33: Close-up view of the contours around (a, b), showing that the gradient is perpendicular to the contours

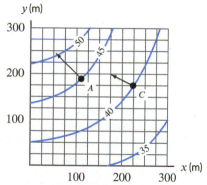

Figure 14.34: A temperature map showing directions and relative magnitudes of two gradient vectors

Examples of Directional Derivatives and Gradient Vectors

Example 6   Explain why the gradient vectors at points A and C in Figure 14.34 have the direction and the relative magnitudes they do.

Solution   The gradient vector points in the direction of greatest increase of the function. This means that in Figure 14.34, the gradient points directly toward warmer temperatures. The magnitude of the gradient vector measures the rate of change. The gradient vector at A is longer than the gradient vector at C because the contours are closer together at A, so the rate of change is larger.

Example 2 on page 810 shows how the contour diagram can tell us the sign of the directional derivative. In the next example we compute the directional derivative in three directions, two that are close to that of the gradient vector and one that is not.

---
[1] This assumes that the same scale is used on both axes.

Example 7   Use the gradient to find the directional derivative of $f(x, y) = x + e^y$ at the point $(1, 1)$ in the direction of the vectors $\vec{i} - \vec{j}, \vec{i} + 2\vec{j}, \vec{i} + 3\vec{j}$.

Solution   In Example 5 we found

$$\operatorname{grad} f(1, 1) = \vec{i} + e\vec{j}.$$

A unit vector in the direction of $\vec{i} - \vec{j}$ is $\vec{s} = (\vec{i} - \vec{j})/\sqrt{2}$, so

$$f_{\vec{s}}(1, 1) = \operatorname{grad} f(1, 1) \cdot \vec{s} = (\vec{i} + e\vec{j}) \cdot \left(\frac{\vec{i} - \vec{j}}{\sqrt{2}}\right) = \frac{1 - e}{\sqrt{2}} \approx -1.215.$$

A unit vector in the direction of $\vec{i} + 2\vec{j}$ is $\vec{v} = (\vec{i} + 2\vec{j})/\sqrt{5}$, so

$$f_{\vec{v}}(1, 1) = \operatorname{grad} f(1, 1) \cdot \vec{v} = (\vec{i} + e\vec{j}) \cdot \left(\frac{\vec{i} + 2\vec{j}}{\sqrt{5}}\right) = \frac{1 + 2e}{\sqrt{5}} \approx 2.879.$$

A unit vector in the direction of $\vec{i} + 3\vec{j}$ is $\vec{w} = (\vec{i} + 3\vec{j})/\sqrt{10}$, so

$$f_{\vec{w}}(1, 1) = \operatorname{grad} f(1, 1) \cdot \vec{w} = (\vec{i} + e\vec{j}) \cdot \left(\frac{\vec{i} + 3\vec{j}}{\sqrt{10}}\right) = \frac{1 + 3e}{\sqrt{10}} \approx 2.895.$$

Now look back at the answers and compare with the value of $\| \operatorname{grad} f \| = \sqrt{1 + e^2} \approx 2.896$. One answer is not close to this value; the other two, $f_{\vec{v}} = 2.879$ and $f_{\vec{w}} = 2.895$, are close but slightly smaller than $\| \operatorname{grad} f \|$. Since $\| \operatorname{grad} f \|$ is the maximum rate of change of f at the point, we have for *any* unit vector $\vec{u}$:

$$f_{\vec{u}}(1, 1) \leq \| \operatorname{grad} f \|.$$

with equality when $\vec{u}$ is in the direction of $\operatorname{grad} f$. Since $e \approx 2.718$, the vectors $\vec{i} + 2\vec{j}$ and $\vec{i} + 3\vec{j}$ both point roughly, but not exactly, in the direction of the gradient vector $\operatorname{grad} f(1, 1) = \vec{i} + e\vec{j}$. Thus, the values of $f_{\vec{v}}$ and $f_{\vec{w}}$ are both close to the value of $\| \operatorname{grad} f \|$. The direction of the vector $\vec{i} - \vec{j}$ is not close to the direction of $\operatorname{grad} f$ and the value of $f_{\vec{s}}$ is not close to the value of $\| \operatorname{grad} f \|$.

Summary for Section 14.4

- The **directional derivative** of a function f in the direction of a **unit vector** $\vec{u}$ at the point (a, b) is denoted $f_{\vec{u}}(a, b)$:
  - $f_{\vec{u}}(a, b)$ measures the rate of change of the function $f(x, y)$ in the $\vec{u}$-direction at (a, b).
  - If $\vec{u} = u_1\vec{i} + u_2\vec{j}$ then

$$f_{\vec{u}}(a, b) = \lim_{h \to 0} \frac{f(a + hu_1, b + hu_2) - f(a, b)}{h}.$$

- The **gradient vector** of a differentiable function f at the point (a, b) is

$$\operatorname{grad} f(a, b) = f_x(a, b)\vec{i} + f_y(a, b)\vec{j}.$$

  The gradient vector is also denoted as ∇f.

- If f is differentiable at (a, b) and $\vec{u} = u_1\vec{i} + u_2\vec{j}$ is a unit vector, then the **directional derivative** can be computed using

$$f_{\vec{u}}(a, b) = f_x(a, b)u_1 + f_y(a, b)u_2 = \operatorname{grad} f(a, b) \cdot \vec{u}.$$

- If f is a differentiable function at the point (a, b) and grad $f(a, b) \neq \vec{0}$, then the **gradient has the following properties**:
  - The direction of grad $f(a, b)$ is
    - Perpendicular to the contour of f through (a, b);
    - In the direction of the maximum rate of increase of f.
  - The magnitude of the gradient vector, $\| \text{grad } f \|$, is
    - The maximum rate of change of f at that point;
    - Large when the contours are close together and small when they are far apart.

Exercises and Problems for Section 14.4   Online Resource: Additional Problems for Section 14.4

EXERCISES

In Exercises **1–14**, find the gradient of the function. Assume the variables are restricted to a domain on which the function is defined.

1. $f(x, y) = \frac{3}{2}x^5 - \frac{4}{7}y^6$

2. $Q = 50K + 100L$

3. $f(m, n) = m^2 + n^2$

4. $z = xe^y$

5. $f(\alpha, \beta) = \sqrt{5\alpha^2 + \beta}$

6. $f(r, h) = \pi r^2 h$

7. $z = (x + y)e^y$

8. $f(K, L) = K^{0.3}L^{0.7}$

9. $f(r, \theta) = r \sin \theta$

10. $f(x, y) = \ln(x^2 + y^2)$

11. $z = \sin(x/y)$

12. $z = \tan^{-1}(x/y)$

13. $f(\alpha, \beta) = \dfrac{2\alpha + 3\beta}{2\alpha - 3\beta}$

14. $z = x\dfrac{e^y}{x + y}$

In Exercises **15–22**, find the gradient at the point.

15. $f(x, y) = x^2 y + 7xy^3$, at $(1, 2)$

16. $f(m, n) = 5m^2 + 3n^4$, at $(5, 2)$

17. $f(r, h) = 2\pi r h + \pi r^2$, at $(2, 3)$

18. $f(x, y) = e^{\sin y}$, at $(0, \pi)$

19. $f(x, y) = \sin(x^2) + \cos y$, at $(\frac{\sqrt{\pi}}{2}, 0)$

20. $f(x, y) = \ln(x^2 + xy)$, at $(4, 1)$

21. $f(x, y) = 1/(x^2 + y^2)$, at $(-1, 3)$

22. $f(x, y) = \sqrt{\tan x + y}$, at $(0, 1)$

In Exercises **23–28**, which of the following vectors gives the direction of the gradient vector at point A on the contour diagram? The scales on the x- and y-axes are the same.

$\vec{i} + \vec{j}$ $\vec{i} - \vec{j}$ $-\vec{i} + 2\vec{j}$ $-2\vec{i} - \vec{j}$

23.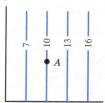

24.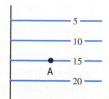

25.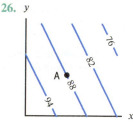

26.

27.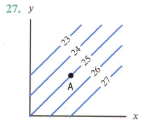

28.

In Exercises **29–34**, use the contour diagram of f in Figure 14.35 to decide if the specified directional derivative is positive, negative, or approximately zero.

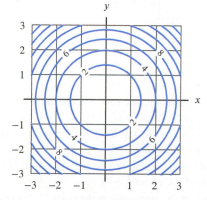

Figure 14.35

29. At point $(-2, 2)$, in direction $\vec{i}$.

30. At point $(0, -2)$, in direction $\vec{j}$.

31. At point $(0, -2)$, in direction $\vec{i} + 2\vec{j}$.

32. At point $(0, -2)$, in direction $\vec{i} - 2\vec{j}$.

33. At point $(-1, 1)$, in direction $\vec{i} + \vec{j}$.

34. At point $(-1, 1)$, in direction $-\vec{i} + \vec{j}$.

■In Exercises 35–42, use the contour diagram of f in Figure 14.35 to find the approximate direction of the gradient vector at the given point.

35. $(-2, 0)$ 36. $(0, -2)$ 37. $(2, 0)$ 38. $(0, 2)$

39. $(-2, 2)$ 40. $(-2, -2)$ 41. $(2, 2)$ 42. $(2, -2)$

■In Exercises 43–44, approximate the directional derivative of f in the direction from P to Q.

43. $P = (10, 12)$, $Q = (10.3, 12.1)$, $f(P) = 50$, $f(Q) = 52$.

44. $P = (-120, 45)$, $Q = (-122, 47)$, $f(P) = 200$, $f(Q) = 205$.

■In Exercises 45–48, find the directional derivative $f_{\vec{u}}(1, 2)$ for the function f with $\vec{u} = (3\vec{i} - 4\vec{j})/5$.

45. $f(x, y) = xy + y^3$ 46. $f(x, y) = 3x - 4y$

47. $f(x, y) = x^2 - y^2$ 48. $f(x, y) = \sin(2x - y)$

49. If $f(x, y) = x^2 y$ and $\vec{v} = 4\vec{i} - 3\vec{j}$, find the directional derivative at the point $(2, 6)$ in the direction of $\vec{v}$.

■In Exercises 50–51, find the differential df from the gradient.

50. grad $f = y\vec{i} + x\vec{j}$

51. grad $f = (2x + 3e^y)\vec{i} + 3xe^y\vec{j}$

■In Exercises 52–53, find grad f from the differential.

52. $df = 2x\,dx + 10y\,dy$

53. $df = (x + 1)ye^x\,dx + xe^x\,dy$

■In Exercises 54–55, assuming P and Q are close, approximate $f(Q)$.

54. $P = (100, 150)$, $Q = (101, 153)$, $f(P) = 2000$, grad $f(P) = 2\vec{i} - 2\vec{j}$.

55. $P = (10, 10)$, $Q = (10.2, 10.1)$, $f(P) = 50$, grad $f(P) = 0.5\vec{i} + \vec{j}$.

56. Where is grad f longer: at a point where contour lines of f are far apart or at a point where contour lines of f are close together?

PROBLEMS

57. A student was asked to find the directional derivative of $f(x, y) = x^2 e^y$ at the point $(1, 0)$ in the direction of $\vec{v} = 4\vec{i} + 3\vec{j}$. The student's answer was

$$f_{\vec{u}}(1, 0) = \operatorname{grad} f(1, 0) \cdot \vec{u} = \frac{8}{5}\vec{i} + \frac{3}{5}\vec{j}.$$

(a) At a glance, how do you know this is wrong?
(b) What is the correct answer?

■In Problems 58–64, find the quantity. Assume that g is a smooth function and that

$$\nabla g(2, 3) = -2\vec{i} + \vec{j} \quad \text{and} \quad \nabla g(2.4, 3) = 4\vec{i}$$

58. $g_y(2.4, 3)$ 59. $g_x(2, 3)$

60. A vector perpendicular to the level curve of g that passes through the point $(2.4, 3)$

61. A vector parallel to the level curve of g that passes through the point $(2, 3)$

62. The slope of the graph of g at the point $(2.4, 3)$ in the direction of the vector $\vec{i} + 3\vec{j}$.

63. The slope of the graph of g at the point $(2, 3)$ in the direction of the vector $\vec{i} + 3\vec{j}$.

64. The greatest slope of the graph of g at the point $(2, 3)$.

65. For $f(x, y) = (x + y)/(1 + x^2)$, find the directional derivative at $(1, -2)$ in the direction of $\vec{v} = 3\vec{i} + 4\vec{j}$.

66. For $g(x, y)$ with $g(5, 10) = 100$ and $g_{\vec{u}}(5, 10) = 0.5$, where $\vec{u}$ is the unit vector in the direction of the vector $\vec{i} + \vec{j}$, estimate $g(5.1, 10.1)$.

67. Let $f(P) = 15$ and $f(Q) = 20$ where $P = (3, 4)$ and $Q = (3.03, 3.96)$. Approximate the directional derivative of f at P in the direction of Q.

68. (a) Give Q, the point at a distance of 0.1 from $P = (4, 5)$ in the direction of $\vec{v} = -\vec{i} + 3\vec{j}$. Give five decimal places in your answer.
    (b) Use P and Q to approximate the directional derivative of $f(x, y) = \sqrt{x + y}$ in the direction of $\vec{v}$.
    (c) Give the exact value for the directional derivative you estimated in part (b).

69. For $f(x, y) = e^x \tan(y) + 2x^2 y$, find the directional derivative at the point $(0, \pi/4)$ in the direction

    (a) $\vec{i} - \vec{j}$ (b) $\vec{i} + \sqrt{3}\vec{j}$

70. Find the rate of change of $f(x, y) = x^2 + y^2$ at the point $(1, 2)$ in the direction of the vector $\vec{u} = 0.6\vec{i} + 0.8\vec{j}$.

71. (a) Let $f(x, y) = (x + y)/(1 + x^2)$. Find the directional derivative of f at $P = (1, -2)$ in the direction of:
    (i) $\vec{v} = 3\vec{i} - 2\vec{j}$ (ii) $\vec{v} = -\vec{i} + 4\vec{j}$
    (b) What is the direction of greatest increase of f at P?

72. Let $f(5, 10) = 200$ and $f(5.2, 9.9) = 197$.

    (a) Approximate the directional derivative at $(5, 10)$ in the direction from $(5, 10)$ toward $(5.2, 9.9)$.
    (b) Approximate $f(Q)$ at the point Q that is distance 0.1 from $(5, 10)$ in the direction of $(5.2, 9.9)$.
    (c) Give coordinates for the point Q.

73. Let $f(100, 100) = 500$ and grad $f(100, 100) = 2\vec{i} + 3\vec{j}$.

   (a) Find the directional derivative of f at the point $(100, 100)$ in the direction $\vec{i} + \vec{j}$.
   (b) Use the directional derivative to approximate $f(102, 102)$.

74. Let grad $f(50, 60) = 0.3\vec{i} + 0.5\vec{j}$. Approximate the directional derivative of f at the point $(50, 60)$ in the direction of the point $(49.5, 62)$.

75. Let $f(x, y) = x^2 y^3$. At the point $(-1, 2)$, find a vector

   (a) In the direction of maximum rate of change.
   (b) In the direction of minimum rate of change.
   (c) In a direction in which the rate of change is zero.

76. Let $f(x, y) = e^{xy}$. At the point $(1, 1)$, find a unit vector

   (a) In the direction of the steepest ascent.
   (b) In the direction of the steepest descent.
   (c) In a direction in which the rate of change is zero.

■ For Problems 77–81 use Figure 14.36, showing level curves of $f(x, y)$, to estimate the directional derivatives.

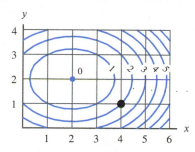

Figure 14.36

77. $f_{\vec{i}}(4, 1)$ 78. $f_{\vec{j}}(4, 1)$
79. $f_{\vec{u}}(4, 1)$ where $\vec{u} = (\vec{i} - \vec{j})/\sqrt{2}$
80. $f_{\vec{u}}(4, 1)$ where $\vec{u} = (-\vec{i} + \vec{j})/\sqrt{2}$
81. $f_{\vec{u}}(4, 1)$ with $\vec{u} = (-2\vec{i} + \vec{j})/\sqrt{5}$
82. The surface $z = g(x, y)$ is in Figure 14.37. What is the sign of each of the following directional derivatives?

   (a) $g_{\vec{u}}(2, 5)$ where $\vec{u} = (\vec{i} - \vec{j})/\sqrt{2}$.
   (b) $g_{\vec{u}}(2, 5)$ where $\vec{u} = (\vec{i} + \vec{j})/\sqrt{2}$.

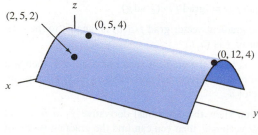

Figure 14.37

83. The table gives values of a differentiable function $f(x, y)$. At the point $(1.2, 0)$, into which quadrant does the gradient vector of f point? Justify your answer.

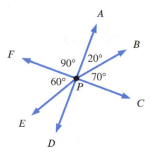

| | | y | |
	-1	0	1
x 1.0	0.7	0.1	-0.5
1.2	4.8	4.2	3.6
1.4	8.9	8.3	7.7

84. The gradient of f at a point P has magnitude 10 and is in the direction of A in Figure 14.38. Find the directional derivatives of f at P in the six directions shown.

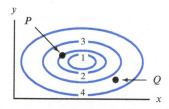

Figure 14.38

85. Figure 14.39 represents the level curves $f(x, y) = c$; the values of f on each curve are marked. In each of the following parts, decide whether the given quantity is positive, negative or zero. Explain your answer.

   (a) The value of $\nabla f \cdot \vec{i}$ at P.
   (b) The value of $\nabla f \cdot \vec{j}$ at P.
   (c) $\partial f / \partial x$ at Q.
   (d) $\partial f / \partial y$ at Q.

Figure 14.39

86. In Figure 14.39, which is larger: $\|\nabla f\|$ at P or $\|\nabla f\|$ at Q? Explain how you know.

87. Let P, Q, R and S be four distinct points in the plane. Let $\vec{u}$ be the unit vector in the direction from P to Q, $\vec{v}$ the unit vector in the direction from P to R, and $\vec{w}$ the unit vector in the direction from P to S. Let $f(x, y)$ be a linear function with $f(P) = 10$, $f(Q) = 7$, $f(R) = 15$, and $f(S) = 10$. List the directional derivatives $f_{\vec{u}}(P)$, $f_{\vec{v}}(P)$, and $f_{\vec{w}}(P)$ in increasing order.

88. Let $f_x(3,1) = -5$ and $f_y(3,1) = 2$. Find a unit vector $\vec{u}$ such that:

(a) $f_{\vec{u}}(3,1) > 0$ (b) $f_{\vec{u}}(3,1) < 0$

(c) $f_{\vec{u}}(3,1) = 0$

89. Let $f(0,0) = -4$ and $f_{\vec{u}}(0,0) = 20$ for a unit vector $\vec{u}$. Suppose that points P and Q in Figure 14.40 are close. Find approximate values of $f(P)$ and $f(Q)$.

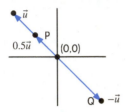

Figure 14.40

In Problems 90–93, check that the point $(2,3)$ lies on the curve. Then, viewing the curve as a contour of $f(x,y)$, use grad $f(2,3)$ to find a vector normal to the curve at $(2,3)$ and an equation for the tangent line to the curve at $(2,3)$.

90. $x^2 + y^2 = 13$ **91.** $xy = 6$

92. $y = x^2 - 1$ **93.** $(y - x)^2 + 2 = xy - 3$

94. The temperature H in °Fahrenheit y miles north of the Canadian border t hours after midnight is given by $H = 30 - 0.05y - 5t$. A moose runs north at a speed of 20 mph. At what rate does the moose perceive the temperature to be changing?

95. At a certain point on a heated plate, the greatest rate of temperature increase, 5° C per meter, is toward the northeast. If an object at this point moves directly north, at what rate is the temperature increasing?

96. An ant is at the point $(1,1,3)$ on the surface of a bowl with equation $z = x^2 + 2y^2$, where x and y are in cm. In what two horizontal directions can the ant move away from the point $(1,1,3)$ so that its initial rate of ascent is 2 vertical cm for each horizontal cm moved? Give your answers as vectors in the plane.

97. Let $T = f(x,y) = 100e^{-(x^2/2)-y^2}$ represent the temperature, in °C, at the point (x,y) with x and y in meters.

(a) Describe the contours of f, and explain their meaning in the context of this problem.

(b) Find the rate at which the temperature changes as you move away from the point $(1,1)$ toward the point $(2,3)$. Give units in your answer.

(c) In what direction would you move away from $(1,1)$ for the temperature to increase as fast as possible?

98. You are climbing a mountain by the steepest route at a slope of 20° when you come upon a trail branching off at a 30° angle from yours. What is the angle of ascent of the branch trail?

99. You are standing at the point $(1,1,3)$ on the hill whose equation is given by $z = 5y - x^2 - y^2$.

(a) If you choose to climb in the direction of steepest ascent, what is your initial rate of ascent relative to the horizontal distance?

(b) If you decide to go straight northwest, will you be ascending or descending? At what rate?

(c) If you decide to maintain your altitude, in what directions can you go?

Strengthen Your Understanding

In Problems 100–102, explain what is wrong with the statement.

100. A function f has a directional derivative given by $f_{\vec{u}}(0,0) = 3\vec{i} + 4\vec{j}$.

101. A function f has gradient grad $f(0,0) = 7$.

102. The gradient vector grad $f(x,y)$ is perpendicular to the contours of f, and the closer together the contours for equally spaced values of f, the shorter the gradient vector.

In Problems 103–104, give an example of:

103. A unit vector $\vec{u}$ such that $f_{\vec{u}}(0,0) < 0$, given that $f_x(0,0) = 2$ and $f_y(0,0) = 3$.

104. A contour diagram of a function with two points in the domain where the gradients are parallel but different lengths.

Are the statements in Problems 105–116 true or false? Give reasons for your answer.

105. If the point (a,b) is on the contour $f(x,y) = k$, then the slope of the line tangent to this contour at (a,b) is $f_y(a,b)/f_x(a,b)$.

106. The gradient vector grad $f(a,b)$ is a vector in 3-space.

107. $\text{grad}(fg) = (\text{grad } f) \cdot (\text{grad } g)$

108. The gradient vector grad $f(a,b)$ is tangent to the contour of f at (a,b).

109. If you know the gradient vector of f at (a,b) then you can find the directional derivative $f_{\vec{u}}(a,b)$ for any unit vector $\vec{u}$.

110. If you know the directional derivative $f_{\vec{u}}(a,b)$ for all unit vectors $\vec{u}$ then you can find the gradient vector of f at (a,b).

111. The directional derivative $f_{\vec{u}}(a,b)$ is parallel to $\vec{u}$.

112. The gradient grad $f(3,4)$ is perpendicular to the vector $3\vec{i} + 4\vec{j}$.

113. If grad $f(1,2) = \vec{i}$, then f decreases in the $-\vec{i}$ direction at $(1,2)$.

114. If grad $f(1,2) = \vec{i}$, then $f(10,2) > f(1,2)$.

115. At the point $(3,0)$, the function $g(x,y) = x^2 + y^2$ has the same maximal rate of increase as that of the function $h(x,y) = 2xy$.

116. If $f(x,y) = e^{x+y}$, then the directional derivative in any direction $\vec{u}$ (with $\|\vec{u}\| = 1$) at the point $(0,0)$ is always less than or equal to $\sqrt{2}$.

14.5 GRADIENTS AND DIRECTIONAL DERIVATIVES IN SPACE

The Gradient Vector and Directional Derivative of a Function of Three Variables

The gradient of a function of three variables is defined in the same way as for two variables:

> **The gradient vector** of a differentiable function $f(x,y,z)$ is
>
> $$\text{grad } f = f_x\vec{i} + f_y\vec{j} + f_z\vec{k}.$$

As in two dimensions, directional derivatives in space give the rate of change of a function in the direction of a unit vector $\vec{u}$. If a function f of three variables is differentiable at the point (a,b,c) and $\vec{u} = u_1\vec{i} + u_2\vec{j} + u_3\vec{k}$, then the directional derivative $f_{\vec{u}}$ is related to the gradient by

$$f_{\vec{u}}(a,b,c) = f_x(a,b,c)u_1 + f_y(a,b,c)u_2 + f_z(a,b,c)u_3 = \text{grad } f(a,b,c) \cdot \vec{u}.$$

Since grad $f(a,b,c) \cdot \vec{u} = \|\text{grad } f(a,b,c)\| \cos\theta$, where θ is the angle between grad $f(a,b,c)$ and $\vec{u}$, the value of $f_{\vec{u}}(a,b,c)$ is largest when $\theta = 0$, that is, when $\vec{u}$ is in the same direction as grad $f(a,b,c)$. In addition, $f_{\vec{u}}(a,b,c) = 0$ when $\theta = \pi/2$, so grad $f(a,b,c)$ is perpendicular to the level surface of f. The properties of gradients in space are similar to those in the plane:

> ### Properties of the Gradient Vector in Space
>
> If f is differentiable at (a,b,c) and $\vec{u}$ is a unit vector, then
>
> $$f_{\vec{u}}(a,b,c) = \text{grad } f(a,b,c) \cdot \vec{u}.$$
>
> If, in addition, grad $f(a,b,c) \neq \vec{0}$, then
> - grad $f(a,b,c)$ is perpendicular to the level surface of f at (a,b,c)
> - grad $f(a,b,c)$ is in the direction of the greatest rate of increase of f
> - $\|\text{grad } f(a,b,c)\|$ is the maximum rate of change of f at (a,b,c).

Example 1   Find the directional derivative of $f(x,y,z) = xy + z$ at the point $(-1,0,1)$ in the direction of the vector $\vec{v} = 2\vec{i} + \vec{k}$.

Solution   The magnitude of $\vec{v}$ is $\|\vec{v}\| = \sqrt{2^2 + 1} = \sqrt{5}$, so a unit vector in the same direction as $\vec{v}$ is

$$\vec{u} = \frac{\vec{v}}{\|\vec{v}\|} = \frac{2}{\sqrt{5}}\vec{i} + 0\vec{j} + \frac{1}{\sqrt{5}}\vec{k}.$$

The partial derivatives of f are $f_x(x, y, z) = y$ and $f_y(x, y, z) = x$ and $f_z(x, y, z) = 1$. Thus,

$$f_{\vec{u}}(-1, 0, 1) = f_x(-1, 0, 1)u_1 + f_y(-1, 0, 1)u_2 + f_z(-1, 0, 1)u_3$$

$$= (0)\left(\frac{2}{\sqrt{5}}\right) + (-1)(0) + (1)\left(\frac{1}{\sqrt{5}}\right) = \frac{1}{\sqrt{5}}.$$

Example 2   Let $f(x, y, z) = x^2 + y^2$ and $g(x, y, z) = -x^2 - y^2 - z^2$. What can we say about the direction of the following vectors?

(a) grad $f(0, 1, 1)$ (b) grad $f(1, 0, 1)$ (c) grad $g(0, 1, 1)$ (d) grad $g(1, 0, 1)$.

Solution   The cylinder $x^2 + y^2 = 1$ in Figure 14.41 is a level surface of f and contains both the points $(0, 1, 1)$ and $(1, 0, 1)$. Since the value of f does not change at all in the z-direction, all the gradient vectors are horizontal. They are perpendicular to the cylinder and point outward because the value of f increases as we move out.

Similarly, the points $(0, 1, 1)$ and $(1, 0, 1)$ also lie on the same level surface of g, namely $g(x, y, z) = -x^2 - y^2 - z^2 = -2$, which is the sphere $x^2 + y^2 + z^2 = 2$. Part of this level surface is shown in Figure 14.42. This time the gradient vectors point inward, since the negative signs mean that the function increases (from large negative values to small negative values) as we move inward.

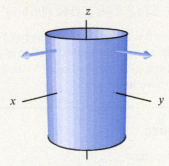

Figure 14.41: The level surface
$f(x, y, z) = x^2 + y^2 = 1$ with two gradient vectors

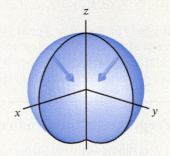

Figure 14.42: The level surface
$g(x, y, z) = -x^2 - y^2 - z^2 = -2$ with two gradient vectors

Example 3   Consider the functions $f(x, y) = 4 - x^2 - 2y^2$ and $g(x, y) = 4 - x^2$. Calculate a vector perpendicular to each of the following:

(a) The level curve of f at the point $(1, 1)$ (b) The surface $z = f(x, y)$ at the point $(1, 1, 1)$
(c) The level curve of g at the point $(1, 1)$ (d) The surface $z = g(x, y)$ at the point $(1, 1, 3)$

Solution   (a) The vector we want is a 2-vector in the plane. Since grad $f = -2x\vec{i} - 4y\vec{j}$, we have

$$\text{grad } f(1, 1) = -2\vec{i} - 4\vec{j}.$$

Any nonzero multiple of this vector is perpendicular to the level curve at the point $(1, 1)$.

(b) In this case we want a 3-vector in space. To find it we rewrite $z = 4 - x^2 - 2y^2$ as the level surface of the function F, where

$$F(x, y, z) = 4 - x^2 - 2y^2 - z = 0.$$

Then

$$\text{grad } F = -2x\vec{i} - 4y\vec{j} - \vec{k},$$

so

$$\operatorname{grad} F(1,1,1) = -2\vec{i} - 4\vec{j} - \vec{k},$$

and $\operatorname{grad} F(1,1,1)$ is perpendicular to the surface $z = 4 - x^2 - 2y^2$ at the point $(1,1,1)$. Notice that $-2\vec{i} - 4\vec{j} - \vec{k}$ is not the only possible answer: any multiple of this vector will do.

(c) We are looking for a 2-vector. Since $\operatorname{grad} g = -2x\vec{i} + 0\vec{j}$, we have

$$\operatorname{grad} g(1,1) = -2\vec{i}.$$

Any multiple of this vector is perpendicular to the level curve also.

(d) We are looking for a 3-vector. We rewrite $z = 4 - x^2$ as the level surface of the function G, where

$$G(x,y,z) = 4 - x^2 - z = 0.$$

Then

$$\operatorname{grad} G = -2x\vec{i} - \vec{k}$$

So

$$\operatorname{grad} G(1,1,3) = -2\vec{i} - \vec{k},$$

and any multiple of $\operatorname{grad} G(1,1,3)$ is perpendicular to the surface $z = 4 - x^2$ at this point.

Example 4   (a) A hiker on the surface $f(x,y) = 4 - x^2 - 2y^2$ at the point $(1,-1,1)$ starts to climb along the path of steepest ascent. What is the relation between the vector $\operatorname{grad} f(1,-1)$ and a vector tangent to the path at the point $(1,-1,1)$ and pointing uphill?

(b) At the point $(1,-1,1)$ on the surface $f(x,y) = 4 - x^2 - 2y^2$, calculate a vector, $\vec{n}$, perpendicular to the surface and a vector, $\vec{T}$, tangent to the curve of steepest ascent.

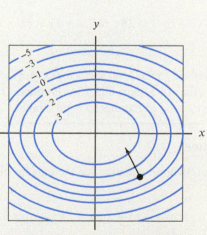

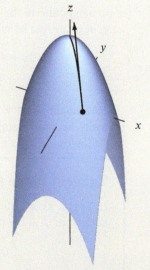

Figure 14.43: Contour diagram for $z = f(x,y) = 4 - x^2 - 2y^2$ showing direction of $\operatorname{grad} f(1,-1)$

Figure 14.44: Graph of $f(x,y) = 4 - x^2 - 2y^2$ showing path of steepest ascent from the point $(1,-1,1)$

Solution   (a) The hiker at the point $(1,-1,1)$ lies directly above the point $(1,-1)$ in the xy-plane. The vector $\operatorname{grad} f(1,-1)$ lies in 2-space, pointing like a compass in the direction in which f increases most rapidly. Therefore, $\operatorname{grad} f(1,-1)$ lies directly under a vector tangent to the hiker's path at $(1,-1,1)$ and pointing uphill. (See Figures 14.43 and 14.44.)

(b) The surface is represented by $F(x, y, z) = 4 - x^2 - 2y^2 - z = 0$. Since grad $F = -2x\vec{i} - 4y\vec{j} - \vec{k}$, a normal, $\vec{n}$, to the surface is given by

$$\vec{n} = \text{grad } F(1, -1, 1) = -2(1)\vec{i} - 4(-1)\vec{j} - \vec{k} = -2\vec{i} + 4\vec{j} - \vec{k}.$$

We take the $\vec{i}$ and $\vec{j}$ components of $\vec{T}$ to be the vector grad $f(1, -1) = -2\vec{i} + 4\vec{j}$. Thus, we have that, for some $a > 0$,

$$\vec{T} = -2\vec{i} + 4\vec{j} + a\vec{k}.$$

We want $\vec{n} \cdot \vec{T} = 0$, so

$$\vec{n} \cdot \vec{T} = (-2\vec{i} + 4\vec{j} - \vec{k}) \cdot (-2\vec{i} + 4\vec{j} + a\vec{k}) = 4 + 16 - a = 0.$$

Thus, $a = 20$ and hence

$$\vec{T} = -2\vec{i} + 4\vec{j} + 20\vec{k}.$$

Example 5   Find the equation of the tangent plane to the sphere $x^2 + y^2 + z^2 = 14$ at the point $(1, 2, 3)$.

Solution   We write the sphere as a level surface as follows:

$$f(x, y, z) = x^2 + y^2 + z^2 = 14.$$

We have

$$\text{grad } f = 2x\vec{i} + 2y\vec{j} + 2z\vec{k},$$

so the vector

$$\text{grad } f(1, 2, 3) = 2\vec{i} + 4\vec{j} + 6\vec{k}$$

is perpendicular to the sphere at the point $(1, 2, 3)$. Since the vector grad $f(1, 2, 3)$ is normal to the tangent plane, the equation of the plane is

$$2x + 4y + 6z = 2 \cdot 1 + 4 \cdot 2 + 6 \cdot 3 = 28 \quad \text{or} \quad x + 2y + 3z = 14.$$

We could also try to find the tangent plane to the level surface $f(x, y, z) = k$ by solving algebraically for z and using the method of Section 14.3, page 802. (See Problem 47.) Solving for z can be difficult or impossible, however, so the method of Example 5 is preferable.

> ### Tangent Plane to a Level Surface
>
> If $f(x, y, z)$ is differentiable at (a, b, c), then an equation for the tangent plane to the level surface of f at the point (a, b, c) is
>
> $$f_x(a, b, c)(x - a) + f_y(a, b, c)(y - b) + f_z(a, b, c)(z - c) = 0.$$

Caution: Scale on the Axis and the Geometric Interpretation of the Gradient

When we interpreted the gradient of a function geometrically (page 813), we tacitly assumed that the units and scales along the x and y axes were the same. If the scales are not the same, the gradient vector may not look perpendicular to the contours. Consider the function $f(x, y) = x^2 + y$ with gradient vector grad $f = 2x\vec{i} + \vec{j}$. Figure 14.45 shows the gradient vector at $(1, 1)$ using the same scales in the x and y directions. As expected, the gradient vector is perpendicular to the contour line. Figure 14.46 shows contours of the same function with unequal scales on the two axes. Notice that the gradient vector no longer appears perpendicular to the contour lines. Thus, we see that the geometric interpretation of the gradient vector requires that the same scale be used on both axes.

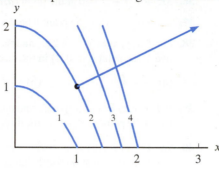

Figure 14.45: The gradient vector with x and y scales equal

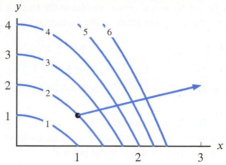

Figure 14.46: The gradient vector with x and y scales unequal

Summary for Section 14.5

If $f(x, y, z)$ is differentiable:

- The **gradient vector** is

$$\operatorname{grad} f = f_x\vec{i} + f_y\vec{j} + f_z\vec{k}.$$

- If $\vec{u}$ is a unit vector, then the **directional derivative** of f can be computed using

$$f_{\vec{u}}(a, b, c) = \operatorname{grad} f(a, b, c) \cdot \vec{u}.$$

If, in addition, $\operatorname{grad} f(a, b, c) \neq \vec{0}$, then

- $\operatorname{grad} f(a, b, c)$ is perpendicular to the level surface of f at (a, b, c)
- $\operatorname{grad} f(a, b, c)$ is in the direction of the greatest rate of increase of f
- $\| \operatorname{grad} f(a, b, c) \|$ is the maximum rate of change of f at (a, b, c).

- An equation for the **tangent plane** to the level surface of f at the point (a, b, c) is

$$f_x(a, b, c)(x - a) + f_y(a, b, c)(y - b) + f_z(a, b, c)(z - c) = 0.$$

Exercises and Problems for Section 14.5   Online Resource: Additional Problems for Section 14.5
EXERCISES

In Exercises 1–12, find the gradient of the function.

1. $f(x, y, z) = x^2$

2. $f(x, y, z) = x^2 + y^3 - z^4$

3. $f(x, y, z) = e^{x+y+z}$

4. $f(x, y, z) = \cos(x + y) + \sin(y + z)$

5. $f(x, y, z) = yz^2/(1 + x^2)$

6. $f(x, y, z) = 1/(x^2 + y^2 + z^2)$

7. $f(x, y, z) = \sqrt{x^2 + y^2 + z^2}$

8. $f(x, y, z) = xe^y \sin z$

9. $f(x, y, z) = xy + \sin(e^z)$

10. $f(x_1, x_2, x_3) = x_1^2 x_2^3 x_3^4$

11. $f(p, q, r) = e^p + \ln q + e^{r^2}$

12. $f(x, y, z) = e^{z^2} + y \ln(x^2 + 5)$

In Exercises 13–18, find the gradient at the point.

13. $f(x, y, z) = zy^2$, at $(1, 0, 1)$

14. $f(x, y, z) = 2x + 3y + 4z$, at $(1, 1, 1)$

15. $f(x, y, z) = x^2 + y^2 - z^4$, at $(3, 2, 1)$

16. $f(x, y, z) = xyz$, at $(1, 2, 3)$

17. $f(x, y, z) = \sin(xy) + \sin(yz)$, at $(1, \pi, -1)$

18. $f(x, y, z) = x \ln(yz)$, at $(2, 1, e)$

In Exercises 19–24, find the directional derivative using $f(x, y, z) = xy + z^2$.

19. At $(1, 2, 3)$ in the direction of $\vec{i} + \vec{j} + \vec{k}$.

20. At $(1, 1, 1)$ in the direction of $\vec{i} + 2\vec{j} + 3\vec{k}$.

21. As you leave the point $(1, 1, 0)$ heading in the direction of the point $(0, 1, 1)$.

22. As you arrive at $(0, 1, 1)$ from the direction of $(1, 1, 0)$.

23. At the point $(2, 3, 4)$ in the direction of a vector making an angle of $3\pi/4$ with $\operatorname{grad} f(2, 3, 4)$.

24. At the point $(2, 3, 4)$ in the direction of the maximum rate of change of f.

In Exercises 25–30, check that the point $(-1, 1, 2)$ lies on the given surface. Then, viewing the surface as a level surface for a function $f(x, y, z)$, find a vector normal to the surface and an equation for the tangent plane to the surface at $(-1, 1, 2)$.

25. $x^2 - y^2 + z^2 = 4$ **26.** $z = x^2 + y^2$

27. $y^2 = z^2 - 3$ **28.** $x^2 - xyz = 3$

29. $\cos(x + y) = e^{xz+2}$ **30.** $y = 4/(2x + 3z)$

In Exercises 31–32, the gradient of f and a point P on the level surface $f(x, y, z) = 0$ are given. Find an equation for the tangent plane to the surface at the point P.

31. $\operatorname{grad} f = yz\vec{i} + xz\vec{j} + xy\vec{k}$, $P = (1, 2, 3)$

32. $\operatorname{grad} f = 2x\vec{i} + z^2\vec{j} + 2yz\vec{k}$, $P = (10, -10, 30)$

In Exercises 33–37, find an equation of the tangent plane to the surface at the given point.

33. $x^2 + y^2 + z^2 = 17$ at the point $(2, 3, 2)$

34. $x^2 + y^2 = 1$ at the point $(1, 0, 0)$

35. $z = 2x + y + 3$ at the point $(0, 0, 3)$

36. $3x^2 - 4xy + z^2 = 0$ at the point (a, a, a), where $a \neq 0$

37. $z = 9/(x + 4y)$ at the point where $x = 1$ and $y = 2$

38. For $f(x, y, z) = 3x^2y^2+2yz$, find the directional derivative at the point $(-1, 0, 4)$ in the direction of

   (a) $\vec{i} - \vec{k}$ **(b)** $-\vec{i} + 3\vec{j} + 3\vec{k}$

39. If $f(x, y, z) = x^2+3xy+2z$, find the directional derivative at the point $(2, 0, -1)$ in the direction of $2\vec{i} + \vec{j} - 2\vec{k}$.

40. **(a)** Let $f(x, y, z) = x^2 + y^2 - xyz$. Find $\operatorname{grad} f$.
   (b) Find the equation for the tangent plane to the surface $f(x, y, z) = 7$ at the point $(2, 3, 1)$.

41. Find the equation of the tangent plane at the point $(3, 2, 2)$ to $z = \sqrt{17 - x^2 - y^2}$.

42. Find the equation of the tangent plane to $z = 8/(xy)$ at the point $(1, 2, 4)$.

43. Find an equation of the tangent plane and of a normal vector to the surface $x = y^3z^7$ at the point $(1, -1, -1)$.

PROBLEMS

44. Let $f(x, y, z)$ represent the temperature in °C at the point (x, y, z) with x, y, z in meters. Let $\vec{v}$ be your velocity in meters per second. Give units and an interpretation of each of the following quantities.

   (a) $\| \operatorname{grad} f \|$ **(b)** $\operatorname{grad} f \cdot \vec{v}$ **(c)** $\| \operatorname{grad} f \| \cdot \| \vec{v} \|$

45. Consider the surface $g(x, y) = 4 - x^2$. What is the relation between $\operatorname{grad} g(-1, -1)$ and a vector tangent to the path of steepest ascent at $(-1, -1, 3)$? Illustrate your answer with a sketch.

46. Match the functions $f(x, y, z)$ in (a)–(d) with the descriptions of their gradients in (I)–(IV).

   (a) $x^2 + y^2 + z^2$ **(b)** $x^2 + y^2$

   (c) $\dfrac{1}{x^2 + y^2 + z^2}$ **(d)** $\dfrac{1}{x^2 + y^2}$

   I Points radially outward from the z-axis.
   II Points radially inward toward the z-axis.
   III Points radially outward from the origin.
   IV Points radially inward toward the origin.

47. Find the equation of the tangent plane at $(2, 3, 1)$ to the surface $x^2 + y^2 - xyz = 7$. Do this in two ways:

   (a) Viewing the surface as the level set of a function of three variables, $F(x, y, z)$.
   (b) Viewing the surface as the graph of a function of two variables $z = f(x, y)$.

48. At what point on the surface $z = 1+x^2+y^2$ is its tangent plane parallel to the following planes?

   (a) $z = 5$ **(b)** $z = 5 + 6x - 10y$

49. Let $g_x(2, 1, 7) = 3$, $g_y(2, 1, 7) = 10$, $g_z(2, 1, 7) = -5$. Find the equation of the tangent plane to $g(x, y, z) = 0$ at the point $(2, 1, 7)$.

50. The vector ∇f at point P and four unit vectors $\vec{u}_1, \vec{u}_2, \vec{u}_3, \vec{u}_4$ are shown in Figure 14.47. Arrange the following quantities in ascending order

$$f_{\vec{u}_1}, \quad f_{\vec{u}_2}, \quad f_{\vec{u}_3}, \quad f_{\vec{u}_4}, \quad \text{the number } 0.$$

The directional derivatives are all evaluated at the point P and the function $f(x, y)$ is differentiable at P.

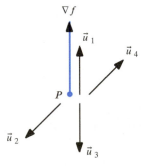

Figure 14.47

51. Let $f(x, y, z) = x^2 + y^2 + z^2$. At the point $(1, 2, 1)$, find the rate of change of f in the direction perpendicular to the plane $x + 2y + 3z = 8$ and moving away from the origin.

52. Let $f(x, y) = \cos x \sin y$ and let S be the surface $z = f(x, y)$.

(a) Find a normal vector to the surface S at the point $(0, \pi/2, 1)$.

(b) What is the equation of the tangent plane to the surface S at the point $(0, \pi/2, 1)$?

53. Let $f(x, y, z) = \sin(x^2 + y^2 + z^2)$.

(a) Describe in words the shape of the level surfaces of f.

(b) Find grad f.

(c) Consider the two vectors $\vec{r} = x\vec{i} + y\vec{j} + z\vec{k}$ and grad f at a point (x, y, z) where $\sin(x^2 + y^2 + z^2) \neq 0$. What is (are) the possible value(s) of the angle between these vectors?

54. Each diagram (I) – (IV) in Figure 14.48 represents the level curves of a function $f(x, y)$. For each function f, consider the point above P on the surface $z = f(x, y)$ and choose from the lists of vectors and equations that follow:

(a) A vector which could be the normal to the surface at that point;

(b) An equation which could be the equation of the tangent plane to the surface at that point.

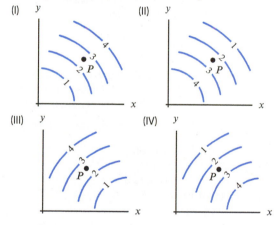

Figure 14.48

Vectors

(E) $2\vec{i} + 2\vec{j} - 2\vec{k}$
(F) $2\vec{i} + 2\vec{j} + 2\vec{k}$
(G) $2\vec{i} - 2\vec{j} + 2\vec{k}$
(H) $-2\vec{i} + 2\vec{j} + 2\vec{k}$

Equations

(J) $x + y + z = 4$
(K) $2x - 2y - 2z = 2$
(L) $-3x - 3y + 3z = 6$
(M) $-\dfrac{x}{2} + \dfrac{y}{2} - \dfrac{z}{2} = -7$

55. (a) What is the shape of the curve in which the following surface cuts the yz-plane:

$$5(x-1)^2 + 2(y+1)^2 + 2(z-3)^2 = 25?$$

(b) Does the curve in part (a) go through the origin?

(c) Find an expression for a vector perpendicular to the surface at the origin.

56. Find the points on the surface $y = 4 + x^2 + z^2$ where the gradient is parallel to $\vec{i} + \vec{j} + \vec{k}$.

57. A particle moves at a speed of 3 units per second perpendicular to the surface $x = 4 + y^2 + z^2$ from the point $(9, 1, 2)$ toward the yz-plane.

(a) What is the particle's velocity vector?

(b) Where is the particle after one second?

58. For the surface $z + 7 = 2x^2 + 3y^2$, where does the tangent plane at the point $(-1, 1, -2)$ meet the three axes?

59. Find a vector perpendicular to the surface $z = 4 - x^2 - y^2$ at the point above the point $(1, 1, 0)$. (The z-axis is vertical.)

60. (a) Where does the surface $x^2 + y^2 - (z-1)^2 = 0$ cut the xy-plane? What is the shape of the curve?

(b) At the points where the surface cuts the xy-plane, do vectors perpendicular to the surface lie in the xy-plane?

61. A unit vector is perpendicular to the surface $z = x^2 - y^2$. At which point on the surface does this unit vector have the largest dot product with the vector $\vec{i} + 2\vec{j} + 3\vec{k}$?

62. The surface S is represented by the equation $F = 0$ where $F(x, y, z) = x^2 - (y/z^2)$.

(a) Find the unit vectors $\vec{u}_1$ and $\vec{u}_2$ pointing in the direction of maximum increase of F at the points $(0, 0, 1)$ and $(1, 1, 1)$ respectively.

(b) Find the tangent plane to S at the points $(0, 0, 1)$ and $(1, 1, 1)$.

(c) Find all points on S where a normal vector is parallel to the xy-plane.

63. Consider the function $f(x, y) = (e^x - x) \cos y$. Suppose S is the surface $z = f(x, y)$.

(a) Find a vector which is perpendicular to the level curve of f through the point $(2, 3)$ in the direction in which f decreases most rapidly.

(b) Suppose $\vec{v} = 5\vec{i} + 4\vec{j} + a\vec{k}$ is a vector in 3-space which is tangent to the surface S at the point P lying on the surface above $(2, 3)$. What is a?

64. (a) Find the tangent plane to the surface $x^2 + y^2 + 3z^2 = 4$ at the point $(0.6, 0.8, 1)$.

(b) Is there a point on the surface $x^2 + y^2 + 3z^2 = 4$ at which the tangent plane is parallel to the plane $8x + 6y + 30z = 1$? If so, find it. If not, explain why not.

65. Your house lies on the surface $z = f(x, y) = 2x^2 - y^2$ directly above the point $(4, 3)$ in the xy-plane.

(a) How high above the xy-plane do you live?

(b) What is the slope of your lawn as you look from your house directly toward the z-axis (that is, along the vector $-4\vec{i} - 3\vec{j}$)?

(c) When you wash your car in the driveway, on this surface above the point $(4, 3)$, which way does the water run off? (Give your answer as a two-dimensional vector.)

(d) What is the equation of the tangent plane to this surface at your house?

66. (a) Sketch the contours of $z = y - \sin x$ for $z = -1, 0, 1, 2$.

(b) A bug starts on the surface at the point $(\pi/2, 1, 0)$ and walks on the surface $z = y - \sin x$ in the direction parallel to the y-axis, in the direction of increasing y. Is the bug walking in a valley or on top of a ridge? Explain.

(c) On the contour $z = 0$ in your sketch for part (a), draw the gradients of z at $x = 0$, $x = \pi/2$, and $x = \pi$.

67. The function $f(x, y, z) = 2x - 3y + z + 10$ gives the temperature, T, in degrees Celsius, at the point (x, y, z).

(a) In words, describe the isothermal surfaces.

(b) Calculate $f_z(0, 0, 0)$ and interpret in terms of temperature.

(c) If you are standing at the point $(0, 0, 0)$, in what direction should you move to increase your temperature the fastest?

(d) Is $z = -2x + 3y + 17$ an isothermal surface? If so, what is the temperature on this isotherm?

68. The concentration of salt in a fluid at (x, y, z) is given by $F(x, y, z) = x^2 + y^4 + x^2 z^2$ mg/cm^3. You are at the point $(-1, 1, 1)$.

(a) In which direction should you move if you want the concentration to increase the fastest?

(b) You start to move in the direction you found in part (a) at a speed of 4 cm/sec. How fast is the concentration changing?

69. The temperature of a gas at the point (x, y, z) is given by $G(x, y, z) = x^2 - 5xy + y^2 z$.

(a) What is the rate of change in the temperature at the point $(1, 2, 3)$ in the direction $\vec{v} = 2\vec{i} + \vec{j} - 4\vec{k}$?

(b) What is the direction of maximum rate of change of temperature at the point $(1, 2, 3)$?

(c) What is the maximum rate of change at the point $(1, 2, 3)$?

70. The temperature at the point (x, y, z) in 3-space is given, in degrees Celsius, by $T(x, y, z) = e^{-(x^2 + y^2 + z^2)}$.

(a) Describe in words the shape of surfaces on which the temperature is constant.

(b) Find grad T.

(c) You travel from the point $(1, 0, 0)$ to the point $(2, 1, 0)$ at a speed of 3 units per second. Find the instantaneous rate of change of the temperature as you leave the point $(1, 0, 0)$. Give units.

71. A spaceship is plunging into the atmosphere of a planet. With coordinates in miles and the origin at the center of the planet, the pressure of the atmosphere at (x, y, z) is

$$P = 5e^{-0.1\sqrt{x^2 + y^2 + z^2}} \text{ atmospheres.}$$

The velocity, in miles/sec, of the spaceship at $(0, 0, 1)$ is $\vec{v} = \vec{i} - 2.5\vec{k}$. At $(0, 0, 1)$, what is the rate of change with respect to time of the pressure on the spaceship?

72. The earth has mass M and is located at the origin in 3-space, while the moon has mass m. Newton's Law of Gravitation states that if the moon is located at the point (x, y, z) then the attractive force exerted by the earth on the moon is given by the vector

$$\vec{F} = -GMm\frac{\vec{r}}{\|\vec{r}\|^3},$$

where $\vec{r} = x\vec{i} + y\vec{j} + z\vec{k}$. Show that $\vec{F} = \text{grad } \varphi$, where φ is the function given by

$$\varphi(x, y, z) = \frac{GMm}{\|\vec{r}\|}.$$

73. Let $\vec{r} = x\vec{i} + y\vec{j} + z\vec{k}$ and $\vec{a}$ be a constant vector. For each of the quantities in (a)–(c), choose the statement in (I)–(V) that describes it. No reasons are needed.

(a) grad$(\vec{r} + \vec{a})$ **(b)** grad$(\vec{r} \cdot \vec{a})$ **(c)** grad$(\vec{r} \times \vec{a})$

I Scalar, independent of $\vec{a}$.
II Scalar, depends on $\vec{a}$.
III Vector, independent of $\vec{a}$.
IV Vector, depends on $\vec{a}$.
V Not defined.

Strengthen Your Understanding

■ In Problems 74–75, explain what is wrong with the statement.

74. The gradient vector grad $f(x, y)$ points in the direction perpendicular to the surface $z = f(x, y)$.

75. The tangent plane at the origin to a surface $f(x, y, z) = 1$ that contains the point $(0, 0, 0)$ has equation

$$f_x(0, 0, 0)x + f_y(0, 0, 0)y + f_z(0, 0, 0)z + 1 = 0.$$

■ In Problems 76–78, give an example of:

76. A surface $z = f(x, y)$ such that the vector $\vec{i} - 2\vec{j} - \vec{k}$ is normal to the tangent plane at the point where $(x, y) = (0, 0)$.

77. A function $f(x, y, z)$ such that grad $f = 2\vec{i} + 3\vec{j} + 4\vec{k}$.

78. Two nonparallel unit vectors $\vec{u}$ and $\vec{v}$ such that $f_{\vec{u}}(0, 0, 0) = f_{\vec{v}}(0, 0, 0) = 0$, where $f(x, y, z) = 2x - 3y$.

■ Are the statements in Problems 79–82 true or false? Give reasons for your answer.

79. An equation for the tangent plane to the surface $z = x^2 + y^3$ at $(1, 1)$ is $z = 2 + 2x(x - 1) + 3y^2(y - 1)$.

80. There is a function $f(x, y)$ which has a tangent plane with equation $z = 0$ at a point (a, b).

81. There is a function with $\|$ grad $f\| = 4$ and $f_{\vec{k}} = 5$ at some point.

82. There is a function with $\|$ grad $f\| = 5$ and $f_{\vec{k}} = -3$ at some point.

14.6   THE CHAIN RULE

Composition of Functions of Many Variables and Rates of Change

The chain rule enables us to differentiate *composite functions*. If we have a function of two variables $z = f(x, y)$ and we substitute $x = g(t)$, $y = h(t)$ into $z = f(x, y)$, then we have a composite function in which z is a function of t:

$$z = f(g(t), h(t)).$$

If, on the other hand, we substitute $x = g(u, v)$, $y = h(u, v)$, then we have a different composite function in which z is a function of u and v:

$$z = f(g(u, v), h(u, v)).$$

The next example shows how to calculate the rate of change of a composite function.

Example 1   Corn production, C, depends on annual rainfall, R, and average temperature, T, so $C = f(R, T)$. Global warming predicts that both rainfall and temperature depend on time. Suppose that according to a particular model of global warming, rainfall is decreasing at 0.2 cm per year and temperature is increasing at 0.1°C per year. Use the fact that at current levels of production, $f_R = 3.3$ and $f_T = -5$ to estimate the current rate of change, dC/dt.

Solution   By local linearity, we know that changes ΔR and ΔT generate a change, ΔC, in C given approximately by

$$\Delta C \approx f_R \Delta R + f_T \Delta T = 3.3\Delta R - 5\Delta T.$$

We want to know how ΔC depends on the time increment, Δt. A change Δt causes changes ΔR and ΔT, which in turn cause a change ΔC. The model of global warming tells us that

$$\frac{dR}{dt} = -0.2 \quad \text{and} \quad \frac{dT}{dt} = 0.1.$$

Thus, a time increment, Δt, generates changes of ΔR and ΔT given by

$$\Delta R \approx -0.2\Delta t \quad \text{and} \quad \Delta T \approx 0.1\Delta t.$$

Substituting for ΔR and ΔT in the expression for ΔC gives us

$$\Delta C \approx 3.3(-0.2\Delta t) - 5(0.1\Delta t) = -1.16\Delta t.$$

Thus,

$$\frac{\Delta C}{\Delta t} \approx -1.16 \quad \text{and, therefore,} \quad \frac{dC}{dt} \approx -1.16.$$

The relationship between ΔC and Δt, which gives the value of dC/dt, is an example of the *chain rule*. The argument in Example 1 leads to more general versions of the chain rule.

The Chain Rule for $z = f(x, y)$, $x = g(t)$, $y = h(t)$

Since $z = f(g(t), h(t))$ is a function of t, we can consider the derivative dz/dt. The chain rule gives dz/dt in terms of the derivatives of f, g, and h. Since dz/dt represents the rate of change of z with t, we look at the change Δz generated by a small change, Δt.

We substitute the local linearizations

$$\Delta x \approx \frac{dx}{dt} \Delta t \quad \text{and} \quad \Delta y \approx \frac{dy}{dt} \Delta t$$

into the local linearization

$$\Delta z \approx \frac{\partial z}{\partial x} \Delta x + \frac{\partial z}{\partial y} \Delta y,$$

yielding

$$\Delta z \approx \frac{\partial z}{\partial x} \frac{dx}{dt} \Delta t + \frac{\partial z}{\partial y} \frac{dy}{dt} \Delta t$$

$$= \left(\frac{\partial z}{\partial x} \frac{dx}{dt} + \frac{\partial z}{\partial y} \frac{dy}{dt} \right) \Delta t.$$

Thus,

$$\frac{\Delta z}{\Delta t} \approx \frac{\partial z}{\partial x} \frac{dx}{dt} + \frac{\partial z}{\partial y} \frac{dy}{dt}.$$

Taking the limit as $\Delta t \to 0$, we get the following result.

If f, g, and h are differentiable and if $z = f(x, y)$, and $x = g(t)$, and $y = h(t)$, then

$$\frac{dz}{dt} = \frac{\partial z}{\partial x} \frac{dx}{dt} + \frac{\partial z}{\partial y} \frac{dy}{dt}.$$

Visualizing the Chain Rule with a Diagram

The diagram in Figure 14.49 provides a way of remembering the chain rule. It shows the chain of dependence: z depends on x and y, which in turn depend on t. Each line in the diagram is labeled with a derivative relating the variables at its ends.

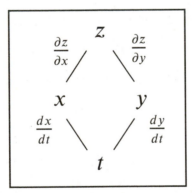

Figure 14.49: Diagram for $z = f(x, y)$, $x = g(t)$, $y = h(t)$. Lines represent dependence of z on x and y, and of x and y on t

The diagram keeps track of how a change in t propagates through the chain of composed functions. There are two paths from t to z, one through x and one through y. For each path, we multiply together the derivatives along the path. Then, to calculate dz/dt, we add the contributions from the two paths.

Example 2   Suppose that $z = f(x, y) = x \sin y$, where $x = t^2$ and $y = 2t + 1$. Let $z = g(t)$. Compute $g'(t)$ directly and using the chain rule.

Solution   Since $z = g(t) = f(t^2, 2t+1) = t^2 \sin(2t+1)$, it is possible to compute $g'(t)$ directly by one-variable methods:

$$g'(t) = t^2 \frac{d}{dt}(\sin(2t+1)) + \left(\frac{d}{dt}(t^2)\right)\sin(2t+1) = 2t^2\cos(2t+1) + 2t\sin(2t+1).$$

The chain rule provides an alternative route to the same answer. We have

$$\frac{dz}{dt} = \frac{\partial z}{\partial x}\frac{dx}{dt} + \frac{\partial z}{\partial y}\frac{dy}{dt} = (\sin y)(2t) + (x\cos y)(2) = 2t\sin(2t+1) + 2t^2\cos(2t+1).$$

Example 3   The capacity, C, of a communication channel, such as a telephone line, to carry information depends on the ratio of the signal strength, S, to the noise, N. For some positive constant k,

$$C = k\ln\left(1 + \frac{S}{N}\right).$$

Suppose that the signal and noise are given as a function of time, t in seconds, by

$$S(t) = 4 + \cos(4\pi t) \qquad N(t) = 2 + \sin(2\pi t).$$

What is dC/dt one second after transmission started? Is the capacity increasing or decreasing at that instant?

Solution   By the chain rule,

$$\frac{dC}{dt} = \frac{\partial C}{\partial S}\frac{dS}{dt} + \frac{\partial C}{\partial N}\frac{dN}{dt}$$

$$= \frac{k}{1 + S/N}\cdot\frac{1}{N}(-4\pi\sin 4\pi t) + \frac{k}{1 + S/N}\left(-\frac{S}{N^2}\right)(2\pi\cos 2\pi t).$$

When $t = 1$, the first term is zero, $S(1) = 5$, and $N(1) = 2$, so

$$\frac{dC}{dt} = \frac{k}{1 + S(1)/N(1)}\left(-\frac{S(1)}{(N(1))^2}\right)\cdot 2\pi = \frac{k}{1 + 5/2}\left(-\frac{5}{4}\right)\cdot 2\pi.$$

Since dC/dt is negative, the capacity is decreasing at time $t = 1$ second.

How to Formulate a General Chain Rule

A diagram can be used to write the chain rule for general compositions.

> To find the rate of change of one variable with respect to another in a chain of composed differentiable functions:
> - Draw a diagram expressing the relationship between the variables, and label each link in the diagram with the derivative relating the variables at its ends.
> - For each path between the two variables, multiply together the derivatives from each step along the path.
> - Add the contributions from each path.

The diagram keeps track of all the ways in which a change in one variable can cause a change in another; the diagram generates all the terms we would get from the appropriate substitutions into the local linearizations.

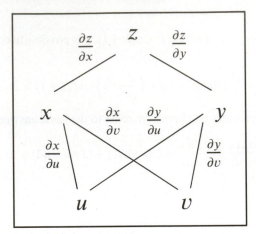

Figure 14.50: Diagram for $z = f(x, y)$, $x = g(u, v)$, $y = h(u, v)$. Lines represent dependence of z on x and y, and of x and y on u and v

For example, we can use Figure 14.50 to find formulas for $\partial z/\partial u$ and $\partial z/\partial v$. Adding the contributions for the two paths from z to u, we get the following results:

If f, g, h are differentiable and if $z = f(x, y)$, with $x = g(u, v)$ and $y = h(u, v)$, then

$$\frac{\partial z}{\partial u} = \frac{\partial z}{\partial x}\frac{\partial x}{\partial u} + \frac{\partial z}{\partial y}\frac{\partial y}{\partial u},$$

$$\frac{\partial z}{\partial v} = \frac{\partial z}{\partial x}\frac{\partial x}{\partial v} + \frac{\partial z}{\partial y}\frac{\partial y}{\partial v}.$$

Example 4   Let $w = x^2 e^y$, $x = 4u$, and $y = 3u^2 - 2v$. Compute $\partial w/\partial u$ and $\partial w/\partial v$ using the chain rule.

Solution   Using the previous result, we have

$$\frac{\partial w}{\partial u} = \frac{\partial w}{\partial x}\frac{\partial x}{\partial u} + \frac{\partial w}{\partial y}\frac{\partial y}{\partial u} = 2xe^y(4) + x^2 e^y(6u) = (8x + 6x^2 u)e^y$$

$$= (32u + 96u^3)e^{3u^2 - 2v}.$$

Similarly,

$$\frac{\partial w}{\partial v} = \frac{\partial w}{\partial x}\frac{\partial x}{\partial v} + \frac{\partial w}{\partial y}\frac{\partial y}{\partial v} = 2xe^y(0) + x^2 e^y(-2) = -2x^2 e^y$$

$$= -32u^2 e^{3u^2 - 2v}.$$

Example 5   A quantity z can be expressed either as a function of x and y, so that $z = f(x, y)$, or as a function of u and v, so that $z = g(u, v)$. The two coordinate systems are related by

$$x = u + v, \quad y = u - v.$$

(a) Use the chain rule to express $\partial z/\partial u$ and $\partial z/\partial v$ in terms of $\partial z/\partial x$ and $\partial z/\partial y$.
(b) Solve the equations in part (a) for $\partial z/\partial x$ and $\partial z/\partial y$.
(c) Show that the expressions we get in part (b) are the same as we get by expressing u and v in terms of x and y and using the chain rule.

Solution

(a) We have $\partial x/\partial u = 1$ and $\partial x/\partial v = 1$, and also $\partial y/\partial u = 1$ and $\partial y/\partial v = -1$. Thus,

$$\frac{\partial z}{\partial u} = \frac{\partial z}{\partial x}(1) + \frac{\partial z}{\partial y}(1) = \frac{\partial z}{\partial x} + \frac{\partial z}{\partial y}$$

and

$$\frac{\partial z}{\partial v} = \frac{\partial z}{\partial x}(1) + \frac{\partial z}{\partial y}(-1) = \frac{\partial z}{\partial x} - \frac{\partial z}{\partial y}.$$

(b) Adding together the equations for $\partial z/\partial u$ and $\partial z/\partial v$, we get

$$\frac{\partial z}{\partial u} + \frac{\partial z}{\partial v} = 2\frac{\partial z}{\partial x}, \quad \text{so} \quad \frac{\partial z}{\partial x} = \frac{1}{2}\frac{\partial z}{\partial u} + \frac{1}{2}\frac{\partial z}{\partial v}.$$

Similarly, subtracting the equations for $\partial z/\partial u$ and $\partial z/\partial v$ yields

$$\frac{\partial z}{\partial y} = \frac{1}{2}\frac{\partial z}{\partial u} - \frac{1}{2}\frac{\partial z}{\partial v}.$$

(c) Alternatively, we can solve the equations

$$x = u + v, \quad y = u - v$$

for u and v, which yields

$$u = \frac{1}{2}x + \frac{1}{2}y, \quad v = \frac{1}{2}x - \frac{1}{2}y.$$

Now we can think of z as a function of u and v, and u and v as functions of x and y, and apply the chain rule again. This gives us

$$\frac{\partial z}{\partial x} = \frac{\partial z}{\partial u}\frac{\partial u}{\partial x} + \frac{\partial z}{\partial v}\frac{\partial v}{\partial x} = \frac{1}{2}\frac{\partial z}{\partial u} + \frac{1}{2}\frac{\partial z}{\partial v}$$

and

$$\frac{\partial z}{\partial y} = \frac{\partial z}{\partial u}\frac{\partial u}{\partial y} + \frac{\partial z}{\partial v}\frac{\partial v}{\partial y} = \frac{1}{2}\frac{\partial z}{\partial u} - \frac{1}{2}\frac{\partial z}{\partial v}.$$

These are the same expressions we got in part (b).

An Application to Physical Chemistry

A chemist investigating the properties of a gas such as carbon dioxide may want to know how the internal energy U of a given quantity of the gas depends on its temperature, T, pressure, P, and volume, V. The three quantities T, P, and V are not independent, however. For instance, according to the ideal gas law, they satisfy the equation

$$PV = kT$$

where k is a constant which depends only upon the quantity of the gas. The internal energy can then be thought of as a function of any two of the three quantities T, P, and V:

$$U = U_1(T, P) = U_2(T, V) = U_3(P, V).$$

The chemist writes, for example, $\left(\frac{\partial U}{\partial T}\right)_P$ to indicate the partial derivative of U with respect to T holding P constant, signifying that for this computation U is viewed as a function of T and P. Thus, we interpret $\left(\frac{\partial U}{\partial T}\right)_P$ as

$$\left(\frac{\partial U}{\partial T}\right)_P = \frac{\partial U_1(T, P)}{\partial T}.$$

If U is to be viewed as a function of T and V, the chemist writes $\left(\frac{\partial U}{\partial T}\right)_V$ for the partial derivative of U with respect to T holding V constant: thus, $\left(\frac{\partial U}{\partial T}\right)_V = \frac{\partial U_2(T,V)}{\partial T}.$

Each of the functions U_1, U_2, U_3 gives rise to one of the following formulas for the differential dU:

$$dU = \left(\frac{\partial U}{\partial T}\right)_P dT + \left(\frac{\partial U}{\partial P}\right)_T dP \qquad \text{corresponds to } U_1,$$

$$dU = \left(\frac{\partial U}{\partial T}\right)_V dT + \left(\frac{\partial U}{\partial V}\right)_T dV \qquad \text{corresponds to } U_2,$$

$$dU = \left(\frac{\partial U}{\partial P}\right)_V dP + \left(\frac{\partial U}{\partial V}\right)_P dV \qquad \text{corresponds to } U_3.$$

All the six partial derivatives appearing in formulas for dU have physical meaning, but they are not all equally easy to measure experimentally. A relationship among the partial derivatives, usually derived from the chain rule, may make it possible to evaluate one of the partials in terms of others that are more easily measured.

Example 6   Suppose a gas satisfies the equation $PV = 2T$ and $P = 3$ when $V = 4$. If $\left(\frac{\partial U}{\partial P}\right)_V = 7$ and $\left(\frac{\partial U}{\partial V}\right)_P = 8$, find the values of $\left(\frac{\partial U}{\partial P}\right)_T$ and $\left(\frac{\partial U}{\partial T}\right)_P$.

Solution   Since we know the values of $\left(\frac{\partial U}{\partial P}\right)_V$ and $\left(\frac{\partial U}{\partial V}\right)_P$, we think of U as a function of P and V and use the function U_3 to write

$$dU = \left(\frac{\partial U}{\partial P}\right)_V dP + \left(\frac{\partial U}{\partial V}\right)_P dV$$
$$dU = 7dP + 8dV.$$

To calculate $\left(\frac{\partial U}{\partial P}\right)_T$ and $\left(\frac{\partial U}{\partial T}\right)_P$, we think of U as a function of T and P. Thus, we want to substitute for dV in terms of dT and dP. Since $PV = 2T$, we have

$$P dV + V dP = 2dT,$$
$$3dV + 4dP = 2dT.$$

Solving gives $dV = (2dT - 4dP)/3$, so

$$dU = 7dP + 8\left(\frac{2dT - 4dP}{3}\right)$$
$$dU = -\frac{11}{3}dP + \frac{16}{3}dT.$$

Comparing with the formula for dU obtained from U_1,

$$dU = \left(\frac{\partial U}{\partial T}\right)_P dT + \left(\frac{\partial U}{\partial P}\right)_T dP,$$

we have

$$\left(\frac{\partial U}{\partial T}\right)_P = \frac{16}{3} \qquad \text{and} \qquad \left(\frac{\partial U}{\partial P}\right)_T = -\frac{11}{3}.$$

In Example 6, we could have substituted for dP instead of dV, leading to values of $\left(\frac{\partial U}{\partial T}\right)_V$ and $\left(\frac{\partial U}{\partial V}\right)_T$. See Problem 41.

In general, if for some particular P, V, and T, we can measure two of the six quantities $\left(\frac{\partial U}{\partial P}\right)_V$, $\left(\frac{\partial U}{\partial V}\right)_P$, $\left(\frac{\partial U}{\partial P}\right)_T$, $\left(\frac{\partial U}{\partial T}\right)_P$, $\left(\frac{\partial U}{\partial V}\right)_T$, $\left(\frac{\partial U}{\partial T}\right)_V$, then we can compute the other four using the relationship between dP, dV, and dT given by the gas law. General formulas for each partial derivative in terms of others can be obtained in the same way. See the following example and Problem 41.

Example 7   Express $\left(\dfrac{\partial U}{\partial T}\right)_P$ in terms of $\left(\dfrac{\partial U}{\partial T}\right)_V$ and $\left(\dfrac{\partial U}{\partial V}\right)_T$ and $\left(\dfrac{\partial V}{\partial T}\right)_P$.

Solution   Since we are interested in the derivatives $\left(\dfrac{\partial U}{\partial T}\right)_V$ and $\left(\dfrac{\partial U}{\partial V}\right)_T$, we think of U as a function of T and V and use the formula

$$dU = \left(\frac{\partial U}{\partial T}\right)_V dT + \left(\frac{\partial U}{\partial V}\right)_T dV \qquad \text{corresponding to } U_2.$$

We want to find a formula for $\left(\dfrac{\partial U}{\partial T}\right)_P$, which means thinking of U as a function of T and P. Thus, we want to substitute for dV. Since V is a function of T and P, we have

$$dV = \left(\frac{\partial V}{\partial T}\right)_P dT + \left(\frac{\partial V}{\partial P}\right)_T dP.$$

Substituting for dV into the formula for dU corresponding to U_2 gives

$$dU = \left(\frac{\partial U}{\partial T}\right)_V dT + \left(\frac{\partial U}{\partial V}\right)_T \left(\left(\frac{\partial V}{\partial T}\right)_P dT + \left(\frac{\partial V}{\partial P}\right)_T dP \right).$$

Collecting the terms containing dT and the terms containing dP gives

$$dU = \left(\left(\frac{\partial U}{\partial T}\right)_V + \left(\frac{\partial U}{\partial V}\right)_T \left(\frac{\partial V}{\partial T}\right)_P \right) dT + \left(\frac{\partial U}{\partial V}\right)_T \left(\frac{\partial V}{\partial P}\right)_T dP.$$

But we also have the formula

$$dU = \left(\frac{\partial U}{\partial T}\right)_P dT + \left(\frac{\partial U}{\partial P}\right)_T dP \qquad \text{corresponding to } U_1.$$

We now have two formulas for dU in terms of dT and dP. The coefficients of dT must be identical, so we conclude

$$\left(\frac{\partial U}{\partial T}\right)_P = \left(\frac{\partial U}{\partial T}\right)_V + \left(\frac{\partial U}{\partial V}\right)_T \left(\frac{\partial V}{\partial T}\right)_P.$$

Example 7 expresses $\left(\dfrac{\partial U}{\partial T}\right)_P$ in terms of three other partial derivatives. Two of them, namely $\left(\dfrac{\partial U}{\partial T}\right)_V$, the constant-volume heat capacity, and $\dfrac{1}{V}\left(\dfrac{\partial V}{\partial T}\right)_P$, the expansion coefficient, can be easily measured experimentally. The third, the internal pressure, $\left(\dfrac{\partial U}{\partial V}\right)_T$, cannot be measured directly but can be related to $\left(\dfrac{\partial P}{\partial T}\right)_V$, which is measurable. Thus, $\left(\dfrac{\partial U}{\partial T}\right)_P$ can be determined indirectly using this identity.

Summary for Section 14.6

- The **chain rule** is a method for writing the partial derivatives of composite functions in terms of the individual functions' partials.
- If f, g, and h are differentiable and if $z = f(x, y)$, and $x = g(t)$, and $y = h(t)$, then

$$\frac{dz}{dt} = \frac{\partial z}{\partial x}\frac{dx}{dt} + \frac{\partial z}{\partial y}\frac{dy}{dt}.$$

- The chain rule can be visualized using a **diagram**:

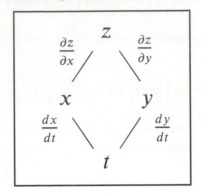

- If f, g, h are differentiable and if $z = f(x, y)$, with $x = g(u, v)$ and $y = h(u, v)$, then we get the **diagram**

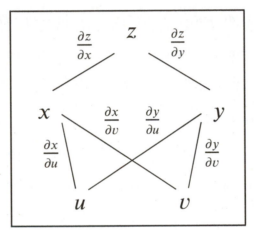

and from this we can read off the chain rule for the various partials:

$$\frac{\partial z}{\partial u} = \frac{\partial z}{\partial x}\frac{\partial x}{\partial u} + \frac{\partial z}{\partial y}\frac{\partial y}{\partial u},$$

$$\frac{\partial z}{\partial v} = \frac{\partial z}{\partial x}\frac{\partial x}{\partial v} + \frac{\partial z}{\partial y}\frac{\partial y}{\partial v}.$$

Exercises and Problems for Section 14.6   Online Resource: Additional Problems for Section 14.6
EXERCISES

For Exercises 1–6, find dz/dt using the chain rule. Assume the variables are restricted to domains on which the functions are defined.

1. $z = xy^2$, $x = e^{-t}$, $y = \sin t$

2. $z = x \sin y + y \sin x$, $x = t^2$, $y = \ln t$

3. $z = \sin(x/y)$, $x = 2t$, $y = 1 - t^2$

4. $z = \ln(x^2 + y^2)$, $x = 1/t$, $y = \sqrt{t}$

5. $z = xe^y$, $x = 2t$, $y = 1 - t^2$

6. $z = (x + y)e^y$, $x = 2t$, $y = 1 - t^2$

For Exercises 7–15, find $\partial z/\partial u$ and $\partial z/\partial v$. The variables are restricted to domains on which the functions are defined.

7. $z = \sin(x/y)$, $x = \ln u$, $y = v$

8. $z = \ln(xy)$, $x = (u^2 + v^2)^2$, $y = (u^3 + v^3)^2$

9. $z = xe^y$, $x = \ln u$, $y = v$

10. $z = (x + y)e^y$, $x = \ln u$, $y = v$

11. $z = xe^y$, $x = u^2 + v^2$, $y = u^2 - v^2$

12. $z = (x + y)e^y$, $x = u^2 + v^2$, $y = u^2 - v^2$

13. $z = xe^{-y} + ye^{-x}$, $x = u \sin v$, $y = v \cos u$

14. $z = \cos(x^2 + y^2)$, $x = u \cos v$, $y = u \sin v$

15. $z = \tan^{-1}(x/y)$, $x = u^2 + v^2$, $y = u^2 - v^2$

PROBLEMS

16. Use the chain rule to find dz/dt, and check the result by expressing z as a function of t and differentiating directly.
$$z = x^3 y^2, \quad x = t^3, \quad y = t^2$$

17. Use the chain rule to find $\partial w/\partial \rho$ and $\partial w/\partial \theta$, given that
$$w = x^2 + y^2 - z^2,$$
and
$$x = \rho \sin \phi \cos \theta, \quad y = \rho \sin \phi \sin \theta, \quad z = \rho \cos \phi.$$

18. Let $z = f(x, y)$ where $x = g(t)$, $y = h(t)$ and f, g, h are all differentiable functions. Given the information in the table, find $\dfrac{\partial z}{\partial t}\Big|_{t=1}$.

$f(3, 10) = 7$	$f(4, 11) = -20$
$f_x(3, 10) = 100$	$f_y(3, 10) = 0.1$
$f_x(4, 11) = 200$	$f_y(4, 11) = 0.2$
$f(3, 4) = -10$	$f(10, 11) = -1$
$g(1) = 3$	$h(1) = 10$
$g'(1) = 4$	$h'(1) = 11$

19. A bison is charging across the plain one morning. His path takes him to location (x, y) at time t where x and y are functions of t and north is in the direction of increasing y. The temperature is always colder farther north. As time passes, the sun rises in the sky, sending out more heat, and a cold front blows in from the east. At time t the air temperature H near the bison is given by $H = f(x, y, t)$. The chain rule expresses the derivative dH/dt as a sum of three terms:
$$\frac{dH}{dt} = \frac{\partial f}{\partial x}\frac{dx}{dt} + \frac{\partial f}{\partial y}\frac{dy}{dt} + \frac{\partial f}{\partial t}.$$

Identify the term that gives the contribution to the change in temperature experienced by the bison that is due to

   (a) The rising sun.
   (b) The coming cold front.
   (c) The bison's change in latitude.

20. The voltage, V (in volts), across a circuit is given by Ohm's law: $V = IR$, where I is the current (in amps) flowing through the circuit and R is the resistance (in ohms). If we place two circuits, with resistance R_1 and R_2, in parallel, then their combined resistance, R, is given by
$$\frac{1}{R} = \frac{1}{R_1} + \frac{1}{R_2}.$$

Suppose the current is 2 amps and increasing at 10^{-2} amp/sec and R_1 is 3 ohms and increasing at 0.5 ohm/sec, while R_2 is 5 ohms and decreasing at 0.1 ohm/sec. Calculate the rate at which the voltage is changing.

21. The air pressure is decreasing at a rate of 2 pascals per kilometer in the eastward direction. In addition, the air pressure is dropping at a constant rate with respect to time everywhere. A ship sailing eastward at 10 km/hour past an island takes barometer readings and records a pressure drop of 50 pascals in 2 hours. Estimate the time rate of change of air pressure on the island. (A pascal is a unit of air pressure.)

22. A steel bar with square cross sections 5 cm by 5 cm and length 3 meters is being heated. For each dimension, the bar expands $13 \cdot 10^{-6}$ meters for each 1°C rise in temperature.[2] What is the rate of change in the volume of the steel bar?

23. Corn production, C, is a function of rainfall, R, and temperature, T. (See Example 1 on page 827.) Figures 14.51 and 14.52 show how rainfall and temperature are predicted to vary with time because of global warming. Suppose we know that $\Delta C \approx 3.3\Delta R - 5\Delta T$. Use this to estimate the change in corn production between the year 2020 and the year 2021. Hence, estimate dC/dt when $t = 2020$.

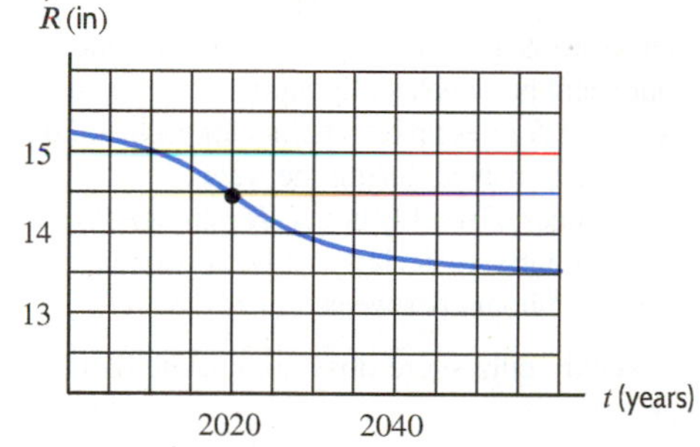

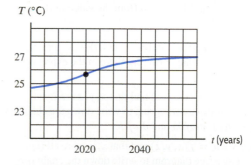

Figure 14.51: Rainfall as a function of time

Figure 14.52: Temperature as a function of time

24. At a point x miles east and y miles north of a campground, the height above sea level is $f(x, y)$ feet. Let
$$\nabla f(10, 2) = -2\vec{i} + \vec{j} \quad \text{and} \quad \vec{v} = \vec{i} + 3\vec{j}.$$

Find the following quantities, including units, as you leave a point 10 miles east and 2 miles north of the campground.

   (a) The slope of the land in the direction of $\vec{v}$.

[2] www.engineeringtoolbox.com, accessed January 2, 2020.

(b) Your vertical speed if you move in the direction of $\vec{v}$ at a speed of 2 miles per hour.

(c) Your vertical speed if you move east at a speed of 2 miles per hour.

25. The function $g(x, y)$ gives the temperature, in degrees Fahrenheit, x miles east and y miles north of a camp-ground. Let $\vec{u}$ be the unit vector in the direction of $\vec{v} = \vec{i} + 3\vec{j}$ and

$$\nabla g(-1, -3) = \vec{i}.$$

A camper located one mile to the west and three miles to the south of the camp starts walking back to camp in the direction $\vec{u}$ at a speed of 2.5 miles/hr. Find the value of the following expressions, and interpret each in everyday terms for the camper.

(a) $g_{\vec{u}}(-1, -3)$ **(b)** $2.5g_{\vec{u}}(-1, -3)$

(c) $2.5g_{\vec{i}}(-1, -3)$

26. Mina's score on her weekly multivariable calculus quiz, S, in points, is a function of the number of hours, H, she spends studying the course materials and the num-ber of problems, P, she solves per week.

- Her score S goes up 2 points for each additional hour spent per week studying the course materials.
- Her score S goes up by 3 points for each additional 5 problems solved during the week.
- The number of weekly hours, H, she spends study-ing the course materials has been decreasing at a rate of 1.5 hours per week.

Mina's weekly quiz score does not change from week to week.

(a) Find the value of dP/dt, where t is time in weeks, include units.

(b) What can Mina learn from the value of the deriva-tive in part (a)?

27. Let $z = g(u, v, w)$ and $u = u(s, t), v = v(s, t), w = w(s, t)$. How many terms are there in the expression for $\partial z/\partial t$?

28. Suppose $w = f(x, y, z)$ and that x, y, z are functions of u and v. Use a tree diagram to write down the chain rule formula for $\partial w/\partial u$ and $\partial w/\partial v$.

29. Suppose $w = f(x, y, z)$ and that x, y, z are all functions of t. Use a tree diagram to write down the chain rule for dw/dt.

30. Let $z = f(t)g(t)$. Use the chain rule applied to $h(x, y) = f(x)g(y)$ to show that $dz/dt = f'(t)g(t) + f(t)g'(t)$. The one-variable product rule for differentiation is a special case of the two-variable chain rule.

31. Let $F(u, v)$ be a function of two variables. Find $f'(x)$ if

(a) $f(x) = F(x, 3)$ **(b)** $f(x) = F(3, x)$

(c) $f(x) = F(x, x)$ **(d)** $f(x) = F(5x, x^2)$

32. The function $g(\rho)$ is graphed in Figure 14.53. Let $\rho = \sqrt{x^2 + y^2 + z^2}$. Define f, a function of x, y, z by $f(x, y, z) = g\left(\sqrt{x^2 + y^2 + z^2}\right)$. Let $\vec{F} = \text{grad } f$.

(a) Describe precisely in words the level surfaces of f.

(b) Give a unit vector in the direction of $\vec{F}$ at the point $(1, 2, 2)$.

(c) Estimate $\|\vec{F}\|$ at the point $(1, 2, 2)$.

(d) Estimate $\vec{F}$ at the point $(1, 2, 2)$.

(e) The points $(1, 2, 2)$ and $(3, 0, 0)$ are both on the sphere $x^2 + y^2 + z^2 = 9$. Estimate $\vec{F}$ at $(3, 0, 0)$.

(f) If P and Q are any two points on the sphere $x^2 + y^2 + z^2 = k^2$:

　(i) Compare the magnitudes of $\vec{F}$ at P and at Q.

　(ii) Describe the directions of $\vec{F}$ at P and at Q.

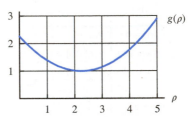

Figure 14.53

■In Problems **33–34**, let $z = f(x, y), x = x(u, v), y = y(u, v)$ and $x(1, 2) = 5, y(1, 2) = 3$, calculate the partial derivative in terms of some of the numbers a, b, c, d, e, k, p, q:

$$f_x(1, 2) = a \quad f_y(1, 2) = c \quad x_u(1, 2) = e \quad y_u(1, 2) = p$$
$$f_x(5, 3) = b \quad f_y(5, 3) = d \quad x_v(1, 2) = k \quad y_v(1, 2) = q$$

33. $z_u(1, 2)$ **34.** $z_v(1, 2)$

■In Problems **35–36**, let $z = f(x, y), x = x(u, v), y = y(u, v)$ and $x(4, 5) = 2, y(4, 5) = 3$. Calculate the partial derivative in terms of $a, b, c, d, e, k, p, q, r, s, t, w$:

$$f_x(4, 5) = a \quad f_y(4, 5) = c \quad x_u(4, 5) = e \quad y_u(4, 5) = p$$
$$f_x(2, 3) = b \quad f_y(2, 3) = d \quad x_v(4, 5) = k \quad y_v(4, 5) = q$$
$$x_u(2, 3) = r \quad y_u(2, 3) = s \quad x_v(2, 3) = t \quad y_v(2, 3) = w$$

35. $z_u(4, 5)$ **36.** $z_v(4, 5)$

■For Problems **37–38**, suppose that $x > 0, y > 0$ and that z can be expressed either as a function of Cartesian coor-dinates (x, y) or as a function of polar coordinates (r, θ), so that $z = f(x, y) = g(r, \theta)$. [Recall that $x = r\cos\theta, y = r\sin\theta, r = \sqrt{x^2 + y^2}$, and, for $x > 0, y > 0, \theta = \arctan(y/x)$.]

37. (a) Use the chain rule to find $\partial z/\partial r$ and $\partial z/\partial\theta$ in terms of $\partial z/\partial x$ and $\partial z/\partial y$.

(b) Solve the equations you have just written down for $\partial z/\partial x$ and $\partial z/\partial y$ in terms of $\partial z/\partial r$ and $\partial z/\partial\theta$.

(c) Show that the expressions you get in part (b) are the same as you would get by using the chain rule to find $\partial z/\partial x$ and $\partial z/\partial y$ in terms of $\partial z/\partial r$ and $\partial z/\partial\theta$.

38. Show that

$$\left(\frac{\partial z}{\partial x}\right)^2 + \left(\frac{\partial z}{\partial y}\right)^2 = \left(\frac{\partial z}{\partial r}\right)^2 + \frac{1}{r^2}\left(\frac{\partial z}{\partial \theta}\right)^2.$$

■ Problems **39–44** are continuations of the physical chemistry example on page 833.

39. Write $\left(\frac{\partial U}{\partial P}\right)_V$ as a partial derivative of one of the functions U_1, U_2, or U_3.

40. Write $\left(\frac{\partial U}{\partial P}\right)_T$ as a partial derivative of one of the functions U_1, U_2, U_3.

41. For the gas in Example 6, find $\left(\frac{\partial U}{\partial T}\right)_V$ and $\left(\frac{\partial U}{\partial V}\right)_T$. [Hint: Use the same method as the example, but substitute for dP instead of dV.]

42. Show that $\left(\frac{\partial T}{\partial V}\right)_P = 1 \bigg/ \left(\frac{\partial V}{\partial T}\right)_P$.

43. Use Example 7 and Problem 42 to show that

$$\left(\frac{\partial U}{\partial V}\right)_P = \left(\frac{\partial U}{\partial V}\right)_T + \frac{\left(\frac{\partial U}{\partial T}\right)_V}{\left(\frac{\partial V}{\partial T}\right)_P}.$$

Strengthen Your Understanding

■ In Problems **45–47**, explain what is wrong with the statement.

45. If $z = f(g(t), h(t))$, then $dz/dt = f(g'(t), h(t)) + f(g(t), h'(t))$.

46. If $C = C(R, T), R = R(x, y), T = T(x, y)$ and $R(0, 2) = 5$, $T(0, 2) = 1$, then $C_x(0, 2) = C_R(0, 2)R_x(0, 2) + C_T(0, 2)T_x(0, 2)$.

47. If $z = f(x, y)$ and $x = g(t), y = h(t)$ with $g(0) = 2$ and $h(0) = 3$, then

$$\frac{dz}{dt}\bigg|_{t=0} = f_x(0, 0)g'(0) + f_y(0, 0)h'(0).$$

■ In Problems **48–52**, give an example of:

48. Functions $x = g(t)$ and $y = h(t)$ such that $(dz/dt)|_{t=0} = 9$, given that $z = x^2 y$.

49. A function $z = f(x, y)$ such that $dz/dt|_{t=0} = 10$, given that $x = e^{2t}$ and $y = \sin t$.

50. Functions z, x and y where you need to follow the diagram in order to answer questions about the derivative of z with respect to the other variables.

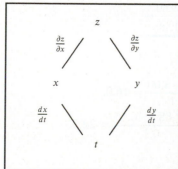

44. In Example 6, we calculated values of $(\partial U/\partial T)_P$ and $(\partial U/\partial P)_T$ using the relationship $PV = 2T$ for a specific gas. In this problem, you will derive general relationships for these two partial derivatives.

(a) Think of V as a function of P and T and write an expression for dV.

(b) Substitute for dV into the following formula for dU (thinking of U as a function of P and V):

$$dU = \left(\frac{\partial U}{\partial P}\right)_V dP + \left(\frac{\partial U}{\partial V}\right)_P dV.$$

(c) Thinking of U as a function of P and T, write an expression for dU.

(d) By comparing coefficients of dP and dT in your answers to parts (b) and (c), show that

$$\left(\frac{\partial U}{\partial T}\right)_P = \left(\frac{\partial U}{\partial V}\right)_P \cdot \left(\frac{\partial V}{\partial T}\right)_P$$
$$\left(\frac{\partial U}{\partial P}\right)_T = \left(\frac{\partial U}{\partial P}\right)_V + \left(\frac{\partial U}{\partial V}\right)_P \cdot \left(\frac{\partial V}{\partial P}\right)_T.$$

51. Functions w, u and v where you need to follow the diagram in order to answer questions about the derivative of w with respect to the other variables.

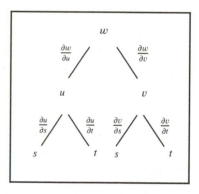

52. Function $z = f(x, y)$ where x and y are functions of one variable, t, for which $\frac{\partial z}{\partial t} = 2$.

53. Let $z = g(u, v)$ and $u = u(x, y, t), v = v(x, y, t)$ and $x = x(t), y = y(t)$. Then the expression for dz/dt has

(a) Three terms   **(b)** Four terms

(c) Six terms   **(d)** Seven terms

(e) Nine terms   **(f)** None of the above

14.7 SECOND-ORDER PARTIAL DERIVATIVES

Since the partial derivatives of a function are themselves functions, we can differentiate them, giving *second-order partial derivatives*. A function $z = f(x, y)$ has two first-order partial derivatives, f_x and f_y, and four second-order partial derivatives.

> **The Second-Order Partial Derivatives of $z = f(x, y)$**
>
> $$\frac{\partial^2 z}{\partial x^2} = f_{xx} = (f_x)_x, \qquad \frac{\partial^2 z}{\partial x \partial y} = f_{yx} = (f_y)_x,$$
>
> $$\frac{\partial^2 z}{\partial y \partial x} = f_{xy} = (f_x)_y, \qquad \frac{\partial^2 z}{\partial y^2} = f_{yy} = (f_y)_y.$$

It is usual to omit the parentheses, writing f_{xy} instead of $(f_x)_y$ and $\dfrac{\partial^2 z}{\partial y \partial x}$ instead of $\dfrac{\partial}{\partial y}\left(\dfrac{\partial z}{\partial x}\right)$.

Example 1 Compute the four second-order partial derivatives of $f(x, y) = xy^2 + 3x^2 e^y$.

Solution From $f_x(x, y) = y^2 + 6xe^y$ we get

$$f_{xx}(x, y) = \frac{\partial}{\partial x}(y^2 + 6xe^y) = 6e^y \quad \text{and} \quad f_{xy}(x, y) = \frac{\partial}{\partial y}(y^2 + 6xe^y) = 2y + 6xe^y.$$

From $f_y(x, y) = 2xy + 3x^2 e^y$ we get

$$f_{yx}(x, y) = \frac{\partial}{\partial x}(2xy + 3x^2 e^y) = 2y + 6xe^y \quad \text{and} \quad f_{yy}(x, y) = \frac{\partial}{\partial y}(2xy + 3x^2 e^y) = 2x + 3x^2 e^y.$$

Observe that $f_{xy} = f_{yx}$ in this example.

Example 2 Use the values of the function $f(x, y)$ in Table 14.7 to estimate $f_{xy}(1, 2)$ and $f_{yx}(1, 2)$.

Table 14.7 *Values of $f(x, y)$*

$y \backslash x$	0.9	1.0	1.1
1.8	4.72	5.83	7.06
2.0	6.48	8.00	9.60
2.2	8.62	10.65	12.88

Solution Since $f_{xy} = (f_x)_y$, we first estimate f_x

$$f_x(1, 2) \approx \frac{f(1.1, 2) - f(1, 2)}{0.1} = \frac{9.60 - 8.00}{0.1} = 16.0,$$

$$f_x(1, 2.2) \approx \frac{f(1.1, 2.2) - f(1, 2.2)}{0.1} = \frac{12.88 - 10.65}{0.1} = 22.3.$$

Thus,

$$f_{xy}(1,2) \approx \frac{f_x(1,2.2) - f_x(1,2)}{0.2} = \frac{22.3 - 16.0}{0.2} = 31.5.$$

Similarly,

$$f_{yx}(1,2) \approx \frac{f_y(1.1,2) - f_y(1,2)}{0.1} \approx \frac{1}{0.1}\left(\frac{f(1.1,2.2) - f(1.1,2)}{0.2} - \frac{f(1,2.2) - f(1,2)}{0.2}\right)$$

$$= \frac{1}{0.1}\left(\frac{12.88 - 9.60}{0.2} - \frac{10.65 - 8.00}{0.2}\right) = 31.5.$$

Observe that in this example also, $f_{xy} = f_{yx}$.

The Mixed Partial Derivatives Are Equal

It is not an accident that the estimates for $f_{xy}(1,2)$ and $f_{yx}(1,2)$ are equal in Example 2, because the same values of the function are used to calculate each one. The fact that $f_{xy} = f_{yx}$ in Examples 1 and 2 corroborates the following general result; Problem 73 (available online) suggests why you might expect it to be true.[3]

> ### Theorem 14.1: Equality of Mixed Partial Derivatives
>
> If f_{xy} and f_{yx} are continuous at (a, b), an interior point of their domain, then
>
> $$f_{xy}(a, b) = f_{yx}(a, b).$$

For most functions f we encounter and most points (a, b) in their domains, not only are f_{xy} and f_{yx} continuous at (a, b), but all their higher-order partial derivatives (such as f_{xxy} or f_{xyyy}) exist and are continuous at (a, b). In that case we say f is *smooth* at (a, b). We say f is smooth on a region R if it is smooth at every point of R.

What Do the Second-Order Partial Derivatives Tell Us?

Example 3   Let us return to the guitar string of Example 4, page 796. The string is 1 meter long and at time t seconds, the point x meters from one end is displaced $f(x, t)$ meters from its rest position, where

$$f(x, t) = 0.003 \sin(\pi x) \sin(2765t).$$

Compute the four second-order partial derivatives of f at the point $(x, t) = (0.3, 1)$ and describe the meaning of their signs in practical terms.

Solution   First we compute $f_x(x, t) = 0.003\pi \cos(\pi x) \sin(2765t)$, from which we get

$$f_{xx}(x, t) = \frac{\partial}{\partial x}(f_x(x, t)) = -0.003\pi^2 \sin(\pi x) \sin(2765t), \qquad \text{so} \qquad f_{xx}(0.3, 1) \approx -0.01;$$

and

$$f_{xt}(x, t) = \frac{\partial}{\partial t}(f_x(x, t)) = (0.003)(2765)\pi \cos(\pi x) \cos(2765t), \qquad \text{so} \qquad f_{xt}(0.3, 1) \approx 14.$$

On page 796 we saw that $f_x(x, t)$ gives the slope of the string at any point and time. Therefore,

---

[3] For a proof, see M. Spivak, *Calculus on Manifolds*, p. 26 (New York: Benjamin, 1965).

$f_{xx}(x,t)$ measures the concavity of the string. The fact that $f_{xx}(0.3,1) < 0$ means the string is concave down at the point $x = 0.3$ when $t = 1$. (See Figure 14.54.)

On the other hand, $f_{xt}(x,t)$ is the rate of change of the slope of the string with respect to time. Thus, $f_{xt}(0.3,1) > 0$ means that at time $t = 1$ the slope at the point $x = 0.3$ is increasing. (See Figure 14.55.)

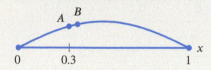

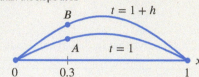

Figure 14.54: Interpretation of $f_{xx}(0.3,1) < 0$: The concavity of the string at $t = 1$

Figure 14.55: Interpretation of $f_{xt}(0.3,1) > 0$: The slope of one point on the string at two different times

Now we compute $f_t(x,t) = (0.003)(2765)\sin(\pi x)\cos(2765t)$, from which we get

$$f_{tx}(x,t) = \frac{\partial}{\partial x}(f_t(x,t)) = (0.003)(2765)\pi\cos(\pi x)\cos(2765t), \quad \text{so} \quad f_{tx}(0.3,1) \approx 14$$

and

$$f_{tt}(x,t) = \frac{\partial}{\partial t}(f_t(x,t)) = -(0.003)(2765)^2\sin(\pi x)\sin(2765t), \quad \text{so} \quad f_{tt}(0.3,1) \approx -7200.$$

On page 796 we saw that $f_t(x,t)$ gives the velocity of the string at any point and time. Therefore, $f_{tx}(x,t)$ and $f_{tt}(x,t)$ will both be rates of change of velocity. That $f_{tx}(0.3,1) > 0$ means that at time $t = 1$ the velocities of points just to the right of $x = 0.3$ are greater than the velocity at $x = 0.3$. (See Figure 14.56.) That $f_{tt}(0.3,1) < 0$ means that the velocity of the point $x = 0.3$ is decreasing at time $t = 1$. Thus, $f_{tt}(0.3,1) = -7200$ m/sec^2 is the acceleration of this point. (See Figure 14.57.)

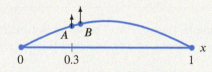

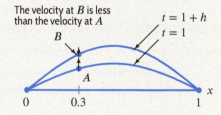

Figure 14.56: Interpretation of $f_{tx}(0.3,1) > 0$: The velocity of different points on the string at $t = 1$

Figure 14.57: Interpretation of $f_{tt}(0.3,1) < 0$: Negative acceleration. The velocity of one point on the string at two different times

Taylor Approximations

We use second derivatives to construct quadratic Taylor approximations. In Section 14.3, we saw how to approximate $f(x,y)$ by a linear function (its local linearization). We now see how to improve this approximation of $f(x,y)$ using a quadratic function.

Linear and Quadratic Approximations Near (0,0)

For a function of one variable, local linearity tells us that the best *linear* approximation is the degree-1 Taylor polynomial

$$f(x) \approx f(a) + f'(a)(x - a) \quad \text{for } x \text{ near } a.$$

A better approximation to $f(x)$ is given by the degree-2 Taylor polynomial:

$$f(x) \approx f(a) + f'(a)(x - a) + \frac{f''(a)}{2}(x - a)^2 \quad \text{for } x \text{ near } a.$$

For a function of two variables the local linearization for (x, y) near (a, b) is

$$f(x, y) \approx L(x, y) = f(a, b) + f_x(a, b)(x - a) + f_y(a, b)(y - b).$$

In the case $(a, b) = (0, 0)$, we have:

> **Taylor Polynomial of Degree 1 Approximating $f(x, y)$ for (x, y) Near (0,0)**
> If f has continuous first-order partial derivatives, then
>
> $$f(x, y) \approx L(x, y) = f(0, 0) + f_x(0, 0)x + f_y(0, 0)y.$$

We get a better approximation to f by using a quadratic polynomial. We choose a quadratic polynomial $Q(x, y)$, with the same partial derivatives as the original function f. You can check that the following Taylor polynomial of degree 2 has this property.

> **Taylor Polynomial of Degree 2 Approximating $f(x, y)$ for (x, y) Near (0,0)**
> If f has continuous second-order partial derivatives, then
>
> $$f(x, y) \approx Q(x, y)$$
> $$= f(0, 0) + f_x(0, 0)x + f_y(0, 0)y + \frac{f_{xx}(0, 0)}{2}x^2 + f_{xy}(0, 0)xy + \frac{f_{yy}(0, 0)}{2}y^2.$$

Example 4   Let $f(x, y) = \cos(2x + y) + 3\sin(x + y)$

(a) Compute the linear and quadratic Taylor polynomials, L and Q, approximating f near $(0, 0)$.
(b) Explain why the contour plots of L and Q for $-1 \le x \le 1$, $-1 \le y \le 1$ look the way they do.

Solution   (a) We have $f(0, 0) = 1$. The derivatives we need are as follows:

$$f_x(x, y) = -2\sin(2x + y) + 3\cos(x + y) \quad \text{so} \quad f_x(0, 0) = 3,$$
$$f_y(x, y) = -\sin(2x + y) + 3\cos(x + y) \quad \text{so} \quad f_y(0, 0) = 3,$$
$$f_{xx}(x, y) = -4\cos(2x + y) - 3\sin(x + y) \quad \text{so} \quad f_{xx}(0, 0) = -4,$$
$$f_{xy}(x, y) = -2\cos(2x + y) - 3\sin(x + y) \quad \text{so} \quad f_{xy}(0, 0) = -2,$$
$$f_{yy}(x, y) = -\cos(2x + y) - 3\sin(x + y) \quad \text{so} \quad f_{yy}(0, 0) = -1.$$

Thus, the linear approximation, $L(x, y)$, to $f(x, y)$ at $(0, 0)$ is given by

$$f(x, y) \approx L(x, y) = f(0, 0) + f_x(0, 0)x + f_y(0, 0)y = 1 + 3x + 3y.$$

The quadratic approximation, $Q(x, y)$, to $f(x, y)$ near $(0, 0)$ is given by

$$f(x, y) \approx Q(x, y)$$
$$= f(0, 0) + f_x(0, 0)x + f_y(0, 0)y + \frac{f_{xx}(0, 0)}{2}x^2 + f_{xy}(0, 0)xy + \frac{f_{yy}(0, 0)}{2}y^2$$
$$= 1 + 3x + 3y - 2x^2 - 2xy - \frac{1}{2}y^2.$$

Notice that the linear terms in $Q(x, y)$ are the same as the linear terms in $L(x, y)$. The quadratic terms in $Q(x, y)$ can be thought of as "correction terms" to the linear approximation.

(b) The contour plots of $f(x, y)$, $L(x, y)$, and $Q(x, y)$ are in Figures 14.58–14.60.

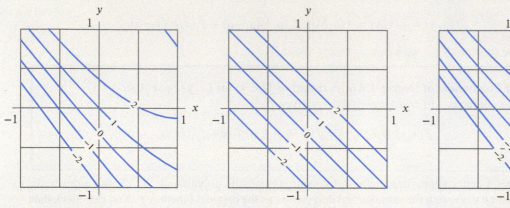

Figure 14.58: Original function, $f(x, y)$ **Figure 14.59**: Linear approximation, $L(x, y)$ **Figure 14.60**: Quadratic approximation, $Q(x, y)$

Notice that the contour plot of Q is more similar to the contour plot of f than is the contour plot of L. Since L is linear, the contour plot of L consists of parallel, equally spaced lines.

An alternative, and much quicker, way to find the Taylor polynomial in the previous example is to use the single-variable approximations. For example, since

$$\cos u = 1 - \frac{u^2}{2!} + \frac{u^4}{4!} + \cdots \quad \text{and} \quad \sin v = v - \frac{v^3}{3!} + \cdots,$$

we can substitute $u = 2x + y$ and $v = x + y$ and expand. We discard terms beyond the second (since we want the quadratic polynomial), getting

$$\cos(2x + y) = 1 - \frac{(2x + y)^2}{2!} + \frac{(2x + y)^4}{4!} + \cdots \approx 1 - \frac{1}{2}(4x^2 + 4xy + y^2) = 1 - 2x^2 - 2xy - \frac{1}{2}y^2$$

and

$$\sin(x + y) = (x + y) - \frac{(x + y)^3}{3!} + \cdots \approx x + y.$$

Combining these results, we get

$$\cos(2x + y) + 3\sin(x + y) \approx 1 - 2x^2 - 2xy - \frac{1}{2}y^2 + 3(x + y) = 1 + 3x + 3y - 2x^2 - 2xy - \frac{1}{2}y^2.$$

Linear and Quadratic Approximations Near (a, b)

The local linearization for a function $f(x, y)$ at a point (a, b) is

> **Taylor Polynomial of Degree 1 Approximating $f(x, y)$ for (x, y) Near (a, b)**
> If f has continuous first-order partial derivatives, then
> $$f(x, y) \approx L(x, y) = f(a, b) + f_x(a, b)(x - a) + f_y(a, b)(y - b).$$

This suggests that a quadratic polynomial approximation $Q(x, y)$ for $f(x, y)$ near a point (a, b) should be written in terms of $(x - a)$ and $(y - b)$ instead of x and y. If we require that $Q(a, b) = f(a, b)$ and that the first- and second-order partial derivatives of Q and f at (a, b) be equal, then we get the following polynomial:

> **Taylor Polynomial of Degree 2 Approximating $f(x, y)$ for (x, y) Near (a, b)**
> If f has continuous second-order partial derivatives, then
>
> $$f(x, y) \approx Q(x, y)$$
> $$= f(a, b) + f_x(a, b)(x - a) + f_y(a, b)(y - b)$$
> $$+ \frac{f_{xx}(a, b)}{2}(x - a)^2 + f_{xy}(a, b)(x - a)(y - b) + \frac{f_{yy}(a, b)}{2}(y - b)^2.$$

These coefficients are derived in exactly the same way as for $(a, b) = (0, 0)$.

Example 5   Find the Taylor polynomial of degree 2 at the point $(1, 2)$ for the function $f(x, y) = \dfrac{1}{xy}$.

Solution   Table 14.8 contains the partial derivatives and their values at the point $(1, 2)$.

Table 14.8   *Partial derivatives of $f(x, y) = 1/(xy)$*

Derivative	Formula	Value at $(1, 2)$	Derivative	Formula	Value at $(1, 2)$
$f(x, y)$	$1/(xy)$	$1/2$	$f_{xx}(x, y)$	$2/(x^3 y)$	1
$f_x(x, y)$	$-1/(x^2 y)$	$-1/2$	$f_{xy}(x, y)$	$1/(x^2 y^2)$	$1/4$
$f_y(x, y)$	$-1/(xy^2)$	$-1/4$	$f_{yy}(x, y)$	$2/(xy^3)$	$1/4$

So, the quadratic Taylor polynomial for f near $(1, 2)$ is

$$\frac{1}{xy} \approx Q(x, y)$$

$$= \frac{1}{2} - \frac{1}{2}(x - 1) - \frac{1}{4}(y - 2) + \frac{1}{2}(1)(x - 1)^2 + \frac{1}{4}(x - 1)(y - 2) + \left(\frac{1}{2}\right)\left(\frac{1}{4}\right)(y - 2)^2$$

$$= \frac{1}{2} - \frac{x - 1}{2} - \frac{y - 2}{4} + \frac{(x - 1)^2}{2} + \frac{(x - 1)(y - 2)}{4} + \frac{(y - 2)^2}{8}.$$

Summary for Section 14.7

- **Second-order partial derivatives** are partial derivatives of partial derivatives.
- A function $z = f(x, y)$ has four **second-order partial derivatives**:

$$\frac{\partial^2 z}{\partial x^2} = f_{xx} = (f_x)_x, \qquad \frac{\partial^2 z}{\partial x \partial y} = f_{yx} = (f_y)_x,$$

$$\frac{\partial^2 z}{\partial y \partial x} = f_{xy} = (f_x)_y, \qquad \frac{\partial^2 z}{\partial y^2} = f_{yy} = (f_y)_y.$$

- The **mixed partials are equal**:
  If f_{xy} and f_{yx} are continuous at (a, b), an interior point of their domain, then

$$f_{xy}(a, b) = f_{yx}(a, b).$$

- If f has continuous first-order partial derivatives, then the **Taylor polynomial of degree** 1 approximating $f(x, y)$ for (x, y) near (a, b) is:

$$f(x, y) \approx L(x, y) = f(a, b) + f_x(a, b)(x - a) + f_y(a, b)(y - b).$$

- If f has continuous second-order partial derivatives, then the **Taylor polynomial of degree** 2 approximating $f(x, y)$ for (x, y) near (a, b) is:

$$f(x, y) \approx Q(x, y)$$
$$= f(a, b) + f_x(a, b)(x - a) + f_y(a, b)(y - b)$$
$$+ \frac{f_{xx}(a, b)}{2}(x - a)^2 + f_{xy}(a, b)(x - a)(y - b) + \frac{f_{yy}(a, b)}{2}(y - b)^2.$$

Exercises and Problems for Section 14.7   Online Resource: Additional Problems for Section 14.7

EXERCISES

In Exercises **1–11**, calculate all four second-order partial derivatives and check that $f_{xy} = f_{yx}$. Assume the variables are restricted to a domain on which the function is defined.

1. $f(x, y) = (x + y)^2$ **2.** $f(x, y) = (x + y)^3$

3. $f(x, y) = 3x^2y + 5xy^3$ **4.** $f(x, y) = e^{2xy}$

5. $f(x, y) = (x + y)e^y$ **6.** $f(x, y) = xe^y$

7. $f(x, y) = \sin(x/y)$ **8.** $f(x, y) = \sqrt{x^2 + y^2}$

9. $f(x, y) = 5x^3y^2 - 7xy^3 + 9x^2 + 11$

10. $f(x, y) = \sin(x^2 + y^2)$

11. $f(x, y) = 3 \sin 2x \cos 5y$

In Exercises **12–19**, find the quadratic Taylor polynomials about $(0, 0)$ for the function.

12. $(y - 1)(x + 1)^2$ **13.** $(x - y + 1)^2$

14. $e^{-2x^2 - y^2}$ **15.** $e^x \cos y$

16. $1/(1 + 2x - y)$ **17.** $\cos(x + 3y)$

18. $\sin 2x + \cos y$ **19.** $\ln(1 + x^2 - y)$

In Exercises **20–21**, find the best quadratic approximation for $f(x, y)$ for (x, y) near $(0, 0)$.

20. $f(x, y) = \ln(1 + x - 2y)$
21. $f(x, y) = \sqrt{1 + 2x - y}$

In Exercises **22–33**, use the level curves of the function $z = f(x, y)$ to decide the sign (positive, negative, or zero) of each of the following partial derivatives at the point P. Assume the x- and y-axes are in the usual positions.

(a) $f_x(P)$ **(b)** $f_y(P)$ **(c)** $f_{xx}(P)$
(d) $f_{yy}(P)$ **(e)** $f_{xy}(P)$

22.

23.

24.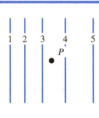

25.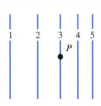

26.

27.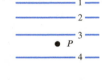

28.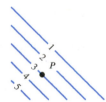

29.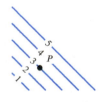

30.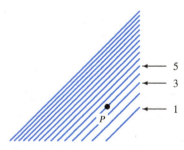

31.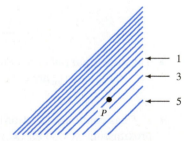

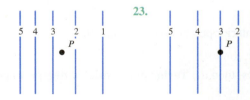

32.

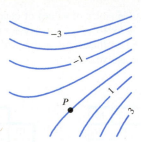

33.

PROBLEMS

In Problems 34–37 estimate the quantity, if possible. If it is not possible, explain why. Assume that g is smooth and

$$\nabla g(2,3) = -7\vec{i} + 3\vec{j}$$
$$\nabla g(2.4,3) = -10.2\vec{i} + 4.2\vec{j}$$

34. $g_{yx}(2,3)$ **35.** $g_{xx}(2,3)$ **36.** $g_{yy}(2,3)$ **37.** $g_{xy}(2,3)$

In Problems 38–41 estimate the quantity, if possible. If it is not possible, explain why. Assume that h is smooth and

$$\nabla h(2.4,3) = -20.4\vec{i} + 8.4\vec{j}$$
$$\nabla h(2.4,2.7) = -22.2\vec{i} + 9\vec{j}$$

38. $h_{yy}(2.4,2.7)$ **39.** $h_{xx}(2.4,3)$

40. $h_{xy}(2.4,2.7)$ **41.** $h_{yx}(2.4,2.7)$

In Problems 42–46, find the linear, $L(x,y)$, and quadratic, $Q(x,y)$, Taylor polynomials valid near $(1,0)$. Compare the values of the approximations $L(0.9,0.2)$ and $Q(0.9,0.2)$ with the exact value of the function $f(0.9,0.2)$.

42. $f(x,y) = \sqrt{x+2y}$ **43.** $f(x,y) = x^2 y$

44. $f(x,y) = xe^{-y}$

45. $F(x,y) = e^x \sin y + e^y \sin x$

46. $f(x,y) = \sin(x-1)\cos y$

In Problems 47–48, show that the function satisfies Laplace's equation, $F_{xx} + F_{yy} = 0$.

47. $F(x,y) = e^{-x} \sin y$

48. $F(x,y) = \arctan(y/x)$

49. If $u(x,t) = e^{at} \sin(bx)$ satisfies the heat equation $u_t = u_{xx}$, find the relationship between a and b.

50. (a) Check that $u(x,t)$ satisfies the heat equation $u_t = u_{xx}$ for $t > 0$ and all x, where

$$u(x,t) = \frac{1}{2\sqrt{\pi t}} e^{-x^2/(4t)}$$

(b) Graph $u(x,t)$ against x for $t = 0.01, 0.1, 1, 10$. These graphs represent the temperature in an infinitely long insulated rod that at $t = 0$ is $0°C$ everywhere except at the origin $x = 0$, and that is infinitely hot at $t = 0$ at the origin.

51. Figure 14.61 shows a graph of $z = f(x,y)$. Is $f_{xx}(0,0)$ positive, zero, or negative? What about $f_{yy}(0,0)$? Give reasons for your answers.

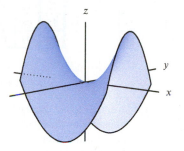

Figure 14.61

52. If $z = f(x) + yg(x)$, what can you say about z_{yy}? Explain your answer.

53. If $z_{xy} = 4y$, what can you say about the value of
(a) z_{yx}?   (b) z_{xyx}?   (c) z_{xyy}?

54. A contour diagram for the smooth function $z = f(x,y)$ is in Figure 14.62.

(a) Is z an increasing or decreasing function of x? Of y?

(b) Is f_x positive or negative? How about f_y?

(c) Is f_{xx} positive or negative? How about f_{yy}?

(d) Sketch the direction of grad f at points P and Q.

(e) Is grad f longer at P or at Q? How do you know?

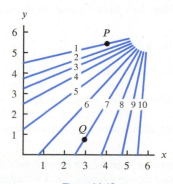

Figure 14.62

■ Problems 55–58 give tables of values of quadratic polynomials $P(x, y) = a + bx + cy + dx^2 + exy + fy^2$. Determine whether each of the coefficients d, e and f of the quadratic terms is positive, negative, or zero.

55.

		x	
	10	12	14
10	35	37	39
y 15	45	47	49
20	55	57	59

56.

		x	
	10	12	14
10	26	36	54
y 15	31	41	59
20	36	46	64

57.

		x	
	10	12	14
10	90	82	74
y 15	75	87	99
20	10	42	74

58.

		x	
	10	12	14
10	13	33	61
y 15	28	28	36
20	93	73	61

59. You are hiking on a level trail going due east and planning to strike off cross country up the mountain to your left. The slope up to the left is too steep now and seems to be gentler the further you go along the trail, so you decide to wait before turning off.

(a) Sketch a topographical contour map that illustrates this story.

(b) What information does the story give about partial derivatives? Define all variables and functions that you use.

(c) What partial derivative influenced your decision to wait before turning?

60. The weekly production, Y, in factories that manufacture a certain item is modeled as a function of the quantity of capital, K, and quantity of labor, L, at the factory. Data shows that hiring a few extra workers increases production. Moreover, for two factories with the same number of workers, hiring a few extra workers increases production more for the factory with more capital. (With more equipment, additional labor can be used more effectively.) What does this tell you about the sign of

(a) $\partial Y/\partial L$?

(b) $\partial^2 Y/(\partial K \partial L)$?

61. Data suggests that human surface area, S, can reasonably be modeled as a function of height, h, and weight, w. In the Dubois model, we have $\partial^2 S/\partial w^2 < 0$ and $\partial^2 S/(\partial h \partial w) > 0$. Two people A and B each gain 1 pound. Which experiences the greater increase in surface area if

(a) They have the same weight but A is taller?

(b) They have the same height, but A is heavier?

62. You plan to buy a used car. You are debating between a 5-year old car and a 10-year old car and thinking about the price. Experts report that the original price matters more when buying a 5-year old car than a 10-year old car. This suggests that we model the average market price, P, in dollars as a function of two variables: the original price, C, in dollars, and the age of the car, A, in years.

(a) Give units for the following partial derivatives and say whether you think they are positive or negative. Explain your reasoning.

(i) $\partial P/\partial A$ (ii) $\partial P/\partial C$

(b) Express the experts' report in terms of partial derivatives.

(c) Using a quadratic polynomial to model P, we have

$$P = a + bC + cA + dC^2 + eCA + fA^2.$$

Which term in this polynomial is most relevant to the experts' report?

63. The tastiness, T, of a soup depends on the volume, V, of the soup in the pot and the quantity, S, of salt in the soup. If you have more soup, you need more salt to make it taste good. Match the three stories (a)–(c) to the three statements (I)–(III) about partial derivatives.

(a) I started adding salt to the soup in the pot. At first the taste improved, but eventually the soup became too salty and continuing to add more salt made it worse.

(b) The soup was too salty, so I started adding unsalted soup. This improved the taste at first, but eventually there was too much soup for the salt, and continuing to add unsalted soup just made it worse.

(c) The soup was too salty, so adding more salt would have made it taste worse. I added a quart of unsalted soup instead. Now it is not salty enough, but I can improve the taste by adding salt.

(I) $\partial^2 T/\partial V^2 < 0$

(II) $\partial^2 T/\partial S^2 < 0$

(III) $\partial^2 T/\partial V \partial S > 0$

64. Figure 14.63 shows the level curves of a function $f(x, y)$ around a maximum or minimum, M. One of the points P and Q has coordinates (x_1, y_1) and the other has coordinates (x_2, y_2). Suppose $b > 0$ and $c > 0$. Consider the two linear approximations to f given by

$$f(x, y) \approx a + b(x - x_1) + c(y - y_1)$$
$$f(x, y) \approx k + m(x - x_2) + n(y - y_2).$$

(a) What is the relationship between the values of a and k?

(b) What are the coordinates of P?

(c) Is M a maximum or a minimum?

(d) What can you say about the sign of the constants m and n?

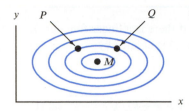

Figure 14.63

65. Consider the function $f(x, y) = (\sin x)(\sin y)$.

    (a) Find the Taylor polynomials of degree 2 for f about the points $(0, 0)$ and $(\pi/2, \pi/2)$.

    (b) Use the Taylor polynomials to sketch the contours of f close to each of the points $(0, 0)$ and $(\pi/2, \pi/2)$.

66. Let $f(x, y) = \sqrt{x + 2y + 1}$.

    (a) Compute the local linearization of f at $(0, 0)$.

    (b) Compute the quadratic Taylor polynomial for f at $(0, 0)$.

    (c) Compare the values of the linear and quadratic approximations in part (a) and part (b) with the true values for $f(x, y)$ at the points $(0.1, 0.1)$, $(-0.1, 0.1)$, $(0.1, -0.1)$, $(-0.1, -0.1)$. Which approximation gives the closest values?

67. Using a computer and your answer to Problem 66, draw the six contour diagrams of $f(x, y) = \sqrt{x + 2y + 1}$ and its linear and quadratic approximations, $L(x, y)$ and $Q(x, y)$, in the two windows $[-0.6, 0.6] \times [-0.6, 0.6]$ and $[-2, 2] \times [-2, 2]$. Explain the shape of the contours, their spacing, and the relationship between the contours of f, L, and Q.

Strengthen Your Understanding

In Problems **68–69**, explain what is wrong with the statement.

68. If $f(x, y) \neq 0$, then the Taylor polynomial of degree 2 approximating $f(x, y)$ near $(0, 0)$ is also nonzero.

69. There is a function $f(x, y)$ with partial derivatives $f_x = xy$ and $f_y = y^2$.

In Problems **70–72**, give an example of:

70. A function $f(x, y)$ such that $f_{xx} \neq 0$, $f_{yy} \neq 0$, and $f_{xy} = 0$.

71. Formulas for two different functions $f(x, y)$ and $g(x, y)$ with the same quadratic approximation near $(0, 0)$.

72. Contour diagrams for two different functions $f(x, y)$ and $g(x, y)$ that have the same quadratic approximations near $(0, 0)$.

14.8 DIFFERENTIABILITY

In Section 14.3 we gave an informal introduction to the concept of differentiability. We called a function $f(x, y)$ *differentiable* at a point (a, b) if it is well approximated by a linear function near (a, b). This section focuses on the precise meaning of the phrase "well approximated." By looking at examples, we shall see that local linearity requires the existence of partial derivatives, but they do not tell the whole story. In particular, existence of partial derivatives at a point is not sufficient to guarantee local linearity at that point.

We begin by discussing the relation between continuity and differentiability. As an illustration, take a sheet of paper, crumple it into a ball and smooth it out again. Wherever there is a crease it would be difficult to approximate the surface by a plane—these are points of nondifferentiability of the function giving the height of the paper above the floor. Yet the sheet of paper models a graph which is continuous—there are no breaks. As in the case of one-variable calculus, continuity does not imply differentiability. But differentiability does *require* continuity: there cannot be linear approximations to a surface at points where there are abrupt changes in height.

Differentiability for Functions of Two Variables

For a function of two variables, as for a function of one variable, we define differentiability at a point in terms of the error and the distance from the point. If the point is (a, b) and a nearby point is $(a + h, b + k)$, the distance between them is $\sqrt{h^2 + k^2}$. (See Figure 14.64.)

A function $f(x, y)$ is **differentiable at the point** (a, b) if there is a linear function $L(x, y) = f(a, b) + m(x - a) + n(y - b)$ such that if the *error* $E(x, y)$ is defined by

$$f(x, y) = L(x, y) + E(x, y),$$

and if $h = x - a, k = y - b$, then the *relative error* $E(a + h, b + k)/\sqrt{h^2 + k^2}$ satisfies

$$\lim_{\substack{h \to 0 \\ k \to 0}} \frac{E(a + h, b + k)}{\sqrt{h^2 + k^2}} = 0.$$

The function f is **differentiable on a region** R if it is differentiable at each point of R. The function $L(x, y)$ is called the *local linearization* of $f(x, y)$ near (a, b).

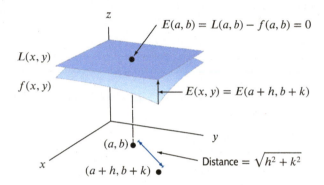

Figure 14.64: Graph of function $z = f(x, y)$ and its local linearization $z = L(x, y)$ near the point (a, b)

Partial Derivatives and Differentiability

In the next example, we show that this definition of differentiability is consistent with our previous notion — that is, that $m = f_x$ and $n = f_y$ and that the graph of $L(x, y)$ is the tangent plane.

Example 1   Show that if f is a differentiable function with local linearization $L(x, y) = f(a, b) + m(x - a) + n(y - b)$, then $m = f_x(a, b)$ and $n = f_y(a, b)$.

Solution   Since f is differentiable, we know that the relative error in $L(x, y)$ tends to 0 as we get close to (a, b). Suppose $h > 0$ and $k = 0$. Then we know that

$$0 = \lim_{h \to 0} \frac{E(a + h, b + k)}{\sqrt{h^2 + k^2}} = \lim_{h \to 0} \frac{E(a + h, b)}{h} = \lim_{h \to 0} \frac{f(a + h, b) - L(a + h, b)}{h}$$

$$= \lim_{h \to 0} \frac{f(a + h, b) - f(a, b) - mh}{h}$$

$$= \lim_{h \to 0} \left(\frac{f(a + h, b) - f(a, b)}{h} \right) - m = f_x(a, b) - m.$$

A similar result holds if $h < 0$, so we have $m = f_x(a, b)$. The result $n = f_y(a, b)$ is found in a similar manner.

The previous example shows that if a function is differentiable at a point, it has partial derivatives there. Therefore, if any of the partial derivatives fail to exist, then the function cannot be differentiable. This is what happens in the following example of a cone.

Example 2   Consider the function $f(x, y) = \sqrt{x^2 + y^2}$. Is f differentiable at the origin?

Solution   If we zoom in on the graph of the function $f(x, y) = \sqrt{x^2 + y^2}$ at the origin, as shown in Figure 14.65, the sharp point remains; the graph never flattens out to look like a plane. Near its vertex, the graph does not look as if is well approximated (in any reasonable sense) by any plane.

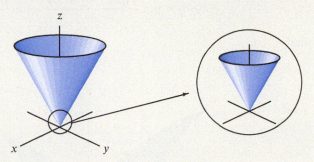

Figure 14.65: The function $f(x, y) = \sqrt{x^2 + y^2}$ is not locally linear at $(0, 0)$: Zooming in around $(0, 0)$ does not make the graph look like a plane

Judging from the graph of f, we would not expect f to be differentiable at $(0, 0)$. Let us check this by trying to compute the partial derivatives of f at $(0, 0)$:

$$f_x(0, 0) = \lim_{h \to 0} \frac{f(h, 0) - f(0, 0)}{h} = \lim_{h \to 0} \frac{\sqrt{h^2 + 0} - 0}{h} = \lim_{h \to 0} \frac{|h|}{h}.$$

Since $|h|/h = \pm 1$, depending on whether h approaches 0 from the left or right, this limit does not exist and so neither does the partial derivative $f_x(0, 0)$. Thus, f cannot be differentiable at the origin. If it were, both of the partial derivatives, $f_x(0, 0)$ and $f_y(0, 0)$, would exist.

Alternatively, we could show directly that there is no linear approximation near $(0, 0)$ that satisfies the small relative error criterion for differentiability. Any plane passing through the point $(0, 0, 0)$ has the form $L(x, y) = mx + ny$ for some constants m and n. If $E(x, y) = f(x, y) - L(x, y)$, then

$$E(x, y) = \sqrt{x^2 + y^2} - mx - ny.$$

Then for f to be differentiable at the origin, we would need to show that

$$\lim_{\substack{h \to 0 \\ k \to 0}} \frac{\sqrt{h^2 + k^2} - mh - nk}{\sqrt{h^2 + k^2}} = 0.$$

Taking $k = 0$ gives

$$\lim_{h \to 0} \frac{|h| - mh}{|h|} = 1 - m \lim_{h \to 0} \frac{h}{|h|}.$$

This limit exists only if $m = 0$ for the same reason as before. But then the value of the limit is 1 and not 0 as required. Thus, we again conclude f is not differentiable.

In Example 2 the partial derivatives f_x and f_y did not exist at the origin and this was sufficient to establish nondifferentiability there. We might expect that if both partial derivatives do exist, then f *is* differentiable. But the next example shows that this not necessarily true: the existence of both partial derivatives at a point is *not* sufficient to guarantee differentiability.

Example 3   Consider the function $f(x, y) = x^{1/3} y^{1/3}$. Show that the partial derivatives $f_x(0,0)$ and $f_y(0,0)$ exist, but that f is not differentiable at $(0, 0)$.

Solution   See Figure 14.66 for the part of the graph of $z = x^{1/3} y^{1/3}$ when $z \geq 0$. We have $f(0, 0) = 0$ and we compute the partial derivatives using the definition:

$$f_x(0,0) = \lim_{h \to 0} \frac{f(h, 0) - f(0, 0)}{h} = \lim_{h \to 0} \frac{0 - 0}{h} = 0,$$

and similarly

$$f_y(0, 0) = 0.$$

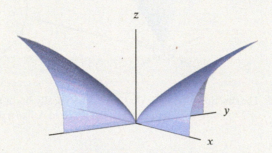

Figure 14.66: Graph of $z = x^{1/3} y^{1/3}$ for $z \geq 0$

So, if there did exist a linear approximation near the origin, it would have to be $L(x, y) = 0$. But we can show that this choice of $L(x, y)$ does not result in the small relative error that is required for differentiability. In fact, since $E(x, y) = f(x, y) - L(x, y) = f(x, y)$, we need to look at the limit

$$\lim_{\substack{h \to 0 \\ k \to 0}} \frac{h^{1/3} k^{1/3}}{\sqrt{h^2 + k^2}}.$$

If this limit exists, we get the same value no matter how h and k approach 0. Suppose we take $k = h > 0$. Then the limit becomes

$$\lim_{h \to 0} \frac{h^{1/3} h^{1/3}}{\sqrt{h^2 + h^2}} = \lim_{h \to 0} \frac{h^{2/3}}{h\sqrt{2}} = \lim_{h \to 0} \frac{1}{h^{1/3}\sqrt{2}}.$$

But this limit does not exist, since small values for h will make the fraction arbitrarily large. So the only possible candidate for a linear approximation at the origin does not have a sufficiently small relative error. Thus, this function is *not* differentiable at the origin, even though the partial derivatives $f_x(0, 0)$ and $f_y(0, 0)$ exist. Figure 14.66 confirms that near the origin the graph of $z = f(x, y)$ is not well approximated by any plane.

In summary,

- If a function is differentiable at a point, then both partial derivatives exist there.
- Having both partial derivatives at a point does not guarantee that a function is differentiable there.

Continuity and Differentiability

We know that differentiable functions of one variable are continuous. Similarly, it can be shown that if a function of two variables is differentiable at a point, then the function is continuous there.

In Example 3 the function f was continuous at the point where it was not differentiable. Example 4 shows that even if the partial derivatives of a function exist at a point, the function is not necessarily continuous at that point if it is not differentiable there.

Example 4　Suppose that f is the function of two variables defined by

$$f(x,y) = \begin{cases} \dfrac{xy}{x^2+y^2}, & (x,y) \neq (0,0), \\ 0, & (x,y) = (0,0). \end{cases}$$

Problem 27 on page 744 showed that $f(x,y)$ is not continuous at the origin. Show that the partial derivatives $f_x(0,0)$ and $f_y(0,0)$ exist. Could f be differentiable at $(0,0)$?

Solution　From the definition of the partial derivative we see that

$$f_x(0,0) = \lim_{h \to 0} \frac{f(h,0) - f(0,0)}{h} = \lim_{h \to 0} \left(\frac{1}{h} \cdot \frac{0}{h^2 + 0^2} \right) = \lim_{h \to 0} \frac{0}{h} = 0,$$

and similarly

$$f_y(0,0) = 0.$$

So, the partial derivatives $f_x(0,0)$ and $f_y(0,0)$ exist. However, f cannot be differentiable at the origin since it is not continuous there.

In summary,

- If a function is differentiable at a point, then it is continuous there.
- Having both partial derivatives at a point does not guarantee that a function is continuous there.

How Do We Know If a Function Is Differentiable?

Can we use partial derivatives to tell us if a function is differentiable? As we see from Examples 3 and 4, it is not enough that the partial derivatives exist. However, the following theorem gives conditions that *do* guarantee differentiability[4]:

Theorem 14.2: Continuity of Partial Derivatives Implies Differentiability

If the partial derivatives, f_x and f_y, of a function f exist and are continuous on a small disk centered at the point (a,b), then f is differentiable at (a,b).

We will not prove this theorem, although it provides a criterion for differentiability which is often simpler to use than the definition. It turns out that the requirement of continuous partial derivatives is more stringent than that of differentiability, so there exist differentiable functions which do not have continuous partial derivatives. However, most functions we encounter will have continuous partial derivatives. The class of functions with continuous partial derivatives is given the name C^1.

Example 5　Show that the function $f(x,y) = \ln(x^2 + y^2)$ is differentiable everywhere in its domain.

Solution　The domain of f is all of 2-space except for the origin. We shall show that f has continuous partial derivatives everywhere in its domain (that is, the function f is in C^1). The partial derivatives are

$$f_x = \frac{2x}{x^2+y^2} \quad \text{and} \quad f_y = \frac{2y}{x^2+y^2}.$$

Since each of f_x and f_y is the quotient of continuous functions, the partial derivatives are continuous everywhere except the origin (where the denominators are zero). Thus, f is differentiable everywhere in its domain.

---

[4]For a proof, see M. Spivak, *Calculus on Manifolds*, p. 31 (New York: Benjamin, 1965).

Most functions built up from elementary functions have continuous partial derivatives, except perhaps at a few obvious points. Thus, in practice, we can often identify functions as being C^1 without explicitly computing the partial derivatives.

Summary for Section 14.8

- A function $f(x, y)$ is **differentiable at the point** (a, b) if there is a linear function $L(x, y) = f(a, b) + m(x - a) + n(y - b)$ such that if the *error* $E(x, y)$ is defined by

$$f(x, y) = L(x, y) + E(x, y),$$

and if $h = x - a, k = y - b$, then the *relative error* $E(a + h, b + k)/\sqrt{h^2 + k^2}$ satisfies

$$\lim_{\substack{h \to 0 \\ k \to 0}} \frac{E(a + h, b + k)}{\sqrt{h^2 + k^2}} = 0.$$

The function f is **differentiable on a region** R if it is differentiable at each point of R. The function $L(x, y)$ is called the *local linearization* of $f(x, y)$ near (a, b).

- **Differentiability** is **not the same** as having **partial derivatives**:
  - If a function is differentiable at a point, then both partial derivatives exist there.
  - Having both partial derivatives at a point does not guarantee that a function is differentiable there.
- If a function is **differentiable** at a point it is **also continuous** at that point.
- A function $f(x, y)$ with both partial derivatives at a point can fail to be continuous at that point.
- If both the partial derivatives of a function $f(x, y)$ **exist and are continuous** on a small disk centered at the point (a, b), then f is **differentiable** at (a, b).

Exercises and Problems for Section 14.8

EXERCISES

In Exercises **1–10**, list the points in the xy-plane, if any, at which the function $z = f(x, y)$ is not differentiable.

1. $z = -\sqrt{x^2 + y^2}$

2. $z = \sqrt{(x + 1)^2 + y^2}$

3. $z = |x| + |y|$

4. $z = |x + 2| - |y - 3|$

5. $z = e^{-(x^2 + y^2)}$

6. $z = x^{1/3} + y^2$

7. $z = |x - 3|^2 + y^3$

8. $z = (\sin x)(\cos |y|)$

9. $z = 4 + \sqrt{(x - 1)^2 + (y - 2)^2}$

10. $z = 1 + \left((x - 1)^2 + (y - 2)^2\right)^2$

PROBLEMS

In Problems **11–14**, a function f is given.
(a) Use a computer to draw a contour diagram for f.
(b) Is f differentiable at all points $(x, y) \neq (0, 0)$?
(c) Do the partial derivatives f_x and f_y exist and are they continuous at all points $(x, y) \neq (0, 0)$?
(d) Is f differentiable at $(0, 0)$?
(e) Do the partial derivatives f_x and f_y exist and are they continuous at $(0, 0)$?

11. $f(x, y) = \begin{cases} \dfrac{x}{y} + \dfrac{y}{x}, & x \neq 0 \text{ and } y \neq 0, \\ 0, & x = 0 \text{ or } y = 0. \end{cases}$

12. $f(x, y) = \begin{cases} \dfrac{2xy}{(x^2 + y^2)^2}, & (x, y) \neq (0, 0), \\ 0, & (x, y) = (0, 0). \end{cases}$

13. $f(x, y) = \begin{cases} \dfrac{x^2 y}{x^4 + y^2}, & (x, y) \neq (0, 0), \\ 0, & (x, y) = (0, 0). \end{cases}$

14. $f(x, y) = \begin{cases} \dfrac{xy}{\sqrt{x^2 + y^2}}, & (x, y) \neq (0, 0), \\ 0, & (x, y) = (0, 0). \end{cases}$

15. Consider the function

$$f(x, y) = \begin{cases} \dfrac{xy^2}{x^2 + y^2}, & (x, y) \neq (0, 0), \\ 0, & (x, y) = (0, 0). \end{cases}$$

(a) Use a computer to draw the contour diagram for f.
(b) Is f differentiable for $(x, y) \neq (0, 0)$?
(c) Show that $f_x(0, 0)$ and $f_y(0, 0)$ exist.
(d) Is f differentiable at $(0, 0)$?
(e) Suppose $x(t) = at$ and $y(t) = bt$, where a and b are constants, not both zero. If $g(t) = f(x(t), y(t))$, show that

$$g'(0) = \frac{ab^2}{a^2 + b^2}.$$

(f) Show that

$$f_x(0, 0)x'(0) + f_y(0, 0)y'(0) = 0.$$

Does the chain rule hold for the composite function $g(t)$ at $t = 0$? Explain.
(g) Show that the directional derivative $f_{\vec{u}}(0, 0)$ exists for each unit vector $\vec{u}$. Does this imply that f is differentiable at $(0, 0)$?

16. Consider the function $f(x, y) = \sqrt{|xy|}$.

(a) Use a computer to draw the contour diagram for f. Does the contour diagram look like that of a plane when we zoom in on the origin?
(b) Use a computer to draw the graph of f. Does the graph look like a plane when we zoom in on the origin?
(c) Is f differentiable for $(x, y) \neq (0, 0)$?
(d) Show that $f_x(0, 0)$ and $f_y(0, 0)$ exist.
(e) Is f differentiable at $(0, 0)$? [Hint: Consider the directional derivative $f_{\vec{u}}(0, 0)$ for $\vec{u} = (\vec{i} + \vec{j})/\sqrt{2}$.]

17. Consider the function

$$f(x, y) = \begin{cases} \dfrac{xy^2}{x^2 + y^4}, & (x, y) \neq (0, 0), \\ 0, & (x, y) = (0, 0). \end{cases}$$

(a) Use a computer to draw the contour diagram for f.
(b) Show that the directional derivative $f_{\vec{u}}(0, 0)$ exists for each unit vector $\vec{u}$.
(c) Is f continuous at $(0, 0)$? Is f differentiable at $(0, 0)$? Explain.

18. Suppose $f(x, y)$ is a function such that $f_x(0, 0) = 0$ and $f_y(0, 0) = 0$, and $f_{\vec{u}}(0, 0) = 3$ for $\vec{u} = (\vec{i} + \vec{j})/\sqrt{2}$.

(a) Is f differentiable at $(0, 0)$? Explain.
(b) Give an example of a function f defined on 2-space which satisfies these conditions. [Hint: The function f does not have to be defined by a single formula valid over all of 2-space.]

19. Consider the following function:

$$f(x, y) = \begin{cases} \dfrac{xy(x^2 - y^2)}{x^2 + y^2}, & (x, y) \neq (0, 0), \\ 0, & (x, y) = (0, 0). \end{cases}$$

The graph of f is shown in Figure 14.67, and the contour diagram of f is shown in Figure 14.68.

(a) Find $f_x(x, y)$ and $f_y(x, y)$ for $(x, y) \neq (0, 0)$.
(b) Show that $f_x(0, 0) = 0$ and $f_y(0, 0) = 0$.
(c) Are the functions f_x and f_y continuous at $(0, 0)$?
(d) Is f differentiable at $(0, 0)$?

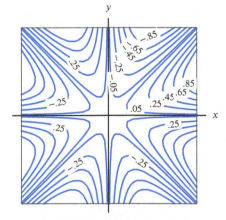

Figure 14.67: Graph of $\dfrac{xy(x^2 - y^2)}{x^2 + y^2}$

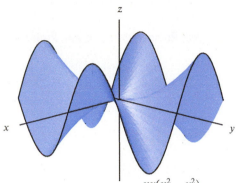

Figure 14.68: Contour diagram of $\dfrac{xy(x^2 - y^2)}{x^2 + y^2}$

20. Suppose a function f is differentiable at the point (a, b). Show that f is continuous at (a, b).

Strengthen Your Understanding

In Problems 21–22, explain what is wrong with the statement.

21. If $f(x, y)$ is continuous at the origin, then it is differentiable at the origin.

22. If the partial derivatives $f_x(0,0)$ and $f_y(0,0)$ both exist, then $f(x, y)$ is differentiable at the origin.

In Problems 23–24, give an example of:

23. A continuous function $f(x, y)$ that is not differentiable at the origin.

24. A continuous function $f(x, y)$ that is not differentiable on the line $x = 1$.

25. Which of the following functions $f(x, y)$ is differentiable at the given point?

(a) $\sqrt{1 - x^2 - y^2}$ at $(0,0)$ (b) $\sqrt{4 - x^2 - y^2}$ at $(2, 0)$

(c) $-\sqrt{x^2 + 2y^2}$ at $(0, 0)$ (d) $-\sqrt{x^2 + 2y^2}$ at $(2, 0)$

Online Resource: Review Problems and Projects

Chapter 15

OPTIMIZATION: LOCAL AND GLOBAL EXTREMA

CONTENTS

15.1   CRITICAL POINTS: LOCAL EXTREMA AND SADDLE POINTS

Functions of several variables, like functions of one variable, can have *local* and *global* extrema. (That is, local and global maxima and minima.) A function has a local extremum at a point where it takes on the largest or smallest value in a small region around the point. Global extrema are the largest or smallest values anywhere on the domain under consideration. (See Figures 15.1 and 15.2.)

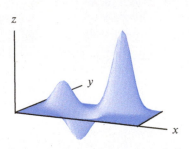

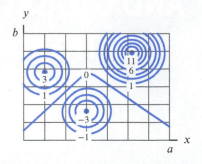

Figure 15.1: Local and global extrema for a function of two variables on $0 \leq x \leq a$, $0 \leq y \leq b$

Figure 15.2: Contour map of the function in Figure 15.1

More precisely, considering only points at which f is defined, we say:

- f has a **local maximum** at the point P_0 if $f(P_0) \geq f(P)$ for all points P near P_0.
- f has a **local minimum** at the point P_0 if $f(P_0) \leq f(P)$ for all points P near P_0.

For example, the function whose contour map is shown in Figure 15.2 has a local minimum value of -3 and local maximum values of 3 and 11 in the rectangle shown.

How Do We Detect a Local Maximum or Minimum?

Recall that if the gradient vector of a function is defined and nonzero, then it points in a direction in which the function increases. Suppose that a function f has a local maximum at a point P_0 which is not on the boundary of the domain. If the vector grad $f(P_0)$ were defined and nonzero, then we could increase f by moving in the direction of grad $f(P_0)$. Since f has a local maximum at P_0, there is no direction in which f is increasing. Thus, if grad $f(P_0)$ is defined, we must have

$$\text{grad } f(P_0) = \vec{0}.$$

Similarly, suppose f has a local minimum at the point P_0. If grad $f(P_0)$ were defined and nonzero, then we could decrease f by moving in the direction opposite to grad $f(P_0)$, and so we must again have grad $f(P_0) = \vec{0}$. Therefore, we make the following definition:

Points where the gradient is either $\vec{0}$ or undefined are called **critical points** of the function.

If a function has a local maximum or minimum at a point P_0, not on the boundary of its domain, then P_0 is a critical point. For a function of two variables, we can also see that the gradient vector must be zero or undefined at a local maximum by looking at its contour diagram and a plot of its gradient vectors. (See Figures 15.3 and 15.4.) Around the maximum the vectors are all pointing inward, perpendicularly to the contours. At the maximum the gradient vector must be zero or undefined. A similar argument shows that the gradient must be zero or undefined at a local minimum.

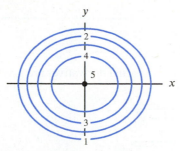

Figure 15.3: Contour diagram around a local maximum of a function

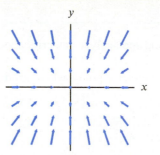

Figure 15.4: Gradients pointing toward the local maximum of the function in Figure 15.3

Finding and Analyzing Critical Points

To find critical points of f we set grad $f = f_x \vec{i} + f_y \vec{j} + f_z \vec{k} = \vec{0}$, which means setting all the partial derivatives of f equal to zero. We must also look for the points where one or more of the partial derivatives is undefined.

Example 1   Find and analyze the critical points of $f(x, y) = x^2 - 2x + y^2 - 4y + 5$.

Solution   To find the critical points, we set both partial derivatives equal to zero:

$$f_x(x, y) = 2x - 2 = 0$$
$$f_y(x, y) = 2y - 4 = 0.$$

Solving these equations gives $x = 1$, $y = 2$. Hence, f has only one critical point, namely $(1, 2)$. To see the behavior of f near $(1, 2)$, look at the values of the function in Table 15.1.

Table 15.1   *Values of $f(x, y)$ near the point $(1, 2)$*

		|	0.8	0.9	1.0	1.1	1.2
					x		
	1.8		0.08	0.05	0.04	0.05	0.08
	1.9		0.05	0.02	0.01	0.02	0.05
y	2.0		0.04	0.01	0.00	0.01	0.04
	2.1		0.05	0.02	0.01	0.02	0.05
	2.2		0.08	0.05	0.04	0.05	0.08

The table suggests that the function has a local minimum value of 0 at $(1, 2)$. We can confirm this by completing the square:

$$f(x, y) = x^2 - 2x + y^2 - 4y + 5 = (x - 1)^2 + (y - 2)^2.$$

Figure 15.5 shows that the graph of f is a paraboloid with vertex at the point $(1, 2, 0)$. It is the same shape as the graph of $z = x^2 + y^2$ (see Figure 12.12 on page 703), except that the vertex has been shifted to $(1, 2)$. So the point $(1, 2)$ is a local minimum of f (as well as a global minimum).

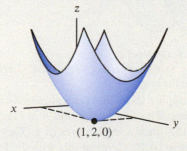

$(1, 2, 0)$

Figure 15.5: The graph of $f(x, y) = x^2 - 2x + y^2 - 4y + 5$ with a local minimum at the point $(1, 2)$

Example 2   Find and analyze any critical points of $f(x, y) = -\sqrt{x^2 + y^2}$.

Solution   We look for points where grad $f = \vec{0}$ or is undefined. The partial derivatives are given by

$$f_x(x, y) = -\frac{x}{\sqrt{x^2 + y^2}},$$

$$f_y(x, y) = -\frac{y}{\sqrt{x^2 + y^2}}.$$

These partial derivatives are never simultaneously zero, but they are undefined at $x = 0$, $y = 0$. Thus, $(0, 0)$ is a critical point and a possible extreme point. The graph of f (see Figure 15.6) is a cone, with vertex at $(0, 0)$. So f has a local and global maximum at $(0, 0)$.

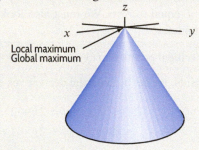

Local maximum
Global maximum

Figure 15.6: Graph of $f(x, y) = -\sqrt{x^2 + y^2}$

Example 3   Find and analyze any critical points of $g(x, y) = x^2 - y^2$.

Solution   To find the critical points, we look for points where both partial derivatives are zero:

$$g_x(x, y) = 2x = 0$$
$$g_y(x, y) = -2y = 0.$$

Solving gives $x = 0$, $y = 0$, so the origin is the only critical point.

Figure 15.7 shows that near the origin g takes on both positive and negative values. Since $g(0, 0) = 0$, the origin is a critical point which is neither a local maximum nor a local minimum. The graph of g looks like a saddle.

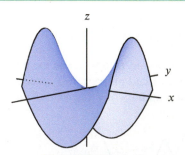

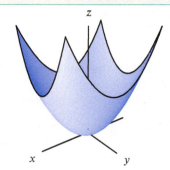

Figure 15.7: Graph of $g(x, y) = x^2 - y^2$, showing saddle shape at the origin

Figure 15.8: Graph of $h(x, y) = x^2 + y^2$, showing minimum at the origin

The previous examples show that critical points can occur at local maxima or minima, or at points which are neither: The functions g and h in Figures 15.7 and 15.8 both have critical points at the origin. Figure 15.9 shows level curves of g. They are hyperbolas showing both positive and negative values of g near $(0, 0)$. Contrast this with the level curves of h near the local minimum in Figure 15.10.

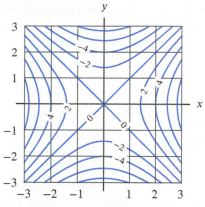

Figure 15.9: Contours of $g(x, y) = x^2 - y^2$, showing a saddle shape at the origin

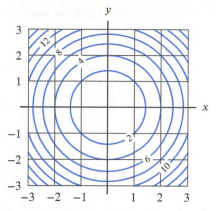

Figure 15.10: Contours of $h(x, y) = x^2 + y^2$, showing a local minimum at the origin

Example 4

Find the local extrema of the function $f(x, y) = 8y^3 + 12x^2 - 24xy$.

Solution

We begin by looking for critical points:

$$f_x(x, y) = 24x - 24y,$$
$$f_y(x, y) = 24y^2 - 24x.$$

Setting these expressions equal to zero gives the system of equations

$$x = y, \qquad x = y^2,$$

which has two solutions, $(0, 0)$ and $(1, 1)$. Are these local maxima, local minima or neither? Figure 15.11 shows contours of f near the points. Notice that $f(1, 1) = -4$ and the contours at nearby points have larger function values. This suggests f has a local minimum at $(1, 1)$.

We have $f(0, 0) = 0$ and the contours near $(0, 0)$ show that f takes both positive and negative values nearby. This suggests that $(0, 0)$ is a critical point which is neither a local maximum nor a local minimum.

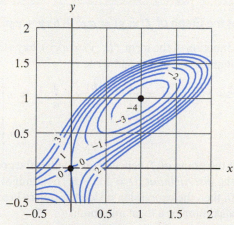

Figure 15.11: Contour diagram of $f(x, y) = 8y^3 + 12x^2 - 24xy$ showing critical points at $(0, 0)$ and $(1, 1)$

Classifying Critical Points

We can see whether a critical point of a function, f, is a local maximum, local minimum, or neither by looking at the contour diagram. There is also an analytic method for making this distinction.

Quadratic Functions of the Form $f(x, y) = ax^2 + bxy + cy^2$

Near most critical points, a function has the same behavior as its quadratic Taylor approximation about that point. Thus, we start by investigating critical points of quadratic functions of the form $f(x, y) = ax^2 + bxy + cy^2$, where a, b and c are constants.

Example 5   Find and analyze the local extrema of the function $f(x, y) = x^2 + xy + y^2$.

Solution   To find critical points, we set

$$f_x(x, y) = 2x + y = 0,$$
$$f_y(x, y) = x + 2y = 0.$$

The only critical point is $(0, 0)$, and the value of the function there is $f(0, 0) = 0$. If f is always positive or zero near $(0, 0)$, then $(0, 0)$ is a local minimum; if f is always negative or zero near $(0, 0)$, it is a local maximum; if f takes both positive and negative values, it is neither. The graph in Figure 15.12 suggests that $(0, 0)$ is a local minimum.

How can we be sure that $(0, 0)$ is a local minimum? We complete the square. Writing

$$f(x, y) = x^2 + xy + y^2 = \left(x + \frac{1}{2}y\right)^2 + \frac{3}{4}y^2,$$

shows that $f(x, y)$ is a sum of two nonnegative terms, so it is always greater than or equal to zero. Thus, the critical point is both a local and a global minimum.

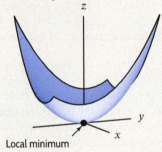

Local minimum

Figure 15.12: Graph of $f(x, y) = x^2 + xy + y^2 = (x + \frac{1}{2}y)^2 + \frac{3}{4}y^2$ showing local minimum at the origin

The Shape of the Graph of $f(x, y) = ax^2 + bxy + cy^2$

In general, a function of the form $f(x, y) = ax^2 + bxy + cy^2$ has one critical point at $(0, 0)$. Assuming $a \neq 0$, we complete the square and write

$$ax^2 + bxy + cy^2 = a\left[x^2 + \frac{b}{a}xy + \frac{c}{a}y^2\right] = a\left[\left(x + \frac{b}{2a}y\right)^2 + \left(\frac{c}{a} - \frac{b^2}{4a^2}\right)y^2\right]$$

$$= a\left[\left(x + \frac{b}{2a}y\right)^2 + \left(\frac{4ac - b^2}{4a^2}\right)y^2\right].$$

The shape of the graph of f depends on whether the coefficient of y^2 is positive, negative, or zero. The sign of the *discriminant*, $D = 4ac - b^2$, determines the sign of the coefficient of y^2.

- If $D > 0$, then the expression inside the square brackets is positive or zero, so the function has a local maximum or a local minimum.
  - If $a > 0$, the function has a local minimum, since the graph is a paraboloid opening upward, like $z = x^2 + y^2$. (See Figure 15.13.)
  - If $a < 0$, the function has a local maximum, since the graph is a paraboloid opening downward, like $z = -x^2 - y^2$. (See Figure 15.14.)
- If $D < 0$, then the function goes up in some directions and goes down in others, like $z = x^2 - y^2$. We say the function has a *saddle point*, that is, a critical point at which the function value increases in some directions but decreases in others. (See Figure 15.15.)
- If $D = 0$, then the quadratic function is $a(x + by/2a)^2$, whose graph is a parabolic cylinder. (See Figure 15.16.)

Figure 15.13: Local minimum: $D > 0$ and $a > 0$ **Figure 15.14:** Local maximum: $D > 0$ and $a < 0$ **Figure 15.15:** Saddle point: $D < 0$ **Figure 15.16:** Parabolic cylinder: $D = 0$

More generally, the graph of $g(x, y) = a(x - x_0)^2 + b(x - x_0)(y - y_0) + c(y - y_0)^2$ has the same shape as the graph of $f(x, y) = ax^2 + bxy + cy^2$, except that the critical point is at (x_0, y_0) rather than $(0, 0)$.[1]

Classifying the Critical Points of a Function

Suppose that f is any function with grad $f(0,0) = \vec{0}$. Its quadratic Taylor polynomial near $(0, 0)$,

$$f(x, y) \approx f(0,0) + f_x(0,0)x + f_y(0,0)y$$
$$+ \frac{1}{2}f_{xx}(0,0)x^2 + f_{xy}(0,0)xy + \frac{1}{2}f_{yy}(0,0)y^2,$$

can be simplified using $f_x(0,0) = f_y(0,0) = 0$, which gives

$$f(x, y) - f(0,0) \approx \frac{1}{2}f_{xx}(0,0)x^2 + f_{xy}(0,0)xy + \frac{1}{2}f_{yy}(0,0)y^2.$$

The discriminant of this quadratic polynomial is

$$D = 4ac - b^2 = 4\left(\frac{1}{2}f_{xx}(0,0)\right)\left(\frac{1}{2}f_{yy}(0,0)\right) - \left(f_{xy}(0,0)\right)^2,$$

which simplifies to

$$D = f_{xx}(0,0)f_{yy}(0,0) - (f_{xy}(0,0))^2.$$

There is a similar formula for D if the critical point is at (x_0, y_0). An analogy with quadratic functions suggests the following test for classifying a critical point of a function of two variables:

Second-Derivative Test for Functions of Two Variables

Suppose (x_0, y_0) is a point where grad $f(x_0, y_0) = \vec{0}$. Let

$$D = f_{xx}(x_0, y_0)f_{yy}(x_0, y_0) - (f_{xy}(x_0, y_0))^2.$$

- If $D > 0$ and $f_{xx}(x_0, y_0) > 0$, then f has a local minimum at (x_0, y_0).
- If $D > 0$ and $f_{xx}(x_0, y_0) < 0$, then f has a local maximum at (x_0, y_0).
- If $D < 0$, then f has a saddle point at (x_0, y_0).
- If $D = 0$, anything can happen: f can have a local maximum, or a local minimum, or a saddle point, or none of these, at (x_0, y_0).

Example 6   Find the local maxima, minima, and saddle points of $f(x, y) = \frac{1}{2}x^2 + 3y^3 + 9y^2 - 3xy + 9y - 9x$.

Solution   Setting the partial derivatives of f to zero gives

$$f_x(x, y) = x - 3y - 9 = 0,$$
$$f_y(x, y) = 9y^2 + 18y - 3x + 9 = 0.$$

Eliminating x gives $9y^2 + 9y - 18 = 0$, with solutions $y = -2$ and $y = 1$. The corresponding values

---

[1]We assumed that $a \neq 0$. If $a = 0$ and $c \neq 0$, the same argument works. If both $a = 0$ and $c = 0$, then $f(x, y) = bxy$, which has a saddle point.

of x are $x = 3$ and $x = 12$, so the critical points of f are $(3, -2)$ and $(12, 1)$. The discriminant is

$$D(x, y) = f_{xx}f_{yy} - f_{xy}^2 = (1)(18y + 18) - (-3)^2 = 18y + 9.$$

Since $D(3, -2) = -36 + 9 < 0$, we know that $(3, -2)$ is a saddle point of f. Since $D(12, 1) = 18 + 9 > 0$ and $f_{xx}(12, 1) = 1 > 0$, we know that $(12, 1)$ is a local minimum of f.

The second-derivative test does not give any information if $D = 0$. However, as the following example illustrates, we may still be able to classify the critical points.

Example 7   Classify the critical points of $f(x, y) = x^4 + y^4$, and $g(x, y) = -x^4 - y^4$, and $h(x, y) = x^4 - y^4$.

Solution   Each of these functions has a critical point at $(0, 0)$. Since all the second partial derivatives are 0 there, each function has $D = 0$. Near the origin, the graphs of f, g and h look like the surfaces in Figures 15.13–15.15, respectively, so f has a local minimum at $(0, 0)$, and g has a local maximum at $(0, 0)$, and h is saddle-shaped at $(0, 0)$.

We can get the same results algebraically. Since $f(0, 0) = 0$ and $f(x, y) > 0$ elsewhere, f has a local minimum at the origin. Since $g(0, 0) = 0$ and $g(x, y) < 0$ elsewhere, g has a local maximum at the origin. Lastly, h is saddle-shaped at the origin since $h(0, 0) = 0$ and, away from the origin, $h(x, y) > 0$ on the x-axis and $h(x, y) < 0$ on the y-axis.

Summary for Section 15.1

- A function has a local maximum or minimum at a point where it takes on the largest or smallest value in a small region around the point:
  - f has a **local maximum** at the point P_0 if $f(P_0) \geq f(P)$ for all points P near P_0.
  - f has a **local minimum** at the point P_0 if $f(P_0) \leq f(P)$ for all points P near P_0.
- Points where the gradient is either $\vec{0}$ or undefined are called **critical points** of the function.
- If a (non-boundary) point is a local maximum or local minimum it **must** be a critical point.
- If a point is a critical point, it **need not** be a local maximum or minimum.
- The **second-derivative test for local extrema:** Suppose (x_0, y_0) is a point where $\operatorname{grad} f(x_0, y_0) = \vec{0}$. Let
$$D = f_{xx}(x_0, y_0)f_{yy}(x_0, y_0) - (f_{xy}(x_0, y_0))^2.$$
  - If $D > 0$ and $f_{xx}(x_0, y_0) > 0$, then f has a local minimum at (x_0, y_0).
  - If $D > 0$ and $f_{xx}(x_0, y_0) < 0$, then f has a local maximum at (x_0, y_0).
  - If $D < 0$, then f has a saddle point at (x_0, y_0).
  - If $D = 0$, anything can happen: f can have a local maximum, or a local minimum, or a saddle point, or none of these, at (x_0, y_0).

Exercises and Problems for Section 15.1

EXERCISES

1. Figures (I)–(VI) show level curves of six functions around a critical point P. Does each function have a local maximum, a local minimum, or a saddle point at P?

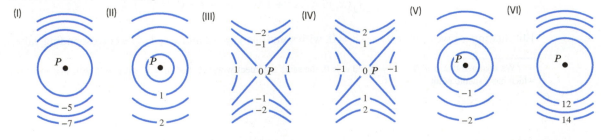

2. Which of the points A, B, C in Figure 15.17 appear to be critical points? Classify those that are critical points.

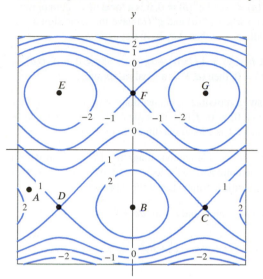

Figure 15.17

3. Which of the points D–G in Figure 15.17 appear to be
   (a) Local maxima?
   (b) Local minima?
   (c) Saddle points?

4. A function $f(x, y)$ has partial derivatives $f_x(1, 2) = 3$, $f_y(1, 2) = 5$. Explain how you know that f does not have a minimum at $(1, 2)$.

5. Assume $f(x, y)$ has a critical point at $(2, 3)$ with $f(2, 3) = 5$. Draw possible cross-sections for $x = 2$ and $y = 3$, and label one value on each axis, if $f(2, 3)$ is:
   (a) A local minimum
   (b) A local maximum
   (c) A saddle point

In Problems 6–13, the function has a critical point at $(0, 0)$. What sort of critical point is it?

6. $f(x, y) = x^2 - \cos y$

7. $f(x, y) = x \sin y$

8. $g(x, y) = x^4 + y^3$

9. $f(x, y) = x^6 + y^6$

10. $k(x, y) = \sin x \sin y$

11. $h(x, y) = \cos x \cos y$

12. $g(x, y) = (x - e^x)(1 - y^2)$

13. $h(x, y) = x^2 - xy + \sin^2 y$

In Problems 14–27, find the critical points and classify them as local maxima, local minima, saddle points, or none of these.

14. $f(x, y) = x^2 - 2xy + 3y^2 - 8y$

15. $f(x, y) = 5 + 6x - x^2 + xy - y^2$

16. $f(x, y) = x^2 - y^2 + 4x + 2y$

17. $f(x, y) = 400 - 3x^2 - 4x + 2xy - 5y^2 + 48y$

18. $f(x, y) = 15 - x^2 + 2y^2 + 6x - 8y$

19. $f(x, y) = x^2 y + 2y^2 - 2xy + 6$

20. $f(x, y) = 2x^3 - 3x^2 y + 6x^2 - 6y^2$

21. $f(x, y) = x^3 - 3x + y^3 - 3y$

22. $f(x, y) = x^3 + y^3 - 3x^2 - 3y + 10$

23. $f(x, y) = x^3 + y^3 - 6y^2 - 3x + 9$

24. $f(x, y) = (x + y)(xy + 1)$

25. $f(x, y) = 8xy - \frac{1}{4}(x + y)^4$

26. $f(x, y) = \sqrt[3]{x^2 + y^2}$

27. $f(x, y) = e^{2x^2 + y^2}$

PROBLEMS

28. Let $f(x, y) = 3x^2 + ky^2 + 9xy$. Determine the values of k (if any) for which the critical point at $(0, 0)$ is:
    (a) A saddle point
    (b) A local maximum
    (c) A local minimum

29. Let $f(x, y) = x^3 + ky^2 - 5xy$. Determine the values of k (if any) for which the critical point at $(0, 0)$ is:
    (a) A saddle point
    (b) A local maximum
    (c) A local minimum

30. Find A and B so that $f(x, y) = x^2 + Ax + y^2 + B$ has a local minimum value of 20 at $(1, 0)$.

31. For $f(x, y) = x^2 + xy + y^2 + ax + by + c$, find values of a, b, and c giving a local minimum at $(2, 5)$ and so that $f(2, 5) = 11$.

32. (a) Find critical points for $f(x, y) = e^{-(x-a)^2 - (y-b)^2}$.
    (b) Find a and b such that the critical point is at $(-1, 5)$.
    (c) For the values of a and b in part (b), is $(-1, 5)$ a local maximum, local minimum, or a saddle point?

33. Let $f(x, y) = kx^2 + y^2 - 4xy$. Determine the values of k (if any) for which the critical point at $(0, 0)$ is:
    (a) A saddle point
    (b) A local maximum
    (c) A local minimum

■ For Problems 34–36, use the contours of f in Figure 15.18.

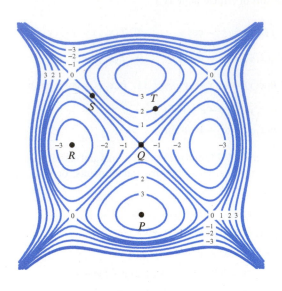

Figure 15.18

34. Decide whether you think each point is a local maximum, local minimum, saddle point, or none of these.

   (a) P (b) Q (c) R (d) S

35. Sketch the direction of ∇f at points surrounding each of P, R, S, and T.

36. At which of P, Q, R, S, or T does $\|\nabla f\|$ seem largest?

■ For Problems 37–40, find critical points and classify them as local maxima, local minima, saddle points, or none of these.

37. $f(x, y) = x^3 + e^{-y^2}$

38. $f(x, y) = \sin x \sin y$

39. $f(x, y) = 1 - \cos x + y^2/2$

40. $f(x, y) = e^x(1 - \cos y)$

41. At the point $(1, 3)$, suppose that $f_x = f_y = 0$ and $f_{xx} > 0, f_{yy} > 0, f_{xy} = 0$.

   (a) What can you conclude about the behavior of the function near the point $(1, 3)$?
   (b) Sketch a possible contour diagram.

42. At the point (a, b), suppose that $f_x = f_y = 0, f_{xx} > 0, f_{yy} = 0, f_{xy} > 0$.

   (a) What can you conclude about the shape of the graph of f near the point (a, b)?
   (b) Sketch a possible contour diagram.

43. Let $h(x, y) = f(x)g(y)$ where $f(0) = g(0) = 0$ and $f'(0) \neq 0, g'(0) \neq 0$. Show that $(0, 0)$ is a saddle point of h.

44. Let $h(x, y) = f(x) + g(y)$. Show that h has a critical point at (a, b) if $f'(a) = g'(b) = 0$, and, assuming $f''(a) \neq 0$ and $g''(b) \neq 0$, it is a local maximum or minimum when $f''(a)$ and $g''(b)$ have the same sign and a saddle point when they have opposite signs.

45. Let $h(x, y) = (f(x))^2 + (g(y))^2$. Show that if $f(a) = g(b) = 0$, then (a, b) is a local minimum.

46. Draw a possible contour diagram of f such that $f_x(-1, 0) = 0, f_y(-1, 0) < 0, f_x(3, 3) > 0, f_y(3, 3) > 0$, and f has a local maximum at $(3, -3)$.

47. Draw a possible contour diagram of a function with a saddle point at $(2, 1)$, a local minimum at $(2, 4)$, and no other critical points. Label the contours.

48. For constants a and b with $ab \neq 0$ and $ab \neq 1$, let

$$f(x, y) = ax^2 + by^2 - 2xy - 4x - 6y.$$

   (a) Find the x- and y-coordinates of the critical point. Your answer will be in terms of a and b.
   (b) If $a = b = 2$, is the critical point a local maximum, a local minimum, or neither? Give a reason for your answer.
   (c) Classify the critical point for all values of a and b with $ab \neq 0$ and $ab \neq 1$.

49. (a) Find the critical point of $f(x, y) = (x^2 - y)(x^2 + y)$.
   (b) Show that at the critical point, the discriminant $D = 0$, so the second-derivative test gives no information about the nature of the critical point.
   (c) Sketch contours near the critical point to determine whether it is a local maximum, a local minimum, a saddle point, or none of these.

50. On a computer, draw contour diagrams for functions

$$f(x, y) = k(x^2 + y^2) - 2xy$$

for $k = -2, -1, 0, 1, 2$. Use these figures to classify the critical point at $(0, 0)$ for each value of k. Explain your observations using the discriminant, D.

51. The behavior of a function can be complicated near a critical point where $D = 0$. Suppose that

$$f(x, y) = x^3 - 3xy^2.$$

Show that there is one critical point at $(0, 0)$ and that $D = 0$ there. Show that the contour for $f(x, y) = 0$ consists of three lines intersecting at the origin and that these lines divide the plane into six regions around the origin where f alternates from positive to negative. Sketch a contour diagram for f near $(0, 0)$. The graph of this function is called a *monkey saddle*.

52. The contour diagrams for four functions $z = f(x, y)$ are in (a)–(d). Each function has a critical point with $z = 0$ at the origin. Graphs (I)–(IV) show the value of z for these four functions on a small circle around the origin, expressed as function of θ, the angle between the positive x-axis and a line through the origin. Match the contour diagrams (a)–(d) with the graphs (I)–(IV). Classify the critical points as local maxima, local minima or saddle points.

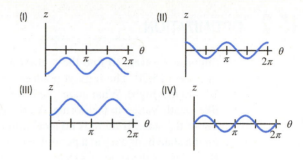

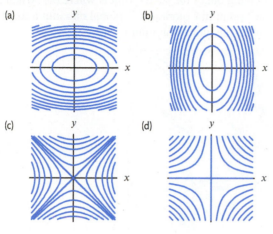

Strengthen Your Understanding

In Problems 53–55, explain what is wrong with the statement.

53. If $f_x = f_y = 0$ at $(1, 3)$, then f has a local maximum or local minimum at $(1, 3)$.

54. For $f(x, y)$, if $D = f_{xx}f_{yy} - (f_{xy})^2 = 0$ at (a, b), then (a, b) is a saddle point.

55. A critical point (a, b) for the function f must be a local minimum if both cross-sections for $x = a$ and $y = b$ are concave up.

In Problems 56–57, give an example of:

56. A nonlinear function having no critical points

57. A function $f(x, y)$ with a local maximum at $(2, -3, 4)$.

Are the statements in Problems 58–69 true or false? Give reasons for your answer.

58. If $f_x(P_0) = f_y(P_0) = 0$, then P_0 is a critical point of f.

59. If $f_x(P_0) = f_y(P_0) = 0$, then P_0 is a local maximum or local minimum of f.

60. If P_0 is a critical point of f, then P_0 is either a local maximum or local minimum of f.

61. If P_0 is a local maximum or local minimum of f, and not on the boundary of the domain of f, then P_0 is a critical point of f.

62. The function whose contour diagram is shown in Figure 15.19 has a saddle point at P.

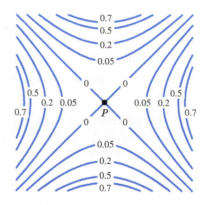

Figure 15.19

63. The function $f(x, y) = \sqrt{x^2 + y^2}$ has a local minimum at the origin.

64. The function $f(x, y) = x^2 - y^2$ has a local minimum at the origin.

65. If f has a local minimum at P_0 then so does the function $g(x, y) = f(x, y) + 5$.

66. If f has a local minimum at P_0 then the function $g(x, y) = -f(x, y)$ has a local maximum at P_0.

67. Every function has at least one local maximum.

68. If P_0 is a local maximum of f, then $f(a, b) \leq f(P_0)$ for all points (a, b) in 2-space.

69. If P_0 is a local maximum of f, then P_0 is also a global maximum of f.

15.2 OPTIMIZATION

Suppose we want to find the highest and the lowest points in Colorado. A contour map is shown in Figure 15.20. The highest point is the top of a mountain peak (point A on the map, Mt. Elbert, 14,440 feet high). What about the lowest point? Colorado does not have large pits without drainage, like Death Valley in California. A drop of rain falling at any point in Colorado will eventually flow out of the state. If there is no local minimum inside the state, where is the lowest point? It must be on the state boundary at a point where a river is flowing out of the state (point B where the Arikaree River leaves the state, 3,315 feet high). The highest point in Colorado is a global maximum for the elevation function in Colorado and the lowest point is the global minimum.

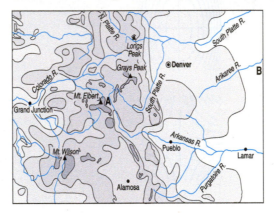

Figure 15.20: The highest and lowest points in the state of Colorado

In general, if we are given a function f defined on a region R, we say:

- f has a **global maximum on R** at the point P_0 if $f(P_0) \geq f(P)$ for all points P in R.
- f has a **global minimum on R** at the point P_0 if $f(P_0) \leq f(P)$ for all points P in R.

The process of finding a global maximum or minimum for a function f on a region R is called *optimization*. If the region R is not stated explicitly, we take it to be the whole xy-plane unless the context of the problem suggests otherwise.

How Do We Find Global Maxima and Minima?

As the Colorado example illustrates, a global extremum can occur either at a critical point inside the region or at a point on the boundary of the region. This is analogous to single-variable calculus, where a function achieves its global extrema on an interval either at a critical point inside the interval or at an endpoint of the interval.

To locate **global maxima and minima** for a function f on a region R:

- Find the critical points of f in the region R.
- Investigate whether the critical points give global maxima or minima.
- If the region R has a boundary, investigate whether f attains a global maximum or minimum on the boundary of R.

Investigating the boundary of a region for possible maxima and minima is the topic of Section 15.1. In this section, we focus on finding global maxima and minima of functions on regions that do not include boundaries.

Not all functions have a global maximum or minimum: it depends on the function and the region. First, we consider applications in which global extrema are expected from practical considerations. At the end of this section, we examine the conditions that lead to global extrema. In general, the fact that a function has a single local maximum or minimum does not guarantee that the point is the global maximum or minimum. (See Problem 38.) An exception is if the function is quadratic, in which case the local maximum or minimum is the global maximum or minimum. (See Example 1 on page 857 and Example 5 on page 860.)

Maximizing Profit and Minimizing Cost

In planning production of an item, a company often chooses the combination of price and quantity that maximizes its profit. We use

$$\text{Profit} = \text{Revenue} - \text{Cost},$$

and, provided the price is constant,

$$\text{Revenue} = \text{Price} \cdot \text{Quantity} = pq.$$

In addition, we need to know how the cost and price depend on quantity.

Example 1   A company manufactures two items which are sold in two separate markets where it has a monopoly. The quantities, q_1 and q_2, demanded by consumers, and the prices, p_1 and p_2 (in dollars), of each item are related by

$$p_1 = 600 - 0.3q_1 \quad \text{and} \quad p_2 = 500 - 0.2q_2.$$

Thus, if the price for either item increases, the demand for it decreases. The company's total production cost is given by

$$C = 16 + 1.2q_1 + 1.5q_2 + 0.2q_1q_2.$$

To maximize its total profit, how much of each product should be produced? What is the maximum profit? [2]

Solution   The total revenue, R, is the sum of the revenues, p_1q_1 and p_2q_2, from each market. Substituting for p_1 and p_2, we get

$$R = p_1q_1 + p_2q_2 = (600 - 0.3q_1)q_1 + (500 - 0.2q_2)q_2$$
$$= 600q_1 - 0.3q_1^2 + 500q_2 - 0.2q_2^2.$$

Thus, the total profit P is given by

$$P = R - C = 600q_1 - 0.3q_1^2 + 500q_2 - 0.2q_2^2 - (16 + 1.2q_1 + 1.5q_2 + 0.2q_1q_2)$$
$$= -16 + 598.8q_1 - 0.3q_1^2 + 498.5q_2 - 0.2q_2^2 - 0.2q_1q_2.$$

Since q_1 and q_2 cannot be negative,[3] the region we consider is the first quadrant with boundary $q_1 = 0$ and $q_2 = 0$.

To maximize P, we look for critical points by setting the partial derivatives equal to 0:

$$\frac{\partial P}{\partial q_1} = 598.8 - 0.6q_1 - 0.2q_2 = 0,$$

$$\frac{\partial P}{\partial q_2} = 498.5 - 0.4q_2 - 0.2q_1 = 0.$$

Since grad P is defined everywhere, the only critical points of P are those where grad $P = \vec{0}$. Thus, solving for q_1, and q_2, we find that

$$q_1 = 699.1 \quad \text{and} \quad q_2 = 896.7.$$

---

[2] Adapted from M. Rosser and P. Lis, *Basic Mathematics for Economists*, 3rd ed. (New York: Routledge, 2016), p. 351.
[3] Restricting prices to be nonnegative further restricts the region but does not alter the solution.

The corresponding prices are

$$p_1 = 390.27 \quad \text{and} \quad p_2 = 320.66.$$

To see whether or not we have found a local maximum, we compute second partial derivatives:

$$\frac{\partial^2 P}{\partial q_1^2} = -0.6, \quad \frac{\partial^2 P}{\partial q_2^2} = -0.4, \quad \frac{\partial^2 P}{\partial q_1 \partial q_2} = -0.2,$$

so,

$$D = \frac{\partial^2 P}{\partial q_1^2} \frac{\partial^2 P}{\partial q_2^2} - \left(\frac{\partial^2 P}{\partial q_1 \partial q_2} \right)^2 = (-0.6)(-0.4) - (-0.2)^2 = 0.2.$$

Therefore we have found a local maximum. The graph of P is a paraboloid opening downward, so $(699.1, 896.7)$ is a global maximum. This point is within the region, so points on the boundary give smaller values of P.

The company should produce 699.1 units of the first item priced at \$390.27 per unit, and 896.7 units of the second item priced at \$320.66 per unit. The maximum profit $P(699.1, 896.7) \approx \$433,000$.

Example 2   A delivery of 480 cubic meters of gravel is to be made to a landfill. The trucker plans to purchase an open-top box in which to transport the gravel in numerous trips. The total cost to the trucker is the cost of the box plus \$80 per trip. The box must have height 2 meters, but the trucker can choose the length and width. The cost of the box is \$100/m^2 for the ends, \$50/m^2 for the sides and \$200/m^2 for the bottom. Notice the tradeoff: A smaller box is cheaper to buy but requires more trips. What size box should the trucker buy to minimize the total cost?[4]

Solution   We first get an algebraic expression for the trucker's cost. Let the length of the box be x meters and the width be y meters; the height is 2 meters. (See Figure 15.21.)

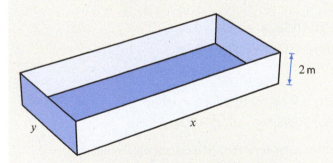

Table 15.2   *Trucker's itemized cost*

Expense	Cost in dollars
Travel: $480/(2xy)$ at \$80/trip	$80 \cdot 480/(2xy)$
Ends: 2 at \$100/m$^2 \cdot 2y$ m^2	$400y$
Sides: 2 at \$50/m$^2 \cdot 2x$ m^2	$200x$
Bottom: 1 at \$200/m$^2 \cdot xy$ m^2	$200xy$

Figure 15.21: The box for transporting gravel

The volume of the box is $2xy$ m^3, so delivery of 480 m^3 of gravel requires $480/(2xy)$ trips. The number of trips is a whole number; however, we treat it as continuous so that we can optimize using derivatives. The trucker's cost is itemized in Table 15.2. The problem is to minimize

$$\text{Total cost} = 80 \cdot \frac{480}{2xy} + 400y + 200x + 200xy = 200 \left(\frac{96}{xy} + 2y + x + xy \right).$$

The length and width of the box must be positive. Thus, the region is the first quadrant but it does not contain the boundary, $x = 0$ and $y = 0$.

[4]Adapted from Claude McMillan, Jr., *Mathematical Programming*, 2nd ed. (New York: Wiley, 1978), pp. 156–157.

Our problem is to minimize

$$f(x, y) = \frac{96}{xy} + 2y + x + xy.$$

The critical points of this function occur where

$$f_x(x, y) = -\frac{96}{x^2 y} + 1 + y = 0$$

$$f_y(x, y) = -\frac{96}{xy^2} + 2 + x = 0.$$

We put the $96/(x^2 y)$ and $96/(xy^2)$ terms on the other side of the the equation, divide, and simplify:

$$\frac{96/(x^2 y)}{96/(xy^2)} = \frac{1+y}{2+x} \quad \text{so} \quad \frac{y}{x} = \frac{1+y}{2+x} \quad \text{giving} \quad 2y = x.$$

Substituting $x = 2y$ in the equation $f_y(x, y) = 0$ gives

$$-\frac{96}{2y \cdot y^2} + 2 + 2y = 0$$

$$y^4 + y^3 - 24 = 0.$$

The only positive solution to this equation is $y = 2$, so the only critical point in the region is $(4, 2)$.
To check that the critical point is a local minimum, we use the second-derivative test. Since

$$D(4, 2) = f_{xx} f_{yy} - (f_{xy})^2 = \frac{192}{4^3 \cdot 2} \cdot \frac{192}{4 \cdot 2^3} - \left(\frac{96}{4^2 \cdot 2^2} + 1\right)^2 = 9 - \frac{25}{4} > 0$$

and $f_{xx}(4, 2) > 0$, the point $(4, 2)$ is a local minimum. Since the value of f increases without bound as x or y increases without bound and as $x \to 0^+$ and $y \to 0^+$, it can be shown that $(4, 2)$ is a global minimum. (See Problem 42.) Thus, the optimal box is 4 meters long and 2 meters wide. With a box of this size, the trucker would need to make 30 trips to haul all of the gravel. This large number lends some credibility to our decision to treat the number of trips as a continuous variable.

Fitting a Line to Data: Least Squares

Suppose we want to fit the "best" line to some data in the plane. We measure the distance from a line to the data points by adding the squares of the vertical distances from each point to the line. The smaller this sum of squares is, the better the line fits the data. The line with the minimum sum of square distances is called the *least squares line*, or the *regression line*. If the data is nearly linear, the least squares line is a good fit; otherwise it may not be. (See Figure 15.22.)

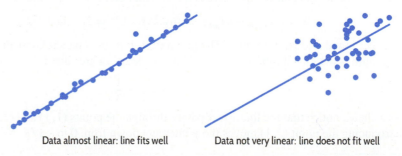

Data almost linear: line fits well   Data not very linear: line does not fit well

Figure 15.22: Fitting lines to data points

Example 3   Find a least squares line for the following data points: $(1, 1)$, $(2, 1)$, and $(3, 3)$.

Solution   Suppose the line has equation $y = b + mx$. If we find b and m then we have found the line. So, for this problem, b and m are the two variables. Any values of m and b are possible, so this is an unconstrained problem. We want to minimize the function $f(b, m)$ that gives the sum of the three squared vertical distances from the points to the line in Figure 15.23.

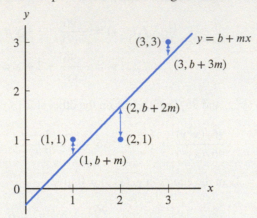

Figure 15.23: The least squares line minimizes the sum of the squares of these vertical distances

The vertical distance from the point $(1, 1)$ to the line is the difference in the y-coordinates $1 - (b + m)$; similarly for the other points. Thus, the sum of squares is

$$f(b, m) = (1 - (b + m))^2 + (1 - (b + 2m))^2 + (3 - (b + 3m))^2.$$

To minimize f we look for critical points. First we differentiate f with respect to b:

$$\frac{\partial f}{\partial b} = -2(1 - (b + m)) - 2(1 - (b + 2m)) - 2(3 - (b + 3m))$$

$$= -2 + 2b + 2m - 2 + 2b + 4m - 6 + 2b + 6m$$

$$= -10 + 6b + 12m.$$

Now we differentiate with respect to m:

$$\frac{\partial f}{\partial m} = 2(1 - (b + m))(-1) + 2(1 - (b + 2m))(-2) + 2(3 - (b + 3m))(-3)$$

$$= -2 + 2b + 2m - 4 + 4b + 8m - 18 + 6b + 18m$$

$$= -24 + 12b + 28m.$$

The equations $\dfrac{\partial f}{\partial b} = 0$ and $\dfrac{\partial f}{\partial m} = 0$ give a system of two linear equations in two unknowns:

$$-10 + 6b + 12m = 0,$$
$$-24 + 12b + 28m = 0.$$

The solution to this pair of equations is the critical point $b = -1/3$ and $m = 1$. Since

$$D = f_{bb}f_{mm} - (f_{mb})^2 = (6)(28) - 12^2 = 24 \quad \text{and} \quad f_{bb} = 6 > 0,$$

we have found a local minimum. The graph of $f(b, m)$ is a parabola opening upward, so the local minimum is the global minimum of f. Thus, the least squares line is

$$y = x - \frac{1}{3}.$$

As a check, notice that the line $y = x$ passes through the points $(1, 1)$ and $(3, 3)$. It is reasonable that introducing the point $(2, 1)$ moves the y-intercept down from 0 to $-1/3$.

The general formulas for the slope and y-intercept of a least squares line are in Project 2 (available online). Many calculators have built-in formulas for b and m, as well as for the *correlation coefficient*, which measures how well the data points fit the least squares line.

How Do We Know Whether a Function Has a Global Maximum or Minimum?

Under what circumstances does a function of two variables have a global maximum or minimum? The next example shows that a function may have both a global maximum and a global minimum on a region, or just one, or neither.

Example 4   Investigate the global maxima and minima of the following functions:

    (a) $h(x, y) = 1 + x^2 + y^2$ on the disk $x^2 + y^2 \leq 1$.
    (b) $f(x, y) = x^2 - 2x + y^2 - 4y + 5$ on the xy-plane.
    (c) $g(x, y) = x^2 - y^2$ on the xy-plane.

Solution   (a) The graph of $h(x, y) = 1 + x^2 + y^2$ is a bowl-shaped paraboloid with a global minimum of 1 at $(0, 0)$, and a global maximum of 2 on the edge of the region, $x^2 + y^2 = 1$.
    (b) The graph of f in Figure 15.5 on page 857 shows that f has a global minimum at the point $(1, 2)$ and no global maximum (because the value of f increases without bound as $x \to \infty$, $y \to \infty$).
    (c) The graph of g in Figure 15.7 on page 858 shows that g has no global maximum because $g(x, y) \to \infty$ as $x \to \infty$ if y is constant. Similarly, g has no global minimum because $g(x, y) \to -\infty$ as $y \to \infty$ if x is constant.

Sometimes a function is guaranteed to have a global maximum and minimum. For example, a continuous function, $h(x)$, of one variable has a global maximum and minimum on every closed interval $a \leq x \leq b$. On a non-closed interval, such as $a \leq x < b$ or $a < x < b$, or on an unbounded interval, such as $a < x < \infty$, h may not have a maximum or minimum value.

What is the situation for functions of two variables? As it turns out, a similar result is true for continuous functions defined on regions which are closed and bounded, analogous to the closed and bounded interval $a \leq x \leq b$. In everyday language we say

> • A **closed** region is one which contains its boundary;
> • A **bounded** region is one which does not stretch to infinity in any direction.

More precise definitions follow. Suppose R is a region in 2-space. A point (x_0, y_0) is a *boundary point* of R if, for every $r > 0$, the circular disk with center (x_0, y_0) and radius r contains both points which are in R and points which are not in R. See Figure 15.24. A point (x_0, y_0) can be a boundary point of the region R without belonging to R. The collection of all the boundary points is the *boundary* of R. The region R is *closed* if it contains its boundary.

A region R in 2-space is *bounded* if the distance between every point (x, y) in R and the origin is less than some constant K. Closed and bounded regions in 3-space are defined in the same way.

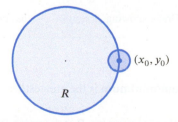

(x_0, y_0)

R

Figure 15.24: Boundary point (x_0, y_0) of R

Example 5   (a) Consider the square $-1 \leq x \leq 1$, $-1 \leq y \leq 1$. Every point in this region is within distance $\sqrt{2}$ of the origin, so the region is bounded. The region's boundary consists of four line segments, all of which belong to the region, so the region is closed.

(b) Consider the first quadrant $x \geq 0$, $y \geq 0$. The boundary of this region consists of the origin, the positive x-axis, and the positive y-axis. All of these belong to the region, so the region is closed. However, the region is not bounded, since there is no upper bound on distances between points in the region and the origin.

(c) The disk $x^2 + y^2 < 1$ is bounded, because each point in the region is within distance 1 of the origin. However, the disk is not closed, because $(1, 0)$ is a boundary point of the region but not included in the region.

(d) The half-plane $y > 0$ is neither closed nor bounded. The origin is a boundary point of this region but is not included in the region.

The reason that closed and bounded regions are useful is the following theorem, which is also true for functions of three or more variables:[5]

Theorem 15.1: Extreme Value Theorem for Multivariable Functions

If f is a continuous function on a closed and bounded region R, then f has a global maximum at some point (x_0, y_0) in R and a global minimum at some point (x_1, y_1) in R.

If f is not continuous or the region R is not closed and bounded, there is no guarantee that f achieves a global maximum or global minimum on R. In Example 4, the function g is continuous but does not achieve a global maximum or minimum in 2-space, a region which is closed but not bounded. Example 6 illustrates what can go wrong when the region is bounded but not closed.

Example 6   Does the function f have a global maximum or minimum on the region R given by $0 < x^2 + y^2 \leq 1$?

$$f(x, y) = \frac{1}{x^2 + y^2}$$

Solution   The region R is bounded, but it is not closed since it does not contain the boundary point $(0, 0)$. We see from the graph of $z = f(x, y)$ in Figure 15.25 that f has a global minimum on the circle $x^2 + y^2 = 1$. However, $f(x, y) \to \infty$ as $(x, y) \to (0, 0)$, so f has no global maximum.

Figure 15.25: Graph showing $f(x, y) = \frac{1}{x^2 + y^2}$ has no global maximum on $0 < x^2 + y^2 \leq 1$

Summary for Section 15.2

- A **global maximum/minimum** is the greatest/least value taken on by a function f anywhere in its domain R:
  - f has a **global maximum on R** at the point P_0 if $f(P_0) \geq f(P)$ for all points P in R.
  - f has a **global minimum on R** at the point P_0 if $f(P_0) \leq f(P)$ for all points P in R.

[5]For a proof, see Walter Rudin, *Principles of Mathematical Analysis*, 3rd ed. (New York: McGraw-Hill, 1976), p. 89.

- To locate **global maxima and minima** for a function f on a region R:
  - Find the critical points of f in the region R.
  - Investigate whether the critical points give global maxima or minima.
  - If the region R has a boundary, investigate whether f attains a global maximum or minimum on the boundary of R.
- A function is guaranteed (by the **Extreme Value Theorem**) to have both a global maximum and global minimum over a **closed** and **bounded** domain:
  - A **closed** region is one which contains its boundary.
  - A **bounded** region is one which does not stretch to infinity in any direction.

Exercises and Problems for Section 15.2

EXERCISES

1. By looking at the weather map in Figure 12.1 on page 694, find the maximum and minimum daily high temperatures in the states of Mississippi, Alabama, Pennsylvania, New York, California, Arizona, and Massachusetts.

■ In Exercises 2–4, estimate the position and approximate value of the global maxima and minima on the closed region shown.

2.

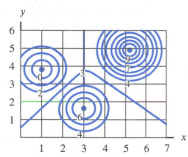

3.

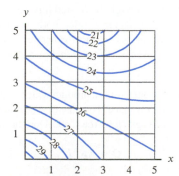

4.

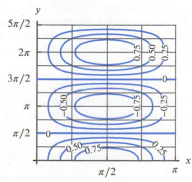

■ In Exercises 5–9, without calculus, find the highest and lowest points (if they exist) on the surface. The z-axis is upward.

5. $x^2 + y^2 + (z - 1)^2 = 49$

6. $(x + 1)^2 + (y - 3)^2 + 2z^2 = 162$

7. $z = (x - 5)^2 + (y - \pi)^2 + 2\pi$

8. $z = 44 - 2x^2 - 2y^2$

9. $x = 4 + y^2 + 2z^2$

10. The surface $z = 27 - x^2 - y^2$ cuts the plane $z = 2$ in a curve. Without calculus, find the point on this curve with the greatest y-coordinate.

■ In Exercises 11–13, find the global maximum and minimum of the function on $-1 \le x \le 1$, $-1 \le y \le 1$, and say whether it occurs on the boundary of the square. [Hint: Use graphs.]

11. $z = x^2 + y^2$ 12. $z = -x^2 - y^2$ 13. $z = x^2 - y^2$

■ In Exercises 14–21, does the function have a global maximum? A global minimum?

14. $f(x, y) = x^2 - 2y^2$ 15. $g(x, y) = x^2 y^2$

16. $h(x, y) = x^3 + y^3$ 17. $f(x, y) = -2x^2 - 7y^2$

18. $f(x, y) = e^{x^2 + y^2}$ 19. $h(x, y) = 1 - y^2 e^{xy}$

20. $f(x, y) = x^2/2 + 3y^3 + 9y^2 - 3x$

21. $g(x, y) = x^2 - \cos(x + y)$

PROBLEMS

22. **(a)** Compute and classify the critical points of $f(x, y) = 2x^2 - 3xy + 8y^2 + x - y$.
  (b) By completing the square, plot the contour diagram of f and show that the local extremum found in part (a) is a global one.

■ In Problems 23–26, find and classify the critical points of the function. Then decide whether the function has global extrema on the xy-plane, and find them if they exist.

23. $f(x, y) = y^2 + 2xy - y - x^3 - x + 2$

24. $f(x, y) = 2x^2 - 4xy + 5y^2 + 9y + 2$

25. $f(x, y) = -x^4 + 2xy^2 - 2y^3$

26. $f(x, y) = xe^{-x^2 - y^2}$

27. A closed rectangular box has volume 32 cm^3. What are the lengths of the edges giving the minimum surface area?

28. A closed rectangular box with faces parallel to the coordinate planes has one bottom corner at the origin and the opposite top corner in the first octant on the plane $3x + 2y + z = 1$. What is the maximum volume of such a box?

29. An international airline has a regulation that each passenger can carry a suitcase having the sum of its width, length and height less than or equal to 135 cm. Find the dimensions of the suitcase of maximum volume that a passenger may carry under this regulation.

30. Design a rectangular milk carton box of width w, length l, and height h which holds 512 cm^3 of milk. The sides of the box cost 1 cent/cm^2 and the top and bottom cost 2 cent/cm^2. Find the dimensions of the box that minimize the total cost of materials used.

31. Find the point on the plane $3x + 2y + z = 1$ that is closest to the origin by minimizing the square of the distance.

32. What is the shortest distance from the surface $xy + 3x + z^2 = 9$ to the origin?

33. For constants a, b, and c, let $f(x, y) = ax + by + c$ be a linear function, and let R be a region in the xy-plane.

  (a) If R is any disk, show that the maximum and minimum values of f on R occur on the boundary of the disk.
  (b) If R is any rectangle, show that the maximum and minimum values of f on R occur at the corners of the rectangle. They may occur at other points of the rectangle as well.
  (c) Use a graph of the plane $z = f(x, y)$ to explain your answers in parts (a) and (b).

34. Two products are manufactured in quantities q_1 and q_2 and sold at prices of p_1 and p_2, respectively. The cost of producing them is given by

$$C = 2q_1^2 + 2q_2^2 + 10.$$

(a) Find the maximum profit that can be made, assuming the prices are fixed.
(b) Find the rate of change of that maximum profit as p_1 increases.

35. A company operates two plants which manufacture the same item and whose total cost functions are

$$C_1 = 8.5 + 0.03q_1^2 \quad \text{and} \quad C_2 = 5.2 + 0.04q_2^2,$$

where q_1 and q_2 are the quantities produced by each plant. The company is a monopoly. The total quantity demanded, $q = q_1 + q_2$, is related to the price, p, by

$$p = 60 - 0.04q.$$

How much should each plant produce in order to maximize the company's profit?[6]

36. The quantity of a product demanded by consumers is a function of its price. The quantity of one product demanded may also depend on the price of other products. For example, if the only chocolate shop in town (a monopoly) sells milk and dark chocolates, the price it sets for each affects the demand of the other. The quantities demanded, q_1 and q_2, of two products depend on their prices, p_1 and p_2, as follows:

$$q_1 = 150 - 2p_1 - p_2$$

$$q_2 = 200 - p_1 - 3p_2.$$

(a) What does the fact that the coefficients of p_1 and p_2 are negative tell you? Give an example of two products that might be related this way.
(b) If one manufacturer sells both products, how should the prices be set to generate the maximum possible revenue? What is that maximum possible revenue?

37. A company manufactures a product which requires capital and labor to produce. The quantity, Q, of the product manufactured is given by the Cobb-Douglas function

$$Q = AK^a L^b,$$

where K is the quantity of capital; L is the quantity of labor used; and A, a, and b are positive constants with $0 < a < 1$ and $0 < b < 1$. One unit of capital costs $\$k$ and one unit of labor costs $\$\ell$. The price of the product is fixed at $\$p$ per unit.

(a) If $a + b < 1$, how much capital and labor should the company use to maximize its profit?
(b) Is there a maximum profit in the case $a + b = 1$? What about $a + b \geq 1$? Explain.

---

[6] Adapted from M. Rosser and P. Lis, *Basic Mathematics for Economists*, 3rd ed. (New York: Routledge, 2016), p. 354.

38. Let $f(x, y) = x^2(y + 1)^3 + y^2$. Show that f has only one critical point, namely $(0, 0)$, and that point is a local minimum but not a global minimum. Contrast this with the case of a function with a single local minimum in one-variable calculus.

39. Find the parabola of the form $y = ax^2 + b$ which best fits the points $(1, 0)$, $(2, 2)$, $(3, 4)$ by minimizing the sum of squares, S, given by

$$S = (a + b)^2 + (4a + b - 2)^2 + (9a + b - 4)^2.$$

40. For the data points $(11, 16)$, $(12, 17)$, $(13, 17)$, and $(16, 20)$, find an expression for $f(b, m)$, the sum of squared errors that are minimized on the least squares line $y = b + mx$. (You need not do the minimization.)

41. Find the least squares line for the data points $(0, 4)$, $(1, 3)$, $(2, 1)$.

42. Let $f(x, y) = 80/(xy) + 20y + 10x + 10xy$ in the region R where $x, y > 0$.

    (a) Explain why $f(x, y) > f(2, 1)$ at every point in R where

    (i) $x > 20$ (ii) $y > 20$

    (iii) $x < 0.01$ and $y \leq 20$

    (iv) $y < 0.01$ and $x \leq 20$

    (b) Explain why f must have a global minimum at a critical point in R.

    (c) Explain why f must have a global minimum in R at the point $(2, 1)$.

43. Let $f(x, y) = 2/x + 3/y + 4x + 5y$ in the region R where $x, y > 0$.

    (a) Explain why f must have a global minimum at some point in R.

    (b) Find the global minimum.

44. (a) The energy, E, required to compress a gas from a fixed initial pressure P_0 to a fixed final pressure P_F through an intermediate pressure p is[7]

$$E = \left(\frac{p}{P_0}\right)^2 + \left(\frac{P_F}{p}\right)^2 - 1.$$

How should p be chosen to minimize the energy?

    (b) Now suppose the compression takes place in two stages with two intermediate pressures, p_1 and p_2. What choices of p_1 and p_2 minimize the energy if

$$E = \left(\frac{p_1}{P_0}\right)^2 + \left(\frac{p_2}{p_1}\right)^2 + \left(\frac{P_F}{p_2}\right)^2 - 2?$$

45. The Dorfman-Steiner rule shows how a company which has a monopoly should set the price, p, of its product and how much advertising, a, it should buy. The price of advertising is p_a per unit. The quantity, q, of the product sold is given by $q = Kp^{-E}a^{\theta}$, where $K > 0$, $E > 1$, and $0 < \theta < 1$ are constants. The cost to the company to make each item is c.

    (a) How does the quantity sold, q, change if the price, p, increases? If the quantity of advertising, a, increases?

    (b) Show that the partial derivatives can be written in the form $\partial q/\partial p = -Eq/p$ and $\partial q/\partial a = \theta q/a$.

    (c) Explain why profit, π, is given by $\pi = pq - cq - p_a a$.

    (d) If the company wants to maximize profit, what must be true of the partial derivatives, $\partial \pi/\partial p$ and $\partial \pi/\partial a$?

    (e) Find $\partial \pi/\partial p$ and $\partial \pi/\partial a$.

    (f) Use your answers to parts (d) and (e) to show that at maximum profit,

$$\frac{p - c}{p} = \frac{1}{E} \quad \text{and} \quad \frac{p - c}{p_a} = \frac{a}{\theta q}.$$

    (g) By dividing your answers in part (f), show that at maximum profit,

$$\frac{p_a a}{pq} = \frac{\theta}{E}.$$

This is the Dorfman-Steiner rule, that the ratio of the advertising budget to revenue does not depend on the price of advertising.

Strengthen Your Understanding

In Problems 46–48, explain what is wrong with the statement.

46. A function having no critical points in a region R cannot have a global maximum in the region.

47. No continuous function has a global minimum on an unbounded region R.

48. If $f(x, y)$ has a local maximum value of 1 at the origin, then the global maximum is 1.

In Problems 49–50, give an example of:

49. A continuous function $f(x, y)$ that has no global maximum and no global minimum on the xy-plane.

50. A function $f(x, y)$ and a region R such that the maximum value of f on R is on the boundary of R.

[7]Adapted from Aris Rutherford, *Discrete Dynamic Programming*, p. 35 (New York: Blaisdell, 1964).

■ Are the statements in Problems **51–59** true or false? Give reasons for your answer.

51. If P_0 is a global maximum of f, where f is defined on all of 2-space, then P_0 is also a local maximum of f.

52. Every function has a global maximum.

53. The region consisting of all points (x, y) satisfying $x^2 + y^2 < 1$ is bounded.

54. The region consisting of all points (x, y) satisfying $x^2 + y^2 < 1$ is closed.

55. The function $f(x, y) = x^2 + y^2$ has a global minimum on the region $x^2 + y^2 < 1$.

56. The function $f(x, y) = x^2 + y^2$ has a global maximum on the region $x^2 + y^2 < 1$.

57. If P and Q are two distinct points in 2-space, and f has a global maximum at P, then f cannot have a global maximum at Q.

58. The function $f(x, y) = \sin(1 + e^{xy})$ must have a global minimum in the square region $0 \le x \le 1, 0 \le y \le 1$.

59. If P_0 is a global minimum of f on a closed and bounded region, then P_0 need not be a critical point of f.

15.3   CONSTRAINED OPTIMIZATION: LAGRANGE MULTIPLIERS

Many, perhaps most, real optimization problems are constrained by external circumstances. For example, a city wanting to build a public transportation system that will serve the greatest possible number of people has only a limited number of tax dollars it can spend on the project. In this section, we see how to find an optimum value under such constraints.

In Section 15.2, we saw how to optimize a function $f(x, y)$ on a region R. If the region R is the entire xy-plane, we have *unconstrained optimization*; if the region R is not the entire xy-plane, that is, if x or y is restricted in some way, then we have *constrained optimization*.

Graphical Approach: Maximizing Production Subject to a Budget Constraint

Suppose we want to maximize production under a budget constraint. Suppose production, f, is a function of two variables, x and y, which are quantities of two raw materials, and that

$$f(x, y) = x^{2/3} y^{1/3}.$$

If x and y are purchased at prices of p_1 and p_2 thousands of dollars per unit, what is the maximum production f that can be obtained with a budget of c thousand dollars?

To maximize f without regard to the budget, we simply increase x and y. However, the budget constraint prevents us from increasing x and y beyond a certain point. Exactly how does the budget constrain us? With prices of p_1 and p_2, the amount spent on x is $p_1 x$ and the amount spent on y is $p_2 y$, so we must have

$$g(x, y) = p_1 x + p_2 y \le c,$$

where $g(x, y)$ is the total cost of the raw materials and c is the budget in thousands of dollars.

Let's look at the case when $p_1 = p_2 = 1$ and $c = 3.78$. Then

$$x + y \le 3.78.$$

Figure 15.26 shows some contours of f and the budget constraint represented by the line $x + y = 3.78$. Any point on or below the line represents a pair of values of x and y that we can afford. A point on the line completely exhausts the budget, while a point below the line represents values of x and y which can be bought without using up the budget. Any point above the line represents a pair of values that we cannot afford.

To maximize f, we find the point which lies on the level curve with the largest possible value of f *and* which lies within the budget. The point must lie on the budget constraint because production is maximized when we spend all the available money. Unless we are at a point where the budget constraint is tangent to a contour of f, we can increase f by moving in some direction along the line representing the budget constraint in Figure 15.26. For example, if we are on the line to the left of the point of tangency, moving right on the constraint will increase f; if we are on the line to the right of the point of tangency, moving left will increase f. Thus, the maximum value of f on the budget constraint occurs at the point where the budget constraint is tangent to the contour $f = 2$.

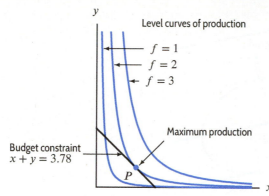

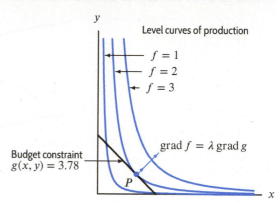

Figure 15.26: Optimal point, P, where budget constraint is tangent to a level of production function

Figure 15.27: At the point, P, of maximum production, the vectors grad f and grad g are parallel

Analytical Solution: Lagrange Multipliers

Figure 15.26 suggests that maximum production is achieved at the point where the budget constraint is tangent to a level curve of the production function. The method of Lagrange multipliers uses this fact in algebraic form. Figure 15.27 shows that at the optimum point, P, the gradient of f and the normal to the budget line $g(x, y) = x + y = 3.78$ are parallel. Thus, at P, grad f and grad g are parallel, so for some scalar λ, called the *Lagrange multiplier*,

$$\text{grad } f = \lambda \text{ grad } g.$$

Calculating the gradients, we find that

$$\left(\frac{2}{3}x^{-1/3}y^{1/3}\right)\vec{i} + \left(\frac{1}{3}x^{2/3}y^{-2/3}\right)\vec{j} = \lambda\left(\vec{i} + \vec{j}\right).$$

Equating components gives

$$\frac{2}{3}x^{-1/3}y^{1/3} = \lambda \quad \text{and} \quad \frac{1}{3}x^{2/3}y^{-2/3} = \lambda.$$

Eliminating λ gives

$$\frac{2}{3}x^{-1/3}y^{1/3} = \frac{1}{3}x^{2/3}y^{-2/3}, \quad \text{which leads to} \quad 2y = x.$$

Since the constraint $x + y = 3.78$ must be satisfied, we have $x = 2.52$ and $y = 1.26$. Then

$$f(2.52, 1.26) = (2.52)^{2/3}(1.26)^{1/3} \approx 2.$$

As before, we see that the maximum value of f is approximately 2. Thus, to maximize production on a budget of $3780, we should use 2.52 units of one raw material and 1.26 units of the other.

Lagrange Multipliers in General

Suppose we want to optimize an *objective function* $f(x, y)$ subject to a *constraint* $g(x, y) = c$. We look for extrema among the points which satisfy the constraint. We make the following definition.

Suppose P_0 is a point satisfying the constraint $g(x, y) = c$.

- f has a **local maximum** at P_0 **subject to the constraint** if $f(P_0) \geq f(P)$ for all points P near P_0 satisfying the constraint.
- f has a **global maximum** at P_0 **subject to the constraint** if $f(P_0) \geq f(P)$ for all points P satisfying the constraint.

Local and global minima are defined similarly.

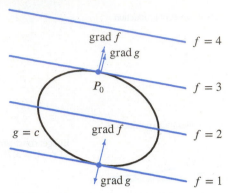

Figure 15.28: Maximum and minimum values
of $f(x, y)$ on $g(x, y) = c$ are at points where
grad f is parallel to grad g

As we saw in the production example, constrained extrema occur at points of tangency of contours of f and g; they can also occur at endpoints of constraints. At a point of tangency, grad f is perpendicular to the constraint and so parallel to grad g. At interior points on the constraint where grad f is not perpendicular to the constraint, the value of f can be increased or decreased by moving along the constraint. Therefore constrained extrema occur only at points where grad f and grad g are parallel or at endpoints of the constraint. (See Figure 15.28.) At points where the gradients are parallel, provided grad $g \neq \vec{0}$, there is a constant λ such that grad $f = \lambda$ grad g.

Optimizing f Subject to the Constraint $g = c$:
If a smooth function, f, has a maximum or minimum subject to a smooth constraint $g = c$ at a point P_0, then either P_0 satisfies the equations

$$\text{grad } f = \lambda \text{ grad } g \quad \text{and} \quad g = c,$$

or P_0 is an endpoint of the constraint, or grad $g(P_0) = \vec{0}$. To investigate whether P_0 is a global maximum or minimum, compare values of f at the points satisfying these three conditions. The number λ is called the **Lagrange multiplier**.

If the set of points satisfying the constraint is closed and bounded, such as a circle or line segment, then there must be a global maximum and minimum of f subject to the constraint. If the constraint is not closed and bounded, such as a line or hyperbola, then there may or may not be a global maximum and minimum.

Example 1   Find the maximum and minimum values of $x + y$ on the circle $x^2 + y^2 = 4$.

Solution   The objective function is

$$f(x, y) = x + y,$$

and the constraint is

$$g(x, y) = x^2 + y^2 = 4.$$

Since grad $f = f_x \vec{i} + f_y \vec{j} = \vec{i} + \vec{j}$ and grad $g = g_x \vec{i} + g_y \vec{j} = 2x\vec{i} + 2y\vec{j}$, the condition grad $f = \lambda$ grad g gives

$$1 = 2\lambda x \quad \text{and} \quad 1 = 2\lambda y,$$

so

$$x = y.$$

We also know that

$$x^2 + y^2 = 4,$$

giving $x = y = \sqrt{2}$ or $x = y = -\sqrt{2}$. The constraint has no endpoints (it's a circle) and $\operatorname{grad} g \neq \vec{0}$ on the circle, so we compare values of f at $(\sqrt{2}, \sqrt{2})$ and $(-\sqrt{2}, -\sqrt{2})$. Since $f(x, y) = x + y$, the maximum value of f is $f(\sqrt{2}, \sqrt{2}) = 2\sqrt{2}$; the minimum value is $f(-\sqrt{2}, -\sqrt{2}) = -2\sqrt{2}$. (See Figure 15.29.)

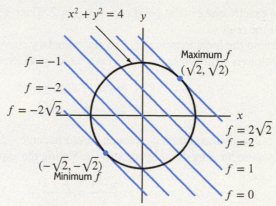

Figure 15.29: Maximum and minimum values of $f(x, y) = x + y$ on the circle $x^2 + y^2 = 4$ are at points where contours of f are tangent to the circle

How to Distinguish Maxima from Minima

There is a second-derivative test[8] for classifying the critical points of constrained optimization problems, but it is more complicated than the test in Section 15.1. However, a graph of the constraint and some contours usually shows which points are maxima, which points are minima, and which are neither.

Optimization with Inequality Constraints

The production problem that we looked at first was to maximize production $f(x, y)$ subject to a budget constraint

$$g(x, y) = p_1 x + p_2 y \leq c.$$

Since the inputs are nonnegative, $x \geq 0$ and $y \geq 0$, we have three inequality constraints, which restrict (x, y) to a region of the plane rather than to a curve in the plane. In principle, we should first check to see whether or not $f(x, y)$ has any critical points in the interior:

$$p_1 x + p_2 y < c, \qquad x > 0 \quad y > 0.$$

However, in the case of a budget constraint, we can see that the maximum of f must occur when the budget is exhausted, so we look for the maximum value of f on the boundary line:

$$p_1 x + p_2 y = c, \qquad x \geq 0 \quad y \geq 0.$$

> **Strategy for Optimizing $f(x, y)$ Subject to the Constraint $g(x, y) \leq c$**
> - Find all points in the region $g(x, y) < c$ where $\operatorname{grad} f$ is zero or undefined.
> - Use Lagrange multipliers to find the local extrema of f on the boundary $g(x, y) = c$.
> - Evaluate f at the points found in the previous two steps and compare the values.

From Section 15.2 we know that if f is continuous on a closed and bounded region, R, then f is guaranteed to attain its global maximum and minimum values on R.

---

[8]See J. E. Marsden and A. J. Tromba, *Vector Calculus*, 6th ed. (New York: W.H. Freeman, 2011), p. 220..

Example 2   Find the maximum and minimum values of $f(x, y) = (x - 1)^2 + (y - 2)^2$ subject to the constraint $x^2 + y^2 \leq 45$.

Solution   First, we look for all critical points of f in the interior of the region. Setting

$$f_x(x, y) = 2(x - 1) = 0$$
$$f_y(x, y) = 2(y - 2) = 0,$$

we find f has exactly one critical point at $x = 1$, $y = 2$. Since $1^2 + 2^2 < 45$, that critical point is in the interior of the region.

Next, we find the local extrema of f on the boundary curve $x^2 + y^2 = 45$. To do this, we use Lagrange multipliers with constraint $g(x, y) = x^2 + y^2 = 45$. Setting grad $f = \lambda$ grad g, we get

$$2(x - 1) = \lambda \cdot 2x,$$
$$2(y - 2) = \lambda \cdot 2y.$$

We can't have $x = 0$ since the first equation would become $-2 = 0$. Similarly, $y \neq 0$. So we can solve each equation for λ by dividing by x and y. Setting the expressions for λ equal gives

$$\frac{x - 1}{x} = \frac{y - 2}{y},$$

so

$$y = 2x.$$

Combining this with the constraint $x^2 + y^2 = 45$, we get

$$5x^2 = 45,$$

so

$$x = \pm 3.$$

Since $y = 2x$, we have possible local extrema at $x = 3$, $y = 6$ and $x = -3$, $y = -6$.

We conclude that the only candidates for the maximum and minimum values of f in the region occur at $(1, 2)$, $(3, 6)$, and $(-3, -6)$. Evaluating f at these three points, we find

$$f(1, 2) = 0, \qquad f(3, 6) = 20, \qquad f(-3, -6) = 80.$$

Therefore, the minimum value of f is 0 at $(1, 2)$ and the maximum value is 80 at $(-3, -6)$.

The Meaning of λ

In the uses of Lagrange multipliers so far, we never found (or needed) the value of λ. However, λ does have a practical interpretation. In the production example, we wanted to maximize

$$f(x, y) = x^{2/3} y^{1/3}$$

subject to the constraint

$$g(x, y) = x + y = 3.78.$$

We solved the equations

$$\frac{2}{3} x^{-1/3} y^{1/3} = \lambda,$$

$$\frac{1}{3} x^{2/3} y^{-2/3} = \lambda,$$

$$x + y = 3.78,$$

to get $x = 2.52$, $y = 1.26$ and $f(2.52, 1.26) \approx 2$. Continuing to find λ gives us

$$\lambda \approx 0.53.$$

Now we do another, apparently unrelated, calculation. Suppose our budget is increased by one, from 3.78 to 4.78, giving a new budget constraint of $x + y = 4.78$. Then the corresponding solution is at $x = 3.19$ and $y = 1.59$ and the new maximum value (instead of $f = 2$) is

$$f = (3.19)^{2/3}(1.59)^{1/3} \approx 2.53.$$

Notice that the amount by which f has increased is 0.53, the value of λ. Thus, in this example, the value of λ represents the extra production achieved by increasing the budget by one—in other words, the extra "bang" you get for an extra "buck" of budget. In fact, this is true in general:

- The value of λ is approximately the increase in the optimum value of f when the budget is increased by 1 unit.

More precisely:

- The value of λ represents the rate of change of the optimum value of f as the budget increases.

An Expression for λ

To interpret λ, we look at how the optimum value of the objective function f changes as the value c of the constraint function g is varied. In general, the optimum point (x_0, y_0) depends on the constraint value c. So, provided x_0 and y_0 are differentiable functions of c, we can use the chain rule to differentiate the optimum value $f(x_0(c), y_0(c))$ with respect to c:

$$\frac{df}{dc} = \frac{\partial f}{\partial x}\frac{dx_0}{dc} + \frac{\partial f}{\partial y}\frac{dy_0}{dc}.$$

At the optimum point (x_0, y_0), we have $f_x = \lambda g_x$ and $f_y = \lambda g_y$, and therefore

$$\frac{df}{dc} = \lambda \left(\frac{\partial g}{\partial x}\frac{dx_0}{dc} + \frac{\partial g}{\partial y}\frac{dy_0}{dc} \right) = \lambda \frac{dg}{dc}.$$

But, as $g(x_0(c), y_0(c)) = c$, we see that $dg/dc = 1$, so $df/dc = \lambda$. Thus, we have the following interpretation of the Lagrange multiplier λ:

> The value of λ is the rate of change of the optimum value of f as c increases (where $g(x, y) = c$). If the optimum value of f is written as $f(x_0(c), y_0(c))$, then
>
> $$\frac{d}{dc}f(x_0(c), y_0(c)) = \lambda.$$

Example 3   The quantity of goods produced according to the function $f(x, y) = x^{2/3}y^{1/3}$ is maximized subject to the budget constraint $x + y \leq 3.78$. The budget is increased to allow for a small increase in production. What is the price of the product if the sale of the additional goods covers the budget increase?

Solution   We know that $\lambda = 0.53$, which tells us that $df/dc = 0.53$. The constraint corresponds to a budget of \$3.78 thousand. Therefore increasing the budget by \$1000 increases production by about 0.53 units. In order to make the increase in budget profitable, the extra goods produced must sell for more than \$1000. Thus, if p is the price of each unit of the good, then $0.53p$ is the revenue from the extra 0.53 units sold. Thus, we need $0.53p \geq 1000$, so $p \geq 1000/0.53 = \$1890$.

The Lagrangian Function

Constrained optimization problems are frequently solved using a *Lagrangian function*, $\mathcal{L}$. For example, to optimize $f(x, y)$ subject to the constraint $g(x, y) = c$, we use the Lagrangian function

$$\mathcal{L}(x, y, \lambda) = f(x, y) - \lambda(g(x, y) - c).$$

To see how the function $\mathcal{L}$ is used, compute the partial derivatives of $\mathcal{L}$:

$$\frac{\partial \mathcal{L}}{\partial x} = \frac{\partial f}{\partial x} - \lambda \frac{\partial g}{\partial x},$$

$$\frac{\partial \mathcal{L}}{\partial y} = \frac{\partial f}{\partial y} - \lambda \frac{\partial g}{\partial y},$$

$$\frac{\partial \mathcal{L}}{\partial \lambda} = -(g(x, y) - c).$$

Notice that if (x_0, y_0) is an extreme point of $f(x, y)$ subject to the constraint $g(x, y) = c$ and λ_0 is the corresponding Lagrange multiplier, then at the point (x_0, y_0, λ_0) we have

$$\frac{\partial \mathcal{L}}{\partial x} = 0 \quad \text{and} \quad \frac{\partial \mathcal{L}}{\partial y} = 0 \quad \text{and} \quad \frac{\partial \mathcal{L}}{\partial \lambda} = 0.$$

In other words, (x_0, y_0, λ_0) is a critical point for the unconstrained Lagrangian function, $\mathcal{L}(x, y, \lambda)$. Thus, the Lagrangian converts a constrained optimization problem to an unconstrained problem.

Example 4

A company has a production function with three inputs x, y, and z given by

$$f(x, y, z) = 50x^{2/5}y^{1/5}z^{1/5}.$$

The total budget is \$24,000 and the company can buy x, y, and z at \$80, \$12, and \$10 per unit, respectively. What combination of inputs will maximize production?[9]

Solution

We need to maximize the objective function

$$f(x, y, z) = 50x^{2/5}y^{1/5}z^{1/5},$$

subject to the constraint

$$g(x, y, z) = 80x + 12y + 10z = 24,000.$$

The method for functions of two variables works for functions of three variables, so we construct the Lagrangian function

$$\mathcal{L}(x, y, z, \lambda) = 50x^{2/5}y^{1/5}z^{1/5} - \lambda(80x + 12y + 10z - 24,000),$$

and solve the system of equations we get from grad $\mathcal{L} = \vec{0}$:

$$\frac{\partial \mathcal{L}}{\partial x} = 20x^{-3/5}y^{1/5}z^{1/5} - 80\lambda = 0,$$

$$\frac{\partial \mathcal{L}}{\partial y} = 10x^{2/5}y^{-4/5}z^{1/5} - 12\lambda = 0,$$

$$\frac{\partial \mathcal{L}}{\partial z} = 10x^{2/5}y^{1/5}z^{-4/5} - 10\lambda = 0,$$

$$\frac{\partial \mathcal{L}}{\partial \lambda} = -(80x + 12y + 10z - 24,000) = 0.$$

Simplifying this system gives

$$\lambda = \frac{1}{4}x^{-3/5}y^{1/5}z^{1/5},$$

$$\lambda = \frac{5}{6}x^{2/5}y^{-4/5}z^{1/5},$$

$$\lambda = x^{2/5}y^{1/5}z^{-4/5},$$

$$80x + 12y + 10z = 24,000.$$

Eliminating z from the first two equations gives $x = 0.3y$. Eliminating x from the second and third equations gives $z = 1.2y$. Substituting for x and z into $80x + 12y + 10z = 24,000$ gives

$$80(0.3y) + 12y + 10(1.2y) = 24,000,$$

---

[9]Adapted from M. Rosser and P. Lis, *Basic Mathematics for Economists*, 3rd ed. (New York: Routledge, 2016), p. 360.

so $y = 500$. Then $x = 150$ and $z = 600$, and $f(150, 500, 600) = 4,622$ units.

The graph of the constraint, $80x + 12y + 10z = 24,000$, is a plane. Since the inputs x, y, z must be nonnegative, the graph is a triangle in the first octant, with edges on the coordinate planes. On the boundary of the triangle, one (or more) of the variables x, y, z is zero, so the function f is zero. Thus production is maximized within the budget using $x = 150$, $y = 500$, and $z = 600$.

Summary for Section 15.3

- Suppose P_0 is a point satisfying the constraint $g(x, y) = c$. Then a **constrained** local/global maxima is defined by:
  - f has a **local maximum** at P_0 **subject to the constraint** if $f(P_0) \geq f(P)$ for all points P near P_0 satisfying the constraint.
  - f has a **global maximum** at P_0 **subject to the constraint** if $f(P_0) \geq f(P)$ for all points P satisfying the constraint.

  Local and global minima are defined similarly.

- Constrained local extrema of a function $f(x, y)$ can occur at points where the contours of f are **tangent** to the constraint curve $g(x, y) = c$.

- Finding constrained extrema using the **Lagrange multiplier** λ:

  If a smooth function, f, has a maximum or minimum subject to a smooth constraint $g(x, y) = c$ at a point P_0, then either P_0 satisfies the equations

  $$\operatorname{grad} f = \lambda \operatorname{grad} g \quad \text{and} \quad g(x, y) = c,$$

  or P_0 is an endpoint of the constraint, or $\operatorname{grad} g(P_0) = \vec{0}$. To investigate whether P_0 is a global maximum or minimum, compare values of f at the points satisfying these three conditions.

- A **strategy** for optimizing $f(x, y)$ subject to the constraint $g(x, y) \leq c$ is:
  - Find all points in the region $g(x, y) < c$ where $\operatorname{grad} f$ is zero or undefined.
  - Use Lagrange multipliers to find the local extrema of f on the boundary $g(x, y) = c$.
  - Evaluate f at the points found in the previous two steps and compare the values.

- An **economic interpretation** of the value of λ in the Lagrange multiplier method: λ represents the rate of change of the optimum value of f as the budget constraint c in $g(x, y) = c$ increases.

Exercises and Problems for Section 15.3   Online Resource: Additional Problems for Section 15.3

EXERCISES

In Exercises 1–18, use Lagrange multipliers to find the maximum and minimum values of f subject to the given constraint, if such values exist.

1. $f(x, y) = x + y$, $\quad x^2 + y^2 = 1$

2. $f(x, y) = x + 3y + 2$, $\quad x^2 + y^2 = 10$

3. $f(x, y) = (x - 1)^2 + (y + 2)^2$, $\quad x^2 + y^2 = 5$

4. $f(x, y) = x^3 + y$, $\quad 3x^2 + y^2 = 4$

5. $f(x, y) = 3x - 2y$, $\quad x^2 + 2y^2 = 44$

6. $f(x, y) = xy$, $\quad 4x^2 + y^2 = 8$

7. $f(x, y) = 2xy$, $\quad 5x + 4y = 100$

8. $f(x_1, x_2) = x_1^2 + x_2^2$, $\quad x_1 + x_2 = 1$

9. $f(x, y) = x^2 + y$, $\quad x^2 - y^2 = 1$

10. $f(x, y, z) = x + 3y + 5z$, $\quad x^2 + y^2 + z^2 = 1$

11. $f(x, y, z) = x^2 - y^2 - 2z$, $\quad x^2 + y^2 = z$

12. $f(x, y, z) = xyz$, $\quad x^2 + y^2 + 4z^2 = 12$

13. $f(x, y) = x^2 + 2y^2$, $\quad x^2 + y^2 \leq 4$

14. $f(x, y) = x + 3y$, $\quad x^2 + y^2 \leq 2$

15. $f(x, y) = xy$, $\quad x^2 + 2y^2 \leq 1$

16. $f(x, y) = x^3 + y$, $\quad x + y \geq 1$

17. $f(x, y) = (x + 3)^2 + (y - 3)^2$, $\quad x^2 + y^2 \leq 2$

18. $f(x, y) = x^2 y + 3y^2 - y$, $\quad x^2 + y^2 \leq 10$

19. For each point marked in Figure 15.30, decide whether:

    (a) The point is a local minimum, maximum, or neither for the function f constrained by the loop.
    (b) The point is a global minimum, maximum, or neither subject to the constraint.

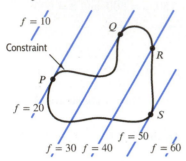

Figure 15.30

■ In Exercises 20–23, a Cobb-Douglas production function $P(K, L)$ and budget $B(K, L)$ are given, where K represents capital and L represents labor. Use Lagrange multipliers to find the values of K and L that maximize production given a budget constraint or minimize budget given a production constraint. Then give the value for λ and its meaning.

20. Maximize production: $P = K^{1/4} L^{3/4}$
    Budget constraint: $B = 2K + L = 40$

21. Maximize production: $P = K^{2/3} L^{1/3}$
    Budget constraint: $B = 10K + 4L = 60$

22. Maximize production: $P = K^{2/5} L^{3/5}$
    Budget constraint: $B = 4K + 5L = 100$

23. Minimize budget: $B = 4K + L$
    Production constraint: $P = K^{1/2} L^{1/2} = 200$

PROBLEMS

24. Find the maximum value of $f(x, y) = x + y - (x - y)^2$ on the triangular region $x \geq 0$, $y \geq 0$, $x + y \leq 1$.

25. For $f(x, y) = x^2 + 6xy$, find the global maximum and minimum on the closed region in the first quadrant bounded by the line $x + y = 4$ and the curve $xy = 3$.

26. (a) Draw contours of $f(x, y) = 2x + y$ for $z = -7, -5, -3, -1, 1, 3, 5, 7$.
    (b) On the same axes, graph the constraint $x^2 + y^2 = 5$.
    (c) Use the graph to approximate the points at which f has a maximum or a minimum value subject to the constraint $x^2 + y^2 = 5$.
    (d) Use Lagrange multipliers to find the maximum and minimum values of $f(x, y) = 2x + y$ subject to $x^2 + y^2 = 5$.

27. Let $f(x, y) = x^\alpha y^{1-\alpha}$ for $0 < \alpha < 1$. Find the value of α such that the maximum value of f on the line $2x + 3y = 6$ occurs at $(1.5, 1)$.

28. Figure 15.31 shows contours of f. Does f have a maximum value subject to the constraint $g(x, y) = c$ for $x \geq 0$, $y \geq 0$? If so, approximately where is it and what is its value? Does f have a minimum value subject to the constraint? If so, approximately where and what?

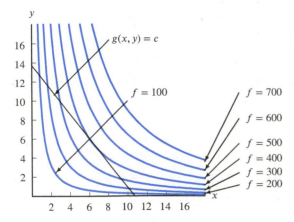

Figure 15.31

29. Each person tries to balance his or her time between leisure and work. The tradeoff is that as you work less your income falls. Therefore each person has *indifference curves* which connect the number of hours of leisure, l, and income, s. If, for example, you are indifferent between 0 hours of leisure and an income of $1125 a week on the one hand, and 10 hours of leisure and an income of $750 a week on the other hand, then the points $l = 0$, $s = 1125$, and $l = 10$, $s = 750$ both lie on the same indifference curve. Table 15.3 gives information on three indifference curves, I, II, and III.

Table 15.3

Weekly income			Weekly leisure hours		
I	II	III	I	II	III
1125	1250	1375	0	20	40
750	875	1000	10	30	50
500	625	750	20	40	60
375	500	625	30	50	70
250	375	500	50	70	90

(a) Graph the three indifference curves.
(b) You have 100 hours a week available for work and leisure combined, and you earn $10/hour. Write an equation in terms of l and s which represents this constraint.
(c) On the same axes, graph this constraint.
(d) Estimate from the graph what combination of leisure hours and income you would choose under these circumstances. Give the corresponding number of hours per week you would work.

30. Figure 15.32 shows ∇f for a function $f(x, y)$ and two curves $g(x, y) = 1$ and $g(x, y) = 2$. Mark the following:

   (a) The point(s) A where f has a local maximum.
   (b) The point(s) B where f has a saddle point.
   (c) The point C where f has a maximum on $g = 1$.
   (d) The point D where f has a minimum on $g = 1$.
   (e) If you used Lagrange multipliers to find C, what would the sign of λ be? Why?

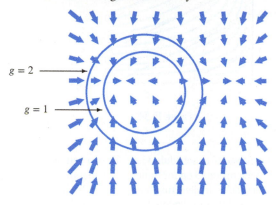

Figure 15.32

31. The point P is a maximum or minimum of the function f subject to the constraint $g(x, y) = x + y = c$, with $x, y \geq 0$. For the graphs (a) and (b), does P give a maximum or a minimum of f? What is the sign of λ? If P gives a maximum, where does the minimum of f occur? If P gives a minimum, where does the maximum of f occur?

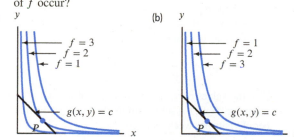

32. Figure 15.33 shows the optimal point (marked with a dot) in three optimization problems with the same constraint. Arrange the corresponding values of λ in increasing order. (Assume λ is positive.)

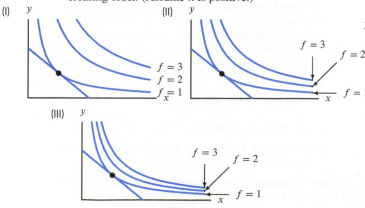

Figure 15.33

33. If the right side of the constraint in Exercise 5 is changed by the small amount Δc, by approximately how much do the maximum and minimum values change?

34. If the right side of the constraint in Exercise 6 is changed by the small amount Δc, by approximately how much do the maximum and minimum values change?

35. The function $P(x, y)$ gives the number of units produced and $C(x, y)$ gives the cost of production.

   (a) A company wishes to maximize production at a fixed cost of \$50,000. What is the objective function f? What is the constraint equation? What is the meaning of λ in this situation?
   (b) A company wishes to minimize costs at a fixed production level of 2000 units. What is the objective function f? What is the constraint equation? What is the meaning of λ in this situation?

36. Design a closed cylindrical container which holds 100 cm^3 and has the minimal possible surface area. What should its dimensions be?

37. A company manufactures x units of one item and y units of another. The total cost in dollars, C, of producing these two items is approximated by the function

$$C = 5x^2 + 2xy + 3y^2 + 800.$$

   (a) If the production quota for the total number of items (both types combined) is 39, find the minimum production cost.
   (b) Estimate the additional production cost or savings if the production quota is raised to 40 or lowered to 38.

38. An international organization must decide how to spend the \$2,000,000 they have been allotted for famine relief in a remote area. They expect to divide the money between buying rice at \$38.5/sack and beans at \$35/sack. The number, P, of people who would be fed if they buy x sacks of rice and y sacks of beans is given by

$$P = 1.1x + y - \frac{xy}{10^8}.$$

What is the maximum number of people that can be fed, and how should the organization allocate its money?

39. The quantity, q, of a product manufactured depends on the number of workers, W, and the amount of capital invested, K, and is given by

$$q = 6W^{3/4}K^{1/4}.$$

Labor costs are \$10 per worker and capital costs are \$20 per unit, and the budget is \$3000.

   (a) What are the optimum number of workers and the optimum number of units of capital?

(b) Show that at the optimum values of W and K, the ratio of the marginal productivity of labor ($\partial q/\partial W$) to the marginal productivity of capital ($\partial q/\partial K$) is the same as the ratio of the cost of a unit of labor to the cost of a unit of capital.

(c) Recompute the optimum values of W and K when the budget is increased by one dollar. Check that increasing the budget by \$1 allows the production of λ extra units of the good, where λ is the Lagrange multiplier.

40. A neighborhood health clinic has a budget of \$600,000 per quarter. The director of the clinic wants to allocate the budget to maximize the number of patient visits, V, which is given as a function of the number of doctors, D, and the number of nurses, N, by

$$V = 1000D^{0.6}N^{0.3}.$$

A doctor gets \$40,000 per quarter; nurses get \$10,000 per quarter.

(a) Set up the director's constrained optimization problem.

(b) Describe, in words, the conditions which must be satisfied by $\partial V/\partial D$ and $\partial V/\partial N$ for V to have an optimum value.

(c) Solve the problem formulated in part (a).

(d) Find the value of the Lagrange multiplier and interpret its meaning in this problem.

(e) At the optimum point, what is the marginal cost of a patient visit (that is, the cost of an additional visit)? Will that marginal cost rise or fall with the number of visits? Why?

41. (a) In Problem 39, does the value of λ change if the budget changes from \$3000 to \$4000?

(b) In Problem 40, does the value of λ change if the budget changes from \$600,000 to \$700,000?

(c) What condition must a Cobb-Douglas production function, $Q = cK^aL^b$, satisfy to ensure that the marginal increase of production (that is, the rate of increase of production with budget) is not affected by the size of the budget?

42. The production function $P(K, L)$ gives the number of pairs of skis produced per week at a factory operating with K units of capital and L units of labor. The contour diagram for P is in Figure 15.34; the parallel lines are budget constraints for budgets, B, in dollars.

(a) On each budget constraint, mark the point that gives the maximum production.

(b) Complete the table, where the budget, B, is in dollars and the maximum production is the number of pairs of skis to be produced each week.

B	2000	4000	6000	8000	10000
M					

(c) Estimate the Lagrange multiplier $\lambda = dM/dB$ at a budget of \$6000. Give units for the multiplier.

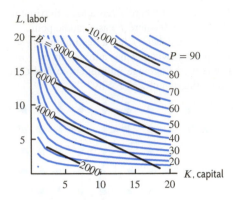

Figure 15.34

43. A doctor wants to schedule visits for two patients who have been operated on for tumors so as to minimize the expected delay in detecting a new tumor. Visits for patients 1 and 2 are scheduled at intervals of x_1 and x_2 weeks, respectively. A total of m visits per week is available for both patients combined.

The recurrence rates for tumors for patients 1 and 2 are judged to be v_1 and v_2 tumors per week, respectively. Thus, $v_1/(v_1+v_2)$ and $v_2/(v_1+v_2)$ are the probabilities that patient 1 and patient 2, respectively, will have the next tumor. It is known that the expected delay in detecting a tumor for a patient checked every x weeks is $x/2$. Hence, the expected detection delay for both patients combined is given by[10]

$$f(x_1,x_2) = \frac{v_1}{v_1+v_2}\cdot\frac{x_1}{2} + \frac{v_2}{v_1+v_2}\cdot\frac{x_2}{2}.$$

Find the values of x_1 and x_2 in terms of v_1 and v_2 that minimize $f(x_1, x_2)$ subject to the fact that m, the number of visits per week, is fixed.

44. What is the value of the Lagrange multiplier in Problem 43? What are the units of λ? What is its practical significance to the doctor?

45. Figure 15.35 shows two weightless springs with spring constants k_1 and k_2 attached between a ceiling and floor without tension or compression. A mass m is placed between the springs which settle into equilibrium as in Figure 15.36. The magnitudes f_1 and f_2 of the forces of the springs on the mass minimize the complementary energy

$$\frac{f_1^2}{2k_1} + \frac{f_2^2}{2k_2}$$

subject to the force balance constraint $f_1 + f_2 = mg$.

(a) Determine f_1 and f_2 by the method of Lagrange multipliers.

---

[10]Adapted from Daniel Kent, Ross Shachter, et al., "Efficient Scheduling of Cystoscopies in Monitoring for Recurrent Bladder Cancer," *Medical Decision Making* (Philadelphia: Hanley and Belfus, 1989).

(b) If you are familiar with Hooke's law, find the meaning of λ.

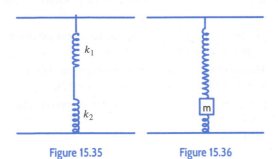

Figure 15.35   **Figure 15.36**

46. (a) If $\sum_{i=1}^{3} x_i = 1$, find the values of x_1, x_2, x_3 making $\sum_{i=1}^{3} x_i^2$ minimum.
(b) Generalize the result of part (a) to find the minimum value of $\sum_{i=1}^{n} x_i^2$ subject to $\sum_{i=1}^{n} x_i = 1$.

47. Let $f(x, y) = ax^2 + bxy + cy^2$. Show that the maximum value of $f(x, y)$ subject to the constraint $x^2 + y^2 = 1$ is equal to λ, the Lagrange multiplier.

48. Find the minimum distance from the point $(1, 2, 10)$ to the paraboloid given by the equation $z = x^2 + y^2$. Give a geometric justification for your answer.

49. A company produces one product from two inputs (for example, capital and labor). Its production function $g(x, y)$ gives the quantity of the product that can be produced with x units of the first input and y units of the second. The *cost function* (or *expenditure function*) is the three-variable function $C(p, q, u)$ where p and q are the unit prices of the two inputs. For fixed p, q, and u, the value $C(p, q, u)$ is the minimum of $f(x, y) = px + qy$ subject to the constraint $g(x, y) = u$.

(a) What is the practical meaning of $C(p, q, u)$?
(b) Find a formula for $C(p, q, u)$ if $g(x, y) = xy$.

50. A *utility function* $U(x, y)$ for two items gives the utility (benefit) to a consumer of x units of item 1 and y units of item 2. The *indirect utility function* is the three-variable function $V(p, q, I)$ where p and q are the unit prices of the two items. For fixed p, q, and I, the value $V(p, q, I)$ is the maximum of $U(x, y)$ subject to the constraint $px + qy = I$.

(a) What is the practical meaning of $V(p, q, I)$?
(b) The Lagrange multiplier λ that arises in the maximization defining V is called the *marginal utility of money*. What is its practical meaning?
(c) Find formulas for $V(p, q, I)$ and λ if $U(x, y) = xy$.

51. The function $h(x, y) = x^2 + y^2 - \lambda(2x + 4y - 15)$ has a minimum value $m(\lambda)$ for each value of λ.

(a) Find $m(\lambda)$.
(b) For which value of λ is $m(\lambda)$ the largest and what is that maximum value?
(c) Find the minimum value of $f(x, y) = x^2 + y^2$ subject to the constraint $2x + 4y = 15$ using the method of Lagrange multipliers and evaluate λ.
(d) Compare your answers to parts (b) and (c).

52. Let f be differentiable and grad $f(2, 1) = -3\vec{i} + 4\vec{j}$. You want to see if $(2, 1)$ is a candidate for the maximum and minimum values of f subject to a constraint satisfied by the point $(2, 1)$.

(a) Show $(2, 1)$ is not a candidate if the constraint is $x^2 + y^2 = 5$.
(b) Show $(2, 1)$ is a candidate if the constraint is $(x - 5)^2 + (y + 3)^2 = 25$. From a sketch of the contours for f near $(2, 1)$ and the constraint, decide whether $(2, 1)$ is a candidate for a maximum or minimum.
(c) Do the same as part (b), but using the constraint $(x + 1)^2 + (y - 5)^2 = 25$.

53. A person's satisfaction from consuming a quantity x_1 of one item and a quantity x_2 of another item is given by

$$S = u(x_1, x_2) = a \ln x_1 + (1 - a) \ln x_2,$$

where a is a constant, $0 < a < 1$. The prices of the two items are p_1 and p_2 respectively, and the budget is b.

(a) Express the maximum satisfaction that can be achieved as a function of p_1, p_2, and b.
(b) Find the amount of money that must be spent to achieve a particular level of satisfaction, c, as a function of p_1, p_2, and c.

Strengthen Your Understanding

In Problems 54–55, explain what is wrong with the statement.

54. The function $f(x, y) = xy$ has a maximum of 2 on the constraint $x + y = 2$.

55. If the level curves of $f(x, y)$ and the level curves of $g(x, y)$ are not tangent at any point on the constraint $g(x, y) = c$, $x \geq 0$, $y \geq 0$, then f has no maximum on the constraint.

In Problems 56–60, give an example of:

56. A function $f(x, y)$ whose maximum subject to the constraint $x^2 + y^2 = 5$ is at $(3, 4)$.

57. A function $f(x, y)$ to be optimized with constraint $x^2 + 2y^2 \leq 1$ such that the minimum value does not change when the constraint is changed to $x^2 + y^2 \leq 1 + c$ for $c > 0$.

58. A function $f(x, y)$ with a minimum at $(1, 1)$ on the constraint $x + y = 2$.

59. A function $f(x, y)$ that has a maximum but no minimum on the constraint $x + y = 4$.

60. A contour diagram of a function f whose maximum value on the constraint $x + 2y = 6$, $x \geq 0$, $y \geq 0$ occurs at one of the endpoints.

■ For Problems **61–62**, use Figure 15.37. The grid lines are one unit apart.

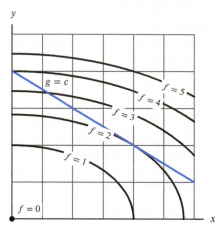

Figure 15.37

61. Find the maximum and minimum values of f on $g = c$. At which points do they occur?

62. Find the maximum and minimum values of f on the triangular region below $g = c$ in the first quadrant.

■ Are the statements in Problems **63–67** true or false? Give reasons for your answer.

63. If $f(x, y)$ has a local maximum at (a, b) subject to the constraint $g(x, y) = c$, then $g(a, b) = c$.

64. If $f(x, y)$ has a local maximum at (a, b) subject to the constraint $g(x, y) = c$, then $\operatorname{grad} f(a, b) = \vec{0}$.

65. The function $f(x, y) = x + y$ has no global maximum subject to the constraint $x - y = 0$.

66. The point $(2, -1)$ is a local minimum of $f(x, y) = x^2 + y^2$ subject to the constraint $x + 2y = 0$.

67. If $\operatorname{grad} f(a, b)$ and $\operatorname{grad} g(a, b)$ point in opposite directions, then (a, b) is a local minimum of $f(x, y)$ constrained by $g(x, y) = c$.

■ In Problems **68–75**, suppose that M and m are the maximum and minimum values of $f(x, y)$ subject to the constraint $g(x, y) = c$ and that (a, b) satisfies $g(a, b) = c$. Decide whether the statements are true or false. Give an explanation for your answer.

68. If $f(a, b) = M$, then $f_x(a, b) = f_y(a, b) = 0$.

69. If $f(a, b) = M$, then $f(a, b) = \lambda g(a, b)$ for some value of λ.

70. If $\operatorname{grad} f(a, b) = \lambda \operatorname{grad} g(a, b)$, then $f(a, b) = M$ or $f(a, b) = m$.

71. If $f(a, b) = M$ and $f_x(a, b)/f_y(a, b) = 5$, then $g_x(a, b)/g_y(a, b) = 5$.

72. If $f(a, b) = m$ and $g_x(a, b) = 0$, then $f_x(a, b) = 0$.

73. Increasing the value of c increases the value of M.

74. Suppose that $f(a, b) = M$ and that $\operatorname{grad} f(a, b) = 3 \operatorname{grad} g(a, b)$. Then increasing the value of c by 0.02 increases the value of M by about 0.06.

75. Suppose that $f(a, b) = m$ and that $\operatorname{grad} f(a, b) = 3 \operatorname{grad} g(a, b)$. Then increasing the value of c by 0.02 decreases the value of m by about 0.06.

Online Resource: Review Problems and Projects

Chapter 16

INTEGRATING FUNCTIONS OF SEVERAL VARIABLES

CONTENTS

16.1   THE DEFINITE INTEGRAL OF A FUNCTION OF TWO VARIABLES

The definite integral of a continuous one-variable function, f, is a limit of Riemann sums:

$$\int_a^b f(x)\, dx = \lim_{\Delta x \to 0} \sum_i f(x_i)\, \Delta x,$$

where x_i is a point in the ith subdivision of the interval $[a, b]$. In this section we extend this definition to functions of two variables. We start by considering how to estimate total population from a two-variable population density.

Population Density of Foxes in England

The fox population in parts of England can be important to public health officials because animals can spread diseases, such as rabies. Biologists use a contour diagram to display the fox population density, D; see Figure 16.1, where D is in foxes per square kilometer.[1] The bold contour is the coastline, which may be thought of as the $D = 0$ contour; clearly the density is zero outside it. We can think of D as a function of position, $D = f(x, y)$ where x and y are in kilometers from the southwest corner of the map.

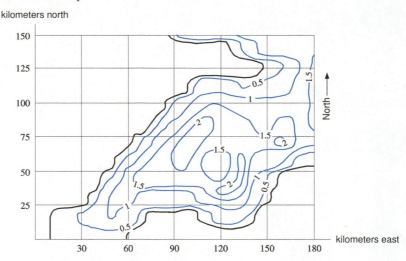

Figure 16.1: Population density of foxes in southwestern England

Example 1   Estimate the total fox population in the region represented by the map in Figure 16.1.

Solution   We subdivide the map into the rectangles shown in Figure 16.1 and estimate the population in each rectangle. For simplicity, we use the population density at the northeast corner of each rectangle. For example, in the bottom left rectangle, the density is 0 at the northeast corner; in the next rectangle to the east (right), the density in the northeast corner is 1. Continuing in this way, we get the estimates in Table 16.1. To estimate the population in a rectangle, we multiply the density by the area of the rectangle, $30 \cdot 25 = 750 \text{ km}^2$. Adding the results, we obtain

$$\text{Estimate of population} = (0.2 + 0.7 + 1.2 + 1.2 + 0.1 + 1.6 + 0.5 + 1.4$$
$$+ 1.1 + 1.6 + 1.5 + 1.8 + 1.5 + 1.3 + 1.1 + 2.0$$
$$+ 1.4 + 1.0 + 1.0 + 0.6 + 1.2)750 = 18{,}000 \text{ foxes.}$$

[1]J. D. Murray et al., "On the Spatial Spread of Rabies Among Foxes", *Proc. R. Soc. Lond. B* 229, pp. 111–150 (1986).

Taking the upper and lower bounds for the population density on each rectangle enables us to find upper and lower estimates for the population. Using the same rectangles, the upper estimate is approximately 35,000 and the lower estimate is 4,000. There is a wide discrepancy between the upper and lower estimates; we could make them closer by taking finer subdivisions.

Table 16.1 *Estimates of population density (northeast corner)*

0.0	0.0	0.2	0.7	1.2	1.2
0.0	0.0	0.0	0.0	0.1	1.6
0.0	0.0	0.5	1.4	1.1	1.6
0.0	0.0	1.5	1.8	1.5	1.3
0.0	1.1	2.0	1.4	1.0	0.0
0.0	1.0	0.6	1.2	0.0	0.0

Definition of the Definite Integral

The sums used to approximate the fox population are Riemann sums. We now define the definite integral for a function f of two variables on a rectangular region. Given a continuous function $f(x, y)$ defined on a region $a \leq x \leq b$ and $c \leq y \leq d$, we subdivide each of the intervals $a \leq x \leq b$ and $c \leq y \leq d$ into n and m equal subintervals respectively, giving nm subrectangles. (See Figure 16.2.)

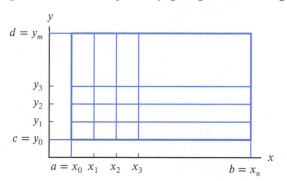

Figure 16.2: Subdivision of a rectangle into nm subrectangles

The area of each subrectangle is $\Delta A = \Delta x \, \Delta y$, where $\Delta x = (b - a)/n$ is the width of each subdivision on the x-axis, and $\Delta y = (d - c)/m$ is the width of each subdivision on the y-axis. To compute the Riemann sum, we multiply the area of each subrectangle by the value of the function at a point in the rectangle and add the resulting numbers. Choosing the maximum value, M_{ij}, of the function on each rectangle and adding for all i, j gives the *upper sum*, $\sum_{i,j} M_{ij} \Delta x \Delta y$.

The *lower sum*, $\sum_{i,j} L_{ij} \Delta x \Delta y$, is obtained by taking the minimum value on each rectangle. If (u_{ij}, v_{ij}) is any point in the ij^{th} subrectangle, any other Riemann sum satisfies

$$\sum_{i,j} L_{ij} \Delta x \Delta y \leq \sum_{i,j} f(u_{ij}, v_{ij}) \, \Delta x \, \Delta y \leq \sum_{i,j} M_{ij} \Delta x \Delta y.$$

We define the definite integral by taking the limit as the numbers of subdivisions, n and m, tend to infinity. By comparing upper and lower sums, as we did for the fox population, it can be shown that the limit exists when the function, f, is continuous. We get the same limit by letting Δx and Δy tend to 0. Thus, we have the following definition:

Suppose the function f is continuous on R, the rectangle $a \leq x \leq b$, $c \leq y \leq d$. If (u_{ij}, v_{ij}) is any point in the ij^{th} subrectangle, we define the **definite integral** of f over R:

$$\int_R f \, dA = \lim_{\Delta x, \Delta y \to 0} \sum_{i,j} f(u_{ij}, v_{ij}) \Delta x \Delta y.$$

Such an integral is called a **double integral**.

The case when R is not rectangular is considered on page 894. Sometimes we think of dA as being the area of an infinitesimal rectangle of length dx and height dy, so that $dA = dx \, dy$. Then we use the notation[2]

$$\int_R f \, dA = \int_R f(x, y) \, dx \, dy.$$

For this definition, we used a particular type of Riemann sum with equal-sized rectangular subdivisions. In a general Riemann sum, the subdivisions do not all have to be the same size.

Interpretation of the Double Integral as Volume

Just as the definite integral of a positive one-variable function can be interpreted as an area, so the double integral of a positive two-variable function can be interpreted as a volume. In the one-variable case we visualize the Riemann sums as the total area of rectangles above the subdivisions. In the two-variable case we get solid bars instead of rectangles. As the number of subdivisions grows, the tops of the bars approximate the surface better, and the volume of the bars gets closer to the volume under the graph of the function. (See Figure 16.3.)

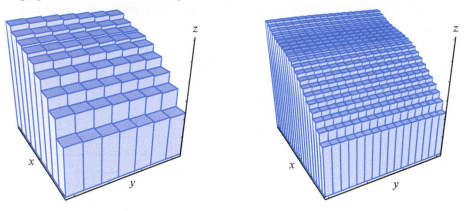

Figure 16.3: Approximating volume under a graph with finer and finer Riemann sums

Thus, we have the following result:

If x, y, z represent length and f is positive, then

$$\text{Volume under graph of } f \text{ above region } R = \int_R f \, dA.$$

Example 2   Let R be the rectangle $0 \leq x \leq 1$ and $0 \leq y \leq 1$. Use Riemann sums to make upper and lower estimates of the volume of the region above R and under the graph of $z = e^{-(x^2 + y^2)}$.

---

[2] Another common notation for the double integral is $\int \int_R f \, dA$.

Solution If R is the rectangle $0 \leq x \leq 1, 0 \leq y \leq 1$, the volume we want is given by

$$\text{Volume} = \int_R e^{-(x^2+y^2)} \, dA.$$

We divide R into 16 subrectangles by dividing each edge into four parts. Figure 16.4 shows that $f(x, y) = e^{-(x^2+y^2)}$ decreases as we move away from the origin. Thus, to get an upper sum we evaluate f on each subrectangle at the corner nearest the origin. For example, in the rectangle $0 \leq x \leq 0.25, 0 \leq y \leq 0.25$, we evaluate f at $(0, 0)$. Using Table 16.2, we find that

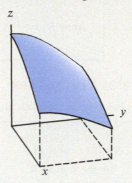

Figure 16.4: Graph of $e^{-(x^2+y^2)}$ above the rectangle R

$$
\begin{aligned}
\text{Upper sum} = (&1 + 0.9394 + 0.7788 + 0.5698 \\
&+ 0.9394 + 0.8825 + 0.7316 + 0.5353 \\
&+ 0.7788 + 0.7316 + 0.6065 + 0.4437 \\
&+ 0.5698 + 0.5353 + 0.4437 + 0.3247)(0.0625) = 0.68.
\end{aligned}
$$

To get a lower sum, we evaluate f at the opposite corner of each rectangle because the surface slopes down in both the x and y directions. This yields a lower sum of 0.44. Thus,

$$0.44 \leq \int_R e^{-(x^2+y^2)} \, dA \leq 0.68.$$

To get a better approximation of the volume under the graph, we use more subdivisions. See Table 16.3.

Table 16.2 *Values of $f(x, y) = e^{-(x^2+y^2)}$ on the rectangle R*

				y		
		0.0	0.25	0.50	0.75	1.00
	0.0	1	0.9394	0.7788	0.5698	0.3679
	0.25	0.9394	0.8825	0.7316	0.5353	0.3456
x	0.50	0.7788	0.7316	0.6065	0.4437	0.2865
	0.75	0.5698	0.5353	0.4437	0.3247	0.2096
	1.00	0.3679	0.3456	0.2865	0.2096	0.1353

Table 16.3 *Riemann sum approximations to $\int_R e^{-(x^2+y^2)} \, dA$*

	Number of subdivisions in x and y directions			
	8	16	32	64
Upper	0.6168	0.5873	0.5725	0.5651
Lower	0.4989	0.5283	0.5430	0.5504

The exact value of the double integral, 0.5577 ..., is trapped between the lower and upper sums. Notice that the lower sum increases and the upper sum decreases as the number of subdivisions increases. However, even with 64 subdivisions, the lower and upper sums agree with the exact value of the integral only in the first decimal place.

Interpretation of the Double Integral as Area

In the special case that $f(x, y) = 1$ for all points (x, y) in the region R, each term in the Riemann sum is of the form $1 \cdot \Delta A = \Delta A$ and the double integral gives the area of the region R:

$$\text{Area}(R) = \int_R 1 \, dA = \int_R dA$$

Interpretation of the Double Integral as Average Value

As in the one-variable case, the definite integral can be used to define the average value of a function:

$$\begin{array}{c} \text{Average value of } f \\ \text{on the region } R \end{array} = \frac{1}{\text{Area of } R} \int_R f \, dA$$

We can rewrite this as

$$\text{Average value} \times \text{Area of } R = \int_R f \, dA.$$

If we interpret the integral as the volume under the graph of f, then we can think of the average value of f as the height of the box with the same volume that is on the same base. (See Figure 16.5.) Imagine that the volume under the graph is made out of wax. If the wax melted within the perimeter of R, then it would end up box-shaped with height equal to the average value of f.

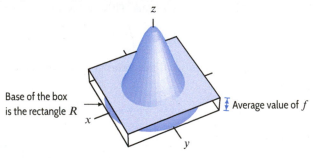

Figure 16.5: Volume and average value

Integral over Regions that Are Not Rectangles

We defined the definite integral $\int_R f(x, y) \, dA$, for a rectangular region R. Now we extend the definition to regions of other shapes, including triangles, circles, and regions bounded by the graphs of piecewise continuous functions.

To approximate the definite integral over a region, R, which is not rectangular, we use a grid of rectangles approximating the region. We obtain this grid by enclosing R in a large rectangle and subdividing that rectangle; we consider just the subrectangles which are inside R.

As before, we pick a point (u_{ij}, v_{ij}) in each subrectangle and form a Riemann sum

$$\sum_{i,j} f(u_{ij}, v_{ij}) \Delta x \Delta y.$$

This time, however, the sum is over only those subrectangles within R. For example, in the case of the fox population we can use the rectangles which are entirely on land. As the subdivisions become finer, the grid approximates the region R more closely. For a function, f, which is continuous on R, we define the definite integral as follows:

$$\int_R f \, dA = \lim_{\Delta x, \Delta y \to 0} \sum_{i,j} f(u_{ij}, v_{ij}) \Delta x \Delta y$$

where the Riemann sum is taken over the subrectangles inside R.

You may wonder why we can leave out the rectangles which cover the edge of R—if we included them, might we get a different value for the integral? The answer is that for any region that we are likely to meet, the area of the subrectangles covering the edge tends to 0 as the grid becomes finer. Therefore, omitting these rectangles does not affect the limit.

Convergence of Upper and Lower Sums to Same Limit

We have said that if f is continuous on the rectangle R, then the difference between upper and lower sums for f converges to 0 as Δx and Δy approach 0. In the following example, we show this in a particular case. The ideas in this example can be used in a general proof.

Example 3   Let $f(x, y) = x^2 y$ and let R be the rectangle $0 \le x \le 1, 0 \le y \le 1$. Show that the difference between upper and lower Riemann sums for f on R converges to 0, as Δx and Δy approach 0.

Solution   The difference between the sums is

$$\sum M_{ij} \Delta x \Delta y - \sum L_{ij} \Delta x \Delta y = \sum (M_{ij} - L_{ij}) \Delta x \Delta y,$$

where M_{ij} and L_{ij} are the maximum and minimum of f on the ij^{th} subrectangle. Since f increases in both the x and y directions, M_{ij} occurs at the corner of the subrectangle farthest from the origin and L_{ij} at the closest. Moreover, since the slopes in the x and y directions don't decrease as x and y increase, the difference $M_{ij} - L_{ij}$ is largest in the subrectangle R_{nm} which is farthest from the origin. Thus,

$$\sum (M_{ij} - L_{ij}) \Delta x \Delta y \le (M_{nm} - L_{nm}) \sum \Delta x \Delta y = (M_{nm} - L_{nm}) \text{Area}(R).$$

Thus, the difference converges to 0 as long as $(M_{nm} - L_{nm})$ does. The maximum M_{nm} of f on the nm^{th} subrectangle occurs at $(1, 1)$, the subrectangle's top right corner, and the minimum L_{nm} occurs at the opposite corner, $(1 - 1/n, 1 - 1/m)$. Substituting into $f(x, y) = x^2 y$ gives

$$M_{nm} - L_{nm} = (1)^2 (1) - \left(1 - \frac{1}{n}\right)^2 \left(1 - \frac{1}{m}\right) = \frac{2}{n} - \frac{1}{n^2} + \frac{1}{m} - \frac{2}{nm} + \frac{1}{n^2 m}.$$

The right-hand side converges to 0 as $n, m \to \infty$, that is, as $\Delta x, \Delta y \to 0$.

Summary for Section 16.1

- The **double integral** of a continuous function over the rectangle R ($a \le x \le b, c \le y \le d$) is defined by the limit of a double Riemann sum. First, we divide R into $n \cdot m$ subrectangles of dimensions $\Delta x = (b - a)/n, \Delta y = (d - c)/m$. Then, if (u_{ij}, v_{ij}) is any point in the ij^{th} subrectangle, we define

$$\int_R f \, dA = \lim_{\Delta x, \Delta y \to 0} \sum_{i,j} f(u_{ij}, v_{ij}) \Delta x \Delta y.$$

- If $f(x, y) = 1$ for all points (x, y) in the region R, then $\int_R 1 \, dA$ gives the **area** of the region R.

- If $f(x, y)$ **is positive** for all points (x, y) in the region R, then $\int_R f \, dA$ gives the **volume under the graph** of f above the region R.

- The **average value** of f on the region R is given by $\dfrac{1}{\text{Area of } R} \displaystyle\int_R f \, dA$.

Exercises and Problems for Section 16.1   Online Resource: Additional Problems for Section 16.1

EXERCISES

1. Table 16.4 gives values of the function $f(x, y)$, which is increasing in x and decreasing in y on the region $R : 0 \le x \le 6, 0 \le y \le 1$. Make the best possible upper and lower estimates of $\int_R f(x, y) \, dA$.

Table 16.4

	x		
	0	3	6
y 0	5	7	10
0.5	4	5	7
1	3	4	6

2. Values of $f(x, y)$ are in Table 16.5. Let R be the rectangle $1 \le x \le 1.2, 2 \le y \le 2.4$. Find Riemann sums which are reasonable over- and underestimates for $\int_R f(x, y) \, dA$ with $\Delta x = 0.1$ and $\Delta y = 0.2$.

Table 16.5

	x		
	1.0	1.1	1.2
y 2.0	5	7	10
2.2	4	6	8
2.4	3	5	4

3. Figure 16.6 shows contours of $g(x, y)$ on the region R, with $5 \le x \le 11$ and $4 \le y \le 10$. Using $\Delta x = \Delta y = 2$, find an overestimate and an underestimate for $\int_R g(x, y) \, dA$.

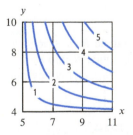

Figure 16.6

4. Figure 16.7 shows contours of $f(x, y)$ on the rectangle R with $0 \le x \le 30$ and $0 \le y \le 15$. Using $\Delta x = 10$ and $\Delta y = 5$, find an overestimate and an underestimate for $\int_R f(x, y) \, dA$.

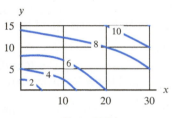

Figure 16.7

5. Figure 16.8 shows a contour plot of population density, people per square kilometer, in a rectangle of land 3 km by 2 km. Estimate the population in the region represented by Figure 16.8.

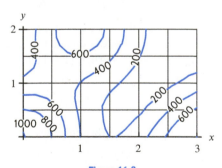

Figure 16.8

■ In Exercises **6–7**, for x and y in meters and R a region on the xy-plane, what does the integral represent? Give units.

6. $\displaystyle\int_R \sigma(x, y) \, dA$, where $\sigma(x, y)$ is bacteria population, in thousands per m^2.

7. $\dfrac{1}{\text{Area of } R} \displaystyle\int_R h(x, y) \, dA$, where $h(x, y)$ is the height of a tent, in meters.

PROBLEMS

In Problems **8–14**, decide (without calculation) whether the integrals are positive, negative, or zero. Let D be the region inside the unit circle centered at the origin, let R be the right half of D, and let B be the bottom half of D.

8. $\int_D 1 \, dA$

9. $\int_R 5x \, dA$

10. $\int_B 5x \, dA$

11. $\int_D (y^3 + y^5) \, dA$

12. $\int_B (y^3 + y^5) \, dA$

13. $\int_D (y - y^3) \, dA$

14. $\int_B (y - y^3) \, dA$

15. Figure 16.9 shows contours of $f(x, y)$. Let R be the square $-0.5 \le x \le 1$, $-0.5 \le y \le 1$. Is the integral $\int_R f \, dA$ positive or negative? Explain your reasoning.

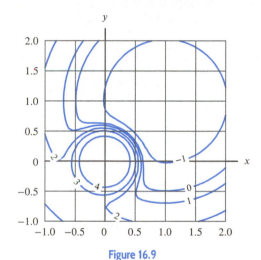

Figure 16.9

16. Table 16.6 gives values of $f(x, y)$, the number of milligrams of mosquito larvae per square meter in a swamp. If x and y are in meters and R is the rectangle $0 \le x \le 8$, $0 \le y \le 6$, estimate $\int_R f(x, y) \, dA$. Give units and interpret your answer.

Table 16.6

		0	4	8
	0	1	3	6
y	3	2	5	9
	6	4	9	15

17. Table 16.7 gives values of $f(x, y)$, the depth of volcanic ash, in meters, after an eruption. If x and y are in kilometers and R is the rectangle $0 \le x \le 100$, $0 \le y \le 100$, estimate the volume of volcanic ash in R in km³.

Table 16.7

		0	50	100
	0	0.82	0.56	0.43
y	50	0.63	0.45	0.3
	100	0.55	0.44	0.26

18. Table 16.8 gives the density of cacti, $f(x, y)$, in a desert region, in thousands of cacti per km². If x and y are in kilometers and R is the square $0 \le x \le 30$, $0 \le y \le 30$, estimate the number of cacti in the region R.

Table 16.8

		0	10	20	30
	0	8.5	8.2	7.9	8.1
	10	9.5	10.6	10.5	10.1
y	20	9.3	10.5	10.4	9.5
	30	8.3	8.6	9.3	9.1

19. Use four subrectangles to approximate the volume of the object whose base is the region $0 \le x \le 4$ and $0 \le y \le 6$, and whose height is given by $f(x, y) = x + y$. Find an overestimate and an underestimate and average the two.

20. Figure 16.10 shows the rainfall, in inches, in Tennessee on May 1–2, 2010.[3] Using three contours (red, yellow, and green), make a rough estimate of how many cubic miles of rain fell on the state during this time.

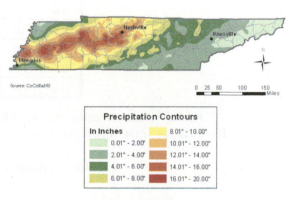

Figure 16.10

---

[3] www.srh.noaa.gov/images/ohx/rainfall/TN_May2010_rainfall_map.png, accessed June 13, 2016.

Strengthen Your Understanding

■ In Problems 21–22, explain what is wrong with the statement.

21. For all f, the integral $\int_R f(x, y)\, dA$ gives the volume of the solid under the graph of f over the region R.

22. If R is a region in the third quadrant where $x < 0, y < 0$, then $\int_R f(x, y)\, dA$ is negative.

■ In Problems 23–24, give an example of:

23. A function $f(x, y)$ and rectangle R such that the Riemann sums obtained using the lower left-hand corner of each subrectangle are an overestimate.

24. A function $f(x, y)$ whose average value over the square $0 \le x \le 1, 0 \le y \le 1$ is negative.

■ Are the statements in Problems 25–34 true or false? Give reasons for your answer.

25. The double integral $\int_R f\, dA$ is always positive.

26. If $f(x, y) = k$ for all points (x, y) in a region R then $\int_R f\, dA = k \cdot \text{Area}(R)$.

27. If R is the rectangle $0 \le x \le 1, 0 \le y \le 1$ then $\int_R e^{xy}\, dA > 3$.

28. If R is the rectangle $0 \le x \le 2, 0 \le y \le 3$ and S is the rectangle $-2 \le x \le 0, -3 \le y \le 0$, then $\int_R f\, dA = -\int_S f\, dA$.

29. Let $\rho(x, y)$ be the population density of a city, in people per km^2. If R is a region in the city, then $\int_R \rho\, dA$ gives the total number of people in the region R.

30. If $\int_R f\, dA = 0$, then $f(x, y) = 0$ at all points of R.

31. If $g(x, y) = kf(x, y)$, where k is constant, then $\int_R g\, dA = k \int_R f\, dA$.

32. If f and g are two functions continuous on a region R, then $\int_R f \cdot g\, dA = \int_R f\, dA \cdot \int_R g\, dA$.

33. If R is the rectangle $0 \le x \le 1, 0 \le y \le 2$ and S is the square $0 \le x \le 1, 0 \le y \le 1$, then $\int_R f\, dA = 2 \int_S f\, dA$.

34. If R is the rectangle $2 \le x \le 4, 5 \le y \le 9$, $f(x, y) = 2x$ and $g(x, y) = x + y$, then the average value of f on R is less than the average value of g on R.

16.2   ITERATED INTEGRALS

In Section 16.1 we approximated double integrals using Riemann sums. In this section we see how to compute double integrals exactly using one-variable integrals.

The Fox Population Again: Expressing a Double Integral as an Iterated Integral

To estimate the fox population, we computed a sum of the form

$$\text{Total population} \approx \sum_{i,j} f(u_{ij}, v_{ij}) \Delta x\, \Delta y,$$

where $1 \le i \le n$ and $1 \le j \le m$ and the values $f(u_{ij}, v_{ij})$ can be arranged as in Table 16.9.

Table 16.9   *Estimates for fox population densities for $n = m = 6$*

0.0	0.0	0.2	0.7	1.2	1.2
0.0	0.0	0.0	0.0	0.1	1.6
0.0	0.0	0.5	1.4	1.1	1.6
0.0	0.0	1.5	1.8	1.5	1.3
0.0	1.1	2.0	1.4	1.0	0.0
0.0	1.0	0.6	1.2	0.0	0.0

For any values of n and m, we can either add across the rows first or add down the columns first. If we add rows first, we can write the sum in the form

$$\text{Total population} \approx \sum_{j=1}^{m} \left(\sum_{i=1}^{n} f(u_{ij}, v_{ij}) \Delta x \right) \Delta y.$$

The inner sum, $\sum_{i=1}^{n} f(u_{ij}, v_{ij})\, \Delta x$, approximates the integral $\int_0^{180} f(x, v_{ij})\, dx$. Thus, we have

$$\text{Total population} \approx \sum_{j=1}^{m} \left(\int_0^{180} f(x, v_{ij})\, dx \right) \Delta y.$$

The outer Riemann sum approximates another integral, this time with integrand $\int_0^{180} f(x, y)\, dx$, which is a function of y. Thus, we can write the total population in terms of nested, or *iterated*, one-variable integrals:

$$\text{Total population} = \int_0^{150} \left(\int_0^{180} f(x, y)\, dx \right) dy.$$

Since the total population is represented by $\int_R f\, dA$, this suggests the method of computing double integrals in the following theorem:[4]

Theorem 16.1: Writing a Double Integral as an Iterated Integral

If R is the rectangle $a \leq x \leq b, c \leq y \leq d$ and f is a continuous function on R, then the integral of f over R exists and is equal to the **iterated integral**

$$\int_R f\, dA = \int_{y=c}^{y=d} \left(\int_{x=a}^{x=b} f(x, y)\, dx \right) dy.$$

The expression $\int_{y=c}^{y=d} \left(\int_{x=a}^{x=b} f(x, y)\, dx \right) dy$ can be written $\int_c^d \int_a^b f(x, y)\, dx\, dy$.

To evaluate the iterated integral, first perform the inside integral with respect to x, holding y constant; then integrate the result with respect to y.

Example 1

A building is 8 meters wide and 16 meters long. It has a flat roof that is 12 meters high at one corner and 10 meters high at each of the adjacent corners. What is the volume of the building?

Solution

If we put the high corner on the z-axis, the long side along the y-axis, and the short side along the x-axis, as in Figure 16.11, then the roof is a plane with z-intercept 12, and x slope $(-2)/8 = -1/4$, and y slope $(-2)/16 = -1/8$. Hence, the equation of the roof is

$$z = 12 - \frac{1}{4}x - \frac{1}{8}y.$$

The volume is given by the double integral

$$\text{Volume} = \int_R \left(12 - \frac{1}{4}x - \frac{1}{8}y\right) dA,$$

where R is the rectangle $0 \leq x \leq 8, 0 \leq y \leq 16$. Setting up an iterated integral, we get

$$\text{Volume} = \int_0^{16} \int_0^8 \left(12 - \frac{1}{4}x - \frac{1}{8}y\right) dx\, dy.$$

The inside integral is

$$\int_0^8 \left(12 - \frac{1}{4}x - \frac{1}{8}y\right) dx = \left.\left(12x - \frac{1}{8}x^2 - \frac{1}{8}xy\right)\right|_{x=0}^{x=8} = 88 - y.$$

Then the outside integral gives

$$\text{Volume} = \int_0^{16} (88 - y)\, dy = \left.\left(88y - \frac{1}{2}y^2\right)\right|_0^{16} = 1280.$$

The volume of the building is 1280 cubic meters.

---

[4]For a proof, see M. Spivak, *Calculus on Manifolds*, pp. 53 and 58 (New York: Benjamin, 1965).

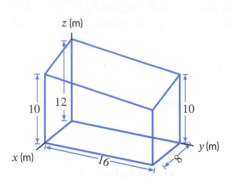

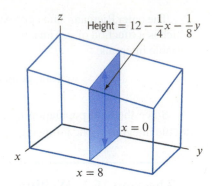

Figure 16.11: A slant-roofed building   **Figure 16.12:** Cross-section of a building

Notice that the inner integral $\int_0^8 (12 - \frac{1}{4}x - \frac{1}{8}y)\,dx$ in Example 1 gives the area of the cross section of the building perpendicular to the y-axis in Figure 16.12.

The iterated integral $\int_0^{16}\int_0^8 (12 - \frac{1}{4}x - \frac{1}{8}y)\,dx\,dy$ thus calculates the volume by adding the volumes of thin cross-sectional slabs.

The Order of Integration

In computing the fox population, we could have chosen to add columns (fixed x) first, instead of the rows. This leads to an iterated integral where x is constant in the inner integral instead of y. Thus,

$$\int_R f(x, y)\,dA = \int_a^b \left(\int_c^d f(x, y)\,dy \right) dx$$

where R is the rectangle $a \leq x \leq b$ and $c \leq y \leq d$.

For any function we are likely to meet, it does not matter in which order we integrate over a rectangular region R; we get the same value for the double integral either way.

$$\int_R f\,dA = \int_c^d \left(\int_a^b f(x, y)\,dx \right) dy = \int_a^b \left(\int_c^d f(x, y)\,dy \right) dx$$

Example 2   Compute the volume of Example 1 as an iterated integral by integrating with respect to y first.

Solution   Rewriting the integral, we have

$$\text{Volume} = \int_0^8 \left(\int_0^{16} (12 - \frac{1}{4}x - \frac{1}{8}y)\,dy \right) dx = \int_0^8 \left(\left. (12y - \frac{1}{4}xy - \frac{1}{16}y^2) \right|_{y=0}^{y=16} \right) dx$$

$$= \int_0^8 (176 - 4x)\,dx = \left. (176x - 2x^2) \right|_0^8 = 1280 \text{ meter}^3.$$

Iterated Integrals Over Non-Rectangular Regions

Example 3   The density at the point (x, y) of a triangular metal plate, as shown in Figure 16.13, is $\delta(x, y)$. Express its mass as an iterated integral.

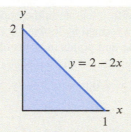

Figure 16.13: A triangular metal plate with density $\delta(x, y)$ at the point (x, y)

Solution   Approximate the triangular region using a grid of small rectangles of sides Δx and Δy. The mass of one rectangle is given by

$$\text{Mass of rectangle} \approx \text{Density} \cdot \text{Area} \approx \delta(x, y)\Delta x \Delta y.$$

Summing over all rectangles gives a Riemann sum which approximates the double integral:

$$\text{Mass} = \int_R \delta(x, y)\, dA,$$

where R is the triangle. We want to compute this integral using an iterated integral.

Think about how the iterated integral over the rectangle $a \le x \le b$, $c \le y \le d$ works:

$$\int_a^b \int_c^d f(x, y)\, dy\, dx.$$

The inside integral with respect to y is along vertical strips which begin at the horizontal line $y = c$ and end at the line $y = d$. There is one such strip for each x between $x = a$ and $x = b$. (See Figure 16.14.)

For the triangular region in Figure 16.13, the idea is the same. The only difference is that the individual vertical strips no longer all go from $y = c$ to $y = d$. The vertical strip that starts at the point $(x, 0)$ ends at the point $(x, 2 - 2x)$, because the top edge of the triangle is the line $y = 2 - 2x$. See Figure 16.15. On this vertical strip, y goes from 0 to $2 - 2x$. Hence, the inside integral is

$$\int_0^{2-2x} \delta(x, y)\, dy.$$

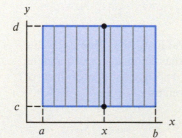

Figure 16.14: Integrating over a rectangle using vertical strips

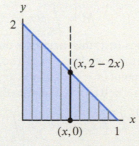

Figure 16.15: Integrating over a triangle using vertical strips

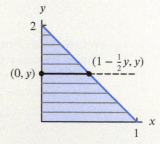

Figure 16.16: Integrating over a triangle using horizontal strips

Finally, since there is a vertical strip for each x between 0 and 1, the outside integral goes from $x = 0$ to $x = 1$. Thus, the iterated integral we want is

$$\text{Mass} = \int_0^1 \int_0^{2-2x} \delta(x, y)\, dy\, dx.$$

We could have chosen to integrate in the opposite order, keeping y fixed in the inner integral instead of x. The limits are formed by looking at horizontal strips instead of vertical ones, and expressing the x-values at the end points in terms of y. See Figure 16.16. To find the right endpoint of the strip, we use the equation of the top edge of the triangle in the form $x = 1 - \frac{1}{2}y$. Thus, a horizontal strip goes from $x = 0$ to $x = 1 - \frac{1}{2}y$. Since there is a strip for every y from 0 to 2, the iterated integral is

$$\text{Mass} = \int_0^2 \int_0^{1-\frac{1}{2}y} \delta(x, y)\, dx\, dy.$$

Limits on Iterated Integrals

- The limits on the outer integral must be constants.
- The limits on the inner integral can involve only the variable in the outer integral. For example, if the inner integral is with respect to x, its limits can be functions of y.

Example 4   Find the mass M of a metal plate R bounded by $y = x$ and $y = x^2$, with density given by $\delta(x, y) = 1 + xy$ kg/meter^2. (See Figure 16.17.)

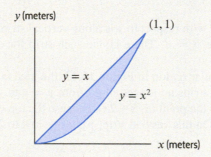

Figure 16.17: A metal plate with density $\delta(x, y)$

Solution   The mass is given by

$$M = \int_R \delta(x, y)\, dA.$$

We integrate along vertical strips first; this means we do the y integral first, which goes from the bottom boundary $y = x^2$ to the top boundary $y = x$. The left edge of the region is at $x = 0$ and the right edge is at the intersection point of $y = x$ and $y = x^2$, which is $(1, 1)$. Thus, the x-coordinate of the vertical strips can vary from $x = 0$ to $x = 1$, and so the mass is given by

$$M = \int_0^1 \int_{x^2}^x \delta(x, y)\, dy\, dx = \int_0^1 \int_{x^2}^x (1 + xy)\, dy\, dx.$$

Calculating the inner integral first gives

$$M = \int_0^1 \int_{x^2}^x (1 + xy)\, dy\, dx = \int_0^1 \left(y + x\frac{y^2}{2} \right) \Big|_{y=x^2}^{y=x} dx$$

$$= \int_0^1 \left(x - x^2 + \frac{x^3}{2} - \frac{x^5}{2} \right) dx = \left(\frac{x^2}{2} - \frac{x^3}{3} + \frac{x^4}{8} - \frac{x^6}{12} \right) \Big|_0^1 = \frac{5}{24} = 0.208 \text{ kg}.$$

Example 5   A semicircular city of radius 3 km borders the ocean on the straight side. Find the average distance from points in the city to the ocean.

Solution   Think of the ocean as everything below the x-axis in the xy-plane and think of the city as the upper half of the circular disk of radius 3 bounded by $x^2 + y^2 = 9$. (See Figure 16.18.)

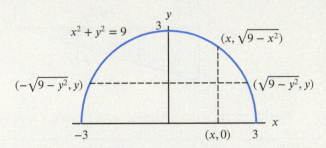

Figure 16.18: The city by the ocean showing a typical vertical strip and a typical horizontal strip

The distance from any point (x, y) in the city to the ocean is the vertical distance to the x-axis, namely y. Thus, we want to compute

$$\text{Average distance} = \frac{1}{\text{Area}(R)} \int_R y \, dA,$$

where R is the region between the upper half of the circle $x^2 + y^2 = 9$ and the x-axis. The area of R is $\pi 3^2/2 = 9\pi/2$.

To compute the integral, let's take the inner integral with respect to y. A vertical strip goes from the x-axis, namely $y = 0$, to the semicircle. The upper limit must be expressed in terms of x, so we solve $x^2 + y^2 = 9$ to get $y = \sqrt{9 - x^2}$. Since there is a strip for every x from -3 to 3, the integral is:

$$\int_R y \, dA = \int_{-3}^{3} \left(\int_0^{\sqrt{9-x^2}} y \, dy \right) dx = \int_{-3}^{3} \left(\frac{y^2}{2} \Big|_{y=0}^{y=\sqrt{9-x^2}} \right) dx$$

$$= \int_{-3}^{3} \frac{1}{2}(9 - x^2) \, dx = \frac{1}{2}\left(9x - \frac{x^3}{3} \right) \Big|_{-3}^{3} = \frac{1}{2}(18 - (-18)) = 18.$$

Therefore, the average distance is $18/(9\pi/2) = 4/\pi = 1.273$ km.

What if we choose the inner integral with respect to x? Then we get the limits by looking at horizontal strips, not vertical, and we solve $x^2 + y^2 = 9$ for x in terms of y. We get $x = -\sqrt{9 - y^2}$ at the left end of the strip and $x = \sqrt{9 - y^2}$ at the right. There is a strip for every y from 0 to 3, so

$$\int_R y \, dA = \int_0^3 \left(\int_{-\sqrt{9-y^2}}^{\sqrt{9-y^2}} y \, dx \right) dy = \int_0^3 \left(yx \Big|_{x=-\sqrt{9-y^2}}^{x=\sqrt{9-y^2}} \right) dy = \int_0^3 2y\sqrt{9 - y^2} \, dy$$

$$= -\frac{2}{3}(9 - y^2)^{3/2} \Big|_0^3 = -\frac{2}{3}(0 - 27) = 18.$$

We get the same result as before. The average distance to the ocean is $(2/(9\pi))18 = 4/\pi = 1.273$ km.

In the examples so far, a region was given and the problem was to determine the limits for an iterated integral. Sometimes the limits are known and we want to determine the region.

Example 6   Sketch the region of integration for the iterated integral $\int_0^6 \int_{x/3}^2 x\sqrt{y^3+1}\,dy\,dx$.

Solution   The inner integral is with respect to y, so we imagine the region built of vertical strips. The bottom of each strip is on the line $y = x/3$, and the top is on the horizontal line $y = 2$. Since the limits of the outer integral are 0 and 6, the whole region is contained between the vertical lines $x = 0$ and $x = 6$. Notice that the lines $y = 2$ and $y = x/3$ meet where $x = 6$. See Figure 16.19.

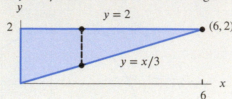

Figure 16.19: The region of integration for Example 6, showing the vertical strip

Reversing the Order of Integration

It is sometimes helpful to reverse the order of integration in an iterated integral. An integral which is difficult or impossible with the integration in one order can be quite straightforward in the other. The next example is such a case.

Example 7   Evaluate $\int_0^6 \int_{x/3}^2 x\sqrt{y^3+1}\,dy\,dx$ using the region sketched in Figure 16.19.

Solution   Since $\sqrt{y^3+1}$ has no elementary antiderivative, we cannot calculate the inner integral symbolically. We try reversing the order of integration. From Figure 16.19, we see that horizontal strips go from $x = 0$ to $x = 3y$ and that there is a strip for every y from 0 to 2. Thus, when we change the order of integration we get

$$\int_0^6 \int_{x/3}^2 x\sqrt{y^3+1}\,dy\,dx = \int_0^2 \int_0^{3y} x\sqrt{y^3+1}\,dx\,dy.$$

Now we can at least do the inner integral because we know the antiderivative of x. What about the outer integral?

$$\int_0^2 \int_0^{3y} x\sqrt{y^3+1}\,dx\,dy = \int_0^2 \left(\frac{x^2}{2}\sqrt{y^3+1}\right)\Big|_{x=0}^{x=3y} dy = \int_0^2 \frac{9y^2}{2}(y^3+1)^{1/2}\,dy$$

$$= (y^3+1)^{3/2}\Big|_0^2 = 27 - 1 = 26.$$

Thus, reversing the order of integration made the integral in the previous problem much easier. Notice that to reverse the order it is essential first to sketch the region over which the integration is being performed.

Summary for Section 16.2

- If R is the rectangle $a \leq x \leq b$, $c \leq y \leq d$ and f is a continuous function on R, then the integral of f over R exists and is equal to the **iterated integral**

$$\int_R f\,dA = \int_c^d \left(\int_a^b f(x,y)\,dx\right) dy.$$

To evaluate the iterated integral, first perform the inside integral with respect to x, holding y constant; then integrate the result with respect to y.

- For most functions f the **order of integration** can be switched giving the same result:

$$\int_R f\, dA = \int_c^d \left(\int_a^b f(x, y)\, dx \right) dy = \int_a^b \left(\int_c^d f(x, y)\, dy \right) dx.$$

- When writing double integrals over **non-rectangular** regions as iterated integrals, the inner integral can only have limits which are functions of the variable of the outer integral; the outer integral's limits must be constants.

- When **reversing** the order of integration in an iterated integral over a non-rectangular region, it is useful to sketch the region of integration first.

Exercises and Problems for Section 16.2 Online Resource: Additional Problems for Section 16.2
EXERCISES

■ In Exercises 1–4, sketch the region of integration.

1. $\displaystyle\int_0^\pi \int_0^x y \sin x \, dy \, dx$ **2.** $\displaystyle\int_0^1 \int_{y^2}^y xy \, dx \, dy$

3. $\displaystyle\int_0^2 \int_0^{y^2} y^2 x \, dx \, dy$ **4.** $\displaystyle\int_0^1 \int_{x-2}^{\cos \pi x} y \, dy \, dx$

■ For Exercises 5–12, evaluate the integral.

5. $\displaystyle\int_0^3 \int_0^4 (4x + 3y)\, dx \, dy$ **6.** $\displaystyle\int_0^2 \int_0^3 (x^2 + y^2)\, dy \, dx$

7. $\displaystyle\int_0^3 \int_0^2 6xy \, dy \, dx$ **8.** $\displaystyle\int_0^1 \int_0^2 x^2 y \, dy \, dx$

9. $\displaystyle\int_0^1 \int_0^1 ye^{xy}\, dx \, dy$ **10.** $\displaystyle\int_0^2 \int_0^y y \, dx \, dy$

11. $\displaystyle\int_0^3 \int_0^y \sin x \, dx \, dy$ **12.** $\displaystyle\int_0^{\pi/2} \int_0^{\sin x} x \, dy \, dx$

■ For Exercises 13–20, sketch the region of integration and evaluate the integral.

13. $\displaystyle\int_1^3 \int_0^4 e^{x+y}\, dy \, dx$ **14.** $\displaystyle\int_0^2 \int_0^x e^{x^2}\, dy \, dx$

15. $\displaystyle\int_1^5 \int_x^{2x} \sin x \, dy \, dx$ **16.** $\displaystyle\int_1^4 \int_{\sqrt{y}}^y x^2 y^3 \, dx \, dy$

17. $\displaystyle\int_1^2 \int_y^{3y} xy \, dx \, dy$ **18.** $\displaystyle\int_0^1 \int_x^{\sqrt{x}} 30x \, dy \, dx$

19. $\displaystyle\int_0^2 \int_0^{2x} xe^{x^3}\, dy \, dx$ **20.** $\displaystyle\int_0^1 \int_1^{1+x^2} \frac{x}{\sqrt{y}}\, dy \, dx$

■ In Exercises 21–26, write $\int_R f\, dA$ as an iterated integral for the shaded region R.

21.

22.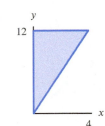

23.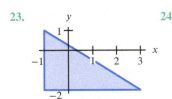

24.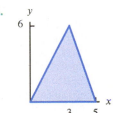

25.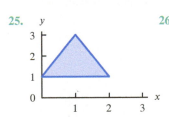

26.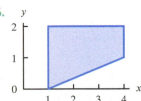

■ For Exercises 27–28, write $\int_R f\, dA$ as an iterated integral in two different ways for the shaded region R.

27. **28.**
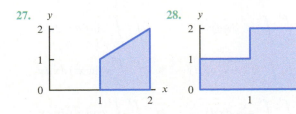

■For Exercises 29–33, evaluate the integral.

29. $\int_R \sqrt{x+y}\, dA$, where R is the rectangle $0 \le x \le 1$, $0 \le y \le 2$.

30. The integral in Exercise 29 using the other order of integration.

31. $\int_R (5x^2 + 1)\sin 3y\, dA$, where R is the rectangle $-1 \le x \le 1, 0 \le y \le \pi/3$.

32. $\int_R xy\, dA$, where R is the triangle $x+y \le 1$, $x \ge 0$, $y \ge 0$.

33. $\int_R (2x+3y)^2\, dA$, where R is the triangle with vertices at $(-1, 0)$, $(0, 1)$, and $(1, 0)$.

PROBLEMS

■In Problems 34–37, integrate $f(x, y) = xy$ over the region R.

34.
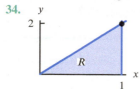

35.
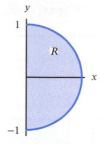

36.
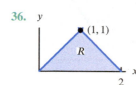

37.
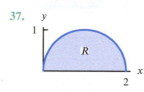

38. (a) Use four subrectangles to approximate the volume of the object whose base is the region $0 \le x \le 4$ and $0 \le y \le 6$, and whose height is given by $f(x, y) = xy$. Find an overestimate and an underestimate and average the two.

(b) Integrate to find the exact volume of the three-dimensional object described in part (a).

■For Problems 39–42, sketch the region of integration, then rewrite the integral with the order of integration reversed.

39. $\int_0^3 \int_{2y}^6 f(x, y)\, dx\, dy$

40. $\int_0^2 \int_0^{\sqrt{4-x^2}} f(x, y)\, dy\, dx$

41. $\int_{-3}^3 \int_0^{9-x^2} f(x, y)\, dy\, dx$

42. $\int_0^2 \int_{y-2}^{2-y} f(x, y)\, dx\, dy$

■In Problems 43–50, evaluate the integral by reversing the order of integration.

43. $\int_0^1 \int_y^1 e^{x^2}\, dx\, dy$

44. $\int_0^1 \int_y^1 \sin(x^2)\ dx\, dy$

45. $\int_0^1 \int_{\sqrt{y}}^1 \sqrt{2+x^3}\, dx\, dy$

46. $\int_0^3 \int_{y^2}^9 y\sin(x^2)\, dx\, dy$

47. $\int_0^1 \int_{e^y}^e \frac{x}{\ln x}\, dx\, dy$

48. $\int_0^1 \int_x^1 \cos(y^2)\, dy\, dx$

49. $\int_0^8 \int_{\sqrt[3]{y}}^2 \frac{1}{1+x^4}\, dx\, dy$

50. $\int_0^1 \int_0^x e^{2y-y^2}\, dy\, dx$

51. Each of the integrals (I)–(VI) takes one of two distinct values. Without evaluating, group them by value.

I. $\int_0^5 \int_0^{10} xy^2\, dx\, dy$ II. $\int_0^5 \int_0^{10} xy^2\, dy\, dx$

III. $\int_0^{10} \int_0^5 xy^2\, dx\, dy$ IV. $\int_0^{10} \int_0^5 xy^2\, dy\, dx$

V. $\int_0^5 \int_0^{10} uv^2\, du\, dv$ VI. $\int_0^5 \int_0^{10} uv^2\, dv\, du$

52. Find the volume under the graph of the function $f(x, y) = 6x^2y$ over the region shown in Figure 16.20.

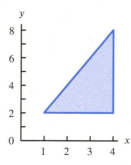

Figure 16.20

53. (a) Find the volume below the surface $z = x^2 + y^2$ and above the xy-plane for $-1 \le x \le 1, -1 \le y \le 1$.

(b) Find the volume above the surface $z = x^2 + y^2$ and below the plane $z = 2$ for $-1 \le x \le 1$, $-1 \le y \le 1$.

54. Compute the integral

$$\int\int_R (2x^2 + y)\, dA,$$

where R is the triangular region with vertices at $(0, 1)$, $(-2, 3)$ and $(2, 3)$.

55. (a) Sketch the region in the xy-plane bounded by the x-axis, $y = x$, and $x + y = 1$.

(b) Express the integral of $f(x, y)$ over this region in terms of iterated integrals in two ways. (In one, use $dx\, dy$; in the other, use $dy\, dx$.)

(c) Using one of your answers to part (b), evaluate the integral exactly with $f(x, y) = x$.

56. Let $f(x, y) = x^2 e^{x^2}$ and let R be the triangle bounded by the lines $x = 3$, $x = y/2$, and $y = x$ in the xy-plane.

  (a) Express $\int_R f \, dA$ as a double integral in two different ways.
  (b) Evaluate one of them.

57. Find the average value of $f(x, y) = x^2 + 4y$ on the rectangle $0 \leq x \leq 3$ and $0 \leq y \leq 6$.

58. Find the average value of $f(x, y) = xy^2$ on the rectangle $0 \leq x \leq 4, 0 \leq y \leq 3$.

59. Figure 16.21 shows two metal plates carrying electrical charges. The charge density (in coulombs per square meter) of each at the point (x, y) is $\sigma(x, y) = 6x + 6$ for x, y in meters.

  (a) Without calculation, decide which plate carries a greater total charge, and explain your reasoning.
  (b) Find the total charge on both plates, and compare to your answer from part (a).

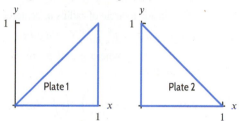

Figure 16.21

60. The population density in people per km² for the trapezoid-shaped town in Figure 16.22 for x, y in kilometers is $\delta(x, y) = 100x + 200y$. Find the town's population.

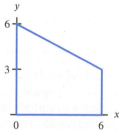

Figure 16.22

61. The quarter-disk-shaped metal plate in Figure 16.23 has radius 3 and density $\delta(x, y) = 2y$ gm/cm², with x, y in cm. Find the mass of the plate.

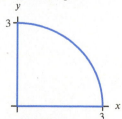

Figure 16.23

In Problems **62–63** set up, but do not evaluate, an iterated integral for the volume of the solid.

62. Under the graph of $f(x, y) = 25 - x^2 - y^2$ and above the xy-plane.

63. Below the graph of $f(x, y) = 25 - x^2 - y^2$ and above the plane $z = 16$.

64. A solid with flat base in the xy-plane is bounded by the vertical planes $y = 0$ and $y - x = 4$, and the slanted plane $2x + y + z = 4$.

  (a) Draw the base of the solid.
  (b) Set up, but do not evaluate, an iterated integral for the volume of the solid.

In Problems **65–69**, find the volume of the solid region.

65. Under the graph of $f(x, y) = xy$ and above the square $0 \leq x \leq 2, 0 \leq y \leq 2$ in the xy-plane.

66. Under the graph of $f(x, y) = x^2 + y^2$ and above the triangle $0 \leq y \leq x, 0 \leq x \leq 1$.

67. Under the graph of $f(x, y) = x + y$ and above the region $y^2 \leq x, 0 \leq x \leq 9, y \geq 0$.

68. Under the graph of $2x + y + z = 4$ in the first octant.

69. The solid region R bounded by the coordinate planes and the graph of $ax + by + cz = 1$. Assume a, b, and $c > 0$.

70. If R is the region $x + y \geq a, x^2 + y^2 \leq a^2$, with $a > 0$, evaluate the integral

$$\int_R xy \, dA.$$

71. The region W lies below the surface $f(x, y) = 2e^{-(x-1)^2 - y^2}$ and above the disk $x^2 + y^2 \leq 4$ in the xy-plane.

  (a) Describe in words the contours of f, using $f(x, y) = 1$ as an example.
  (b) Write an integral giving the area of the cross-section of W in the plane $x = 1$.
  (c) Write an iterated double integral giving the volume of W.

72. Find the average distance to the x-axis for points in the region in the first quadrant bounded by the x-axis and the graph of $y = x - x^2$.

73. Give the contour diagram of a function f whose average value on the square $0 \leq x \leq 1, 0 \leq y \leq 1$ is

  (a) Greater than the average of the values of f at the four corners of the square.
  (b) Less than the average of the values of f at the four corners of the square.

74. The function $f(x, y) = ax + by$ has an average value of 20 on the rectangle $0 \le x \le 2, 0 \le y \le 3$.

   (a) What can you say about the constants a and b?

   (b) Find two different choices for f that have average value 20 on the rectangle, and give their contour diagrams on the rectangle.

75. The function $f(x, y) = ax^2 + bxy + cy^2$ has an average value of 20 on the square $0 \le x \le 2, 0 \le y \le 2$.

   (a) What can you say about the constants a, b, and c?

   (b) Find two different choices for f that have average value 20 on the square, and give their contour diagrams on the square.

Strengthen Your Understanding

■ In Problems 76–77, explain what is wrong with the statement.

76. $\int_0^1 \int_0^x f(x, y) \, dy \, dx = \int_0^1 \int_0^1 f(x, y) \, dx \, dy$

77. $\int_0^1 \int_0^y xy \, dx \, dy = \int_0^y \int_0^1 xy \, dy \, dx$

■ In Problems 78–80, give an example of:

78. An iterated double integral, with limits of integration, giving the volume of a cylinder standing vertically with a circular base in the xy-plane.

79. A nonconstant function, f, whose integral is 4 over the triangular region with vertices $(0, 0)$, $(1, 0)$, $(1, 1)$.

80. A double integral representing the volume of a triangular prism of base area 6.

■ Are the statements in Problems 81–88 true or false? Give reasons for your answer.

81. The iterated integral $\int_0^1 \int_5^{12} f \, dx \, dy$ is computed over the rectangle $0 \le x \le 1, 5 \le y \le 12$.

82. If R is the region inside the triangle with vertices $(0, 0), (1, 1)$ and $(0, 2)$, then the double integral $\int_R f \, dA$ can be evaluated by an iterated integral of the form $\int_0^2 \int_0^1 f \, dx \, dy$.

83. The region of integration of the iterated integral $\int_1^2 \int_{x^2}^{x^3} f \, dy \, dx$ lies completely in the first quadrant (that is, $x \ge 0, y \ge 0$).

84. If the limits a, b, c and d in the iterated integral $\int_a^b \int_c^d f \, dy \, dx$ are all positive, then the value of $\int_a^b \int_c^d f \, dy \, dx$ is also positive.

85. If $f(x, y)$ is a function of y only, then $\int_a^b \int_0^1 f \, dx \, dy = \int_a^b f \, dy$.

86. If R is the region inside a circle of radius a, centered at the origin, then $\int_R f \, dA = \int_{-a}^a \int_0^{\sqrt{a^2 - x^2}} f \, dy \, dx$.

87. If $f(x, y) = g(x) \cdot h(y)$, where g and h are single-variable functions, then

$$\int_a^b \int_c^d f \, dy \, dx = \left(\int_a^b g(x) \, dx \right) \cdot \left(\int_c^d h(y) \, dy \right).$$

88. If $f(x, y) = g(x) + h(y)$, where g and h are single-variable functions, then

$$\int_a^b \int_c^d f \, dx \, dy = \left(\int_a^b g(x) \, dx \right) + \left(\int_c^d h(y) \, dy \right).$$

16.3 TRIPLE INTEGRALS

A continuous function of three variables can be integrated over a solid region W in 3-space in the same way as a function of two variables is integrated over a flat region in 2-space. Again, we start with a Riemann sum. First we subdivide W into smaller regions, then we multiply the volume of each region by a value of the function in that region, and then we add the results. For example, if W is the box $a \le x \le b, c \le y \le d, p \le z \le q$, then we subdivide each side into n, m, and l pieces, thereby chopping W into nml smaller boxes, as shown in Figure 16.24.

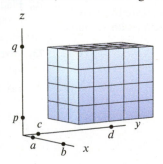

Figure 16.24: Subdividing a three-dimensional box

The volume of each smaller box is

$$\Delta V = \Delta x \Delta y \Delta z,$$

where $\Delta x = (b - a)/n$, and $\Delta y = (d - c)/m$, and $\Delta z = (q - p)/l$. Using this subdivision, we pick a point $(u_{ijk}, v_{ijk}, w_{ijk})$ in the ijk^{th} small box and construct a Riemann sum

$$\sum_{i,j,k} f(u_{ijk}, v_{ijk}, w_{ijk}) \, \Delta V.$$

If f is continuous, as Δx, Δy, and Δz approach 0, this Riemann sum approaches the definite integral, $\int_W f \, dV$, called a *triple integral*, which is defined as

$$\int_W f \, dV = \lim_{\Delta x, \Delta y, \Delta z \to 0} \sum_{i,j,k} f(u_{ijk}, v_{ijk}, w_{ijk}) \, \Delta x \, \Delta y \, \Delta z.$$

As in the case of a double integral, we can evaluate this integral as an iterated integral:

Triple integral as an iterated integral:

$$\int_W f \, dV = \int_p^q \left(\int_c^d \left(\int_a^b f(x, y, z) \, dx \right) dy \right) dz,$$

where y and z are treated as constants in the innermost (dx) integral, and z is treated as a constant in the middle (dy) integral. Other orders of integration are possible.

Example 1 A cube C has sides of length 4 cm and is made of a material of variable density. If one corner is at the origin and the adjacent corners are on the positive x, y, and z axes, then the density at the point (x, y, z) is $\delta(x, y, z) = 1 + xyz$ gm/cm^3. Find the mass of the cube.

Solution Consider a small piece ΔV of the cube, small enough so that the density remains close to constant over the piece. Then

$$\text{Mass of small piece} = \text{Density} \cdot \text{Volume} \approx \delta(x, y, z) \, \Delta V.$$

To get the total mass, we add the masses of the small pieces and take the limit as $\Delta V \to 0$. Thus, the mass is the triple integral

$$M = \int_C \delta \, dV = \int_0^4 \int_0^4 \int_0^4 (1 + xyz) \, dx \, dy \, dz = \int_0^4 \int_0^4 \left(x + \frac{1}{2} x^2 yz \right) \Big|_{x=0}^{x=4} dy \, dz$$

$$= \int_0^4 \int_0^4 (4 + 8yz) \, dy \, dz = \int_0^4 \left(4y + 4y^2 z \right) \Big|_{y=0}^{y=4} dz = \int_0^4 (16 + 64z) \, dz = 576 \text{ gm}.$$

Example 2   Express the volume of the building described in Example 1 on page 899 as a triple integral.

Solution   The building is given by $0 \leq x \leq 8$, $0 \leq y \leq 16$, and $0 \leq z \leq 12 - x/4 - y/8$. (See Figure 16.25.) To find its volume, divide it into small cubes of volume $\Delta V = \Delta x \, \Delta y \, \Delta z$ and add. First, make a vertical stack of cubes above the point $(x, y, 0)$. This stack goes from $z = 0$ to $z = 12 - x/4 - y/8$, so

$$\text{Volume of vertical stack} \approx \sum_z \Delta V = \sum_z \Delta x \, \Delta y \, \Delta z = \left(\sum_z \Delta z \right) \Delta x \, \Delta y.$$

Next, line up these stacks parallel to the y-axis to form a slice from $y = 0$ to $y = 16$. So

$$\text{Volume of slice} \approx \left(\sum_y \sum_z \Delta z \, \Delta y \right) \Delta x.$$

Finally, line up the slices along the x-axis from $x = 0$ to $x = 8$ and add up their volumes, to get

$$\text{Volume of building} \approx \sum_x \sum_y \sum_z \Delta z \, \Delta y \, \Delta x.$$

Thus, in the limit,

$$\text{Volume of building} = \int_0^8 \int_0^{16} \int_0^{12-x/4-y/8} 1 \, dz \, dy \, dx.$$

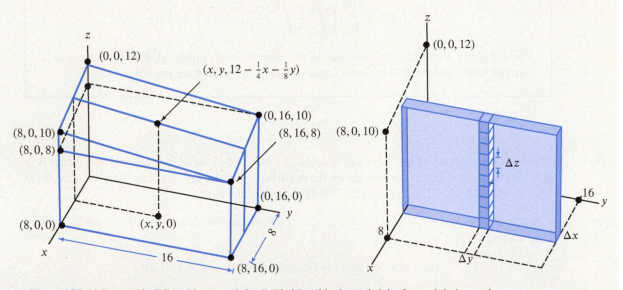

Figure 16.25: Volume of building (shown to left) divided into blocks and slabs for a triple integral

The preceding examples show that the triple integral has interpretations similar to the double integral:

- If $\rho(x, y, z)$ is density, then $\displaystyle\int_W \rho \, dV$ is the total quantity in the solid region W.

- $\displaystyle\int_W 1 \, dV$ is the volume of the solid region W.

Example 3   Set up an iterated integral to compute the mass of the solid cone bounded by $z = \sqrt{x^2 + y^2}$ and $z = 3$, if the density is given by $\delta(x, y, z) = z$.

Solution   We break the cone in Figure 16.26 into small cubes of volume $\Delta V = \Delta x \, \Delta y \, \Delta z$, on which the density is approximately constant, and approximate the mass of each cube by $\delta(x, y, z) \, \Delta x \, \Delta y \, \Delta z$. Stacking the cubes vertically above the point $(x, y, 0)$, starting on the cone at height $z = \sqrt{x^2 + y^2}$ and going up to $z = 3$, tells us that the inner integral is

$$\int_{\sqrt{x^2+y^2}}^{3} \delta(x, y, z) \, dz = \int_{\sqrt{x^2+y^2}}^{3} z \, dz.$$

There is a stack for every point in the xy-plane in the shadow of the cone. The cone $z = \sqrt{x^2 + y^2}$ intersects the horizontal plane $z = 3$ in the circle $x^2 + y^2 = 9$, so there is a stack for all (x, y) in the region $x^2 + y^2 \leq 9$. Lining up the stacks parallel to the y-axis gives a slice from $y = -\sqrt{9 - x^2}$ to $y = \sqrt{9 - x^2}$, for each fixed value of x. Thus, the limits on the middle integral are

$$\int_{-\sqrt{9-x^2}}^{\sqrt{9-x^2}} \int_{\sqrt{x^2+y^2}}^{3} z \, dz \, dy.$$

Finally, there is a slice for each x between -3 and 3, so the integral we want is

$$\text{Mass} = \int_{-3}^{3} \int_{-\sqrt{9-x^2}}^{\sqrt{9-x^2}} \int_{\sqrt{x^2+y^2}}^{3} z \, dz \, dy \, dx.$$

Notice that setting up the limits on the two outer integrals was just like setting up the limits for a double integral over the region $x^2 + y^2 \leq 9$.

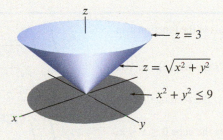

Figure 16.26: The cone $z = \sqrt{x^2 + y^2}$ with its shadow on the xy-plane

As the previous example illustrates, for a region W contained between two surfaces, the innermost limits correspond to these surfaces. The middle and outer limits ensure that we integrate over the "shadow" of W in the xy-plane.

Limits on Triple Integrals

- The limits for the outer integral are constants.
- The limits for the middle integral can involve only one variable (that in the outer integral).
- The limits for the inner integral can involve two variables (those on the two outer integrals).

Summary for Section 16.3

- A **triple integral** of a continuous function $f(x, y, z)$ over a solid rectangular box W with $a \leq x \leq b, c \leq y \leq d, p \leq z \leq q$, can be evaluated using the iterated integral

$$\int_W f \, dV = \int_p^q \left(\int_c^d \left(\int_a^b f(x, y, z) \, dx \right) dy \right) dz,$$

where y and z are treated as constants in the innermost (dx) integral, and z is treated as a constant in the middle (dy) integral. The five other orders of integration give the same result.

- The **volume** of the solid region W is given by the triple integral $\int_W 1 \, dV$.

- If $\rho(x, y, z)$ is **density**, then $\int_W \rho \, dV$ is the **total quantity** in the solid region W.

- Limits on a **triple integrated integral** satisfy:

  o The limits for the outer integral are constants.
  o The limits for the middle integral can involve only one variable (that in the outer integral).
  o The limits for the inner integral can involve two variables (those on the two outer integrals).

- If a solid region W is **above** the graph of $z = g(x, y)$ and **below** $z = h(x, y)$ with (x, y) lying inside the region R in the xy-plane, then

$$\int_W f \, dV = \int \left(\int \left(\int_{g(x,y)}^{h(x,y)} f(x, y, z) \, dz \right) dy \right) dx,$$

where the limits on the outer two iterated integrals are the same as those for a double integral over R.

Exercises and Problems for Section 16.3   Online Resource: Additional Problems for Section 16.3

EXERCISES

■ In Exercises 1–4, find the triple integrals of the function over the region W.

1. $f(x, y, z) = x^2 + 5y^2 - z$, W is the rectangular box $0 \leq x \leq 2, -1 \leq y \leq 1, 2 \leq z \leq 3$.

2. $h(x, y, z) = ax + by + cz$, W is the rectangular box $0 \leq x \leq 1, 0 \leq y \leq 1, 0 \leq z \leq 2$.

3. $f(x, y, z) = \sin x \cos(y + z)$, W is the cube $0 \leq x \leq \pi$, $0 \leq y \leq \pi, 0 \leq z \leq \pi$.

4. $f(x, y, z) = e^{-x-y-z}$, W is the rectangular box with corners at $(0, 0, 0)$, $(a, 0, 0)$, $(0, b, 0)$, and $(0, 0, c)$.

■ Sketch the region of integration in Exercises 5–13.

5. $\int_0^1 \int_{-1}^1 \int_0^{\sqrt{1-x^2}} f(x, y, z) \, dz \, dx \, dy$

6. $\int_0^1 \int_{-1}^1 \int_0^{\sqrt{1-z^2}} f(x, y, z) \, dy \, dz \, dx$

7. $\int_0^1 \int_{-1}^1 \int_{-\sqrt{1-x^2}}^{\sqrt{1-x^2}} f(x, y, z) \, dz \, dx \, dy$

8. $\int_{-1}^1 \int_0^1 \int_{-\sqrt{1-z^2}}^{\sqrt{1-z^2}} f(x, y, z) \, dy \, dz \, dx$

9. $\int_{-1}^1 \int_{-\sqrt{1-x^2}}^{\sqrt{1-x^2}} \int_0^{\sqrt{1-x^2-z^2}} f(x, y, z) \, dy \, dz \, dx$

10. $\int_0^1 \int_{-\sqrt{1-z^2}}^{\sqrt{1-z^2}} \int_0^{\sqrt{1-x^2-z^2}} f(x, y, z) \, dy \, dx \, dz$

11. $\int_0^1 \int_0^{\sqrt{1-y^2}} \int_{-\sqrt{1-x^2-y^2}}^{\sqrt{1-x^2-y^2}} f(x, y, z) \, dz \, dx \, dy$

12. $\int_0^1 \int_{-\sqrt{1-z^2}}^{\sqrt{1-z^2}} \int_{-\sqrt{1-y^2-z^2}}^{\sqrt{1-y^2-z^2}} f(x, y, z) \, dx \, dy \, dz$

13. $\int_0^1 \int_0^{\sqrt{1-z^2}} \int_{-\sqrt{1-x^2-z^2}}^{\sqrt{1-x^2-z^2}} f(x, y, z) \, dy \, dx \, dz$

■ In Exercises 14–15, for x, y and z in meters, what does the integral over the solid region E represent? Give units.

14. $\int_E 1 \, dV$

15. $\int_E \delta(x, y, z) \, dV$, where $\delta(x, y, z)$ is density, in kg/m^3.

PROBLEMS

In Problems 16–20, decide whether the integrals are positive, negative, or zero. Let S be the solid sphere $x^2 + y^2 + z^2 \leq 1$, and T be the top half of this sphere (with $z \geq 0$), and B be the bottom half (with $z \leq 0$), and R be the right half of the sphere (with $x \geq 0$), and L be the left half (with $x \leq 0$).

16. $\int_T e^z \, dV$

17. $\int_B e^z \, dV$

18. $\int_S \sin z \, dV$

19. $\int_T \sin z \, dV$

20. $\int_R \sin z \, dV$

Let W be the solid cone bounded by $z = \sqrt{x^2 + y^2}$ and $z = 2$. For Problems 21–29, decide (without calculating its value) whether the integral is positive, negative, or zero.

21. $\int_W y \, dV$

22. $\int_W x \, dV$

23. $\int_W z \, dV$

24. $\int_W xy \, dV$

25. $\int_W xyz \, dV$

26. $\int_W (z - 2) \, dV$

27. $\int_W \sqrt{x^2 + y^2} \, dV$

28. $\int_W e^{-xyz} \, dV$

29. $\int_W (z - \sqrt{x^2 + y^2}) \, dV$

In Problems 30–34, let W be the solid cylinder bounded by $x^2 + y^2 = 1$, $z = 0$, and $z = 2$. Decide (without calculating its value) whether the integral is positive, negative, or zero.

30. $\int_W x \, dV$

31. $\int_W z \, dV$

32. $\int_W (x^2 + y^2 - 2) \, dV$

33. $\int_W (z - 1) \, dV$

34. $\int_W e^{-y} \, dV$

35. Find the volume of the region bounded by the planes $z = 3y$, $z = y$, $y = 1$, $x = 1$, and $x = 2$.

36. Find the volume of the region bounded by $z = x^2$, $0 \leq x \leq 5$, and the planes $y = 0$, $y = 3$, and $z = 0$.

37. Find the volume of the region in the first octant bounded by the coordinate planes and the surface $x + y + z = 2$.

38. A trough with triangular cross-section lies along the x-axis for $0 \leq x \leq 10$. The slanted sides are given by $z = y$ and $z = -y$ for $0 \leq z \leq 1$ and the ends by $x = 0$ and $x = 10$, where x, y, z are in meters. The trough contains a sludge whose density at the point (x, y, z) is $\delta = e^{-3x}$ kg per m^3.

    (a) Express the total mass of sludge in the trough in terms of triple integrals.
    (b) Find the mass.

39. Find the volume of the region bounded by $z = x + y$, $z = 10$, and the planes $x = 0$, $y = 0$.

In Problems 40–45, write a triple integral, including limits of integration, that gives the specified volume.

40. Between $z = x + y$ and $z = 1 + 2x + 2y$ and above $0 \leq x \leq 1$, $0 \leq y \leq 2$.

41. Between the paraboloid $z = x^2 + y^2$ and the sphere $x^2 + y^2 + z^2 = 4$ and above the disk $x^2 + y^2 \leq 1$.

42. Between $2x + 2y + z = 6$ and $3x + 4y + z = 6$ and above $x + y \leq 1$, $x \geq 0$, $y \geq 0$.

43. Under the sphere $x^2 + y^2 + z^2 = 9$ and above the region between $y = x$ and $y = 2x - 2$ in the xy-plane in the first quadrant.

44. Between the top portion of the sphere $x^2 + y^2 + z^2 = 9$ and the plane $z = 2$.

45. Under the sphere $x^2 + y^2 + z^2 = 4$ and above the region $x^2 + y^2 \leq 4$, $0 \leq x \leq 1$, $0 \leq y \leq 2$ in the xy-plane.

In Problems 46–49, write limits of integration for the integral $\int_W f(x, y, z) \, dV$ where W is the quarter or half sphere or cylinder shown.

46.
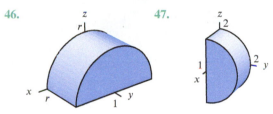

47.

48.
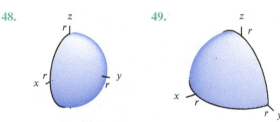

49.

50. Find the volume of the region between the plane $z = x$ and the surface $z = x^2$, and the planes $y = 0$, and $y = 3$.

51. Find the volume of the region bounded by $z = x + y$, $0 \leq x \leq 5$, $0 \leq y \leq 5$, and the planes $x = 0$, $y = 0$, and $z = 0$.

52. Find the volume of the pyramid with base in the plane $z = -6$ and sides formed by the three planes $y = 0$ and $y - x = 4$ and $2x + y + z = 4$.

53. Find the volume between the planes $z = 1 + x + y$ and $x + y + z = 1$ and above the triangle $x + y \leq 1$, $x \geq 0$, $y \geq 0$ in the xy-plane.

54. Find the mass of a triangular-shaped solid bounded by the planes $z = 1 + x$, $z = 1 - x$, $z = 0$, and with $0 \leq y \leq 3$. The density is $\delta = 10 - z$ gm/cm^3, and x, y, z are in cm.

55. Find the mass of the solid bounded by the xy-plane, yz-plane, xz-plane, and the plane $(x/3) + (y/2) + (z/6) = 1$, if the density of the solid is given by $\delta(x, y, z) = x + y$.

56. Find the mass of the pyramid with base in the plane $z = -6$ and sides formed by the three planes $y = 0$ and $y - x = 4$ and $2x + y + z = 4$, if the density of the solid is given by $\delta(x, y, z) = y$.

57. Let E be the solid pyramid bounded by the planes $x + z = 6$, $x - z = 0$, $y + z = 6$, $y - z = 0$, and above the plane $z = 0$ (see Figure 16.27). The density at any point in the pyramid is given by $\delta(x, y, z) = z$ grams per cm^3, where x, y, and z are measured in cm.

(a) Explain in practical terms what the triple integral $\int_E z \, dV$ represents.

(b) In evaluating the integral from part (a), how many separate triple integrals would be required if we chose to integrate in the z-direction first?

(c) Evaluate the triple integral from part (a) by integrating in a well-chosen order.

Figure 16.27

58. (a) What is the equation of the plane passing through the points $(1, 0, 0)$, $(0, 1, 0)$, and $(0, 0, 1)$?

(b) Find the volume of the region bounded by this plane and the planes $x = 0$, $y = 0$, and $z = 0$.

◼Problems 59–61 refer to Figure 16.28, which shows triangular portions of the planes $2x + 4y + z = 4$, $3x - 2y = 0$, $z = 2$, and the three coordinate planes $x = 0$, $y = 0$, and $z = 0$. For each solid region E, write down an iterated integral for the triple integral $\int_E f(x, y, z) \, dV$.

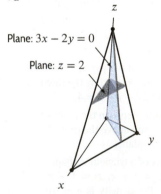

Plane: $3x - 2y = 0$

Plane: $z = 2$

Figure 16.28

59. E is the region bounded by $y = 0$, $z = 0$, $3x - 2y = 0$, and $2x + 4y + z = 4$.

60. E is the region bounded by $x = 0$, $y = 0$, $z = 0$, $z = 2$, and $2x + 4y + z = 4$.

61. E is the region bounded by $x = 0$, $z = 0$, $3x - 2y = 0$, and $2x + 4y + z = 4$.

62. Figure 16.29 shows part of a spherical ball of radius 5 cm. Write an iterated triple integral which represents the volume of this region.

2 cm

Figure 16.29

63. A solid region D is a half cylinder of radius 1 lying horizontally with its rectangular base in the xy-plane and its axis along the y-axis from $y = 0$ to $y = 10$. (The region is above the xy-plane.)

(a) What is the equation of the curved surface of this half cylinder?

(b) Write the limits of integration of the integral $\int_D f(x, y, z) \, dV$ in Cartesian coordinates.

64. Set up, but do not evaluate, an iterated integral for the volume of the solid formed by the intersections of the cylinders $x^2 + z^2 = 1$ and $y^2 + z^2 = 1$.

◼Problems 65–67 refer to Figure 16.30, which shows E, the region in the first octant bounded by the parabolic cylinder $z = 6y^2$ and the elliptical cylinder $x^2 + 3y^2 = 12$. For the given order of integration, write an iterated integral equivalent to the triple integral $\int_E f(x, y, z) \, dV$.

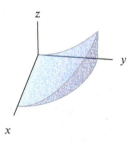

Figure 16.30

65. $dz \, dx \, dy$ 66. $dx \, dz \, dy$ 67. $dy \, dz \, dx$

◼Problems 68–71 refer to Figure 16.31, which shows E, the region in the first octant bounded by the planes $z = 5$ and $5x + 3z = 15$ and the elliptical cylinder $4x^2 + 9y^2 = 36$. For the given order of integration, write an iterated integral equivalent to the triple integral $\int_E f(x, y, z) \, dV$.

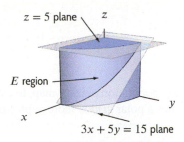

z = 5 plane

E region

$3x + 5y = 15$ plane

Figure 16.31

68. $dz\,dy\,dx$

69. $dz\,dx\,dy$

70. $dy\,dz\,dx$

71. $dy\,dx\,dz$

■Problems 72–74 refer to Figure 16.32, which shows E, the region in the first octant bounded by the plane $x + y = 2$ and the parabolic cylinder $z = 4 - x^2$. For the given order of integration, write an iterated integral, or sum of integrals, equivalent to the triple integral $\int_E f(x, y, z)\,dV$.

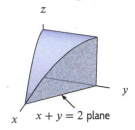

$x + y = 2$ plane

Figure 16.32

72. $dz\,dy\,dx$ 73. $dy\,dz\,dx$ 74. $dy\,dx\,dz$

Strengthen Your Understanding

■In Problems 77–78, explain what is wrong with the statement.

77. Let S be the solid sphere $x^2 + y^2 + z^2 \leq 1$ and let U be the upper half of S where $z \geq 0$. Then $\int_S f(x, y, z)\,dV = 2\int_U f(x, y, z)\,dV$.

78. $\int_0^1 \int_0^x \int_0^y f(x, y, z)\,dz\,dy\,dx = \int_0^1 \int_y^1 \int_0^x f(x, y, z)\,dz\,dx\,dy$

■In Problems 79–80, give an example of:

79. A function f such that $\int_R f\,dV = 7$, where R is the cylinder $x^2 + y^2 \leq 4, 0 \leq z \leq 3$.

80. A nonconstant function $f(x, y, z)$ such that if B is the region enclosed by the sphere of radius 1 centered at the origin, the integral $\int_B f(x, y, z)\,dx\,dy\,dz$ is zero.

■Are the statements in Problems 81–90 true or false? Give reasons for your answer.

81. If $\rho(x, y, z)$ is mass density of a material in 3-space, then $\int_W \rho(x, y, z)\,dV$ gives the volume of the solid region W.

82. The region of integration of the triple iterated integral $\int_0^1 \int_0^1 \int_0^x f\,dz\,dy\,dx$ lies above a square in the xy-plane and below a plane.

■Problems 75–76 concern the *center of mass*, the point at which the mass of a solid body in motion can be considered to be concentrated. If the object has density $\rho(x, y, z)$ at the point (x, y, z) and occupies a region W, then the coordinates $(\bar{x}, \bar{y}, \bar{z})$ of the center of mass are given by

$$\bar{x} = \frac{1}{m} \int_W x\rho\,dV \quad \bar{y} = \frac{1}{m} \int_W y\rho\,dV \quad \bar{z} = \frac{1}{m} \int_W z\rho\,dV$$

where $m = \int_W \rho\,dV$ is the total mass of the body.

75. A solid is bounded below by the square $z = 0, 0 \leq x \leq 1, 0 \leq y \leq 1$ and above by the surface $z = x + y + 1$. Find the total mass and the coordinates of the center of mass if the density is 1 gm/cm^3 and x, y, z are measured in centimeters.

76. Find the center of mass of the tetrahedron that is bounded by the xy, yz, xz planes and the plane $x + 2y + 3z = 1$. Assume the density is 1 gm/cm^3 and x, y, z are in centimeters.

83. If W is the unit ball $x^2 + y^2 + z^2 \leq 1$ then an iterated integral over W is $\int_0^1 \int_0^{\sqrt{1-x^2}} \int_0^{\sqrt{1-x^2-y^2}} f\,dz\,dy\,dx$.

84. The iterated integrals $\int_0^1 \int_0^{1-x} \int_0^{1-x-y} f\,dz\,dy\,dx$ and $\int_0^1 \int_0^{1-z} \int_0^{1-y-z} f\,dx\,dy\,dz$ are equal.

85. The iterated integrals $\int_{-1}^1 \int_0^1 \int_0^{1-x^2} f\,dz\,dy\,dx$ and $\int_0^1 \int_0^1 \int_{-\sqrt{1-z}}^{\sqrt{1-z}} f\,dx\,dy\,dz$ are equal.

86. If W is a rectangular solid in 3-space, then $\int_W f\,dV = \int_a^b \int_c^d \int_e^k f\,dz\,dy\,dx$, where a, b, c, d, e, and k are constants.

87. If W is the unit cube $0 \leq x \leq 1, 0 \leq y \leq 1, 0 \leq z \leq 1$ and $\int_W f\,dV = 0$, then $f = 0$ everywhere in the unit cube.

88. If $f > g$ at all points in the solid region W, then $\int_W f\,dV > \int_W g\,dV$.

89. If W_1 and W_2 are solid regions with volume$(W_1) >$ volume(W_2) then $\int_{W_1} f\,dV > \int_{W_2} f\,dV$.

90. Both double and triple integrals can be used to compute volume.

16.4   DOUBLE INTEGRALS IN POLAR COORDINATES

Integration in Polar Coordinates

We started this chapter by putting a rectangular grid on the fox population density map, to estimate the total population using a Riemann sum. However, sometimes a polar grid is more appropriate.

Example 1   A biologist studying insect populations around a circular lake divides the area into the polar sectors of radii 2, 3, and 4 km in Figure 16.33. The approximate population density in each sector is shown in millions per square km. Estimate the total insect population around the lake.

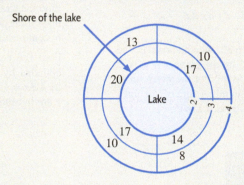

Figure 16.33: An insect-infested lake showing the insect population density by sector

Solution   To get the estimate, we multiply the population density in each sector by the area of that sector. Unlike the rectangles in a rectangular grid, the sectors in this grid do not all have the same area. The inner sectors have area

$$\frac{1}{4}(\pi 3^2 - \pi 2^2) = \frac{5\pi}{4} \approx 3.93 \text{ km}^2,$$

and the outer sectors have area

$$\frac{1}{4}(\pi 4^2 - \pi 3^2) = \frac{7\pi}{4} \approx 5.50 \text{ km}^2,$$

so we estimate

$$\text{Population} \approx (20)(3.93) + (17)(3.93) + (14)(3.93) + (17)(3.93)$$
$$+ (13)(5.50) + (10)(5.50) + (8)(5.50) + (10)(5.50)$$
$$= 492.74 \text{ million insects}.$$

What Is dA in Polar Coordinates?

The previous example used a polar grid rather than a rectangular grid. A rectangular grid is constructed from vertical and horizontal lines of the form $x = k$ (a constant) and $y = l$ (another constant). In polar coordinates, $r = k$ gives a circle of radius k centered at the origin and $\theta = l$ gives a ray emanating from the origin (at angle l with the x-axis). A polar grid is built out of these circles and rays. Suppose we want to integrate $f(r, \theta)$ over the region R in Figure 16.34.

Choosing (r_{ij}, θ_{ij}) in the ij^{th} bent rectangle in Figure 16.34 gives a Riemann sum:

$$\sum_{i,j} f(r_{ij}, \theta_{ij}) \, \Delta A.$$

To calculate the area ΔA, look at Figure 16.35. If Δr and $\Delta \theta$ are small, the shaded region is approximately a rectangle with sides $r \, \Delta \theta$ and Δr, so

$$\Delta A \approx r \Delta \theta \Delta r.$$

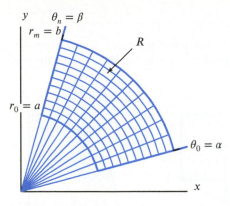

Figure 16.34: Dividing up a region using a polar grid

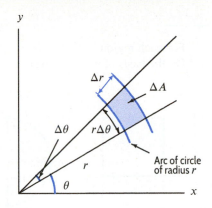

Figure 16.35: Calculating area ΔA in polar coordinates

Thus, the Riemann sum is approximately

$$\sum_{i,j} f(r_{ij}, \theta_{ij}) \, r_{ij} \, \Delta\theta \, \Delta r.$$

If we take the limit as Δr and $\Delta\theta$ approach 0, we obtain

$$\int_R f \, dA = \int_\alpha^\beta \int_a^b f(r, \theta) \, r \, dr \, d\theta.$$

> When computing integrals in polar coordinates, use $x = r\cos\theta$, $y = r\sin\theta$, $x^2 + y^2 = r^2$. Put $dA = r \, dr \, d\theta$ or $dA = r \, d\theta \, dr$.

Example 2   Compute the integral of $f(x, y) = 1/(x^2 + y^2)^{3/2}$ over the region R shown in Figure 16.36.

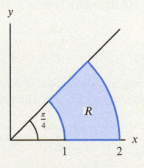

Figure 16.36: Integrate f over the polar region

Solution   The region R is described by the inequalities $1 \le r \le 2$, $0 \le \theta \le \pi/4$. In polar coordinates, $r = \sqrt{x^2 + y^2}$, so we can write f as

$$f(x, y) = \frac{1}{(x^2 + y^2)^{3/2}} = \frac{1}{(r^2)^{3/2}} = \frac{1}{r^3}.$$

Then

$$\int_R f \, dA = \int_0^{\pi/4} \int_1^2 \frac{1}{r^3} r \, dr \, d\theta = \int_0^{\pi/4} \left(\int_1^2 r^{-2} \, dr \right) d\theta$$

$$= \int_0^{\pi/4} -\frac{1}{r} \Big|_{r=1}^{r=2} d\theta = \int_0^{\pi/4} \frac{1}{2} \, d\theta = \frac{\pi}{8}.$$

Example 3   For each region in Figure 16.37, decide whether to integrate using polar or Cartesian coordinates. On the basis of its shape, write an iterated integral of an arbitrary function $f(x, y)$ over the region.

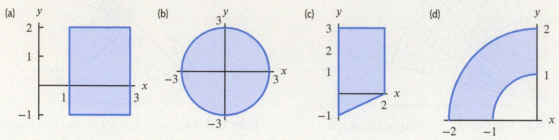

Figure 16.37

Solution   (a) Since this is a rectangular region, Cartesian coordinates are likely to be a better choice. The rectangle is described by the inequalities $1 \le x \le 3$ and $-1 \le y \le 2$, so the integral is

$$\int_{-1}^{2} \int_{1}^{3} f(x, y)\, dx\, dy.$$

(b) A circle is best described in polar coordinates. The radius is 3, so r goes from 0 to 3, and to describe the whole circle, θ goes from 0 to 2π. The integral is

$$\int_{0}^{2\pi} \int_{0}^{3} f(r \cos \theta, r \sin \theta)\, r\, dr\, d\theta.$$

(c) The bottom boundary of this trapezoid is the line $y = (x/2) - 1$ and the top is the line $y = 3$, so we use Cartesian coordinates. If we integrate with respect to y first, the lower limit of the integral is $(x/2) - 1$ and the upper limit is 3. The x limits are $x = 0$ to $x = 2$. So the integral is

$$\int_{0}^{2} \int_{(x/2)-1}^{3} f(x, y)\, dy\, dx.$$

(d) This is another polar region: it is a piece of a ring in which r goes from 1 to 2. Since it is in the second quadrant, θ goes from $\pi/2$ to π. The integral is

$$\int_{\pi/2}^{\pi} \int_{1}^{2} f(r \cos \theta, r \sin \theta)\, r\, dr\, d\theta.$$

Summary for Section 16.4

- When computing $\int_{R} f(x, y)\, dA$ in **polar coordinates**:
  - Convert the integrand to r, θ by using $x = r \cos \theta$, $y = r \sin \theta$, $x^2 + y^2 = r^2$.
  - Put $dA = r\, dr\, d\theta$ or $dA = r\, d\theta\, dr$.

Exercises and Problems for Section 16.4   Online Resource: Additional Problems for Section 16.4
EXERCISES

For the regions R in Exercises **1–4**, write $\int_R f\, dA$ as an iterated integral in polar coordinates.

1.
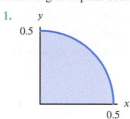

2.
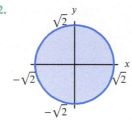

3.
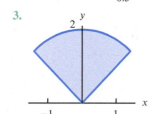

4.
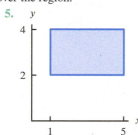

In Exercises **5–8**, choose rectangular or polar coordinates to set up an iterated integral of an arbitrary function $f(x,y)$ over the region.

5.
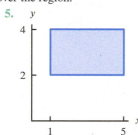

6.
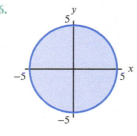

7.

8.
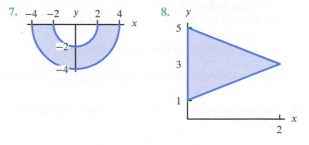

Sketch the region of integration in Exercises **9–15**.

9. $\displaystyle\int_0^4 \int_{-\pi/2}^{\pi/2} f(r,\theta)\, r\, d\theta\, dr$

10. $\displaystyle\int_{\pi/2}^{\pi} \int_0^1 f(r,\theta)\, r\, dr\, d\theta$

11. $\displaystyle\int_0^{2\pi} \int_1^2 f(r,\theta)\, r\, dr\, d\theta$

12. $\displaystyle\int_{\pi/6}^{\pi/3} \int_0^1 f(r,\theta)\, r\, dr\, d\theta$

13. $\displaystyle\int_0^{\pi/4} \int_0^{1/\cos\theta} f(r,\theta)\, r\, dr\, d\theta$

14. $\displaystyle\int_3^4 \int_{3\pi/4}^{3\pi/2} f(r,\theta)\, r\, d\theta\, dr$

15. $\displaystyle\int_{\pi/4}^{\pi/2} \int_0^{2/\sin\theta} f(r,\theta)\, r\, dr\, d\theta$

PROBLEMS

In Problems **16–18**, evaluate the integral.

16. $\int_R \sqrt{x^2+y^2}\, dx\, dy$ where R is $4 \le x^2+y^2 \le 9$.

17. $\int_R \sin(x^2+y^2)\, dA$, where R is the disk of radius 2 centered at the origin.

18. $\int_R (x^2-y^2)\, dA$, where R is the first quadrant region between the circles of radius 1 and radius 2.

Convert the integrals in Problems **19–21** to polar coordinates and evaluate.

19. $\displaystyle\int_{-1}^0 \int_{-\sqrt{1-x^2}}^{\sqrt{1-x^2}} x\, dy\, dx$

20. $\displaystyle\int_0^{\sqrt{6}} \int_{-x}^x dy\, dx$

21. $\displaystyle\int_0^{\sqrt{2}} \int_y^{\sqrt{4-y^2}} xy\, dx\, dy$

Problems **22–26** concern Figure 16.38, which shows regions R_1, R_2, and R_3 contained in the semicircle $x^2+y^2=4$ with $y \ge 0$.

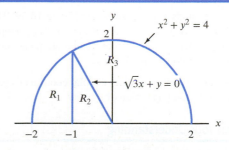

Figure 16.38

22. In Cartesian coordinates, write $\int_{R_1} 2y\, dA$ as an iterated integral in two different ways and then evaluate it.

23. In Cartesian coordinates, write $\int_{R_2} 2y\, dA$ as an iterated integral in two different ways.

24. Evaluate $\int_{R_3} (x^2+y^2)\, dA$.

25. Evaluate $\int_R 12y\, dA$, where R is the region formed by combining the regions R_1 and R_2.

26. Evaluate $\int_S x\, dA$, where S is the region formed by combining the regions R_2 and R_3.

27. Consider the integral $\int_0^3 \int_{x/3}^1 f(x,y)\,dy\,dx$.

   (a) Sketch the region R over which the integration is being performed.
   (b) Rewrite the integral with the order of integration reversed.
   (c) Rewrite the integral in polar coordinates.

28. Describe the region of integration for $\int_{\pi/4}^{\pi/2} \int_{1/\sin\theta}^{4/\sin\theta} f(r,\theta)r\,dr\,d\theta$.

29. Evaluate the integral by converting it into Cartesian coordinates:
$$\int_0^{\pi/6} \int_0^{2/\cos\theta} r\,dr\,d\theta.$$

30. (a) Sketch the region of integration of
$$\int_0^1 \int_{\sqrt{1-x^2}}^{\sqrt{4-x^2}} x\,dy\,dx + \int_1^2 \int_0^{\sqrt{4-x^2}} x\,dy\,dx$$

   (b) Evaluate the quantity in part (a).

31. Find the volume of the region between the graph of $f(x,y) = 25 - x^2 - y^2$ and the xy plane.

32. Find the volume of an ice cream cone bounded by the hemisphere $z = \sqrt{8 - x^2 - y^2}$ and the cone $z = \sqrt{x^2 + y^2}$.

33. (a) For $a > 0$, find the volume under the graph of $z = e^{-(x^2+y^2)}$ above the disk $x^2 + y^2 \le a^2$.
   (b) What happens to the volume as $a \to \infty$?

34. A circular metal disk of radius 3 lies in the xy-plane with its center at the origin. At a distance r from the origin, the density of the metal per unit area is $\delta = \dfrac{1}{r^2 + 1}$.

   (a) Write a double integral giving the total mass of the disk. Include limits of integration.
   (b) Evaluate the integral.

35. A city surrounds a bay as shown in Figure 16.39. The population density of the city (in thousands of people per square km) is $\delta(r,\theta)$, where r and θ are polar coordinates and distances are in km.

   (a) Set up an iterated integral in polar coordinates giving the total population of the city.

   (b) The population density decreases the farther you live from the shoreline of the bay; it also decreases the farther you live from the ocean. Which of the following functions best describes this situation?
      (i) $\delta(r,\theta) = (4-r)(2 + \cos\theta)$
      (ii) $\delta(r,\theta) = (4-r)(2 + \sin\theta)$
      (iii) $\delta(r,\theta) = (r+4)(2 + \cos\theta)$

   (c) Estimate the population using your answers to parts (a) and (b).

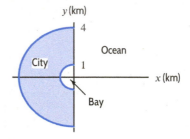

Figure 16.39

36. A disk of radius 5 cm has density 10 gm/cm² at its center and density 0 at its edge, and its density is a linear function of the distance from the center. Find the mass of the disk.

37. Electric charge is distributed over the xy-plane, with density inversely proportional to the distance from the origin. Show that the total charge inside a circle of radius R centered at the origin is proportional to R. What is the constant of proportionality?

38. (a) Graph $r = 1/(2\cos\theta)$ for $-\pi/2 \le \theta \le \pi/2$ and $r = 1$.
   (b) Write an iterated integral representing the area inside the curve $r = 1$ and to the right of $r = 1/(2\cos\theta)$. Evaluate the integral.

39. (a) Sketch the circles $r = 2\cos\theta$ for $-\pi/2 \le \theta \le \pi/2$ and $r = 1$.
   (b) Write an iterated integral representing the area inside the circle $r = 2\cos\theta$ and outside the circle $r = 1$. Evaluate the integral.

Strengthen Your Understanding

In Problems 40–44, explain what is wrong with the statement.

40. If R is the region bounded by $x = 1$, $y = 0$, $y = x$, then in polar coordinates $\int_R x\,dA = \int_0^{\pi/4} \int_0^1 r^2 \cos\theta\,dr\,d\theta$.

41. If R is the region $x^2 + y^2 \le 4$, then $\int_R (x^2 + y^2)\,dA = \int_0^{2\pi} \int_0^2 r^2\,dr\,d\theta$.

42. $\int_0^1 \int_0^1 \sqrt{x^2 + y^2}\,dy\,dx = \int_0^{\pi/2} \int_0^1 r^2\,dr\,d\theta$

43. $\int_1^2 \int_0^{\sqrt{4-x^2}} 1\,dy\,dx = \int_0^{\pi/2} \int_1^2 r\,dr\,d\theta$

44. $\int_0^1 \int_0^\pi r\,dr\,d\theta = \int_0^\pi \int_0^1 r\,dr\,d\theta$

In Problems 45–48, give an example of:

45. A region R of integration in the first quadrant which suggests the use of polar coordinates.

46. An integrand $f(x,y)$ that suggests the use of polar coordinates.

47. A function $f(x, y)$ such that $\int_R f(x, y) \, dy \, dx$ in polar coordinates has an integrand without a factor of r.

48. A region R such that $\int_R f(x, y) \, dA$ must be broken into two integrals in Cartesian coordinates, but only needs one integral in polar coordinates.

49. Which of the following integrals give the area of the unit circle?

(a) $\displaystyle\int_{-1}^{1} \int_{-\sqrt{1-x^2}}^{\sqrt{1-x^2}} dy \, dx$ **(b)** $\displaystyle\int_{-1}^{1} \int_{-\sqrt{1-x^2}}^{\sqrt{1-x^2}} x \, dy \, dx$

(c) $\displaystyle\int_{0}^{2\pi} \int_{0}^{1} r \, dr \, d\theta$ **(d)** $\displaystyle\int_{0}^{2\pi} \int_{0}^{1} dr \, d\theta$

(e) $\displaystyle\int_{0}^{1} \int_{0}^{2\pi} r \, d\theta \, dr$ **(f)** $\displaystyle\int_{0}^{1} \int_{0}^{2\pi} d\theta \, dr$

■ Are the statements in Problems **50–55** true or false? Give reasons for your answer.

50. The integral $\int_0^{2\pi} \int_0^1 dr \, d\theta$ gives the area of the unit circle.

51. The quantity $8 \int_5^7 \int_0^{\pi/4} r \, d\theta \, dr$ gives the area of a ring with radius between 5 and 7.

52. Let R be the region inside the semicircle $x^2 + y^2 = 9$ with $y \geq 0$. Then $\int_R (x+y) \, dA = \int_0^\pi \int_0^3 r \, dr \, d\theta$.

53. The integrals $\int_0^\pi \int_0^1 r^2 \cos\theta \, dr \, d\theta$ and $2\int_0^{\pi/2} \int_0^1 r^2 \cos\theta \, dr \, d\theta$ are equal.

54. The integral $\int_0^{\pi/4} \int_0^{1/\cos\theta} r \, dr \, d\theta$ gives the area of the region $0 \leq x \leq 1, 0 \leq y \leq x$.

55. The integral $\int_0^{2\pi} \int_0^1 r^3 \, dr \, d\theta$ gives the area of the unit circle.

16.5 INTEGRALS IN CYLINDRICAL AND SPHERICAL COORDINATES

Some double integrals are easier to evaluate in polar, rather than Cartesian, coordinates. Similarly, some triple integrals are easier in non-Cartesian coordinates.

Cylindrical Coordinates

The cylindrical coordinates of a point (x, y, z) in 3-space are obtained by representing the x and y coordinates in polar coordinates and letting the z-coordinate be the z-coordinate of the Cartesian coordinate system. (See Figure 16.40.)

Relation Between Cartesian and Cylindrical Coordinates

Each point in 3-space is represented using $0 \leq r < \infty, 0 \leq \theta \leq 2\pi, -\infty < z < \infty$.

$$x = r\cos\theta,$$
$$y = r\sin\theta,$$
$$z = z.$$

As with polar coordinates in the plane, note that $x^2 + y^2 = r^2$.

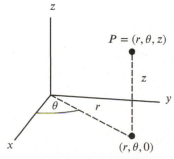

Figure 16.40: Cylindrical coordinates: (r, θ, z)

A useful way to visualize cylindrical coordinates is to sketch the surfaces obtained by setting one of the coordinates equal to a constant. See Figures 16.41–16.43.

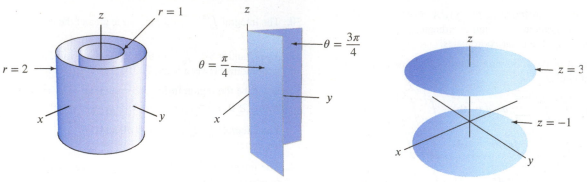

Figure 16.41: The surfaces $r = 1$ and $r = 2$

Figure 16.42: The surfaces $\theta = \pi/4$ and $\theta = 3\pi/4$

Figure 16.43: The surfaces $z = -1$ and $z = 3$

Setting $r = c$ (where c is constant) gives a cylinder around the z-axis whose radius is c. Setting $\theta = c$ gives a half-plane perpendicular to the xy plane, with one edge along the z-axis, making an angle c with the x-axis. Setting $z = c$ gives a horizontal plane $|c|$ units from the xy-plane. We call these *fundamental surfaces*.

The regions that can most easily be described in cylindrical coordinates are those regions whose boundaries are such fundamental surfaces (for example, vertical cylinders, or wedge-shaped parts of vertical cylinders).

Example 1   Describe in cylindrical coordinates a wedge of cheese cut from a cylinder 4 cm high and 6 cm in radius; this wedge subtends an angle of $\pi/6$ at the center. (See Figure 16.44.)

Solution   The wedge is described by the inequalities $0 \leq r \leq 6$, and $0 \leq z \leq 4$, and $0 \leq \theta \leq \pi/6$.

Integration in Cylindrical Coordinates

To integrate a double integral $\int_R f \, dA$ in polar coordinates, we had to express the area element dA in terms of polar coordinates: $dA = r \, dr \, d\theta$. To evaluate a triple integral $\int_W f \, dV$ in cylindrical coordinates, we need to express the volume element dV in cylindrical coordinates.

In Figure 16.45, consider the volume element ΔV bounded by fundamental surfaces. The area of the base is $\Delta A \approx r \, \Delta r \, \Delta \theta$. Since the height is Δz, the volume element is given approximately by $\Delta V \approx r \, \Delta r \, \Delta \theta \, \Delta z$.

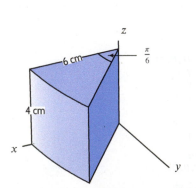

Figure 16.44: A wedge of cheese

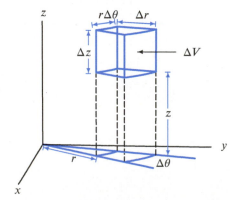

Figure 16.45: Volume element in cylindrical coordinates

> When computing integrals in cylindrical coordinates, put $dV = r\,dr\,d\theta\,dz$. Other orders of integration are also possible.

Example 2   Find the mass of the wedge of cheese in Example 1, if its density is 1.2 grams/cm^3.

Solution   If the wedge is W, its mass is

$$\int_W 1.2\,dV.$$

In cylindrical coordinates this integral is

$$\int_0^4 \int_0^{\pi/6} \int_0^6 1.2\,r\,dr\,d\theta\,dz = \int_0^4 \int_0^{\pi/6} 0.6r^2 \Big|_0^6 d\theta\,dz = 21.6 \int_0^4 \int_0^{\pi/6} d\theta\,dz$$

$$= 21.6\left(\frac{\pi}{6}\right)4 = 45.239 \text{ grams}.$$

Example 3   A water tank in the shape of a hemisphere has radius a; its base is its plane face. Find the volume, V, of water in the tank as a function of h, the depth of the water.

Solution   In Cartesian coordinates, a sphere of radius a has the equation $x^2 + y^2 + z^2 = a^2$. (See Figure 16.46.) In cylindrical coordinates, $r^2 = x^2 + y^2$, so this becomes

$$r^2 + z^2 = a^2.$$

Thus, if we want to describe the amount of water in the tank in cylindrical coordinates, we let r go from 0 to $\sqrt{a^2 - z^2}$, we let θ go from 0 to 2π, and we let z go from 0 to h, giving

$$\begin{aligned}
\text{Volume of water} &= \int_W 1\,dV = \int_0^{2\pi} \int_0^h \int_0^{\sqrt{a^2-z^2}} r\,dr\,dz\,d\theta = \int_0^{2\pi} \int_0^h \frac{r^2}{2}\Big|_{r=0}^{r=\sqrt{a^2-z^2}} dz\,d\theta \\
&= \int_0^{2\pi} \int_0^h \frac{1}{2}(a^2 - z^2)\,dz\,d\theta = \int_0^{2\pi} \frac{1}{2}\left(a^2 z - \frac{z^3}{3}\right)\Big|_{z=0}^{z=h} d\theta \\
&= \int_0^{2\pi} \frac{1}{2}\left(a^2 h - \frac{h^3}{3}\right) d\theta = \pi\left(a^2 h - \frac{h^3}{3}\right).
\end{aligned}$$

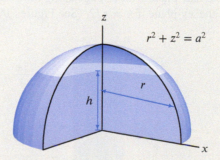

Figure 16.46: Hemispherical water tank with radius a and water of depth h

Spherical Coordinates

In Figure 16.47, the point P has coordinates (x, y, z) in the Cartesian coordinate system. We define spherical coordinates ρ, ϕ, and θ for P as follows: $\rho = \sqrt{x^2 + y^2 + z^2}$ is the distance of P from the origin; ϕ is the angle between the positive z-axis and the line through the origin and the point P; and θ is the same as in cylindrical coordinates.

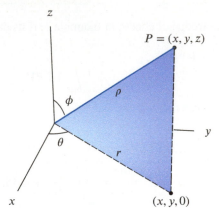

Figure 16.47: Spherical coordinates: (ρ, ϕ, θ)

In cylindrical coordinates,

$$x = r\cos\theta, \quad \text{and} \quad y = r\sin\theta, \quad \text{and} \quad z = z.$$

From Figure 16.47 we have $z = \rho\cos\phi$ and $r = \rho\sin\phi$, giving the following relationship:

Relation Between Cartesian and Spherical Coordinates

Each point in 3-space is represented using $0 \le \rho < \infty$, $0 \le \phi \le \pi$, and $0 \le \theta \le 2\pi$.

$$x = \rho\sin\phi\cos\theta$$
$$y = \rho\sin\phi\sin\theta$$
$$z = \rho\cos\phi.$$

Also, $\rho^2 = x^2 + y^2 + z^2$.

This system of coordinates is useful when there is spherical symmetry with respect to the origin, either in the region of integration or in the integrand. The fundamental surfaces in spherical coordinates are $\rho = k$ (a constant), which is a sphere of radius k centered at the origin, $\theta = k$ (a constant), which is the half-plane with its edge along the z-axis, and $\phi = k$ (a constant), which is a cone if $k \ne \pi/2$ and the xy-plane if $k = \pi/2$. (See Figures 16.48–16.50.)

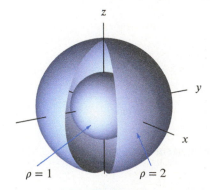

Figure 16.48: The surfaces $\rho = 1$ and $\rho = 2$

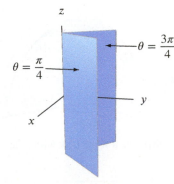

Figure 16.49: The surfaces $\theta = \pi/4$ and $\theta = 3\pi/4$

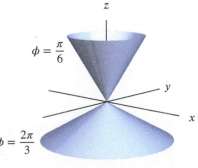

Figure 16.50: The surfaces $\phi = \pi/6$ and $\phi = 2\pi/3$

Integration in Spherical Coordinates

To use spherical coordinates in triple integrals we need to express the volume element, dV, in spherical coordinates. From Figure 16.51, we see that the volume element can be approximated by a box with curved edges. One edge has length $\Delta\rho$. The edge parallel to the xy-plane is an arc of a circle made from rotating the cylindrical radius $r (= \rho\sin\phi)$ through an angle $\Delta\theta$, and so has length $\rho\sin\phi\,\Delta\theta$. The remaining edge comes from rotating the radius ρ through an angle $\Delta\phi$, and so has length $\rho\,\Delta\phi$. Therefore, $\Delta V \approx \Delta\rho(\rho\,\Delta\phi)(\rho\sin\phi\,\Delta\theta) = \rho^2\sin\phi\,\Delta\rho\,\Delta\phi\,\Delta\theta$.

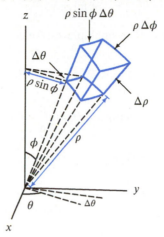

Figure 16.51: Volume element in spherical coordinates

Thus:

> When computing integrals in spherical coordinates, put $dV = \rho^2\sin\phi\,d\rho\,d\phi\,d\theta$. Other orders of integration are also possible.

Example 4

Use spherical coordinates to derive the formula for the volume of a ball of radius a.

Solution

In spherical coordinates, a ball of radius a is described by the inequalities $0 \le \rho \le a$, $0 \le \theta \le 2\pi$, and $0 \le \phi \le \pi$. Note that θ goes from 0 to 2π, whereas ϕ goes from 0 to π. We find the volume by integrating the constant density function 1 over the ball:

$$\text{Volume} = \int_R 1\,dV = \int_0^{2\pi}\int_0^{\pi}\int_0^a \rho^2\sin\phi\,d\rho\,d\phi\,d\theta = \int_0^{2\pi}\int_0^{\pi}\frac{1}{3}a^3\sin\phi\,d\phi\,d\theta$$

$$= \frac{1}{3}a^3\int_0^{2\pi}-\cos\phi\Big|_0^{\pi}d\theta = \frac{2}{3}a^3\int_0^{2\pi}d\theta = \frac{4\pi a^3}{3}.$$

Example 5

Find the magnitude of the gravitational force exerted by a solid hemisphere of radius a and constant density δ on a unit mass located at the center of the base of the hemisphere.

Solution

Assume the base of the hemisphere rests on the xy-plane with center at the origin. (See Figure 16.52.) Newton's law of gravitation says that the force between two masses m_1 and m_2 at a distance r apart is $F = Gm_1m_2/r^2$, where G is the gravitational constant.

In this example, symmetry shows that the net component of the force on the particle at the origin due to the hemisphere is in the z direction only. Any force in the x or y direction from some part of the hemisphere is canceled by the force from another part of the hemisphere directly opposite the first.

To compute the net z-component of the gravitational force, we imagine a small piece of the hemisphere with volume ΔV, located at spherical coordinates (ρ, θ, ϕ). This piece has mass $\delta\Delta V$

and exerts a force of magnitude F on the unit mass at the origin. The z-component of this force is given by its projection onto the z-axis, which can be seen from the figure to be $F \cos \phi$. The distance from the mass $\delta \Delta V$ to the unit mass at the origin is the spherical coordinate ρ. Therefore, the z-component of the force due to the small piece ΔV is

$$\begin{array}{c} z\text{-component} \\ \text{of force} \end{array} = \frac{G(\delta \Delta V)(1)}{\rho^2} \cos \phi.$$

Adding the contributions of the small pieces, we get a vertical force with magnitude

$$F = \int_0^{2\pi} \int_0^{\pi/2} \int_0^a \left(\frac{G\delta}{\rho^2} \right)(\cos \phi)\rho^2 \sin \phi \, d\rho \, d\phi \, d\theta = \int_0^{2\pi} \int_0^{\pi/2} G\delta(\cos \phi \sin \phi)\rho \Big|_{\rho=0}^{\rho=a} d\phi \, d\theta$$

$$= \int_0^{2\pi} \int_0^{\pi/2} G\delta a \cos \phi \sin \phi \, d\phi \, d\theta = \int_0^{2\pi} G\delta a \left(-\frac{(\cos \phi)^2}{2} \right) \Big|_{\phi=0}^{\phi=\pi/2} d\theta$$

$$= \int_0^{2\pi} G\delta a \left(\frac{1}{2} \right) d\theta = G\delta a\pi.$$

The integral in this example is improper because the region of integration contains the origin, where the force is undefined. However, it can be shown that the result is nevertheless correct.

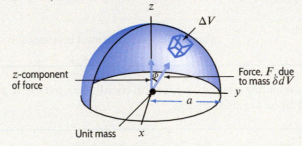

Figure 16.52: Gravitational force of hemisphere on mass at origin

Summary for Section 16.5

- In **cylindrical coordinates**, each point in 3-space is represented using $0 \le r < \infty, 0 \le \theta \le 2\pi$, $-\infty < z < \infty$:

$$x = r \cos \theta,$$
$$y = r \sin \theta,$$
$$z = z$$

with $x^2 + y^2 = r^2$.

- When **computing triple integrals in cylindrical coordinates**, convert the integrand to be in terms of r, θ, z and put $dV = r \, dr \, d\theta \, dz$.

- In **spherical coordinates**, each point in 3-space is represented using $0 \le \rho < \infty, 0 \le \phi \le \pi$, and $0 \le \theta \le 2\pi$:

$$x = \rho \sin \phi \cos \theta$$
$$y = \rho \sin \phi \sin \theta$$
$$z = \rho \cos \phi$$

with $\rho^2 = x^2 + y^2 + z^2$.

- When **computing triple integrals in spherical coordinates**, convert the integrand to be in terms of ρ, θ, ϕ and put $dV = \rho^2 \sin \phi \, d\rho \, d\phi \, d\theta$.

Exercises and Problems for Section 16.5   Online Resource: Additional Problems for Section 16.5
EXERCISES

1. Match the equations in (a)–(f) with one of the surfaces in (I)–(VII).

   (a) $x = 5$ (b) $x^2 + z^2 = 7$ (c) $\rho = 5$

   (d) $z = 1$ (e) $r = 3$ (f) $\theta = 2\pi$

   (I)   Cylinder, centered on x-axis.
   (II)   Cylinder, centered on y-axis.
   (III)   Cylinder, centered on z-axis.
   (IV)   Plane, perpendicular to the x-axis.
   (V)   Plane, perpendicular to the y-axis.
   (VI)   Plane, perpendicular to the z-axis.
   (VII)   Sphere.

■ In Exercises 2–7, find an equation for the surface.

2. The vertical plane $y = x$ in cylindrical coordinates.

3. The top half of the sphere $x^2 + y^2 + z^2 = 1$ in cylindrical coordinates.

4. The cone $z = \sqrt{x^2 + y^2}$ in cylindrical coordinates.

5. The cone $z = \sqrt{x^2 + y^2}$ in spherical coordinates.

6. The plane $z = 10$ in spherical coordinates.

7. The plane $z = 4$ in spherical coordinates.

8. (a) In words, what are the shapes of the surfaces

   $$z = \sqrt{x^2 + y^2} \quad \text{and} \quad x^2 + y^2 + z^2 = 1?$$

   (b) Write the equations of these surfaces in cylindrical coordinates.
   (c) In words, describe the intersection of the surfaces.
   (d) Write the equation of the intersection in cylindrical coordinates.
   (e) Write the equation of the intersection in Cartesian coordinates.

9. (a) In words, what are the shapes of the surfaces

   $$z = \sqrt{x^2 + y^2} \quad \text{and} \quad z = 6 - \sqrt{x^2 + y^2}?$$

   (b) Write the equations of these surfaces in cylindrical coordinates.
   (c) In words, describe the intersection of the surfaces.
   (d) Write the equation of the intersection in cylindrical coordinates.
   (e) Write the equation of the intersection in Cartesian coordinates.

■ In Exercises 10–11, evaluate the triple integrals in cylindrical coordinates over the region W.

10. $f(x, y, z) = \sin(x^2 + y^2)$, W is the solid cylinder with height 4 and with base of radius 1 centered on the z axis at $z = -1$.

11. $f(x, y, z) = x^2 + y^2 + z^2$, W is the region $0 \le r \le 4$, $\pi/4 \le \theta \le 3\pi/4$, $-1 \le z \le 1$.

■ In Exercises 12–13, evaluate the triple integrals in spherical coordinates.

12. $f(\rho, \theta, \phi) = \sin \phi$, over the region $0 \le \theta \le 2\pi$, $0 \le \phi \le \pi/4$, $1 \le \rho \le 2$.

13. $f(x, y, z) = 1/(x^2 + y^2 + z^2)^{1/2}$ over the bottom half of the sphere of radius 5 centered at the origin.

■ For Exercises 14–20, choose coordinates and set up a triple integral, including limits of integration, for a density function f over the region.

14.

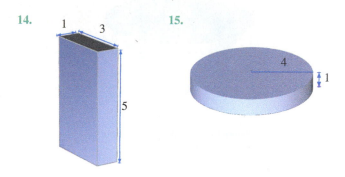

15.

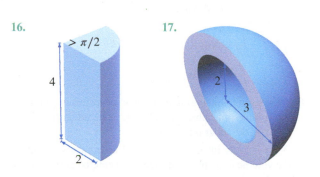

16.

17.

18. A piece of a sphere; angle at the center is $\pi/3$.

19.

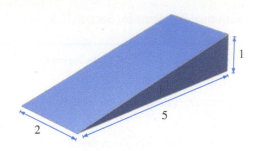

20.

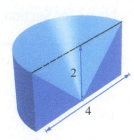

PROBLEMS

In Problems **21–23**, if W is the region in Figure 16.53, what are the limits of integration?

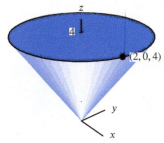

Figure 16.53: Cone with flat top, symmetric about z-axis

21. $\displaystyle\int_?^? \int_?^? \int_?^? f(r, \theta, z) r\, dz\, dr\, d\theta$

22. $\displaystyle\int_?^? \int_?^? \int_?^? g(\rho, \phi, \theta)\rho^2 \sin\phi\, d\rho\, d\phi\, d\theta$

23. $\displaystyle\int_?^? \int_?^? \int_?^? h(x, y, z)\, dz\, dy\, dx$

24. Write a triple integral in cylindrical coordinates giving the volume of a sphere of radius K centered at the origin. Use the order $dz\, dr\, d\theta$.

25. Write a triple integral in spherical coordinates giving the volume of a sphere of radius K centered at the origin. Use the order $d\theta\, d\rho\, d\phi$.

In Problems **26–28**, for the regions W shown, write the limits of integration for $\int_W dV$ in the following coordinates:

(a) Cartesian **(b)** Cylindrical **(c)** Spherical

26.

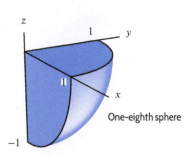

One-eighth sphere

27.

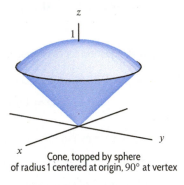

Cone, topped by sphere
of radius 1 centered at origin, 90° at vertex

28.

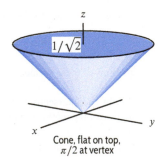

Cone, flat on top,
$\pi/2$ at vertex

29. Write a triple integral representing the volume above the cone $z = \sqrt{x^2 + y^2}$ and below the sphere of radius 2 centered at the origin. Include limits of integration but do not evaluate. Use:

(a) Cylindrical coordinates
(b) Spherical coordinates

30. Write a triple integral representing the volume of the region between spheres of radius 1 and 2, both centered at the origin. Include limits of integration but do not evaluate. Use:

(a) Spherical coordinates.
(b) Cylindrical coordinates. Write your answer as the difference of two integrals.

In Problems 31–36, write a triple integral including limits of integration that gives the specified volume.

31. Under $\rho = 3$ and above $\phi = \pi/3$.

32. Under $\rho = 3$ and above $z = r$.

33. The region between $z = 5$ and $z = 10$, with $2 \leq x^2 + y^2 \leq 3$ and $0 \leq \theta \leq \pi$.

34. Between the cone $z = \sqrt{x^2 + y^2}$ and the first quadrant of the xy-plane, with $x^2 + y^2 \leq 7$.

35. The cap of the solid sphere $x^2 + y^2 + z^2 \leq 10$ cut off by the plane $z = 1$.

36. Below the cone $z = r$, above the xy-plane, and inside the sphere $x^2 + y^2 + z^2 = 8$.

37. (a) Write an integral (including limits of integration) representing the volume of the region inside the cone $z = \sqrt{3(x^2 + y^2)}$ and below the plane $z = 1$.
    (b) Evaluate the integral.

38. Find the volume between the cone $z = \sqrt{x^2 + y^2}$ and the plane $z = 10 + x$ above the disk $x^2 + y^2 \leq 1$.

39. Find the volume between the cone $x = \sqrt{y^2 + z^2}$ and the sphere $x^2 + y^2 + z^2 = 4$.

40. The sphere of radius 2 centered at the origin is sliced horizontally at $z = 1$. What is the volume of the cap above the plane $z = 1$?

41. Suppose W is the region outside the cylinder $x^2 + y^2 = 1$ and inside the sphere $x^2 + y^2 + z^2 = 2$. Calculate

$$\int_W (x^2 + y^2)\,dV.$$

42. Write and evaluate a triple integral representing the volume of a slice of the cylindrical cake of height 2 and radius 5 between the planes $\theta = \pi/6$ and $\theta = \pi/3$.

43. Write a triple integral representing the volume of the cone in Figure 16.54 and evaluate it.

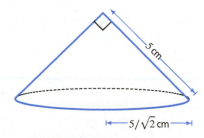

Figure 16.54

44. Find the average distance from the origin of

    (a) The points in the interval $|x| \leq 12$.
    (b) The points in the plane in the disc $r \leq 12$.
    (c) The points in space in the ball $\rho \leq 12$.

In Problems 45–46, without performing the integration, decide whether the integral is positive, negative, or zero.

45. W_1 is the unit ball, $x^2 + y^2 + z^2 \leq 1$.

    (a) $\int_{W_1} \sin\phi\,dV$ (b) $\int_{W_1} \cos\phi\,dV$

46. W_2 is $0 \leq z \leq \sqrt{1 - x^2 - y^2}$, the top half of the unit ball.

    (a) $\int_{W_2} (z^2 - z)\,dV$ (b) $\int_{W_2} (-xz)\,dV$

47. The insulation surrounding a pipe of length l is the region between two cylinders with the same axis. The inner cylinder has radius a, the outer radius of the pipe, and the insulation has thickness h. Write a triple integral, including limits of integration, giving the volume of the insulation. Evaluate the integral.

48. Assume p, q, r are positive constants. Find the volume contained between the coordinate planes and the plane

$$\frac{x}{p} + \frac{y}{q} + \frac{z}{r} = 1.$$

49. A cone stands with its flat base on a table. The cone's circular base has radius a; the vertex (tip) is at a height of h above the center of the base. Write a triple integral, including limits of integration, representing the volume of the cone. Evaluate the integral.

50. A half-melon is approximated by the region between two concentric spheres, one of radius a and the other of radius b, with $0 < a < b$. Write a triple integral, including limits of integration, giving the volume of the half-melon. Evaluate the integral.

51. A bead is made by drilling a cylindrical hole of radius 1 mm through a sphere of radius 5 mm. See Figure 16.55.

    (a) Set up a triple integral in cylindrical coordinates representing the volume of the bead.
    (b) Evaluate the integral.

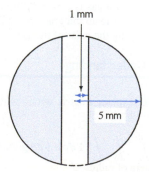

Figure 16.55

52. A pile of hay is in the region $0 \leq z \leq 2-x^2-y^2$, where x, y, z are in meters. At height z, the density of the hay is $\delta = (2 - z)$ kg/m^3.

    (a) Write an integral representing the mass of hay in the pile.
    (b) Evaluate the integral.

53. Find the mass M of the solid region W given in spherical coordinates by $0 \leq \rho \leq 3, 0 \leq \theta < 2\pi, 0 \leq \phi \leq \pi/4$. The density, $\delta(P)$, at any point P is given by the distance of P from the origin.

54. Write an integral representing the mass of a sphere of radius 3 if the density of the sphere at any point is twice the distance of that point from the center of the sphere.

55. A sphere has density at each point proportional to the square of the distance of the point from the z-axis. The density is 2 gm/cm^3 at a distance of 2 cm from the axis. What is the mass of the sphere if it is centered at the origin and has radius 3 cm?

56. The density of a solid sphere at any point is proportional to the square of the distance of the point to the center of the sphere. What is the ratio of the mass of a sphere of radius 1 to a sphere of radius 2?

57. A spherical shell centered at the origin has an inner radius of 6 cm and an outer radius of 7 cm. The density, δ, of the material increases linearly with the distance from the center. At the inner surface, $\delta = 9$ gm/cm^3; at the outer surface, $\delta = 11$ gm/cm^3.

    (a) Using spherical coordinates, write the density, δ, as a function of radius, ρ.
    (b) Write an integral giving the mass of the shell.
    (c) Find the mass of the shell.

58. (a) Write an iterated integral which represents the mass of a solid ball of radius a. The density at each point in the ball is k times the distance from that point to a fixed plane passing through the center of the ball.
    (b) Evaluate the integral.

59. In the region under $z = 4 - x^2 - y^2$ and above the xy-plane the density of a gas is $\delta = e^{-x-y}$gm/cm^3, where x, y, z are in cm. Write an integral, with limits of integration, representing the mass of the gas.

60. The density, δ, of the cylinder $x^2 + y^2 \leq 4, 0 \leq z \leq 3$ varies with the distance, r, from the z-axis:

$$\delta = 1 + r \text{ gm/cm}^3.$$

Find the mass of the cylinder if x, y, z are in cm.

61. The density of material at a point in a solid cylinder is proportional to the distance of the point from the z-axis. What is the ratio of the mass of the cylinder $x^2+y^2 \leq 1$, $0 \leq z \leq 2$ to the mass of the cylinder $x^2 + y^2 \leq 9$, $0 \leq z \leq 2$?

62. Electric charge is distributed throughout 3-space, with density proportional to the distance from the xy-plane. Show that the total charge inside a cylinder of radius R and height h, sitting on the xy-plane and centered along the z-axis, is proportional to $R^2 h^2$.

63. Electric charge is distributed throughout 3-space with density inversely proportional to the distance from the origin. Show that the total charge inside a sphere of radius R is proportional to R^2.

■ For Problems 64–67, use the definition of center of mass given on page 915. Assume x, y, z are in cm.

64. Let C be a solid cone with both height and radius 1 and contained between the surfaces $z = \sqrt{x^2 + y^2}$ and $z = 1$. If C has constant mass density of 1 gm/cm^3, find the z-coordinate of C's center of mass.

65. The density of the cone C in Problem 64 is given by $\delta(z) = z^2$ gm/cm^3. Find

    (a) The mass of C.
    (b) The z-coordinate of C's center of mass.

66. For $a > 0$, consider the family of solids bounded below by the paraboloid $z = a(x^2 + y^2)$ and above by the plane $z = 1$. If the solids all have constant mass density 1 gm/cm^3, show that the z-coordinate of the center of mass is $2/3$ and so independent of the parameter a.

67. Find the location of the center of mass of a hemisphere of radius a and density b gm/cm^3.

Strengthen Your Understanding

■ In Problems 68–70, explain what is wrong with the statement.

68. The integral $\int_0^{2\pi} \int_0^{\pi} \int_0^1 1 \, d\rho \, d\phi \, d\theta$ gives the volume inside the sphere of radius 1.

69. Changing the order of integration gives

$$\int_0^{2\pi} \int_0^{\pi/4} \int_0^{2/\cos\phi} \rho^2 \sin\phi \, d\rho \, d\phi \, d\theta$$

$$= \int_0^{2/\cos\phi} \int_0^{\pi/4} \int_0^{2\pi} \rho^2 \sin\phi \, d\theta \, d\phi \, d\rho.$$

70. The volume of a cylinder of height and radius 1 is

$$\int_0^{2\pi} \int_0^1 \int_0^1 1 \, dz \, dr \, d\theta.$$

In Problems **71–72**, give an example of:

71. An integral in spherical coordinates that gives the volume of a hemisphere.

72. An integral for which it is more convenient to use spherical coordinates than to use Cartesian coordinates.

73. Which of the following integrals give the volume of the unit sphere?

(a) $\int_0^{2\pi} \int_0^{2\pi} \int_0^1 1\, d\rho\, d\theta\, d\phi$

(b) $\int_0^{\pi} \int_0^{2\pi} \int_0^1 1\, d\rho\, d\theta\, d\phi$

(c) $\int_0^{\pi} \int_0^{2\pi} \int_0^1 \rho^2 \sin\phi\, d\rho\, d\theta\, d\phi$

(d) $\int_0^{\pi} \int_0^{2\pi} \int_0^1 \rho^2 \sin\phi\, d\rho\, d\phi\, d\theta$

(e) $\int_0^{\pi} \int_0^{2\pi} \int_0^1 \rho\, d\rho\, d\phi\, d\theta$

16.6 APPLICATIONS OF INTEGRATION TO PROBABILITY

To represent how a quantity such as height or weight is distributed throughout a population, we use a density function. To study two or more quantities at the same time and see how they are related, we use a multivariable density function.

Density Functions

Distribution of Weight and Height in Expectant Mothers

Table 16.10 shows the distribution of weight and height in a survey of expectant mothers. The histogram in Figure 16.56 is constructed so that the volume of each bar represents the percentage in the corresponding weight and height range. For example, the bar representing the mothers who weighed 60–70 kg and were 160–165 cm tall has base of area 10 kg · 5 cm = 50 kg cm. The volume of this bar is 12%, so its height is 12%/50 kg cm = 0.24%/ kg cm. Notice that the units on the vertical axis are % per kg cm, so the volume of a bar is a %. The total volume is 100% = 1.

Table 16.10 *Distribution of weight and height in a survey of expectant mothers, in %*

	45-50 kg	50-60 kg	60-70 kg	70-80 kg	80-105 kg	Totals by height
150-155 cm	2	4	4	2	1	13
155-160 cm	0	12	8	2	1	23
160-165 cm	1	7	12	4	3	27
165-170 cm	0	8	12	6	2	28
170-180 cm	0	1	3	4	1	9
Totals by weight	3	32	39	18	8	100

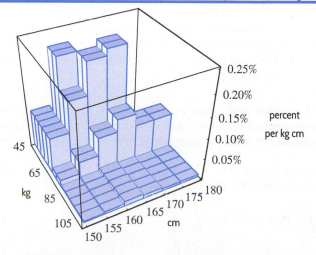

Figure 16.56: Histogram representing the data in Table 16.10

Example 1   Find the percentage of mothers in the survey with height between 170 and 180 cm.

Solution   We add the percentages across the row corresponding to the 170–180 cm height range; this is equivalent to adding the volumes of the corresponding rectangular solids in the histogram.

$$\text{Percentage of mothers} = 0 + 1 + 3 + 4 + 1 = 9\%.$$

Smoothing the Histogram

If we group the data using narrower weight and height groups (and a larger sample), we can draw a smoother histogram and get finer estimates. In the limit, we replace the histogram with a smooth surface, in such a way that the volume under the surface above a rectangle is the percentage of mothers in that rectangle. We define a *density function*, $p(w, h)$, to be the function whose graph is the smooth surface. It has the property that

$$
\begin{array}{ccc}
\text{Fraction of sample with} & & \text{Volume under graph of } p \\
\text{weight between } a \text{ and } b \text{ and} & = & \text{over the rectangle} \\
\text{height between } c \text{ and } d & & a \le w \le b, c \le h \le d
\end{array}
= \int_a^b \int_c^d p(w, h)\, dh\, dw.
$$

This density also gives the probability that a mother is in these height and weight groups.

Joint Probability Density Functions

We generalize this idea to represent any two characteristics, x and y, distributed throughout a population.

A function $p(x, y)$ is called a **joint probability density function**, or **pdf**, for x and y if

$$
\begin{array}{ccc}
\text{Probability that member of} & & \text{Volume under graph of } p \\
\text{population has } x \text{ between } a \text{ and } b & = & \text{above the rectangle} \\
\text{and } y \text{ between } c \text{ and } d & & a \le x \le b, c \le y \le d
\end{array}
= \int_a^b \int_c^d p(x, y)\, dy\, dx,
$$

where

$$\int_{-\infty}^{\infty} \int_{-\infty}^{\infty} p(x, y)\, dy\, dx = 1 \quad \text{and} \quad p(x, y) \ge 0 \text{ for all } x \text{ and } y.$$

The probability that x falls in an interval of width Δx around x_0 and y falls in an interval of width Δy around y_0 is approximately $p(x_0, y_0)\Delta x \Delta y$.

A joint density function need not be continuous, as in Example 2. In addition, as in Example 4, the integrals involved may be improper and must be computed by methods similar to those used for improper one-variable integrals.

Example 2   Let $p(x, y)$ be defined on the square $0 \le x \le 1, 0 \le y \le 1$ by $p(x, y) = x + y$; let $p(x, y) = 0$ if (x, y) is outside this square. Check that p is a joint density function. In terms of the distribution of x and y in the population, what does it mean that $p(x, y) = 0$ outside the square?

Solution   First, we have $p(x, y) \ge 0$ for all x and y. To check that p is a joint density function, we show that the total volume under the graph is 1:

$$\int_{-\infty}^{\infty} \int_{-\infty}^{\infty} p(x, y)\, dy\, dx = \int_0^1 \int_0^1 (x + y)\, dy\, dx$$

$$= \int_0^1 \left(xy + \frac{y^2}{2}\right)\Bigg|_0^1 dx = \int_0^1 \left(x + \frac{1}{2}\right) dx = \left(\frac{x^2}{2} + \frac{x}{2}\right)\Bigg|_0^1 = 1.$$

The fact that $p(x, y) = 0$ outside the square means that the variables x and y never take values outside the interval $[0, 1]$; that is, the value of x and y for any individual in the population is always between 0 and 1.

Example 3   Two variables x and y are distributed in a population according to the density function of Example 2. Find the fraction of the population with $x \leq 1/2$, the fraction with $y \leq 1/2$, and the fraction with both $x \leq 1/2$ and $y \leq 1/2$.

Solution   The fraction with $x \leq 1/2$ is the volume under the graph to the left of the line $x = 1/2$:

$$\int_0^{1/2} \int_0^1 (x + y)\, dy\, dx = \int_0^{1/2} \left(xy + \frac{y^2}{2} \right) \Big|_0^1 dx = \int_0^{1/2} \left(x + \frac{1}{2} \right) dx$$

$$= \left(\frac{x^2}{2} + \frac{x}{2} \right) \Big|_0^{1/2} = \frac{1}{8} + \frac{1}{4} = \frac{3}{8}.$$

Since the function and the regions of integration are symmetric in x and y, the fraction with $y \leq 1/2$ is also $3/8$. Finally, the fraction with both $x \leq 1/2$ and $y \leq 1/2$ is

$$\int_0^{1/2} \int_0^{1/2} (x + y)\, dy\, dx = \int_0^{1/2} \left(xy + \frac{y^2}{2} \right) \Big|_0^{1/2} dx = \int_0^{1/2} \left(\frac{1}{2}x + \frac{1}{8} \right) dx$$

$$= \left(\frac{1}{4}x^2 + \frac{1}{8}x \right) \Big|_0^{1/2} = \frac{1}{16} + \frac{1}{16} = \frac{1}{8}.$$

Recall that a one-variable density function $p(x)$ is a function such that $p(x) \geq 0$ for all x, and $\int_{-\infty}^{\infty} p(x)\, dx = 1$.

Example 4   Let p_1 and p_2 be one-variable density functions for x and y, respectively. Check that $p(x, y) = p_1(x)p_2(y)$ is a joint density function.

Solution   Since both p_1 and p_2 are density functions, they are nonnegative everywhere. Thus, their product $p_1(x)p_2(x) = p(x, y)$ is nonnegative everywhere. Now we must check that the volume under the graph of p is 1. Since $\int_{-\infty}^{\infty} p_2(y)\, dy = 1$ and $\int_{-\infty}^{\infty} p_1(x)\, dx = 1$, we have

$$\int_{-\infty}^{\infty} \int_{-\infty}^{\infty} p(x, y)\, dy\, dx = \int_{-\infty}^{\infty} \int_{-\infty}^{\infty} p_1(x)p_2(y)\, dy\, dx = \int_{-\infty}^{\infty} p_1(x) \left(\int_{-\infty}^{\infty} p_2(y)\, dy \right) dx$$

$$= \int_{-\infty}^{\infty} p_1(x)(1)\, dx = \int_{-\infty}^{\infty} p_1(x)\, dx = 1.$$

Example 5   A machine in a factory is set to produce components 10 cm long and 5 cm in diameter. In fact, there is a slight variation from one component to the next. A component is usable if its length and diameter deviate from the correct values by less than 0.1 cm. With the length, x, in cm and the diameter, y, in cm, the probability density function is

$$p(x, y) = \frac{50\sqrt{2}}{\pi} e^{-100(x-10)^2} e^{-50(y-5)^2}.$$

What is the probability that a component is usable? (See Figure 16.57.)

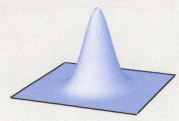

Figure 16.57: The density function $p(x, y) = \frac{50\sqrt{2}}{\pi} e^{-100(x-10)^2} e^{-50(y-5)^2}$

Solution   We know that

$$
\begin{array}{l}
\text{Probability that } x \text{ and } y \text{ satisfy} \\
x_0 - \Delta x \leq x \leq x_0 + \Delta x \\
y_0 - \Delta y \leq y \leq y_0 + \Delta y
\end{array}
= \frac{50\sqrt{2}}{\pi} \int_{y_0-\Delta y}^{y_0+\Delta y} \int_{x_0-\Delta x}^{x_0+\Delta x} e^{-100(x-10)^2} e^{-50(y-5)^2} \, dx \, dy.
$$

Thus,

$$
\begin{array}{l}
\text{Probability that} \\
\text{component is usable}
\end{array}
= \frac{50\sqrt{2}}{\pi} \int_{4.9}^{5.1} \int_{9.9}^{10.1} e^{-100(x-10)^2} e^{-50(y-5)^2} \, dx \, dy.
$$

The double integral must be evaluated numerically. This yields

$$
\begin{array}{l}
\text{Probability that} \\
\text{component is usable}
\end{array}
= \frac{50\sqrt{2}}{\pi}(0.02556) = 0.57530.
$$

Thus, there is a 57.530% chance that the component is usable.

Summary for Section 16.6

- A function $p(x, y)$ is called a **joint probability density function**, or **pdf**, for x and y if

$$
\begin{array}{l}
\text{Probability that member of} \\
\text{population has } x \text{ between } a \text{ and } b \\
\text{and } y \text{ between } c \text{ and } d
\end{array}
=
\begin{array}{l}
\text{Volume under graph of } p \\
\text{above the rectangle} \\
a \leq x \leq b, c \leq y \leq d
\end{array}
= \int_a^b \int_c^d p(x, y) \, dy \, dx,
$$

where

$$
\int_{-\infty}^{\infty} \int_{-\infty}^{\infty} p(x, y) \, dy \, dx = 1 \quad \text{and} \quad p(x, y) \geq 0 \text{ for all } x \text{ and } y.
$$

- If $p(x, y)$ is a joint probability density function, then the **probability** that x falls in an interval of width Δx around x_0 and y falls in an interval of width Δy around y_0 is approximately $p(x_0, y_0)\Delta x \Delta y$.

Exercises and Problems for Section 16.6

EXERCISES

In Exercises 1–6, check whether p is a joint density function. Assume $p(x, y) = 0$ outside the region R.

1. $p(x, y) = 1/2$, where R is $4 \leq x \leq 5, -2 \leq y \leq 0$

2. $p(x, y) = 1$, where R is $0 \leq x \leq 1, 0 \leq y \leq 2$

3. $p(x, y) = x + y$, where R is $-1 \leq x \leq 1, 0 \leq y \leq 1$

4. $p(x, y) = 6(y - x)$, where R is $0 \leq x \leq y \leq 2$

5. $p(x, y) = (2/\pi)(1 - x^2 - y^2)$, where R is $x^2 + y^2 \leq 1$

6. $p(x, y) = xye^{-x-y}$, where R is $x \geq 0, y \geq 0$

■ In Exercises **7–10**, a joint probability density function is given by $p(x, y) = xy/4$ in R, the rectangle $0 \leq x \leq 2$, $0 \leq y \leq 2$, and $p(x, y) = 0$ else. Find the probability that a point (x, y) satisfies the given conditions.

7. $x \leq 1$ and $y \leq 1$

8. $x \geq 1$ and $y \geq 1$

9. $x \geq 1$ and $y \leq 1$

10. $1/3 \leq x \leq 1$

■ In Exercises **11–14**, a joint probability density function is given by $p(x, y) = 0.005x + 0.025y$ in R, the rectangle $0 \leq x \leq 10, 0 \leq y \leq 2$, and $p(x, y) = 0$ else. Find the probability that a point (x, y) satisfies the given conditions.

11. $x \leq 4$

12. $y \geq 1$

13. $x \leq 4$ and $y \geq 1$

14. $x \geq 5$ and $y \geq 1$

■ In Exercises **15–22**, let p be the joint density function such that $p(x, y) = xy$ in R, the rectangle $0 \leq x \leq 2, 0 \leq y \leq 1$, and $p(x, y) = 0$ outside R. Find the fraction of the population satisfying the given constraints.

15. $x \geq 3$

16. $x = 1$

17. $x + y \leq 3$

18. $-1 \leq x \leq 1$

19. $x \geq y$

20. $x + y \leq 1$

21. $0 \leq x \leq 1, 0 \leq y \leq 1/2$

22. Within a distance 1 from the origin

PROBLEMS

23. Let x and y have joint density function

$$p(x, y) = \begin{cases} \frac{2}{3}(x + 2y) & \text{for } 0 \leq x \leq 1, 0 \leq y \leq 1, \\ 0 & \text{otherwise.} \end{cases}$$

Find the probability that

(a) $x > 1/3$.

(b) $x < (1/3) + y$.

24. The joint density function for x, y is given by

$$f(x, y) = \begin{cases} kxy & \text{for } 0 \leq x \leq y \leq 1, \\ 0 & \text{otherwise.} \end{cases}$$

(a) Determine the value of k.

(b) Find the probability that (x, y) lies in the shaded region in Figure 16.58.

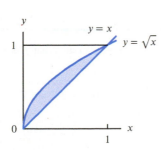

Figure 16.58

25. A joint density function is given by

$$f(x, y) = \begin{cases} kx^2 & \text{for } 0 \leq x \leq 2 \text{ and } 0 \leq y \leq 1, \\ 0 & \text{otherwise.} \end{cases}$$

(a) Find the value of the constant k.

(b) Find the probability that (x, y) satisfies $x + y \leq 2$.

(c) Find the probability that (x, y) satisfies $x \leq 1$ and $y \leq 1/2$.

26. A point is chosen at random from the region S in the xy-plane containing all points (x, y) such that $-1 \leq x \leq 1, -2 \leq y \leq 2$ and $x - y \geq 0$ ("at random" means that the density function is constant on S).

(a) Determine the joint density function for x and y.

(b) If T is a subset of S with area α, then find the probability that a point (x, y) is in T.

27. A probability density function on a square has constant values in different triangular regions as shown in Figure 16.59. Find the probability that

(a) $x \geq 2$

(b) $y \geq x$

(c) $y \geq x$ and $x \geq 2$

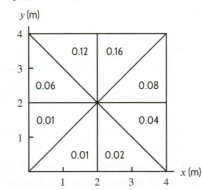

Figure 16.59: Probability density on a square (per m²)

28. A health insurance company wants to know what proportion of its policies are going to cost the company a lot of money because the insured people are over 65 and sick. In order to compute this proportion, the company defines a *disability index*, x, with $0 \leq x \leq 1$, where $x = 0$ represents perfect health and $x = 1$ represents total disability. In addition, the company uses a density function, $f(x, y)$, defined in such a way that the quantity

$$f(x, y) \Delta x \Delta y$$

approximates the fraction of the population with disability index between x and $x+\Delta x$, and aged between y and $y+\Delta y$. The company knows from experience that a policy no longer covers its costs if the insured person is over 65 and has a disability index exceeding 0.8. Write an expression for the fraction of the company's policies held by people meeting these criteria.

29. The probability that a radioactive substance will decay at time t is modeled by the density function

$$p(t) = \lambda e^{-\lambda t}$$

for $t \geq 0$, and $p(t) = 0$ for $t < 0$. The positive constant λ depends on the material, and is called the decay rate.

(a) Check that p is a density function.
(b) Two materials with decay rates λ and μ decay independently of each other; their joint density function is the product of the individual density functions. Write the joint density function for the probability that the first material decays at time t and the second at time s.
(c) Find the probability that the first substance decays before the second.

30. Figure 16.60 represents a baseball field, with the bases at $(1,0)$, $(1,1)$, $(0,1)$, and home plate at $(0,0)$. The outer bound of the outfield is a piece of a circle about the origin with radius 4. When a ball is hit by a batter we record the spot on the field where the ball is caught. Let $p(r,\theta)$ be a function in the plane that gives the density of the distribution of such spots. Write an expression that represents the probability that a hit is caught in

(a) The right field (region R).
(b) The center field (region C).

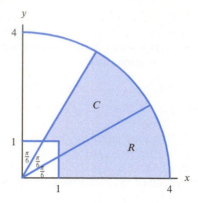

Figure 16.60

31. Two independent random numbers x and y between 0 and 1 have joint density function

$$p(x, y) = \begin{cases} 1 & \text{if } 0 \leq x, y \leq 1 \\ 0 & \text{otherwise.} \end{cases}$$

This problem concerns the average $z = (x+y)/2$, which has a one-variable probability density function of its own.

(a) Find $F(t)$, the probability that $z \leq t$. Treat separately the cases $t \leq 0$, $0 < t \leq 1/2$, $1/2 < t \leq 1$, $1 < t$. Note that $F(t)$ is the cumulative distribution function of z.
(b) Find and graph the probability density function of z.
(c) Are x and y more likely to be near 0, 1/2, or 1? What about z?

Strengthen Your Understanding

■ In Problems 32–33, explain what is wrong with the statement.

32. If $p_1(x, y)$ and $p_2(x, y)$ are joint density functions, then $p_1(x, y) + p_2(x, y)$ is a joint density function.

33. If $p(w, h)$ is the probability density function of the weight and height of mothers discussed in Section 16.6, then the probability that a mother weighs 60 kg and has a height of 170 cm is $p(60, 170)$.

■ In Problems 34–35, give an example of:

34. Values for a, b, c and d such that f is a joint density function:

$$f(x, y) = \begin{cases} 1 & \text{for } a \leq x \leq b \text{ and } c \leq y \leq d, \\ 0 & \text{otherwise} \end{cases}$$

35. A one-variable function $g(y)$ such that f is a joint density function:

$$f(x, y) = \begin{cases} g(y) & \text{for } 0 \leq x \leq 2 \text{ and } 0 \leq y \leq 1, \\ 0 & \text{otherwise} \end{cases}$$

■ For Problems 36–39, let $p(x, y)$ be a joint density function for x and y. Are the following statements true or false?

36. $\int_a^b \int_{-\infty}^{\infty} p(x, y)\, dy\, dx$ is the probability that $a \leq x \leq b$.

37. $0 \leq p(x, y) \leq 1$ for all x.

38. $\int_a^b p(x, y)\, dx$ is the probability that $a \leq x \leq b$.

39. $\int_{-\infty}^{\infty} \int_{-\infty}^{\infty} p(x, y)\, dy\, dx = 1$.

Online Resource: Review Problems and Projects

Chapter 17

PARAMETERIZATION AND VECTOR FIELDS

CONTENTS

17.1 PARAMETERIZED CURVES

A curve in the plane may be parameterized by a pair of equations of the form $x = f(t)$, $y = g(t)$. As the parameter t changes, the point (x, y) traces out the curve. In this section we find parametric equations for curves in three dimensions, and we see how to write parametric equations using position vectors.

Parametric Equations in Three Dimensions

We describe motion in the plane by giving parametric equations for x and y in terms of t. To describe a motion in 3-space parametrically, we need a third equation giving z in terms of t.

Example 1   Find parametric equations for the curve $y = x^2$ in the xy-plane.

Solution   A possible parameterization in two dimensions is $x = t$, $y = t^2$. Since the curve is in the xy-plane, the z-coordinate is zero, so a parameterization in three dimensions is

$$x = t, \quad y = t^2, \quad z = 0.$$

Example 2   Find parametric equations for a particle that starts at $(0, 3, 0)$ and moves around a circle as shown in Figure 17.1.

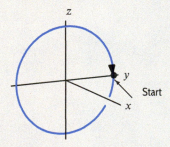

Figure 17.1: Circle of radius 3 in the yz-plane, centered at origin

Solution   Since the motion is in the yz-plane, we have $x = 0$ at all times t. Looking at the yz-plane from the positive x-direction, we see motion around a circle of radius 3 in the clockwise direction. Thus,

$$x = 0, \quad y = 3 \cos t, \quad z = -3 \sin t.$$

Example 3   Describe in words the motion given parametrically by

$$x = \cos t, \quad y = \sin t, \quad z = t.$$

Solution   The particle's x- and y-coordinates give circular motion in the xy-plane, while the z-coordinate increases steadily. Thus, the particle traces out a rising spiral, like a coiled spring. (See Figure 17.2.) This curve is called a *helix*.

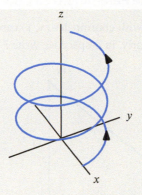

Figure 17.2: The helix $x = \cos t$, $y = \sin t$, $z = t$

Example 4   Find parametric equations for the line parallel to the vector $2\vec{i} + 3\vec{j} + 4\vec{k}$ and through the point $(1, 5, 7)$.

Solution   Let's imagine a particle at the point $(1, 5, 7)$ at time $t = 0$ and moving through a displacement of $2\vec{i} + 3\vec{j} + 4\vec{k}$ for each unit of time, t. When $t = 0$, $x = 1$ and x increases by 2 units for every unit of time. Thus, at time t, the x-coordinate of the particle is given by

$$x = 1 + 2t.$$

Similarly, the y-coordinate starts at $y = 5$ and increases at a rate of 3 units for every unit of time. The z-coordinate starts at $y = 7$ and increases by 4 units for every unit of time. Thus, the parametric equations of the line are

$$x = 1 + 2t, \quad y = 5 + 3t, \quad z = 7 + 4t.$$

We can generalize the previous example as follows:

Parametric Equations of a Line through the point (x_0, y_0, z_0) and parallel to the vector $a\vec{i} + b\vec{j} + c\vec{k}$ are

$$x = x_0 + at, \quad y = y_0 + bt, \quad z = z_0 + ct.$$

Notice that the coordinates x, y, and z are linear functions of the parameter t.

Example 5   (a) Describe in words the curve given by the parametric equations $x = 3 + t$, $y = 2t$, $z = 1 - t$.
              (b) Find parametric equations for the line through the points $(1, 2, -1)$ and $(3, 3, 4)$.

Solution   (a) The curve is a line through the point $(3, 0, 1)$ and parallel to the vector $\vec{i} + 2\vec{j} - \vec{k}$.
              (b) The line is parallel to the vector between the points $P = (1, 2, -1)$ and $Q = (3, 3, 4)$.

$$\vec{PQ} = (3 - 1)\vec{i} + (3 - 2)\vec{j} + (4 - (-1))\vec{k} = 2\vec{i} + \vec{j} + 5\vec{k}.$$

Thus, using the point P, the parametric equations are

$$x = 1 + 2t, \quad y = 2 + t, \quad z = -1 + 5t.$$

Using the point Q gives the equations $x = 3 + 2t$, $y = 3 + t$, $z = 4 + 5t$, which represent the same line. The point where $t = 0$ in the second equations is given by $t = 1$ in the first equations.

Using Position Vectors to Write Parameterized Curves as Vector-Valued Functions

A point in the plane with coordinates (x, y) can be represented by the position vector $\vec{r} = x\vec{i} + y\vec{j}$ in Figure 17.3. Similarly, in 3-space we write $\vec{r} = x\vec{i} + y\vec{j} + z\vec{k}$. (See Figure 17.4.)

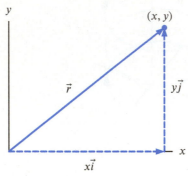

Figure 17.3: Position vector $\vec{r}$ for the point (x, y)

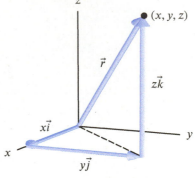

Figure 17.4: Position vector $\vec{r}$ for the point (x, y, z)

We can write the parametric equations $x = f(t)$, $y = g(t)$, $z = h(t)$ as a single vector equation

$$\vec{r}(t) = f(t)\vec{i} + g(t)\vec{j} + h(t)\vec{k}$$

called a *parameterization*. As the parameter t varies, the point with position vector $\vec{r}(t)$ traces out a curve in 3-space. For example, the circular motion in the plane

$$x = \cos t, y = \sin t \quad \text{can be written as} \quad \vec{r} = (\cos t)\vec{i} + (\sin t)\vec{j}$$

and the helix in 3-space

$$x = \cos t, y = \sin t, z = t \quad \text{can be written as} \quad \vec{r} = (\cos t)\vec{i} + (\sin t)\vec{j} + t\vec{k}.$$

See Figure 17.5.

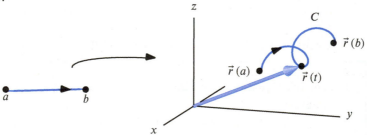

Figure 17.5: The parameterization sends the interval, $a \leq t \leq b$, to the curve, C, in 3-space

---

Example 6 Use vectors to give a parameterization for the circle of radius $\frac{1}{2}$ centered at the point $(-1, 2)$.

Solution The circle of radius 1 centered at the origin is parameterized by the vector-valued function

$$\vec{r}_1(t) = \cos t\vec{i} + \sin t\vec{j}, \quad 0 \leq t \leq 2\pi.$$

The point $(-1, 2)$ has position vector $\vec{r}_0 = -\vec{i} + 2\vec{j}$. The position vector, $\vec{r}(t)$, of a point on the circle of radius $\frac{1}{2}$ centered at $(-1, 2)$ is found by adding $\frac{1}{2}\vec{r}_1$ to $\vec{r}_0$. (See Figures 17.6 and 17.7.) Thus,

$$\vec{r}(t) = \vec{r}_0 + \frac{1}{2}\vec{r}_1(t) = -\vec{i} + 2\vec{j} + \frac{1}{2}(\cos t\vec{i} + \sin t\vec{j}) = (-1 + \frac{1}{2}\cos t)\vec{i} + (2 + \frac{1}{2}\sin t)\vec{j},$$

or, equivalently,

$$x = -1 + \frac{1}{2}\cos t, \quad y = 2 + \frac{1}{2}\sin t, \quad 0 \leq t \leq 2\pi.$$

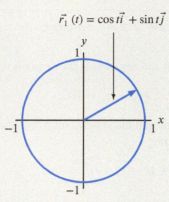

Figure 17.6: The circle $x^2 + y^2 = 1$ parameterized by $\vec{r}_1(t) = \cos t\vec{i} + \sin t\vec{j}$

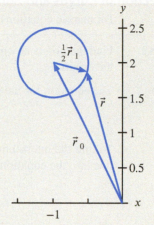

Figure 17.7: The circle of radius $\frac{1}{2}$ and center $(-1, 2)$ parameterized by $\vec{r}(t) = \vec{r}_0 + \frac{1}{2}\vec{r}_1(t)$

Parametric Equation of a Line

Consider a straight line in the direction of a vector $\vec{v}$ passing through the point (x_0, y_0, z_0) with position vector $\vec{r}_0$. We start at $\vec{r}_0$ and move up and down the line, adding different multiples of $\vec{v}$ to $\vec{r}_0$. (See Figure 17.8.)

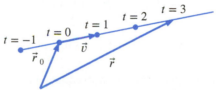

Figure 17.8: The line $\vec{r}(t) = \vec{r}_0 + t\vec{v}$

In this way, every point on the line can be written as $\vec{r}_0 + t\vec{v}$, which yields the following:

Parametric Equation of a Line in Vector Form

The line through the point with position vector $\vec{r}_0 = x_0\vec{i} + y_0\vec{j} + z_0\vec{k}$ in the direction of the vector $\vec{v} = a\vec{i} + b\vec{j} + c\vec{k}$ has parametric equation

$$\vec{r}(t) = \vec{r}_0 + t\vec{v}.$$

Example 7   (a) Find parametric equations for the line passing through the points $(2, -1, 3)$ and $(-1, 5, 4)$.
(b) Represent the line segment from $(2, -1, 3)$ to $(-1, 5, 4)$ parametrically.

Solution   (a) The line passes through $(2, -1, 3)$ and is parallel to the displacement vector $\vec{v} = -3\vec{i} + 6\vec{j} + \vec{k}$ from $(2, -1, 3)$ to $(-1, 5, 4)$. Thus, the parametric equation is

$$\vec{r}(t) = 2\vec{i} - \vec{j} + 3\vec{k} + t(-3\vec{i} + 6\vec{j} + \vec{k}).$$

(b) In the parameterization in part (a), $t = 0$ corresponds to the point $(2, -1, 3)$ and $t = 1$ corresponds to the point $(-1, 5, 4)$. So the parameterization of the segment is

$$\vec{r}(t) = 2\vec{i} - \vec{j} + 3\vec{k} + t(-3\vec{i} + 6\vec{j} + \vec{k}), \qquad 0 \le t \le 1.$$

Intersection of Curves and Surfaces

Parametric equations for a curve enable us to find where a curve intersects a surface.

Example 8   Find the points at which the line $x = t$, $y = 2t$, $z = 1 + t$ pierces the sphere of radius 10 centered at the origin.

Solution   The equation for the sphere of radius 10 and centered at the origin is

$$x^2 + y^2 + z^2 = 100.$$

To find the intersection points of the line and the sphere, substitute the parametric equations of the line into the equation of the sphere, giving

$$t^2 + 4t^2 + (1 + t)^2 = 100,$$

so

$$6t^2 + 2t - 99 = 0,$$

which has the two solutions at approximately $t = -4.23$ and $t = 3.90$. Using the parametric equation for the line, $(x, y, z) = (t, 2t, 1 + t)$, we see that the line cuts the sphere at the two points

$$(x, y, z) = (-4.23, 2(-4.23), 1 + (-4.23)) = (-4.23, -8.46, -3.23),$$

and

$$(x, y, z) = (3.90, 2(3.90), 1 + 3.90) = (3.90, 7.80, 4.90).$$

We can also use parametric equations to find the intersection of two curves.

Example 9   Two particles move through space, with equations $\vec{r}_1(t) = t\vec{i} + (1 + 2t)\vec{j} + (3 - 2t)\vec{k}$ and $\vec{r}_2(t) = (-2 - 2t)\vec{i} + (1 - 2t)\vec{j} + (1 + t)\vec{k}$. Do the particles ever collide? Do their paths cross?

Solution   To see if the particles collide, we must find out if they pass through the same point at the same time t. So we must find a solution to the vector equation $\vec{r}_1(t) = \vec{r}_2(t)$, which is the same as finding a common solution to the three scalar equations

$$t = -2 - 2t, \qquad 1 + 2t = 1 - 2t, \qquad 3 - 2t = 1 + t.$$

Separately, the solutions are $t = -2/3$, $t = 0$, and $t = 2/3$, so there is no common solution, and the particles don't collide. To see if their paths cross, we find out if they pass through the same point at two possibly different times, t_1 and t_2. So we solve the equations

$$t_1 = -2 - 2t_2, \qquad 1 + 2t_1 = 1 - 2t_2, \qquad 3 - 2t_1 = 1 + t_2.$$

We solve the first two equations simultaneously and get $t_1 = 2$, $t_2 = -2$. Since these values also satisfy the third equation, the paths cross. The position of the first particle at time $t = 2$ is the same as the position of the second particle at time $t = -2$, namely the point $(2, 5, -1)$.

Example 10   Are the lines $x = -1 + t$, $y = 1 + 2t$, $z = 5 - t$ and $x = 2 + 2t$, $y = 4 + t$, $z = 3 + t$ parallel? Do they intersect?

Solution   In vector form the lines are parameterized by

$$\vec{r} = -\vec{i} + \vec{j} + 5\vec{k} + t(\vec{i} + 2\vec{j} - \vec{k})$$
$$\vec{r} = 2\vec{i} + 4\vec{j} + 3\vec{k} + t(2\vec{i} + \vec{j} + \vec{k})$$

Their direction vectors $\vec{i} + 2\vec{j} - \vec{k}$ and $2\vec{i} + \vec{j} + \vec{k}$ are not multiples of each other, so the lines are not parallel. To find out if they intersect, we see if they pass through the same point at two possibly different times, t_1 and t_2:

$$-1 + t_1 = 2 + 2t_2, \qquad 1 + 2t_1 = 4 + t_2, \qquad 5 - t_1 = 3 + t_2.$$

The first two equations give $t_1 = 1$, $t_2 = -1$. Since these values do not satisfy the third equation, the paths do not cross, and so the lines do not intersect.

The next example shows how to tell if two different parameterizations give the same line.

Example 11   Show that the following two lines are the same:

$$\vec{r} = -\vec{i} - \vec{j} + \vec{k} + t(3\vec{i} + 6\vec{j} - 3\vec{k})$$
$$\vec{r} = \vec{i} + 3\vec{j} - \vec{k} + t(-\vec{i} - 2\vec{j} + \vec{k})$$

Solution   The direction vectors of the two lines, $3\vec{i} + 6\vec{j} - 3\vec{k}$ and $-\vec{i} - 2\vec{j} + \vec{k}$, are multiples of each other, so the lines are parallel. To see if they are the same, we pick a point on the first line and see if it is on the second line. For example, the point on the first line with $t = 0$ has position vector $-\vec{i} - \vec{j} + \vec{k}$. Solving

$$\vec{i} + 3\vec{j} - \vec{k} + t(-\vec{i} - 2\vec{j} + \vec{k}) = -\vec{i} - \vec{j} + \vec{k},$$

we get $t = 2$, so the two lines have a point in common. Thus, they are the same line, parameterized in two different ways.

Summary for Section 17.1

- **Parametric equations in three dimensions:** To describe motion in 3-space parametrically, we need three equations: $x = f(t)$, $y = g(t)$, and $z = h(t)$.
- **Parametric equations of a line** through the point (x_0, y_0, z_0) and parallel to the vector $a\vec{i} + b\vec{j} + c\vec{k}$ are
$$x = x_0 + at, \qquad y = y_0 + bt, \qquad z = z_0 + ct.$$
- **Vector form of parameterized curves:** We can write the parametric equations $x = f(t)$, $y = g(t)$, $z = h(t)$ as a single vector equation
$$\vec{r}(t) = f(t)\vec{i} + g(t)\vec{j} + h(t)\vec{k}.$$
- **Parametric equation of a line in vector form:** The line through the point with position vector $\vec{r}_0 = x_0\vec{i} + y_0\vec{j} + z_0\vec{k}$ in the direction of the vector $\vec{v} = a\vec{i} + b\vec{j} + c\vec{k}$ has parametric equation
$$\vec{r}(t) = \vec{r}_0 + t\vec{v}.$$

Exercises and Problems for Section 17.1   Online Resource: Additional Problems for Section 17.1
EXERCISES

In Exercises 1–6, find a parameterization for the curve.

1.

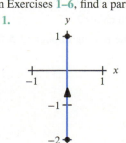

2.

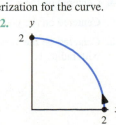

3.

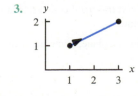

4.

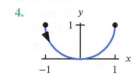

5.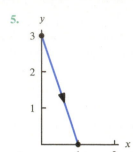

6.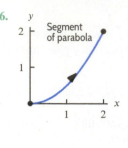

■ In Exercises **7–17**, find parametric equations for the line.

7. The line in the direction of the vector $\vec{i} - \vec{k}$ and through the point $(0, 1, 0)$.

8. The line in the direction of the vector $\vec{i} + 2\vec{j} - \vec{k}$ and through the point $(3, 0, -4)$.

9. The line parallel to the z-axis passing through the point $(1, 0, 0)$.

10. The line in the direction of the vector $5\vec{j} + 2\vec{k}$ and through the point $(5, -1, 1)$.

11. The line in the direction of the vector $3\vec{i} - 3\vec{j} + \vec{k}$ and through the point $(1, 2, 3)$.

12. The line in the direction of the vector $2\vec{i} + 2\vec{j} - 3\vec{k}$ and through the point $(-3, 4, -2)$.

13. The line through $(-3, -2, 1)$ and $(-1, -3, -1)$.

14. The line through the points $(1, 5, 2)$ and $(5, 0, -1)$.

15. The line through the points $(2, 3, -1)$ and $(5, 2, 0)$.

16. The line through $(3, -2, 2)$ and intersecting the y-axis at $y = 2$.

17. The line intersecting the x-axis at $x = 3$ and the z-axis at $z = -5$.

■ In Exercises **18–34**, find a parameterization for the curve.

18. A line segment between $(2, 1, 3)$ and $(4, 3, 2)$.

19. A circle of radius 3 centered on the z-axis and lying in the plane $z = 5$.

20. A line perpendicular to the plane $z = 2x - 3y + 7$ and through the point $(1, 1, 6)$.

21. The circle of radius 2 in the xy-plane, centered at the origin, clockwise.

22. The circle of radius 2 parallel to the xy-plane, centered at the point $(0, 0, 1)$, and traversed counterclockwise when viewed from below.

23. The circle of radius 2 in the xz-plane, centered at the origin.

24. The circle of radius 3 parallel to the xy-plane, centered at the point $(0, 0, 2)$.

25. The circle of radius 3 in the yz-plane, centered at the point $(0, 0, 2)$.

26. The circle of radius 5 parallel to the yz-plane, centered at the point $(-1, 0, -2)$.

27. The curve $x = y^2$ in the xy-plane.

28. The curve $y = x^3$ in the xy-plane.

29. The curve $x = -3z^2$ in the xz-plane.

30. The curve in which the plane $z = 2$ cuts the surface $z = \sqrt{x^2 + y^2}$.

31. The curve $y = 4 - 5x^4$ through the point $(0, 4, 4)$, parallel to the xy-plane.

32. The ellipse of major diameter 5 parallel to the y-axis and minor diameter 2 parallel to the z-axis, centered at $(0, 1, -2)$.

33. The ellipse of major diameter 6 along the x-axis and minor diameter 4 along the y-axis, centered at the origin.

34. The ellipse of major diameter 3 parallel to the x-axis and minor diameter 2 parallel to the z-axis, centered at $(0, 1, -2)$.

■ In Exercises **35–42**, find a parametric equation for the curve segment.

35. Line from $(-1, 2, -3)$ to $(2, 2, 2)$.

36. Line from $P_0 = (-1, -3)$ to $P_1 = (5, 2)$.

37. Line from $P_0 = (1, -3, 2)$ to $P_1 = (4, 1, -3)$.

38. Semicircle from $(0, 0, 5)$ to $(0, 0, -5)$ in the yz-plane with $y \geq 0$.

39. Semicircle from $(1, 0, 0)$ to $(-1, 0, 0)$ in the xy-plane with $y \geq 0$.

40. Graph of $y = \sqrt{x}$ from $(1, 1)$ to $(16, 4)$.

41. Arc of a circle of radius 5 from $P = (0, 0)$ to $Q = (10, 0)$.

42. Quarter-ellipse from $(4, 0, 3)$ to $(0, -3, 3)$ in the plane $z = 3$.

■ In Exercises **43–46**, find parametric equations for a helix satisfying the given conditions.

43. Centered on the z-axis, with radius 10.

44. Centered on the x-axis, with radius 5.

45. Centered on the y-axis, with radius 2.

46. Centered on the vertical line passing through $(3, 5, 0)$, with radius 1.

PROBLEMS

In Problems **47–51**, parameterize the line through $P = (2, 5)$ and $Q = (12, 9)$ so that the points P and Q correspond to the given parameter values.

47. $t = 0$ and 1 **48.** $t = 0$ and 5

49. $t = 20$ and 30 **50.** $t = 10$ and 11

51. $t = 0$ and -1

52. At the point where $t = -1$, find an equation for the plane perpendicular to the line

$$x = 5 - 3t, \quad y = 5t - 7, \quad \frac{z}{t} = 6.$$

53. Determine whether the following line is parallel to the plane $2x - 3y + 5z = 5$:

$$x = 5 + 7t, \quad y = 4 + 3t, \quad z = -3 - 2t.$$

54. Show that the equations $x = 3 + t$, $y = 2t$, $z = 1 - t$ satisfy the equations $x + y + 3z = 6$ and $x - y - z = 2$. What does this tell you about the curve parameterized by these equations?

55. (a) Explain why the line of intersection of two planes must be parallel to the cross product of a normal vector to the first plane and a normal vector to the second.
    (b) Find a vector parallel to the line of intersection of the two planes $x + 2y - 3z = 7$ and $3x - y + z = 0$.
    (c) Find parametric equations for the line in part (b).

56. Find an equation for the plane containing the point $(2, 3, 4)$ and the line $x = 1 + 2t$, $y = 3 - t$, $z = 4 + t$.

57. (a) Find an equation for the line perpendicular to the plane $2x - 3y = z$ and through the point $(1, 3, 7)$.
    (b) Where does the line cut the plane?
    (c) What is the distance between the point $(1, 3, 7)$ and the plane?

58. If possible, find a value of a making the given parameterized line perpendicular to the plane $ax + 8y = 4z - 1$.

    (a) $\vec{r}_1(t) = (9 - t)\vec{i} + 4t\vec{j} + (7 - 2t)\vec{k}$
    (b) $\vec{r}_2(t) = (1 + 2t)\vec{i} + (3t - 1)\vec{j} + (-1 + t)\vec{k}$

59. Consider two points P_0 and P_1 in 3-space.

    (a) Show that the line segment from P_0 to P_1 can be parameterized by

$$\vec{r}(t) = (1 - t)\overrightarrow{OP_0} + t\overrightarrow{OP_1}, \quad 0 \le t \le 1.$$

    (b) What is represented by the parametric equation

$$\vec{r}(t) = t\overrightarrow{OP_0} + (1 - t)\overrightarrow{OP_1}, \quad 0 \le t \le 1?$$

60. (a) Find a vector parallel to the line of intersection of the planes $2x - y - 3z = 0$ and $x + y + z = 1$.
    (b) Show that the point $(1, -1, 1)$ lies on both planes.
    (c) Find parametric equations for the line of intersection.

61. Find the intersection of the line $x = 5 + 7t$, $y = 4 + 3t$, $z = -3 - 2t$ and the plane $2x - 3y + 5z = -7$.

In Problems **62–65**, are the lines L_1 and L_2 the same line?

62. L_1: $x = 5 + t$, $y = 3 - 2t$, $z = 5t$
     L_2: $x = 5 + 2t$, $y = 3 - 4t$, $z = 10t$

63. L_1: $x = 2 + 3t$, $y = 1 + 4t$, $z = 6 - t$
     L_2: $x = 2 + 6t$, $y = 4 + 3t$, $z = 3 - 2t$

64. L_1: $x = 2 + 3t$, $y = 1 + 4t$, $z = 6 - t$
     L_2: $x = 5 + 6t$, $y = 5 + 8t$, $z = 5 - 2t$

65. L_1: $x = 1 + 2t$, $y = 1 - 3t$, $z = 1 + t$
     L_2: $x = 1 - 4t$, $y = 6t$, $z = 4 - 2t$

In Problems **66–68** two parameterized lines are given. Are they the same line?

66. $\vec{r}_1(t) = (5 - 3t)\vec{i} + 2t\vec{j} + (7 + t)\vec{k}$
     $\vec{r}_2(t) = (5 - 6t)\vec{i} + 4t\vec{j} + (7 + 3t)\vec{k}$

67. $\vec{r}_1(t) = (5 - 3t)\vec{i} + (1 + t)\vec{j} + 2t\vec{k}$
     $\vec{r}_2(t) = (2 + 6t)\vec{i} + (2 - 2t)\vec{j} + (2 - 4t)\vec{k}$

68. $\vec{r}_1(t) = (5 - 3t)\vec{i} + (1 + t)\vec{j} + 2t\vec{k}$
     $\vec{r}_2(t) = (2 + 6t)\vec{i} + (2 - 2t)\vec{j} + (3 - 4t)\vec{k}$

69. If it exists, find the value of c for which the lines $l(t) = (c + t, 1 + t, 5 + t)$ and $m(s) = (s, 1 - s, 3 + s)$ intersect.

70. (a) Where does the line $\vec{r} = 2\vec{i} + 5\vec{j} + t(3\vec{i} + \vec{j} + 2\vec{k})$ cut the plane $x + y + z = 1$?
    (b) Find a vector perpendicular to the line and lying in the plane.
    (c) Find an equation for the line that passes through the point of intersection of the line and plane, is perpendicular to the line, and lies in the plane.

In Problems **71–74**, find parametric equations for the line.

71. The line of intersection of the planes $x - y + z = 3$ and $2x + y - z = 5$.

72. The line of intersection of the planes $x + y + z = 3$ and $x - y + 2z = 2$.

73. The line perpendicular to the surface $z = x^2 + y^2$ at the point $(1, 2, 5)$.

74. The line through the point $(-4, 2, 3)$ and parallel to a line in the yz-plane which makes a $45°$ angle with the positive y-axis and the positive z-axis.

75. Is the point $(-3, -4, 2)$ visible from the point $(4, 5, 0)$ if there is an opaque ball of radius 1 centered at the origin?

76. Two particles are traveling through space. At time t the first particle is at the point $(-1 + t, 4 - t, -1 + 2t)$ and the second particle is at $(-7 + 2t, -6 + 2t, -1 + t)$.

   (a) Describe the two paths in words.
   (b) Do the two particles collide? If so, when and where?
   (c) Do the paths of the two particles cross? If so, where?

77. Match the parameterizations with their graphs in Figure 17.9.

   (a) $x = 2\cos 4\pi t$, $y = 2\sin 4\pi t$, $z = t$
   (b) $x = 2\cos 4\pi t$, $y = \sin 4\pi t$, $z = t$
   (c) $x = 0.5t\cos 4\pi t$, $y = 0.5t\sin 4\pi t$, $z = t$
   (d) $x = 2\cos 4\pi t$, $y = 2\sin 4\pi t$, $z = 0.5t^3$

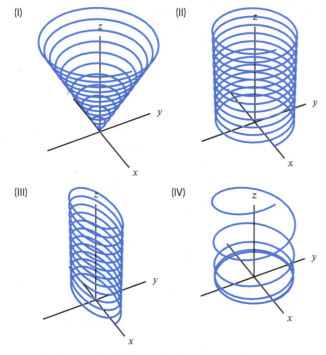

(I) (II) (III) (IV)

Figure 17.9

■ In Problems 78–81, find c so that one revolution about the z-axis of the helix gives an increase of Δz in the z-coordinate.

78. $x = 2\cos t$, $y = 2\sin t$, $z = ct$, $\Delta z = 15$
79. $x = 2\cos t$, $y = 2\sin t$, $z = ct$, $\Delta z = 50$
80. $x = 2\cos 3t$, $y = 2\sin 3t$, $z = ct$, $\Delta z = 10$
81. $x = 2\cos \pi t$, $y = 2\sin \pi t$, $z = ct$, $\Delta z = 20$

82. For $t > 0$, a particle moves along the curve $x = a + b\sin kt$, $y = a + b\cos kt$, where a, b, k are positive constants.

   (a) Describe the motion in words.

(b) What is the effect on the curve of the following changes?

   (i) Increasing b
   (ii) Increasing a
   (iii) Increasing k
   (iv) Setting a and b equal

83. In the Atlantic Ocean off the coast of Newfoundland, Canada, the temperature and salinity (saltiness) vary throughout the year. Figure 17.10 shows a parametric curve giving the average temperature, T (in °C) and salinity (in grams of salt per kg of water) for t in months, with $t = 1$ corresponding to mid-January.[1]

   (a) Why does the parameterized curve form a loop?
   (b) When is the water temperature highest?
   (c) When is the water saltiest?
   (d) Estimate dT/dt at $t = 6$, and give the units. What is the meaning of your answer for seawater?

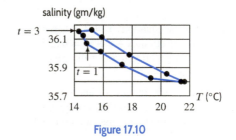

Figure 17.10

84. A light shines on the helix of Example 3 on page 938 from far down each axis. Sketch the shadow the helix casts on each of the coordinate planes: xy, xz, and yz.

85. The paraboloid $z = x^2 + y^2$ and the plane $z = 2x + 4y + 4$ intersect in a curve in 3-space.

   (a) Show that the shadow of the intersection in the xy-plane is a circle and find its center and radius.
   (b) Parameterize the circle in the xy-plane.
   (c) Parameterize the intersection of the paraboloid and the plane in 3-space.

86. The cone $z = \sqrt{x^2 + y^2}$ and the paraboloid $z = 6 - x^2 - y^2$ intersect in a curve in 3-space.

   (a) Find an equation for the horizontal plane that contains the intersection of the paraboloid and the cone.
   (b) Show that the shadow of the intersection in the xy-plane is a circle and find its center and radius.
   (c) Parameterize the intersection of the paraboloid and the cone in 3-space.

[1] Based on http://www.vub.ac.be. Accessed November, 2011.

87. For a positive constant a and $t \geq 0$, the parametric equations I-V represent the curves described in (a)-(e). Match each description (a)-(e) with its parametric equations and write an equation involving only x and y for the curve.

   (a) Line through the origin.
   (b) Line not through the origin.
   (c) Hyperbola opening along x-axis.
   (d) Circle traversed clockwise.
   (e) Circle traversed counterclockwise.

   I. $x = a \sin t, y = a \cos t$ II. $x = a \sin t, y = a \sin t$
   III. $x = a \cos t, y = a \sin t$ IV. $x = a \cos^2 t, y = a \sin^2 t$
   V. $x = a/\cos t, y = a \tan t$

88. **(a)** Find a parametric equation for the line through the point $(2, 1, 3)$ and in the direction of $a\vec{i} + b\vec{j} + c\vec{k}$.
   (b) Find conditions on a, b, c so that the line you found in part (a) goes through the origin. Give a reason for your answer.

89. Consider the line $x = 5 - 2t, y = 3 + 7t, z = 4t$ and the plane $ax + by + cz = d$. All the following questions have many possible answers. Find values of a, b, c, d such that:

   (a) The plane is perpendicular to the line.
   (b) The plane is perpendicular to the line and through the point $(5, 3, 0)$.
   (c) The line lies in the plane.

90. Explain the significance of the constants $\alpha > 0$ and $\beta > 0$ in the family of helices given by $\vec{r} = \alpha \cos t\vec{i} + \alpha \sin t\vec{j} + \beta t\vec{k}$.

91. Find parametric equations of the line passing through the points $(1, 2, 3)$, $(3, 5, 7)$ and calculate the shortest distance from the line to the origin.

92. Show that for a fixed value of θ, the line parameterized by $x = \cos \theta + t \sin \theta$, $y = \sin \theta - t \cos \theta$ and $z = t$ lies on the graph of the hyperboloid $x^2 + y^2 = z^2 + 1$.

93. A line has equation $\vec{r} = \vec{a} + t\vec{b}$ where $\vec{r} = x\vec{i} + y\vec{j} + z\vec{k}$ and $\vec{a}$ and $\vec{b}$ are constant vectors such that $\vec{a} \neq \vec{0}$, $\vec{b} \neq \vec{0}$, $\vec{b}$ not parallel or perpendicular to $\vec{a}$. For each of the planes (a)–(c), pick the equation (i)–(ix) which represents it. Explain your choice.

   (a) A plane perpendicular to the line and through the origin.
   (b) A plane perpendicular to the line and not through the origin.
   (c) A plane containing the line.

   (i) $\vec{a} \cdot \vec{r} = ||\vec{b}||$ (ii) $\vec{b} \cdot \vec{r} = ||\vec{a}||$
   (iii) $\vec{a} \cdot \vec{r} = \vec{b} \cdot \vec{r}$ (iv) $(\vec{a} \times \vec{b}) \cdot (\vec{r} - \vec{a}) = 0$
   (v) $\vec{r} - \vec{a} = \vec{b}$ (vi) $\vec{a} \cdot \vec{r} = 0$
   (vii) $\vec{b} \cdot \vec{r} = 0$ (viii) $\vec{a} + \vec{r} = \vec{b}$
   (ix) $(\vec{a} \times \vec{b}) \cdot (\vec{r} - \vec{b}) = ||\vec{a}||$

94. **(a)** Find a parametric equation for the line through the point $(1, 5, 2)$ and in the direction of the vector $2\vec{i} + 3\vec{j} - \vec{k}$.
   (b) By minimizing the square of the distance from a point on the line to the origin, find the exact point on the line which is closest to the origin.

95. Figure 17.11 shows the parametric curve $x = x(t), y = y(t)$ for $a \leq t \leq b$.

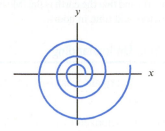

Figure 17.11

   (a) Match a graph to each of the parametric curves given, for the same t values, by
     (i) $(-x(t), -y(t))$ (ii) $(-x(t), y(t))$
     (iii) $(x(t) + 1, y(t))$ (iv) $(x(t) + 1, y(t) + 1)$

(A) 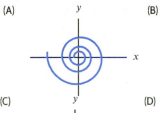 (B)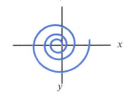

(C) 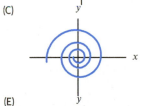 (D)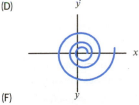

(E)  (F)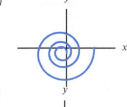

(G) 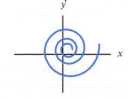 (H)

   (b) Which of the following could be the formulas for the functions $x(t), y(t)$?
     (i) $x = 10 \cos t$ $y = 10 \sin t$
     (ii) $x = (10 + 8t) \cos t$ $y = (10 + 8t) \sin t$
     (iii) $x = e^{t^2/200} \cos t$ $y = e^{t^2/200} \sin t$
     (iv) $x = (10 - 8t) \cos t$ $y = (10 - 8t) \sin t$
     (v) $x = 10 \cos(t^2 + t)$ $y = 10 \sin(t^2 + t)$

96. A plane from Denver, Colorado, (altitude 1650 meters) flies to Bismark, North Dakota (altitude 550 meters). It travels at 650 km/hour at a constant height of 8000 meters above the line joining Denver and Bismark. Bismark is about 850 km in the direction 60° north of east from Denver. Find parametric equations describing the plane's motion. Assume the origin is at sea level beneath Denver, that the x-axis points east and the y-axis points north, and that the earth is flat. Measure distances in kilometers and time in hours.

97. The vector $\vec{n}$ is perpendicular to the plane P_1. The vector $\vec{v}$ is parallel to the line L.

   (a) If $\vec{n} \cdot \vec{v} = 0$, what does this tell you about the directions of P_1 and L? (Are they parallel? Perpendicular? Or is it impossible to tell?)

   (b) Suppose $\vec{n} \times \vec{v} \neq \vec{0}$. The plane P_2 has normal $\vec{n} \times \vec{v}$. What can you say about the directions of

   (i) P_1 and P_2?   (ii) L and P_2?

Strengthen Your Understanding

■ In Problems 98–99, explain what is wrong with the statement.

98. The curve parameterized by $\vec{r}_1(t) = \vec{r}(t-2)$, defined for all t, is a shift in the $\vec{i}$-direction of the curve parameterized by $\vec{r}(t)$.

99. All points of the curve $\vec{r}(t) = R\cos t\vec{i} + R\sin t\vec{j} + t\vec{k}$ are the same distance, R, from the origin.

■ In Problems 100–102, give an example of:

100. Parameterizations of two different circles that have the same center and equal radii.

101. Parameterizations of two different lines that intersect at the point $(1, 2, 3)$.

102. A parameterization of the line $x = t, y = 2t, z = 3+4t$ that is not given by linear functions.

■ Are the statements in Problems 103–114 true or false? Give reasons for your answer.

103. The parametric curve $x = 3t+2, y = -2t$ for $0 \leq t \leq 5$ passes through the origin.

104. The parametric curve $x = t^2, y = t^4$ for $0 \leq t \leq 1$ is a parabola.

105. A parametric curve $x = g(t), y = h(t)$ for $a \leq t \leq b$ is always the graph of a function $y = f(x)$.

106. The parametric curve $x = (3t+2)^2, y = (3t+2)^2 - 1$ for $0 \leq t \leq 3$ is a line.

107. The parametric curve $x = -\sin t, y = -\cos t$ for $0 \leq t \leq 2\pi$ traces out a unit circle counterclockwise as t increases.

108. A parameterization of the graph of $y = \ln x$ for $x > 0$ is given by $x = e^t, y = t$ for $-\infty < t < \infty$.

109. Both $x = -t + 1, y = 2t$ and $x = 2s, y = -4s + 2$ describe the same line.

110. The line of intersection of the two planes $z = x + y$ and $z = 1 - x - y$ can be parameterized by $x = t, y = \frac{1}{2} - t, z = \frac{1}{2}$.

111. The two lines given by $x = t, y = 2+t, z = 3+t$ and $x = 2s, y = 1-s, z = s$ do not intersect.

112. The line parameterized by $x = 1, y = 2t, z = 3+t$ is parallel to the x-axis.

113. The equation $\vec{r}(t) = 3t\vec{i} + (6t+1)\vec{j}$ parameterizes a line.

114. The lines parameterized by $\vec{r}_1(t) = t\vec{i} + (-2t+1)\vec{j}$ and $\vec{r}_2(t) = (2t+5)\vec{i} + (-t)\vec{j}$ are parallel.

17.2   MOTION, VELOCITY, AND ACCELERATION

In this section we see how to find the vector quantities of velocity and acceleration from a parametric equation for the motion of an object.

The Velocity Vector

The velocity of a moving particle can be represented by a vector with the following properties:

> The **velocity vector** of a moving object is a vector $\vec{v}$ such that:
> - The magnitude of $\vec{v}$ is the speed of the object.
> - The direction of $\vec{v}$ is the direction of motion.
>
> Thus, the speed of the object is $\|\vec{v}\|$ and the velocity vector is tangent to the object's path.

Example 1   A child is sitting on a Ferris wheel of diameter 10 meters, making one revolution every 2 minutes. Find the speed of the child and draw velocity vectors at two different times.

Solution   The child moves at a constant speed around a circle of radius 5 meters, completing one revolution every 2 minutes. One revolution around a circle of radius 5 is a distance of 10π, so the child's speed is $10\pi/2 = 5\pi \approx 15.7$ m/min. Hence, the magnitude of the velocity vector is 15.7 m/min. The direction of motion is tangent to the circle, and hence perpendicular to the radius at that point. Figure 17.12 shows the direction of the vector at two different times.

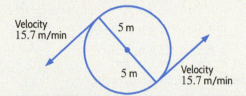

Figure 17.12: Velocity vectors of a child on a Ferris wheel (note that vectors would be in opposite direction if viewed from the other side)

Computing the Velocity

We find the velocity, as in one-variable calculus, by taking a limit. If the position vector of the particle is $\vec{r}(t)$ at time t, then the displacement vector between its positions at times t and $t + \Delta t$ is $\Delta \vec{r} = \vec{r}(t + \Delta t) - \vec{r}(t)$. (See Figure 17.13.) Over this interval,

$$\text{Average velocity} = \frac{\Delta \vec{r}}{\Delta t}.$$

In the limit as Δt goes to zero we have the instantaneous velocity at time t:

The **velocity vector**, $\vec{v}(t)$, of a moving object with position vector $\vec{r}(t)$ at time t is

$$\vec{v}(t) = \lim_{\Delta t \to 0} \frac{\Delta \vec{r}}{\Delta t} = \lim_{\Delta t \to 0} \frac{\vec{r}(t + \Delta t) - \vec{r}(t)}{\Delta t},$$

whenever the limit exists. We use the notation $\vec{v} = \dfrac{d\vec{r}}{dt} = \vec{r}\,'(t)$.

Notice that the direction of the velocity vector $\vec{r}\,'(t)$ in Figure 17.13 is approximated by the direction of the vector $\Delta \vec{r}$ and that the approximation gets better as $\Delta t \to 0$.

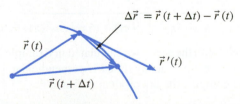

Figure 17.13: The change, $\Delta \vec{r}$, in the position vector for a particle moving on a curve and the velocity vector $\vec{v} = \vec{r}\,'(t)$

The Components of the Velocity Vector

If we represent a curve parametrically by $x = f(t)$, $y = g(t)$, $z = h(t)$, then we can write its position vector as: $\vec{r}(t) = f(t)\vec{i} + g(t)\vec{j} + h(t)\vec{k}$. Now we can compute the velocity vector:

$$\vec{v}(t) = \lim_{\Delta t \to 0} \frac{\vec{r}(t + \Delta t) - \vec{r}(t)}{\Delta t}$$

$$= \lim_{\Delta t \to 0} \frac{(f(t + \Delta t)\vec{i} + g(t + \Delta t)\vec{j} + h(t + \Delta t)\vec{k}) - (f(t)\vec{i} + g(t)\vec{j} + h(t)\vec{k})}{\Delta t}$$

$$= \lim_{\Delta t \to 0} \left(\frac{f(t + \Delta t) - f(t)}{\Delta t}\vec{i} + \frac{g(t + \Delta t) - g(t)}{\Delta t}\vec{j} + \frac{h(t + \Delta t) - h(t)}{\Delta t}\vec{k} \right)$$

$$= f'(t)\vec{i} + g'(t)\vec{j} + h'(t)\vec{k}$$

$$= \frac{dx}{dt}\vec{i} + \frac{dy}{dt}\vec{j} + \frac{dz}{dt}\vec{k}.$$

Thus, we have the following result:

> The **components of the velocity vector** of a particle moving in space with position vector $\vec{r}(t) = f(t)\vec{i} + g(t)\vec{j} + h(t)\vec{k}$ at time t are given by
>
> $$\vec{v}(t) = f'(t)\vec{i} + g'(t)\vec{j} + h'(t)\vec{k} = \frac{dx}{dt}\vec{i} + \frac{dy}{dt}\vec{j} + \frac{dz}{dt}\vec{k}.$$

Example 2   Find the components of the velocity vector for the child on the Ferris wheel in Example 1 using a coordinate system which has its origin at the center of the Ferris wheel and which makes the rotation counterclockwise.

Solution   The Ferris wheel has radius 5 meters and completes 1 revolution counterclockwise every 2 minutes. The motion is parameterized by an equation of the form

$$\vec{r}(t) = 5\cos(\omega t)\vec{i} + 5\sin(\omega t)\vec{j},$$

where ω is chosen to make the period 2 minutes. Since the period of $\cos(\omega t)$ and $\sin(\omega t)$ is $2\pi/\omega$, we must have

$$\frac{2\pi}{\omega} = 2, \quad \text{so} \quad \omega = \pi.$$

Thus, the motion is described by the equation

$$\vec{r}(t) = 5\cos(\pi t)\vec{i} + 5\sin(\pi t)\vec{j},$$

where t is in minutes. The velocity is given by

$$\vec{v} = \frac{dx}{dt}\vec{i} + \frac{dy}{dt}\vec{j} = -5\pi\sin(\pi t)\vec{i} + 5\pi\cos(\pi t)\vec{j}.$$

To check, we calculate the magnitude of $\vec{v}$,

$$\|\vec{v}\| = \sqrt{(-5\pi)^2\sin^2(\pi t) + (5\pi)^2\cos^2(\pi t)} = 5\pi\sqrt{\sin^2(\pi t) + \cos^2(\pi t)} = 5\pi \approx 15.7,$$

which agrees with the speed we calculated in Example 1. To see that the direction is correct, we must show that the vector $\vec{v}$ at any time t is perpendicular to the position vector of the child at time t. To do this, we compute the dot product of $\vec{v}$ and $\vec{r}$:

$$\vec{v} \cdot \vec{r} = (-5\pi\sin(\pi t)\vec{i} + 5\pi\cos(\pi t)\vec{j}) \cdot (5\cos(\pi t)\vec{i} + 5\sin(\pi t)\vec{j})$$

$$= -25\pi\sin(\pi t)\cos(\pi t) + 25\pi\cos(\pi t)\sin(\pi t) = 0.$$

So the velocity vector, $\vec{v}$, is perpendicular to $\vec{r}$ and hence tangent to the circle. The direction is counterclockwise, since in the first quadrant, x is decreasing while y is increasing. (See Figure 17.14.)

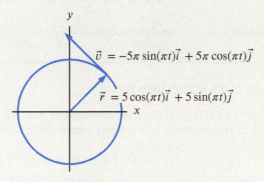

Figure 17.14: Velocity and radius vector of motion around a circle

Velocity Vectors and Tangent Lines

Since the velocity vector is tangent to the path of motion, it can be used to find parametric equations for the tangent line, if there is one.

Example 3   Find the tangent line at the point $(1, 1, 2)$ to the curve defined by the parametric equation

$$\vec{r}(t) = t^2\vec{i} + t^3\vec{j} + 2t\vec{k}.$$

Solution   At time $t = 1$ the particle is at the point $(1, 1, 2)$ with position vector $\vec{r}_0 = \vec{i} + \vec{j} + 2\vec{k}$. The velocity vector at time t is $\vec{r}\,'(t) = 2t\vec{i} + 3t^2\vec{j} + 2\vec{k}$, so at time $t = 1$ the velocity is $\vec{v} = \vec{r}\,'(1) = 2\vec{i} + 3\vec{j} + 2\vec{k}$. The tangent line passes through $(1, 1, 2)$ in the direction of $\vec{v}$, so it has the parametric equation

$$\vec{r}(t) = \vec{r}_0 + t\vec{v} = (\vec{i} + \vec{j} + 2\vec{k}) + t(2\vec{i} + 3\vec{j} + 2\vec{k}).$$

The Acceleration Vector

Just as the velocity of a particle moving in 2-space or 3-space is a vector quantity, so is the rate of change of the velocity of the particle, namely its acceleration. Figure 17.15 shows a particle at time t with velocity vector $\vec{v}(t)$ and then a little later at time $t + \Delta t$. The vector $\Delta\vec{v} = \vec{v}(t + \Delta t) - \vec{v}(t)$ is the change in velocity and points approximately in the direction of the acceleration. So,

$$\text{Average acceleration} = \frac{\Delta\vec{v}}{\Delta t}.$$

In the limit as $\Delta t \to 0$, we have the instantaneous acceleration at time t:

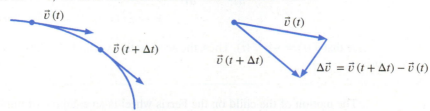

Figure 17.15: Computing the difference between two velocity vectors

The **acceleration vector** of an object moving with velocity $\vec{v}(t)$ at time t is

$$\vec{a}(t) = \lim_{\Delta t \to 0} \frac{\Delta \vec{v}}{\Delta t} = \lim_{\Delta t \to 0} \frac{\vec{v}(t + \Delta t) - \vec{v}(t)}{\Delta t},$$

if the limit exists. We use the notation $\vec{a} = \dfrac{d\vec{v}}{dt} = \dfrac{d^2\vec{r}}{dt^2} = \vec{r}\,''(t)$.

Components of the Acceleration Vector

If we represent a curve in space parametrically by $x = f(t)$, $y = g(t)$, $z = h(t)$, we can express the acceleration in components. The velocity vector $\vec{v}(t)$ is given by

$$\vec{v}(t) = f'(t)\vec{i} + g'(t)\vec{j} + h'(t)\vec{k}.$$

From the definition of the acceleration vector, we have

$$\vec{a}(t) = \lim_{\Delta t \to 0} \frac{\vec{v}(t + \Delta t) - \vec{v}(t)}{\Delta t} = \frac{d\vec{v}}{dt}.$$

Using the same method to compute $d\vec{v}/dt$ as we used to compute $d\vec{r}/dt$ on page 950, we obtain

The **components of the acceleration vector**, $\vec{a}(t)$, at time t of a particle moving in space with position vector $\vec{r}(t) = f(t)\vec{i} + g(t)\vec{j} + h(t)\vec{k}$ at time t are given by

$$\vec{a}(t) = f''(t)\vec{i} + g''(t)\vec{j} + h''(t)\vec{k} = \frac{d^2x}{dt^2}\vec{i} + \frac{d^2y}{dt^2}\vec{j} + \frac{d^2z}{dt^2}\vec{k}.$$

Motion in a Circle and Along a Line

We now consider the velocity and acceleration vectors for two basic motions: uniform motion around a circle, and motion along a straight line.

Example 4   Find the acceleration vector for the child on the Ferris wheel in Examples 1 and 2.

Solution   The child's position vector is given by $\vec{r}(t) = 5\cos(\pi t)\vec{i} + 5\sin(\pi t)\vec{j}$. In Example 2 we saw that the velocity vector is

$$\vec{v}(t) = \frac{dx}{dt}\vec{i} + \frac{dy}{dt}\vec{j} = -5\pi\sin(\pi t)\vec{i} + 5\pi\cos(\pi t)\vec{j}.$$

Thus, the acceleration vector is

$$\vec{a}(t) = \frac{d^2x}{dt^2}\vec{i} + \frac{d^2y}{dt^2}\vec{j} = -(5\pi)\cdot\pi\cos(\pi t)\vec{i} - (5\pi)\cdot\pi\sin(\pi t)\vec{j}$$
$$= -5\pi^2\cos(\pi t)\vec{i} - 5\pi^2\sin(\pi t)\vec{j}.$$

Notice that $\vec{a}(t) = -\pi^2\vec{r}(t)$. Thus, the acceleration vector is a multiple of $\vec{r}(t)$ and points toward the origin.

The motion of the child on the Ferris wheel is an example of uniform circular motion, whose properties follow. (See Problem 45.)

Uniform Circular Motion: For a particle whose motion is described by

$$\vec{r}(t) = R\cos(\omega t)\vec{i} + R\sin(\omega t)\vec{j},$$

- Motion is in a circle of radius R with period $2\pi/|\omega|$.
- Velocity, $\vec{v}$, is tangent to the circle and speed is constant $\|\vec{v}\| = |\omega|R$.
- Acceleration, $\vec{a}$, points toward the center of the circle with $\|\vec{a}\| = \|\vec{v}\|^2/R$.

In uniform circular motion, the acceleration vector is perpendicular to the velocity vector, $\vec{v}$, because $\vec{v}$ does not change in magnitude, only in direction. There is no acceleration in the direction of $\vec{v}$.

We now look at straight-line motion in which the velocity vector always has the same direction but its magnitude changes. In straight-line motion, the acceleration vector points in the same direction as the velocity vector if the speed is increasing and in the opposite direction to the velocity vector if the speed is decreasing.

Example 5

Consider the motion given by the vector equation

$$\vec{r}(t) = 2\vec{i} + 6\vec{j} + (t^3 + t)(4\vec{i} + 3\vec{j} + \vec{k}).$$

Show that this is straight-line motion in the direction of the vector $4\vec{i} + 3\vec{j} + \vec{k}$ and relate the acceleration vector to the velocity vector.

Solution

The velocity vector is

$$\vec{v} = (3t^2 + 1)(4\vec{i} + 3\vec{j} + \vec{k}).$$

Since $(3t^2 + 1)$ is a positive scalar, the velocity vector $\vec{v}$ always points in the direction of the vector $4\vec{i} + 3\vec{j} + \vec{k}$. In addition,

$$\text{Speed} = \|\vec{v}\| = (3t^2 + 1)\sqrt{4^2 + 3^2 + 1^2} = \sqrt{26}(3t^2 + 1).$$

Notice that the speed is decreasing until $t = 0$, then starts increasing. The acceleration vector is

$$\vec{a} = 6t(4\vec{i} + 3\vec{j} + \vec{k}).$$

For $t > 0$, the acceleration vector points in the same direction as $4\vec{i} + 3\vec{j} + \vec{k}$, which is the same direction as $\vec{v}$. This makes sense because the object is speeding up. For $t < 0$, the acceleration vector $6t(4\vec{i} + 3\vec{j} + \vec{k})$ points in the opposite direction to $\vec{v}$ because the object is slowing down.

Motion in a Straight Line: For a particle whose motion is described by

$$\vec{r}(t) = \vec{r}_0 + f(t)\vec{v},$$

- Motion is along a straight line through the point with position vector $\vec{r}_0$ parallel to $\vec{v}$.
- Velocity, $\vec{v}$, and acceleration, $\vec{a}$, are parallel to the line.

If $f(t) = t$, then we have $\vec{r}(t) = \vec{r}_0 + t\vec{v}$, the equation of a line obtained on page 941.

The Length of a Curve

The speed of a particle is the magnitude of its velocity vector:

$$\text{Speed} = \|\vec{v}\| = \sqrt{\left(\frac{dx}{dt}\right)^2 + \left(\frac{dy}{dt}\right)^2 + \left(\frac{dz}{dt}\right)^2}.$$

As in one dimension, we can find the distance traveled by a particle along a curve by integrating its speed. Thus,

$$\text{Distance traveled} = \int_a^b \|\vec{v}(t)\| \, dt.$$

If the particle never stops or reverses its direction as it moves along the curve, the distance it travels will be the same as the length of the curve. This suggests the following formula, which is justified in Problem 71 (available online):

> If the curve C is given parametrically for $a \leq t \leq b$ by smooth functions and if the velocity vector $\vec{v}$ is not $\vec{0}$ for $a < t < b$, then
>
> $$\text{Length of } C = \int_a^b \|\vec{v}\| \, dt.$$

Example 6

Find the circumference of the ellipse given by the parametric equations

$$x = 2\cos t, \quad y = \sin t, \quad 0 \leq t \leq 2\pi.$$

Solution

The circumference of this curve is given by an integral which must be calculated numerically:

$$\text{Circumference} = \int_0^{2\pi} \sqrt{\left(\frac{dx}{dt}\right)^2 + \left(\frac{dy}{dt}\right)^2} \, dt = \int_0^{2\pi} \sqrt{(-2\sin t)^2 + (\cos t)^2} \, dt$$

$$= \int_0^{2\pi} \sqrt{4\sin^2 t + \cos^2 t} \, dt = 9.69.$$

Since the ellipse is inscribed in a circle of radius 2 and circumscribes a circle of radius 1, we would expect the length of the ellipse to be between $2\pi(2) \approx 12.57$ and $2\pi(1) \approx 6.28$, so the value of 9.69 is reasonable.

Summary for Section 17.2

- The **velocity vector** of a moving object is a vector $\vec{v}$ such that:
  - The magnitude of $\vec{v}$ is the speed of the object.
  - The direction of $\vec{v}$ is the direction of motion and is tangent to the object's path.
- The **components of the velocity vector**, $\vec{v}(t)$, of a particle moving with position vector $\vec{r}(t) = f(t)\vec{i} + g(t)\vec{j} + h(t)\vec{k}$ at time t are given by

$$\vec{v}(t) = f'(t)\vec{i} + g'(t)\vec{j} + h'(t)\vec{k} = \frac{dx}{dt}\vec{i} + \frac{dy}{dt}\vec{j} + \frac{dz}{dt}\vec{k}.$$

- The **components of the acceleration vector**, $\vec{a}(t)$, at time t of a particle moving with position vector $\vec{r}(t) = f(t)\vec{i} + g(t)\vec{j} + h(t)\vec{k}$ at time t are given by

$$\vec{a}(t) = f''(t)\vec{i} + g''(t)\vec{j} + h''(t)\vec{k} = \frac{d^2x}{dt^2}\vec{i} + \frac{d^2y}{dt^2}\vec{j} + \frac{d^2z}{dt^2}\vec{k}.$$

- **Length of a curve:** If the curve C is given parametrically for $a \leq t \leq b$ by smooth functions and if the velocity vector $\vec{v}$ is not $\vec{0}$ for $a < t < b$, then

$$\text{Length of } C = \int_a^b \|\vec{v}\| \, dt.$$

Exercises and Problems for Section 17.2   Online Resource: Additional Problems for Section 17.2

EXERCISES

In Exercises 1–6, find the velocity and acceleration vectors.

1. $x = 2 + 3t,\ y = 4 + t,\ z = 1 - t$

2. $x = 2 + 3t^2,\ y = 4 + t^2,\ z = 1 - t^2$

3. $x = t,\ y = t^2,\ z = t^3$

4. $x = t,\ y = t^3 - t$

5. $x = 3\cos t,\ y = 4\sin t$

6. $x = 3\cos(t^2),\ y = 3\sin(t^2),\ z = t^2$

In Exercises 7–12, find the velocity $\vec{v}(t)$ and speed $\|\vec{v}(t)\|$. Find any times at which the particle stops.

7. $x = t,\ y = t^2,\ z = t^3$

8. $x = \cos 3t,\ y = \sin 5t$

9. $x = 3t^2,\ y = t^3 + 1$

10. $x = (t - 1)^2,\ y = 2,\ z = 2t^3 - 3t^2$

11. $x = 3\sin(t^2) - 1,\ y = 3\cos(t^2)$

12. $x = 3\sin^2 t,\ y = \cos t - 1,\quad z = t^2$

In Exercises 13–16, find the length of the curve.

13. $x = 3 + 5t,\ y = 1 + 4t,\ z = 3 - t$ for $1 \leq t \leq 2$. Check by calculating the length by another method.

14. $x = \cos 3t,\ y = \sin 5t$ for $0 \leq t \leq 2\pi$.

15. $x = \cos(e^t),\ y = \sin(e^t)$ for $0 \leq t \leq 1$. Check by calculating the length by another method.

16. $\vec{r}(t) = 2t\vec{i} + \ln t\vec{j} + t^2\vec{k}$ for $1 \leq t \leq 2$.

In Exercises 17–18, find the velocity and acceleration vectors of the uniform circular motion and check that they are perpendicular. Check that the speed and magnitude of the acceleration are constant.

17. $x = 3\cos(2\pi t),\ y = 3\sin(2\pi t),\ z = 0$

18. $x = 2\pi,\ y = 2\sin(3t),\ z = 2\cos(3t)$

In Exercises 19–20, find the velocity and acceleration vectors of the straight-line motion. Check that the acceleration vector points in the same direction as the velocity vector if the speed is increasing and in the opposite direction if the speed is decreasing.

19. $x = 2 + t^2,\ y = 3 - 2t^2,\ z = 5 - t^2$

20. $x = -2t^3 - 3t + 1,\ y = 4t^3 + 6t - 5,\ z = 6t^3 + 9t - 2$

21. Find parametric equations for the tangent line at $t = 2$ for Exercise 10.

PROBLEMS

22. A particle passes through the point $P = (5, 4, -2)$ at time $t = 4$, moving with constant velocity $\vec{v} = 2\vec{i} - 3\vec{j} + \vec{k}$. Find a parametric equation for its motion.

In Problems 23–24, find all values of t for which the particle is moving parallel to the x-axis and to the y-axis. Determine the end behavior and graph the particle's path.

23. $x = t^2 - 6t,\quad y = t^3 - 3t$

24. $x = t^3 - 12t,\quad y = t^2 + 10t$

25. The table gives x and y coordinates of a particle in the plane at time t. Assuming that the particle moves smoothly and that the points given show all the major features of the motion, estimate the following quantities:

    (a) The velocity vector and speed at time $t = 2$.
    (b) Any times when the particle is moving parallel to the y-axis.
    (c) Any times when the particle has come to a stop.

t	0	0.5	1.0	1.5	2.0	2.5	3.0	3.5	4.0
x	1	4	6	7	6	3	2	3	5
y	3	2	3	5	8	10	11	10	9

26. A particle starts at the point $P = (3, 2, -5)$ and moves along a straight line toward $Q = (5, 7, -2)$ at a speed of 5 cm/sec. Let x, y, z be measured in centimeters.

    (a) Find the particle's velocity vector.
    (b) Find parametric equations for the particle's motion.

27. A particle moves at a constant speed along a line from the point $P = (2, -1, 5)$ at time $t = 0$ to the point $Q = (5, 3, -1)$. Find parametric equations for the particle's motion if:

    (a) The particle takes 5 seconds to move from P to Q.
    (b) The speed of the particle is 5 units per second.

28. A particle travels along the line $x = 1 + t$, $y = 5 + 2t$, $z = -7 + t$, where t is in seconds and x, y, z are in meters.

    (a) When and where does the particle hit the plane $x + y + z = 1$?
    (b) How fast is the particle going when it hits the plane? Give units.

29. A stone is thrown from a rooftop at time $t = 0$ seconds. Its position at time t is given by

$$\vec{r}(t) = 10t\vec{i} - 5t\vec{j} + (6.4 - 4.9t^2)\vec{k}.$$

The origin is at the base of the building, which is standing on flat ground. Distance is measured in meters. The vector $\vec{i}$ points east, $\vec{j}$ points north, and $\vec{k}$ points up.

(a) How high is the rooftop above the ground?
(b) At what time does the stone hit the ground?
(c) How fast is the stone moving when it hits the ground?
(d) Where does the stone hit the ground?
(e) What is the stone's acceleration when it hits the ground?

30. A child wanders slowly down a circular staircase from the top of a tower. With x, y, z in feet and the origin at the base of the tower, her position t minutes from the start is given by

$$x = 10\cos t, \quad y = 10\sin t, \quad z = 90 - 5t.$$

(a) How tall is the tower?
(b) When does the child reach the bottom?
(c) What is her speed at time t?
(d) What is her acceleration at time t?

31. The origin is on flat ground and the z-axis points upward. For time $0 \leq t \leq 10$ in seconds and distance in centimeters, a particle moves along a path given by

$$\vec{r} = 2t\vec{i} + 3t\vec{j} + (100 - (t - 5)^2)\vec{k}.$$

(a) When is the particle at the highest point? What is that point?
(b) When in the interval $0 \leq t \leq 10$ is the particle moving fastest? What is its speed at that moment?
(c) When in the interval $0 \leq t \leq 10$ is the particle moving slowest? What is its speed at that moment?

32. The function $w = f(x, y, z)$ has grad $f(7, 2, 5) = 4\vec{i} - 3\vec{j} + \vec{k}$. A particle moves along the curve $\vec{r}(t)$, arriving at the point $(7, 2, 5)$ with velocity $2\vec{i} + 3\vec{j} + 6\vec{k}$ when $t = 0$. Find the rate of change of w with respect to time at $t = 0$.

33. Suppose x measures horizontal distance in meters, and y measures distance above the ground in meters. At time $t = 0$ in seconds, a projectile starts from a point h meters above the origin with speed v meters/sec at an angle θ to the horizontal. Its path is given by

$$x = (v\cos\theta)t, \quad y = h + (v\sin\theta)t - \frac{1}{2}gt^2.$$

Using this information about a general projectile, analyze the motion of a ball which travels along the path

$$x = 20t, \quad y = 2 + 25t - 4.9t^2.$$

(a) When does the ball hit the ground?

(b) Where does the ball hit the ground?
(c) At what height above the ground does the ball start?
(d) What is the value of g, the acceleration due to gravity?
(e) What are the values of v and θ?

34. A particle is moving on a path in the xz-plane given by $x = 20t$, $z = 5t - 0.5t^2$, where z is the height of the particle above the ground in meters, x is the horizontal distance in meters, and t is time in seconds.

(a) What is the equation of the path in terms of x and z only?
(b) When is the particle at ground level?
(c) What is the velocity of the particle at time t?
(d) What is the speed of the particle at time t?
(e) Is the speed ever 0?
(f) When is the particle at the highest point?

35. The base of a 20-meter tower is at the origin; the base of a 20-meter tree is at $(0, 20, 0)$. The ground is flat and the z-axis points upward. The following parametric equations describe the motion of six projectiles each launched at time $t = 0$ in seconds.

  (I) $\vec{r}(t) = (20 + t^2)\vec{k}$
 (II) $\vec{r}(t) = 2t^2\vec{j} + 2t^2\vec{k}$
(III) $\vec{r}(t) = 20\vec{i} + 20\vec{j} + (20 - t^2)\vec{k}$
(IV) $\vec{r}(t) = 2t\vec{j} + (20 - t^2)\vec{k}$
 (V) $\vec{r}(t) = (20 - 2t)\vec{i} + 2t\vec{j} + (20 - t)\vec{k}$
(VI) $\vec{r}(t) = t\vec{i} + t\vec{j} + t\vec{k}$

(a) Which projectile is launched from the top of the tower and goes downward? When and where does it hit the ground?
(b) Which projectile hits the top of the tree? When? From where is it launched?
(c) Which projectile is not launched from somewhere on the tower and hits the tree? Where and when does it hit the tree?

36. A particle moves on a circle of radius 5 cm, centered at the origin, in the xy-plane (x and y measured in centimeters). It starts at the point $(0, 5)$ and moves counterclockwise, going once around the circle in 8 seconds.

(a) Write a parameterization for the particle's motion.
(b) What is the particle's speed? Give units.

37. A particle moves along a curve with velocity vector $\vec{v}(t) = -\sin t\vec{i} + \cos t\vec{j}$. At time $t = 0$ the particle is at $(2, 3)$.

(a) Find the displacement vector for the particle from time $t = 0$ to $t = \pi$.
(b) Find the position of the particle at time $t = \pi$.
(c) Find the distance traveled by the particle from time $t = 0$ to time $t = \pi$.

38. Determine the position vector $\vec{r}(t)$ for a rocket which is launched from the origin at time $t = 0$ seconds, reaches its highest point of $(x, y, z) = (1000, 3000, 10,000)$, where x, y, z are in meters, and after the launch is subject only to the acceleration due to gravity, 9.8 m/sec^2.

39. Emily is standing on the outer edge of a merry-go-round, 10 meters from the center. The merry-go-round completes one full revolution every 20 seconds. As Emily passes over a point P on the ground, she drops a ball from 3 meters above the ground.

    (a) How fast is Emily going?
    (b) How far from P does the ball hit the ground? (The acceleration due to gravity is 9.8 m/sec^2.)
    (c) How far from Emily does the ball hit the ground?

40. A point P moves in a circle of radius a. Show that $\vec{r}(t)$, the position vector of P, and its velocity vector $\vec{r}\,'(t)$ are perpendicular.

41. A wheel of radius 1 meter rests on the x-axis with its center on the y-axis. There is a spot on the rim at the point $(1, 1)$. See Figure 17.16. At time $t = 0$ the wheel starts rolling on the x-axis in the direction shown at a rate of 1 radian per second.

    (a) Find parametric equations describing the motion of the center of the wheel.
    (b) Find parametric equations describing the motion of the spot on the rim. Plot its path.

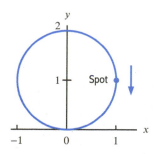

Figure 17.16

42. Suppose $\vec{r}(t) = \cos t\,\vec{i} + \sin t\,\vec{j} + 2t\,\vec{k}$ represents the position of a particle on a helix, where z is the height of the particle above the ground.

    (a) Is the particle ever moving downward? When?
    (b) When does the particle reach a point 10 units above the ground?
    (c) What is the velocity of the particle when it is 10 units above the ground?
    (d) When it is 10 units above the ground, the particle leaves the helix and moves along the tangent. Find parametric equations for this tangent line.

43. Show that the helix $\vec{r} = \alpha \cos t\,\vec{i} + \alpha \sin t\,\vec{j} + \beta t\,\vec{k}$ is parameterized with constant speed.

44. An ant crawls along the radius from the center to the edge of a circular disk of radius 1 meter, moving at a constant rate of 1 cm/sec. Meanwhile, the disk is turning counterclockwise about its center at 1 revolution per second.

    (a) Parameterize the motion of the ant.
    (b) Find the velocity and speed of the ant.
    (c) Determine the acceleration and magnitude of the acceleration of the ant.

45. The motion of a particle is given by $\vec{r}(t) = R\cos(\omega t)\vec{i} + R\sin(\omega t)\vec{j}$, with $R > 0$, $\omega > 0$.

    (a) Show that the particle moves on a circle and find the radius, direction, and period.
    (b) Determine the velocity vector of the particle and its direction and speed.
    (c) What are the direction and magnitude of the acceleration vector of the particle?

46. You bicycle along a straight flat road with a safety light attached to one foot. Your bike moves at a speed of 25 km/hr and your foot moves in a circle of radius 20 cm centered 30 cm above the ground, making one revolution per second.

    (a) Find parametric equations for x and y which describe the path traced out by the light, where y is distance (in cm) above the ground.
    (b) Sketch the light's path.
    (c) How fast (in revolutions/sec) would your foot have to be rotating if an observer standing at the side of the road sees the light moving backward?

47. How do the motions of objects A and B differ, if A has position vector $\vec{r}_A(t)$ and B has position vector $\vec{r}_B(t) = \vec{r}_A(2t)$ for $t \geq 0$. Illustrate your answer with $\vec{r}_A(t) = t\vec{i} + t^2\vec{j}$.

48. At time $t = 0$ an object is moving with velocity vector $\vec{v} = 2\vec{i} + \vec{j}$ and acceleration vector $\vec{a} = \vec{i} + \vec{j}$. Can it be in uniform circular motion about some point in the plane?

49. Figure 17.17 shows the velocity and acceleration vectors of an object in uniform circular motion about a point in the plane at a particular moment. Is it moving round the circle in the clockwise or counterclockwise direction?

Figure 17.17

50. Let $\vec{v}(t)$ be the velocity of a particle moving in the plane. Let $s(t)$ be the magnitude of $\vec{v}$ and let $\theta(t)$ be the angle of $\vec{v}(t)$ with the positive x-axis at time t, so that $\vec{v} = s\cos\theta\,\vec{i} + s\sin\theta\,\vec{j}$.

Let $\vec{T}$ be the unit vector in the direction of $\vec{v}$, and let $\vec{N}$ be the unit vector in the direction of $\vec{k} \times \vec{v}$, perpendicular to $\vec{v}$. Show that the acceleration $\vec{a}(t)$ is given

by

$$\vec{a} = \frac{ds}{dt}\vec{T} + s\frac{d\theta}{dt}\vec{N}.$$

This shows how to separate the acceleration into the sum of one component, $\dfrac{ds}{dt}\vec{T}$, due to changing speed and a perpendicular component, $s\dfrac{d\theta}{dt}\vec{N}$, due to changing direction of the motion.

Strengthen Your Understanding

In Problems 51–53, explain what is wrong with the statement.

51. When a particle moves around a circle its velocity and acceleration are always orthogonal.

52. A particle with position $\vec{r}(t)$ at time t has acceleration equal to 3 m/sec^2 at time $t = 0$.

53. A parameterized curve $\vec{r}(t)$, $A \leq t \leq B$, has length $B - A$.

In Problems 54–55, give an example of:

54. A function $\vec{r}(t)$ such that the particle with position $\vec{r}(t)$ at time t has velocity $\vec{v} = \vec{i} + 2\vec{j}$ and acceleration $\vec{a} = 4\vec{i} + 6\vec{k}$ at $t = 0$.

55. An interval $a \leq t \leq b$ corresponding to a piece of the helix $\vec{r}(t) = \cos t\,\vec{i} + \sin t\,\vec{j} + t\vec{k}$ of length 10.

Are the statements in Problems 56–63 true or false? Give reasons for your answer.

56. A particle whose motion in the plane is given by $\vec{r}(t) = t^2\vec{i} + (1-t)\vec{j}$ has the same velocity at $t = 1$ and $t = -1$.

57. A particle whose motion in the plane is given by $\vec{r}(t) = t^2\vec{i} + (1-t)\vec{j}$ has the same speed at $t = 1$ and $t = -1$.

58. If a particle is moving along a parameterized curve $\vec{r}(t)$ then the acceleration vector at any point is always perpendicular to the velocity vector at that point.

59. If a particle is moving along a parameterized curve $\vec{r}(t)$ then the acceleration vector at a point cannot be parallel to the velocity vector at that point.

60. If $\vec{r}(t)$ for $a \leq t \leq b$ is a parameterized curve, then $\vec{r}(-t)$ for $a \leq t \leq b$ is the same curve traced backward.

61. If $\vec{r}(t)$ for $a \leq t \leq b$ is a parameterized curve C and the speed $\|\vec{v}(t)\| = 1$, then the length of C is $b - a$.

62. If a particle moves with motion $\vec{r}(t) = 3t\vec{i} + 2t\vec{j} + t\vec{k}$, then the particle stops at the origin.

63. If a particle moves with constant speed, the path of the particle must be a line.

For Problems 64–67, decide if the statement is true or false for all smooth parameterized curves $\vec{r}(t)$ and all values of t for which $\vec{r}\,'(t) \neq \vec{0}$.

64. The vector $\vec{r}\,'(t)$ is tangent to the curve at the point with position vector $\vec{r}(t)$.

65. $\vec{r}\,'(t) \times \vec{r}(t) = \vec{0}$

66. $\vec{r}\,'(t) \cdot \vec{r}(t) = 0$

67. $\vec{r}\,''(t) = -\omega^2\vec{r}(t)$

17.3   VECTOR FIELDS

Introduction to Vector Fields

A *vector field* is a function that assigns a vector to each point in the plane or in 3-space. One example of a vector field is the gradient of a function $f(x, y)$; at each point (x, y) the vector grad $f(x, y)$ points in the direction of maximum rate of increase of f. In this section we look at other vector fields representing velocities and forces.

Velocity Vector Fields

Figure 17.18 shows the flow of a part of the Gulf Stream, a current in the Atlantic Ocean.[2] It is an example of a *velocity vector field*: each vector shows the velocity of the current at that point. The current is fastest where the velocity vectors are longest in the middle of the stream. Beside the stream are eddies where the water flows round and round in circles.

---

[2]Based on data supplied by Avijit Gangopadhyay of the Jet Propulsion Laboratory.

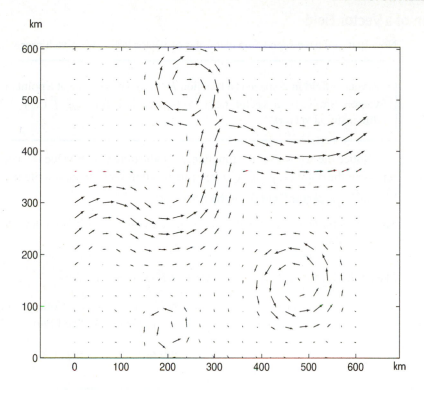

Figure 17.18: The velocity vector field of the Gulf Stream

Force Fields

Another physical quantity represented by a vector is force. When we experience a force, sometimes it results from direct contact with the object that supplies the force (for example, a push). Many forces, however, can be felt at all points in space. For example, the earth exerts a gravitational pull on all other masses. Such forces can be represented by vector fields.

Figure 17.19 shows the gravitational force exerted by the earth on a mass of one kilogram at different points in space. This is a sketch of the vector field in 3-space. You can see that the vectors all point toward the earth (which is not shown in the diagram) and that the vectors farther from the earth are smaller in magnitude.

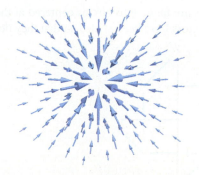

Figure 17.19: The gravitational field of the earth

Definition of a Vector Field

Now that you have seen some examples of vector fields, we give a more formal definition.

> A **vector field** in 2-space is a function $\vec{F}(x, y)$ whose value at a point (x, y) is a 2-dimensional vector. Similarly, a vector field in 3-space is a function $\vec{F}(x, y, z)$ whose values are 3-dimensional vectors.

Notice the arrow over the function, $\vec{F}$, indicating that its value is a vector, not a scalar. We often represent the point (x, y) or (x, y, z) by its position vector $\vec{r}$ and write the vector field as $\vec{F}(\vec{r})$.

Visualizing a Vector Field Given by a Formula

Since a vector field is a function that assigns a vector to each point, a vector field can often be given by a formula.

Example 1   Sketch the vector field in 2-space given by $\vec{F}(x, y) = -y\vec{i} + x\vec{j}$.

Solution   Table 17.1 shows the value of the vector field at a few points. Notice that each value is a vector. To plot the vector field, we plot $\vec{F}(x, y)$ with its tail at (x, y). (See Figure 17.20.)

Table 17.1   *Values of $\vec{F}(x, y) = -y\vec{i} + x\vec{j}$*

		y		
		-1	0	1
	-1	$\vec{i} - \vec{j}$	$-\vec{j}$	$-\vec{i} - \vec{j}$
x	0	$\vec{i}$	$\vec{0}$	$-\vec{i}$
	1	$\vec{i} + \vec{j}$	$\vec{j}$	$-\vec{i} + \vec{j}$

Now we look at the formula. The magnitude of the vector at (x, y) is the distance from (x, y) to the origin, since

$$\|\vec{F}(x, y)\| = \| - y\vec{i} + x\vec{j}\| = \sqrt{x^2 + y^2}.$$

Therefore, all the vectors at a fixed distance from the origin (that is, on a circle centered at the origin) have the same magnitude. The magnitude gets larger as we move farther from the origin.

What about the direction? Figure 17.20 suggests that at each point (x, y) the vector $\vec{F}(x, y)$ is perpendicular to the position vector $\vec{r} = x\vec{i} + y\vec{j}$. We confirm this using the dot product:

$$\vec{r} \cdot \vec{F}(x, y) = (x\vec{i} + y\vec{j}) \cdot (-y\vec{i} + x\vec{j}) = 0.$$

Thus, the vectors are tangent to circles centered at the origin and get longer as we go out. In Figure 17.21, the vectors have been scaled so that they do not obscure each other.

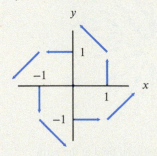

Figure 17.20: The value $\vec{F}(x, y)$ is placed at the point (x, y)

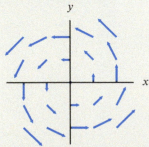

Figure 17.21: The vector field $\vec{F}(x, y) = -y\vec{i} + x\vec{j}$, vectors scaled smaller to fit in diagram

Example 2   Sketch the vector fields in 2-space given by   (a) $\vec{F}(x, y) = x\vec{j}$ (b) $\vec{G}(x, y) = x\vec{i}$.

Solution   (a) The vector $x\vec{j}$ is parallel to the y-direction, pointing up when x is positive and down when x is negative. Also, the larger $|x|$ is, the longer the vector. The vectors in the field are constant along vertical lines since the vector field does not depend on y. (See Figure 17.22.)

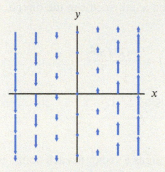

Figure 17.22: The vector field $\vec{F}(x, y) = x\vec{j}$ **Figure 17.23**: The vector field $\vec{F}(x, y) = x\vec{i}$

(b) This is similar to the previous example, except that the vector $x\vec{i}$ is parallel to the x-direction, pointing to the right when x is positive and to the left when x is negative. Again, the larger $|x|$ is the longer the vector, and the vectors are constant along vertical lines, since the vector field does not depend on y. (See Figure 17.23.)

Example 3   Describe the vector field in 3-space given by $\vec{F}(\vec{r}) = \vec{r}$, where $\vec{r} = x\vec{i} + y\vec{j} + z\vec{k}$.

Solution   The notation $\vec{F}(\vec{r}) = \vec{r}$ means that the value of $\vec{F}$ at the point (x, y, z) with position vector $\vec{r}$ is the vector $\vec{r}$ with its tail at (x, y, z). Thus, the vector field points outward. See Figure 17.24. Note that the lengths of the vectors have been scaled down so as to fit into the diagram.

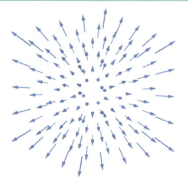

Figure 17.25: Force exerted on mass m by mass M

Figure 17.24: The vector field $\vec{F}(\vec{r}) = \vec{r}$

Finding a Formula for a Vector Field

Example 4   Newton's Law of Gravitation states that the magnitude of the gravitational force exerted by an object of mass M on an object of mass m is proportional to M and m and inversely proportional to the square of the distance between them. The direction of the force is from m to M along the line connecting them. (See Figure 17.25.) Find a formula for the vector field $\vec{F}(\vec{r})$ that represents the gravitational force, assuming M is located at the origin and m is located at the point with position vector $\vec{r}$.

Solution   Since the mass m is located at $\vec{r}$, Newton's law says that the magnitude of the force is given by

$$\|\vec{F}(\vec{r})\| = \frac{GMm}{\|\vec{r}\|^2},$$

where G is the universal gravitational constant. A unit vector in the direction of the force is $-\vec{r}/\|\vec{r}\|$, where the negative sign indicates that the direction of force is toward the origin (gravity is attractive). By taking the product of the magnitude of the force and a unit vector in the direction of the force, we obtain an expression for the force vector field:

$$\vec{F}(\vec{r}) = \frac{GMm}{\|\vec{r}\|^2}\left(-\frac{\vec{r}}{\|\vec{r}\|}\right) = \frac{-GMm\vec{r}}{\|\vec{r}\|^3}.$$

We have already seen a picture of this vector field in Figure 17.19.

Gradient Vector Fields

The gradient of a scalar function f is a function that assigns a vector to each point, and is therefore a vector field. It is called the *gradient field* of f. Many vector fields in physics are gradient fields.

Example 5   Sketch the gradient field of the functions in Figures 17.26–17.28.

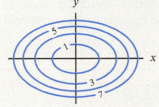

Figure 17.26: The contour map of $f(x, y) = x^2 + 2y^2$

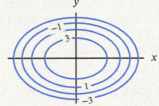

Figure 17.27: The contour map of $g(x, y) = 5 - x^2 - 2y^2$

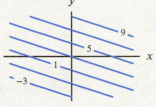

Figure 17.28: The contour map of $h(x, y) = x + 2y + 3$

Solution   See Figures 17.29–17.31. For a function $f(x, y)$, the gradient vector of f at a point is perpendicular to the contours in the direction of increasing f and its magnitude is the rate of change in that direction. The rate of change is large when the contours are close together and small when they are far apart. Notice that in Figure 17.29 the vectors all point outward, away from the local minimum of f, and in Figure 17.30 the vectors of grad g all point inward, toward the local maximum of g. Since h is a linear function, its gradient is constant, so grad h in Figure 17.31 is a constant vector field.

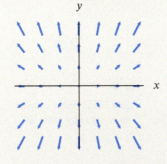

Figure 17.29: grad f

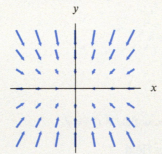

Figure 17.30: grad g

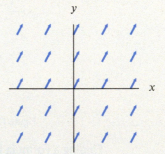

Figure 17.31: grad h

Summary for Section 17.3

- A **vector field in 2-space** is a function $\vec{F}(x, y)$ whose value at (x, y) is a 2-dimensional vector.
- A **vector field in 3-space** is a function $\vec{F}(x, y, z)$ whose value at (x, y, z) is a 3-dimensional vector.
- The gradient of a scalar function f is called the **gradient field** of f.

Exercises and Problems for Section 17.3

EXERCISES

■ For Exercises 1–6, find formulas for the vector fields. (There are many possible answers.)

1.

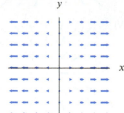

y, *x*

2.

y, *x*

3.

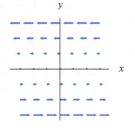

y, *x*

4.

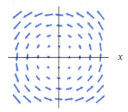

y, *x*

5.

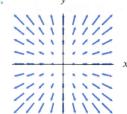

y, *x*

6.

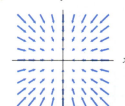

y, *x*

13. $\vec{F}(x, y) = -y\vec{j}$

14. $\vec{F}(\vec{r}) = 2\vec{r}$

15. $\vec{F}(\vec{r}) = \vec{r}/\|\vec{r}\|$

16. $\vec{F}(\vec{r}) = -\vec{r}/\|\vec{r}\|^3$

17. $\vec{F} = y\vec{i} - x\vec{j}$

18. $\vec{F}(x, y) = 2x\vec{i} + x\vec{j}$

19. $\vec{F}(x, y) = (x + y)\vec{i} + (x - y)\vec{j}$

20. Match vector fields $\vec{A} - \vec{D}$ in the tables with vector fields (I)–(IV) in Figure 17.32.

Vector field $\vec{A}$

$y \backslash x$	-1	1
-1	$\vec{i} + \vec{j}$	$\vec{i} + \vec{j}$
1	$\vec{i} + \vec{j}$	$\vec{i} + \vec{j}$

Vector field $\vec{B}$

$y \backslash x$	-1	1
-1	$-\vec{i} - \vec{j}$	$-\vec{i} - \vec{j}$
1	$\vec{i} + \vec{j}$	$\vec{i} + \vec{j}$

Vector field $\vec{C}$

$y \backslash x$	-1	1
-1	$-2\vec{i} + \vec{j}$	$2\vec{i} + \vec{j}$
1	$-2\vec{i} + \vec{j}$	$2\vec{i} + \vec{j}$

Vector field $\vec{D}$

$y \backslash x$	-1	1
-1	$\vec{i} + \vec{j}$	$-\vec{i} - \vec{j}$
1	$-\vec{i} + \vec{j}$	$\vec{i} - \vec{j}$

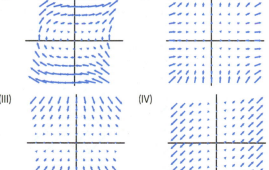

(I)   (II)   (III)   (IV)

Figure 17.32

■ In Exercises 7–10, assume $x, y > 0$ and decide if

(a) The vector field is parallel to the x-axis, parallel to the y-axis, or neither.

(b) As x increases, the length increases, decreases, or neither.

(c) As y increases, the length increases, decreases, or neither.

Assume $x, y > 0$.

7. $\vec{F} = x\vec{j}$

8. $\vec{F} = y\vec{i} + \vec{j}$

9. $\vec{F} = (x + e^{1-y})\vec{i}$

10. $\text{grad}(x^4 + e^{3y})$

■ Sketch the vector fields in Exercises 11–19 in the xy-plane.

11. $\vec{F}(x, y) = 2\vec{i} + 3\vec{j}$

12. $\vec{F}(x, y) = y\vec{i}$

21. For each description of a vector field in (a)-(d), choose one or more of the vector fields I-IX.

(a) Pointing radially outward, increasing in length away from the origin.

(b) Pointing in a circular direction around the origin, remaining the same length.

(c) Pointing towards the origin, increasing in length farther from the origin.

(d) Pointing clockwise around the origin.

I. $\dfrac{x\vec{i} + y\vec{j}}{\sqrt{x^2 + y^2}}$ II. $\dfrac{-y\vec{i} + x\vec{j}}{\sqrt{x^2 + y^2}}$ III. $\vec{r}$

IV. $-\vec{r}$ V. $-y\vec{i} + x\vec{j}$ VI. $y\vec{i} - x\vec{j}$

VII. $y\vec{i} + x\vec{j}$ VIII. $\dfrac{\vec{r}}{\|\vec{r}\|^3}$ IX. $-\dfrac{\vec{r}}{\|\vec{r}\|^3}$

22. Each vector field in Figures (I)–(IV) represents the force on a particle at different points in space as a result of another particle at the origin. Match up the vector fields with the descriptions below.

  (a) A repulsive force whose magnitude decreases as distance increases, such as between electric charges of the same sign.
  (b) A repulsive force whose magnitude increases as distance increases.
  (c) An attractive force whose magnitude decreases as distance increases, such as gravity.
  (d) An attractive force whose magnitude increases as distance increases.

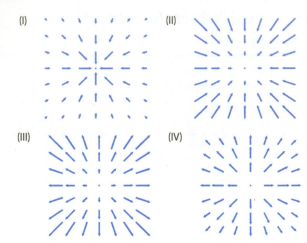

PROBLEMS

■ In Problems **23–27**, give an example of a vector field $\vec{F}(x, y)$ in 2-space with the stated properties.

23. $\vec{F}$ is constant

24. $\vec{F}$ has a constant direction but $\|\vec{F}\|$ is not constant

25. $\|\vec{F}\|$ is constant but $\vec{F}$ is not constant

26. Neither $\|\vec{F}\|$ nor the direction of $\vec{F}$ is constant

27. $\vec{F}$ is perpendicular to $\vec{G} = (x+y)\vec{i} + (1+y^2)\vec{j}$ at every point

28. Match the level curves in (I)–(IV) with the gradient fields in (A)–(D). All figures use the same square window.

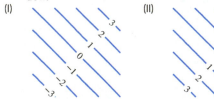

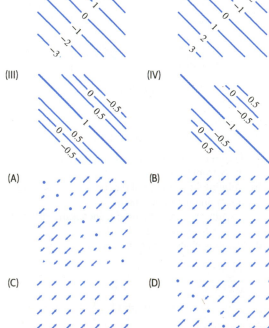

■ Problems **29–30** concern the vector fields $\vec{F} = x\vec{i} + y\vec{j}$, $\vec{G} = -y\vec{i} + x\vec{j}$, and $\vec{H} = x\vec{i} - y\vec{j}$.

29. Match $\vec{F}, \vec{G}, \vec{H}$ with their sketches in (I)–(III).

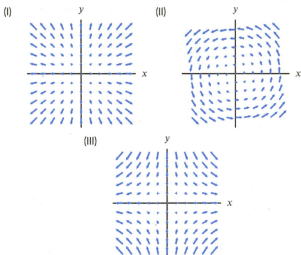

30. Match the vector fields with their sketches, (I)–(IV).

  (a) $\vec{F} + \vec{G}$ **(b)** $\vec{F} + \vec{H}$ **(c)** $\vec{G} + \vec{H}$ **(d)** $-\vec{F} + \vec{G}$

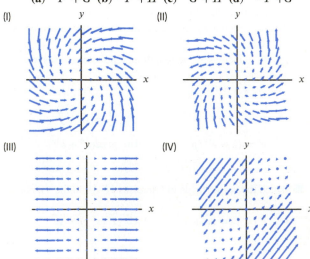

31. Match vector fields (a)–(f) with their graphs (I)–(VI).

(a) $-y\vec{i} + x\vec{j}$ (b) $x\vec{i}$

(c) $y\vec{j}$ (d) $z\vec{k}$

(e) $2\vec{k}$ (f) $\vec{r}$

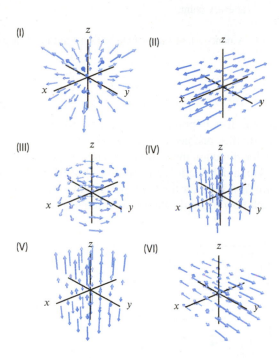

In Problems 32–34, write formulas for vector fields with the given properties.

32. All vectors are parallel to the x-axis; all vectors on a vertical line have the same magnitude.

33. All vectors point toward the origin and have constant length.

34. All vectors are of unit length and perpendicular to the position vector at that point.

35. (a) Let $\vec{F} = x\vec{i} + (x + y)\vec{j} + (x - y + z)\vec{k}$. Find a point at which $\vec{F}$ is parallel to l, the line $x = 5 + t$, $y = 6 - 2t$, $z = 7 - 3t$.

(b) Find a point at which $\vec{F}$ and l are perpendicular.

(c) Give an equation for and describe in words the set of all points at which $\vec{F}$ and l are perpendicular.

In Problems 36–37, let $\vec{F} = x\vec{i} + y\vec{j}$ and $\vec{G} = -y\vec{i} + x\vec{j}$.

36. Sketch the vector field $\vec{L} = a\vec{F} + \vec{G}$ if:

(a) $a = 0$ (b) $a > 0$ (c) $a < 0$

37. Sketch the vector field $\vec{L} = \vec{F} + b\vec{G}$ if:

(a) $b = 0$ (b) $b > 0$ (c) $b < 0$

38. In the middle of a wide, steadily flowing river there is a fountain that spouts water horizontally in all directions. The river flows in the $\vec{i}$-direction in the xy-plane and the fountain is at the origin.

(a) If $A > 0$, $K > 0$, explain why the following expression could represent the velocity field for the combined flow of the river and the fountain:

$$\vec{v} = A\vec{i} + K(x^2 + y^2)^{-1}(x\vec{i} + y\vec{j}).$$

(b) What is the significance of the constants A and K?

(c) Using a computer, sketch the vector field $\vec{v}$ for $K = 1$ and $A = 1$ and $A = 2$, and for $A = 0.2$, $K = 2$.

39. Figures 17.33 and 17.34 show the gradient of the functions $z = f(x, y)$ and $z = g(x, y)$.

(a) For each function, draw a rough sketch of the level curves, showing possible z-values.

(b) The xz-plane cuts each of the surfaces $z = f(x, y)$ and $z = g(x, y)$ in a curve. Sketch each of these curves, making clear how they are similar and how they are different from one another.

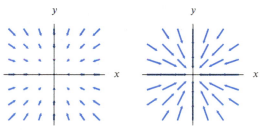

Figure 17.33: Gradient of $z = f(x, y)$ **Figure 17.34**: Gradient of $z = g(x, y)$

40. Let $\vec{F} = u\vec{i} + v\vec{j}$ be a vector field in 2-space with magnitude $F = \|\vec{F}\|$.

(a) Let $\vec{T} = (1/F)\vec{F}$. Show that $\vec{T}$ is the unit vector in the direction of $\vec{F}$. See Figure 17.35.

(b) Let $\vec{N} = (1/F)(\vec{k} \times \vec{F}) = (1/F)(-v\vec{i} + u\vec{j})$. Show that $\vec{N}$ is the unit vector pointing to the left of and at right angles to $\vec{F}$. See Figure 17.35.

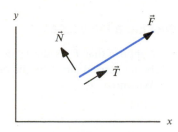

Figure 17.35

Strengthen Your Understanding

■ In Problems **41–42**, explain what is wrong with the statement.

41. A plot of the vector field $\vec{G}(x, y, z) = \vec{F}(2x, 2y, 2z)$ can be obtained from a plot of the vector field $\vec{F}(x, y, z)$ by doubling the lengths of all the arrows.

42. A vector field $\vec{F}$ is defined by the formula $\vec{F}(x, y, z) = x^2 - yz$.

■ In Problems **43–44**, give an example of:

43. A nonconstant vector field that is parallel to $\vec{i} + \vec{j} + \vec{k}$ at every point.

44. A nonconstant vector field with magnitude 1 at every point.

17.4   THE FLOW OF A VECTOR FIELD

When an iceberg is spotted in the North Atlantic, it is important to be able to predict where the iceberg is likely to be a day or a week later. To do this, one needs to know the velocity vector field of the ocean currents, that is, how fast and in what direction the water is moving at each point.

In this section we use differential equations to find the path of an object in a fluid flow. This path is called a flow line. Figure 17.36 shows several flow lines for the Gulf Stream velocity vector field in Figure 17.18 on page 959. The arrows on each flow line indicate the direction of flow.

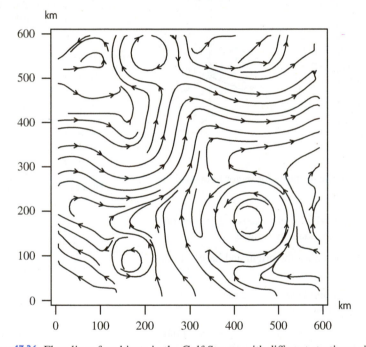

Figure 17.36: Flow lines for objects in the Gulf Stream with different starting points

How Do We Find a Flow Line?

Suppose that $\vec{F}$ is the velocity vector field of water on the surface of a creek and imagine a seed being carried along by the current. We want to know the position vector $\vec{r}(t)$ of the seed at time t. We know

$$\begin{array}{c} \text{Velocity of seed} \\ \text{at time } t \end{array} = \begin{array}{c} \text{Velocity of current at seed's position} \\ \text{at time } t; \end{array}$$

that is,

$$\vec{r}'(t) = \vec{F}(\vec{r}(t)).$$

We make the following definition:

A **flow line** of a vector field $\vec{v} = \vec{F}(\vec{r})$ is a path $\vec{r}(t)$ whose velocity vector equals $\vec{v}$. Thus,

$$\vec{r}'(t) = \vec{v} = \vec{F}(\vec{r}(t)).$$

The **flow** of a vector field is the family of all of its flow lines.

A flow line is also called an *integral curve* or a *streamline*. We define flow lines for any vector field, as it turns out to be useful to study the flow of fields (for example, electric and magnetic) that are not velocity fields.

After resolving $\vec{F}$ and $\vec{r}$ into components, $\vec{F} = F_1\vec{i} + F_2\vec{j}$ and $\vec{r}(t) = x(t)\vec{i} + y(t)\vec{j}$, the definition of a flow line tells us that $x(t)$ and $y(t)$ satisfy the system of differential equations

$$x'(t) = F_1(x(t), y(t)) \quad \text{and} \quad y'(t) = F_2(x(t), y(t)).$$

Solving these differential equations gives a parameterization of the flow line.

Example 1   Find the flow line of the constant velocity field $\vec{v} = 3\vec{i} + 4\vec{j}$ cm/sec that passes through the point $(1, 2)$ at time $t = 0$.

Solution   Let $\vec{r}(t) = x(t)\vec{i} + y(t)\vec{j}$ be the position in cm of a particle at time t, where t is in seconds. We have

$$x'(t) = 3 \quad \text{and} \quad y'(t) = 4.$$

Thus,

$$x(t) = 3t + x_0 \quad \text{and} \quad y(t) = 4t + y_0.$$

Since the path passes the point $(1, 2)$ at $t = 0$, we have $x_0 = 1$ and $y_0 = 2$ and so

$$x(t) = 3t + 1 \quad \text{and} \quad y(t) = 4t + 2.$$

Thus, the path is the line given parametrically by

$$\vec{r}(t) = (3t + 1)\vec{i} + (4t + 2)\vec{j}.$$

(See Figure 17.37.) To find an explicit equation for the path, eliminate t between these expressions to get

$$\frac{x - 1}{3} = \frac{y - 2}{4} \quad \text{or} \quad y = \frac{4}{3}x + \frac{2}{3}.$$

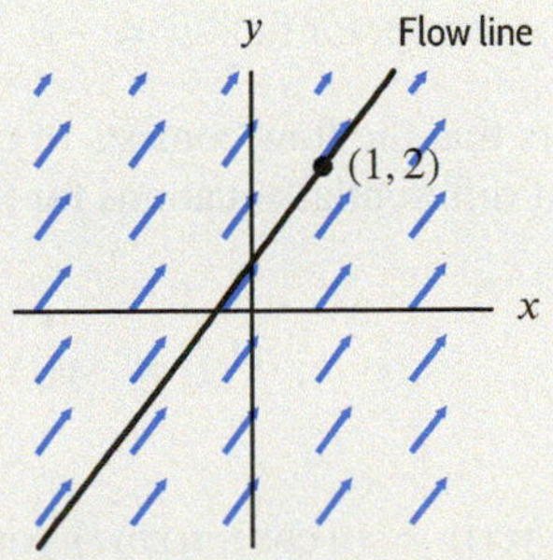

Figure 17.37: Vector field $\vec{F} = 3\vec{i} + 4\vec{j}$ with the flow line through $(1, 2)$

Example 2   The velocity of a flow at the point (x, y) is $\vec{F}(x, y) = \vec{i} + x\vec{j}$. Find the path of motion of an object in the flow that is at the point $(-2, 2)$ at time $t = 0$.

Solution   Figure 17.38 shows this field. Since $\vec{r}'(t) = \vec{F}(\vec{r}(t))$, we are looking for the flow line that satisfies the system of differential equations

$$x'(t) = 1, \quad y'(t) = x(t) \qquad \text{satisfying } x(0) = -2 \text{ and } y(0) = 2.$$

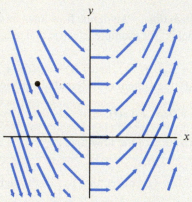

Figure 17.38: The velocity field
$\vec{v} = \vec{i} + x\vec{j}$

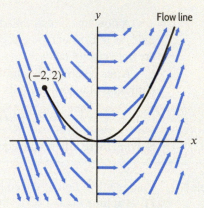

Figure 17.39: A flow line of the
velocity field $\vec{v} = \vec{i} + x\vec{j}$

Solving for $x(t)$ first, we get $x(t) = t + x_0$, where x_0 is a constant of integration. Thus, $y'(t) = t + x_0$, so $y(t) = \frac{1}{2}t^2 + x_0 t + y_0$, where y_0 is also a constant of integration. Since $x(0) = x_0 = -2$ and $y(0) = y_0 = 2$, the path of motion is given by

$$x(t) = t - 2, \quad y(t) = \frac{1}{2}t^2 - 2t + 2,$$

or, equivalently,

$$\vec{r}(t) = (t - 2)\vec{i} + (\tfrac{1}{2}t^2 - 2t + 2)\vec{j}.$$

The graph of this flow line in Figure 17.39 looks like a parabola. We check this by seeing that an explicit equation for the path is $y = \frac{1}{2}x^2$.

Example 3   Determine the flow of the vector field $\vec{v} = -y\vec{i} + x\vec{j}$.

Solution   Figure 17.40 suggests that the flow consists of concentric counterclockwise circles, centered at the origin. The system of differential equations for the flow is

$$x'(t) = -y(t) \qquad y'(t) = x(t).$$

The equations $(x(t), y(t)) = (a \cos t, a \sin t)$ parameterize a family of counterclockwise circles of radius a, centered at the origin. We check that this family satisfies the system of differential equations:

$$x'(t) = -a \sin t = -y(t) \quad \text{and} \quad y'(t) = a \cos t = x(t).$$

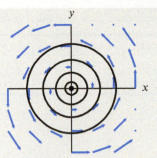

Figure 17.40: The flow of the
vector field $\vec{v} = -y\vec{i} + x\vec{j}$

Approximating Flow Lines Numerically

Often it is not possible to find formulas for the flow lines of a vector field. However, we can approximate them numerically by Euler's method for solving differential equations. Since the flow lines $\vec{r}(t) = x(t)\vec{i} + y(t)\vec{j}$ of a vector field $\vec{v} = \vec{F}(x, y)$ satisfy the differential equation $\vec{r}'(t) = \vec{F}(\vec{r}(t))$, we have

$$\vec{r}(t + \Delta t) \approx \vec{r}(t) + (\Delta t)\vec{r}'(t)$$
$$= \vec{r}(t) + (\Delta t)\vec{F}(\vec{r}(t)) \quad \text{for } \Delta t \text{ near } 0.$$

To approximate a flow line, we start at a point $\vec{r}_0 = \vec{r}(0)$ and estimate the position $\vec{r}_1$ of a particle at time Δt later:

$$\vec{r}_1 = \vec{r}(\Delta t) \approx \vec{r}(0) + (\Delta t)\vec{F}(\vec{r}(0))$$
$$= \vec{r}_0 + (\Delta t)\vec{F}(\vec{r}_0).$$

We then repeat the same procedure starting at $\vec{r}_1$, and so on. The general formula for getting from one point to the next is

$$\vec{r}_{n+1} = \vec{r}_n + (\Delta t)\vec{F}(\vec{r}_n).$$

The points with position vectors $\vec{r}_0, \vec{r}_1, \ldots$ trace out the path, as shown in the next example.

Example 4   Use Euler's method to approximate the flow line through $(1, 2)$ for the vector field $\vec{v} = y^2\vec{i} + 2x^2\vec{j}$.

Solution   The flow is determined by the differential equations $\vec{r}'(t) = \vec{v}$, or equivalently

$$x'(t) = y^2, \qquad y'(t) = 2x^2.$$

We use Euler's method with $\Delta t = 0.02$, giving

$$\vec{r}_{n+1} = \vec{r}_n + 0.02\,\vec{v}(x_n, y_n)$$
$$= x_n\vec{i} + y_n\vec{j} + 0.02(y_n^2\vec{i} + 2x_n^2\vec{j}),$$

or equivalently
$$x_{n+1} = x_n + 0.02 y_n^2, \qquad y_{n+1} = y_n + 0.02 \cdot 2x_n^2.$$

When $t = 0$, we have $(x_0, y_0) = (1, 2)$. Then

$$x_1 = x_0 + 0.02 \cdot y_0^2 = 1 + 0.02 \cdot 2^2 = 1.08,$$
$$y_1 = y_0 + 0.02 \cdot 2x_0^2 = 2 + 0.02 \cdot 2 \cdot 1^2 = 2.04.$$

So after one step $x(0.02) \approx 1.08$ and $y(0.02) \approx 2.04$. Similarly, $x(0.04) = x(2\Delta t) \approx 1.16$, $y(0.04) = y(2\Delta t) \approx 2.08$ and so on. Farther values along the flow line are given in Table 17.2 and plotted in Figure 17.41.

Table 17.2   *Approximated flow line starting at $(1, 2)$ for the vector field $\vec{v} = y^2\vec{i} + 2x^2\vec{j}$*

t	0	0.02	0.04	0.06	0.08	0.1	0.12	0.14	0.16	0.18
x	1	1.08	1.16	1.25	1.34	1.44	1.54	1.65	1.77	1.90
y	2	2.04	2.08	2.14	2.20	2.28	2.36	2.45	2.56	2.69

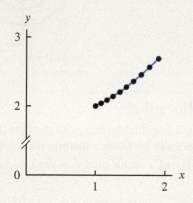

Figure 17.41: Euler's method solution to $x' = y^2$, $y' = 2x^2$

Summary for Section 17.4

- A **flow line** of a vector field $\vec{v} = \vec{F}(\vec{r})$ is a path $\vec{r}(t)$ whose velocity vector equals $\vec{v}$.
- The **flow** of a vector field is the family of all of its flow lines.
- Suppose $\vec{F} = F_1\vec{i} + F_2\vec{j}$ has flow line given parametrically by $\vec{r}(t) = x(t)\vec{i} + y(t)\vec{j}$. Then $x(t)$ and $y(t)$ is a solution to the system of differential equations

$$x'(t) = F_1(x(t), y(t)) \quad \text{and} \quad y'(t) = F_2(x(t), y(t)).$$

- **Approximating flow lines by Euler's method:** The general formula for getting from point $\vec{r}_n$ to the next point $\vec{r}_{n+1}$ on a flow line of a vector field $\vec{F}$ by Euler's method is

$$\vec{r}_{n+1} = \vec{r}_n + (\Delta t)\vec{F}(\vec{r}_n).$$

Exercises and Problems for Section 17.4

EXERCISES

In Exercises 1–3, sketch the vector field and its flow.

1. $\vec{v} = 2\vec{j}$ **2.** $\vec{v} = 3\vec{i}$ **3.** $\vec{v} = 3\vec{i} - 2\vec{j}$

In Exercises 4–9, sketch the vector field and the flow. Then find the system of differential equations associated with the vector field and check that the flow satisfies the system.

4. $\vec{v} = x\vec{i}$; $x(t) = ae^t$, $y(t) = b$

5. $\vec{v} = x\vec{j}$; $x(t) = a$, $y(t) = at + b$

6. $\vec{v} = x\vec{i} + y\vec{j}$; $x(t) = ae^t$, $y(t) = be^t$

7. $\vec{v} = x\vec{i} - y\vec{j}$; $x(t) = ae^t$, $y(t) = be^{-t}$

8. $\vec{v} = y\vec{i} - x\vec{j}$; $x(t) = a\sin t$, $y(t) = a\cos t$

9. $\vec{v} = y\vec{i} + x\vec{j}$; $x(t) = a(e^t + e^{-t})$, $y(t) = a(e^t - e^{-t})$

10. Use a computer or calculator with Euler's method to approximate the flow line through $(1, 2)$ for the vector field $\vec{v} = y^2\vec{i} + 2x^2\vec{j}$ using 5 steps with $\Delta t = 0.1$.

PROBLEMS

For Problems **11–14**, find the region of the Gulf Stream velocity field in Figure 17.18 on page 959 represented by the given table of velocity vectors (in cm/sec).

11.

$35\vec{i} + 131\vec{j}$	$48\vec{i} + 92\vec{j}$	$47\vec{i} + \vec{j}$
$-32\vec{i} + 132\vec{j}$	$-44\vec{i} + 92\vec{j}$	$-42\vec{i} + \vec{j}$
$-51\vec{i} + 73\vec{j}$	$-119\vec{i} + 84\vec{j}$	$-128\vec{i} + 6\vec{j}$

12.

$10\vec{i} - 3\vec{j}$	$11\vec{i} + 16\vec{j}$	$20\vec{i} + 75\vec{j}$
$53\vec{i} - 7\vec{j}$	$58\vec{i} + 23\vec{j}$	$64\vec{i} + 80\vec{j}$
$119\vec{i} - 8\vec{j}$	$121\vec{i} + 31\vec{j}$	$114\vec{i} + 66\vec{j}$

13.

$97\vec{i} - 41\vec{j}$	$72\vec{i} - 24\vec{j}$	$54\vec{i} - 10\vec{j}$
$134\vec{i} - 49\vec{j}$	$131\vec{i} - 44\vec{j}$	$129\vec{i} - 18\vec{j}$
$103\vec{i} - 36\vec{j}$	$122\vec{i} - 30\vec{j}$	$131\vec{i} - 17\vec{j}$

14.

$-95\vec{i} - 60\vec{j}$	$18\vec{i} - 48\vec{j}$	$82\vec{i} - 22\vec{j}$
$-29\vec{i} + 48\vec{j}$	$76\vec{i} + 63\vec{j}$	$128\vec{i} - 16\vec{j}$
$26\vec{i} + 105\vec{j}$	$49\vec{i} + 119\vec{j}$	$88\vec{i} + 13\vec{j}$

15. $\vec{F}(x, y)$ and $\vec{G}(x, y) = 2\vec{F}(x, y)$ are two vector fields. Illustrating your answer with $\vec{F}(x, y) = -y\vec{i} + x\vec{j}$, describe the graphical difference between:

(a) The vector fields   **(b)** Their flows

16. Match the vector fields (a)–(f) with their flow lines (I)–(VI). Put arrows on the flow lines indicating the direction of flow.

(a) $y\vec{i} + x\vec{j}$ **(b)** $-y\vec{i} + x\vec{j}$
(c) $x\vec{i} + y\vec{j}$ **(d)** $-y\vec{i} + (x + y/10)\vec{j}$
(e) $-y\vec{i} + (x - y/10)\vec{j}$ **(f)** $(x - y)\vec{i} + (x - y)\vec{j}$

(I)

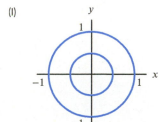

(II)

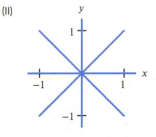

(III)

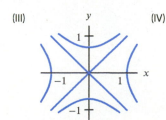

(IV)

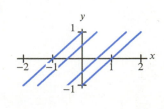

(V)

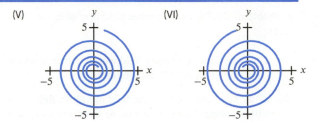

(VI)

17. Show that the acceleration $\vec{a}$ of an object flowing in a velocity field $\vec{F}(x, y) = u(x, y)\vec{i} + v(x, y)\vec{j}$ is given by $\vec{a} = (u_x u + u_y v)\vec{i} + (v_x u + v_y v)\vec{j}$.

18. A velocity vector field $\vec{v} = -H_y\vec{i} + H_x\vec{j}$ is based on the partial derivatives of a smooth function $H(x, y)$. Explain why

(a) $\vec{v}$ is perpendicular to grad H.
(b) the flow lines of $\vec{v}$ are along the level curves of H.

In Problems **19–21**, show that every flow line of the vector field $\vec{v}$ lies on a level curve for the function $f(x, y)$.

19. $\vec{v} = x\vec{i} - y\vec{j}$, $f(x, y) = xy$

20. $\vec{v} = y\vec{i} + x\vec{j}$, $f(x, y) = x^2 - y^2$

21. $\vec{v} = ay\vec{i} + bx\vec{j}$, $f(x, y) = bx^2 - ay^2$

22. A velocity vector field, $\vec{F}(x, y) = (x + 2y)\vec{i} + xy\vec{j}$, in meters per second, has x and y in meters. For an object starting at $(2, 1)$, use Euler's method with $\Delta t = 0.01$ sec to approximate its position 0.01 sec later.

23. A solid metal ball has its center at the origin of a fixed set of axes. The ball rotates once every 24 hours around the z-axis. The direction of rotation is counterclockwise when viewed from above. Let $\vec{v}(x, y, z)$ be the velocity vector of the particle of metal at the point (x, y, z) inside the ball. Time is in hours and x, y, z are in meters.

(a) Find a formula for the vector field $\vec{v}$. Give units for your answer.
(b) Describe in words the flow lines of $\vec{v}$.

24. (a) Show that $h(t) = e^{-2at}(x^2 + y^2)$ is constant along any flow line of $\vec{v} = (ax - y)\vec{i} + (x + ay)\vec{j}$.
(b) Show that points moving with the flow that are on the unit circle centered at the origin at time 0 are on the circle of radius e^{at} centered at the origin at time t.

Strengthen Your Understanding

In Problems 25–26, explain what is wrong with the statement.

25. The flow lines of a vector field whose components are linear functions are all straight lines.

26. If the flow lines of a vector field are all straight lines with the same slope pointing in the same direction, then the vector field is constant.

In Problems 27–28, give an example of:

27. A vector field $\vec{F}(x, y, z)$ such that the path $\vec{r}(t) = t\vec{i} + t^2\vec{j} + t^3\vec{k}$ is a flow line.

28. A vector field whose flow lines are rays from the origin.

Are the statements in Problems 29–38 true or false? Give reasons for your answer.

29. The flow lines for $\vec{F}(x, y) = x\vec{j}$ are parallel to the y-axis.

30. The flow lines of $\vec{F}(x, y) = y\vec{i} - x\vec{j}$ are hyperbolas.

31. The flow lines of $\vec{F}(x, y) = x\vec{i}$ are parabolas.

32. The vector field in Figure 17.42 has a flow line which lies in the first and third quadrants.

Figure 17.42

33. The vector field in Figure 17.42 has a flow line on which both x and y tend to infinity.

34. If $\vec{F}$ is a gradient vector field, $\vec{F}(x, y) = \nabla f(x, y)$, then the flow lines for $\vec{F}$ are the contours for f.

35. If the flow lines for the vector field $\vec{F}(\vec{r})$ are all concentric circles centered at the origin, then $\vec{F}(\vec{r}) \cdot \vec{r} = 0$ for all $\vec{r}$.

36. If the flow lines for the vector field $\vec{F}(x, y)$ are all straight lines parallel to the constant vector $\vec{v} = 3\vec{i} + 5\vec{j}$, then $\vec{F}(x, y) = \vec{v}$.

37. No flow line for the vector field $\vec{F}(x, y) = x\vec{i} + 2\vec{j}$ has a point where the y-coordinate reaches a relative maximum.

38. The vector field $\vec{F}(x, y) = e^x\vec{i} + y\vec{j}$ has a flow line that crosses the x-axis.

Online Resource: Review Problems and Projects

Chapter 18

LINE INTEGRALS

CONTENTS

18.1   THE IDEA OF A LINE INTEGRAL

Imagine that you are rowing on a river with a noticeable current. At times you may be working against the current and at other times you may be moving with it. At the end you have a sense of whether, overall, you were helped or hindered by the current. The line integral, defined in this section, measures the extent to which a curve in a vector field is, overall, going with the vector field or against it.

Orientation of a Curve

A curve can be traced out in two directions, as shown in Figure 18.1. We need to choose one direction before we can define a line integral.

> A curve is said to be **oriented** if we have chosen a direction of travel on it.

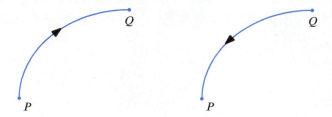

Figure 18.1: A curve with two different orientations represented by arrowheads

Definition of the Line Integral

Consider a vector field $\vec{F}$ and an oriented curve C. We begin by dividing C into n small, almost straight pieces along which $\vec{F}$ is approximately constant. Each piece can be represented by a displacement vector $\Delta\vec{r}_i = \vec{r}_{i+1} - \vec{r}_i$ and the value of $\vec{F}$ at each point of this small piece of C is approximately $\vec{F}(\vec{r}_i)$. See Figures 18.2 and 18.3.

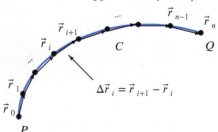

Figure 18.2: The curve C, oriented from P to Q, approximated by straight line segments represented by displacement vectors
$$\Delta\vec{r}_i = \vec{r}_{i+1} - \vec{r}_i$$

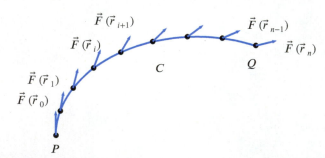

Figure 18.3: The vector field $\vec{F}$ evaluated at the points with position vector $\vec{r}_i$ on the curve C oriented from P to Q

Returning to our initial example, the vector field $\vec{F}$ represents the current and the oriented curve C is the path of the person rowing the boat. We wish to determine to what extent the vector field $\vec{F}$ helps or hinders motion along C. Since the dot product can be used to measure to what extent two vectors point in the same or opposing directions, we form the dot product $\vec{F}(\vec{r}_i) \cdot \Delta\vec{r}_i$ for each point with position vector $\vec{r}_i$ on C. Summing over all such pieces, we get a Riemann sum:

$$\sum_{i=0}^{n-1} \vec{F}(\vec{r}_i) \cdot \Delta\vec{r}_i.$$

We define the line integral, written $\int_C \vec{F} \cdot d\vec{r}$, by taking the limit as $\|\Delta\vec{r}_i\| \to 0$. Provided the limit exists, we make the following definition:

> The **line integral** of a vector field $\vec{F}$ along an oriented curve C is
>
> $$\int_C \vec{F} \cdot d\vec{r} = \lim_{\|\Delta\vec{r}_i\| \to 0} \sum_{i=0}^{n-1} \vec{F}(\vec{r}_i) \cdot \Delta\vec{r}_i.$$

How Does the Limit Defining a Line Integral Work?

The limit in the definition of a line integral exists if $\vec{F}$ is continuous on the curve C and if C is made by joining end to end a finite number of smooth curves. (A vector field $\vec{F} = F_1\vec{i} + F_2\vec{j} + F_3\vec{k}$ is *continuous* if F_1, F_2, and F_3 are continuous, and a *smooth curve* is one that can be parameterized by smooth functions.) We subdivide the curve using a parameterization that goes from one end of the curve to the other, in the forward direction, without retracing any portion of the curve. A subdivision of the parameter interval gives a subdivision of the curve. All the curves we consider in this book are *piecewise smooth* in this sense. Section 18.2 shows how to use a parameterization to compute a line integral.

Example 1   Find the line integral of the constant vector field $\vec{F} = \vec{i} + 2\vec{j}$ along the path from $(1, 1)$ to $(10, 10)$ shown in Figure 18.4.

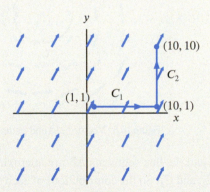

Figure 18.4: The constant vector field $\vec{F} = \vec{i} + 2\vec{j}$ and the path from $(1, 1)$ to $(10, 10)$

Solution   Let C_1 be the horizontal segment of the path going from $(1, 1)$ to $(10, 1)$. When we break this path into pieces, each piece $\Delta\vec{r}$ is horizontal, so $\Delta\vec{r} = \Delta x\vec{i}$ and $\vec{F} \cdot \Delta\vec{r} = (\vec{i} + 2\vec{j}) \cdot \Delta x\vec{i} = \Delta x$. Hence,

$$\int_{C_1} \vec{F} \cdot d\vec{r} = \int_{x=1}^{x=10} dx = 9.$$

Similarly, along the vertical segment C_2, we have $\Delta\vec{r} = \Delta y\vec{j}$ and $\vec{F} \cdot \Delta\vec{r} = (\vec{i} + 2\vec{j}) \cdot \Delta y\vec{j} = 2\Delta y$, so

$$\int_{C_2} \vec{F} \cdot d\vec{r} = \int_{y=1}^{y=10} 2\, dy = 18.$$

Thus,

$$\int_C \vec{F} \cdot d\vec{r} = \int_{C_1} \vec{F} \cdot d\vec{r} + \int_{C_2} \vec{F} \cdot d\vec{r} = 9 + 18 = 27.$$

What Does the Line Integral Tell Us?

Remember that for any two vectors $\vec{u}$ and $\vec{v}$, the dot product $\vec{u} \cdot \vec{v}$ is positive if $\vec{u}$ and $\vec{v}$ point roughly in the same direction (that is, if the angle between them is less than $\pi/2$). The dot product is zero if $\vec{u}$ is perpendicular to $\vec{v}$ and is negative if they point roughly in opposite directions (that is, if the angle between them is greater than $\pi/2$).

The line integral of $\vec{F}$ adds the dot products of $\vec{F}$ and $\Delta \vec{r}$ along the path. If $||\vec{F}||$ is constant, the line integral gives a positive number if $\vec{F}$ is mostly pointing in the same direction as C, and a negative number if $\vec{F}$ is mostly pointing in the opposite direction to C. The line integral is zero if $\vec{F}$ is perpendicular to the path at all points or if the positive and negative contributions cancel out. In general, the line integral of a vector field $\vec{F}$ along a curve C measures the extent to which C is going with $\vec{F}$ or against it.

Example 2   The vector field $\vec{F}$ and the oriented curves C_1, C_2, C_3, C_4 are shown in Figure 18.5. The curves C_1 and C_3 are the same length. Which of the line integrals $\int_{C_i} \vec{F} \cdot d\vec{r}$, for $i = 1, 2, 3, 4$, are positive? Which are negative? Arrange these line integrals in ascending order.

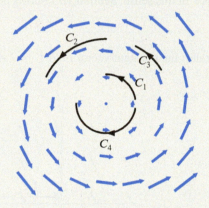

Figure 18.5: Vector field and paths C_1, C_2, C_3, C_4

Solution   The vector field $\vec{F}$ and the line segments $\Delta \vec{r}$ are approximately parallel and in the same direction for the curves C_1, C_2, and C_3. So the contributions of each term $\vec{F} \cdot \Delta \vec{r}$ are positive for these curves. Thus, $\int_{C_1} \vec{F} \cdot d\vec{r}$, $\int_{C_2} \vec{F} \cdot d\vec{r}$, and $\int_{C_3} \vec{F} \cdot d\vec{r}$ are each positive. For the curve C_4, the vector field and the line segments are in opposite directions, so each term $\vec{F} \cdot \Delta \vec{r}$ is negative, and therefore the integral $\int_{C_4} \vec{F} \cdot d\vec{r}$ is negative.

Since the magnitude of the vector field is smaller along C_1 than along C_3, and these two curves are the same length, we have

$$\int_{C_1} \vec{F} \cdot d\vec{r} < \int_{C_3} \vec{F} \cdot d\vec{r}.$$

In addition, the magnitude of the vector field is the same along C_2 and C_3, but the curve C_2 is longer than the curve C_3. Thus,

$$\int_{C_3} \vec{F} \cdot d\vec{r} < \int_{C_2} \vec{F} \cdot d\vec{r}.$$

Putting these results together with the fact that $\int_{C_4} \vec{F} \cdot d\vec{r}$ is negative, we have

$$\int_{C_4} \vec{F} \cdot d\vec{r} < \int_{C_1} \vec{F} \cdot d\vec{r} < \int_{C_2} \vec{F} \cdot d\vec{r} < \int_{C_3} \vec{F} \cdot d\vec{r}.$$

Interpretations of the Line Integral

Work

Recall from Section 13.3 that if a constant force $\vec{F}$ acts on an object while it moves along a straight line through a displacement $\vec{d}$, the work done by the force on the object is

$$\text{Work done} = \vec{F} \cdot \vec{d}.$$

Now suppose we want to find the work done by gravity on an object moving far above the surface of the earth. Since the force of gravity varies with distance from the earth and the path may not be straight, we can't use the formula $\vec{F} \cdot \vec{d}$. We approximate the path by line segments which are small enough that the force is approximately constant on each one. Suppose the force at a point with position vector $\vec{r}$ is $\vec{F}(\vec{r})$, as in Figures 18.2 and 18.3. Then

$$\begin{array}{c}\text{Work done by force } \vec{F}(\vec{r}_i) \\ \text{over small displacement } \Delta\vec{r}_i\end{array} \quad \approx \quad \vec{F}(\vec{r}_i) \cdot \Delta\vec{r}_i,$$

and so,

$$\begin{array}{c}\text{Total work done by force} \\ \text{along oriented curve } C\end{array} \quad \approx \quad \sum_i \vec{F}(\vec{r}_i) \cdot \Delta\vec{r}_i.$$

Taking the limit as $\|\Delta\vec{r}_i\| \to 0$, we get

$$\begin{array}{c}\text{Work done by force } \vec{F}(\vec{r}) \\ \text{along curve } C\end{array} = \lim_{\|\Delta\vec{r}_i\|\to 0} \sum_i \vec{F}(\vec{r}_i) \cdot \Delta\vec{r}_i = \int_C \vec{F} \cdot d\vec{r}.$$

Example 3   A mass lying on a flat table is attached to a spring whose other end is fastened to the wall. (See Figure 18.6.) The spring is extended 20 cm beyond its rest position and released. If the axes are as shown in Figure 18.6, when the spring is extended by a distance of x, by Hooke's Law, the force exerted by the spring on the mass is given by

$$\vec{F}(x) = -kx\vec{i},$$

where k is a positive constant that depends on the properties of the particular spring.

Suppose the mass moves back to the rest position. How much work is done by the force exerted by the spring?

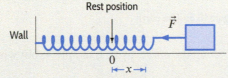

Figure 18.6: Force on mass due to an extended spring

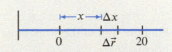

Figure 18.7: Dividing up the interval $0 \le x \le 20$ in order to calculate the work done

Solution   The path from $x = 20$ to $x = 0$ is divided as shown in Figure 18.7, with a typical segment represented by

$$\Delta\vec{r} = \Delta x\vec{i}.$$

Since we are moving from $x = 20$ to $x = 0$, the quantity Δx will be negative. The work done by the force as the mass moves through this segment is approximated by

$$\text{Work done} \approx \vec{F} \cdot \Delta\vec{r} = (-kx\vec{i}\,) \cdot (\Delta x \vec{i}\,) = -kx\,\Delta x.$$

Thus, we have

$$\text{Total work done} \approx \sum -kx\,\Delta x.$$

In the limit, as $\|\Delta x\| \to 0$, this sum becomes an ordinary definite integral. Since the path starts at $x = 20$, this is the lower limit of integration; $x = 0$ is the upper limit. Thus, we get

$$\text{Total work done} = \int_{x=20}^{x=0} -kx\,dx = -\frac{kx^2}{2}\bigg|_{20}^{0} = \frac{k(20)^2}{2} = 200k.$$

Note that the work done is positive, since the force acts in the direction of motion.

Example 3 shows how a line integral over a path parallel to the x-axis reduces to a one-variable integral. Section 18.2 shows how to convert *any* line integral into a one-variable integral.

Example 4   A particle with position vector $\vec{r}$ is subject to a force, $\vec{F}$, due to gravity. What is the *sign* of the work done by $\vec{F}$ as the particle moves along the path C_1, a radial line through the center of the earth, starting 8000 km from the center and ending 10,000 km from the center? (See Figure 18.8.)

Solution   We divide the path into small radial segments, $\Delta\vec{r}$, pointing away from the center of the earth and parallel to the gravitational force. The vectors $\vec{F}$ and $\Delta\vec{r}$ point in opposite directions, so each term $\vec{F} \cdot \Delta\vec{r}$ is negative. Adding all these negative quantities and taking the limit results in a negative value for the total work. Thus, the work done by gravity is negative. The negative sign indicates that we would have to do work *against* gravity to move the particle along the path C_1.

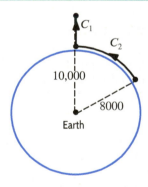

Figure 18.8: The earth

Example 5   Find the sign of the work done by gravity along the curve C_1 in Example 4, but with the opposite orientation.

Solution   Tracing a curve in the opposite direction changes the sign of the line integral because all the segments $\Delta\vec{r}$ change direction, and so every term $\vec{F} \cdot \Delta\vec{r}$ changes sign. Thus, the result will be the negative of the answer found in Example 4. Therefore, the work done by gravity as a particle moves along C_1 toward the center of the earth is positive.

Example 6   Find the work done by gravity as a particle moves along C_2, an arc of a circle 8000 km long at a distance of 8000 km from the center of the earth. (See Figure 18.8.)

Solution   Since C_2 is everywhere perpendicular to the gravitational force, $\vec{F} \cdot \Delta \vec{r} = 0$ for all $\Delta \vec{r}$ along C_2. Thus,

$$\text{Work done} = \int_{C_2} \vec{F} \cdot d\vec{r} = 0,$$

so the work done is zero. This is why satellites can remain in orbit without expending any fuel, once they have attained the correct altitude and velocity.

Circulation

The velocity vector field for the Gulf Stream on page 959 shows distinct eddies or regions where the water circulates. We can measure this circulation using a *closed curve*, that is, one that starts and ends at the same point.

> If C is an oriented closed curve, the line integral of a vector field $\vec{F}$ around C is called the **circulation** of $\vec{F}$ around C.

Circulation is a measure of the net tendency of the vector field to point around the curve C. To emphasize that C is closed, the circulation is sometimes denoted $\oint_C \vec{F} \cdot d\vec{r}$, with a small circle on the integral sign.

Example 7   Describe the rotation of the vector fields in Figures 18.9 and 18.10. Find the sign of the circulation of the vector fields around the indicated paths.

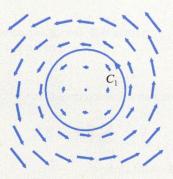

Figure 18.9: A circulating flow

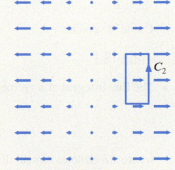

Figure 18.10: A flow with zero circulation

Solution   Consider the vector field in Figure 18.9. If you think of this as representing the velocity of water flowing in a pond, you see that the water is circulating. The line integral around C_1, measuring the circulation around C_1, is positive, because the vectors of the field are all pointing in the direction of the path. By way of contrast, look at the vector field in Figure 18.10. Here the line integral around C_2 is zero because the vertical portions of the path are perpendicular to the field and the contributions from the two horizontal portions cancel out. This means that there is no net tendency for the water to circulate around C_2.

It turns out that the vector field in Figure 18.10 has the property that its circulation around *any* closed path is zero. Water moving according to this vector field has no tendency to circulate around any point, and a leaf dropped into the water will not spin. We'll look at such special fields again later when we introduce the notion of the *curl* of a vector field.

Properties of Line Integrals

Line integrals share some basic properties with ordinary one-variable integrals:

For a scalar constant λ, vector fields $\vec{F}$ and $\vec{G}$, and oriented curves C, C_1, and C_2,

1. $\displaystyle\int_C \lambda\vec{F} \cdot d\vec{r} = \lambda \int_C \vec{F} \cdot d\vec{r}.$ 2. $\displaystyle\int_C (\vec{F} + \vec{G}) \cdot d\vec{r} = \int_C \vec{F} \cdot d\vec{r} + \int_C \vec{G} \cdot d\vec{r}.$

3. $\displaystyle\int_{-C} \vec{F} \cdot d\vec{r} = -\int_C \vec{F} \cdot d\vec{r}.$ 4. $\displaystyle\int_{C_1+C_2} \vec{F} \cdot d\vec{r} = \int_{C_1} \vec{F} \cdot d\vec{r} + \int_{C_2} \vec{F} \cdot d\vec{r}.$

Properties 3 and 4 are concerned with the curve C over which the line integral is taken. If C is an oriented curve, then $-C$ is the same curve traversed in the opposite direction, that is, with the opposite orientation. (See Figure 18.11.) Property 3 holds because if we integrate along $-C$, the vectors $\Delta\vec{r}$ point in the opposite direction and the dot products $\vec{F} \cdot \Delta\vec{r}$ are the negatives of what they were along C.

If C_1 and C_2 are oriented curves with C_1 ending where C_2 begins, we construct a new oriented curve, called $C_1 + C_2$, by joining them together. (See Figure 18.12.) Property 4 is the analogue for line integrals of the property for definite integrals which says that

$$\int_a^b f(x)\,dx = \int_a^c f(x)\,dx + \int_c^b f(x)\,dx.$$

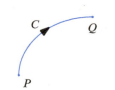 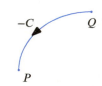

Figure 18.11: A curve, C, and its opposite, $-C$

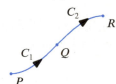 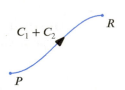

Figure 18.12: Joining two curves, C_1, and C_2, to make a new one, $C_1 + C_2$

Summary for Section 18.1

- The **line integral** of a vector field $\vec{F}$ along an oriented curve C is

$$\int_C \vec{F} \cdot d\vec{r} = \lim_{\|\Delta\vec{r}_i\| \to 0} \sum_{i=0}^{n-1} \vec{F}(\vec{r}_i) \cdot \Delta\vec{r}_i,$$

  where $\Delta\vec{r}$ denotes a small displacement vector along C.

- **Interpreting line integrals:** The line integral of $\vec{F}$ along an oriented curve C is:
  - Positive if $\vec{F}$ is mostly pointing in the same direction as C.
  - Negative if $\vec{F}$ is mostly pointing in the opposite direction to C.
  - Zero if $\vec{F}$ is perpendicular to the C at all points or if the positive and negative contributions cancel out.

- The **work** done by force $\vec{F}$ along curve C is $\int_C \vec{F} \cdot d\vec{r}$.

- If C is an oriented closed curve, then $\int_C \vec{F} \cdot d\vec{r}$ is called the **circulation** of $\vec{F}$ around C.

- **Properties of line integrals:** For a scalar constant λ, vector fields $\vec{F}$ and $\vec{G}$, and oriented curves C, C_1, and C_2,

1. $\displaystyle\int_C \lambda\vec{F} \cdot d\vec{r} = \lambda \int_C \vec{F} \cdot d\vec{r}.$ 2. $\displaystyle\int_C (\vec{F} + \vec{G}) \cdot d\vec{r} = \int_C \vec{F} \cdot d\vec{r} + \int_C \vec{G} \cdot d\vec{r}.$

3. $\displaystyle\int_{-C} \vec{F} \cdot d\vec{r} = -\int_C \vec{F} \cdot d\vec{r}.$ 4. $\displaystyle\int_{C_1+C_2} \vec{F} \cdot d\vec{r} = \int_{C_1} \vec{F} \cdot d\vec{r} + \int_{C_2} \vec{F} \cdot d\vec{r}.$

Exercises and Problems for Section 18.1   Online Resource: Additional Problems for Section 18.1
EXERCISES

In Exercises 1–6, say whether you expect the line integral of the pictured vector field over the given curve to be positive, negative, or zero.

1.

2.
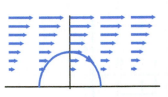

3.
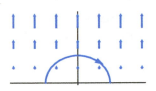

4.
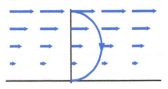

5.
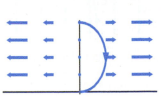

6.
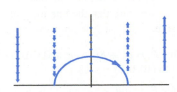

In Exercises 7–15, calculate the line integral of the vector field along the line between the given points.

7. $\vec{F} = x\vec{j}$, from $(1, 0)$ to $(3, 0)$

8. $\vec{F} = x\vec{j}$, from $(2, 0)$ to $(2, 5)$

9. $\vec{F} = 6\vec{i} - 7\vec{j}$, from $(0, 0)$ to $(7, 6)$

10. $\vec{F} = 6\vec{i} + y^2\vec{j}$, from $(3, 0)$ to $(7, 0)$

11. $\vec{F} = 3\vec{i} + 4\vec{j}$, from $(0, 6)$ to $(0, 13)$

12. $\vec{F} = x\vec{i}$, from $(2, 0)$ to $(6, 0)$

13. $\vec{F} = x\vec{i} + y\vec{j}$, from $(2, 0)$ to $(6, 0)$

14. $\vec{F} = \vec{r} = x\vec{i} + y\vec{j}$, from $(2, 2)$ to $(6, 6)$

15. $\vec{F} = x\vec{i} + 6\vec{j} - \vec{k}$, from $(0, -2, 0)$ to $(0, -10, 0)$

In Exercises 16–18, find $\int_C \vec{F} \cdot d\vec{r}$ for the given $\vec{F}$ and C.

16. $\vec{F} = 5\vec{i} + 7\vec{j}$, and C is the x-axis from $(-1, 0)$ to $(-9, 0)$.

17. $\vec{F} = x^2\vec{i} + y^2\vec{j}$, and C is the x-axis from $(2, 0)$ to $(3, 0)$.

18. $\vec{F} = 6x\vec{i} + (x + y^2)\vec{j}$; C is the y-axis from $(0, 3)$ to $(0, 5)$.

In Exercises 19–23, calculate the line integral.

19. $\displaystyle\int_C (2\vec{j} + 3\vec{k}) \cdot d\vec{r}$ where C is the y-axis from the origin to the point $(0, 10, 0)$.

20. $\int_C (2x\vec{i} + 3y\vec{j}) \cdot d\vec{r}$, where C is the line from $(1, 0, 0)$ to $(1, 0, 5)$.

21. $\displaystyle\int_C ((2y + 7)\vec{i} + 3x\vec{j}) \cdot d\vec{r}$, where C is the line from $(1, 0, 0)$ to $(5, 0, 0)$.

22. $\displaystyle\int_C (x\vec{i} + y\vec{j} + z\vec{k}) \cdot d\vec{r}$ where C is the unit circle in the xy-plane, oriented counterclockwise.

23. $\displaystyle\int_C (3z\vec{i} + 4x^2\vec{j} - xy\vec{k}) \cdot d\vec{r}$, where C is the line from $(2, 1, 3)$ to $(2, 1, 8)$.

In Exercises 24–27, find the work done by the force $\vec{F}$ along the curve C.

24. $\vec{F} = 3\vec{i} - x\vec{j}$, C is the line from $(2, 6)$ to $(9, 6)$.

25. $\vec{F} = y^3\vec{i} + 2xy\vec{j}$, C is the line from $(-1, 0)$ to $(-1, 3)$.

26. $\vec{F} = 7\vec{i} - 5\vec{j}$, C is the line from $(2, -2)$ to $(1, 6)$.

27. $\vec{F} = -x\vec{i} - y\vec{j}$, C is the upper half of the unit circle from $(1, 0)$ to $(-1, 0)$.

PROBLEMS

In Problems 28–31, let C_1 be the line from $(0,0)$ to $(0,1)$; let C_2 be the line from $(1,0)$ to $(0,1)$; let C_3 be the semicircle in the upper half plane from $(-1,0)$ to $(1,0)$. Do the line integrals of the vector field along each of the paths C_1, C_2, and C_3 appear to be positive, negative, or zero?

28.

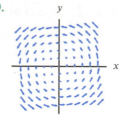

29.

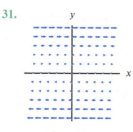

30.

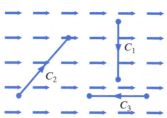

31.

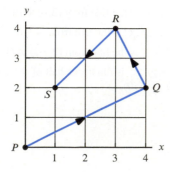

32. Consider the vector field $\vec{F}$ shown in Figure 18.13, together with the paths C_1, C_2, and C_3. Arrange the line integrals $\int_{C_1} \vec{F} \cdot d\vec{r}$, $\int_{C_2} \vec{F} \cdot d\vec{r}$ and $\int_{C_3} \vec{F} \cdot d\vec{r}$ in ascending order.

Figure 18.13

33. Compute $\int_C \vec{F} \cdot d\vec{r}$, where C is the oriented curve in Figure 18.14 and $\vec{F}$ is a vector field constant on each of the three straight segments of C:

$$\vec{F} = \begin{cases} \vec{i} & \text{on } PQ \\ 2\vec{i} - \vec{j} & \text{on } QR \\ 3\vec{i} + \vec{j} & \text{on } RS. \end{cases}$$

Figure 18.14

34. An object moves along the curve C in Figure 18.15 while being acted on by the force field $\vec{F}(x,y) = y\vec{i} + x^2\vec{j}$.

   (a) Evaluate $\vec{F}$ at the points $(0,-1)$, $(1,-1)$, $(2,-1)$, $(3,-1)$, $(4,-1)$, $(4,0)$, $(4,1)$, $(4,2)$, $(4,3)$.
   (b) Make a sketch showing the force field along C.
   (c) Find the work done by $\vec{F}$ on the object.

Figure 18.15

35. Let $\vec{F}$ be the constant force field $\vec{j}$ in Figure 18.16. On which of the paths C_1, C_2, C_3 is zero work done by $\vec{F}$? Explain.

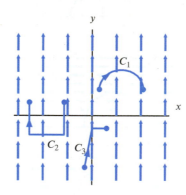

Figure 18.16

In Problems 36–40, give conditions on one or more of the constants a, b, c to ensure that the line integral $\int_C \vec{F} \cdot d\vec{r}$ has the given sign.

36. Positive for $\vec{F} = a\vec{i} + b\vec{j} + c\vec{k}$ and C is the line from the origin to $(10,0,0)$.

37. Positive for $\vec{F} = ay\vec{i} + c\vec{k}$ and C is the unit circle in the xy-plane, centered at the origin and oriented counterclockwise when viewed from above.

38. Negative for $\vec{F} = b\vec{j} + c\vec{k}$ and C is the parabola $y = x^2$ in the xy-plane from the origin to $(3,9,0)$.

39. Positive for $\vec{F} = ay\vec{i} - ax\vec{j} + (c-1)\vec{k}$ and C is the line segment from the origin to $(1,1,1)$.

40. Negative for $\vec{F} = a\vec{i} + b\vec{j} - \vec{k}$ and C is the line segment from $(1,2,3)$ to $(1,2,c)$.

41. (a) For each of the vector fields, $\vec{F}$, shown in Figure 18.17, sketch a curve for which the integral $\int_C \vec{F} \cdot d\vec{r}$ is positive.
   (b) For which of the vector fields is it possible to make your answer to part (a) a closed curve?

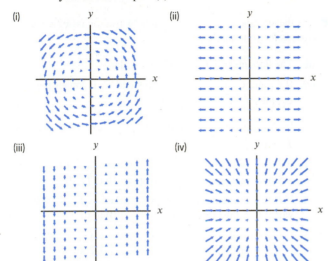

(i)

(ii)

(iii)

(iv)

Figure 18.17

■ For Problems 42–46, say whether you expect the given vector field to have positive, negative, or zero circulation around the closed curve $C = C_1 + C_2 + C_3 + C_4$ in Figure 18.18. The segments C_1 and C_3 are circular arcs centered at the origin; C_2 and C_4 are radial line segments. You may find it helpful to sketch the vector field.

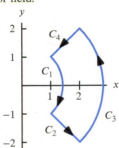

Figure 18.18

42. $\vec{F}(x, y) = x\vec{i} + y\vec{j}$

43. $\vec{F}(x, y) = -y\vec{i} + x\vec{j}$

44. $\vec{F}(x, y) = y\vec{i} - x\vec{j}$

45. $\vec{F}(x, y) = x^2\vec{i}$

46. $\vec{F}(x, y) = -\dfrac{y}{x^2 + y^2}\vec{i} + \dfrac{x}{x^2 + y^2}\vec{j}$

■ In Problems 47–50, C_1 and C_2 are oriented curves, and C_1 ends where C_2 begins. Find the integral given that $\int_{C_1} \vec{F} \cdot d\vec{r} = 8$, $\int_{C_1} \vec{G} \cdot d\vec{r} = 3$, $\int_{C_2} \vec{F} \cdot d\vec{r} = -5$, and $\int_{C_2} \vec{G} \cdot d\vec{r} = 15$.

47. $\displaystyle\int_{C_1} \left(\vec{F} + \vec{G} \right) \cdot d\vec{r}$

48. $\displaystyle\int_{C_2} 3\vec{G} \cdot d\vec{r}$

49. $\displaystyle\int_{C_1 + C_2} 2\vec{F} \cdot d\vec{r}$

50. $\displaystyle\int_{C_1 + C_2} \left(\vec{G} - \vec{F} \right) \cdot d\vec{r}$

51. A force $\vec{F}$ moves an object in a line from $(1, 1)$ to $(2, 4)$ with force $\vec{F} = 2\vec{i} + 3\vec{j}$, and then along a line from $(2, 4)$ to $(3, 3)$ with force $\vec{F} = \vec{i} - \vec{j}$. How much work does the force do on the object in total?

52. Find the work done by a constant force $\vec{F}$ moving an object through straight line displacement $\vec{r}$ if
   (a) $\vec{F}$ is in the same direction as $\vec{r}$, $\|\vec{F}\| = 5$ newtons and $\|\vec{r}\| = 3$ meters.
   (b) $\vec{F}$ and $\vec{r}$ are perpendicular.
   (c) $\vec{F} = 4\vec{i} + \vec{j} + 4\vec{k}$ pounds and $\vec{r} = \vec{i} + \vec{j} + \vec{k}$ foot.

53. A horizontal square has sides of 1000 km running north-south and east-west. A wind blows from the east and decreases in magnitude toward the north at a rate of 6 meter/sec for every 500 km. Compute the circulation of the wind counterclockwise around the square.

54. Let $\vec{F} = x\vec{i} + y\vec{j}$ and let C_1 be the line joining $(1, 0)$ to $(0, 2)$ and let C_2 be the line joining $(0, 2)$ to $(-1, 0)$. Is $\int_{C_1} \vec{F} \cdot d\vec{r} = -\int_{C_2} \vec{F} \cdot d\vec{r}$? Explain.

55. The vector field $\vec{F}$ has $\|\vec{F}\| \leq 7$ everywhere and C is the circle of radius 1 centered at the origin. What is the largest possible value of $\int_C \vec{F} \cdot d\vec{r}$? The smallest possible value? What conditions lead to these values?

56. Along a curve C, a vector field $\vec{F}$ is everywhere tangent to C in the direction of orientation and has constant magnitude $\|\vec{F}\| = m$. Use the definition of the line integral to explain why

$$\int_C \vec{F} \cdot d\vec{r} = m \cdot \text{Length of } C.$$

57. Explain why the following statement is true: Whenever the line integral of a vector field around every closed curve is zero, the line integral along a curve with fixed endpoints has a constant value independent of the path taken between the endpoints.

58. Explain why the converse to the statement in Problem 57 is also true: Whenever the line integral of a vector field depends only on endpoints and not on paths, the circulation around every closed curve is zero.

■ In Problems 59–60, use the fact that the force of gravity on a particle of mass m at the point with position vector $\vec{r}$ is

$$\vec{F} = -\frac{GMm\vec{r}}{\|\vec{r}\|^3},$$

where G is a constant and M is the mass of the earth.

59. Calculate the work done by the force of gravity on a particle of mass m as it moves radially from 8000 km to 10,000 km from the center of the earth.

60. Calculate the work done by the force of gravity on a particle of mass m as it moves radially from 8000 km from the center of the earth to infinitely far away.

Strengthen Your Understanding

■ In Problems 61–62, explain what is wrong with the statement.

61. If $\vec{F}$ is a vector field and C is an oriented curve, then $\int_{-C} \vec{F} \cdot d\vec{r}$ must be less than zero.

62. It is possible that for a certain vector field $\vec{F}$ and oriented path C, we have $\int_C \vec{F} \cdot d\vec{r} = 2\vec{i} - 3\vec{j}$.

■ In Problems 63–64, give an example of:

63. A nonzero vector field $\vec{F}$ such that $\int_C \vec{F} \cdot d\vec{r} = 0$, where C is the straight line curve from $(0,0)$ to $(1,1)$.

64. Two oriented curves C_1 and C_2 in the plane such that, for $\vec{F}(x, y) = x\vec{j}$, we have $\int_{C_1} \vec{F} \cdot d\vec{r} > 0$ and $\int_{C_2} \vec{F} \cdot d\vec{r} < 0$.

■ Are the statements in Problems 65–67 true or false? Explain why or give a counterexample.

65. $\int_C \vec{F} \cdot d\vec{r}$ is a vector.

66. Suppose C_1 is the unit square joining the points $(0,0)$, $(1,0)$, $(1,1)$, $(0,1)$ oriented clockwise and C_2 is the same square but traversed twice in the opposite direction. If $\int_{C_1} \vec{F} \cdot d\vec{r} = 3$, then $\int_{C_2} \vec{F} \cdot d\vec{r} = -6$.

67. The line integral of $\vec{F} = x\vec{i} + y\vec{j} = \vec{r}$ along the semicircle $x^2 + y^2 = 1$, $y \geq 0$, oriented counterclockwise, is zero.

■ Are the statements in Problems 68–74 true or false? Give reasons for your answer.

68. The line integral $\int_C \vec{F} \cdot d\vec{r}$ is a scalar.

69. If C_1 and C_2 are oriented curves and C_1 is longer than C_2, then $\int_{C_1} \vec{F} \cdot d\vec{r} > \int_{C_2} \vec{F} \cdot d\vec{r}$.

70. If C is an oriented curve and $\int_C \vec{F} \cdot d\vec{r} = 0$, then $\vec{F} = \vec{0}$.

71. If $\vec{F} = \vec{i}$ is a vector field in 2-space, then $\int_C \vec{F} \cdot d\vec{r} > 0$, where C is the oriented line from $(0,0)$ to $(1,0)$.

72. If $\vec{F} = \vec{i}$ is a vector field in 2-space, then $\int_C \vec{F} \cdot d\vec{r} > 0$, where C is the oriented line from $(0,0)$ to $(0,1)$.

73. If C_1 is the upper semicircle $x^2 + y^2 = 1$, $y \geq 0$ and C_2 is the lower semicircle $x^2 + y^2 = 1$, $y \leq 0$, both oriented counterclockwise, then for any vector field $\vec{F}$, we have $\int_{C_1} \vec{F} \cdot d\vec{r} = -\int_{C_2} \vec{F} \cdot d\vec{r}$.

74. The work done by the force $\vec{F} = -y\vec{i} + x\vec{j}$ on a particle moving clockwise around the boundary of the square $-1 \leq x \leq 1$, $-1 \leq y \leq 1$ is positive.

18.2 COMPUTING LINE INTEGRALS OVER PARAMETERIZED CURVES

The goal of this section is to show how to use a parameterization of a curve to convert a line integral into an ordinary one-variable integral.

Using a Parameterization to Evaluate a Line Integral

Recall the definition of the line integral,

$$\int_C \vec{F} \cdot d\vec{r} = \lim_{\|\Delta\vec{r}_i\| \to 0} \sum \vec{F}(\vec{r}_i) \cdot \Delta\vec{r}_i,$$

where the $\vec{r}_i$ are the position vectors of points subdividing the curve into short pieces. Now suppose we have a smooth parameterization, $\vec{r}(t)$, of C for $a \leq t \leq b$, so that $\vec{r}(a)$ is the position vector of the starting point of the curve and $\vec{r}(b)$ is the position vector of the end. Then we can divide C into n pieces by dividing the interval $a \leq t \leq b$ into n pieces, each of size $\Delta t = (b-a)/n$. See Figures 18.19 and 18.20.

At each point $\vec{r}_i = \vec{r}(t_i)$ we want to compute

$$\vec{F}(\vec{r}_i) \cdot \Delta\vec{r}_i.$$

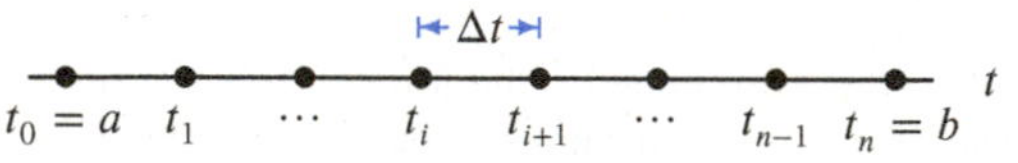

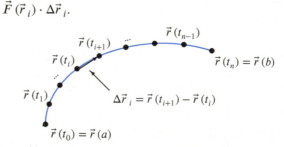

Figure 18.19: Subdivision of the interval $a \leq t \leq b$

Figure 18.20: Corresponding subdivision of the parameterized path C

Since $t_{i+1} = t_i + \Delta t$, the displacement vectors $\Delta \vec{r}_i$ are given by

$$
\begin{aligned}
\Delta \vec{r}_i &= \vec{r}(t_{i+1}) - \vec{r}(t_i) \\
&= \vec{r}(t_i + \Delta t) - \vec{r}(t_i) \\
&= \frac{\vec{r}(t_i + \Delta t) - \vec{r}(t_i)}{\Delta t} \cdot \Delta t \\
&\approx \vec{r}\,'(t_i)\Delta t,
\end{aligned}
$$

where we use the facts that Δt is small and that $\vec{r}(t)$ is differentiable to obtain the last approximation. Therefore,

$$
\int_C \vec{F} \cdot d\vec{r} \approx \sum \vec{F}(\vec{r}_i) \cdot \Delta \vec{r}_i \approx \sum \vec{F}(\vec{r}(t_i)) \cdot \vec{r}\,'(t_i)\,\Delta t.
$$

Notice that $\vec{F}(\vec{r}(t_i)) \cdot \vec{r}\,'(t_i)$ is the value at t_i of a one-variable function of t, so this last sum is really a one-variable Riemann sum. In the limit as $\Delta t \to 0$, we get a definite integral:

$$
\lim_{\Delta t \to 0} \sum \vec{F}(\vec{r}(t_i)) \cdot \vec{r}\,'(t_i)\,\Delta t = \int_a^b \vec{F}(\vec{r}(t)) \cdot \vec{r}\,'(t)\,dt.
$$

Thus, we have the following result:

If $\vec{r}(t)$, for $a \le t \le b$, is a smooth parameterization of an oriented curve C and $\vec{F}$ is a vector field which is continuous on C, then

$$
\int_C \vec{F} \cdot d\vec{r} = \int_a^b \vec{F}(\vec{r}(t)) \cdot \vec{r}\,'(t)\,dt.
$$

In words: To compute the line integral of $\vec{F}$ over C, take the dot product of $\vec{F}$ evaluated on C with the velocity vector, $\vec{r}\,'(t)$, of the parameterization of C, then integrate along the curve.

Even though we assumed that C is smooth, we can use the same formula to compute line integrals over curves which are *piecewise smooth*, such as the boundary of a rectangle. If C is piecewise smooth, we apply the formula to each one of the smooth pieces and add the results by applying property 4 on page 980.

Example 1   Compute $\int_C \vec{F} \cdot d\vec{r}$ where $\vec{F} = (x+y)\vec{i} + y\vec{j}$ and C is the quarter unit circle, oriented counterclockwise as shown in Figure 18.21.

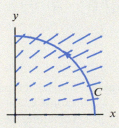

Figure 18.21: The vector field $\vec{F} = (x+y)\vec{i} + y\vec{j}$ and the quarter circle C

Solution   Since most of the vectors in $\vec{F}$ along C point generally in a direction opposite to the orientation of C, we expect our answer to be negative. The first step is to parameterize C by

$$\vec{r}(t) = x(t)\vec{i} + y(t)\vec{j} = \cos t\vec{i} + \sin t\vec{j}, \quad 0 \le t \le \frac{\pi}{2}.$$

Substituting the parameterization into $\vec{F}$, we get $\vec{F}(x(t), y(t)) = (\cos t + \sin t)\vec{i} + \sin t\vec{j}$. The vector $\vec{r}'(t) = x'(t)\vec{i} + y'(t)\vec{j} = -\sin t\vec{i} + \cos t\vec{j}$. Then

$$\int_C \vec{F} \cdot d\vec{r} = \int_0^{\pi/2} ((\cos t + \sin t)\vec{i} + \sin t\vec{j}) \cdot (-\sin t\vec{i} + \cos t\vec{j})\,dt$$

$$= \int_0^{\pi/2} (-\cos t \sin t - \sin^2 t + \sin t \cos t)\,dt$$

$$= \int_0^{\pi/2} -\sin^2 t\,dt = -\frac{\pi}{4} \approx -0.7854.$$

So the answer is negative, as expected.

Example 2   Consider the vector field $\vec{F} = x\vec{i} + y\vec{j}$.

(a) Suppose C_1 is the line segment joining $(1, 0)$ to $(0, 2)$ and C_2 is a part of a parabola with its vertex at $(0, 2)$, joining the same points in the same order. (See Figure 18.22.) Verify that

$$\int_{C_1} \vec{F} \cdot d\vec{r} = \int_{C_2} \vec{F} \cdot d\vec{r}.$$

(b) If C is the triangle shown in Figure 18.23, show that $\int_C \vec{F} \cdot d\vec{r} = 0$.

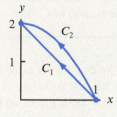

Figure 18.22

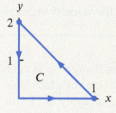

Figure 18.23

Solution   (a) We parameterize C_1 by $\vec{r}(t) = (1 - t)\vec{i} + 2t\vec{j}$ with $0 \le t \le 1$. Then $\vec{r}'(t) = -\vec{i} + 2\vec{j}$, so

$$\int_{C_1} \vec{F} \cdot d\vec{r} = \int_0^1 \vec{F}(1 - t, 2t) \cdot (-\vec{i} + 2\vec{j})\,dt = \int_0^1 ((1 - t)\vec{i} + 2t\vec{j}) \cdot (-\vec{i} + 2\vec{j})\,dt$$

$$= \int_0^1 (5t - 1)\,dt = \frac{3}{2}.$$

To parameterize C_2, we use the fact that it is part of a parabola with vertex at $(0, 2)$, so its equation is of the form $y = -kx^2 + 2$ for some k. Since the parabola crosses the x-axis at $(1, 0)$, we find that $k = 2$ and $y = -2x^2 + 2$. Therefore, we use the parameterization $\vec{r}(t) = t\vec{i} + (-2t^2 + 2)\vec{j}$ with $0 \le t \le 1$, which has $\vec{r}' = \vec{i} - 4t\vec{j}$. This traces out C_2 in reverse, since $t = 0$ gives $(0, 2)$, and $t = 1$ gives $(1, 0)$. Thus, we make $t = 0$ the upper limit of integration and $t = 1$ the lower limit:

$$\int_{C_2} \vec{F} \cdot d\vec{r} = \int_1^0 \vec{F}(t, -2t^2 + 2) \cdot (\vec{i} - 4t\vec{j}) \, dt = -\int_0^1 (t\vec{i} + (-2t^2 + 2)\vec{j}) \cdot (\vec{i} - 4t\vec{j}) \, dt$$

$$= -\int_0^1 (8t^3 - 7t) \, dt = \frac{3}{2}.$$

So the line integrals of $\vec{F}$ along C_1 and C_2 have the same value.

(b) We break $\int_C \vec{F} \cdot d\vec{r}$ into three pieces, one of which we have already computed (namely, the piece connecting $(1,0)$ to $(0,2)$, where the line integral has value $3/2$). The piece running from $(0,2)$ to $(0,0)$ can be parameterized by $\vec{r}(t) = (2 - t)\vec{j}$ with $0 \le t \le 2$. The piece running from $(0,0)$ to $(1,0)$ can be parameterized by $\vec{r}(t) = t\vec{i}$ with $0 \le t \le 1$. Then

$$\int_C \vec{F} \cdot d\vec{r} = \frac{3}{2} + \int_0^2 \vec{F}(0, 2 - t) \cdot (-\vec{j}) \, dt + \int_0^1 \vec{F}(t, 0) \cdot \vec{i} \, dt$$

$$= \frac{3}{2} + \int_0^2 (2 - t)\vec{j} \cdot (-\vec{j}) \, dt + \int_0^1 t\vec{i} \cdot \vec{i} \, dt$$

$$= \frac{3}{2} + \int_0^2 (t - 2) \, dt + \int_0^1 t \, dt = \frac{3}{2} + (-2) + \frac{1}{2} = 0.$$

Example 3   Let C be the closed curve consisting of the upper half-circle of radius 1 and the line forming its diameter along the x-axis, oriented counterclockwise. (See Figure 18.24.) Find $\int_C \vec{F} \cdot d\vec{r}$ where $\vec{F}(x, y) = -y\vec{i} + x\vec{j}$.

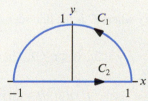

Figure 18.24: The curve $C = C_1 + C_2$ for Example 3

Solution   We write $C = C_1 + C_2$, where C_1 is the half-circle and C_2 is the line, and compute $\int_{C_1} \vec{F} \cdot d\vec{r}$ and $\int_{C_2} \vec{F} \cdot d\vec{r}$ separately. We parameterize C_1 by $\vec{r}(t) = \cos t\vec{i} + \sin t\vec{j}$, with $0 \le t \le \pi$. Then

$$\int_{C_1} \vec{F} \cdot d\vec{r} = \int_0^\pi (-\sin t\vec{i} + \cos t\vec{j}) \cdot (-\sin t\vec{i} + \cos t\vec{j}) \, dt$$

$$= \int_0^\pi (\sin^2 t + \cos^2 t) \, dt = \int_0^\pi 1 \, dt = \pi.$$

For C_2, we have $\int_{C_2} \vec{F} \cdot d\vec{r} = 0$, since the vector field $\vec{F}$ has no $\vec{i}$ component along the x-axis (where $y = 0$) and is therefore perpendicular to C_2 at all points.

Finally, we can write

$$\int_C \vec{F} \cdot d\vec{r} = \int_{C_1} \vec{F} \cdot d\vec{r} + \int_{C_2} \vec{F} \cdot d\vec{r} = \pi + 0 = \pi.$$

It is no accident that the result for $\int_{C_1} \vec{F} \cdot d\vec{r}$ is the same as the length of the curve C_1. See Problem 56 on page 983 and Problem 41 on page 991.

The next example illustrates the computation of a line integral over a path in 3-space.

Example 4   A particle travels along the helix C given by $\vec{r}(t) = \cos t\vec{i} + \sin t\vec{j} + 2t\vec{k}$ and is subject to a force $\vec{F} = x\vec{i} + z\vec{j} - xy\vec{k}$. Find the total work done on the particle by the force for $0 \le t \le 3\pi$.

Solution   The work done is given by a line integral, which we evaluate using the given parameterization:

$$\text{Work done} = \int_C \vec{F} \cdot d\vec{r} = \int_0^{3\pi} \vec{F}(\vec{r}(t)) \cdot \vec{r}\,'(t)\,dt$$

$$= \int_0^{3\pi} (\cos t\vec{i} + 2t\vec{j} - \cos t \sin t\vec{k}) \cdot (-\sin t\vec{i} + \cos t\vec{j} + 2\vec{k})\,dt$$

$$= \int_0^{3\pi} (-\cos t \sin t + 2t \cos t - 2\cos t \sin t)\,dt$$

$$= \int_0^{3\pi} (-3\cos t \sin t + 2t \cos t)\,dt = -4.$$

The Differential Notation $\int_C P\,dx + Q\,dy + R\,dz$

There is an alternative notation for line integrals that is often useful. For the vector field $\vec{F} = P(x, y, z)\vec{i} + Q(x, y, z)\vec{j} + R(x, y, z)\vec{k}$ and an oriented curve C, if we write $d\vec{r} = dx\vec{i} + dy\vec{j} + dz\vec{k}$ we have

$$\int_C \vec{F} \cdot d\vec{r} = \int_C P(x, y, z)dx + Q(x, y, z)dy + R(x, y, z)dz.$$

Example 5   Evaluate $\int_C xy\,dx - y^2\,dy$ where C is the line segment from $(0, 0)$ to $(2, 6)$.

Solution   We parameterize C by $x = t, y = 3t, 0 \le t \le 2$. Thus, $dx = dt, dy = 3dt$, so

$$\int_C xy\,dx - y^2\,dy = \int_0^2 t(3t)dt - (3t)^2(3dt) = \int_0^2 (-24t^2)\,dt = -64.$$

Line integrals can be expressed either using vectors or using differentials. If the independent variables are distances, then visualizing a line integral in terms of dot products can be useful. However, if the independent variables are, for example, temperature and volume, then the dot product does not have physical meaning, so differentials are more natural.

Independence of Parameterization

Since there are many different ways of parameterizing a given oriented curve, you may be wondering what happens to the value of a given line integral if you choose another parameterization. The answer is that the choice of parameterization makes no difference. Since we initially defined the line integral without reference to any particular parameterization, this is exactly as we would expect.

Example 6   Consider the oriented path which is a straight-line segment L running from $(0, 0)$ to $(1, 1)$. Calculate the line integral of the vector field $\vec{F} = (3x - y)\vec{i} + x\vec{j}$ along L using each of the parameterizations

(a) $A(t) = (t, t), \quad 0 \le t \le 1,$ (b) $D(t) = (e^t - 1, e^t - 1), \quad 0 \le t \le \ln 2.$

Solution   The line L has equation $y = x$. Both $A(t)$ and $D(t)$ give a parameterization of L: each has both coordinates equal and each begins at $(0,0)$ and ends at $(1,1)$. Now let's calculate the line integral of the vector field $\vec{F} = (3x - y)\vec{i} + x\vec{j}$ using each parameterization.

(a) Using $A(t)$, we get

$$\int_L \vec{F} \cdot d\vec{r} = \int_0^1 ((3t - t)\vec{i} + t\vec{j}) \cdot (\vec{i} + \vec{j}) \, dt = \int_0^1 3t \, dt = \frac{3t^2}{2}\Big|_0^1 = \frac{3}{2}.$$

(b) Using $D(t)$, we get

$$\int_L \vec{F} \cdot d\vec{r} = \int_0^{\ln 2} \left(\left(3(e^t - 1) - (e^t - 1)\right)\vec{i} + (e^t - 1)\vec{j} \right) \cdot (e^t\vec{i} + e^t\vec{j}) \, dt$$

$$= \int_0^{\ln 2} 3(e^{2t} - e^t) \, dt = 3\left(\frac{e^{2t}}{2} - e^t \right)\Big|_0^{\ln 2} = \frac{3}{2}.$$

The fact that both answers are the same illustrates that the value of a line integral is independent of the parameterization of the path. Problems 59–61 (available online) give another way of seeing this.

Summary for Section 18.2

- **Using a parameterization to evaluate a line integral:** If $\vec{r}(t)$, for $a \le t \le b$, is a smooth parameterization of an oriented curve C and $\vec{F}$ is a vector field which is continuous on C, then

$$\int_C \vec{F} \cdot d\vec{r} = \int_a^b \vec{F}(\vec{r}(t)) \cdot \vec{r}'(t) \, dt.$$

- **Differential notation:** For the vector field $\vec{F} = P(x, y, z)\vec{i} + Q(x, y, z)\vec{j} + R(x, y, z)\vec{k}$ and an oriented curve C, we have

$$\int_C \vec{F} \cdot d\vec{r} = \int_C P(x, y, z)dx + Q(x, y, z)dy + R(x, y, z)dz.$$

Exercises and Problems for Section 18.2   Online Resource: Additional Problems for Section 18.2

EXERCISES

In Exercises 1–3, write $\int_C \vec{F} \cdot d\vec{r}$ in the form $\int_a^b g(t) \, dt$. (Give a formula for g and numbers for a and b. You do not need to evaluate the integral.)

1. $\vec{F} = y\vec{i} + x\vec{j}$ and C is the semicircle from $(0, 1)$ to $(0, -1)$ with $x > 0$.

2. $\vec{F} = x\vec{i} + z^2\vec{k}$ and C is the line from $(0, 1, 0)$ to $(2, 3, 2)$.

3. $\vec{F} = (\cos x)\vec{i} + (\cos y)\vec{j} + (\cos z)\vec{k}$ and C is the unit circle in the plane $z = 10$, centered on the z-axis and oriented counterclockwise when viewed from above.

In Exercises 4–8, find the line integral.

4. $\int_C (3\vec{i} + (y + 5)\vec{j}) \cdot d\vec{r}$ where C is the line from $(0, 0)$ to $(0, 3)$.

5. $\int_C (2x\vec{i} + 3y\vec{j}) \cdot d\vec{r}$ where C is the line from $(1, 0, 0)$ to $(5, 0, 0)$.

6. $\int_C (2y^2\vec{i} + x\vec{j}) \cdot d\vec{r}$ where C is the line segment from $(3, 1)$ to $(0, 0)$.

7. $\int_C (x\vec{i} + y\vec{j}) \cdot d\vec{r}$ where C is the semicircle with center at $(2, 0)$ and going from $(3, 0)$ to $(1, 0)$ in the region $y > 0$.

8. Find $\int_C ((x^2 + y)\vec{i} + y^3\vec{j}) \cdot d\vec{r}$ where C consists of the three line segments from $(4, 0, 0)$ to $(4, 3, 0)$ to $(0, 3, 0)$ to $(0, 3, 5)$.

■In Exercises 9–23, find $\int_C \vec{F} \cdot d\vec{r}$ for the given $\vec{F}$ and C.

9. $\vec{F} = 2\vec{i} + \vec{j}$; C is the x-axis from $x = 10$ to $x = 7$.

10. $\vec{F} = 3\vec{j} - \vec{i}$; C is the line $y = x$ from $(1, 1)$ to $(5, 5)$.

11. $\vec{F} = x\vec{i} + y\vec{j}$ and C is the line from $(0, 0)$ to $(3, 3)$.

12. $\vec{F} = y\vec{i} - x\vec{j}$ and C is the right-hand side of the unit circle, starting at $(0, 1)$.

13. $\vec{F} = x^2\vec{i} + y^2\vec{j}$ and C is the line from the point $(1, 2)$ to the point $(3, 4)$.

14. $\vec{F} = -y \sin x\vec{i} + \cos x\vec{j}$ and C is the parabola $y = x^2$ between $(0, 0)$ and $(2, 4)$.

15. $\vec{F} = y^3\vec{i} + x^2\vec{j}$ and C is the line from $(0, 0)$ to $(3, 2)$.

16. $\vec{F} = 2y\vec{i} - (\sin y)\vec{j}$ counterclockwise around the unit circle C starting at the point $(1, 0)$.

17. $\vec{F} = \ln y\vec{i} + \ln x\vec{j}$ and C is the curve given parametrically by $(2t, t^3)$, for $2 \le t \le 4$.

18. $\vec{F} = x\vec{i} + 6\vec{j} - \vec{k}$, and C is the line $x = y = z$ from $(0, 0, 0)$ to $(2, 2, 2)$.

19. $\vec{F} = (2x - y + 4)\vec{i} + (5y + 3x - 6)\vec{j}$ and C is the triangle with vertices $(0, 0), (3, 0), (3, 2)$ traversed counterclockwise.

20. $\vec{F} = x\vec{i} + 2zy\vec{j} + x\vec{k}$ and C is $\vec{r} = t\vec{i} + t^2\vec{j} + t^3\vec{k}$ for $1 \le t \le 2$.

21. $\vec{F} = x^3\vec{i} + y^2\vec{j} + z\vec{k}$ and C is the line from the origin to the point $(2, 3, 4)$.

22. $\vec{F} = -y\vec{i} + x\vec{j} + 5\vec{k}$ and C is the helix $x = \cos t, y = \sin t, z = t$, for $0 \le t \le 4\pi$.

23. $\vec{F} = e^y\vec{i} + \ln(x^2 + 1)\vec{j} + \vec{k}$ and C is the circle of radius 2 centered at the origin in the yz-plane in Figure 18.25.

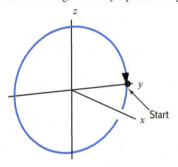

Figure 18.25

24. Every line integral can be written in both vector notation and differential notation. For example,

$$\int_C (2x\vec{i} + (x + y)\vec{j}) \cdot d\vec{r} = \int_C 2x\, dx + (x + y)\, dy.$$

(a) Express $\int_C 3\, dx + xy\, dy$ in vector notation.

(b) Express $\int_C (100 \cos x\vec{i} + e^y \sin x\vec{j}) \cdot d\vec{r}$ in differential notation.

■In Exercises 25–26, express the line integral $\int_C \vec{F} \cdot d\vec{r}$ in differential notation.

25. $\vec{F} = 3x\vec{i} - y \sin x\vec{j}$

26. $\vec{F} = y^2\vec{i} + z^2\vec{j} + (x^2 - 5)\vec{k}$

■In Exercises 27–28, find $\vec{F}$ so that the line integral equals $\int_C \vec{F} \cdot d\vec{r}$.

27. $\int_C (x + 2y)dx + x^2y\, dy$

28. $\int_C e^{-3y}\, dx - yz(\sin x)\, dy + (y + z)\, dz$

■Evaluate the line integrals in Exercises 29–34.

29. $\int_C y\, dx + x\, dy$ where C is the parameterized path $x = t^2, y = t^3, 1 \le t \le 5$.

30. $\int_C dx + y\, dy + z\, dz$ where C is one turn of the helix $x = \cos t, y = \sin t, z = 3t, 0 \le t \le 2\pi$.

31. $\int_C 3y\, dx + 4x\, dy$ where C is the straight-line path from $(1, 3)$ to $(5, 9)$.

32. $\int_C x\, dx + z\, dy - y\, dz$ where C is the circle of radius 3 in the yz-plane centered at the origin, oriented counterclockwise when viewed from the positive y-axis.

33. $\int_C (x + y)\, dx + x^2\, dy$ where C is the path $x = t^2, y = t, 0 \le t \le 10$.

34. $\int_C x\, dy$ where C is the quarter circle centered at the origin going counterclockwise from $(2, 0)$ to $(0, 2)$.

PROBLEMS

35. Evaluate the line integral of $\vec{F} = (3x - y)\vec{i} + x\vec{j}$ over two different paths from $(0, 0)$ to $(1, 1)$.

   (a) The path (t, t^2), with $0 \le t \le 1$
   (b) The path (t^2, t), with $0 \le t \le 1$

36. Curves C_1 and C_2 are parameterized as follows:

   C_1 is $(x(t), y(t)) = (0, t)$ for $-1 \le t \le 1$

   C_2 is $(x(t), y(t)) = (\cos t, \sin t)$ for $\frac{\pi}{2} \le t \le \frac{3\pi}{2}$.

   (a) Sketch C_1 and C_2 with arrows showing their orientation.

(b) Suppose $\vec{F} = (x + 3y)\vec{i} + y\vec{j}$. Calculate $\int_C \vec{F} \cdot d\vec{r}$, where C is the curve given by $C = C_1 + C_2$.

37. Calculate the line integral of $\vec{F} = (3x - y)\vec{i} + x\vec{j}$ along the line segment L from $(0, 0)$ to $(1, 1)$ using each of the parameterizations

   (a) $B(t) = (2t, 2t), \quad 0 \le t \le 1/2$
   (b) $C(t) = \left(\dfrac{t^2 - 1}{3}, \dfrac{t^2 - 1}{3}\right), \quad 1 \le t \le 2$

In Problems 38–39, evaluate the line integral using a short-cut available in two special situations.

- If $\vec{F}$ is perpendicular to C at every point of C, then

$$\int_C \vec{F} \cdot d\vec{r} = 0.$$

- If $\vec{F}$ is tangent to C at every point of C and has constant magnitude on C, then

$$\int_C \vec{F} \cdot d\vec{r} = \pm \|\vec{F}\| \times \text{Length of } C.$$

Choose the $+$ sign if $\vec{F}$ points in the direction of integration and choose the $-$ sign if $\vec{F}$ points in the direction opposite to the direction of integration.

38. $\int_C (x\vec{i} + y\vec{j}) \cdot d\vec{r}$, where C is the circle of radius 10 centered at the origin, oriented counterclockwise.

39. $\int_C (-y\vec{i} + x\vec{j}) \cdot d\vec{r}$, where C is the circle of radius 10 centered at the origin, oriented counterclockwise.

40. Justify the following without parameterizing the paths.

    (a) $\int_C (x^2 + \cos y) \, dy = 0$ where C is the straight line path from $(10, 5)$ to $(20, 5)$.

    (b) $\int_C (x^2 + \cos y) \, dx = 0$ where C is the straight line path from $(10, 5)$ to $(10, 50)$.

41. Let $\vec{F} = -y\vec{i} + x\vec{j}$ and let C be the unit circle oriented counterclockwise.

    (a) Show that $\vec{F}$ has a constant magnitude of 1 on C.

    (b) Show that $\vec{F}$ is always tangent to the circle C.

    (c) Show that $\int_C \vec{F} \cdot d\vec{r} = \text{Length of } C$.

42. A spiral staircase in a building is in the shape of a helix of radius 5 meters. Between two floors of the building, the stairs make one full revolution and climb by 4 meters. A person carries a bag of groceries up two floors. The combined mass of the person and the groceries is $m = 70$ kg and the gravitational force is mg downward, where $g = 9.8$ m/sec^2 is the acceleration due to gravity. Calculate the work done by the person against gravity.

43. If C is $\vec{r} = (t+1)\vec{i} + 2t\vec{j} + 3t\vec{k}$ for $0 \leq t \leq 1$, we know $\int_C \vec{F}(\vec{r}) \cdot d\vec{r} = 5$. Find the value of the integrals:

    (a) $\int_1^0 \vec{F}((t+1)\vec{i} + 2t\vec{j} + 3t\vec{k}) \cdot (\vec{i} + 2\vec{j} + 3\vec{k}) \, dt$

    (b) $\int_0^1 \vec{F}((t^2+1)\vec{i} + 2t^2\vec{j} + 3t^2\vec{k}) \cdot (2t\vec{i} + 4t\vec{j} + 6t\vec{k}) \, dt$

    (c) $\int_{-1}^1 \vec{F}((t^2+1)\vec{i} + 2t^2\vec{j} + 3t^2\vec{k}) \cdot (2t\vec{i} + 4t\vec{j} + 6t\vec{k}) \, dt$

Strengthen Your Understanding

In Problems 44–45, explain what is wrong with the statement.

44. For the vector field $\vec{F} = x\vec{i} - y\vec{j}$ and oriented path C parameterized by $x = \cos t$, $y = \sin t$, $0 \leq t \leq \pi/2$, we have

$$\int_C \vec{F} \cdot d\vec{r} = \int_0^{\pi/2} (\cos t\vec{i} - \sin t\vec{j}) \cdot (\cos t\vec{i} + \sin t\vec{j}) \, dt.$$

45. If $\int_C \vec{F} \cdot d\vec{r} = 0$, then $\vec{F}$ is perpendicular to C at every point on C.

In Problems 46–47, give an example of:

46. A vector field $\vec{F}$ such that, for the parameterized path $\vec{r}(t) = 3\cos t\vec{i} + 3\sin t\vec{j}$, $-\pi/2 \leq t \leq \pi/2$, the integral $\int_C \vec{F} \cdot d\vec{r}$ can be computed geometrically, without using the parameterization.

47. A parameterized path C such that, for the vector field $\vec{F}(x, y) = \sin y\vec{i}$, the integral $\int_C \vec{F} \cdot d\vec{r}$ is nonzero and can be computed geometrically, without using the parameterization.

Are the statements in Problems 48–56 true or false? Give reasons for your answer.

48. If C_1 and C_2 are oriented curves with C_2 beginning where C_1 ends, then $\int_{C_1+C_2} \vec{F} \cdot d\vec{r} > \int_{C_1} \vec{F} \cdot d\vec{r}$.

49. The line integral $\int_C 4\vec{i} \cdot d\vec{r}$ over the curve C parameterized by $\vec{r}(t) = t\vec{i} + t^2\vec{j}$, for $0 \leq t \leq 2$, is positive.

50. If C_1 is the curve parameterized by $\vec{r}_1(t) = \cos t\vec{i} + \sin t\vec{j}$, with $0 \leq t \leq \pi$, and C_2 is the curve parameterized by $\vec{r}_2(t) = \cos t\vec{i} - \sin t\vec{j}$, $0 \leq t \leq \pi$, then for any vector field $\vec{F}$ we have $\int_{C_1} \vec{F} \cdot d\vec{r} = \int_{C_2} \vec{F} \cdot d\vec{r}$.

51. If C_1 is the curve parameterized by $\vec{r}_1(t) = \cos t\vec{i} + \sin t\vec{j}$, with $0 \leq t \leq \pi$, and C_2 is the curve parameterized by $\vec{r}_2(t) = \cos(2t)\vec{i} + \sin(2t)\vec{j}$, $0 \leq t \leq \frac{\pi}{2}$, then for any vector field $\vec{F}$ we have $\int_{C_1} \vec{F} \cdot d\vec{r} = \int_{C_2} \vec{F} \cdot d\vec{r}$.

52. If C is the curve parameterized by $\vec{r}(t)$, for $a \leq t \leq b$ with $\vec{r}(a) = \vec{r}(b)$, then $\int_C \vec{F} \cdot d\vec{r} = 0$ for any vector field $\vec{F}$. (Note that C starts and ends at the same place.)

53. If C_1 is the line segment from $(0, 0)$ to $(1, 0)$ and C_2 is the line segment from $(0, 0)$ to $(2, 0)$, then for any vector field $\vec{F}$, we have $\int_{C_2} \vec{F} \cdot d\vec{r} = 2\int_{C_1} \vec{F} \cdot d\vec{r}$.

54. If C is a circle of radius a, centered at the origin and oriented counterclockwise, then $\int_C (2x\vec{i} + y\vec{j}) \cdot d\vec{r} = 0$.

55. If C is a circle of radius a, centered at the origin and oriented counterclockwise, then $\int_C (2y\vec{i} + x\vec{j}) \cdot d\vec{r} = 0$.

56. If C_1 is the curve parameterized by $\vec{r}_1(t) = t\vec{i} + t^2\vec{j}$, with $0 \leq t \leq 2$, and C_2 is the curve parameterized by $\vec{r}_2(t) = (2-t)\vec{i} + (2-t)^2\vec{j}$, $0 \leq t \leq 2$, then for any vector field $\vec{F}$ we have $\int_{C_1} \vec{F} \cdot d\vec{r} = -\int_{C_2} \vec{F} \cdot d\vec{r}$.

57. If C_1 is the path parameterized by $\vec{r}_1(t) = (t, t)$, $0 \le t \le 1$, and if C_2 is the path parameterized by $\vec{r}_2(t) = (1 - t, 1 - t)$, $0 \le t \le 1$, and if $\vec{F} = x\vec{i} + y\vec{j}$, which of the following is true?

(a) $\int_{C_1} \vec{F} \cdot d\vec{r} > \int_{C_2} \vec{F} \cdot d\vec{r}$

(b) $\int_{C_1} \vec{F} \cdot d\vec{r} < \int_{C_2} \vec{F} \cdot d\vec{r}$

(c) $\int_{C_1} \vec{F} \cdot d\vec{r} = \int_{C_2} \vec{F} \cdot d\vec{r}$

58. If C_1 is the path parameterized by $\vec{r}_1(t) = (t, t)$, for $0 \le t \le 1$, and if C_2 is the path parameterized by $\vec{r}_2(t) = (\sin t, \sin t)$, for $0 \le t \le 1$, and if $\vec{F} = x\vec{i} + y\vec{j}$, which of the following is true?

(a) $\int_{C_1} \vec{F} \cdot d\vec{r} > \int_{C_2} \vec{F} \cdot d\vec{r}$

(b) $\int_{C_1} \vec{F} \cdot d\vec{r} < \int_{C_2} \vec{F} \cdot d\vec{r}$

(c) $\int_{C_1} \vec{F} \cdot d\vec{r} = \int_{C_2} \vec{F} \cdot d\vec{r}$

18.3   GRADIENT FIELDS AND PATH-INDEPENDENT FIELDS

For a function, f, of one variable, the Fundamental Theorem of Calculus tells us that the definite integral of a rate of change, f', gives the total change in f:

$$\int_a^b f'(t)\, dt = f(b) - f(a).$$

What about functions of two or more variables? The quantity that describes the rate of change is the gradient vector field. If we know the gradient of a function f, can we compute the total change in f between two points? The answer is yes, using a line integral.

Finding the Total Change in f from grad f: The Fundamental Theorem

To find the change in f between two points P and Q, we choose a smooth path C from P to Q, then divide the path into many small pieces. See Figure 18.26.

First we estimate the change in f as we move through a displacement $\Delta \vec{r}_i$ from $\vec{r}_i$ to $\vec{r}_{i+1}$. Suppose $\vec{u}$ is a unit vector in the direction of $\Delta \vec{r}_i$. Then the change in f is given by

$$f(\vec{r}_{i+1}) - f(\vec{r}_i) \approx \text{Rate of change of } f \times \text{Distance moved in direction of } \vec{u}$$
$$= f_{\vec{u}}(\vec{r}_i)\|\Delta \vec{r}_i\|$$
$$= \text{grad } f \cdot \vec{u}\, \|\Delta \vec{r}_i\|$$
$$= \text{grad } f \cdot \Delta \vec{r}_i. \qquad \text{since } \Delta \vec{r}_i = \|\Delta \vec{r}_i\|\vec{u}$$

Therefore, summing over all pieces of the path, the total change in f is given by

$$\text{Total change} = f(Q) - f(P) \approx \sum_{i=0}^{n-1} \text{grad } f(\vec{r}_i) \cdot \Delta \vec{r}_i.$$

In the limit as $\|\Delta \vec{r}_i\|$ approaches zero, this suggests the following result:

Theorem 18.1: The Fundamental Theorem of Calculus for Line Integrals

Suppose C is a piecewise smooth oriented path with starting point P and ending point Q. If f is a function whose gradient is continuous on the path C, then

$$\int_C \text{grad } f \cdot d\vec{r} = f(Q) - f(P).$$

Notice that there are many different paths from P to Q. (See Figure 18.27.) However, the value of the line integral $\int_C \text{grad } f \cdot d\vec{r}$ depends only on the endpoints of C; it does not depend on where

C goes in between. Problem 88 (available online) shows how the Fundamental Theorem for Line Integrals can be derived from the one-variable Fundamental Theorem of Calculus.

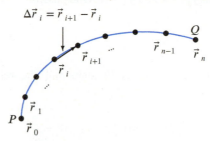

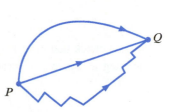

Figure 18.26: Subdivision of the path from P to Q. We estimate the change in f along $\Delta \vec{r}_i$

Figure 18.27: There are many different paths from P to Q: all give the same value of $\int_C \text{grad } f \cdot d\vec{r}$

Example 1   Suppose that grad f is everywhere perpendicular to the curve joining P and Q shown in Figure 18.28.

(a) Explain why you expect the path joining P and Q to be a contour.

(b) Using a line integral, show that $f(P) = f(Q)$.

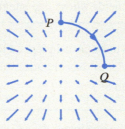

Figure 18.28: The gradient vector field of the function f

Solution   (a) The gradient of f is everywhere perpendicular to the path from P to Q, as you expect along a contour.

(b) Consider the path from P to Q shown in Figure 18.28 and evaluate the line integral

$$\int_C \text{grad } f \cdot d\vec{r} = f(Q) - f(P).$$

Since grad f is everywhere perpendicular to the path, the line integral is 0. Thus, $f(Q) = f(P)$.

Example 2   Consider the vector field $\vec{F} = x\vec{i} + y\vec{j}$. In Example 2 on page 986 we calculated $\int_{C_1} \vec{F} \cdot d\vec{r}$ and $\int_{C_2} \vec{F} \cdot d\vec{r}$ over the oriented curves shown in Figure 18.29 and found they were the same. Find a scalar function f with grad $f = \vec{F}$. Hence, find an easy way to calculate the line integrals, and explain why we could have expected them to be the same.

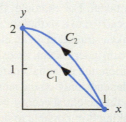

Figure 18.29: Find the line integral of $\vec{F} = x\vec{i} + y\vec{j}$ over the curves C_1 and C_2

Solution   We want to find a function $f(x, y)$ for which $f_x = x$ and $f_y = y$. One possibility for f is

$$f(x, y) = \frac{x^2}{2} + \frac{y^2}{2}.$$

You can check that $\operatorname{grad} f = x\vec{i} + y\vec{j}$. Now we can use the Fundamental Theorem to compute the line integral. Since $\vec{F} = \operatorname{grad} f$ we have

$$\int_{C_1} \vec{F} \cdot d\vec{r} = \int_{C_1} \operatorname{grad} f \cdot d\vec{r} = f(0, 2) - f(1, 0) = \frac{3}{2}.$$

Notice that the calculation looks exactly the same for C_2. Since the value of the integral depends only on the values of f at the endpoints, it is the same no matter what path we choose.

Path-Independent, or Conservative, Vector Fields

In the previous example, the line integral was independent of the path taken between the two (fixed) endpoints. We give vector fields whose line integrals have this property a special name.

> A vector field $\vec{F}$ is said to be **path-independent**, or **conservative**, if for any two points P and Q, the line integral $\int_C \vec{F} \cdot d\vec{r}$ has the same value along any piecewise smooth path C from P to Q lying in the domain of $\vec{F}$.

If, on the other hand, the line integral $\int_C \vec{F} \cdot d\vec{r}$ does depend on the path C joining P to Q, then $\vec{F}$ is said to be a *path-dependent* vector field.

Now suppose that $\vec{F}$ is any continuous gradient field, so $\vec{F} = \operatorname{grad} f$. If C is a path from P to Q, the Fundamental Theorem for Line Integrals tells us that

$$\int_C \vec{F} \cdot d\vec{r} = f(Q) - f(P).$$

Since the right-hand side of this equation does not depend on the path, but only on the endpoints of the path, the vector field $\vec{F}$ is path-independent. Thus, we have the following important result:

> If $\vec{F}$ is a continuous gradient vector field, then $\vec{F}$ is path-independent.

Why Do We Care About Path-Independent, or Conservative, Vector Fields?

Many of the fundamental vector fields of nature are path-independent—for example, the gravitational field and the electric field of particles at rest. The fact that the gravitational field is path-independent means that the work done by gravity when an object moves depends only on the starting and ending points and not on the path taken. For example, the work done by gravity (computed by the line integral) on a bicycle being carried to a sixth floor apartment is the same whether it is carried up the stairs in a zig-zag path or taken straight up in an elevator.

When a vector field is path-independent, we can define the *potential energy* of a body. When the body moves to another position, the potential energy changes by an amount equal to the work done by the vector field, which depends only on the starting and ending positions. If the work done had not been path-independent, the potential energy would depend both on the body's current position *and* on how it got there, making it impossible to define a useful potential energy.

Project 1 (available online) explains why path-independent force vector fields are also called *conservative* vector fields: When a particle moves under the influence of a conservative vector field, the *total energy* of the particle is *conserved*. It turns out that the force field is obtained from the gradient of the potential energy function.

Path-Independent Fields and Gradient Fields

We have seen that every gradient field is path-independent. What about the converse? That is, given a path-independent vector field $\vec{F}$, can we find a function f such that $\vec{F} = \text{grad } f$? The answer is yes, provided that $\vec{F}$ is continuous.

How to Construct f from $\vec{F}$

First, notice that there are many different choices for f, since we can add a constant to f without changing grad f. If we pick a fixed starting point P, then by adding or subtracting a constant to f, we can ensure that $f(P) = 0$. For any other point Q, we define $f(Q)$ by the formula

$$f(Q) = \int_C \vec{F} \cdot d\vec{r}, \quad \text{where } C \text{ is any path from } P \text{ to } Q.$$

Since $\vec{F}$ is path-independent, it does not matter which path we choose from P to Q. On the other hand, if $\vec{F}$ is not path-independent, then different choices might give different values for $f(Q)$, so f would not be a function (a function has to have a single value at each point).

We still have to show that the gradient of the function f really is $\vec{F}$; we do this on page 997. However, by constructing a function f in this manner, we have the following result:

Theorem 18.2: Path-Independent Fields Are Gradient Fields

If $\vec{F}$ is a continuous path-independent vector field on an open region R, then $\vec{F} = \text{grad } f$ for some f defined on R.

Combining Theorems 18.1 and 18.2, we have

A continuous vector field $\vec{F}$ defined on an open region is path-independent if and only if $\vec{F}$ is a gradient vector field.

The function f is sufficiently important that it is given a special name:

If a vector field $\vec{F}$ is of the form $\vec{F} = \text{grad } f$ for some scalar function f, then f is called a **potential function** for the vector field $\vec{F}$.

Warning

Physicists use the convention that a function ϕ is a potential function for a vector field $\vec{F}$ if $\vec{F} = -\text{grad } \phi$. See Problem 89 (available online).

Example 3   Show that the vector field $\vec{F}(x, y) = y \cos x \vec{i} + (\sin x + y)\vec{j}$ is path-independent.

Solution   Let's suppose $\vec{F}$ does have a potential function f, so that $\vec{F} = \text{grad } f$. This means

$$\frac{\partial f}{\partial x} = y \cos x \qquad \text{and} \qquad \frac{\partial f}{\partial y} = \sin x + y.$$

Integrating the expression for $\partial f / \partial x$ with respect to x shows that

$$f(x, y) = y \sin x + C(y) \qquad \text{where } C(y) \text{ is a function of } y \text{ only.}$$

The constant of integration here is an arbitrary function $C(y)$ of y, since $\partial(C(y))/\partial x = 0$. Differentiating this expression for $f(x, y)$ with respect to y and using $\partial f / \partial y = \sin x + y$ gives

$$\frac{\partial f}{\partial y} = \sin x + C'(y) = \sin x + y.$$

Thus, we must have $C'(y) = y$, so $g(y) = y^2/2 + A$, where A is some constant. Thus,

$$f(x, y) = y \sin x + \frac{y^2}{2} + A$$

is a potential function for $\vec{F}$. Therefore, $\vec{F}$ is path-independent.

Example 4   The gravitational field, $\vec{F}$, of an object of mass M is given by

$$\vec{F} = -\frac{GM}{r^3} \vec{r}.$$

Show that $\vec{F}$ is a gradient field by finding f, a potential function for $\vec{F}$.

Solution   The vector $\vec{F}$ points directly in toward the origin. If $\vec{F} = \text{grad } f$, then $\vec{F}$ must be perpendicular to the level surfaces of f, so the level surfaces of f must be spheres. Also, if $\text{grad } f = \vec{F}$, then $\| \text{grad } f \| = \| \vec{F} \| = GM/r^2$ is the rate of change of f in the direction toward the origin. Now, differentiating with respect to r gives the rate of change in a radially outward direction. Thus, if we write $w = f(x, y, z)$, we have

$$\frac{dw}{dr} = -\frac{GM}{r^2} = GM \left(-\frac{1}{r^2} \right) = GM \frac{d}{dr} \left(\frac{1}{r} \right).$$

So for the potential function, let's try

$$w = \frac{GM}{r} \qquad \text{or} \qquad f(x, y, z) = \frac{GM}{\sqrt{x^2 + y^2 + z^2}}.$$

We check that f is the potential function by calculating

$$f_x = \frac{\partial}{\partial x} \frac{GM}{\sqrt{x^2 + y^2 + z^2}} = \frac{-GMx}{(x^2 + y^2 + z^2)^{3/2}},$$

$$f_y = \frac{\partial}{\partial y} \frac{GM}{\sqrt{x^2 + y^2 + z^2}} = \frac{-GMy}{(x^2 + y^2 + z^2)^{3/2}},$$

$$f_z = \frac{\partial}{\partial z} \frac{GM}{\sqrt{x^2 + y^2 + z^2}} = \frac{-GMz}{(x^2 + y^2 + z^2)^{3/2}}.$$

So

$$\text{grad } f = f_x \vec{i} + f_y \vec{j} + f_z \vec{k} = \frac{-GM}{(x^2 + y^2 + z^2)^{3/2}} (x \vec{i} + y \vec{j} + z \vec{k}) = \frac{-GM}{r^3} \vec{r} = \vec{F}.$$

Our computations show that $\vec{F}$ is a gradient field and that $f = GM/r$ is a potential function for $\vec{F}$.

Path-independent vector fields are rare, but often important. Section 18.4 gives a method for determining whether a vector field has the property.

Why Path-Independent Vector Fields Are Gradient Fields: Showing grad $f = \vec{F}$

Suppose $\vec{F}$ is a path-independent vector field. On page 995 we defined the function f, which we hope will satisfy grad $f = \vec{F}$, as follows:

$$f(x_0, y_0) = \int_C \vec{F} \cdot d\vec{r},$$

where C is a path from a fixed starting point P to a point $Q = (x_0, y_0)$. This integral has the same value for any path from P to Q because $\vec{F}$ is path-independent. Now we show why grad $f = \vec{F}$. We consider vector fields in 2-space; the argument in 3-space is essentially the same.

First, we write the line integral in terms of the components $\vec{F}(x, y) = F_1(x, y)\vec{i} + F_2(x, y)\vec{j}$ and the components $d\vec{r} = dx\vec{i} + dy\vec{j}$:

$$f(x_0, y_0) = \int_C F_1(x, y)\, dx + F_2(x, y)\, dy.$$

We want to compute the partial derivatives of f, that is, the rate of change of f at (x_0, y_0) parallel to the axes. To do this easily, we choose a path which reaches the point (x_0, y_0) on a horizontal or vertical line segment. Let C' be a path from P which stops short of Q at a fixed point (a, b) and let L_x and L_y be the paths shown in Figure 18.30. Then we can split the line integral into three pieces. Since $d\vec{r} = \vec{j}\, dy$ on L_y and $d\vec{r} = \vec{i}\, dx$ on L_x, we have:

$$f(x_0, y_0) = \int_{C'} \vec{F} \cdot d\vec{r} + \int_{L_y} \vec{F} \cdot d\vec{r} + \int_{L_x} \vec{F} \cdot d\vec{r} = \int_{C'} \vec{F} \cdot d\vec{r} + \int_b^{y_0} F_2(a, y)\, dy + \int_a^{x_0} F_1(x, y_0)\, dx.$$

The first two integrals do not involve x_0. Thinking of x_0 as a variable and differentiating with respect to it gives

$$f_{x_0}(x_0, y_0) = \frac{\partial}{\partial x_0} \int_{C'} \vec{F} \cdot d\vec{r} + \frac{\partial}{\partial x_0} \int_b^{y_0} F_2(a, y)dy + \frac{\partial}{\partial x_0} \int_a^{x_0} F_1(x, y_0)dx$$
$$= 0 + 0 + F_1(x_0, y_0) = F_1(x_0, y_0),$$

and thus

$$f_x(x, y) = F_1(x, y).$$

A similar calculation for y using the path from P to Q shown in Figure 18.31 gives

$$f_{y_0}(x_0, y_0) = F_2(x_0, y_0).$$

Therefore, as we claimed,

$$\text{grad } f = f_x\vec{i} + f_y\vec{j} = F_1\vec{i} + F_2\vec{j} = \vec{F}.$$

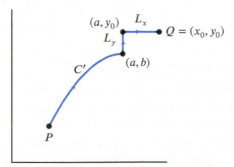

Figure 18.30: The path $C' + L_y + L_x$ is used to show $f_x = F_1$

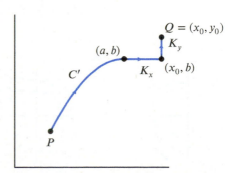

Figure 18.31: The path $C' + K_x + K_y$ is used to show $f_y = F_2$

Summary for Section 18.3

- **The Fundamental Theorem for Line Integrals:** Suppose C is a piecewise smooth oriented path with starting point P and ending point Q, and f is a function whose gradient is continuous on C. Then

$$\int_C \operatorname{grad} f \cdot d\vec{r} = f(Q) - f(P).$$

- A vector field $\vec{F}$ is said to be **path-independent**, or **conservative**, if for any two points P and Q, the line integral $\int_C \vec{F} \cdot d\vec{r}$ has the same value along any piecewise smooth path C from P to Q lying in the domain of $\vec{F}$.

- A continuous vector field $\vec{F}$ defined on an open region is **path-independent** if and only if $\vec{F}$ is a **gradient vector field**.

- Suppose $\vec{F}$ is a gradient vector field. Then a scalar function f is called a **potential function** of $\vec{F}$ if $\vec{F} = \operatorname{grad} f$.

Exercises and Problems for Section 18.3   Online Resource: Additional Problems for Section 18.3
EXERCISES

1. If $\vec{F} = \operatorname{grad}(x^2 + y^4)$, find $\int_C \vec{F} \cdot d\vec{r}$ where C is the quarter of the circle $x^2 + y^2 = 4$ in the first quadrant, oriented counterclockwise.

2. If $\vec{F} = \operatorname{grad}(\sin(xy) + e^z)$, find $\int_C \vec{F} \cdot d\vec{r}$ where C consists of a line from $(0,0,0)$ to $(0,0,1)$, followed by a line to $(0, \sqrt{2}, 3)$, followed by a line to $(\sqrt{2}, \sqrt{5}, 2)$.

■In Exercises **3–6**, let C be a square of side 2, centered at the origin with sides on the lines $x = \pm 1$, $y = \pm 1$ and traversed counterclockwise. What is the sign of the line integral of the vector fields around the curve C? Does the vector field appear to be path-independent?

3.
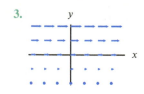

4.
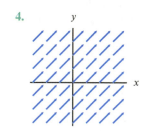

5.
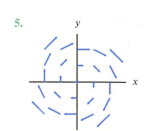

6.
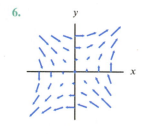

■In Exercises **7–12**, does the vector field appear to be path-independent (conservative)?

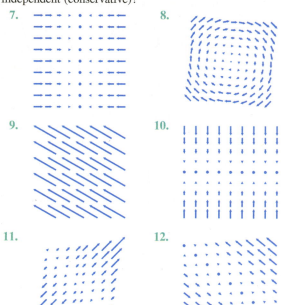

13. Find f if $\operatorname{grad} f = 2xy\vec{i} + x^2\vec{j}$.

14. Find f if $\operatorname{grad} f = 2xy\vec{i} + (x^2 + 8y^3)\vec{j}$.

15. Find f if $\operatorname{grad} f = (yze^{xyz} + z^2 \cos(xz^2))\vec{i} + xze^{xyz}\vec{j} + (xye^{xyz} + 2xz \cos(xz^2))\vec{k}$.

16. Let $f(x, y, z) = x^2 + 2y^3 + 3z^4$ and $\vec{F} = \operatorname{grad} f$. Find $\int_C \vec{F} \cdot d\vec{r}$ where C consists of four line segments from $(4, 0, 0)$ to $(4, 3, 0)$ to $(0, 3, 0)$ to $(0, 3, 5)$ to $(0, 0, 5)$.

In Exercises **17–25**, use the Fundamental Theorem of Line Integrals to calculate $\int_C \vec{F} \cdot d\vec{r}$ exactly.

17. $\vec{F} = 3x^2\vec{i} + 4y^3\vec{j}$ around the top of the unit circle from $(1, 0)$ to $(-1, 0)$.

18. $\vec{F} = (x+2)\vec{i} + (2y+3)\vec{j}$ and C is the line from $(1, 0)$ to $(3, 1)$.

19. $\vec{F} = 2\sin(2x+y)\vec{i} + \sin(2x+y)\vec{j}$ along the path consisting of a line from $(\pi, 0)$ to $(2, 5)$ followed by a line to $(5\pi, 0)$ followed by a quarter circle to $(0, 5\pi)$.

20. $\vec{F} = 2x\vec{i} - 4y\vec{j} + (2z-3)\vec{k}$ and C is the line from $(1, 1, 1)$ to $(2, 3, -1)$.

21. $\vec{F} = x^{2/3}\vec{i} + e^{7y}\vec{j}$ and C is the unit circle oriented clockwise.

22. $\vec{F} = x^{2/3}\vec{i} + e^{7y}\vec{j}$ and C is the quarter of the unit circle in the first quadrant, traced counterclockwise from $(1, 0)$ to $(0, 1)$.

23. $\vec{F} = ye^{xy}\vec{i} + xe^{xy}\vec{j} + (\cos z)\vec{k}$ along the curve consisting of a line from $(0, 0, \pi)$ to $(1, 1, \pi)$ followed by the parabola $z = \pi x^2$ in the plane $y = 1$ to the point $(3, 1, 9\pi)$.

24. $\vec{F} = y\sin(xy)\vec{i} + x\sin(xy)\vec{j}$ and C is the parabola $y = 2x^2$ from $(1, 2)$ to $(3, 18)$.

25. $\vec{F} = 2xy^2ze^{x^2y^2z}\vec{i} + 2x^2yze^{x^2y^2z}\vec{j} + x^2y^2e^{x^2y^2z}\vec{k}$ and C is the circle of radius 1 in the plane $z = 1$, centered on the z-axis, starting at $(1, 0, 1)$ and oriented counterclockwise viewed from above.

PROBLEMS

26. Let $\vec{v} = \text{grad}(x^2 + y^2)$. Consider the path C which is a line between any two of the following points:

$(0,0); (5,0); (-5,0); (0,6); (0,-6); (5,4); (-3,-5).$

Suppose you want to choose the path C in order to maximize $\int_C \vec{v} \cdot d\vec{r}$. What point should be the start of C? What point should be the end of C? Explain your answer.

27. Let $\vec{F} = \text{grad}(2x^2 + 3y^2)$. Which one of the three paths PQ, QR, and RS in Figure 18.32 should you choose as C in order to maximize $\int_C \vec{F} \cdot d\vec{r}$?

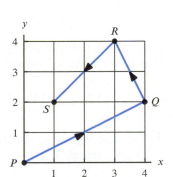

Figure 18.32

28. Compute $\int_C \left(\cos(xy)e^{\sin(xy)}(y\vec{i} + x\vec{j}) + \vec{k} \right) \cdot d\vec{r}$ where C is the line from $(\pi, 2, 5)$ to $(0.5, \pi, 7)$.

29. The vector field $\vec{F}(x, y) = x\vec{i} + y\vec{j}$ is path-independent. Compute geometrically the line integrals over the three paths A, B, and C shown in Figure 18.33 from $(1, 0)$ to $(0, 1)$ and check that they are equal. Here A is a portion of a circle, B is a line, and C consists of two line segments meeting at a right angle.

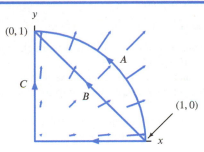

Figure 18.33

30. The vector field $\vec{F}(x, y) = x\vec{i} + y\vec{j}$ is path-independent. Compute algebraically the line integrals over the three paths A, B, and C shown in Figure 18.34 from $(0, 0)$ to $(1, 1)$ and check that they are equal. Here A is a line segment, B is part of the graph of $f(x) = x^2$, and C consists of two line segments meeting at a right angle.

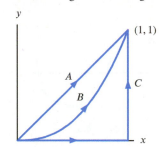

Figure 18.34

In Problems **31–34**, decide whether the vector field could be a gradient vector field. Justify your answer.

31. $\vec{F}(x, y) = x\vec{i}$

32. $\vec{G}(x, y) = (x^2 - y^2)\vec{i} - 2xy\vec{j}$

33. $\vec{F}(\vec{r}) = \vec{r}/\|\vec{r}\|^3$, where $\vec{r} = x\vec{i} + y\vec{j} + z\vec{k}$

34. $\vec{F}(x, y, z) = \dfrac{-z}{\sqrt{x^2 + z^2}}\vec{i} + \dfrac{y}{\sqrt{x^2 + z^2}}\vec{j} + \dfrac{x}{\sqrt{x^2 + z^2}}\vec{k}$

35. Find a potential function for $\vec{F} = 5y\vec{i} + (5x+y)\vec{j}$. Use it to evaluate the integral $\int_C \vec{F} \cdot d\vec{r}$ on the path C if

    (a) C runs from $(10,0)$ to $(0,-10)$ along the circle of radius 10 centered at the origin.

    (b) C runs from $(10,0)$ to $(0,-10)$ along a straight line.

36. Suppose C is a path that begins and ends at the same point $P = (15, 20)$. What, if anything, can you say about $\int_C (p(x,y)\vec{i} + q(x,y)\vec{j}) \cdot d\vec{r}$?

    (a) With no assumptions about p and q.

    (b) If $p(x,y)\vec{i} + q(x,y)\vec{j}$ has a potential function.

37. Let $\vec{F} = -y\vec{i} + x\vec{j}$ and let C be the circle of radius 5 centered at the origin, oriented counterclockwise.

    (a) Evaluate $\int_C \vec{F} \cdot d\vec{r}$.

    (b) Give a potential function for $\vec{F}$ or explain why there are none.

38. Let $\vec{F} = y\vec{i}$.

    (a) Evaluate $\int_{C_1} \vec{F} \cdot d\vec{r}$ if C_1 is the straight line path from $(0,0)$ to $(1,0)$.

    (b) Evaluate $\int_{C_2} \vec{F} \cdot d\vec{r}$ if C_2 is the path along three edges of a square, from $(0,0)$ to $(0,1)$ to $(1,1)$ to $(1,0)$.

    (c) Does $\vec{F}$ have a potential function? Either give one or explain why there are none.

39. If $df = p\,dx + q\,dy$ for smooth f, explain why

$$\frac{\partial p}{\partial y} = \frac{\partial q}{\partial x}.$$

40. If $\vec{F}(x,y,z) = 2xe^{x^2+yz}\vec{i} + ze^{x^2+yz}\vec{j} + ye^{x^2+yz}\vec{k}$, find exactly the line integral of $\vec{F}$ along the curve consisting of the two half circles in the plane $z = 0$ in Figure 18.35.

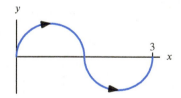

Figure 18.35

41. Let grad $f = 2xe^{x^2} \sin y\vec{i} + e^{x^2} \cos y\vec{j}$. Find the change in f between $(0,0)$ and $(1, \pi/2)$:

    (a) By computing a line integral.

    (b) By computing f.

42. Let C be the quarter of the unit circle centered at the origin, traversed counterclockwise starting on the negative x-axis. Find the exact values of

    (a) $\displaystyle\int_C (2\pi x\vec{i} + y^2\vec{j}) \cdot d\vec{r}$ **(b)** $\displaystyle\int_C (-2y\vec{i} + x\vec{j}) \cdot d\vec{r}$

■For the vector fields in Problems **43–46**, find the line integral along the curve C from the origin along the x-axis to the point $(3,0)$ and then counterclockwise around the circumference of the circle $x^2 + y^2 = 9$ to the point $(3/\sqrt{2}, 3/\sqrt{2})$.

43. $\vec{F} = x\vec{i} + y\vec{j}$

44. $\vec{H} = -y\vec{i} + x\vec{j}$

45. $\vec{F} = y(x+1)^{-1}\vec{i} + \ln(x+1)\vec{j}$

46. $\vec{G} = (ye^{xy} + \cos(x+y))\vec{i} + (xe^{xy} + \cos(x+y))\vec{j}$

47. Let C be the helix $x = \cos t$, $y = \sin t$, $z = t$ for $0 \le t \le 1.25\pi$. Find $\int_C \vec{F} \cdot d\vec{r}$ exactly for

$$\vec{F} = yz^2 e^{xyz^2}\vec{i} + xz^2 e^{xyz^2}\vec{j} + 2xyz e^{xyz^2}\vec{k}.$$

48. Let $\vec{F} = 2x\vec{i} + 2y\vec{j} + 2z\vec{k}$ and $\vec{G} = (2x+y)\vec{i} + 2y\vec{j} + 2z\vec{k}$. Let C be the line from the origin to the point $(1,5,9)$. Find $\int_C \vec{F} \cdot d\vec{r}$ and use the result to find $\int_C \vec{G} \cdot d\vec{r}$.

49. **(a)** If $\vec{F} = ye^x\vec{i} + e^x\vec{j}$, explain how the Fundamental Theorem of Calculus for Line Integrals enables you to calculate $\int_C \vec{F} \cdot d\vec{r}$ where C is any curve going from the point $(1,2)$ to the point $(3,7)$. Explain why it does not matter how the curve goes.

    (b) If C is the line from the point $(1,2)$ to the point $(3,7)$, calculate the line integral in part (a) without using the Fundamental Theorem.

50. Calculate the line integral $\int_C \vec{F} \cdot d\vec{r}$ exactly, where C is the curve from P to Q in Figure 18.36 and

$$\vec{F} = \sin\left(\frac{x}{2}\right)\sin\left(\frac{y}{2}\right)\vec{i} - \cos\left(\frac{x}{2}\right)\cos\left(\frac{y}{2}\right)\vec{j}.$$

The curves PR, RS and SQ are trigonometric functions of period 2π and amplitude 1.

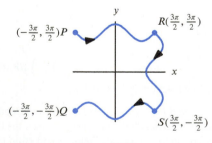

Figure 18.36

51. The domain of $f(x, y)$ is the xy-plane; values of f are in Table 18.1. Find $\int_C \text{grad } f \cdot d\vec{r}$, where C is

   (a) A line from $(0, 2)$ to $(3, 4)$.

   (b) A circle of radius 1 centered at $(1, 2)$ traversed counterclockwise.

Table 18.1

$y \backslash^x$	0	1	2	3	4
0	53	57	59	58	56
1	56	58	59	59	57
2	57	58	59	60	59
3	59	60	61	62	61
4	62	63	65	66	69

52. Figure 18.37 shows the vector field $\vec{F}(x, y) = x\vec{j}$.

   (a) Find paths C_1, C_2, and C_3 from P to Q such that

$$\int_{C_1} \vec{F} \cdot d\vec{r} = 0, \quad \int_{C_2} \vec{F} \cdot d\vec{r} > 0, \quad \int_{C_3} \vec{F} \cdot d\vec{r} < 0.$$

   (b) Is $\vec{F}$ a gradient field? Explain.

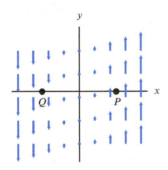

Figure 18.37

53. (a) Figure 18.38 shows level curves of $f(x, y)$. Sketch a vector at P in the direction of grad f.

   (b) Is the length of grad f at P longer, shorter, or the same length as the length of grad f at Q?

   (c) If C is a curve going from P to Q, find $\int_C \text{grad } f \cdot d\vec{r}$.

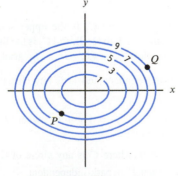

Figure 18.38

54. Consider the line integrals, $\int_{C_i} \vec{F} \cdot d\vec{r}$, for $i = 1, 2, 3, 4$, where C_i is the path from P_i to Q_i shown in Figure 18.39 and $\vec{F} = \text{grad } f$. Level curves of f are also shown in Figure 18.39.

   (a) Which of the line integral(s) is (are) zero?

   (b) Arrange the four line integrals in ascending order (from least to greatest).

   (c) Two of the nonzero line integrals have equal and opposite values. Which are they? Which is negative and which is positive?

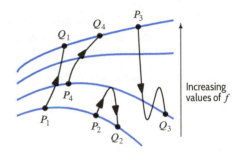

Figure 18.39

55. Consider the vector field $\vec{F}$ shown in Figure 18.40.

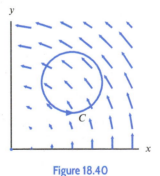

Figure 18.40

   (a) Is $\int_C \vec{F} \cdot d\vec{r}$ positive, negative, or zero?

   (b) From your answer to part (a), can you determine whether or not $\vec{F} = \text{grad } f$ for some function f?

   (c) Which of the following formulas best fits $\vec{F}$?

$$\vec{F}_1 = \frac{x}{x^2 + y^2}\vec{i} + \frac{y}{x^2 + y^2}\vec{j},$$

$$\vec{F}_2 = -y\vec{i} + x\vec{j},$$

$$\vec{F}_3 = \frac{-y}{(x^2 + y^2)^2}\vec{i} + \frac{x}{(x^2 + y^2)^2}\vec{j}.$$

56. If $\vec{F}$ is a path-independent vector field, with
$\int_{(0,0)}^{(1,0)} \vec{F} \cdot d\vec{r} = 5.1$ and $\int_{(1,0)}^{(1,1)} \vec{F} \cdot d\vec{r} = 3.2$ and
$\int_{(0,1)}^{(1,1)} \vec{F} \cdot d\vec{r} = 4.7$, find

$$\int_{(0,1)}^{(0,0)} \vec{F} \cdot d\vec{r}.$$

57. The path C is a line segment of length 10 in the plane starting at $(2,1)$. For $f(x,y) = 3x + 4y$, consider

$$\int_C \text{grad } f \cdot d\vec{r}.$$

(a) Where should the other end of the line segment C be placed to maximize the value of the integral?

(b) What is the maximum value of the integral?

58. Let $\vec{r} = x\vec{i} + y\vec{j} + z\vec{k}$ and $\vec{a} = a_1\vec{i} + a_2\vec{j} + a_3\vec{k}$, a constant vector.

(a) Find $\text{grad}(\vec{r} \cdot \vec{a})$.

(b) Let C be a path from the origin to the point with position vector $\vec{r}_0$. Find $\int_C \text{grad}(\vec{r} \cdot \vec{a}) \cdot d\vec{r}$.

(c) If $||\vec{r}_0|| = 10$, what is the maximum possible value of $\int_C \text{grad}(\vec{r} \cdot \vec{a}) \cdot d\vec{r}$? Explain.

59. The force exerted by gravity on a refrigerator of mass m is $\vec{F} = -mg\vec{k}$.

(a) Find the work done against this force in moving from the point $(1,0,0)$ to the point $(1,0,2\pi)$ along the curve $x = \cos t, y = \sin t, z = t$ by calculating a line integral.

(b) Is $\vec{F}$ conservative (that is, path-independent)? Give a reason for your answer.

60. A particle subject to a force $\vec{F}(x,y) = y\vec{i} - x\vec{j}$ moves clockwise along the arc of the unit circle, centered at the origin, that begins at $(-1,0)$ and ends at $(0,1)$.

(a) Find the work done by $\vec{F}$. Explain the sign of your answer.

(b) Is $\vec{F}$ path-independent? Explain.

Strengthen Your Understanding

■ In Problems 61–63, explain what is wrong with the statement.

61. If $\vec{F}$ is a gradient field and C is an oriented path from point P to point Q, then $\int_C \vec{F} \cdot d\vec{r} = \vec{F}(Q) - \vec{F}(P)$.

62. Given any vector field $\vec{F}$ and a point P, the function $f(Q) = \int_C \vec{F} \cdot d\vec{r}$, where C is a path from P to Q, is a potential function for $\vec{F}$.

63. If a vector field $\vec{F}$ is not a gradient vector field, then $\int_C \vec{F} \cdot d\vec{r}$ can't be evaluated.

■ In Problems 64–65, give an example of:

64. A vector field $\vec{F}$ such that $\int_C \vec{F} \cdot d\vec{r} = 100$, for every oriented path C from $(0,0)$ to $(1,2)$.

65. A path-independent vector field.

■ In Problems 66–69, each of the statements is *false*. Explain why or give a counterexample.

66. If $\int_C \vec{F} \cdot d\vec{r} = 0$ for one particular closed path C, then $\vec{F}$ is path-independent.

67. $\int_C \vec{F} \cdot d\vec{r}$ is the total change in $\vec{F}$ along C.

68. If the vector fields $\vec{F}$ and $\vec{G}$ have $\int_C \vec{F} \cdot d\vec{r} = \int_C \vec{G} \cdot d\vec{r}$ for a particular path C, then $\vec{F} = \vec{G}$.

69. If the total change of a function f along a curve C is zero, then C must be a contour of f.

■ Are the statements in Problems 70–80 true or false? Give reasons for your answer.

70. The fact that the line integral of a vector field $\vec{F}$ is zero around the unit circle $x^2 + y^2 = 1$ means that $\vec{F}$ must be a gradient vector field.

71. If C is the line segment that starts at $(0,0)$ and ends at (a,b) then $\int_C (x\vec{i} + y\vec{j}) \cdot d\vec{r} = \frac{1}{2}(a^2 + b^2)$.

72. The circulation of any vector field $\vec{F}$ around any closed curve C is zero.

73. If $\vec{F} = \text{grad } f$, then $\vec{F}$ is path-independent.

74. If $\vec{F}$ is path-independent, then $\int_{C_1} \vec{F} \cdot d\vec{r} = \int_{C_2} \vec{F} \cdot d\vec{r}$, where C_1 and C_2 are any paths.

75. The line integral $\int_C \vec{F} \cdot d\vec{r}$ is the total change of $\vec{F}$ along C.

76. If $\vec{F}$ is path-independent, then there is a potential function for $\vec{F}$.

77. If $f(x,y) = e^{\cos(xy)}$, and C_1 is the upper semicircle $x^2 + y^2 = 1$ from $(-1,0)$ to $(1,0)$, and C_2 is the line from $(-1,0)$ to $(1,0)$, then $\int_{C_1} \text{grad } f \cdot d\vec{r} = \int_{C_2} \text{grad } f \cdot d\vec{r}$.

78. If $\vec{F}$ is path-independent and C is any closed curve, then $\int_C \vec{F} \cdot d\vec{r} = 0$.

79. The vector field $\vec{F}(x,y) = y^2\vec{i} + k\vec{j}$, where k is constant, is a gradient field.

80. If $\int_C \vec{F} \cdot d\vec{r} = 0$, where C is any circle of the form $x^2 + y^2 = a^2$, then $\vec{F}$ is path-independent.

18.4   PATH-DEPENDENT VECTOR FIELDS AND GREEN'S THEOREM

Suppose we are given a vector field but are not told whether it is path-independent. How can we tell if it has a potential function, that is, if it is a gradient field?

How to Tell If a Vector Field Is Path-Dependent Using Line Integrals

One way to decide if a vector field is path-dependent is to find two paths with the same endpoints such that the line integrals of the vector field along the two paths have different values.

Example 1   Is the vector field $\vec{G}$ shown in Figure 18.41 path-independent? At any point $\vec{G}$ has magnitude equal to the distance from the origin and direction perpendicular to the line joining the point to the origin.

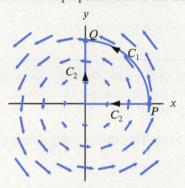

Figure 18.41: Is this vector field
path-independent?

Solution   We choose $P = (1, 0)$ and $Q = (0, 1)$ and two paths between them: C_1, a quarter circle of radius 1, and C_2, formed by parts of the x- and y-axes. (See Figure 18.41.)

Along C_1, the line integral $\int_{C_1} \vec{G} \cdot d\vec{r} > 0$, since $\vec{G}$ points in the direction of the curve.

Along C_2, however, we have $\int_{C_2} \vec{G} \cdot d\vec{r} = 0$, since $\vec{G}$ is perpendicular to C_2 everywhere.

Thus, $\vec{G}$ is not path-independent.

Path-Dependent Fields and Circulation

Notice that the vector field in the previous example has nonzero circulation around the origin. What can we say about the circulation of a general path-independent vector field $\vec{F}$ around a closed curve, C? Suppose C is a *simple* closed curve, that is, a closed curve that does not cross itself. If P and Q are any two points on the path, then we can think of C (oriented as shown in Figure 18.42) as made up of the path C_1 followed by $-C_2$. Since $\vec{F}$ is path-independent, we know that

$$\int_{C_1} \vec{F} \cdot d\vec{r} = \int_{C_2} \vec{F} \cdot d\vec{r}.$$

Thus, we see that the circulation around C is zero:

$$\int_{C} \vec{F} \cdot d\vec{r} = \int_{C_1} \vec{F} \cdot d\vec{r} + \int_{-C_2} \vec{F} \cdot d\vec{r} = \int_{C_1} \vec{F} \cdot d\vec{r} - \int_{C_2} \vec{F} \cdot d\vec{r} = 0.$$

If the closed curve C does cross itself, we break it into simple closed curves as shown in Figure 18.43 and apply the same argument to each one.

Now suppose we know that the line integral around any closed curve is zero. For any two points, P and Q, with two paths, C_1 and C_2, between them, create a closed curve, C, as in Figure 18.42. Since the circulation around this closed curve, C, is zero, the line integrals along the two paths, C_1 and C_2, are equal.[1] Thus, $\vec{F}$ is path-independent.

---

[1] A similar argument is used in Problems 57 and 58 on page 983.

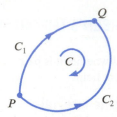

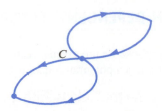

Figure 18.42: A simple closed curve C broken into two pieces, C_1 and C_2

Figure 18.43: A curve C which crosses itself can be broken into simple closed curves

Thus, we have the following result:

> A vector field is path-independent if and only if $\displaystyle\int_C \vec{F} \cdot d\vec{r} = 0$ for every closed curve C.

Hence, to see if a field is *path-dependent*, we look for a closed path with nonzero circulation. For instance, the vector field in Example 1 has nonzero circulation around a circle around the origin, showing it is path-dependent.

How to Tell If a Vector Field Is Path-Dependent Algebraically: The Curl

Example 2   Does the vector field $\vec{F} = 2xy\vec{i} + xy\vec{j}$ have a potential function? If so, find it.

Solution   Let's suppose $\vec{F}$ does have a potential function, f, so $\vec{F} = \text{grad } f$. This means that

$$\frac{\partial f}{\partial x} = 2xy \quad \text{and} \quad \frac{\partial f}{\partial y} = xy.$$

Integrating the expression for $\partial f/\partial x$ shows that we must have

$$f(x, y) = x^2 y + C(y) \qquad \text{where } C(y) \text{ is a function of } y.$$

Differentiating this expression for $f(x, y)$ with respect to y and using the fact that $\partial f/\partial y = xy$, we get

$$\frac{\partial f}{\partial y} = x^2 + C'(y) = xy.$$

Thus, we must have

$$C'(y) = xy - x^2.$$

But this expression for $C'(y)$ is impossible because $C'(y)$ is a function of y alone. This argument shows that there is no potential function for the vector field $\vec{F}$.

Is there an easier way to see that a vector field has no potential function, other than by trying to find the potential function and failing? The answer is yes. First we look at a 2-dimensional vector field $\vec{F} = F_1\vec{i} + F_2\vec{j}$. If $\vec{F}$ is a gradient field, then there is a potential function f such that

$$\vec{F} = F_1\vec{i} + F_2\vec{j} = \frac{\partial f}{\partial x}\vec{i} + \frac{\partial f}{\partial y}\vec{j}.$$

Thus,

$$F_1 = \frac{\partial f}{\partial x} \quad \text{and} \quad F_2 = \frac{\partial f}{\partial y}.$$

Let us assume that f has continuous second partial derivatives. Then, by the equality of mixed partial derivatives,

$$\frac{\partial F_1}{\partial y} = \frac{\partial^2 f}{\partial y \partial x} = \frac{\partial^2 f}{\partial x \partial y} = \frac{\partial F_2}{\partial x}.$$

Thus, we have the following result:

If $\vec{F}(x, y) = F_1 \vec{i} + F_2 \vec{j}$ is a gradient vector field with continuous partial derivatives, then

$$\frac{\partial F_2}{\partial x} - \frac{\partial F_1}{\partial y} = 0.$$

If $\vec{F}(x, y) = F_1 \vec{i} + F_2 \vec{j}$ is an arbitrary vector field, then we define the 2-dimensional or scalar **curl** of the vector field $\vec{F}$ to be

$$\frac{\partial F_2}{\partial x} - \frac{\partial F_1}{\partial y}.$$

Notice that we now know that if $\vec{F}$ is a gradient field, then its curl is 0. We do not (yet) know whether the converse is true. (That is: If the curl is 0, does $\vec{F}$ have to be a gradient field?) However, the curl already enables us to show that a vector field is *not* a gradient field.

Example 3   Show that $\vec{F} = 2xy\vec{i} + xy\vec{j}$ cannot be a gradient vector field.

Solution   We have $F_1 = 2xy$ and $F_2 = xy$. Since $\partial F_1 / \partial y = 2x$ and $\partial F_2 / \partial x = y$, in this case

$$\partial F_2 / \partial x - \partial F_1 / \partial y \neq 0$$

so $\vec{F}$ cannot be a gradient field.

Green's Theorem

We now have two ways of seeing that a vector field $\vec{F}$ in the plane is path-dependent. We can evaluate $\int_C \vec{F} \cdot d\vec{r}$ for some closed curve and find it is not zero, or we can show that $\partial F_2 / \partial x - \partial F_1 / \partial y \neq 0$. It's natural to think that

$$\int_C \vec{F} \cdot d\vec{r} \quad \text{and} \quad \frac{\partial F_2}{\partial x} - \frac{\partial F_1}{\partial y}$$

might be related. The relation is called Green's Theorem.

Theorem 18.3: Green's Theorem

Suppose C is a piecewise smooth simple closed curve that is the boundary of a region R in the plane and oriented so that the region is on the left as we move around the curve. See Figure 18.44. Suppose $\vec{F} = F_1 \vec{i} + F_2 \vec{j}$ is a smooth vector field on a region[2] containing R and C. Then

$$\int_C \vec{F} \cdot d\vec{r} = \int_R \left(\frac{\partial F_2}{\partial x} - \frac{\partial F_1}{\partial y} \right) dx\, dy.$$

The online supplement at www.WileyPLUS.com contains a proof of Green's Theorem with different, but equivalent, conditions on the region R.

---

[2]The region is an open region containing R and C.

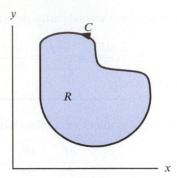

Figure 18.44: Boundary C oriented with R on the left

We first prove Green's Theorem in the case where the region R is the rectangle $a \leq x \leq b, c \leq y \leq d$. Figure 18.45 shows the boundary of R divided into four curves.

On C_1, where $y = c$ and $dy = 0$, we have $d\vec{r} = dx\vec{i}$ and thus

$$\int_{C_1} \vec{F} \cdot d\vec{r} = \int_a^b F_1(x, c)\, dx.$$

Similarly, on C_3 where $y = d$ we have

$$\int_{C_3} \vec{F} \cdot d\vec{r} = \int_b^a F_1(x, d)\, dx = -\int_a^b F_1(x, d)\, dx.$$

Hence

$$\int_{C_1 + C_3} \vec{F} \cdot d\vec{r} = \int_a^b F_1(x, c)\, dx - \int_a^b F_1(x, d)\, dx = -\int_a^b (F_1(x, d) - F_1(x, c))\, dx.$$

By the Fundamental Theorem of Calculus,

$$F_1(x, d) - F_1(x, c) = \int_c^d \frac{\partial F_1}{\partial y}\, dy$$

and therefore

$$\int_{C_1 + C_3} \vec{F} \cdot d\vec{r} = -\int_a^b \int_c^d \frac{\partial F_1}{\partial y}\, dy\, dx = -\int_c^d \int_a^b \frac{\partial F_1}{\partial y}\, dx\, dy.$$

Along the curve C_2, where $x = b$, and the curve C_4, where $x = a$, we get, by a similar argument,

$$\int_{C_2 + C_4} \vec{F} \cdot d\vec{r} = \int_c^d (F_2(b, y) - F_2(a, y))\, dy = \int_c^d \int_a^b \frac{\partial F_2}{\partial x}\, dx\, dy.$$

Adding the line integrals over $C_1 + C_3$ and $C_2 + C_4$, we get

$$\int_C \vec{F} \cdot d\vec{r} = \int_R \left(\frac{\partial F_2}{\partial x} - \frac{\partial F_1}{\partial y} \right) dx\, dy.$$

If R is not a rectangle, we subdivide it into small rectangular pieces as shown in Figure 18.46. The contribution to the integral of the non-rectangular pieces can be made as small as we like by making the subdivision fine enough. The double integrals over each piece add up to the double integral over the whole region R. Figure 18.47 shows how the circulations around adjacent pieces cancel along the common edge, so the circulations around all the pieces add up to the circulation around the boundary C. Since Green's Theorem holds for the rectangular pieces, it holds for the whole region R.

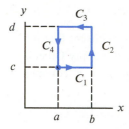

Figure 18.45: The boundary of a rectangle broken into C_1, C_2, C_3, C_4

Figure 18.46: Region R bounded by a closed curve C and split into many small regions, ΔR

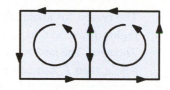

Figure 18.47: Two adjacent small closed curves

Example 4   Use Green's Theorem to evaluate $\int_C \left(y^2 \vec{i} + x \vec{j} \right) \cdot d\vec{r}$ where C is the counterclockwise path around the perimeter of the rectangle $0 \leq x \leq 2$, $0 \leq y \leq 3$.

Solution   We have $F_1 = y^2$ and $F_2 = x$. By Green's Theorem,

$$\int_C \left(y^2 \vec{i} + x \vec{j} \right) \cdot d\vec{r} = \int_R \left(\frac{\partial F_2}{\partial x} - \frac{\partial F_1}{\partial y} \right) dx\, dy = \int_0^3 \int_0^2 (1 - 2y)\, dx\, dy = -12.$$

The Curl Test for Vector Fields in the Plane

We already know that if $\vec{F} = F_1 \vec{i} + F_2 \vec{j}$ is a gradient field with continuous partial derivatives, then

$$\frac{\partial F_2}{\partial x} - \frac{\partial F_1}{\partial y} = 0.$$

Now we show that the converse is true if the domain of $\vec{F}$ has no holes in it. This means that we assume that

$$\frac{\partial F_2}{\partial x} - \frac{\partial F_1}{\partial y} = 0$$

and show that $\vec{F}$ is path-independent. If C is any oriented simple closed curve in the domain of $\vec{F}$ and R is the region inside C, then

$$\int_R \left(\frac{\partial F_2}{\partial x} - \frac{\partial F_1}{\partial y} \right) dx\, dy = 0$$

since the integrand is identically 0. Therefore, by Green's Theorem,

$$\int_C \vec{F} \cdot d\vec{r} = \int_R \left(\frac{\partial F_2}{\partial x} - \frac{\partial F_1}{\partial y} \right) dx\, dy = 0.$$

Thus, $\vec{F}$ is path-independent and therefore a gradient field. This argument is valid for every closed curve, C, provided the region R is entirely in the domain of $\vec{F}$. Thus, we have the following result:

The Curl Test for Vector Fields in 2-Space

Suppose $\vec{F} = F_1 \vec{i} + F_2 \vec{j}$ is a vector field with continuous partial derivatives such that
- The domain of $\vec{F}$ has the property that every closed curve in it encircles a region that lies entirely within the domain. In particular, the domain of $\vec{F}$ has no holes.
- $\dfrac{\partial F_2}{\partial x} - \dfrac{\partial F_1}{\partial y} = 0.$

Then $\vec{F}$ is path-independent, so $\vec{F}$ is a gradient field and has a potential function.

Why Are Holes in the Domain of the Vector Field Important?

The reason for assuming that the domain of the vector field $\vec{F}$ has no holes is to ensure that the region R inside C is actually contained in the domain of $\vec{F}$. Otherwise, we cannot apply Green's Theorem. The next two examples show that if $\partial F_2/\partial x - \partial F_1/\partial y = 0$ but the domain of $\vec{F}$ contains a hole, then $\vec{F}$ can either be path-independent or be path-dependent.

Example 5   Let $\vec{F}$ be the vector field given by $\vec{F}(x, y) = \dfrac{-y\vec{i} + x\vec{j}}{x^2 + y^2}$.

(a) Calculate $\dfrac{\partial F_2}{\partial x} - \dfrac{\partial F_1}{\partial y}$. Does the curl test imply that $\vec{F}$ is path-independent?

(b) Calculate $\displaystyle\int_C \vec{F} \cdot d\vec{r}$, where C is the unit circle centered at the origin and oriented counterclockwise. Is $\vec{F}$ a path-independent vector field?

(c) Explain why the answers to parts (a) and (b) do not contradict Green's Theorem.

Solution   (a) Taking partial derivatives, we have

$$\frac{\partial F_2}{\partial x} = \frac{\partial}{\partial x}\left(\frac{x}{x^2 + y^2}\right) = \frac{1}{x^2 + y^2} - \frac{x \cdot 2x}{(x^2 + y^2)^2} = \frac{y^2 - x^2}{(x^2 + y^2)^2}.$$

Similarly,

$$\frac{\partial F_1}{\partial y} = \frac{\partial}{\partial y}\left(\frac{-y}{x^2 + y^2}\right) = \frac{-1}{x^2 + y^2} + \frac{y \cdot 2y}{(x^2 + y^2)^2} = \frac{y^2 - x^2}{(x^2 + y^2)^2}.$$

Thus,

$$\frac{\partial F_2}{\partial x} - \frac{\partial F_1}{\partial y} = 0.$$

Since $\vec{F}$ is undefined at the origin, the domain of $\vec{F}$ contains a hole. Therefore, the curl test does not apply.

(b) See Figure 18.49. On the unit circle, $\vec{F}$ is tangent to the circle and $||\vec{F}|| = 1$. Thus,[3]

$$\int_C \vec{F} \cdot d\vec{r} = ||\vec{F}|| \cdot \text{Length of curve} = 1 \cdot 2\pi = 2\pi.$$

Since the line integral around the closed curve C is nonzero, $\vec{F}$ is not path-independent. We observe that $\vec{F} = \text{grad}(\arctan(y/x))$ where $x \neq 0$. However, $\arctan(y/x)$ is not defined anywhere on the y-axis. Thus Theorem 18.1 (Fundamental Theorem of Line Integrals) does not hold on the unit circle, so we cannot use it to get path independence.[4]

(c) The domain of $\vec{F}$ is the "punctured plane," as shown in Figure 18.48. Since $\vec{F}$ is not defined at the origin, which is inside C, Green's Theorem does not apply. In this case

$$2\pi = \int_C \vec{F} \cdot d\vec{r} \neq \int_R \left(\frac{\partial F_2}{\partial x} - \frac{\partial F_1}{\partial y}\right) dx\,dy = 0.$$

---

[3] See Problem 56 on page 983.

[4] It can be shown that $\arctan(y/x)$ is θ from polar coordinates. The fact that θ increases by 2π each time we wind once around the origin counterclockwise explains why $\vec{F}$ is not path-independent.

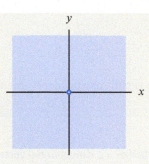

Figure 18.48: The domain of $\vec{F}(x, y) = \frac{-y\vec{i} + x\vec{j}}{x^2 + y^2}$ is the plane minus the origin

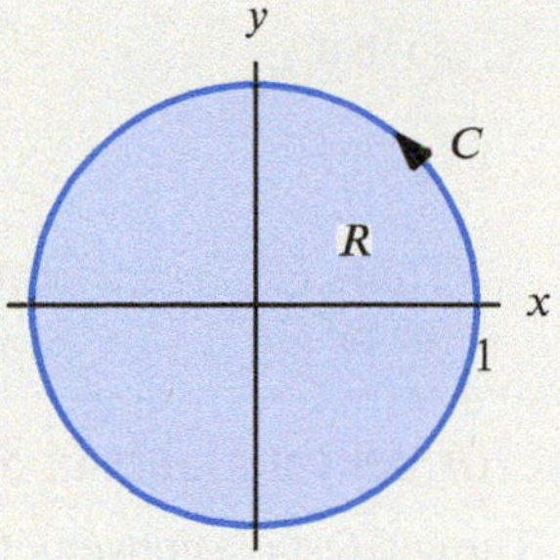

Figure 18.49: The region R is *not* contained in the domain of $\vec{F}(x, y) = \frac{-y\vec{i} + x\vec{j}}{x^2 + y^2}$

Although the vector field $\vec{F}$ in the last example was not defined at the origin, this by itself does not prevent the vector field from being path-independent, as we see in the following example.

Example 6   Consider the vector field $\vec{F}$ given by $\vec{F}(x, y) = \dfrac{x\vec{i} + y\vec{j}}{x^2 + y^2}$.

(a) Calculate $\dfrac{\partial F_2}{\partial x} - \dfrac{\partial F_1}{\partial y}$. Does the curl test imply that $\vec{F}$ is path-independent?

(b) Explain how we know that $\displaystyle\int_C \vec{F} \cdot d\vec{r} = 0$, where C is the unit circle centered at the origin and oriented counterclockwise. Does this imply that $\vec{F}$ is path-independent?

(c) Check that $f(x, y) = \frac{1}{2}\ln(x^2 + y^2)$ is a potential function for $\vec{F}$. Does this imply that $\vec{F}$ is path-independent?

Solution   (a) Taking partial derivatives, we have

$$\frac{\partial F_2}{\partial x} = \frac{\partial}{\partial x}\left(\frac{y}{x^2 + y^2}\right) = \frac{-2xy}{(x^2 + y^2)^2}, \quad \text{and} \quad \frac{\partial F_1}{\partial y} = \frac{\partial}{\partial y}\left(\frac{x}{x^2 + y^2}\right) = \frac{-2xy}{(x^2 + y^2)^2}.$$

Therefore,

$$\frac{\partial F_2}{\partial x} - \frac{\partial F_1}{\partial y} = 0.$$

This does *not* imply that $\vec{F}$ is path-independent: The domain of $\vec{F}$ contains a hole since $\vec{F}$ is undefined at the origin. Thus, the curl test does not apply.

(b) Since $\vec{F}(x, y) = x\vec{i} + y\vec{j} = \vec{r}$ on the unit circle C, the field $\vec{F}$ is everywhere perpendicular to C. Thus

$$\int_C \vec{F} \cdot d\vec{r} = 0.$$

The fact that $\int_C \vec{F} \cdot d\vec{r} = 0$ when C is the unit circle does *not* imply that $\vec{F}$ is path-independent. To be sure that $\vec{F}$ is path-independent, we would have to show that $\int_C \vec{F} \cdot d\vec{r} = 0$ for *every* closed curve C in the domain of $\vec{F}$, not just the unit circle.

(c) To check that grad $f = \vec{F}$, we differentiate f:

$$f_x = \frac{1}{2}\frac{\partial}{\partial x}\ln(x^2 + y^2) = \frac{1}{2}\frac{2x}{x^2 + y^2} = \frac{x}{x^2 + y^2},$$

and

$$f_y = \frac{1}{2}\frac{\partial}{\partial y}\ln(x^2 + y^2) = \frac{1}{2}\frac{2y}{x^2 + y^2} = \frac{y}{x^2 + y^2},$$

so that

$$\text{grad } f = \frac{x\vec{i} + y\vec{j}}{x^2 + y^2} = \vec{F}.$$

Thus, $\vec{F}$ is a gradient field and therefore is path-independent—even though $\vec{F}$ is undefined at the origin.

The Curl Test for Vector Fields in 3-Space

The curl test is a convenient way of deciding whether a 2-dimensional vector field is path-independent. Fortunately, there is an analogous test for 3-dimensional vector fields, although we cannot justify it until Chapter 20.

If $\vec{F}(x, y, z) = F_1\vec{i} + F_2\vec{j} + F_3\vec{k}$ is a vector field on 3-space we define a new vector field, curl $\vec{F}$, on 3-space by

$$\text{curl } \vec{F} = \left(\frac{\partial F_3}{\partial y} - \frac{\partial F_2}{\partial z}\right)\vec{i} + \left(\frac{\partial F_1}{\partial z} - \frac{\partial F_3}{\partial x}\right)\vec{j} + \left(\frac{\partial F_2}{\partial x} - \frac{\partial F_1}{\partial y}\right)\vec{k}.$$

The vector field curl $\vec{F}$ can be used to determine whether the vector field $\vec{F}$ is path-independent.

The Curl Test for Vector Fields in 3-Space

Suppose $\vec{F}$ is a vector field on 3-space with continuous partial derivatives such that
- The domain of $\vec{F}$ has the property that every closed curve in it can be contracted to a point in a smooth way, staying at all times within the domain.
- curl $\vec{F} = \vec{0}$.

Then $\vec{F}$ is path-independent, so $\vec{F}$ is a gradient field and has a potential function.

For the 2-dimensional curl test, the domain of $\vec{F}$ must have no holes. This meant that if $\vec{F}$ was defined on a simple closed curve C, then it was also defined at all points inside C. One way to test for holes is to try to "lasso" them with a closed curve. If every closed curve in the domain can be pulled to a point without hitting a hole, that is, without straying outside the domain, then the domain has no holes. In 3-space, we need the same condition to be satisfied: we must be able to pull every closed curve to a point, like a lasso, without straying outside the domain.

Example 7   Decide if the following vector fields are path-independent and whether or not the curl test applies.

(a) $\vec{F} = \dfrac{x\vec{i} + y\vec{j} + z\vec{k}}{(x^2 + y^2 + z^2)^{3/2}}$

(b) $\vec{G} = \dfrac{-y\vec{i} + x\vec{j}}{x^2 + y^2} + z^2\vec{k}$

Solution   (a) Suppose $f = -(x^2 + y^2 + z^2)^{-1/2}$. Then $f_x = x(x^2 + y^2 + z^2)^{-3/2}$ and f_y and f_z are similar, so grad $f = \vec{F}$. Thus, $\vec{F}$ is a gradient field and therefore path-independent. Calculations show curl $\vec{F} = \vec{0}$. The domain of $\vec{F}$ is all of 3-space minus the origin, and any closed curve in the domain can be pulled to a point without leaving the domain. Thus, the curl test applies.

(b) Let C be the circle $x^2 + y^2 = 1$, $z = 0$ traversed counterclockwise when viewed from the positive z-axis. Since $z = 0$ on the curve C, the vector field $\vec{G}$ reduces to the vector field in Example 5 and is everywhere tangent to C and of magnitude 1, so

$$\int_C \vec{G} \cdot d\vec{r} = \|\vec{G}\| \cdot \text{Length of curve} = 1 \cdot 2\pi = 2\pi.$$

Since the line integral around this closed curve is nonzero, $\vec{G}$ is path-dependent. Computations show curl $\vec{G} = \vec{0}$. However, the domain of $\vec{G}$ is all of 3-space minus the z-axis, and it does not satisfy the curl test domain criterion. For example, the circle, C, is lassoed around the z-axis, and cannot be pulled to a point without hitting the z-axis. Thus, the curl test does not apply.

Summary for Section 18.4

- **The 2-dimensional curl:** If $\vec{F}(x, y) = F_1\vec{i} + F_2\vec{j}$ is a 2-dimensional vector field, then we define the 2-dimensional or scalar curl of $\vec{F}$ to be $\dfrac{\partial F_2}{\partial x} - \dfrac{\partial F_1}{\partial y}$.

- **Green's Theorem:** Suppose C is a piecewise smooth simple closed curve that is the boundary of a region R in the plane and oriented so that the region is on the left as we move around the curve. Suppose $\vec{F} = F_1\vec{i} + F_2\vec{j}$ is a smooth vector field on a region containing R and C. Then

$$\int_C \vec{F} \cdot d\vec{r} = \int_R \left(\frac{\partial F_2}{\partial x} - \frac{\partial F_1}{\partial y} \right) dx\, dy.$$

- **The curl test for vector fields in 2-space:** The vector field $\vec{F} = F_1\vec{i} + F_2\vec{j}$ is a path-independent vector field if both
  - The domain of $\vec{F}$ has no holes, and
  - $\dfrac{\partial F_2}{\partial x} - \dfrac{\partial F_1}{\partial y} = 0.$

- **The curl of a 3-dimensional vector field:** If $\vec{F}(x, y, z) = F_1\vec{i} + F_2\vec{j} + F_3\vec{k}$, we define the curl vector field by

$$\text{curl}\, \vec{F} = \left(\frac{\partial F_3}{\partial y} - \frac{\partial F_2}{\partial z} \right)\vec{i} + \left(\frac{\partial F_1}{\partial z} - \frac{\partial F_3}{\partial x} \right)\vec{j} + \left(\frac{\partial F_2}{\partial x} - \frac{\partial F_1}{\partial y} \right)\vec{k}.$$

- **The curl test for vector fields in 3-space:** The vector field $\vec{F} = F_1\vec{i} + F_2\vec{j} + F_3\vec{k}$ is a path-independent vector field if both
  - The domain of $\vec{F}$ has the property that every closed curve in it can be contracted to a point, and
  - $\text{curl}\, \vec{F} = \vec{0}.$

Exercises and Problems for Section 18.4

EXERCISES

In Exercises 1–10, decide if the given vector field is the gradient of a function f. If so, find f. If not, explain why not.

1. $y\vec{i} - x\vec{j}$

2. $2xy\vec{i} + x^2\vec{j}$

3. $y\vec{i} + y\vec{j}$

4. $2xy\vec{i} + 2xy\vec{j}$

5. $(x^2 + y^2)\vec{i} + 2xy\vec{j}$

6. $(2xy^3 + y)\vec{i} + (3x^2y^2 + x)\vec{j}$

7. $\dfrac{\vec{i}}{x} + \dfrac{\vec{j}}{y} + \dfrac{\vec{k}}{z}$

8. $\dfrac{\vec{i}}{x} + \dfrac{\vec{j}}{y} + \dfrac{\vec{k}}{xy}$

9. $2x\cos(x^2 + z^2)\vec{i} + \sin(x^2 + z^2)\vec{j} + 2z\cos(x^2 + z^2)\vec{k}$

10. $\dfrac{y}{x^2 + y^2}\vec{i} - \dfrac{x}{x^2 + y^2}\vec{j}$

In Exercises 11–14, use Green's Theorem to calculate the circulation of $\vec{F}$ around the curve, oriented counterclockwise.

11. $\vec{F} = y\vec{i} - x\vec{j}$ around the unit circle.

12. $\vec{F} = xy\vec{j}$ around the square $0 \le x \le 1, 0 \le y \le 1$.

13. $\vec{F} = (2x^2 + 3y)\vec{i} + (2x + 3y^2)\vec{j}$ around the triangle with vertices $(2, 0), (0, 3), (-2, 0)$.

14. $\vec{F} = 3y\vec{i} + xy\vec{j}$ around the unit circle.

15. Use Green's Theorem to evaluate $\int_C \left(y^2\vec{i} + x\vec{j} \right) \cdot d\vec{r}$ where C is the counterclockwise path around the perimeter of the rectangle $0 \le x \le 2, 0 \le y \le 3$.

16. If C goes counterclockwise around the perimeter of the rectangle R with vertices $(10, 10), (30, 10), (30, 20)$, and $(10, 20)$, use Green's Theorem to evaluate

$$\int_C -y\, dx + x\, dy.$$

17. Calculate $\int_C ((3x + 5y)\vec{i} + (2x + 7y)\vec{j}) \cdot d\vec{r}$ where C is the circular path with center (a, b) and radius m, oriented counterclockwise. Use Green's Theorem.

PROBLEMS

18. Find the line integral of $\vec{F} = e^{x^2}\vec{i} + (7x + 1)\vec{j}$ around the closed curve C consisting of the two line segments and the circular arc in Figure 18.50.

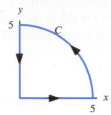

Figure 18.50

19. (a) Sketch $\vec{F} = y\vec{i}$ and determine the sign of the circulation of $\vec{F}$ around the unit circle centered at the origin and oriented counterclockwise.
(b) Use Green's Theorem to compute the circulation in part (a) exactly.

20. Let $\vec{F} = (\sin x)\vec{i} + (x + y)\vec{j}$. Find the line integral of $\vec{F}$ around the perimeter of the rectangle with corners $(3, 0)$, $(3, 5)$, $(-1, 5)$, $(-1, 0)$, traversed in that order.

21. Find $\displaystyle\int_C ((\sin(x^2)\cos y)\vec{i} + (\sin(y^2) + e^x)\vec{j}) \cdot d\vec{r}$ where C is the square of side 1 in the first quadrant of the xy-plane, with one vertex at the origin and sides along the axes, and oriented counterclockwise when viewed from above.

In Problems 22–25, find the line integral of $\vec{F}$ around C, the boundary of the region between $y = x^2$ and $y = x^3$ in the first quadrant, oriented counterclockwise.

22. $\vec{F} = 4\vec{i} + 5\vec{j}$

23. $\vec{F} = (x^2 - y)\vec{i} + (2x - e^y)\vec{j}$

24. $\vec{F} = (y^2 + e^x)\vec{i} + xy\vec{j}$

25. $\vec{F} = xy^2\vec{i} + (x - y)\vec{j}$

In Problems 26–27 find the line integral of $\vec{F}$ around the closed curve in Figure 18.51. The arc is part of a circle.

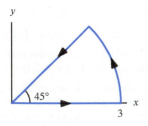

Figure 18.51

26. $\vec{F} = (x - y)\vec{i} + x\vec{j}$

27. $\vec{F} = (x + y)\vec{i} + \sin y\vec{j}$

28. Let $\vec{F} = 2xe^y\vec{i} + x^2e^y\vec{j}$ and $\vec{G} = (x - y)\vec{i} + (x + y)\vec{j}$. Let C be the line from $(0, 0)$ to $(2, 4)$. Find exactly:

  (a) $\displaystyle\int_C \vec{F} \cdot d\vec{r}$ **(b)** $\displaystyle\int_C \vec{G} \cdot d\vec{r}$

29. Let $\vec{F} = y\vec{i} + x\vec{j}$ and $\vec{G} = 3y\vec{i} - 3x\vec{j}$. In Figure 18.52, the curve C_2 is the semicircle centered at the origin from $(-1, 1)$ to $(1, -1)$ and C_1 is the line segment from $(-1, 1)$ to $(1, -1)$, and $C = C_2 - C_1$. Find the following line integrals:

  (a) $\displaystyle\int_{C_1} \vec{F} \cdot d\vec{r}$ **(b)** $\displaystyle\int_C \vec{F} \cdot d\vec{r}$

  (c) $\displaystyle\int_{C_2} \vec{F} \cdot d\vec{r}$ **(d)** $\displaystyle\int_{C_2} \vec{G} \cdot d\vec{r}$

  (e) $\displaystyle\int_C \vec{G} \cdot d\vec{r}$ **(f)** $\displaystyle\int_{C_1} \vec{G} \cdot d\vec{r}$

  (g) $\displaystyle\int_C (\vec{F} + \vec{G}) \cdot d\vec{r}$

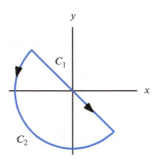

Figure 18.52

30. Calculate $\displaystyle\int_C \left((x^2 - y)\vec{i} + (y^2 + x)\vec{j}\right) \cdot d\vec{r}$ if:

  (a) C is the circle $(x - 5)^2 + (y - 4)^2 = 9$ oriented counterclockwise.
  (b) C is the circle $(x - a)^2 + (y - b)^2 = R^2$ in the xy-plane oriented counterclockwise.

31. Let C_1 be the curve consisting of the circle of radius 2, centered at the origin and oriented counterclockwise, and C_2 be the curve consisting of the line segment from $(0, 0)$ to $(1, 1)$ followed by the line segment from $(1, 1)$ to $(3, 1)$. Let $\vec{F} = 2xy^2\vec{i} + (2yx^2 + 2y)\vec{j}$ and let $\vec{G} = (y + x)\vec{i} + (y - x)\vec{j}$. Compute the following line integrals.

  (a) $\displaystyle\int_{C_1} \vec{F} \cdot d\vec{r}$ **(b)** $\displaystyle\int_{C_2} \vec{F} \cdot d\vec{r}$

  (c) $\displaystyle\int_{C_1} \vec{G} \cdot d\vec{r}$ **(d)** $\displaystyle\int_{C_2} \vec{G} \cdot d\vec{r}$

32. Prove that Green's Theorem is true when the integrand of the line integral has a potential function.

33. Consider the following parametric equations:

$$C_1 \ : \ \vec{r}(t) = t\cos(2\pi t)\,\vec{i} + t\sin(2\pi t)\,\vec{k}\,, 0 \le t \le 2$$
$$C_2 \ : \ \vec{r}(t) = t\cos(2\pi t)\,\vec{i} + t\,\vec{j} + t\sin(2\pi t)\,\vec{k}\,, 0 \le t \le 2$$

(a) Describe, in words, the motion of a particle moving along each of the paths.

(b) Evaluate $\int_{C_2} \vec{F} \cdot d\vec{r}$, for the vector field $\vec{F} = yz\,\vec{i} + z(x+1)\,\vec{j} + (xy+y+1)\,\vec{k}$.

(c) Find a nonzero vector field $\vec{G}$ such that:

$$\int_{C_1} \vec{G} \cdot d\vec{r} = \int_{C_2} \vec{G} \cdot d\vec{r}.$$

Explain how you reasoned to find $\vec{G}$.

(d) Find two different, nonzero vector fields $\vec{H}_1$, $\vec{H}_2$ such that:

$$\int_{C_1} \vec{H}_1 \cdot d\vec{r} = \int_{C_1} \vec{H}_2 \cdot d\vec{r}.$$

Explain how you reasoned to find the two fields.

34. Show that the line integral of $\vec{F} = x\vec{j}$ around a closed curve in the xy-plane, oriented as in Green's Theorem, measures the area of the region enclosed by the curve.

▪ In Problems **35–37**, use the result of Problem 34 to calculate the area of the region within the parameterized curves. In each case, sketch the curve.

35. The ellipse $x^2/a^2 + y^2/b^2 = 1$ parameterized by $x = a\cos t, y = b\sin t$, for $0 \le t \le 2\pi$.

36. The hypocycloid $x^{2/3} + y^{2/3} = a^{2/3}$ parameterized by $x = a\cos^3 t, y = a\sin^3 t, \ 0 \le t \le 2\pi$.

37. The folium of Descartes, $x^3 + y^3 = 3xy$, parameterized by $x = \dfrac{3t}{1+t^3}, y = \dfrac{3t^2}{1+t^3}$, for $0 \le t < \infty$.

38. The vector field $\vec{F}$ is defined on the disk D of radius 5 centered at the origin in the plane:

$$\vec{F} = (-y^3 + y\sin(xy))\vec{i} + (4x(1-y^2) + x\sin(xy))\vec{j}.$$

Consider the line integral $\int_C \vec{F} \cdot d\vec{r}$, where C is some closed curve contained in D. For which C is the value of this integral the largest? [Hint: Assume C is a closed curve, made up of smooth pieces and never crossing itself, and oriented counterclockwise.]

39. Example 1 on page 1003 showed that the vector field in Figure 18.53 could not be a gradient field by showing that it is not path-independent. Here is another way to see the same thing. Suppose that the vector field were the gradient of a function f. Draw and label a diagram showing what the contours of f would have to look like, and explain why it would not be possible for f to have a single value at any given point.

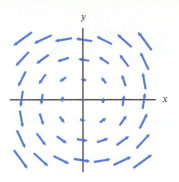

Figure 18.53

40. Repeat Problem 39 for the vector field in Problem 52 on page 1001.

41. (a) By finding potential functions, show that each of the vector fields $\vec{F}, \vec{G}, \vec{H}$ is a gradient field on some domain (not necessarily the whole plane).

(b) Find the line integrals of $\vec{F}, \vec{G}, \vec{H}$ around the unit circle in the xy-plane, centered at the origin, and traversed counterclockwise.

(c) For which of the three vector fields can Green's Theorem be used to calculate the line integral in part (b)? Why or why not?

$$\vec{F} = y\vec{i} + x\vec{j}, \ \vec{G} = \frac{y\vec{i} - x\vec{j}}{x^2 + y^2}, \ \vec{H} = \frac{x\vec{i} + y\vec{j}}{(x^2 + y^2)^{1/2}}$$

42. (a) For which of the following can you use Green's Theorem to evaluate the integral? Explain.

I $\displaystyle\int_C (x^2 + y^2)\,dx + (x^2 + y^2)\,dy$ where C is the curve defined by $y = x, y = x^2, 0 \le x \le 1$ with counterclockwise orientation.

II $\displaystyle\int_C \frac{1}{\sqrt{x^2 + y^2}}\,dx - \frac{1}{\sqrt{x^2 + y^2}}\,dy$ where C is the unit circle centered at the origin, oriented counterclockwise.

III $\displaystyle\int_C \vec{F} \cdot d\vec{r}$ where $\vec{F} = x\vec{i} + y\vec{j}$ where C is the line segment from the origin to $(1, 1)$.

(b) Use Green's Theorem to evaluate the integrals in part (a) that can be done that way.

43. Arrange the line integrals L_1, L_2, L_3 in ascending order, where

$$L_i = \int_{C_i} (-x^2 y\vec{i} + (xy^2 - x)\vec{j}) \cdot d\vec{r}.$$

The points A, B, D lie on the unit circle and C_i is one of the curves shown in Figure 18.54.

C_1: Line segment A to B

C_2: Line segment A to D followed by line segment D to B

C_3: Semicircle ADB

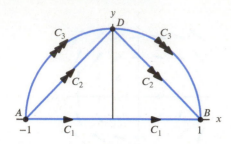

Figure 18.54

44. For all x, y, let $\vec{F} = F_1(x,y)\vec{i} + F_2(x,y)\vec{j}$ satisfy

$$\frac{\partial F_2}{\partial x} - \frac{\partial F_1}{\partial y} = 3.$$

(a) Calculate $\int_{C_1} \vec{F} \cdot d\vec{r}$ where C_1 is the unit circle in the xy-plane centered at the origin, oriented counterclockwise.

(b) Calculate $\int_{C_2} \vec{F} \cdot d\vec{r}$ where C_2 is the boundary of the rectangle of $4 \leq x \leq 7$, $5 \leq y \leq 7$, oriented counterclockwise.

(c) Let C_3 be the circle of radius 7 centered at the point $(10, 2)$; let C_4 be the circle of radius 8 centered at the origin; let C_5 be the square of side 14 centered at $(7, 7)$ with sides parallel to the axes; C_3, C_4, C_5 are all oriented counterclockwise. Arrange the integrals $\int_{C_3} \vec{F} \cdot d\vec{r}$, $\int_{C_4} \vec{F} \cdot d\vec{r}$, $\int_{C_5} \vec{F} \cdot d\vec{r}$ in increasing order.

45. Let $\vec{F} = (3x^2y + y^3 + e^x)\vec{i} + (e^{y^2} + 12x)\vec{j}$. Consider the line integral of $\vec{F}$ around the circle of radius a, centered at the origin and traversed counterclockwise.

(a) Find the line integral for $a = 1$.

(b) For which value of a is the line integral a maximum? Explain.

46. Let

$$\vec{F}(x, y) = \frac{-y\vec{i} + x\vec{j}}{x^2 + y^2}$$

and let oriented curves C_1 and C_2 be as in Figure 18.55. The curve C_2 is an arc of the unit circle centered at the origin. Show that

(a) The curl of $\vec{F}$ is zero.

(b) $\int_{C_1} \vec{F} \cdot d\vec{r} = \int_{C_2} \vec{F} \cdot d\vec{r}$.

(c) $\int_{C_1} \vec{F} \cdot d\vec{r} = \theta$, the angle at the origin subtended by the oriented curve C_1.

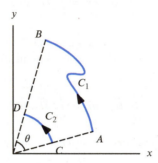

Figure 18.55

47. The electric field $\vec{E}$, at the point with position vector $\vec{r}$ in 3-space, due to a charge q at the origin is given by

$$\vec{E}(\vec{r}) = q \frac{\vec{r}}{\|\vec{r}\|^3}.$$

(a) Compute curl $\vec{E}$. Is $\vec{E}$ a path-independent vector field? Explain.

(b) Find a potential function φ for $\vec{E}$, if possible.

Strengthen Your Understanding

In Problems 48–49, explain what is wrong with the statement.

48. If $\int_C \vec{F} \cdot d\vec{r} = 0$ for a specific closed path C, then $\vec{F}$ must be path-independent.

49. Let $\vec{F} = F_1(x,y)\vec{i} + F_2(x,y)\vec{j}$ with

$$\frac{\partial F_2}{\partial x} - \frac{\partial F_1}{\partial y} = 3$$

and let C be the path consisting of line segments from $(0, 0)$ to $(1, 1)$ to $(2, 0)$. Then

$$\int_C \vec{F} \cdot d\vec{r} = 3.$$

In Problems 50–52, give an example of:

50. A function $Q(x, y)$ such that $\vec{F} = xy\vec{i} + Q(x,y)\vec{j}$ is a gradient field.

51. Two oriented curves, C_1 and C_2, from $(1, 0)$ to $(0, 1)$ such that if

$$\vec{F}(x, y) = \frac{-y\vec{i} + x\vec{j}}{x^2 + y^2},$$

then

$$\int_{C_1} \vec{F} \cdot d\vec{r} \neq \int_{C_2} \vec{F} \cdot d\vec{r}.$$

[Note that the scalar curl of $\vec{F}$ is 0 where $\vec{F}$ is defined.]

52. A vector field that is not a gradient field.

■Are the statements in Problems 53–60 true or false? Give reasons for your answer.

53. If $f(x)$ and $g(y)$ are continuous one-variable functions, then the vector field $\vec{F} = f(x)\vec{i} + g(y)\vec{j}$ is path-independent.

54. If $\vec{F} = \operatorname{grad} f$, and C is the perimeter of a square of side length a oriented counterclockwise and surrounding the region R, then

$$\int_C \vec{F} \cdot d\vec{r} = \int_R f \, dA.$$

55. If $\vec{F}$ and $\vec{G}$ are both path-independent vector fields, then $\vec{F} + \vec{G}$ is path-independent.

Online Resource: Review Problems and Projects

56. If $\vec{F}$ and $\vec{G}$ are both path-dependent vector fields, then $\vec{F} + \vec{G}$ is path-dependent.

57. The vector field $\vec{F}(\vec{r}) = \vec{r}$ in 3-space is path-independent.

58. A constant vector field $\vec{F} = a\vec{i} + b\vec{j}$ is path-independent.

59. If $\vec{F}$ is path-independent and k is a constant, then the vector field $k\vec{F}$ is path-independent.

60. If $\vec{F}$ is path-independent and $h(x, y)$ is a scalar function, then the vector field $h(x, y)\vec{F}$ is path-independent.

Chapter 19

FLUX INTEGRALS AND DIVERGENCE

CONTENTS

19.1   THE IDEA OF A FLUX INTEGRAL

Flow Through a Surface

Imagine water flowing through a fishing net stretched across a stream. Suppose we want to measure the flow rate of water through the net, that is, the volume of fluid that passes through the surface per unit time.

Example 1   A flat square surface of area A, in m^2, is immersed in a fluid. The fluid flows with constant velocity $\vec{v}$, in m/sec, perpendicular to the square. Write an expression for the rate of flow in m^3/sec.

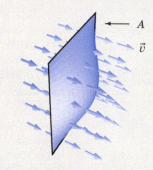

Figure 19.1: Fluid flowing perpendicular to a surface

Solution   In one second a given particle of water moves a distance of $\|\vec{v}\|$ in the direction perpendicular to the square. Thus, the entire body of water moving through the square in one second is a box of length $\|\vec{v}\|$ and cross-sectional area A. So the box has volume $\|\vec{v}\|A$ m^3, and

$$\text{Flow rate} = \|\vec{v}\|A \text{ m}^3/\text{sec}.$$

This flow rate is called the *flux* of the fluid through the surface. We can also compute the flux of vector fields, such as electric and magnetic fields, where no flow is actually taking place. If the vector field is constant and perpendicular to the surface, and if the surface is flat, as in Example 1, the flux is obtained by multiplying the speed by the area.

Next we find the flux of a constant vector field through a flat surface that is not perpendicular to the vector field, using a dot product. In general, we break a surface into small pieces which are approximately flat and where the vector field is approximately constant, leading to a flux integral.

Orientation of a Surface

Before computing the flux of a vector field through a surface, we need to decide which direction of flow through the surface is the positive direction; this is described as choosing an orientation.[1]

> At each point on a smooth surface there are two unit normals, one in each direction. **Choosing an orientation** means picking one of these normals at every point of the surface in a continuous way. The unit normal vector in the direction of the orientation is denoted by $\vec{n}$. For a closed surface (that is, the boundary of a solid region), we choose the **outward orientation** unless otherwise specified.

We say the flux through a piece of surface is positive if the flow is in the direction of the orientation and negative if it is in the opposite direction. (See Figure 19.2.)

---

[1] Although we will not study them, there are a few surfaces for which this cannot be done. See page 1024.

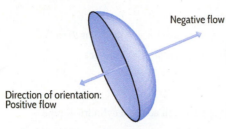

Negative flow

Direction of orientation:
Positive flow

Figure 19.2: An oriented surface showing
directions of positive and negative flow

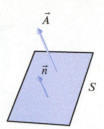

Figure 19.3: Area vector $\vec{A} = \vec{n} A$ of flat surface
with area A and orientation $\vec{n}$

The Area Vector

The flux through a flat surface depends both on the area of the surface and its orientation. Thus, it is useful to represent its area by a vector as shown in Figure 19.3.

> The **area vector** of a flat, oriented surface is a vector $\vec{A}$ such that
> - The magnitude of $\vec{A}$ is the area of the surface.
> - The direction of $\vec{A}$ is the direction of the orientation vector $\vec{n}$.

The Flux of a Constant Vector Field Through a Flat Surface

Suppose the velocity vector field, $\vec{v}$, of a fluid is constant and $\vec{A}$ is the area vector of a flat surface. The flux through this surface is the volume of fluid that flows through in one unit of time. The skewed box in Figure 19.4 has cross-sectional area $\|\vec{A}\|$ and height $\|\vec{v}\|\cos\theta$, so its volume is $(\|\vec{v}\|\cos\theta)\|\vec{A}\| = \vec{v}\cdot\vec{A}$. Thus, we have the following result:

> If $\vec{v}$ is constant and $\vec{A}$ is the area vector of a flat surface, then
>
> $$\text{Flux through surface} = \vec{v}\cdot\vec{A}.$$

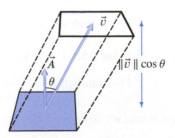

Figure 19.4: Flux of $\vec{v}$ through a surface with area vector $\vec{A}$ is the volume of this skewed box

Example 2   Water is flowing down a cylindrical pipe 2 cm in radius with a velocity of 3 cm/sec. Find the flux of the velocity vector field through the ellipse-shaped region shown in Figure 19.5. The normal to the ellipse makes an angle of θ with the direction of flow and the area of the ellipse is $4\pi/(\cos\theta)$ cm^2.

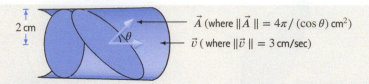

Figure 19.5: Flux through ellipse-shaped region across a cylindrical pipe

Solution   There are two ways to approach this problem. One is to use the formula we just derived, which gives

$$\text{Flux through ellipse} = \vec{v} \cdot \vec{A} = \|\vec{v}\|\|\vec{A}\|\cos\theta = 3(\text{Area of ellipse})\cos\theta$$
$$= 3\left(\frac{4\pi}{\cos\theta}\right)\cos\theta = 12\pi \text{ cm}^3/\text{sec}.$$

The second way is to notice that the flux through the ellipse is equal to the flux through the circle perpendicular to the pipe in Figure 19.5. Since the flux is the rate at which water is flowing down the pipe, we have

$$\text{Flux through circle} = \begin{array}{c}\text{Velocity}\\\text{of water}\end{array} \times \begin{array}{c}\text{Area of}\\\text{circle}\end{array} = \left(3\,\frac{\text{cm}}{\text{sec}}\right)(\pi 2^2 \text{ cm}^2) = 12\pi \text{ cm}^3/\text{sec}.$$

The Flux Integral

If the vector field, $\vec{F}$, is not constant or the surface, S, is not flat, we divide the surface into a patchwork of small, almost flat pieces. (See Figure 19.6.) For a particular patch with area ΔA, we pick a unit orientation vector $\vec{n}$ at a point on the patch and define the area vector of the patch, $\Delta\vec{A}$, as

$$\Delta\vec{A} = \vec{n}\,\Delta A.$$

(See Figure 19.7.) If the patches are small enough, we can assume that $\vec{F}$ is approximately constant on each piece. Then we know that

$$\text{Flux through patch} \approx \vec{F} \cdot \Delta\vec{A},$$

so, adding the fluxes through all the small pieces, we have

$$\text{Flux through whole surface} \approx \sum \vec{F} \cdot \Delta\vec{A}.$$

As each patch becomes smaller and $\|\Delta\vec{A}\| \to 0$, the approximation gets better and we get

$$\text{Flux through } S = \lim_{\|\Delta\vec{A}\|\to 0} \sum \vec{F} \cdot \Delta\vec{A}.$$

Thus, provided the limit exists, we make the following definition:

The **flux integral** of the vector field $\vec{F}$ through the oriented surface S is

$$\int_S \vec{F} \cdot d\vec{A} = \lim_{\|\Delta\vec{A}\|\to 0} \sum \vec{F} \cdot \Delta\vec{A}.$$

If S is a closed surface oriented outward, we describe the flux through S as the flux out of S.

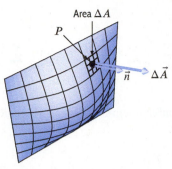

Area ΔA

P

$\vec{n}$ $\Delta \vec{A}$

Figure 19.6: Surface S divided into small, almost flat pieces, showing a typical orientation vector $\vec{n}$ and area vector $\Delta \vec{A}$

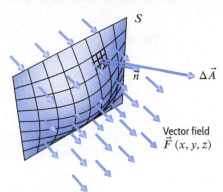

S

$\vec{n}$ $\Delta \vec{A}$

Vector field $\vec{F}(x, y, z)$

Figure 19.7: Flux of a vector field through a curved surface S

In computing a flux integral, we have to divide the surface up in a reasonable way, or the limit might not exist. In practice this problem seldom arises; however, one way to avoid it is to define flux integrals by the method used to compute them shown in Section 21.3.

Flux and Fluid Flow

If $\vec{v}$ is the velocity vector field of a fluid, we have

$$\begin{array}{c} \text{Rate fluid flows} \\ \text{through surface } S \end{array} = \begin{array}{c} \text{Flux of } \vec{v} \\ \text{through } S \end{array} = \int_S \vec{v} \cdot d\vec{A}$$

The rate of fluid flow is measured in units of volume per unit time.

Example 3   Find the flux of the vector field $\vec{B}(x, y, z)$ shown in Figure 19.8 through the square S of side 2 shown in Figure 19.9, oriented in the $\vec{j}$ direction, where

$$\vec{B}(x, y, z) = \frac{-y\vec{i} + x\vec{j}}{x^2 + y^2}.$$

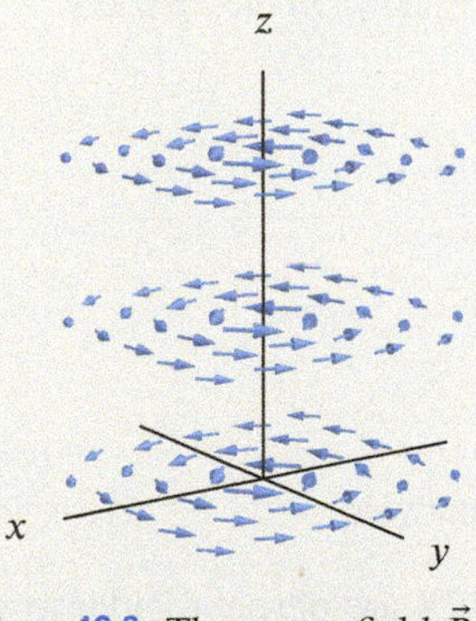

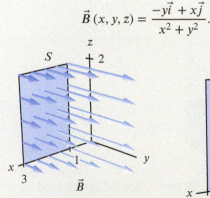

Figure 19.8: The vector field $\vec{B}$ in planes $z = 0$, $z = 1$, $z = 2$, where
$$\vec{B}(x, y, z) = \frac{-y\vec{i} + x\vec{j}}{x^2 + y^2}$$

Figure 19.9: Flux of $\vec{B}$ through the square S of side 2 in xy-plane and oriented in $\vec{j}$ direction

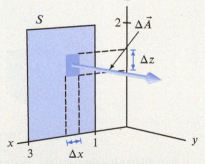

Figure 19.10: A small patch of surface with area $\|\Delta \vec{A}\| = \Delta x \Delta z$

Solution   Consider a small rectangular patch with area vector $\Delta \vec{A}$ in S, with sides Δx and Δz so that $\|\Delta \vec{A}\| = \Delta x \, \Delta z$. Since $\Delta \vec{A}$ points in the $\vec{j}$ direction, we have $\Delta \vec{A} = \vec{j} \, \Delta x \Delta z$. (See Figure 19.10.)

At the point $(x, 0, z)$ in S, substituting $y = 0$ into $\vec{B}$ gives $\vec{B}(x, 0, z) = (1/x)\vec{j}$. Thus, we have

$$\text{Flux through small patch} \approx \vec{B} \cdot \Delta \vec{A} = \left(\frac{1}{x}\vec{j}\right) \cdot (\vec{j} \, \Delta x \Delta z) = \frac{1}{x} \Delta x \, \Delta z.$$

Therefore,

$$\text{Flux through surface} = \int_S \vec{B} \cdot d\vec{A} = \lim_{\|\Delta \vec{A}\| \to 0} \sum \vec{B} \cdot \Delta \vec{A} = \lim_{\substack{\Delta x \to 0 \\ \Delta z \to 0}} \sum \frac{1}{x} \Delta x \, \Delta z.$$

This last expression is a Riemann sum for the double integral $\int_R \frac{1}{x} \, dA$, where R is the square $1 \leq x \leq 3$, $0 \leq z \leq 2$. Thus,

$$\text{Flux through surface} = \int_S \vec{B} \cdot d\vec{A} = \int_R \frac{1}{x} \, dA = \int_0^2 \int_1^3 \frac{1}{x} \, dx \, dz = 2 \ln 3.$$

The result is positive since the vector field is passing through the surface in the positive direction.

Example 4   Each of the vector fields in Figure 19.11 consists entirely of vectors parallel to the xy-plane, and is constant in the z direction (that is, the vector field looks the same in any plane parallel to the xy-plane). For each one, say whether you expect the flux through a closed surface surrounding the origin to be positive, negative, or zero. In part (a) the surface is a closed cube with faces perpendicular to the axes; in parts (b) and (c) the surface is a closed cylinder. In each case we choose the outward orientation. (See Figure 19.12.)

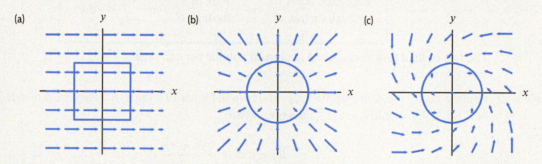

Figure 19.11: Flux of a vector field through the closed surfaces whose cross-sections are shown in the xy-plane

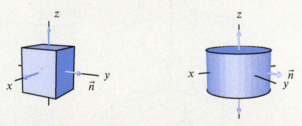

Figure 19.12: The closed cube and closed cylinder, both oriented outward

Solution   (a) Since the vector field appears to be parallel to the faces of the cube which are perpendicular to the y- and z-axes, we expect the flux through these faces to be zero. The fluxes through the two faces perpendicular to the x-axis appear to be equal in magnitude and opposite in sign, so we expect the net flux to be zero.

(b) Since the top and bottom of the cylinder are parallel to the flow, the flux through them is zero. On the curved surface of the cylinder, $\vec{v}$ and $\Delta \vec{A}$ appear to be everywhere parallel and in the same direction, so we expect each term $\vec{v} \cdot \Delta \vec{A}$ to be positive, and therefore the flux integral $\int_S \vec{v} \cdot d\vec{A}$ to be positive.

(c) As in part (b), the flux through the top and bottom of the cylinder is zero. In this case $\vec{v}$ and $\Delta \vec{A}$ are not parallel on the round surface of the cylinder, but since the fluid appears to be flowing inward as well as swirling, we expect each term $\vec{v} \cdot \Delta \vec{A}$ to be negative, and therefore the flux integral to be negative.

Calculating Flux Integrals Using $d\vec{A} = \vec{n}\, dA$

For a small patch of surface ΔS with unit normal $\vec{n}$ and area ΔA, the area vector is $\Delta\vec{A} = \vec{n}\,\Delta A$. The next example shows how we can use this relationship to compute a flux integral.

Example 5   An electric charge q is placed at the origin in 3-space. The resulting electric field $\vec{E}(\vec{r})$ at the point with position vector $\vec{r}$ is given by

$$\vec{E}(\vec{r}) = q\frac{\vec{r}}{\|\vec{r}\|^3}, \qquad \vec{r} \neq \vec{0}.$$

Find the flux of $\vec{E}$ out of the sphere of radius R centered at the origin. (See Figure 19.13.)

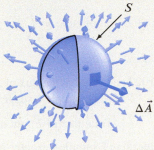

Figure 19.13: Flux of $\vec{E} = q\vec{r}/\|\vec{r}\|^3$ through the surface of a sphere of radius R centered at the origin

Solution   This vector field points radially outward from the origin in the same direction as $\vec{n}$. Thus, since $\vec{n}$ is a unit vector,

$$\vec{E}\cdot\Delta\vec{A} = \vec{E}\cdot\vec{n}\,\Delta A = \|\vec{E}\|\,\Delta A.$$

On the sphere, $\|\vec{E}\| = q/R^2$, so

$$\int_S \vec{E}\cdot d\vec{A} = \lim_{\|\Delta\vec{A}\|\to 0}\sum\vec{E}\cdot\Delta\vec{A} = \lim_{\Delta A\to 0}\sum\frac{q}{R^2}\Delta A = \frac{q}{R^2}\lim_{\Delta A\to 0}\sum\Delta A.$$

The last sum approximates the surface area of the sphere. In the limit as the subdivisions get finer, we have

$$\lim_{\Delta A\to 0}\sum\Delta A = \text{Surface area of sphere}.$$

Thus, the flux is given by

$$\int_S \vec{E}\cdot d\vec{A} = \frac{q}{R^2}\lim_{\Delta A\to 0}\sum\Delta A = \frac{q}{R^2}\cdot(\text{Surface area of sphere}) = \frac{q}{R^2}(4\pi R^2) = 4\pi q.$$

This result is known as Gauss's law.

To compute a flux with an integral instead of Riemann sums, we often write $d\vec{A} = \vec{n}\, dA$, as in the next example.

Example 6   Let S be the surface of the cube bounded by the six planes $x = \pm 1$, $y = \pm 1$, and $z = \pm 1$. Compute the flux of the electric field $\vec{E}$ of the previous example outward through S.

Solution   It is enough to compute the flux of $\vec{E}$ through a single face, say the top face S_1 defined by $z = 1$, where $-1 \leq x \leq 1$ and $-1 \leq y \leq 1$. By symmetry, the flux of $\vec{E}$ through the other five faces of S must be the same.

On the top face, S_1, we have $d\vec{A} = \vec{n}\, dA = \vec{k}\, dx\, dy$ and

$$\vec{E}(x, y, 1) = q\frac{x\vec{i} + y\vec{j} + \vec{k}}{(x^2 + y^2 + 1)^{3/2}}.$$

The corresponding flux integral is given by

$$\int_{S_1} \vec{E} \cdot d\vec{A} = q \int_{-1}^{1}\int_{-1}^{1} \frac{x\vec{i} + y\vec{j} + \vec{k}}{(x^2 + y^2 + 1)^{3/2}} \cdot \vec{k} \, dx \, dy = q \int_{-1}^{1}\int_{-1}^{1} \frac{1}{(x^2 + y^2 + 1)^{3/2}} \, dx \, dy.$$

Computing this integral numerically shows that

$$\text{Flux through top face} = \int_{S_1} \vec{E} \cdot d\vec{A} \approx 2.0944q.$$

Thus,

$$\text{Total flux of } \vec{E} \text{ out of cube} = \int_S \vec{E} \cdot d\vec{A} \approx 6(2.0944q) = 12.5664q.$$

Example 5 on page 1023 showed that the flux of $\vec{E}$ through a sphere of radius R centered at the origin is $4\pi q$. Since $4\pi \approx 12.5664$, Example 6 suggests that

$$\text{Total flux of } \vec{E} \text{ out of cube} = 4\pi q.$$

By computing the flux integral in Example 6 exactly, it is possible to verify that the flux of $\vec{E}$ through the cube and the sphere are exactly equal. When we encounter the Divergence Theorem in Chapter 20 we will see why this is so.

Notes on Orientation

Two difficulties can occur in choosing an orientation. The first is that if the surface is not smooth, it may not have a normal vector at every point. For example, a cube does not have a normal vector along its edges. When we have a surface, such as a cube, which is made of a finite number of smooth pieces, we choose an orientation for each piece separately. The best way to do this is usually clear. For example, on the cube we choose the outward orientation on each face. (See Figure 19.14.)

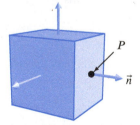

Figure 19.14: The orientation vector field $\vec{n}$ on the cube surface determined by the choice of unit normal vector at the point P

Figure 19.15: The Möbius strip is an example of a non-orientable surface

The second difficulty is that there are some surfaces which cannot be oriented at all, such as the *Möbius strip* in Figure 19.15.

Summary for Section 19.1

- The **area vector** of a flat, oriented surface is a vector $\vec{A}$ such that
  - The magnitude of $\vec{A}$ is the area of the surface.
  - The direction of $\vec{A}$ is the direction of the orientation vector $\vec{n}$.
- **Special case:** If $\vec{v}$ is constant and $\vec{A}$ is the area vector of a flat surface, then

$$\text{Flux through surface} = \vec{v} \cdot \vec{A}.$$

- The **flux integral** of the vector field $\vec{F}$ through the oriented surface S is

$$\int_S \vec{F} \cdot d\vec{A} = \lim_{\|\Delta \vec{A}\| \to 0} \sum \vec{F} \cdot \Delta \vec{A},$$

where $\Delta \vec{A}$ denotes an area vector of a small patch of S.

- **Flux and fluid flow:** If $\vec{v}$ is the velocity vector field of a fluid, we have

$$\begin{array}{ccc} \text{Rate fluid flows} \\ \text{through surface } S \end{array} = \begin{array}{c} \text{Flux of } \vec{v} \\ \text{through } S \end{array} = \int_S \vec{v} \cdot d\vec{A}.$$

Exercises and Problems for Section 19.1   Online Resource: Additional Problems for Section 19.1
EXERCISES

■ In Exercises 1–4, find the area vector of the oriented flat surface.

1. The triangle with vertices $(0,0,0)$, $(0,2,0)$, $(0,0,3)$ oriented in the negative x direction.

2. The circular disc of radius 5 in the xy-plane, oriented upward.

3. $y = 10$, $0 \le x \le 5$, $0 \le z \le 3$, oriented away from the xz-plane.

4. $y = -10$, $0 \le x \le 5$, $0 \le z \le 3$, oriented away from the xz-plane.

5. Find an oriented flat surface with area vector $150\vec{j}$.

■ In Exercises 6–9, for each of the surfaces in (a)–(e), say whether the flux of $\vec{F}$ through the surface is positive, negative, or zero. The normal vector shows the orientation.

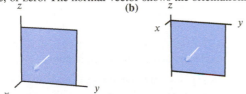

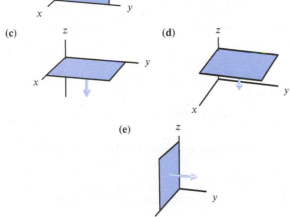

6. $\vec{F}(x,y,z) = \vec{i} + 2\vec{j} + \vec{k}$.

7. $\vec{F}(x,y,z) = z\vec{i}$.

8. $\vec{F}(x,y,z) = -z\vec{i} + x\vec{k}$.

9. $\vec{F}(\vec{r}) = \vec{r}$.

10. Let S be the disk of radius 3 perpendicular the the y-axis, centered at $(0,6,0)$ and oriented away from the origin. Is $\int_S (x\vec{i} + y\vec{j}) \cdot d\vec{A}$ a vector or a scalar?

11. Compute $\int_S (4\vec{i} + 5\vec{k}) \cdot d\vec{A}$, where S is the square of side length 3 perpendicular to the z-axis, centered at $(0,0,-2)$ and oriented

    (a) Toward the origin.   (b) Away from the origin.

12. Compute $\int_S (2\vec{i} + 3\vec{k}) \cdot d\vec{A}$, where S is the disk of radius 4 perpendicular to the x-axis, centered at $(5,0,0)$ and oriented

    (a) Toward the origin.   (b) Away from the origin.

■ In Exercises 13–16, compute the flux of $\vec{v} = \vec{i} + 2\vec{j} - 3\vec{k}$ through the rectangular region with the orientation shown.

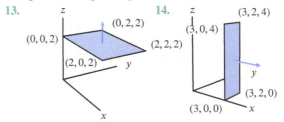

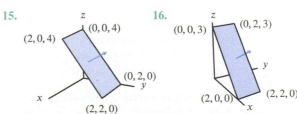

■ For Exercises 17–20 find the flux of the constant vector field $\vec{v} = \vec{i} - \vec{j} + 3\vec{k}$ through the given surface.

17. A disk of radius 2 in the xy-plane oriented upward.

18. A triangular plate of area 4 in the yz-plane oriented in the positive x-direction.

19. A square plate of area 4 in the yz-plane oriented in the positive x-direction.

20. The triangular plate with vertices $(1,0,0)$, $(0,1,0)$, $(0,0,1)$, oriented away from the origin.

■ In Exercises 21–23, find the flux of $\vec{H} = 2\vec{i} + 3\vec{j} + 5\vec{k}$ through the surface S.

21. S is the cylinder $x^2 + y^2 = 1$, closed at the ends by the planes $z = 0$ and $z = 1$ and oriented outward.

22. S is the disk of radius 1 in the plane $x = 2$ oriented in the positive x-direction.

23. S is the disk of radius 1 in the plane $x + y + z = 1$ oriented in upward.

■ Find the flux of the vector fields in Exercises 24–26 out of the closed box $0 \le x \le 1, 0 \le y \le 2, 0 \le z \le 3$.

24. $\vec{F} = 3\vec{i} + 2\vec{j} + \vec{k}$ 25. $\vec{G} = x\vec{i}$

26. $\vec{H} = zx\vec{k}$

■ In Exercises 27–30, calculate the flux integral.

27. $\int_S (x\vec{i} + 4\vec{j}) \cdot d\vec{A}$ where S is the disk of radius 5 perpendicular to the x-axis, centered at $(3, 0, 0)$ and oriented toward the origin.

28. $\int_S \vec{r} \cdot d\vec{A}$ where S is the sphere of radius 3 centered at the origin.

29. $\int_S (\sin x\, \vec{i} + (y^2 + z^2)\vec{j} + y^2\vec{k}) \cdot d\vec{A}$ where S is a disk of radius π in the plane $x = 3\pi/2$, oriented in the positive x-direction.

30. $\int_S (5\vec{i} + 5\vec{j} + 5\vec{k}) \cdot d\vec{A}$ where S is a disk of radius 3 in the plane $x + y + z = 1$, oriented upward.

■ In Exercises 31–34, calculate the flux integral using a shortcut arising from two special cases:

• If $\vec{F}$ is tangent at every point of S, then $\int_S \vec{F} \cdot d\vec{A} = 0$.

• If $\vec{F}$ is perpendicular at every point of S and has constant magnitude on S, then

$$\int_S \vec{F} \cdot d\vec{A} = \pm \|\vec{F}\| \cdot \text{Area of } S.$$

Choose the positive sign if $\vec{F}$ points in the same direction as the orientation of S; choose the negative sign if $\vec{F}$ points in the direction opposite the orientation of S.

31. $\int_S (x\vec{i} + y\vec{j}) \cdot d\vec{A}$, where S is the cylinder of radius 10, centered on the z-axis between $z = 0$ and $z = 10$ and oriented away from the z-axis.

32. $\int_S (-y\vec{i} + x\vec{j}) \cdot d\vec{A}$, where S is the cylinder of radius 10, centered on the z-axis between $z = 0$ and $z = 10$ and oriented away from the z-axis.

33. $\int_S (x\vec{i} + y\vec{j} + z\vec{k}) \cdot d\vec{A}$, where S is the sphere of radius 20, centered at the origin and oriented outward.

34. $\int_S (-y\vec{i} + x\vec{j}) \cdot d\vec{A}$, where S is the sphere of radius 20, centered at the origin and oriented outward.

■ In Exercises 35–57, calculate the flux of the vector field through the surface.

35. $\vec{F} = 2\vec{i} + 3\vec{j}$ through the square of side π in the xy-plane, oriented upward.

36. $\vec{F} = 2\vec{i} + 3\vec{j}$ through the unit disk in the yz-plane, centered at the origin and oriented in the positive x-direction.

37. $\vec{F} = x\vec{i} + y\vec{j} + z\vec{k}$ through the square of side 1.6 centered at $(2, 5, 8)$, parallel to the xz-plane and oriented away from the origin.

38. $\vec{F} = z\vec{k}$ through a square of side $\sqrt{14}$ in a horizontal plane 2 units below the xy-plane and oriented downward.

39. $\vec{F} = -y\vec{i} + x\vec{j}$ and S is the square plate in the yz-plane with corners at $(0, 1, 1)$, $(0, -1, 1)$, $(0, 1, -1)$, and $(0, -1, -1)$, oriented in the positive x-direction.

40. $\vec{F} = 7\vec{i} + 6\vec{j} + 5\vec{k}$ and S is a disk of radius 2 in the yz-plane, centered at the origin and oriented in the positive x-direction.

41. $\vec{F} = x\vec{i} + 2y\vec{j} + 3z\vec{k}$ and S is a square of side 2 in the plane $y = 3$, oriented in the positive y-direction.

42. $\vec{F} = 7\vec{i} + 6\vec{j} + 5\vec{k}$ and S is a sphere of radius π centered at the origin.

43. $\vec{F} = -5\vec{r}$ through the sphere of radius 2 centered at the origin.

44. $\vec{F} = x\vec{i} + y\vec{j} + (z^2 + 3)\vec{k}$ and S is the rectangle $z = 4$, $0 \le x \le 2$, $0 \le y \le 3$, oriented in the positive z-direction.

45. $\vec{F} = 6\vec{i} + 7\vec{j}$ through a triangle of area 10 in the plane $x + y = 5$, oriented in the positive x-direction.

46. $\vec{F} = 6\vec{i} + x^2\vec{j} - \vec{k}$, through the square of side 4 in the plane $y = 3$, centered on the y-axis, with sides parallel to the x and z axes, and oriented in the positive y-direction.

47. $\vec{F} = (x+3)\vec{i} + (y+5)\vec{j} + (z+7)\vec{k}$ through the rectangle $x = 4$, $0 \le y \le 2$, $0 \le z \le 3$, oriented in the positive x-direction.

48. $\vec{F} = 7\vec{r}$ through the sphere of radius 3 centered at the origin.

49. $\vec{F} = -3\vec{r}$ through the sphere of radius 2 centered at the origin.

50. $\vec{F} = 2z\vec{i} + x\vec{j} + x\vec{k}$ through the rectangle $x = 4$, $0 \le y \le 2$, $0 \le z \le 3$, oriented in the positive x-direction.

51. $\vec{F} = \vec{i} + 2\vec{j}$ through a square of side 2 lying in the plane $x + y + z = 1$, oriented away from the origin.

52. $\vec{F} = (x^2 + y^2)\vec{k}$ through the disk of radius 3 in the xy-plane, centered at the origin and oriented upward.

53. $\vec{F} = \cos(x^2 + y^2)\vec{k}$ through the disk $x^2 + y^2 \le 9$ oriented upward in the plane $z = 1$.

54. $\vec{F} = e^{y^2+z^2}\vec{i}$ through the disk of radius 2 in the yz-plane, centered at the origin and oriented in the positive x-direction.

55. $\vec{F} = -y\vec{i} + x\vec{j}$ through the disk in the xy-plane with radius 2, oriented upward and centered at the origin.

56. $\vec{F} = \vec{r}$ through the disk of radius 2 parallel to the xy-plane oriented upward and centered at $(0, 0, 2)$.

57. $\vec{F} = (2 - x)\vec{i}$ through the cube whose vertices include the points $(0, 0, 0)$, $(3, 0, 0)$, $(0, 3, 0)$, $(0, 0, 3)$, and oriented outward.

PROBLEMS

58. Let B be the surface of a box centered at the origin, with edges parallel to the axes and in the planes $x = \pm 1$, $y = \pm 1$, $z = \pm 1$, and let S be the sphere of radius 1 centered at origin.

   (a) Indicate whether the following flux integrals are positive, negative, or zero. No reasons needed.

     (a) $\int_B x\vec{i} \cdot d\vec{A}$ **(b)** $\int_B y\vec{i} \cdot d\vec{A}$
     (c) $\int_S |x|\vec{i} \cdot d\vec{A}$ **(d)** $\int_S (y-x)\vec{i} \cdot d\vec{A}$

   (b) Explain with reasons how you know which flux integral is greater:

$$\int_S x\vec{i} \cdot d\vec{A} \quad \text{or} \quad \int_B x\vec{i} \cdot d\vec{A} \text{ ?}$$

59. Suppose that $\vec{E}$ is a uniform electric field on 3-space, so $\vec{E}(x, y, z) = a\vec{i} + b\vec{j} + c\vec{k}$, for all points (x, y, z), where a, b, c are constants. Show, with the aid of symmetry, that the flux of $\vec{E}$ through each of the following closed surfaces S is zero:

   (a) S is the cube bounded by the planes $x = \pm 1$, $y = \pm 1$, and $z = \pm 1$.
   (b) S is the sphere $x^2 + y^2 + z^2 = 1$.
   (c) S is the cylinder bounded by $x^2 + y^2 = 1$, $z = 0$, and $z = 2$.

60. Water is flowing down a cylindrical pipe of radius 2 cm; its speed is $(3 - (3/4)r^2)$ cm/sec at a distance r cm from the center of the pipe. Find the flux through the circular cross section of the pipe, oriented so that the flux is positive.

61. (a) What do you think will be the electric flux through the cylindrical surface that is placed as shown in the constant electric field in Figure 19.16? Why?
   (b) What if the cylinder is placed upright, as shown in Figure 19.17? Explain.

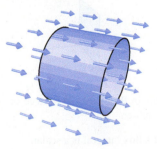

Figure 19.16

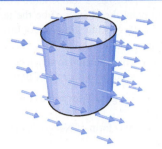

Figure 19.17

62. Let S be part of a cylinder centered on the y-axis. Explain why the three vectors fields $\vec{F}$, $\vec{G}$, and $\vec{H}$ have the same flux through S. Do not compute the flux.

$$\vec{F} = x\vec{i} + 2yz\vec{k}$$
$$\vec{G} = x\vec{i} + y\sin x\vec{j} + 2yz\vec{k}$$
$$\vec{H} = x\vec{i} + \cos(x^2 + z)\vec{j} + 2yz\vec{k}$$

63. Find the flux of $\vec{F} = \vec{r}/\|\vec{r}\|^3$ out of the sphere of radius R centered at the origin.

64. Find the flux of $\vec{F} = \vec{r}/\|r\|^2$ out of the sphere of radius R centered at the origin.

65. Consider the flux of the vector field $\vec{F} = \vec{r}/\|\vec{r}\|^p$ for $p \geq 0$ out of the sphere of radius 2 centered at the origin.

   (a) For what value of p is the flux a maximum?
   (b) What is that maximum value?

66. Let S be the cube with side length 2, faces parallel to the coordinate planes, and centered at the origin.

   (a) Calculate the total flux of the constant vector field $\vec{v} = -\vec{i} + 2\vec{j} + \vec{k}$ out of S by computing the flux through each face separately.
   (b) Calculate the flux out of S for any constant vector field $\vec{v} = a\vec{i} + b\vec{j} + c\vec{k}$.
   (c) Explain why the answers to parts (a) and (b) make sense.

67. Let S be the tetrahedron with vertices at the origin and at $(1, 0, 0)$, $(0, 1, 0)$ and $(0, 0, 1)$.

   (a) Calculate the total flux of the constant vector field $\vec{v} = -\vec{i} + 2\vec{j} + \vec{k}$ out of S by computing the flux through each face separately.
   (b) Calculate the flux out of S in part (a) for any constant vector field $\vec{v}$.
   (c) Explain why the answers to parts (a) and (b) make sense.

68. Let $P(x, y, z)$ be the pressure at the point (x, y, z) in a fluid. Let $\vec{F}(x, y, z) = P(x, y, z)\vec{k}$. Let S be the surface of a body submerged in the fluid. If S is oriented inward, show that $\int_S \vec{F} \cdot d\vec{A}$ is the buoyant force on the body, that is, the force upward on the body due to the pressure of the fluid surrounding it. [Hint: $\vec{F} \cdot d\vec{A} = P(x, y, z)\vec{k} \cdot d\vec{A} = (P(x, y, z) \, d\vec{A}) \cdot \vec{k}$.]

69. A region of 3-space has a temperature which varies from point to point. Let $T(x, y, z)$ be the temperature at a point (x, y, z). Newton's law of cooling says that $\operatorname{grad} T$ is proportional to the heat flow vector field, $\vec{F}$, where $\vec{F}$ points in the direction in which heat is flowing and has magnitude equal to the rate of flow of heat.

   (a) Suppose $\vec{F} = k \operatorname{grad} T$ for some constant k. What is the sign of k?
   (b) Explain why this form of Newton's law of cooling makes sense.
   (c) Let W be a region of space bounded by the surface S. Explain why

$$\begin{array}{c} \text{Rate of heat} \\ \text{loss from } W \end{array} = k \int_S (\operatorname{grad} T) \cdot d\vec{A}.$$

70. The z-axis carries a constant electric charge density of λ units of charge per unit length, with $\lambda > 0$. The resulting electric field is $\vec{E}$.

   (a) Sketch the electric field, $\vec{E}$, in the xy-plane, given

$$\vec{E}(x, y, z) = 2\lambda \frac{x\vec{i} + y\vec{j}}{x^2 + y^2}.$$

   (b) Compute the flux of $\vec{E}$ outward through the cylinder $x^2 + y^2 = R^2$, for $0 \le z \le h$.

71. An infinitely long straight wire lying along the z-axis carries an electric current I flowing in the $\vec{k}$ direction. Ampère's Law in magnetostatics says that the current gives rise to a magnetic field $\vec{B}$ given by

$$\vec{B}(x, y, z) = \frac{I}{2\pi} \frac{-y\vec{i} + x\vec{j}}{x^2 + y^2}.$$

   (a) Sketch the field $\vec{B}$ in the xy-plane.
   (b) Let S_1 be the disk with center at $(0, 0, h)$, radius a, and parallel to the xy-plane, oriented in the $\vec{k}$ direction. What is the flux of $\vec{B}$ through S_1? Does your answer seem reasonable?
   (c) Let S_2 be the rectangle given by $x = 0$, $a \le y \le b$, $0 \le z \le h$, and oriented in the $-\vec{i}$ direction. What is the flux of $\vec{B}$ through S_2? Does your answer seem reasonable?

Strengthen Your Understanding

■ In Problems 72–73, explain what is wrong with the statement.

72. For a certain vector field $\vec{F}$ and oriented surface S, we have $\int_S \vec{F} \cdot d\vec{A} = 2\vec{i} - 3\vec{j} + \vec{k}$.

73. If S is a region in the xy-plane oriented upwards then $\int_S \vec{F} \cdot d\vec{A} > 0$.

■ In Problems 74–75, give an example of:

74. A nonzero vector field $\vec{F}$ such that $\int_S \vec{F} \cdot d\vec{A} = 0$, where S is the triangular surface with corners $(1, 0, 0)$, $(0, 1, 0)$, $(0, 0, 1)$, oriented away from the origin.

75. A nonconstant vector field $\vec{F}(x, y, z)$ and an oriented surface S such that $\int_S \vec{F} \cdot d\vec{A} = 1$.

76. For each of the surfaces in (a)–(e), pick the vector field $\vec{F}_1, \vec{F}_2, \vec{F}_3, \vec{F}_4, \vec{F}_5$, with the largest flux through the surface. The surfaces are all squares of the same size. Note that the orientation is shown.

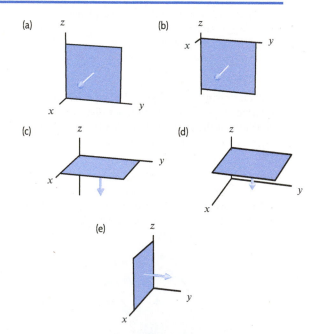

$$\vec{F}_1 = 2\vec{i} - 3\vec{j} - 4\vec{k}$$
$$\vec{F}_2 = \vec{i} - 2\vec{j} + 7\vec{k}$$
$$\vec{F}_3 = -7\vec{i} + 5\vec{j} + 6\vec{k}$$
$$\vec{F}_4 = -11\vec{i} + 4\vec{j} - 5\vec{k}$$
$$\vec{F}_5 = -5\vec{i} + 3\vec{j} + 5\vec{k}$$

■ Are the statements in Problems 77–86 true or false? Give reasons for your answer.

77. The value of a flux integral is a scalar.

78. The area vector $\vec{A}$ of a flat, oriented surface is parallel to the surface.

79. If S is the unit sphere centered at the origin, oriented outward and the flux integral $\int_S \vec{F} \cdot d\vec{A}$ is zero, then $\vec{F} = \vec{0}$.

80. The flux of the vector field $\vec{F} = \vec{i}$ through the plane $x = 0$, with $0 \le y \le 1, 0 \le z \le 1$, oriented in the $\vec{i}$ direction is positive.

81. If S is the unit sphere centered at the origin, oriented outward and $\vec{F} = x\vec{i} + y\vec{j} + z\vec{k} = \vec{r}$, then the flux integral $\int_S \vec{F} \cdot d\vec{A}$ is positive.

82. If S is the cube bounded by the six planes $x = \pm 1, y = \pm 1, z = \pm 1$, oriented outward, and $\vec{F} = \vec{k}$, then

$\int_S \vec{F} \cdot d\vec{A} = 0$.

83. If S is an oriented surface in 3-space, and $-S$ is the same surface, but with the opposite orientation, then $\int_S \vec{F} \cdot d\vec{A} = -\int_{-S} \vec{F} \cdot d\vec{A}$.

84. If S_1 is a rectangle with area 1 and S_2 is a rectangle with area 2, then $2\int_{S_1} \vec{F} \cdot d\vec{A} = \int_{S_2} \vec{F} \cdot d\vec{A}$.

85. If $\vec{F} = 2\vec{G}$, then $\int_S \vec{F} \cdot d\vec{A} = 2\int_S \vec{G} \cdot d\vec{A}$.

86. If $\int_S \vec{F} \cdot d\vec{A} > \int_S \vec{G} \cdot d\vec{A}$ then $||\vec{F}|| > ||\vec{G}||$ at all points on the surface S.

19.2 FLUX INTEGRALS FOR GRAPHS, CYLINDERS, AND SPHERES

In Section 19.1 we computed flux integrals in certain simple cases. In this section we see how to compute flux through surfaces that are graphs of functions, through cylinders, and through spheres.

Flux of a Vector Field Through the Graph of $z = f(x, y)$

Suppose S is the graph of the differentiable function $z = f(x, y)$, oriented upward, and that $\vec{F}$ is a smooth vector field. In Section 19.1 we subdivided the surface into small pieces with area vector $\Delta\vec{A}$ and defined the flux of $\vec{F}$ through S as follows:

$$\int_S \vec{F} \cdot d\vec{A} = \lim_{||\Delta\vec{A}|| \to 0} \sum \vec{F} \cdot \Delta\vec{A}.$$

How do we divide S into small pieces? One way is to use the cross sections of f with x or y constant and take the patches in a wire frame representation of the surface. So we must calculate the area vector of one of these patches, which is approximately a parallelogram.

The Area Vector of a Coordinate Patch

According to the geometric definition of the cross product on page 775, the vector $\vec{v} \times \vec{w}$ has magnitude equal to the area of the parallelogram formed by $\vec{v}$ and $\vec{w}$ and direction perpendicular to this parallelogram and determined by the right-hand rule. Thus, we have

$$\boxed{\text{Area vector of parallelogram} = \vec{A} = \vec{v} \times \vec{w}.}$$

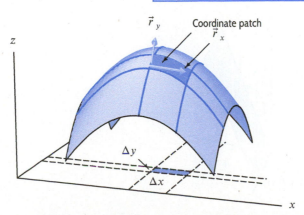

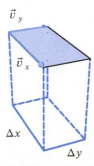

Figure 19.18: Surface showing coordinate patch and tangent vectors $\vec{r}_x$ and $\vec{r}_y$

Figure 19.19: Parallelogram-shaped patch in the tangent plane to the surface

Consider the patch of surface above the rectangular region with sides Δx and Δy in the xy-plane shown in Figure 19.18. We approximate the area vector, $\Delta \vec{A}$, of this patch by the area vector of the corresponding patch on the tangent plane to the surface. See Figure 19.19. This patch is the parallelogram determined by the vectors $\vec{v}_x$ and $\vec{v}_y$, so its area vector is given by

$$\Delta \vec{A} \approx \vec{v}_x \times \vec{v}_y.$$

To find $\vec{v}_x$ and $\vec{v}_y$, notice that a point on the surface has position vector $\vec{r} = x\vec{i} + y\vec{j} + f(x, y)\vec{k}$. Thus, a cross section of S with y constant has tangent vector

$$\vec{r}_x = \frac{\partial \vec{r}}{\partial x} = \vec{i} + f_x \vec{k},$$

and a cross section with x constant has tangent vector

$$\vec{r}_y = \frac{\partial \vec{r}}{\partial y} = \vec{j} + f_y \vec{k}.$$

The vectors $\vec{r}_x$ and $\vec{v}_x$ are parallel because they are both tangent to the surface and parallel to the xz-plane. Since the x-component of $\vec{r}_x$ is $\vec{i}$ and the x-component of $\vec{v}_x$ is $(\Delta x)\vec{i}$, we have $\vec{v}_x = (\Delta x)\vec{r}_x$. Similarly, we have $\vec{v}_y = (\Delta y)\vec{r}_y$. So the upward-pointing area vector of the parallelogram is

$$\Delta \vec{A} \approx \vec{v}_x \times \vec{v}_y = (\vec{r}_x \times \vec{r}_y) \, \Delta x \, \Delta y = \left(-f_x \vec{i} - f_y \vec{j} + \vec{k} \right) \Delta x \, \Delta y.$$

This is our approximation for the area vector $\Delta \vec{A}$ on the surface. Replacing $\Delta \vec{A}$, Δx, and Δy by $d\vec{A}$, dx and dy, we write

$$d\vec{A} = \left(-f_x \vec{i} - f_y \vec{j} + \vec{k} \right) dx \, dy.$$

The Flux of $\vec{F}$ Through a Surface Given by a Graph of $z = f(x, y)$

Suppose the surface S is the part of the graph of $z = f(x, y)$ above[2] a region R in the xy-plane, and suppose S is oriented upward. The flux of $\vec{F}$ through S is

$$\int_S \vec{F} \cdot d\vec{A} = \int_R \vec{F}(x, y, f(x, y)) \cdot \left(-f_x \vec{i} - f_y \vec{j} + \vec{k} \right) dx \, dy.$$

Example 1   Compute $\int_S \vec{F} \cdot d\vec{A}$ where $\vec{F}(x, y, z) = z\vec{k}$ and S is the rectangular plate with corners $(0, 0, 0)$, $(1, 0, 0)$, $(0, 1, 3)$, $(1, 1, 3)$, oriented upward. See Figure 19.20.

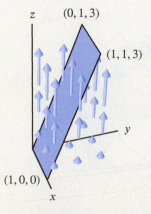

Figure 19.20: The vector field $\vec{F} = z\vec{k}$ on the rectangular surface S

---

[2]The formula is also correct when the graph is below the region R.

Solution   We find the equation for the plane S in the form $z = f(x, y)$. Since f is linear, with x-slope equal to 0 and y-slope equal to 3, and $f(0, 0) = 0$, we have

$$z = f(x, y) = 0 + 0x + 3y = 3y.$$

Thus, we have

$$d\vec{A} = (-f_x\vec{i} - f_y\vec{j} + \vec{k})\,dx\,dy = (0\vec{i} - 3\vec{j} + \vec{k})\,dx\,dy = (-3\vec{j} + \vec{k})\,dx\,dy.$$

The flux integral is therefore

$$\int_S \vec{F} \cdot d\vec{A} = \int_0^1 \int_0^1 3y\vec{k} \cdot (-3\vec{j} + \vec{k})\,dx\,dy = \int_0^1 \int_0^1 3y\,dx\,dy = 1.5.$$

Surface Area of a Graph

Since the magnitude of an area vector is area, we can find area of a surface by integrating the magnitude $\|d\vec{A}\|$. If a surface is the graph of a function $z = f(x, y)$, we have

$$\|d\vec{A}\| = \|-f_x\vec{i} - f_y\vec{j} + \vec{k}\|\,dx\,dy = \sqrt{(f_x)^2 + (f_y)^2 + 1}\,dx\,dy.$$

Thus we have the following result:

> Suppose a surface S is the part of the graph $z = f(x, y)$ where (x, y) is in a region R in the xy-plane. Then
> $$\text{Area of } S = \int_R \sqrt{(f_x)^2 + (f_y)^2 + 1}\,dx\,dy.$$

Example 2   Find the area of the surface $z = f(x, y)$ where $0 \le x \le 4, 0 \le y \le 5$, when:

(a) $f(x, y) = 2x + 3y + 4$ (b) $f(x, y) = x^2 + y^2$

Solution   (a) Since $f_x = 2$ and $f_y = 3$, we have

$$\text{Area} = \int_0^5 \int_0^4 \sqrt{2^2 + 3^2 + 1}\,dx\,dy = 20\sqrt{14}.$$

(b) Since $f_x = 2x$ and $f_y = 2y$. we have

$$\text{Area} = \int_0^5 \int_0^4 \sqrt{4x^2 + 4y^2 + 1}\,dx\,dy = 140.089.$$

Surface area integrals can often only be evaluated numerically.

Flux of a Vector Field Through a Cylindrical Surface

Consider the cylinder of radius R centered on the z-axis illustrated in Figure 19.21 and oriented away from the z-axis. The coordinate patch in Figure 19.22 has surface area given by

$$\Delta A \approx R \, \Delta\theta \, \Delta z.$$

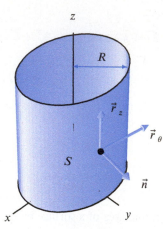

Figure 19.21: Outward-oriented cylinder

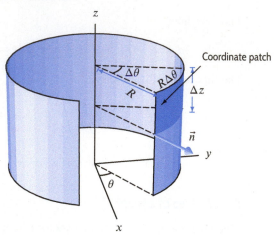

Figure 19.22: Coordinate patch with area $\Delta\vec{A}$ on surface of a cylinder

The outward unit normal $\vec{n}$ points in the direction of $x\vec{i} + y\vec{j}$, so

$$\vec{n} = \frac{x\vec{i} + y\vec{j}}{\|x\vec{i} + y\vec{j}\|} = \frac{R\cos\theta\vec{i} + R\sin\theta\vec{j}}{R} = \cos\theta\vec{i} + \sin\theta\vec{j}.$$

Therefore, the area vector of the coordinate patch is approximated by

$$\Delta\vec{A} = \vec{n}\,\Delta A \approx \left(\cos\theta\vec{i} + \sin\theta\vec{j}\right) R\,\Delta z\,\Delta\theta.$$

Replacing $\Delta\vec{A}$, Δz, and $\Delta\theta$ by $d\vec{A}$, dz, and $d\theta$, we write

$$d\vec{A} = \left(\cos\theta\vec{i} + \sin\theta\vec{j}\right) R\,dz\,d\theta.$$

This gives the following result:

The Flux of a Vector Field Through a Cylinder

The flux of $\vec{F}$ through the cylindrical surface S, of radius R and oriented away from the z-axis, is given by

$$\int_S \vec{F} \cdot d\vec{A} = \int_T \vec{F}(R, \theta, z) \cdot \left(\cos\theta\vec{i} + \sin\theta\vec{j}\right) R\,dz\,d\theta,$$

where T is the θz-region corresponding to S.

Example 3   Compute $\int_S \vec{F} \cdot d\vec{A}$ where $\vec{F}(x, y, z) = y\vec{j}$ and S is the part of the cylinder of radius 2 centered on the z-axis with $x \geq 0$, $y \geq 0$, and $0 \leq z \leq 3$. The surface is oriented toward the z-axis.

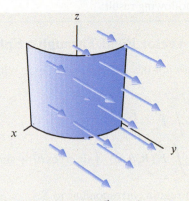

Figure 19.23: The vector field $\vec{F} = y\vec{j}$ on the surface S

Solution

In cylindrical coordinates, we have $R = 2$ and $\vec{F} = y\vec{j} = 2\sin\theta\vec{j}$. Since the orientation of S is toward the z-axis, the flux through S is given by

$$\int_S \vec{F} \cdot d\vec{A} = -\int_T 2\sin\theta\vec{j} \cdot (\cos\theta\vec{i} + \sin\theta\vec{j})2\,dz\,d\theta = -4\int_0^{\pi/2}\int_0^3 \sin^2\theta\,dz\,d\theta = -3\pi.$$

Flux of a Vector Field Through a Spherical Surface

Consider the piece of the sphere of radius R centered at the origin, oriented outward, as illustrated in Figure 19.24. The coordinate patch in Figure 19.24 has surface area given by

$$\Delta A \approx R^2 \sin\phi\,\Delta\phi\,\Delta\theta.$$

The outward unit normal $\vec{n}$ points in the direction of $\vec{r} = x\vec{i} + y\vec{j} + z\vec{k}$, so

$$\vec{n} = \frac{\vec{r}}{\|\vec{r}\|} = \sin\phi\cos\theta\vec{i} + \sin\phi\sin\theta\vec{j} + \cos\phi\vec{k}.$$

Therefore, the area vector of the coordinate patch is approximated by

$$\Delta\vec{A} \approx \vec{n}\,\Delta A = \frac{\vec{r}}{\|\vec{r}\|}\Delta A = \left(\sin\phi\cos\theta\vec{i} + \sin\phi\sin\theta\vec{j} + \cos\phi\vec{k}\right)R^2\sin\phi\,\Delta\phi\,\Delta\theta.$$

Replacing $\Delta\vec{A}$, $\Delta\phi$, and $\Delta\theta$ by $d\vec{A}$, $d\phi$, and $d\theta$, we write

$$d\vec{A} = \frac{\vec{r}}{\|\vec{r}\|}\,dA = \left(\sin\phi\cos\theta\vec{i} + \sin\phi\sin\theta\vec{j} + \cos\phi\vec{k}\right)R^2\sin\phi\,d\phi\,d\theta.$$

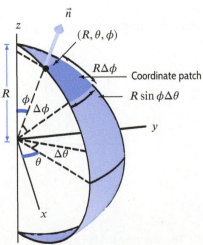

Figure 19.24: Coordinate patch with area $\Delta\vec{A}$ on surface of a sphere

Thus, we obtain the following result:

> ## The Flux of a Vector Field Through a Sphere
>
> The flux of $\vec{F}$ through the spherical surface S, with radius R and oriented away from the origin, is given by
>
> $$\int_S \vec{F} \cdot d\vec{A} = \int_S \vec{F} \cdot \frac{\vec{r}}{\|\vec{r}\|}\, dA$$
>
> $$= \int_T \vec{F}(R, \theta, \phi) \cdot \left(\sin\phi\cos\theta\vec{i} + \sin\phi\sin\theta\vec{j} + \cos\phi\vec{k} \right) R^2 \sin\phi\, d\phi\, d\theta,$$
>
> where T is the $\theta\phi$-region corresponding to S.

Example 4   Find the flux of $\vec{F} = z\vec{k}$ through S, the upper hemisphere of radius 2 centered at the origin, oriented outward.

Solution   The hemisphere S is parameterized by spherical coordinates θ and ϕ, with $0 \le \theta \le 2\pi$ and $0 \le \phi \le \pi/2$. Since $R = 2$ and $\vec{F} = z\vec{k} = 2\cos\phi\vec{k}$, the flux is

$$\int_S \vec{F} \cdot d\vec{A} = \int_S 2\cos\phi\vec{k} \cdot (\sin\phi\cos\theta\vec{i} + \sin\phi\sin\theta\vec{j} + \cos\phi\vec{k})4\sin\phi\, d\phi\, d\theta$$

$$= \int_0^{2\pi} \int_0^{\pi/2} 8\sin\phi\cos^2\phi\, d\phi\, d\theta = 2\pi\left(8\left(\frac{-\cos^3\phi}{3}\right)\Big|_{\phi=0}^{\pi/2}\right) = \frac{16\pi}{3}.$$

Example 5   The magnetic field $\vec{B}$ due to an *ideal magnetic dipole*, $\vec{\mu}$, located at the origin is a multiple of

$$\vec{B}(\vec{r}) = -\frac{\vec{\mu}}{\|\vec{r}\|^3} + \frac{3(\vec{\mu} \cdot \vec{r})\vec{r}}{\|\vec{r}\|^5}.$$

Figure 19.25 shows a sketch of $\vec{B}$ in the plane $z = 0$ for the dipole $\vec{\mu} = \vec{i}$. Notice that $\vec{B}$ is similar to the magnetic field of a bar magnet with its north pole at the tip of the vector $\vec{i}$ and its south pole at the tail of the vector $\vec{i}$.

Compute the flux of $\vec{B}$ outward through the sphere S with center at the origin and radius R.

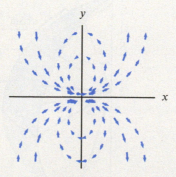

Figure 19.25: The magnetic field of a dipole, $\vec{i}$, at the origin: $\vec{B} = -\dfrac{\vec{i}}{\|\vec{r}\|^3} + \dfrac{3(\vec{i} \cdot \vec{r})\vec{r}}{\|\vec{r}\|^5}$

Solution   Since $\vec{i} \cdot \vec{r} = x$ and $\|\vec{r}\| = R$ on the sphere of radius R, we have

$$\int_S \vec{B} \cdot d\vec{A} = \int_S \left(-\frac{\vec{i}}{\|\vec{r}\|^3} + \frac{3(\vec{i} \cdot \vec{r})\vec{r}}{\|\vec{r}\|^5} \right) \cdot \frac{\vec{r}}{\|\vec{r}\|} \, dA = \int_S \left(-\frac{\vec{i} \cdot \vec{r}}{\|\vec{r}\|^4} + \frac{3(\vec{i} \cdot \vec{r})\|\vec{r}\|^2}{\|\vec{r}\|^6} \right) dA$$

$$= \int_S \frac{2\vec{i} \cdot \vec{r}}{\|\vec{r}\|^4} \, dA = \int_S \frac{2x}{\|\vec{r}\|^4} \, dA = \frac{2}{R^4} \int_S x \, dA.$$

But the sphere S is centered at the origin. Thus, the contribution to the integral from each positive x-value is canceled by the contribution from the corresponding negative x-value; so $\int_S x \, dA = 0$. Therefore,

$$\int_S \vec{B} \cdot d\vec{A} = \frac{2}{R^4} \int_S x \, dA = 0.$$

Summary for Section 19.2

- **The flux through a graph:** Suppose the surface S is the part of the graph of $z = f(x, y)$ above a region R in the xy-plane that is oriented upward. Then

$$\int_S \vec{F} \cdot d\vec{A} = \int_R \vec{F}(x, y, f(x, y)) \cdot \left(-f_x \vec{i} - f_y \vec{j} + \vec{k} \right) dx \, dy.$$

- **Surface area of a graph:** Suppose a surface S is the part of the graph $z = f(x, y)$, where (x, y) is in a region R in the xy-plane. Then

$$\text{Area of } S = \int_R \sqrt{(f_x)^2 + (f_y)^2 + 1} \, dx \, dy.$$

- **The flux through a cylinder:** Suppose the cylindrical surface S has radius R, is oriented away from the z-axis, and has corresponding θz-region T. Then

$$\int_S \vec{F} \cdot d\vec{A} = \int_T \vec{F}(R, \theta, z) \cdot \left(\cos \theta \vec{i} + \sin \theta \vec{j} \right) R \, dz \, d\theta.$$

- **The flux through a spherical surface:** Suppose the spherical surface S has radius R, is oriented away from the origin, and has corresponding $\theta \phi$-region T. Then

$$\int_S \vec{F} \cdot d\vec{A} = \int_T \vec{F}(R, \theta, \phi) \cdot \left(\sin \phi \cos \theta \vec{i} + \sin \phi \sin \theta \vec{j} + \cos \phi \vec{k} \right) R^2 \sin \phi \, d\phi \, d\theta.$$

Exercises and Problems for Section 19.2

EXERCISES

In Exercises **1–4**, find the area vector $d\vec{A}$ for the surface $z = f(x, y)$, oriented upward.

1. $f(x, y) = 3x - 5y$

2. $f(x, y) = 8x + 7y$

3. $f(x, y) = 2x^2 - 3y^2$

4. $f(x, y) = xy + y^2$

In Exercises **5–8**, write an iterated integral for the flux of $\vec{F}$ through the surface S, which is the part of the graph of $z = f(x, y)$ corresponding to the region R, oriented upward. Do not evaluate the integral.

5. $\vec{F}(x, y, z) = 10\vec{i} + 20\vec{j} + 30\vec{k}$
   $f(x, y) = 2x - 3y$
   $R: -2 \leq x \leq 3, 0 \leq y \leq 5$

6. $\vec{F}(x, y, z) = z\vec{i} + x\vec{j} + y\vec{k}$
   $f(x, y) = 50 - 4x + 10y$
   $R: 0 \leq x \leq 4, 0 \leq y \leq 8$

7. $\vec{F}(x, y, z) = yz\vec{i} + xy\vec{j} + xy\vec{k}$
   $f(x, y) = \cos x + \sin 2y$
   R: Triangle with vertices $(0, 0)$, $(0, 5)$, $(5, 0)$

8. $\vec{F}(x, y, z) = \cos(x + 2y)\vec{j}$
   $f(x, y) = xe^{3y}$
   R: Quarter disk of radius 5 centered at the origin, in quadrant I

■ In Exercises 9–12, compute the flux of $\vec{F}$ through the surface S, which is the part of the graph of $z = f(x, y)$ corresponding to region R, oriented upward.

9. $\vec{F}(x, y, z) = 3\vec{i} - 2\vec{j} + 6\vec{k}$
   $f(x, y) = 4x - 2y$
   $R: 0 \leq x \leq 5, 0 \leq y \leq 10$

10. $\vec{F}(x, y, z) = \vec{i} - 2\vec{j} + z\vec{k}$
    $f(x, y) = xy$
    $R: 0 \leq x \leq 10, 0 \leq y \leq 10$

11. $\vec{F}(x, y, z) = \cos y\vec{i} + z\vec{j} + \vec{k}$
    $f(x, y) = x^2 + 2y$
    $R: 0 \leq x \leq 1, 0 \leq y \leq 1$

12. $\vec{F}(x, y, z) = x\vec{i} + z\vec{k}$
    $f(x, y) = x + y + 2$
    R: Triangle with vertices $(-1, 0), (1, 0), (0, 1)$

■ In Exercises 13–16, write an iterated integral for the flux of $\vec{F}$ through the cylindrical surface S centered on the z-axis, oriented away from the z-axis. Do not evaluate the integral.

13. $\vec{F}(x, y, z) = \vec{i} + 2\vec{j} + 3\vec{k}$
    S: radius 10, $x \geq 0, y \geq 0, 0 \leq z \leq 5$

14. $\vec{F}(x, y, z) = x\vec{i} + 2y\vec{j} + 3z\vec{k}$
    S: radius 10, $0 \leq z \leq 5$

15. $\vec{F}(x, y, z) = z^2\vec{i} + e^x\vec{j} + \vec{k}$
    S: radius 6, inside sphere of radius 10

16. $\vec{F}(x, y, z) = x^2yz\vec{j} + z^3\vec{k}$
    S: radius 2, between the xy-plane and the paraboloid $z = x^2 + y^2$

■ In Exercises 17–20, compute the flux of $\vec{F}$ through the cylindrical surface S centered on the z-axis, oriented away from the z-axis.

17. $\vec{F}(x, y, z) = z\vec{j} + 6x\vec{k}$
    S: radius 5, $y \geq 0, 0 \leq z \leq 20$

18. $\vec{F}(x, y, z) = y\vec{i} + xz\vec{k}$
    S: radius 10, $x \geq 0, y \geq 0, 0 \leq z \leq 3$

19. $\vec{F}(x, y, z) = xyz\vec{j} + xe^z\vec{k}$
    S: radius 2, $0 \leq y \leq x, 0 \leq z \leq 10$

20. $\vec{F}(x, y, z) = xy\vec{i} + 2z\vec{j}$
    S: radius 1, $x \geq 0, 0 \leq y \leq 1/2, 0 \leq z \leq 2$

■ In Exercises 21–24, write an iterated integral for the flux of $\vec{F}$ through the spherical surface S centered at the origin, oriented away from the origin. Do not evaluate the integral.

21. $\vec{F}(x, y, z) = \vec{i} + 2\vec{j} + 3\vec{k}$
    S: radius 10, $z \geq 0$

22. $\vec{F}(x, y, z) = x\vec{i} + 2y\vec{j} + 3z\vec{k}$
    S: radius 5, entire sphere

23. $\vec{F}(x, y, z) = z^2\vec{i}$
    S: radius 2, $x \geq 0$

24. $\vec{F}(x, y, z) = e^x\vec{k}$
    S: radius 3, $y \geq 0, z \leq 0$

■ In Exercises 25–27, compute the flux of $\vec{F}$ through the spherical surface S centered at the origin, oriented away from the origin.

25. $\vec{F}(x, y, z) = z\vec{i}$
    S: radius 20, $x \geq 0, y \geq 0, z \geq 0$

26. $\vec{F}(x, y, z) = y\vec{i} - x\vec{j} + z\vec{k}$
    S: radius 4, entire sphere

27. $\vec{F}(x, y, z) = x\vec{i} + y\vec{j}$
    S: radius 1, above the cone $\phi = \pi/4$.

■ In Exercises 28–29, compute the flux of $\vec{v} = z\vec{k}$ through the rectangular region with the orientation shown.

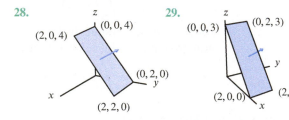

28.   29.

PROBLEMS

■ In Problems 30–46 compute the flux of the vector field $\vec{F}$ through the surface S.

30. $\vec{F} = z\vec{k}$ and S is the portion of the plane $x + y + z = 1$ that lies in the first octant, oriented upward.

31. $\vec{F} = (x - y)\vec{i} + z\vec{j} + 3x\vec{k}$ and S is the part of the plane $z = x + y$ above the rectangle $0 \leq x \leq 2, 0 \leq y \leq 3$, oriented upward.

32. $\vec{F} = 2x\vec{j} + y\vec{k}$ and S is the part of the surface $z = -y + 1$ above the square $0 \leq x \leq 1, 0 \leq y \leq 1$, oriented upward.

33. $\vec{F} = -y\vec{j} + z\vec{k}$ and S is the part of the surface $z = y^2 + 5$ over the rectangle $-2 \leq x \leq 1, 0 \leq y \leq 1$, oriented upward.

34. $\vec{F} = \ln(x^2)\vec{i} + e^x\vec{j} + \cos(1 - z)\vec{k}$ and S is the part of the surface $z = -y + 1$ above the square $0 \leq x \leq 1$, $0 \leq y \leq 1$, oriented upward.

35. $\vec{F} = 5\vec{i} + 7\vec{j} + z\vec{k}$ and S is a closed cylinder of radius 3 centered on the z-axis, with $-2 \leq z \leq 2$, and oriented outward.

36. $\vec{F} = x\vec{i} + y\vec{j} + z\vec{k}$ and S is a closed cylinder of radius 2 centered on the y-axis, with $-3 \leq y \leq 3$, and oriented outward.

37. $\vec{F} = 3x\vec{i} + y\vec{j} + z\vec{k}$ and S is the part of the surface $z = -2x - 4y + 1$, oriented upward, with (x, y) in the triangle R with vertices $(0, 0)$, $(0, 2)$, $(1, 0)$.

38. $\vec{F} = x\vec{i} + y\vec{j}$ and S is the part of the surface $z = 25 - (x^2 + y^2)$ above the disk of radius 5 centered at the origin, oriented upward.

39. $\vec{F} = \cos(x^2 + y^2)\vec{k}$ and S is as in Exercise 38.

40. $\vec{F} = \vec{r}$ and S is the part of the plane $x + y + z = 1$ above the rectangle $0 \leq x \leq 2, 0 \leq y \leq 3$, oriented downward.

41. $\vec{F} = \vec{r}$ and S is the part of the surface $z = x^2 + y^2$ above the disk $x^2 + y^2 \leq 1$, oriented downward.

42. $\vec{F} = xz\vec{i} + y\vec{k}$ and S is the hemisphere $x^2 + y^2 + z^2 = 9, z \geq 0$, oriented upward.

43. $\vec{F} = -xz\vec{i} - yz\vec{j} + z^2\vec{k}$ and S is the cone $z = \sqrt{x^2 + y^2}$ for $0 \leq z \leq 6$, oriented upward.

44. $\vec{F} = yz^4\vec{i} - xz^4\vec{j} + e^{z^2}\vec{k}$ and S is the cone $z = \sqrt{x^2 + y^2}$ for $1 \leq z \leq 2$, oriented upward.

45. $\vec{F} = y\vec{i} + \vec{j} - xz\vec{k}$ and S is the surface $y = x^2 + z^2$, with $x^2 + z^2 \leq 1$, oriented in the positive y-direction.

46. $\vec{F} = x^2\vec{i} + y^2\vec{j} + z^2\vec{k}$ and S is the oriented triangular surface shown in Figure 19.26.

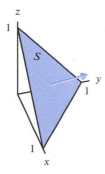

Figure 19.26

In Problems **47–50** find the area of the surface $z = f(x, y)$ over the region R in the xy-plane.

47. $f(x, y) = 50 + 5x - y, R: -5 \leq x \leq 5, 0 \leq y \leq 10$

48. $f(x, y) = 50 + 5x - y, R$: circle of radius 3 centered at the origin

49. $f(x, y) = xe^y, R: 0 \leq x \leq 1, 0 \leq y \leq 1$

50. $f(x, y) = (\sin x)(\sin y), R: 0 \leq x \leq \pi/2, 0 \leq y \leq \pi/2$

51. Let S be the hemisphere $x^2 + y^2 + z^2 = a^2$ of radius a, where $z \geq 0$.

   (a) Express the surface area of S as an integral in Cartesian coordinates.
   (b) Change variables to express the area integral in polar coordinates.
   (c) Find the area of S.

In Problems **52–53**, compute the flux of $\vec{F}$ through the cylindrical surface in Figure 19.27, oriented away from the z-axis.

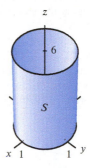

Figure 19.27

52. $\vec{F} = x\vec{i} + y\vec{j}$

53. $\vec{F} = xz\vec{i} + yz\vec{j} + z^3\vec{k}$

In Problems **54–57**, compute the flux of $\vec{F}$ through the spherical surface, S.

54. $\vec{F} = z\vec{k}$ and S is the upper hemisphere of radius 2 centered at the origin, oriented outward.

55. $\vec{F} = y\vec{i} - x\vec{j} + z\vec{k}$ and S is the spherical cap given by $x^2 + y^2 + z^2 = 1, z \geq 0$, oriented upward.

56. $\vec{F} = z^2\vec{k}$ and S is the upper hemisphere of the sphere $x^2 + y^2 + z^2 = 25$, oriented away from the origin.

57. $\vec{F} = x\vec{i} + y\vec{j} + z\vec{k}$ and S is the surface of the sphere $x^2 + y^2 + z^2 = a^2$, oriented outward.

58. Compute the flux of $\vec{F} = x\vec{i} + y\vec{j} + z\vec{k}$ over the quarter cylinder S given by $x^2 + y^2 = 1, 0 \leq x \leq 1, 0 \leq y \leq 1$, $0 \leq z \leq 1$, oriented outward.

59. Compute the flux of $\vec{F} = x\vec{i} + \vec{j} + \vec{k}$ through the surface S given by $x = \sin y \sin z$, with $0 \leq y \leq \pi/2$, $0 \leq z \leq \pi/2$, oriented in the direction of increasing x.

60. Compute the flux of $\vec{F} = (x + z)\vec{i} + \vec{j} + z\vec{k}$ through the surface S given by $y = x^2 + z^2$, with $0 \leq y \leq 1$, $x \geq 0, z \geq 0$, oriented toward the xz-plane.

61. Let $\vec{F} = (xze^{yz})\vec{i} + xz\vec{j} + (5 + x^2 + y^2)\vec{k}$. Calculate the flux of $\vec{F}$ through the disk $x^2 + y^2 \leq 1$ in the xy-plane, oriented upward.

62. Let $\vec{H} = (e^{xy}+3z+5)\vec{i} + (e^{xy}+5z+3)\vec{j} + (3z+e^{xy})\vec{k}$. Calculate the flux of $\vec{H}$ through the square of side 2 with one vertex at the origin, one edge along the positive y-axis, one edge in the xz-plane with $x > 0$, $z > 0$, and the normal $\vec{n} = \vec{i} - \vec{k}$.

63. The vector field, $\vec{F}$, in Figure 19.28 depends only on z; that is, it is of the form $g(z)\vec{k}$, where g is an increasing function. The integral $\int_S \vec{F} \cdot d\vec{A}$ represents the flux of $\vec{F}$ through this rectangle, S, oriented upward. In each of the following cases, how does the flux change?

(a) The rectangle is twice as wide in the x-direction, with new corners at the origin, $(2,0,0)$, $(2,1,3)$, $(0,1,3)$.

(b) The rectangle is moved so that its corners are at $(1,0,0)$, $(2,0,0)$, $(2,1,3)$, $(1,1,3)$.

(c) The orientation is changed to downward.

(d) The rectangle is tripled in size, so that its new corners are at the origin, $(3,0,0)$, $(3,3,9)$, $(0,3,9)$.

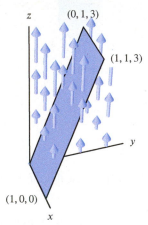

Figure 19.28

64. Electric charge is distributed in space with density (in coulomb/m³) given in spherical coordinates by

$$\delta(\rho, \phi, \theta) = \begin{cases} \delta_0 \text{ (a constant)} & \rho \le a \\ 0 & \rho > a. \end{cases}$$

(a) Describe the charge distribution in words.

(b) Find the electric field $\vec{E}$ due to δ. Assume that $\vec{E}$ can be written in spherical coordinates as $\vec{E} = E(\rho)\vec{e}_\rho$, where $\vec{e}_\rho$ is the unit outward normal to the sphere of radius ρ. In addition, $\vec{E}$ satisfies Gauss's Law for any simple closed surface S enclosing a volume W:

$$\int_S \vec{E} \cdot d\vec{A} = k \int_W \delta\, dV, \quad k \text{ a constant.}$$

65. Electric charge is distributed in space with density (in coulomb/m³) given in cylindrical coordinates by

$$\delta(r, \theta, z) = \begin{cases} \delta_0 \text{ (a constant)} & \text{if } r \le a \\ 0 & \text{if } r > a \end{cases}$$

(a) Describe the charge distribution in words.

(b) Find the electric field $\vec{E}$ due to δ. Assume that $\vec{E}$ can be written in cylindrical coordinates as $\vec{E} = E(r)\vec{e}_r$, where $\vec{e}_r$ is the unit outward vector to the cylinder of radius r, and that $\vec{E}$ satisfies Gauss's Law (see Problem 64).

Strengthen Your Understanding

In Problems 66–67, explain what is wrong with the statement.

66. Flux outward through the cone, given in cylindrical coordinates by $z = r$, can be computed using the formula $d\vec{A} = \left(\cos\theta\vec{i} + \sin\theta\vec{j}\right) R\, dz\, d\theta$.

67. For the surface $z = f(x, y)$ oriented upward, the formula

$$d\vec{A} = \vec{n}\, dA = \left(-f_x\vec{i} - f_y\vec{j} + \vec{k}\right) dx\, dy$$

gives $\vec{n} = -f_x\vec{i} - f_y\vec{j} + \vec{k}$ and $dA = dx\, dy$.

In Problems 68–69, give an example of:

68. A function $f(x, y)$ such that, for the surface $z = f(x, y)$ oriented upwards, we have $d\vec{A} = (\vec{i} + \vec{j} + \vec{k})\, dx\, dy$.

69. An oriented surface S on the cylinder of radius 10 centered on the z-axis such that $\int_S \vec{F} \cdot d\vec{A} = 600$, where $\vec{F} = x\vec{i} + y\vec{j}$.

Are the statements in Problems 70–72 true or false? Give reasons for your answer.

70. If S is the part of the graph of $z = f(x, y)$ above $a \le x \le b, c \le y \le d$, then S has surface area $\int_a^b \int_c^d \sqrt{(f_x)^2 + (f_y)^2 + 1}\, dx\, dy$.

71. If $\vec{A}(x, y)$ is the area vector for $z = f(x, y)$ oriented upward and $\vec{B}(x, y)$ is the area vector for $z = -f(x, y)$ oriented upward, then $\vec{A}(x, y) = -\vec{B}(x, y)$.

72. If S is the sphere $x^2 + y^2 + z^2 = 1$ oriented outward and $\int_S \vec{F} \cdot d\vec{A} = 0$, then $\vec{F}(x, y, z)$ is perpendicular to $x\vec{i} + y\vec{j} + z\vec{k}$ at every point of S.

19.3   THE DIVERGENCE OF A VECTOR FIELD

Imagine that the vector fields in Figures 19.29 and 19.30 are velocity vector fields describing the flow of a fluid.[3] Figure 19.29 suggests outflow from the origin; for example, it could represent the expanding cloud of matter in the big-bang theory of the origin of the universe. We say that the origin is a *source*. Figure 19.30 suggests flow into the origin; in this case we say that the origin is a *sink*.

In this section we use the flux out of a closed surface surrounding a point to measure the outflow per unit volume there, also called the *divergence*, or *flux density*.

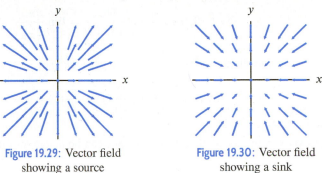

Figure 19.29: Vector field showing a source

Figure 19.30: Vector field showing a sink

Definition of Divergence

To measure the outflow per unit volume of a vector field at a point, we calculate the flux out of a small sphere centered at the point, divide by the volume enclosed by the sphere, then take the limit of this flux-to-volume ratio as the sphere contracts around the point.

> ### Geometric Definition of Divergence
>
> The **divergence**, or **flux density**, of a smooth vector field $\vec{F}$, written **div** $\vec{F}$, is a scalar-valued function defined by
>
> $$\operatorname{div} \vec{F}(x, y, z) = \lim_{\text{Volume} \to 0} \frac{\int_S \vec{F} \cdot d\vec{A}}{\text{Volume of } S}.$$
>
> Here S is a sphere centered at (x, y, z), oriented outward, that contracts down to (x, y, z) in the limit. The limit can be computed using other shapes as well, such as the cubes in Example 2.

In Cartesian coordinates, the divergence can also be calculated using the following formula. We show that these definitions are equivalent later in the section.

> ### Cartesian Coordinate Definition of Divergence
>
> If $\vec{F} = F_1\vec{i} + F_2\vec{j} + F_3\vec{k}$, then
>
> $$\operatorname{div} \vec{F} = \frac{\partial F_1}{\partial x} + \frac{\partial F_2}{\partial y} + \frac{\partial F_3}{\partial z}.$$

The dot product formula gives an easy way to remember the Cartesian coordinate definition and suggests another common notation for div $\vec{F}$, namely $\nabla \cdot \vec{F}$. Using $\nabla = \frac{\partial}{\partial x}\vec{i} + \frac{\partial}{\partial y}\vec{j} + \frac{\partial}{\partial z}\vec{k}$, we can write

$$\operatorname{div} \vec{F} = \nabla \cdot \vec{F} = \left(\frac{\partial}{\partial x}\vec{i} + \frac{\partial}{\partial y}\vec{j} + \frac{\partial}{\partial z}\vec{k}\right) \cdot (F_1\vec{i} + F_2\vec{j} + F_3\vec{k}) = \frac{\partial F_1}{\partial x} + \frac{\partial F_2}{\partial y} + \frac{\partial F_3}{\partial z}.$$

---

[3] Although not all vector fields represent physically realistic fluid flows, it is useful to think of them in this way.

Example 1   Calculate the divergence of $\vec{F}(\vec{r}) = \vec{r}$ at the origin

(a) Using the geometric definition.
(b) Using the Cartesian coordinate definition.

Solution   (a) Using the method of Example 5 on page 1023, we can calculate the flux of $\vec{F}$ out of the sphere of radius a, centered at the origin; it is $4\pi a^3$. So we have

$$\operatorname{div}\vec{F}(0,0,0) = \lim_{a \to 0} \frac{\text{Flux}}{\text{Volume}} = \lim_{a \to 0} \frac{4\pi a^3}{\frac{4}{3}\pi a^3} = \lim_{a \to 0} 3 = 3.$$

(b) In Cartesian coordinates, $\vec{F}(x, y, z) = x\vec{i} + y\vec{j} + z\vec{k}$, so

$$\operatorname{div}\vec{F} = \frac{\partial}{\partial x}(x) + \frac{\partial}{\partial y}(y) + \frac{\partial}{\partial z}(z) = 1 + 1 + 1 = 3.$$

The next example shows that the divergence can be negative if there is net inflow to a point.

Example 2   (a) Using the geometric definition, find the divergence of $\vec{v} = -x\vec{i}$ at: (i) $(0,0,0)$ (ii) $(2,2,0)$.
(b) Confirm that the coordinate definition gives the same results.

Solution   (a) (i) The vector field $\vec{v} = -x\vec{i}$ is parallel to the x-axis and is shown in the xy-plane in Figure 19.31. To compute the flux density at $(0,0,0)$, we use a cube S_1, centered at the origin with edges parallel to the axes, of length $2c$. Then the flux through the faces perpendicular to the y- and z-axes is zero (because the vector field is parallel to these faces). On the faces perpendicular to the x-axis, the vector field and the outward normal are parallel but point in opposite directions. On the face at $x = c$, where $\vec{v} = -c\vec{i}$ and $\Delta\vec{A} = \|\vec{A}\|\vec{i}$, we have

$$\vec{v} \cdot \Delta\vec{A} = -c\|\Delta\vec{A}\|.$$

On the face at $x = -c$, where $\vec{v} = c\vec{i}$ and $\Delta\vec{A} = -\|\vec{A}\|\vec{i}$, the dot product is still negative:

$$\vec{v} \cdot \Delta\vec{A} = -c\|\Delta\vec{A}\|.$$

Therefore, the flux through the cube is given by

$$\int_{S_1} \vec{v} \cdot d\vec{A} = \int_{\text{Face } x=-c} \vec{v} \cdot d\vec{A} + \int_{\text{Face } x=c} \vec{v} \cdot d\vec{A}$$

$$= -c \cdot \text{Area of one face} + (-c) \cdot \text{Area of other face} = -2c(2c)^2 = -8c^3.$$

Thus,

$$\operatorname{div}\vec{v}(0,0,0) = \lim_{\text{Volume} \to 0} \frac{\int_S \vec{v} \cdot d\vec{A}}{\text{Volume of cube}} = \lim_{c \to 0}\left(\frac{-8c^3}{(2c)^3}\right) = -1.$$

Since the vector field points inward toward the yz-plane, it makes sense that the divergence is negative at the origin.

(ii) Take S_2 to be a cube as before, but centered this time at the point $(2,2,0)$. See Figure 19.31. As before, the flux through the faces perpendicular to the y- and z-axes is zero. On the face at $x = 2 + c$,

$$\vec{v} \cdot \Delta\vec{A} = -(2 + c)\|\Delta\vec{A}\|.$$

On the face at $x = 2 - c$ with outward normal, the dot product is positive, and

$$\vec{v} \cdot \Delta\vec{A} = (2 - c)\|\Delta\vec{A}\|.$$

Therefore, the flux through the cube is given by

$$\int_{S_2} \vec{v} \cdot d\vec{A} = \int_{\text{Face } x=2-c} \vec{v} \cdot d\vec{A} + \int_{\text{Face } x=2+c} \vec{v} \cdot d\vec{A}$$

$$= (2-c) \cdot \text{Area of one face} - (2+c) \cdot \text{Area of other face} = -2c(2c)^2 = -8c^3.$$

Then, as before,

$$\text{div } \vec{v}\,(2, 2, 0) = \lim_{\text{Volume} \to 0} \frac{\int_S \vec{v} \cdot d\vec{A}}{\text{Volume of cube}} = \lim_{c \to 0} \left(\frac{-8c^3}{(2c)^3} \right) = -1.$$

Although the vector field is flowing away from the point $(2, 2, 0)$ on the left, this outflow is smaller in magnitude than the inflow on the right, so the net outflow is negative.

(b) Since $\vec{v} = -x\vec{i} + 0\vec{j} + 0\vec{k}$, the formula gives

$$\text{div } \vec{v} = \frac{\partial}{\partial x}(-x) + \frac{\partial}{\partial y}(0) + \frac{\partial}{\partial z}(0) = -1 + 0 + 0 = -1.$$

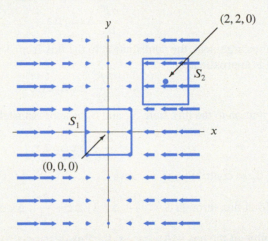

Figure 19.31: Vector field $\vec{v} = -x\vec{i}$ in the xy-plane

Why Do the Two Definitions of Divergence Give the Same Result?

The geometric definition defines div $\vec{F}$ as the flux density of $\vec{F}$. To see why the coordinate definition is also the flux density, imagine computing the flux out of a small box-shaped surface S at (x_0, y_0, z_0), with sides of length Δx, Δy, and Δz parallel to the axes. On S_1 (the back face of the box shown in Figure 19.32, where $x = x_0$), the outward normal is in the negative x-direction, so $d\vec{A} = -dy\,dz\,\vec{i}$. Assuming $\vec{F}$ is approximately constant on S_1, we have

$$\int_{S_1} \vec{F} \cdot d\vec{A} = \int_{S_1} \vec{F} \cdot (-\vec{i})\,dy\,dz \approx -F_1(x_0, y_0, z_0) \int_{S_1} dy\,dz$$

$$= -F_1(x_0, y_0, z_0) \cdot \text{Area of } S_1 = -F_1(x_0, y_0, z_0)\,\Delta y\,\Delta z.$$

On S_2, the face where $x = x_0 + \Delta x$, the outward normal points in the positive x-direction, so $d\vec{A} = dy\,dz\,\vec{i}$. Therefore,

$$\int_{S_2} \vec{F} \cdot d\vec{A} = \int_{S_2} \vec{F} \cdot \vec{i}\,dy\,dz \approx F_1(x_0 + \Delta x, y_0, z_0) \int_{S_2} dy\,dz$$

$$= F_1(x_0 + \Delta x, y_0, z_0) \cdot \text{Area of } S_2 = F_1(x_0 + \Delta x, y_0, z_0)\,\Delta y\,\Delta z.$$

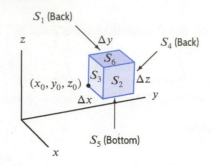

Figure 19.32: Box used to find div $\vec{F}$ at (x_0, y_0, z_0)

Thus,

$$\int_{S_1} \vec{F} \cdot d\vec{A} + \int_{S_2} \vec{F} \cdot d\vec{A} \approx F_1(x_0 + \Delta x, y_0, z_0)\Delta y \Delta z - F_1(x_0, y_0, z_0)\Delta y \Delta z$$

$$= \frac{F_1(x_0 + \Delta x, y_0, z_0) - F_1(x_0, y_0, z_0)}{\Delta x}\Delta x \Delta y \Delta z$$

$$\approx \frac{\partial F_1}{\partial x}\Delta x \Delta y \Delta z.$$

By an analogous argument, the contribution to the flux from S_3 and S_4 (the surfaces perpendicular to the y-axis) is approximately

$$\frac{\partial F_2}{\partial y} \Delta x \, \Delta y \, \Delta z,$$

and the contribution to the flux from S_5 and S_6 is approximately

$$\frac{\partial F_3}{\partial z} \Delta x \, \Delta y \, \Delta z.$$

Thus, adding these contributions, we have

$$\text{Total flux through } S \approx \frac{\partial F_1}{\partial x} \Delta x \, \Delta y \, \Delta z + \frac{\partial F_2}{\partial y} \Delta x \, \Delta y \, \Delta z + \frac{\partial F_3}{\partial z} \Delta x \, \Delta y \, \Delta z.$$

Since the volume of the box is $\Delta x \, \Delta y \, \Delta z$, the flux density is

$$\frac{\text{Total flux through } S}{\text{Volume of box}} \approx \frac{\dfrac{\partial F_1}{\partial x}\Delta x \Delta y \Delta z + \dfrac{\partial F_2}{\partial y}\Delta x \Delta y \Delta z + \dfrac{\partial F_3}{\partial z}\Delta x \Delta y \Delta z}{\Delta x \Delta y \Delta z}$$

$$= \frac{\partial F_1}{\partial x} + \frac{\partial F_2}{\partial y} + \frac{\partial F_3}{\partial z}.$$

Divergence-Free Vector Fields

A vector field $\vec{F}$ is said to be *divergence free* or *solenoidal* if $\text{div}\vec{F} = 0$ everywhere that $\vec{F}$ is defined.

Example 3   Figure 19.33 shows, for three values of the constant p, the vector field

$$\vec{E} = \frac{\vec{r}}{\|\vec{r}\|^p} \qquad \vec{r} = x\vec{i} + y\vec{j} + z\vec{k}, \, \vec{r} \neq \vec{0}.$$

(a) Find a formula for div $\vec{E}$.
(b) Is there a value of p for which $\vec{E}$ is divergence-free? If so, find it.

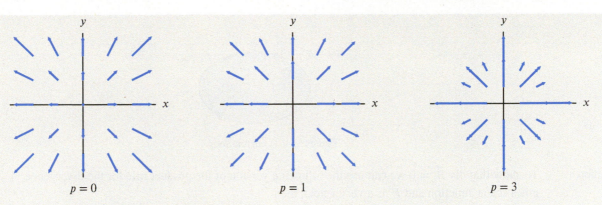

$p = 0$ $p = 1$ $p = 3$

Figure 19.33: The vector field $\vec{E}\,(\vec{r}) = \vec{r}\,/\|\vec{r}\,\|^p$ for $p = 0, 1,$ and 3 in the xy-plane

Solution   (a) The components of $\vec{E}$ are

$$\vec{E} = \frac{x}{(x^2 + y^2 + z^2)^{p/2}}\vec{i} + \frac{y}{(x^2 + y^2 + z^2)^{p/2}}\vec{j} + \frac{z}{(x^2 + y^2 + z^2)^{p/2}}\vec{k}.$$

We compute the partial derivatives

$$\frac{\partial}{\partial x}\left(\frac{x}{(x^2 + y^2 + z^2)^{p/2}}\right) = \frac{1}{(x^2 + y^2 + z^2)^{p/2}} - \frac{px^2}{(x^2 + y^2 + z^2)^{(p/2)+1}}$$

$$\frac{\partial}{\partial y}\left(\frac{y}{(x^2 + y^2 + z^2)^{p/2}}\right) = \frac{1}{(x^2 + y^2 + z^2)^{p/2}} - \frac{py^2}{(x^2 + y^2 + z^2)^{(p/2)+1}}$$

$$\frac{\partial}{\partial z}\left(\frac{z}{(x^2 + y^2 + z^2)^{p/2}}\right) = \frac{1}{(x^2 + y^2 + z^2)^{p/2}} - \frac{pz^2}{(x^2 + y^2 + z^2)^{(p/2)+1}}.$$

So

$$\text{div}\,\vec{E} = \frac{3}{(x^2 + y^2 + z^2)^{p/2}} - \frac{p(x^2 + y^2 + z^2)}{(x^2 + y^2 + z^2)^{(p/2)+1}}$$

$$= \frac{3 - p}{(x^2 + y^2 + z^2)^{p/2}} = \frac{3 - p}{\|\vec{r}\,\|^p}.$$

(b) The divergence is zero when $p = 3$, so $\vec{F}\,(\vec{r}) = \vec{r}\,/\|\vec{r}\,\|^3$ is a divergence-free vector field. Notice that the divergence is zero even though the vectors point outward from the origin.

Magnetic Fields

An important class of divergence-free vector fields is the magnetic fields. One of Maxwell's Laws of Electromagnetism is that the magnetic field $\vec{B}$ satisfies

$$\text{div}\,\vec{B} = 0.$$

Example 4   An infinitesimal current loop, similar to that shown in Figure 19.34, is called a *magnetic dipole*. Its magnitude is described by a constant vector $\vec{\mu}$, called the dipole moment. The magnetic field due to a magnetic dipole with moment $\vec{\mu}$ is a multiple of

$$\vec{B} = -\frac{\vec{\mu}}{\|\vec{r}\,\|^3} + \frac{3(\vec{\mu} \cdot \vec{r}\,)\vec{r}}{\|\vec{r}\,\|^5}, \qquad \vec{r} \neq \vec{0}.$$

Show that div $\vec{B} = 0$.

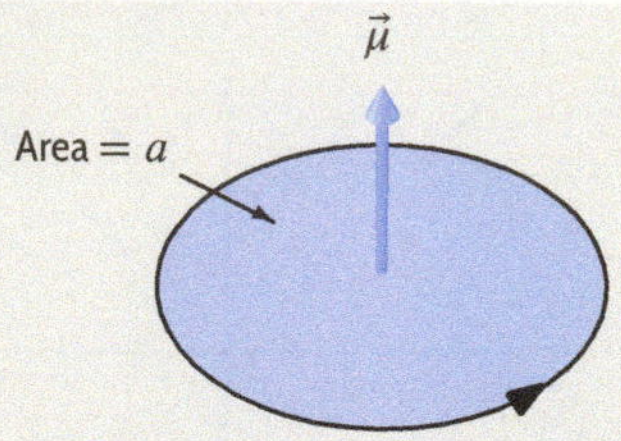

Figure 19.34: A current loop

Solution   To show that div $\vec{B} = 0$ we can use the following version of the product rule for the divergence: if g is a scalar function and $\vec{F}$ is a vector field, then

$$\operatorname{div}(g\vec{F}) = (\operatorname{grad} g) \cdot \vec{F} + g \operatorname{div} \vec{F}.$$

(See Problem 37 on page 1047.) Thus, since $\vec{\mu}$ is constant and div $\vec{\mu} = 0$, we have

$$\operatorname{div}\left(\frac{\vec{\mu}}{\|\vec{r}\|^3}\right) = \operatorname{div}\left(\frac{1}{\|\vec{r}\|^3}\vec{\mu}\right) = \operatorname{grad}\left(\frac{1}{\|\vec{r}\|^3}\right) \cdot \vec{\mu} + \left(\frac{1}{\|\vec{r}\|^3}\right) 0$$

and

$$\operatorname{div}\left(\frac{(\vec{\mu} \cdot \vec{r})\vec{r}}{\|\vec{r}\|^5}\right) = \operatorname{div}\left(\vec{\mu} \cdot \vec{r}\,\frac{\vec{r}}{\|\vec{r}\|^5}\right) = \operatorname{grad}(\vec{\mu} \cdot \vec{r}) \cdot \frac{\vec{r}}{\|\vec{r}\|^5} + (\vec{\mu} \cdot \vec{r})\operatorname{div}\left(\frac{\vec{r}}{\|\vec{r}\|^5}\right).$$

From Problems 83 and 84 of Section 14.5 (available online) and Example 3 on page 1042, we have

$$\operatorname{grad}\left(\frac{1}{\|\vec{r}\|^3}\right) = \frac{-3\vec{r}}{\|\vec{r}\|^5}, \qquad \operatorname{grad}(\vec{\mu} \cdot \vec{r}) = \vec{\mu}, \qquad \operatorname{div}\left(\frac{\vec{r}}{\|\vec{r}\|^5}\right) = \frac{-2}{\|\vec{r}\|^5}.$$

Putting these results together gives

$$\operatorname{div}\vec{B} = -\operatorname{grad}\left(\frac{1}{\|\vec{r}\|^3}\right) \cdot \vec{\mu} + 3\operatorname{grad}(\vec{\mu} \cdot \vec{r}) \cdot \frac{\vec{r}}{\|\vec{r}\|^5} + 3(\vec{\mu} \cdot \vec{r})\operatorname{div}\left(\frac{\vec{r}}{\|\vec{r}\|^5}\right)$$

$$= \frac{3\vec{r} \cdot \vec{\mu}}{\|\vec{r}\|^5} + \frac{3\vec{\mu} \cdot \vec{r}}{\|\vec{r}\|^5} - \frac{6\vec{\mu} \cdot \vec{r}}{\|\vec{r}\|^5}$$

$$= 0.$$

Summary for Section 19.3

- **Geometric definition of divergence:** The divergence of a smooth vector field $\vec{F}$ is a scalar-valued function defined by

$$\operatorname{div}\vec{F}(x, y, z) = \lim_{\text{Volume} \to 0} \frac{\int_S \vec{F} \cdot d\vec{A}}{\text{Volume of } S}.$$

Here S is a sphere centered at (x, y, z), oriented outward, that contracts down to (x, y, z).

- **Cartesian-coordinate definition of divergence:** If $\vec{F} = F_1\vec{i} + F_2\vec{j} + F_3\vec{k}$, then

$$\operatorname{div}\vec{F} = \frac{\partial F_1}{\partial x} + \frac{\partial F_2}{\partial y} + \frac{\partial F_3}{\partial z}.$$

- A vector field $\vec{F}$ is said to be **divergence-free** if $\operatorname{div}\vec{F} = 0$ everywhere that $\vec{F}$ is defined.

Exercises and Problems for Section 19.3 Online Resource: Additional Problems for Section 19.3

EXERCISES

■ In Exercises 1–2, is the quantity a vector or a scalar? Calculate it.

1. $\operatorname{div}\left((x^2 + y)\vec{i} + (xye^z)\vec{j} - \ln(x^2 + y^2)\vec{k}\right)$

2. $\operatorname{div}\left((2\sin(xy) + \tan z)\vec{i} + (\tan y)\vec{j} + (e^{x^2+y^2})\vec{k}\right)$

3. Which of the following two vector fields, sketched in the xy-plane, appears to have the greater divergence at the origin? The scales are the same on each.

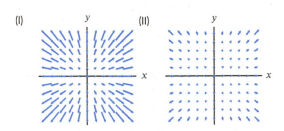

■ In Exercises 4–12, find the divergence of the vector field. (Note: $\vec{r} = x\vec{i} + y\vec{j} + z\vec{k}$.)

4. $\vec{F}(x, y) = -y\vec{i} + x\vec{j}$

5. $\vec{F}(x, y) = -x\vec{i} + y\vec{j}$

6. $\vec{F}(x, y, z) = (-x + y)\vec{i} + (y + z)\vec{j} + (-z + x)\vec{k}$

7. $\vec{F}(x, y) = (x^2 - y^2)\vec{i} + 2xy\vec{j}$

8. $\vec{F}(x, y, z) = 3x^2\vec{i} - \sin(xz)(\vec{i} + \vec{k})$

9. $\vec{F} = \left(\ln(x^2 + 1)\vec{i} + (\cos y)\vec{j} + (xye^z)\vec{k}\right)$

10. $\vec{F}(\vec{r}) = \vec{a} \times \vec{r}$

11. $\vec{F}(x, y) = \dfrac{-y\vec{i} + x\vec{j}}{x^2 + y^2}$

12. $\vec{F}(\vec{r}) = \dfrac{\vec{r} - \vec{r}_0}{\|\vec{r} - \vec{r}_0\|}, \quad \vec{r} \neq \vec{r}_0$

13. For each of the following vector fields, sketched in the xy-plane, decide if the divergence is positive, zero, or negative at the indicated point.

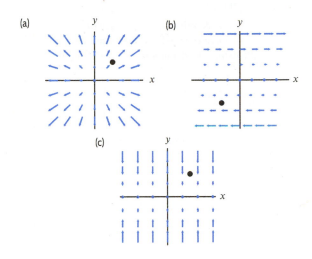

PROBLEMS

14. Draw two vector fields that have positive divergence everywhere.

15. Draw two vector fields that have negative divergence everywhere.

16. Draw two vector fields that have zero divergence everywhere.

17. A small sphere of radius 0.1 surrounds the point $(2, 3, -1)$. The flux of a vector field $\vec{G}$ into this sphere is 0.00004π. Estimate $\operatorname{div} \vec{G}$ at the point $(2, 3, -1)$.

18. A smooth vector field $\vec{F}$ has $\operatorname{div} \vec{F}(1, 2, 3) = 5$. Estimate the flux of $\vec{F}$ out of a small sphere of radius 0.01 centered at the point $(1, 2, 3)$.

19. Let $\vec{F}$ be a vector field with $\operatorname{div} \vec{F} = x^2 + y^2 - z$.

    (a) Estimate $\int_S \vec{F} \cdot d\vec{A}$ where S is
        (i) A sphere of radius 0.1 centered at $(2, 0, 0)$.
        (ii) A box of side 0.2 with edges parallel to the axes and centered at $(0, 0, 10)$.

    (b) The point $(2, 0, 0)$ is called a *source* for the vector field $\vec{F}$; the point $(0, 0, 10)$ is called a *sink*. Explain the reason for these names using your answer to part (a).

20. The flux of $\vec{F}$ out of a small sphere of radius 0.1 centered at $(4, 5, 2)$ is 0.0125. Estimate:

    (a) $\operatorname{div} \vec{F}$ at $(4, 5, 2)$
    (b) The flux of $\vec{F}$ out of a sphere of radius 0.2 centered at $(4, 5, 2)$.

21. (a) Find the flux of $\vec{F} = 2x\vec{i} - 3y\vec{j} + 5z\vec{k}$ through a cube with four of its corners at the points $(a, b, c), (a + w, b, c), (a, b + w, c), (a, b, c + w)$ and edge length w. See Figure 19.35.
    (b) Use the geometric definition and part (a) to find $\operatorname{div} \vec{F}$ at the point (a, b, c).
    (c) Find $\operatorname{div} \vec{F}$ using partial derivatives.

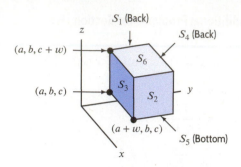

Figure 19.35

22. Suppose $\vec{F} = (3x+2)\vec{i} + 4x\vec{j} + (5x+1)\vec{k}$. Use the method of Exercise 21 to find div $\vec{F}$ at the point (a, b, c) by two different methods.

23. Use the geometric definition of divergence to find div $\vec{v}$ at the origin, where $\vec{v} = -2\vec{r}$. Check that you get the same result using the definition in Cartesian coordinates.

24. (a) Let $f(x, y) = axy + ax^2y + y^3$. Find div grad f.
    (b) Choose a so that div grad $f = 0$ for all x, y.

25. Let $\vec{F} = (9a^2x + 10ay^2)\vec{i} + (10z^3 - 6ay)\vec{j} - (3z + 10x^2 + 10y^2)\vec{k}$. Find the value(s) of a making div $\vec{F}$

    (a) 0   (b) A minimum

26. Let $\vec{F}(\vec{r}) = \vec{r}/\|\vec{r}\|^3$ (in 3-space), $\vec{r} \neq \vec{0}$.

    (a) Calculate div $\vec{F}$.
    (b) Sketch $\vec{F}$. Does $\vec{F}$ appear to have nonzero divergence? Does this agree with your answer to part (a)?

27. The vector field $\vec{F}(\vec{r}) = \vec{r}/\|\vec{r}\|^3$ is not defined at the origin. Nevertheless, we can attempt to use the flux definition to compute div $\vec{F}$ at the origin. What is the result?

28. Let $\vec{F}(x, y, z) = z\vec{k}$.

    (a) Calculate div $\vec{F}$.
    (b) Sketch $\vec{F}$. Does $\vec{F}$ appear to have nonzero divergence? Does this agree with your answer to part (a)?

29. The divergence of a magnetic vector field $\vec{B}$ must be zero everywhere. Which of the following vector fields cannot be a magnetic vector field?

    (a) $\vec{B}(x, y, z) = -y\vec{i} + x\vec{j} + (x+y)\vec{k}$
    (b) $\vec{B}(x, y, z) = -z\vec{i} + y\vec{j} + x\vec{k}$
    (c) $\vec{B}(x, y, z) = (x^2 - y^2 - x)\vec{i} + (y - 2xy)\vec{j}$

■ Problems 30–31 involve electric fields. Electric charge produces a vector field $\vec{E}$, called the electric field, which represents the force on a unit positive charge placed at the point. Two positive or two negative charges repel one another, whereas two charges of opposite sign attract one another. The divergence of $\vec{E}$ is proportional to the density of the electric charge (that is, the charge per unit volume), with a positive constant of proportionality.

30. A certain distribution of electric charge produces the electric field shown in Figure 19.36. Where are the charges that produced this electric field concentrated? Which concentrations are positive and which are negative?

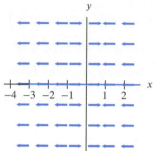

Figure 19.36

31. The electric field at the point $\vec{r}$ as a result of a point charge at the origin is $\vec{E}(\vec{r}) = k\vec{r}/\|\vec{r}\|^3$.

    (a) Calculate div $\vec{E}$ for $\vec{r} \neq \vec{0}$.
    (b) Calculate the limit suggested by the geometric definition of div $\vec{E}$ at the point $(0, 0, 0)$.
    (c) Explain what your answers mean in terms of charge density.

32. Due to roadwork ahead, the traffic on a highway slows linearly from 55 miles/hour to 15 miles/hour over a 2000-foot stretch of road, then crawls along at 15 miles/hour for 5000 feet, then speeds back up linearly to 55 miles/hour in the next 1000 feet, after which it moves steadily at 55 miles/hour.

    (a) Sketch a velocity vector field for the traffic flow.
    (b) Write a formula for the velocity vector field $\vec{v}$ (miles/hour) as a function of the distance x feet from the initial point of slowdown. (Take the direction of motion to be $\vec{i}$ and consider the various sections of the road separately.)
    (c) Compute div $\vec{v}$ at $x = 1000, 5000, 7500, 10{,}000$. Be sure to include the proper units.

33. The velocity field $\vec{v}$ in Problem 32 does not give a complete description of the traffic flow, for it takes no account of the spacing between vehicles. Let ρ be the density (cars/mile) of highway, where we assume that ρ depends only on x.

    (a) Using your highway experience, arrange in ascending order: $\rho(0), \rho(1000), \rho(5000)$.
    (b) What are the units and interpretation of the vector field $\rho\vec{v}$?
    (c) Would you expect $\rho\vec{v}$ to be constant? Why? What does this mean for div$(\rho\vec{v})$?
    (d) Determine $\rho(x)$ if $\rho(0) = 75$ cars/mile and $\rho\vec{v}$ is constant.
    (e) If the highway has two lanes, find the approximate number of feet between cars at $x = 0, 1000$, and 5000.

34. For $\vec{r} = x\vec{i} + y\vec{j} + z\vec{k}$, an arbitrary function $f(x, y, z)$, and an arbitrary vector field $\vec{F}(x, y, z)$, which of the following is a vector field and which is a constant vector field?

   (a) grad f (b) $(\text{div}\,\vec{F}\,)\vec{i}$ (c) $(\text{div}\,\vec{r}\,)\vec{i}$

   (d) $(\text{div}\,\vec{i}\,)\vec{F}$ (e) $\text{grad}(\text{div}\,\vec{F}\,)$

35. Let $\vec{r} = x\vec{i} + y\vec{j} + z\vec{k}$ and $\vec{c} = c_1\vec{i} + c_2\vec{j} + c_3\vec{k}$, a constant vector; let S be a sphere of radius R centered at the origin. Find

   (a) $\text{div}(\vec{r} \times \vec{c}\,)$ (b) $\int_S (\vec{r} \times \vec{c}\,) \cdot d\vec{A}$

36. Show that if $\vec{a}$ is a constant vector and $f(x, y, z)$ is a function, then $\text{div}(f\vec{a}\,) = (\text{grad}\,f) \cdot \vec{a}$.

37. Show that if $g(x, y, z)$ is a scalar-valued function and $\vec{F}(x, y, z)$ is a vector field, then

$$\text{div}(g\vec{F}\,) = (\text{grad}\,g) \cdot \vec{F} + g\,\text{div}\,\vec{F}.$$

38. If $f(x, y, z)$ and $g(x, y, z)$ are functions with continuous second partial derivatives, show that

$$\text{div}(\text{grad}\,f \times \text{grad}\,g) = 0.$$

■ In Problems 39–41, use Problems 37 and 38 to find the divergence of the vector field. The vectors $\vec{a}$ and $\vec{b}$ are constant.

39. $\vec{F} = \dfrac{1}{\|\vec{r}\|^p}\vec{a} \times \vec{r}$ **40.** $\vec{B} = \dfrac{1}{x^a}\vec{r}$

41. $\vec{G} = (\vec{b} \cdot \vec{r}\,)\vec{a} \times \vec{r}$

Strengthen Your Understanding

■ In Problems 44–46, explain what is wrong with the statement.

44. $\text{div}(2x\vec{i}\,) = 2\vec{i}$.

45. For $\vec{F}(x, y, z) = (x^2 + y)\vec{i} + (2y + z)\vec{j} - z^2\vec{k}$ we have $\text{div}\,\vec{F} = 2x\vec{i} + 2\vec{j} - 2z\vec{k}$.

46. The divergence of $f(x, y, z) = x^2 + yz$ is given by $\text{div}\,f(x, y, z) = 2x + z + y$.

■ In Problems 47–49, give an example of:

47. A vector field $\vec{F}(x, y, z)$ whose divergence is a nonzero constant.

48. A nonzero vector field $\vec{F}(x, y, z)$ whose divergence is zero.

49. A vector field that is not divergence free.

■ Are the statements in Problems 50–62 true or false? Give reasons for your answer.

50. $\text{div}(\vec{F} + \vec{G}\,) = \text{div}\,\vec{F} + \text{div}\,\vec{G}$

51. $\text{grad}(\vec{F} \cdot \vec{G}\,) = \vec{F}(\text{div}\,\vec{G}\,) + (\text{div}\,\vec{F}\,)\vec{G}$

52. $\text{div}\,\vec{F}$ is a scalar whose value can vary from point to point.

42. A vector field, $\vec{v}$, in the plane is a *point source* at the origin if its direction is away from the origin at every point, its magnitude depends only on the distance from the origin, and its divergence is zero away from the origin.

   (a) Explain why a point source at the origin must be of the form $\vec{v} = \left(f(x^2 + y^2)\right)(x\vec{i} + y\vec{j}\,)$ for some positive function f.

   (b) Show that $\vec{v} = K(x^2 + y^2)^{-1}(x\vec{i} + y\vec{j}\,)$ is a point source at the origin if $K > 0$.

   (c) What is the magnitude $\|\vec{v}\|$ of the source in part (b) as a function of the distance from its center?

   (d) Sketch the vector field $\vec{v} = (x^2 + y^2)^{-1}(x\vec{i} + y\vec{j}\,)$.

   (e) Show that $\phi = \dfrac{K}{2}\log(x^2 + y^2)$ is a potential function for the source in part (b).

43. A vector field, $\vec{v}$, in the plane is a *point sink* at the origin if its direction is toward the origin at every point, its magnitude depends only on the distance from the origin, and its divergence is zero away from the origin.

   (a) Explain why a point sink at the origin must be of the form $\vec{v} = \left(f(x^2 + y^2)\right)(x\vec{i} + y\vec{j}\,)$ for some negative function f.

   (b) Show that $\vec{v} = K(x^2 + y^2)^{-1}(x\vec{i} + y\vec{j}\,)$ is a point sink at the origin if $K < 0$.

   (c) Determine the magnitude $\|\vec{v}\|$ of the sink in part (b) as a function of the distance from its center.

   (d) Sketch $\vec{v} = -(x^2 + y^2)^{-1}(x\vec{i} + y\vec{j}\,)$.

   (e) Show that $\phi = \dfrac{K}{2}\log(x^2 + y^2)$ is a potential function for the sink in part (b).

53. If $\vec{F}$ is a vector field in 3-space, then $\text{div}\,\vec{F}$ is also a vector field.

54. A constant vector field $\vec{F} = a\vec{i} + b\vec{j} + c\vec{k}$ has zero divergence.

55. If a vector field $\vec{F}$ in 3-space has zero divergence then $\vec{F} = a\vec{i} + b\vec{j} + c\vec{k}$ where a, b and c are constants.

56. If $\vec{F}$ is a vector field in 3-space, and f is a scalar function, then $\text{div}(f\vec{F}\,) = f\,\text{div}\,\vec{F}$.

57. If $\vec{F}$ is a vector field in 3-space, and $\vec{F} = \text{grad}\,f$, then $\text{div}\,\vec{F} = 0$.

58. If $\vec{F}$ is a vector field in 3-space, then $\text{grad}(\text{div}\,\vec{F}\,) = \vec{0}$.

59. The field $\vec{F}(\vec{r}\,) = \vec{r}$ is divergence free.

60. If $f(x, y, z)$ is any given continuous scalar function, then there is at least one vector field $\vec{F}$ such that $\text{div}\,\vec{F} = f$.

61. If $\vec{F}$ and $\vec{G}$ are vector fields satisfying $\text{div}\,\vec{F} = \text{div}\,\vec{G}$ then $\vec{F} = \vec{G}$.

62. There exist a scalar function f and a vector field $\vec{F}$ satisfying $\text{div}(\text{grad}\,f) = \text{grad}(\text{div}\,\vec{F}\,)$.

19.4   THE DIVERGENCE THEOREM

The Divergence Theorem is a multivariable analogue of the Fundamental Theorem of Calculus; it says that the integral of the flux density over a solid region equals the flux integral through the boundary of the region.

The Boundary of a Solid Region

The boundary, S, of a solid region, W, may be thought of as the skin between the interior of the region and the space around it. For example, the boundary of a solid ball is a spherical surface, the boundary of a solid cube is its six faces, and the boundary of a solid cylinder is a tube sealed at both ends by disks. (See Figure 19.37). A surface which is the boundary of a solid region is called a *closed surface*. We assume a closed surface is oriented outward unless otherwise specified.

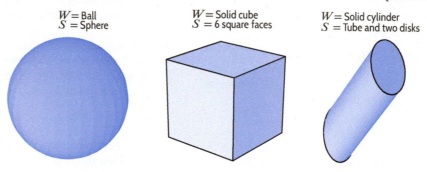

$W = $ Ball
$S = $ Sphere

$W = $ Solid cube
$S = $ 6 square faces

$W = $ Solid cylinder
$S = $ Tube and two disks

Figure 19.37: Several solid regions and their boundaries

Calculating the Flux from the Flux Density

Consider a solid region W in 3-space whose boundary is the closed surface S. There are two ways to find the total flux of a vector field $\vec{F}$ out of W. One is to calculate the flux of $\vec{F}$ through S:

$$\text{Flux out of } W = \int_S \vec{F} \cdot d\vec{A}.$$

Another way is to use div $\vec{F}$, which gives the flux density at any point in W. We subdivide W into small boxes, as shown in Figure 19.38. Then, for a small box of volume ΔV,

$$\text{Flux out of box} \approx \text{Flux density} \cdot \text{Volume} = \text{div } \vec{F} \ \Delta V.$$

What happens when we add the fluxes out of all the boxes? Consider two adjacent boxes, as shown in Figure 19.39. The flux through the shared wall is counted twice, once out of the box on each side. When we add the fluxes, these two contributions cancel, so we get the flux out of the solid region formed by joining the two boxes. Continuing in this way, we find that

$$\text{Flux out of } W = \sum \text{Flux out of small boxes} \approx \sum \text{div } \vec{F} \ \Delta V.$$

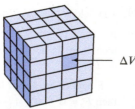

Figure 19.38: Subdivision of region into small boxes

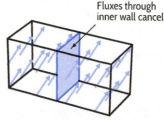

Fluxes through inner wall cancel

Figure 19.39: Adding the flux out of adjacent boxes

We have approximated the flux by a Riemann sum. As the subdivision gets finer, the sum approaches an integral, so

$$\text{Flux out of } W = \int_W \text{div}\,\vec{F}\;dV.$$

We have calculated the flux in two ways, as a flux integral and as a volume integral. Therefore, these two integrals must be equal. This result holds even if W is not a rectangular solid. Thus, we have the following result.[4]

Theorem 19.1: The Divergence Theorem

If W is a solid region whose boundary S is a piecewise smooth surface, and if $\vec{F}$ is a smooth vector field on a solid region[5] containing W and S, then

$$\int_S \vec{F}\cdot d\vec{A} = \int_W \text{div}\,\vec{F}\;dV,$$

where S is given the outward orientation.

Example 1 Use the Divergence Theorem to calculate the flux of the vector field $\vec{F}(\vec{r}) = \vec{r}$ through the sphere of radius a centered at the origin.

Solution In Example 5 on page 1023 we computed the flux using the definition of a flux integral, giving

$$\int_S \vec{r}\cdot d\vec{A} = 4\pi a^3.$$

Now we use $\text{div}\,\vec{F} = \text{div}(x\vec{i} + y\vec{j} + z\vec{k}) = 3$ and the Divergence Theorem:

$$\int_S \vec{r}\cdot d\vec{A} = \int_W \text{div}\,\vec{F}\;dV = \int_W 3\,dV = 3\cdot\frac{4}{3}\pi a^3 = 4\pi a^3.$$

Example 2 Use the Divergence Theorem to calculate the flux of the vector field

$$\vec{F}(x,y,z) = (x^2+y^2)\vec{i} + (y^2+z^2)\vec{j} + (x^2+z^2)\vec{k}$$

through the cube in Figure 19.40.

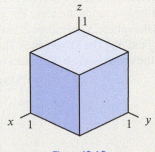

Figure 19.40

---

[4]A proof of the Divergence Theorem using the coordinate definition of the divergence can be found in the online supplement at www.WileyPLUS.com.
[5]The region containing W and S is open.

Solution   The divergence of $\vec{F}$ is div $\vec{F} = 2x + 2y + 2z$. Since div $\vec{F}$ is positive everywhere in the first quadrant, the flux through S is positive. By the Divergence Theorem,

$$\int_S \vec{F} \cdot d\vec{A} = \int_0^1 \int_0^1 \int_0^1 2(x + y + z)\,dx\,dy\,dz = \int_0^1 \int_0^1 \left.(x^2 + 2x(y + z))\right|_0^1 dy\,dz$$

$$= \int_0^1 \int_0^1 1 + 2(y + z)\,dy\,dz = \int_0^1 \left.(y + y^2 + 2yz)\right|_0^1 dz$$

$$= \int_0^1 (2 + 2z)\,dz = \left.(2z + z^2)\right|_0^1 = 3.$$

The Divergence Theorem and Divergence-Free Vector Fields

An important application of the Divergence Theorem is the study of divergence-free vector fields.

Example 3   In Example 3 on page 1042 we saw that the following vector field is divergence free:

$$\vec{F}(\vec{r}) = \frac{\vec{r}}{\|\vec{r}\|^3}, \qquad \vec{r} \neq \vec{0}.$$

Calculate $\int_S \vec{F} \cdot d\vec{A}$, using the Divergence Theorem if possible, for the following surfaces:

(a) S_1 is the sphere of radius a centered at the origin.
(b) S_2 is the sphere of radius a centered at the point $(2a, 0, 0)$.

Solution   (a) We cannot use the Divergence Theorem directly because $\vec{F}$ is not defined everywhere inside the sphere (it is not defined at the origin). Since $\vec{F}$ points outward everywhere on S_1, the flux out of S_1 is positive. On S_1,

$$\vec{F} \cdot d\vec{A} = \|\vec{F}\|\,dA = \frac{a}{a^3}\,dA,$$

so

$$\int_{S_1} \vec{F} \cdot d\vec{A} = \frac{1}{a^2} \int_{S_1} dA = \frac{1}{a^2}(\text{Area of } S_1) = \frac{1}{a^2} 4\pi a^2 = 4\pi.$$

Notice that the flux is not zero, although div $\vec{F}$ is zero everywhere it is defined.

(b) Let W be the solid region enclosed by S_2. Since div $\vec{F} = 0$ everywhere in W, we can use the Divergence Theorem in this case, giving

$$\int_{S_2} \vec{F} \cdot d\vec{A} = \int_W \text{div } \vec{F}\,dV = \int_W 0\,dV = 0.$$

The Divergence Theorem applies to any solid region W and its boundary S, even in cases where the boundary consists of two or more surfaces. For example, if W is the solid region between the sphere S_1 of radius 1 and the sphere S_2 of radius 2, both centered at the same point, then the boundary of W consists of both S_1 and S_2. The Divergence Theorem requires the outward orientation, which on S_2 points away from the center and on S_1 points toward the center. (See Figure 19.41.)

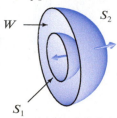

Figure 19.41: Cutaway view of the region W between two spheres, showing orientation vectors

Example 4 Let S_1 be the sphere of radius 1 centered at the origin and let S_2 be the ellipsoid $x^2 + y^2 + 4z^2 = 16$, both oriented outward. For

$$\vec{F}(\vec{r}) = \frac{\vec{r}}{\|\vec{r}\|^3}, \qquad \vec{r} \neq \vec{0},$$

show that

$$\int_{S_1} \vec{F} \cdot d\vec{A} = \int_{S_2} \vec{F} \cdot d\vec{A}.$$

Solution The ellipsoid contains the sphere; let W be the solid region between them. Since W does not contain the origin, div $\vec{F}$ is defined and equal to zero everywhere in W. Thus, if S is the boundary of W, then

$$\int_S \vec{F} \cdot d\vec{A} = \int_W \text{div}\,\vec{F}\,dV = 0.$$

But S consists of S_2 oriented outward and S_1 oriented inward, so

$$0 = \int_S \vec{F} \cdot d\vec{A} = \int_{S_2} \vec{F} \cdot d\vec{A} - \int_{S_1} \vec{F} \cdot d\vec{A},$$

and thus

$$\int_{S_2} \vec{F} \cdot d\vec{A} = \int_{S_1} \vec{F} \cdot d\vec{A}.$$

In Example 3 we showed that $\int_{S_1} \vec{F} \cdot d\vec{A} = 4\pi$, so $\int_{S_2} \vec{F} \cdot d\vec{A} = 4\pi$ also. Note that it would have been more difficult to compute the integral over the ellipsoid directly.

Electric Fields

The electric field produced by a positive point charge q placed at the origin is

$$\vec{E} = q\frac{\vec{r}}{\|\vec{r}\|^3}.$$

Using Example 3, we see that the flux of the electric field through any sphere centered at the origin is $4\pi q$. In fact, using the idea of Example 4, we can show that the flux of $\vec{E}$ through any closed surface containing the origin is $4\pi q$. See Problems 41 and 42 on page 1054. This is a special case of Gauss's Law, which states that the flux of an electric field through any closed surface is proportional to the total charge enclosed by the surface. Carl Friedrich Gauss (1777–1855) also discovered the Divergence Theorem, which is sometimes called Gauss's Theorem.

Summary for Section 19.4

- The **boundary** of a solid region is the skin between the interior of the region and the space around it.
- **The Divergence Theorem** If W is a solid region with boundary S given the outward orientation, and if $\vec{F}$ is a smooth vector field on a solid region containing W and S, then

$$\int_S \vec{F} \cdot d\vec{A} = \int_W \text{div}\,\vec{F}\,dV.$$

- If $\vec{F}$ is a **divergence-free** vector field, and if S is the boundary of solid region W, then by the Divergence Theorem

$$\int_S \vec{F} \cdot d\vec{A} = \int_W \text{div}\,\vec{F}\,dV = 0.$$

Exercises and Problems for Section 19.4

EXERCISES

■For Exercises 1–5, compute the flux integral $\int_S \vec{F} \cdot d\vec{A}$ in two ways, if possible, directly and using the Divergence Theorem. In each case, S is closed and oriented outward.

1. $\vec{F}(\vec{r}) = \vec{r}$ and S is the cube enclosing the volume $0 \le x \le 2, 0 \le y \le 2$, and $0 \le z \le 2$.

2. $\vec{F} = y\vec{j}$ and S is a closed vertical cylinder of height 2, with its base a circle of radius 1 on the xy-plane, centered at the origin.

3. $\vec{F} = x^2\vec{i} + 2y^2\vec{j} + 3z^2\vec{k}$ and S is the surface of the box with faces $x = 1, x = 2, y = 0, y = 1, z = 0, z = 1$.

4. $\vec{F} = (z^2 + x)\vec{i} + (x^2 + y)\vec{j} + (y^2 + z)\vec{k}$ and S is the closed cylinder $x^2 + z^2 = 1$, with $0 \le y \le 1$, oriented outward.

5. $\vec{F} = -z\vec{i} + x\vec{k}$ and S is a square pyramid with height 3 and base on the xy-plane of side length 1.

■In Exercises 6–12, find the flux of the vector field out of the closed box $0 \le x \le 2, 0 \le y \le 3, 0 \le z \le 4$.

6. $\vec{F} = 4\vec{i} + 7\vec{j} - \vec{k}$

7. $\vec{G} = y\vec{i} + z\vec{k}$

8. $\vec{H} = xy\vec{i} + z\vec{j} + y\vec{k}$

9. $\vec{J} = xy^2\vec{j} + x\vec{k}$

10. $\vec{N} = e^z\vec{i} + \sin(xy)\vec{k}$

11. $\vec{M} = (3x + 4y)\vec{i} + (4y + 5z)\vec{j} + (5z + 3x)\vec{k}$

12. $\vec{M}$ where div $\vec{M} = xy + 5$

PROBLEMS

13. Find the flux of $\vec{F} = z\vec{i} + y\vec{j} + x\vec{k}$ out of a sphere of radius 3 centered at the origin.

14. Find the flux of $\vec{F} = xy\vec{i} + yz\vec{j} + zx\vec{k}$ out of a sphere of radius 1 centered at the origin.

15. Find the flux of $\vec{F} = x^3\vec{i} + y^3\vec{j} + z^3\vec{k}$ through the closed surface bounding the solid region $x^2 + y^2 \le 4$, $0 \le z \le 5$, oriented outward.

16. The region W lies between the spheres $x^2 + y^2 + z^2 = 4$ and $x^2 + y^2 + z^2 = 9$ and within the cone $z = \sqrt{x^2 + y^2}$ with $z \ge 0$; its boundary is the closed surface, S, oriented outward. Find the flux of $\vec{F} = x^3\vec{i} + y^3\vec{j} + z^3\vec{k}$ out of S.

17. For $\vec{F} = (2x + \sin z)\vec{i} + (xz - y)\vec{j} + (e^x + 2z)\vec{k}$, find the flux of $\vec{F}$ out of the closed silo-shaped region within the cylinder $x^2 + y^2 = 1$, below the hemisphere $z = 1 + \sqrt{1 - x^2 - y^2}$, and above the xy-plane.

18. Find the flux of $\vec{F}$ through the closed cylinder of radius 2, centered on the z-axis, with $3 \le z \le 7$, if $\vec{F} = (x + 3e^{yz})\vec{i} + (\ln(x^2z^2 + 1) + y)\vec{j} + z\vec{k}$.

19. Find the flux of $\vec{F} = e^{y^2z^2}\vec{i} + (\tan(0.001x^2z^2) + y^2)\vec{j} + (\ln(1 + x^2y^2) + z^2)\vec{k}$ out of the closed box $0 \le x \le 5$, $0 \le y \le 4, 0 \le z \le 3$.

20. Find the flux of $\vec{F} = x^2\vec{i} + z\vec{j} + y\vec{k}$ out of the closed cone $x = \sqrt{y^2 + z^2}$, with $0 \le x \le 1$.

21. Find the flux of $\vec{F} = (e^y + xz)\vec{i} + (e^z + x)\vec{j} + (e^x + y)\vec{k}$ out of the cube of side 2 bounded by the planes $x = -1$, $x = 1, y = -1, y = 1, z = 2$, and the xy-plane.

22. Find the flux of $\vec{F} = y^2\vec{i} + x^3z^3\vec{j} + 3xz\vec{k}$ out of the closed region in the first quadrant bounded by the planes $x = y, x = 1, z = 4$, and the coordinate planes.

23. For $\vec{F} = (ye^z)\vec{i} + (xz^2 + 2y)\vec{j} + (3z - 5x^3)\vec{k}$, find the flux of $\vec{F}$ out of the closed tetrahedron bounded by the plane $x + y + z = 2$ and the coordinate planes.

24. For $\vec{F} = 5xz^2\vec{i} + (2xy - e^z)\vec{j} + (x^2 - 2xz)\vec{k}$, find the flux of $\vec{F}$ out of the closed surface bounded by the cone $z = \sqrt{x^2 + y^2}$ and the sphere $x^2 + y^2 + z^2 = 1$.

25. Suppose $\vec{F}$ is a vector field with div $\vec{F} = 10$. Find the flux of $\vec{F}$ out of a cylinder of height a and radius a, centered on the z-axis and with base in the xy-plane.

26. Let $\vec{F} = (5x + 7y)\vec{i} + (7y + 9z)\vec{j} + (9z + 11x)\vec{k}$, and let Q_i be the flux of $\vec{F}$ through the surfaces S_i for $i = 1$–4. Arrange Q_i in ascending order, where

   (a) S_1 is the sphere of radius 2 centered at the origin
   (b) S_2 is the cube of side 2 centered at the origin and with sides parallel to the axes
   (c) S_3 is the sphere of radius 1 centered at the origin
   (d) S_4 is a pyramid with all four corners lying on S_3

27. A cone has its tip at the point $(0, 0, 5)$, and its base is the disk $D, x^2 + y^2 \le 1$, in the xy-plane. The surface of the cone is the curved and slanted face, S, oriented upward, and the flat base, D, oriented downward. The flux of the constant vector field $\vec{F} = a\vec{i} + b\vec{j} + c\vec{k}$ through S is given by

$$\int_S \vec{F} \cdot d\vec{A} = 3.22.$$

Is it possible to calculate $\int_D \vec{F} \cdot d\vec{A}$? If so, give the answer. If not, explain what additional information you would need to be able to make this calculation.

28. If V is a volume surrounded by a closed surface S, show that $\frac{1}{3}\int_S \vec{r} \cdot d\vec{A} = V$.

29. A vector field $\vec{F}$ satisfies div $\vec{F} = 0$ everywhere. Show that $\int_S \vec{F} \cdot d\vec{A} = 0$ for every closed surface S.

30. Let S be the cube in the first quadrant with side 2, one corner at the origin and edges parallel to the axes. Let

$$\vec{F}_1 = (xy^2 + 3xz^2)\vec{i} + (3x^2y + 2yz^2)\vec{j} + 3zy^2\vec{k}$$

$$\vec{F}_2 = (xy^2 + 5e^{yz})\vec{i} + (yz^2 + 7\sin(xz))\vec{j} + (x^2z + \cos(xy))\vec{k}$$

$$\vec{F}_3 = \left(xz^2 + \frac{x^3}{3}\right)\vec{i} + \left(yz^2 + \frac{y^3}{3}\right)\vec{j} + \left(zy^2 + \frac{z^3}{3}\right)\vec{k}.$$

Arrange the flux integrals of $\vec{F}_1$, $\vec{F}_2$, $\vec{F}_3$ out of S in increasing order.

31. Let div $\vec{F} = 2(6 - x)$ and $0 \le a, b, c \le 10$.

   (a) Find the flux of $\vec{F}$ out of the rectangular box given by $0 \le x \le a, 0 \le y \le b, 0 \le z \le c$.

   (b) For what values of a, b, c is the flux largest? What is that largest flux?

32. **(a)** Find div$(\vec{r}/||\vec{r}||^2)$ where $\vec{r} = x\vec{i} + y\vec{j}$ for $\vec{r} \ne \vec{0}$.

   (b) Can you use the Divergence Theorem to compute the flux of $\vec{r}/||\vec{r}||^2$ out of a closed cylinder of radius 1, length 2, centered at the origin, and with its axis along the z-axis?

   (c) Compute the flux of $\vec{r}/||\vec{r}||^2$ out of the cylinder in part (b).

   (d) Find the flux of $\vec{r}/||\vec{r}||^2$ out of a closed cylinder of radius 2, length 2, centered at the origin, and with its axis along the z-axis.

33. Let $\vec{r} = x\vec{i} + y\vec{j} + z\vec{k}$ and let $\vec{F}$ be the vector field given by

$$\vec{F} = \frac{\vec{r}}{||\vec{r}||^3}.$$

   (a) Find the flux of $\vec{F}$ out of the sphere $x^2 + y^2 + z^2 = 1$ oriented outward.

   (b) Calculate div $\vec{F}$. Show your work and simplify your answer completely.

   (c) Use your answers to parts (a) and (b) to calculate the flux out of a box of side 10 centered at the origin and with sides parallel to the coordinate planes. (The box is also oriented outward.)

In Problems **34–35**, find the flux of $\vec{F} = \vec{r}/||\vec{r}||^3$ through the surface. [Hint: Use the method of Problem 33.]

34. S is the ellipsoid $x^2 + 2y^2 + 3z^2 = 6$.

35. S is the closed cylinder $y^2 + z^2 = 4, -2 \le x \le 2$.

36. **(a)** Let div $\vec{F} = x^2 + y^2 + z^2 + 3$. Calculate $\int_{S_1} \vec{F} \cdot d\vec{A}$ where S_1 is the sphere of radius 1 centered at the origin.

   (b) Let S_2 be the sphere of radius 2 centered at the origin; let S_3 be the sphere of radius 3 centered at the origin; let S_4 be the box of side 6 centered at the origin with edges parallel to the axes. Without calculating them, arrange the following integrals in increasing order:

$$\int_{S_2} \vec{F} \cdot d\vec{A}, \quad \int_{S_3} \vec{F} \cdot d\vec{A}, \quad \int_{S_4} \vec{F} \cdot d\vec{A}.$$

37. Suppose div $\vec{F} = xyz^2$.

   (a) Find div $\vec{F}$ at the point $(1, 2, 1)$. [Note: You are given div $\vec{F}$, not $\vec{F}$.]

   (b) Using your answer to part (a), but no other information about the vector field $\vec{F}$, estimate the flux out of a small box of side 0.2 centered at the point $(1, 2, 1)$ and with edges parallel to the axes.

   (c) Without computing the vector field $\vec{F}$, calculate the exact flux out of the box.

38. Suppose div $\vec{F} = x^2 + y^2 + 3$. Find a surface S such that $\int_S \vec{F} \cdot d\vec{A}$ is negative, or explain why no such surface exists.

39. As a result of radioactive decay, heat is generated uniformly throughout the interior of the earth at a rate of 30 watts per cubic kilometer. (A watt is a rate of heat production.) The heat then flows to the earth's surface where it is lost to space. Let $\vec{F}(x, y, z)$ denote the rate of flow of heat measured in watts per square kilometer. By definition, the flux of $\vec{F}$ across a surface is the quantity of heat flowing through the surface per unit of time.

   (a) What is the value of div $\vec{F}$? Include units.

   (b) Assume the heat flows outward symmetrically. Verify that $\vec{F} = \alpha\vec{r}$, where $\vec{r} = x\vec{i} + y\vec{j} + z\vec{k}$ and α is a suitable constant, satisfies the given conditions. Find α.

   (c) Let $T(x, y, z)$ denote the temperature inside the earth. Heat flows according to the equation $\vec{F} = -k \operatorname{grad} T$, where k is a constant. Explain why this makes sense physically.

   (d) If T is in °C, then $k = 30,000$ watts/km°C. Assuming the earth is a sphere with radius 6400 km and surface temperature 20°C, what is the temperature at the center?

40. If a surface S is submerged in an incompressible fluid, a force $\vec{F}$ is exerted on one side of the surface by the pressure in the fluid. If the z-axis is vertical, with the positive direction upward and the fluid level at $z = 0$, then the component of force in the direction of a unit vector $\vec{u}$ is given by the following:

$$\vec{F} \cdot \vec{u} = -\int_S z\delta g\vec{u} \cdot d\vec{A},$$

where δ is the density of the fluid (mass/volume), g is the acceleration due to gravity, and the surface is oriented away from the side on which the force is exerted. In this problem we consider a totally submerged closed surface enclosing a volume V. We are interested in the force of the liquid on the external surface, so S is oriented inward. Use the Divergence Theorem to show that:

   (a) The force in the $\vec{i}$ and $\vec{j}$ directions is zero.

   (b) The force in the $\vec{k}$ direction is $\delta g V$, the weight of the volume of fluid with the same volume as V. This is *Archimedes' Principle*.

41. According to Coulomb's Law, the electrostatic field $\vec{E}$ at the point $\vec{r}$ due to a charge q at the origin is given by

$$\vec{E}(\vec{r}) = q\frac{\vec{r}}{\|\vec{r}\|^3}.$$

(a) Compute div $\vec{E}$.

(b) Let S_a be the sphere of radius a centered at the origin and oriented outward. Show that the flux of $\vec{E}$ through S_a is $4\pi q$.

(c) Could you have used the Divergence Theorem in part (b)? Explain why or why not.

(d) Let S be an arbitrary, closed, outward-oriented surface surrounding the origin. Show that the flux of $\vec{E}$ through S is again $4\pi q$. [Hint: Apply the Divergence Theorem to the solid region lying between a small sphere S_a and the surface S.]

42. According to Coulomb's Law, the electric field $\vec{E}$ at the point $\vec{r}$ due to a charge q at the point $\vec{r}_0$ is given by

$$\vec{E}(\vec{r}) = q\frac{(\vec{r} - \vec{r}_0)}{\|\vec{r} - \vec{r}_0\|^3}.$$

Suppose S is a closed, outward-oriented surface and that $\vec{r}_0$ does not lie on S. Use Problem 41 to show that

$$\int_S \vec{E} \cdot d\vec{A} = \begin{cases} 4\pi q & \text{if } q \text{ lies inside } S, \\ 0 & \text{if } q \text{ lies outside } S. \end{cases}$$

Strengthen Your Understanding

■ In Problems 43–44, explain what is wrong with the statement.

43. The flux integral $\int_S \vec{F} \cdot d\vec{A}$ can be evaluated using the Divergence Theorem, where $\vec{F} = 2x\vec{i} - 3\vec{j}$ and S is the triangular surface with corners $(1, 0, 0)$, $(0, 1, 0)$, $(0, 0, 1)$ oriented away from the origin.

44. If S is the boundary of a solid region W, where S is oriented outward, and $\vec{F}$ is a vector field, then

$$\int_S \text{div } \vec{F} \, d\vec{A} = \int_W \vec{F} \, dV.$$

■ In Problems 45–46, give an example of:

45. A surface S that is the boundary of a solid region such that $\int_S \vec{F} \cdot d\vec{A} = 0$ if $\vec{F}(x, y, z) = y\vec{i} + xz\vec{j} + y^2\vec{k}$.

46. A vector field $\vec{F}$ such that the flux of $\vec{F}$ out of a sphere of radius 1 centered at the origin is 3.

■ Are the statements in Problems 47–51 true or false? The smooth vector field $\vec{F}$ is defined everywhere in 3-space and has constant divergence equal to 4.

47. The field $\vec{F}$ has a net inflow per unit volume at the point $(-3, 4, 0)$.

48. The vector field $\vec{F}$ could be $\vec{F} = x\vec{i} + (3y)\vec{j} + (y - 5x)\vec{k}$.

49. The vector field $\vec{F}$ could be a constant field.

50. The flux of $\vec{F}$ through a circle of radius 5 lying anywhere on the xy-plane and oriented upward is $4(\pi 5^2)$.

51. The flux of $\vec{F}$ through a closed cylinder of radius 1 centered along the y-axis, $0 \leq y \leq 3$ and oriented outward is $4(3\pi)$.

■ Are the statements in Problems 52–59 true or false? Give reasons for your answer.

52. $\int_S \vec{F} \cdot d\vec{A} = \text{div } \vec{F}$.

53. If $\vec{F}$ is a divergence-free vector field in 3-space and S is a closed surface oriented inward, then $\int_S \vec{F} \cdot d\vec{A} = 0$.

54. If $\vec{F}$ is a vector field in 3-space satisfying div $\vec{F} = 1$, and S is a closed surface oriented outward, then $\int_S \vec{F} \cdot d\vec{A}$ is equal to the volume enclosed by S.

55. Let W be the solid region between the sphere S_1 of radius 1 and S_2 of radius 2, both centered at the origin and oriented outward. If $\vec{F}$ is a vector field in 3-space, then $\int_W \text{div } \vec{F} \, dV = \int_{S_2} \vec{F} \cdot d\vec{A} - \int_{S_1} \vec{F} \cdot d\vec{A}$.

56. Let S_1 be the square $0 \leq x \leq 1, 0 \leq y \leq 1, z = 0$ oriented downward and let S_2 be the square $0 \leq x \leq 1, 0 \leq y \leq 1, z = 1$ oriented upward. If $\vec{F}$ is a vector field, then $\int_W \text{div } \vec{F} \, dV = \int_{S_2} \vec{F} \cdot d\vec{A} + \int_{S_1} \vec{F} \cdot d\vec{A}$, where W is the solid cube $0 \leq x \leq 1, 0 \leq y \leq 1, 0 \leq z \leq 1$.

57. Let S_1 be the square $0 \leq x \leq 1, 0 \leq y \leq 1, z = 0$ oriented downward and let S_2 be the square $0 \leq x \leq 1, 0 \leq y \leq 1, z = 1$ oriented upward. If $\vec{F} = \cos(xyz)\vec{k}$, then $\int_W \text{div } \vec{F} \, dV = \int_{S_2} \vec{F} \cdot d\vec{A} + \int_{S_1} \vec{F} \cdot d\vec{A}$, where W is the solid cube $0 \leq x \leq 1, 0 \leq y \leq 1, 0 \leq z \leq 1$.

58. If S is a sphere of radius 1, centered at the origin, oriented outward, and $\vec{F}$ is a vector field satisfying $\int_S \vec{F} \cdot d\vec{A} = 0$, then div $\vec{F} = 0$ at all points inside S.

59. Let S_h be the surface consisting of a cylinder of height h, closed at the top. The curved sides are $x^2 + y^2 = 1$, for $0 \leq z \leq h$, and the top $x^2 + y^2 \leq 1$, for $z = h$, oriented outward. If $\vec{F}$ is divergence free, then $\int_{S_h} \vec{F} \cdot d\vec{A}$ is independent of the height h.

Online Resource: Review Problems and Projects

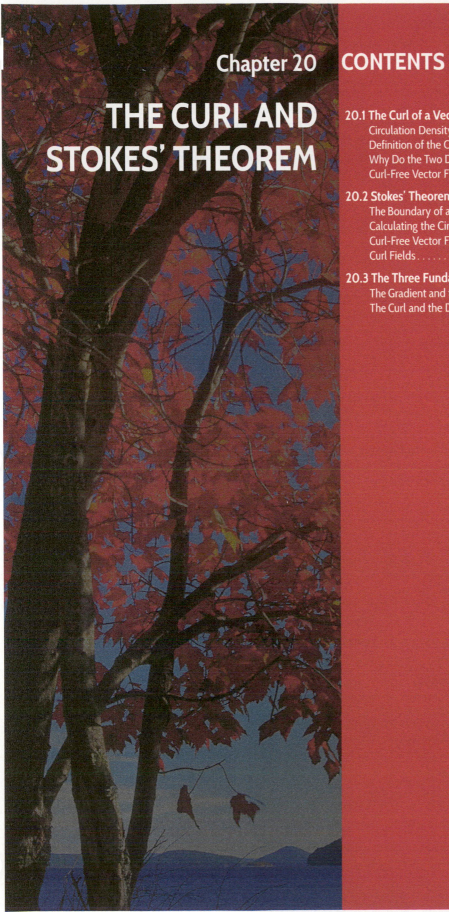

Chapter 20

THE CURL AND STOKES' THEOREM

CONTENTS

20.1   THE CURL OF A VECTOR FIELD

The divergence is a scalar derivative which measures the outflow of a vector field per unit volume. Now we introduce a vector derivative, the curl, which measures the circulation of a vector field.

Imagine holding the paddle-wheel in Figure 20.1 in the flow shown by Figure 20.2. The speed at which the paddle-wheel spins measures the strength of circulation. Notice that the angular velocity depends on the direction in which the stick is pointing. If the stick is pointing horizontally the paddle-wheel does not spin; if the stick is vertical, the paddle-wheel spins.

Figure 20.1: A device for measuring circulation

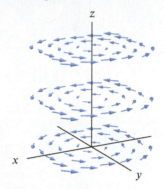

Figure 20.2: A vector field (in the planes $z = 1$, $z = 2$, $z = 3$) with circulation about the z-axis

Circulation Density

We measure the strength of the circulation using a closed curve. Suppose C is a circle with center $P = (x, y, z)$ in the plane perpendicular to $\vec{n}$, traversed in the direction determined from $\vec{n}$ by the right-hand rule. (See Figures 20.3 and 20.4.)

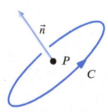

Figure 20.3: Direction of C relates to direction of $\vec{n}$ by the right-hand rule

Figure 20.4: When the thumb points in the direction of $\vec{n}$, the fingers curl in the forward direction around C

We make the following definition:

> The **circulation density** of a smooth vector field $\vec{F}$ at (x, y, z) around the direction of the unit vector $\vec{n}$ is defined, provided the limit exists, to be
>
> $$\text{circ}_{\vec{n}} \vec{F}(x, y, z) = \lim_{\text{Area} \to 0} \frac{\text{Circulation around } C}{\text{Area inside } C} = \lim_{\text{Area} \to 0} \frac{\int_C \vec{F} \cdot d\vec{r}}{\text{Area inside } C}.$$
>
> The circle C is in the plane perpendicular to $\vec{n}$ and oriented by the right-hand rule.

We can use other closed curves for C, such as rectangles, that lie in a plane perpendicular to $\vec{n}$ and include the point (x, y, z).

The circulation density determines the angular velocity of the paddle-wheel in Figure 20.1 provided you could make one sufficiently small and light and insert it without disturbing the flow.

Example 1 Consider the vector field $\vec{F}$ in Figure 20.2. Suppose that $\vec{F}$ is parallel to the xy-plane and that at a distance r from the z-axis it has magnitude $2r$. Calculate $\text{circ}_{\vec{n}} \vec{F}$ at the origin for

(a) $\vec{n} = \vec{k}$ (b) $\vec{n} = -\vec{k}$ (c) $\vec{n} = \vec{i}$.

Solution (a) Take a circle C of radius a in the xy-plane, centered at the origin, traversed in a direction determined from $\vec{k}$ by the right-hand rule. Then, since $\vec{F}$ is tangent to C everywhere and points in the forward direction around C, we have

$$\text{Circulation around } C = \int_C \vec{F} \cdot d\vec{r} = \|\vec{F}\| \cdot \text{Circumference of } C = 2a(2\pi a) = 4\pi a^2.$$

Thus, the circulation density is

$$\text{circ}_{\vec{k}} \vec{F} = \lim_{a \to 0} \frac{\text{Circulation around } C}{\text{Area inside } C} = \lim_{a \to 0} \frac{4\pi a^2}{\pi a^2} = 4.$$

(b) If $\vec{n} = -\vec{k}$ the circle is traversed in the opposite direction, so the line integral changes sign. Thus,

$$\text{circ}_{-\vec{k}} \vec{F} = -4.$$

(c) The circulation around $\vec{i}$ is calculated using circles in the yz-plane. Since $\vec{F}$ is everywhere perpendicular to such a circle C,

$$\int_C \vec{F} \cdot d\vec{r} = 0.$$

Thus, we have

$$\text{circ}_{\vec{i}} \vec{F} = \lim_{a \to 0} \frac{\int_C \vec{F} \cdot d\vec{r}}{\pi a^2} = \lim_{a \to 0} \frac{0}{\pi a^2} = 0.$$

Definition of the Curl

Example 1 shows that the circulation density of a vector field can be positive, negative, or zero, depending on the direction. We assume that there is one direction in which the circulation density is greatest and define a single vector quantity that incorporates all these different circulation densities. We give two definitions, one geometric and one algebraic, which turn out to lead to the same result.

Geometric Definition of Curl

The curl of a smooth vector field $\vec{F}$, written $\text{curl}\,\vec{F}$, is the vector field with the following properties:

- The direction of $\text{curl}\,\vec{F}(x, y, z)$ is the direction $\vec{n}$ for which $\text{circ}_{\vec{n}} \vec{F}(x, y, z)$ is the greatest.
- The magnitude of $\text{curl}\,\vec{F}(x, y, z)$ is the circulation density of $\vec{F}$ around that direction.

If the circulation density is zero around every direction, then we define the curl to be $\vec{0}$.

Cartesian Coordinate Definition of Curl

If $\vec{F} = F_1\vec{i} + F_2\vec{j} + F_3\vec{k}$, then

$$\text{curl}\,\vec{F} = \left(\frac{\partial F_3}{\partial y} - \frac{\partial F_2}{\partial z}\right)\vec{i} + \left(\frac{\partial F_1}{\partial z} - \frac{\partial F_3}{\partial x}\right)\vec{j} + \left(\frac{\partial F_2}{\partial x} - \frac{\partial F_1}{\partial y}\right)\vec{k}.$$

The cross-product formula gives an easy way to remember the Cartesian coordinate definition and suggests another common notation for curl $\vec{F}$, namely $\nabla \times \vec{F}$. Using $\nabla = \frac{\partial}{\partial x}\vec{i} + \frac{\partial}{\partial y}\vec{j} + \frac{\partial}{\partial z}\vec{k}$, we can write

$$\text{curl } \vec{F} = \nabla \times \vec{F} = \begin{vmatrix} \vec{i} & \vec{j} & \vec{k} \\ \frac{\partial}{\partial x} & \frac{\partial}{\partial y} & \frac{\partial}{\partial z} \\ F_1 & F_2 & F_3 \end{vmatrix}.$$

Example 2   For each field in Figure 20.5, use the sketch and the geometric definition to decide whether the curl at the origin appears to point up or down, or to be the zero vector. Then check your answer using the coordinate definition of curl and the formulas in the caption of Figure 20.5. Note that the vector fields have no $\vec{k}$-components and are independent of z.

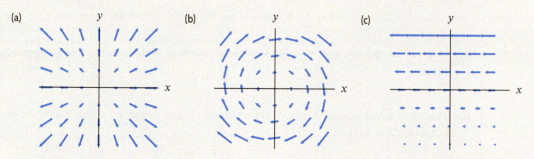

Figure 20.5: Sketches in the xy-plane of (a) $\vec{F} = x\vec{i} + y\vec{j}$ (b) $\vec{F} = y\vec{i} - x\vec{j}$ (c) $\vec{F} = -(y+1)\vec{i}$

Solution

(a) This vector field shows no rotation, and the circulation around any circle in the xy-plane centered at the origin appears to be zero, so we suspect that the circulation density around $\vec{k}$ is zero. The coordinate definition of curl gives

$$\text{curl } \vec{F} = \left(\frac{\partial(0)}{\partial y} - \frac{\partial y}{\partial z}\right)\vec{i} + \left(\frac{\partial x}{\partial z} - \frac{\partial(0)}{\partial x}\right)\vec{j} + \left(\frac{\partial y}{\partial x} - \frac{\partial x}{\partial y}\right)\vec{k} = \vec{0}.$$

(b) This vector field appears to be rotating around the z-axis. By the right-hand rule, the circulation density around $\vec{k}$ is negative, so we expect the z-component of the curl to point down. The coordinate definition gives

$$\text{curl } \vec{F} = \left(\frac{\partial(0)}{\partial y} - \frac{\partial(-x)}{\partial z}\right)\vec{i} + \left(\frac{\partial y}{\partial z} - \frac{\partial(0)}{\partial x}\right)\vec{j} + \left(\frac{\partial(-x)}{\partial x} - \frac{\partial y}{\partial y}\right)\vec{k} = -2\vec{k}.$$

(c) At first glance, you might expect this vector field to have zero curl, as all the vectors are parallel to the x-axis. However, if you find the circulation around the curve C in Figure 20.6, the sides contribute nothing (they are perpendicular to the vector field), the bottom contributes a negative quantity (the curve is in the opposite direction to the vector field), and the top contributes a larger positive quantity (the curve is in the same direction as the vector field and the magnitude of the vector field is larger at the top than at the bottom). Thus, the circulation around C is positive and hence we expect the curl to be nonzero and point up. The coordinate definition gives

$$\text{curl } \vec{F} = \left(\frac{\partial(0)}{\partial y} - \frac{\partial(0)}{\partial z}\right)\vec{i} + \left(\frac{\partial(-(y+1))}{\partial z} - \frac{\partial(0)}{\partial x}\right)\vec{j} + \left(\frac{\partial(0)}{\partial x} - \frac{\partial(-(y+1))}{\partial y}\right)\vec{k} = \vec{k}.$$

Another way to see that the curl is nonzero in this case is to imagine the vector field representing the velocity of moving water. A boat sitting in the water tends to rotate, as the water moves faster on one side than the other.

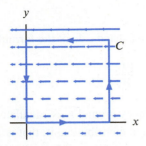

Figure 20.6: Rectangular curve in xy-plane

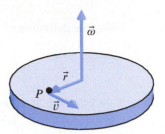

Figure 20.7: Rotating flywheel

Example 3   A flywheel is rotating with angular velocity $\vec{\omega}$ and the velocity of a point P with position vector $\vec{r}$ is given by $\vec{v} = \vec{\omega} \times \vec{r}$. (See Figure 20.7.) Calculate curl $\vec{v}$.

Solution   If $\vec{\omega} = \omega_1\vec{i} + \omega_2\vec{j} + \omega_3\vec{k}$, using the determinant notation introduced in Section 13.4, we have

$$\vec{v} = \vec{\omega} \times \vec{r} = \begin{vmatrix} \vec{i} & \vec{j} & \vec{k} \\ \omega_1 & \omega_2 & \omega_3 \\ x & y & z \end{vmatrix} = (\omega_2 z - \omega_3 y)\vec{i} + (\omega_3 x - \omega_1 z)\vec{j} + (\omega_1 y - \omega_2 x)\vec{k}.$$

The curl formula can also be written using a determinant:

$$\text{curl}\,\vec{v} = \begin{vmatrix} \vec{i} & \vec{j} & \vec{k} \\ \frac{\partial}{\partial x} & \frac{\partial}{\partial y} & \frac{\partial}{\partial z} \\ \omega_2 z - \omega_3 y & \omega_3 x - \omega_1 z & \omega_1 y - \omega_2 x \end{vmatrix}$$

$$= \left(\frac{\partial}{\partial y}(\omega_1 y - \omega_2 x) - \frac{\partial}{\partial z}(\omega_3 x - \omega_1 z) \right)\vec{i} + \left(\frac{\partial}{\partial z}(\omega_2 z - \omega_3 y) - \frac{\partial}{\partial x}(\omega_1 y - \omega_2 x) \right)\vec{j}$$

$$+ \left(\frac{\partial}{\partial x}(\omega_3 x - \omega_1 z) - \frac{\partial}{\partial y}(\omega_2 z - \omega_3 y) \right)\vec{k}$$

$$= 2\omega_1\vec{i} + 2\omega_2\vec{j} + 2\omega_3\vec{k} = 2\vec{\omega}.$$

Thus, as we would expect, curl $\vec{v}$ is parallel to the axis of rotation of the flywheel (namely, the direction of $\vec{\omega}$) and the magnitude of curl $\vec{v}$ is larger the faster the flywheel is rotating (that is, the larger the magnitude of $\vec{\omega}$).

Why Do the Two Definitions of Curl Give the Same Result?

Using Green's Theorem in Cartesian coordinates, we can show that for curl $\vec{F}$ defined in Cartesian coordinates,

$$\boxed{\text{curl}\,\vec{F} \cdot \vec{n} = \text{circ}_{\vec{n}}\,\vec{F}.}$$

This shows that curl $\vec{F}$ defined in Cartesian coordinates satisfies the geometric definition, since the left-hand side takes its maximum value when $\vec{n}$ points in the same direction as curl $\vec{F}$, and in that case its value is $\|\,\text{curl}\,\vec{F}\,\|$.

The following example justifies this formula in a specific case.

Example 4   Use the definition of curl in Cartesian coordinates and Green's Theorem to show that

$$\left(\text{curl}\,\vec{F} \right) \cdot \vec{k} = \text{circ}_{\vec{k}}\,\vec{F}.$$

Solution   Using the definition of curl in Cartesian coordinates, the left-hand side of the formula is

$$\left(\operatorname{curl}\vec{F}\,\right)\cdot\vec{k} = \frac{\partial F_2}{\partial x} - \frac{\partial F_1}{\partial y}.$$

Now let's look at the right-hand side. The circulation density around $\vec{k}$ is calculated using circles perpendicular to $\vec{k}$; hence, the $\vec{k}$-component of $\vec{F}$ does not contribute to it; that is, the circulation density of $\vec{F}$ around $\vec{k}$ is the same as the circulation density of $F_1\vec{i} + F_2\vec{j}$ around $\vec{k}$. But in any plane perpendicular to $\vec{k}$, z is constant, so in that plane F_1 and F_2 are functions of x and y alone. Thus, $F_1\vec{i} + F_2\vec{j}$ can be thought of as a two-dimensional vector field on the horizontal plane through the point (x, y, z) where the circulation density is being calculated. Let C be a circle in this plane, with radius a and centered at (x, y, z), and let R be the region enclosed by C. Green's Theorem says that

$$\int_C (F_1\vec{i} + F_2\vec{j}\,) \cdot d\vec{r} = \int_R \left(\frac{\partial F_2}{\partial x} - \frac{\partial F_1}{\partial y} \right) dA.$$

When the circle is small, $\partial F_2/\partial x - \partial F_1/\partial y$ is approximately constant on R, so

$$\int_R \left(\frac{\partial F_2}{\partial x} - \frac{\partial F_1}{\partial y} \right) dA \approx \left(\frac{\partial F_2}{\partial x} - \frac{\partial F_1}{\partial y} \right) \cdot \text{Area of } R = \left(\frac{\partial F_2}{\partial x} - \frac{\partial F_1}{\partial y} \right) \pi a^2.$$

Thus, taking a limit as the radius of the circle goes to zero, we have

$$\operatorname{circ}_{\vec{k}} \vec{F}(x, y, z) = \lim_{a \to 0} \frac{\int_C (F_1\vec{i} + F_2\vec{j}\,) \cdot d\vec{r}}{\pi a^2} = \lim_{a \to 0} \frac{\int_R \left(\frac{\partial F_2}{\partial x} - \frac{\partial F_1}{\partial y} \right) dA}{\pi a^2} = \frac{\partial F_2}{\partial x} - \frac{\partial F_1}{\partial y}.$$

Curl-Free Vector Fields

A vector field is said to be *curl free* or *irrotational* if $\operatorname{curl}\vec{F} = \vec{0}$ everywhere that $\vec{F}$ is defined.

Example 5   Figure 20.8 shows the vector field $\vec{B}$ for three values of the constant p, where $\vec{B}$ is defined on 3-space by

$$\vec{B} = \frac{-y\vec{i} + x\vec{j}}{(x^2 + y^2)^{p/2}}.$$

(a) Find a formula for curl $\vec{B}$.
(b) Is there a value of p for which $\vec{B}$ is curl free? If so, find it.

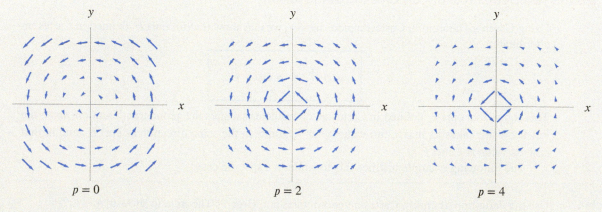

Figure 20.8: The vector field $\vec{B}(\vec{r}) = (-y\vec{i} + x\vec{j}\,)/(x^2 + y^2)^{p/2}$ for $p = 0, 2,$ and 4

Solution   (a) We can use the following version of the product rule for curl. If ϕ is a scalar function and $\vec{F}$ is a vector field, then

$$\operatorname{curl}(\phi \vec{F}) = \phi \operatorname{curl} \vec{F} + (\operatorname{grad} \phi) \times \vec{F}.$$

(See Problem 32 on page 1063.) We write $\vec{B} = \phi \vec{F} = \dfrac{1}{(x^2 + y^2)^{p/2}}(-y\vec{i} + x\vec{j})$. Then

$$\operatorname{curl} \vec{F} = \operatorname{curl}(-y\vec{i} + x\vec{j}) = 2\vec{k}$$

$$\operatorname{grad} \phi = \operatorname{grad}\left(\frac{1}{(x^2 + y^2)^{p/2}}\right) = \frac{-p}{(x^2 + y^2)^{(p/2)+1}}(x\vec{i} + y\vec{j}).$$

Thus, we have

$$\operatorname{curl} \vec{B} = \frac{1}{(x^2 + y^2)^{p/2}} \operatorname{curl}(-y\vec{i} + x\vec{j}) + \operatorname{grad}\left(\frac{1}{(x^2 + y^2)^{p/2}}\right) \times (-y\vec{i} + x\vec{j})$$

$$= \frac{1}{(x^2 + y^2)^{p/2}} 2\vec{k} + \frac{-p}{(x^2 + y^2)^{(p/2)+1}}(x\vec{i} + y\vec{j}) \times (-y\vec{i} + x\vec{j})$$

$$= \frac{1}{(x^2 + y^2)^{p/2}} 2\vec{k} + \frac{-p}{(x^2 + y^2)^{(p/2)+1}}(x^2 + y^2)\vec{k}$$

$$= \frac{2 - p}{(x^2 + y^2)^{p/2}} \vec{k}.$$

(b) The curl is zero when $p = 2$. Thus, when $p = 2$ the vector field is curl free:

$$\vec{B} = \frac{-y\vec{i} + x\vec{j}}{x^2 + y^2}.$$

Summary for Section 20.1

- The **circulation density** of $\vec{F}$ at (x, y, z) around the direction given by $\vec{n}$ is

$$\operatorname{circ}_{\vec{n}} \vec{F}(x, y, z) = \lim_{\text{Area} \to 0} \frac{\text{Circulation around } C}{\text{Area inside } C} = \lim_{\text{Area} \to 0} \frac{\int_C \vec{F} \cdot d\vec{r}}{\text{Area inside } C},$$

  where the circle C is in the plane perpendicular to $\vec{n}$ and oriented by the right-hand rule.

- **Geometric definition of the curl:** The curl of a smooth vector field $\vec{F}$ is the vector field with the properties:
  - The direction of $\operatorname{curl} \vec{F}(x, y, z)$ is the direction $\vec{n}$ for which $\operatorname{circ}_{\vec{n}} \vec{F}(x, y, z)$ is the greatest.
  - The magnitude of $\operatorname{curl} \vec{F}(x, y, z)$ is the circulation density of $\vec{F}$ around that direction.

- **Cartesian-coordinate definition of the curl:** If $\vec{F} = F_1\vec{i} + F_2\vec{j} + F_3\vec{k}$, then

$$\operatorname{curl} \vec{F} = \left(\frac{\partial F_3}{\partial y} - \frac{\partial F_2}{\partial z}\right)\vec{i} + \left(\frac{\partial F_1}{\partial z} - \frac{\partial F_3}{\partial x}\right)\vec{j} + \left(\frac{\partial F_2}{\partial x} - \frac{\partial F_1}{\partial y}\right)\vec{k}.$$

- A vector field is **curl free** or **irrotational** if $\operatorname{curl} \vec{F} = \vec{0}$ everywhere that $\vec{F}$ is defined.

Exercises and Problems for Section 20.1 Online Resource: Additional Problems for Section 20.1
EXERCISES

In Exercises 1–5, is the quantity a vector or a scalar? Calculate it.

1. $\operatorname{curl}(z\vec{i} - x\vec{j} + y\vec{k})$
2. $\operatorname{circ}_{\vec{i}}(z\vec{i} - x\vec{j} + y\vec{k})$
3. $\operatorname{curl}(-2z\vec{i} - z\vec{j} + xy\vec{k})$
4. $\operatorname{circ}_{\vec{k}}(-2z\vec{i} - z\vec{j} + xy\vec{k})$
5. $\operatorname{curl}(x\vec{i} + y\vec{j} + z\vec{k})$

In Exercises 6–13, compute the curl of the vector field.

6. $\vec{F} = 3x\vec{i} - 5z\vec{j} + y\vec{k}$
7. $\vec{F} = (x^2 - y^2)\vec{i} + 2xy\vec{j}$
8. $\vec{F} = (-x + y)\vec{i} + (y + z)\vec{j} + (-z + x)\vec{k}$
9. $\vec{F} = 2yz\vec{i} + 3xz\vec{j} + 7xy\vec{k}$
10. $\vec{F} = x^2\vec{i} + y^3\vec{j} + z^4\vec{k}$
11. $\vec{F} = e^x\vec{i} + \cos y\vec{j} + e^{z^2}\vec{k}$
12. $\vec{F} = (x + yz)\vec{i} + (y^2 + xzy)\vec{j} + (zx^3y^2 + x^7y^6)\vec{k}$
13. $\vec{F}(\vec{r}) = \vec{r}/\|\vec{r}\|$

In Exercises 14–17, does the vector field appear to have nonzero curl at the origin? The vector field is shown in the xy-plane; it has no z-component and is independent of z.

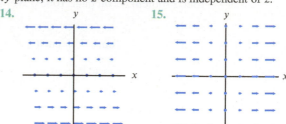

14.

15.

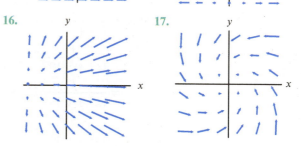

16.

17.

PROBLEMS

18. Let $\vec{F}$ be the vector field in Figure 20.2 on page 1056. It is rotating counterclockwise around the z-axis when viewed from above. At a distance r from the z-axis, $\vec{F}$ has magnitude $2r$.

    (a) Find a formula for $\vec{F}$.
    (b) Find $\operatorname{curl}\vec{F}$ using the coordinate definition and relate your answer to circulation density.

19. Use the geometric definition to find the curl of the vector field $\vec{F}(\vec{r}) = \vec{r}$. Check your answer using the coordinate definition.

20. A smooth vector field $\vec{G}$ has curl $\vec{G}(0,0,0) = 2\vec{i} - 3\vec{j} + 5\vec{k}$. Estimate the circulation around a circle of radius 0.01 centered at the origin in each of the following planes:

    (a) xy-plane, oriented counterclockwise when viewed from the positive z-axis.
    (b) yz-plane, oriented counterclockwise when viewed from the positive x-axis.
    (c) xz-plane, oriented counterclockwise when viewed from the positive y-axis.

21. Three small circles, C_1, C_2, and C_3, each with radius 0.1 and centered at the origin, are in the xy-, yz-, and xz-planes, respectively. The circles are oriented counterclockwise when viewed from the positive z-, x-, and y-axes, respectively. A vector field, $\vec{F}$, has circulation around C_1 of 0.02π, around C_2 of 0.5π, and around C_3 of 3π. Estimate curl $\vec{F}$ at the origin.

22. Using your answers to Exercises 10–11, make a conjecture about a particular form of the vector field $\vec{F} \neq \vec{0}$ that has curl $\vec{F} = \vec{0}$. What form? Show why your conjecture is true.

23. (a) Find curl $\vec{G}$ if $\vec{G} = (ay^3 + be^z)\vec{i} + (cz + dx^2)\vec{j} + (e\sin x + fy)\vec{k}$ and a, b, c, d, e, f are constants.
    (b) If curl $\vec{G}$ is everywhere parallel to the yz-plane, what can you say about the constants a–f?
    (c) If curl $\vec{G}$ is everywhere parallel to the z-axis, what can you say about the constants a–f?

24. Figure 20.9 gives a sketch of the velocity vector field $\vec{F} = y\vec{i} + x\vec{j}$ in the xy-plane.

    (a) What is the direction of rotation of a thin twig placed at the origin along the x-axis?
    (b) What is the direction of rotation of a thin twig placed at the origin along the y-axis?
    (c) Compute curl $\vec{F}$.

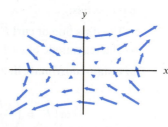

Figure 20.9

25. A tornado is formed when a tube of air circling a horizontal axis is tilted up vertically by the updraft from a thunderstorm. If t is time, this process can be modeled by the wind velocity field

$$\vec{F}(t, x, y, z) = (\cos t \vec{j} + \sin t \vec{k}) \times \vec{r} \quad \text{and} \quad 0 \leq t \leq \frac{\pi}{2}.$$

Determine the direction of curl $\vec{F}$:

(a) At $t = 0$ (b) At $t = \pi/2$
(c) For $0 < t < \pi/2$

26. A large fire becomes a firestorm when the nearby air acquires a circulatory motion. The associated updraft has the effect of bringing more air to the fire, causing it to burn faster. Records show that a firestorm developed during the Chicago Fire of 1871 and during the Second World War bombing of Hamburg, Germany, but there was no firestorm during the Great Fire of London in 1666. Explain how a firestorm could be identified using the curl of a vector field.

27. A vortex that rotates at constant angular velocity ω about the z-axis has velocity vector field $\vec{v} = \omega(-y\vec{i} + x\vec{j})$.

(a) Sketch the vector field with $\omega = 1$ and the vector field with $\omega = -1$.
(b) Determine the speed $\|\vec{v}\|$ of the vortex as a function of the distance from its center.

(c) Compute div $\vec{v}$ and curl $\vec{v}$.
(d) Compute the circulation of $\vec{v}$ counterclockwise about the circle of radius R in the xy-plane, centered at the origin.

28. A central vector field is one of the form $\vec{F} = f(r)\vec{r}$ where f is any function of $r = \|\vec{r}\|$. Show that any central vector field is irrotational.

29. Show that curl $(\vec{F} + \vec{C}) = $ curl $\vec{F}$ for a constant vector field $\vec{C}$.

30. If $\vec{F}$ is any vector field whose components have continuous second partial derivatives, show div curl $\vec{F} = 0$.

31. We have seen that the Fundamental Theorem of Calculus for Line Integrals implies $\int_C$ grad $f \cdot d\vec{r} = 0$ for any smooth closed path C and any smooth function f.

(a) Use the geometric definition of curl to deduce that curl grad $f = \vec{0}$.
(b) Show that curl grad $f = \vec{0}$ using the coordinate definition.

32. Show that curl $(\phi\vec{F}) = \phi$ curl $\vec{F} + (\text{grad } \phi) \times \vec{F}$ for a scalar function ϕ and a vector field $\vec{F}$.

33. Show that if $\vec{F} = f$ grad g for some scalar functions f and g, then curl $\vec{F}$ is everywhere perpendicular to $\vec{F}$.

Strengthen Your Understanding

■ In Problems 34–35, explain what is wrong with the statement.

34. A vector field $\vec{F}$ has curl given by curl $\vec{F} = 2x - 3y$.

35. If all the vectors of a vector field $\vec{F}$ are parallel, then curl $\vec{F} = \vec{0}$.

■ In Problems 36–37, give an example of:

36. A vector field $\vec{F}(x, y, z)$ such that curl $\vec{F} = \vec{0}$.

37. A vector field $\vec{F}(x, y, z)$ such that curl $\vec{F} = \vec{j}$.

■ In Problems 38–46, is the statement true or false? Assume $\vec{F}$ and $\vec{G}$ are smooth vector fields and f is a smooth function on 3-space. Explain.

38. The circulation density, $\text{circ}_{\vec{n}} \vec{F}(x, y, z)$, is a scalar.

39. curl grad $f = 0$.

40. If $\vec{F}$ is a vector field with div $\vec{F} = 0$ and curl $\vec{F} = \vec{0}$, then $\vec{F} = \vec{0}$.

41. If $\vec{F}$ and $\vec{G}$ are vector fields, then curl$(\vec{F} + \vec{G}) = $ curl$\vec{F}$ + curl$\vec{G}$.

42. If $\vec{F}$ and $\vec{G}$ are vector fields, then curl$(\vec{F} \cdot \vec{G}) = $ curl$\vec{F} \cdot$ curl$\vec{G}$.

43. If $\vec{F}$ and $\vec{G}$ are vector fields, then curl$(\vec{F} \times \vec{G}) = $ (curl$\vec{F}) \times$ (curl$\vec{G}$).

44. curl$(f\vec{G}) = (\text{grad } f) \times \vec{G} + f(\text{curl } \vec{G})$.

45. For any vector field $\vec{F}$, the curl of $\vec{F}$ is perpendicular at every point to $\vec{F}$.

46. If $\vec{F}$ is as shown in Figure 20.10, then curl $\vec{F} \cdot \vec{j} > 0$.

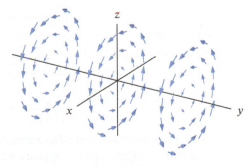

Figure 20.10

47. Of the following vector fields, which ones have a curl which is parallel to one of the axes? Which axis?

(a) $y\vec{i} - x\vec{j} + z\vec{k}$ (b) $y\vec{i} + z\vec{j} + x\vec{k}$ (c) $-z\vec{i} + y\vec{j} + x\vec{k}$
(d) $x\vec{i} + z\vec{j} - y\vec{k}$ (e) $z\vec{i} + x\vec{j} + y\vec{k}$

20.2 STOKES' THEOREM

The Divergence Theorem says that the integral of the flux density over a solid region is equal to the flux through the surface bounding the region. Similarly, Stokes' Theorem says that the integral of the circulation density over a surface is equal to the circulation around the boundary of the surface.

The Boundary of a Surface

The *boundary* of a surface S is the curve or curves running around the edge of S (like the hem around the edge of a piece of cloth). An orientation of S determines an orientation for its boundary, C, as follows. Pick a positive normal vector $\vec{n}$ on S, near C, and use the right-hand rule to determine a direction of travel around $\vec{n}$. This in turn determines a direction of travel around the boundary C. See Figure 20.11. Another way of describing the orientation on C is that someone walking along C in the forward direction, body upright in the direction of the positive normal on S, would have the surface on their left. Notice that the boundary can consist of two or more curves, as the surface on the right in Figure 20.11 shows.

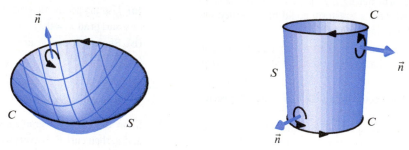

Figure 20.11: Two oriented surfaces and their boundaries

Calculating the Circulation from the Circulation Density

Consider a closed, oriented curve C in 3-space. We can find the circulation of a vector field $\vec{F}$ around C by calculating the line integral:

$$\begin{array}{c} \text{Circulation} \\ \text{around } C \end{array} = \int_C \vec{F} \cdot d\vec{r}.$$

If C is the boundary of an oriented surface S, there is another way to calculate the circulation using curl $\vec{F}$. We subdivide S into pieces as shown on the surface on the left in Figure 20.11. If $\vec{n}$ is a positive unit normal vector to a piece of surface with area ΔA, then $\Delta \vec{A} = \vec{n}\,\Delta A$. In addition, $\text{circ}_{\vec{n}}\,\vec{F}$ is the circulation density of $\vec{F}$ around $\vec{n}$, so

$$\begin{array}{c} \text{Circulation of } \vec{F} \text{ around} \\ \text{boundary of the piece} \end{array} \approx \left(\text{circ}_{\vec{n}}\,\vec{F} \right) \Delta A = ((\text{curl }\vec{F}) \cdot \vec{n})\Delta A = (\text{curl}\vec{F}) \cdot \Delta \vec{A}.$$

Next we add up the circulations around all the small pieces. The line integral along the common edge of a pair of adjacent pieces appears with opposite sign in each piece, so it cancels out. (See Figure 20.12.) When we add up all the pieces the internal edges cancel and we are left with the circulation around C, the boundary of the entire surface. Thus,

$$\begin{array}{c} \text{Circulation} \\ \text{around } C \end{array} = \sum \begin{array}{c} \text{Circulation around} \\ \text{boundary of pieces} \end{array} \approx \sum \text{curl }\vec{F} \cdot \Delta \vec{A}.$$

Taking the limit as $\Delta A \to 0$, we get

$$\begin{array}{c} \text{Circulation} \\ \text{around } C \end{array} = \int_S \text{curl }\vec{F} \cdot d\vec{A}.$$

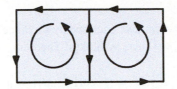

Figure 20.12: Two adjacent pieces of the surface

We have expressed the circulation as a line integral around C and as a flux integral over S; thus, the two integrals must be equal. Hence we have[1]

Theorem 20.1: Stokes' Theorem

If S is a smooth oriented surface with piecewise smooth, oriented boundary C, and if $\vec{F}$ is a smooth vector field on a solid region[2] containing S and C, then

$$\int_C \vec{F} \cdot d\vec{r} = \int_S \operatorname{curl} \vec{F} \cdot d\vec{A}.$$

The orientation of C is determined from the orientation of S according to the right-hand rule.

Example 1   Let $\vec{F}(x, y, z) = -2y\vec{i} + 2x\vec{j}$. Use Stokes' Theorem to find $\int_C \vec{F} \cdot d\vec{r}$, where C is a circle.

(a) Parallel to the yz-plane, of radius a, centered at a point on the x-axis, with either orientation.
(b) Parallel to the xy-plane, of radius a, centered at a point on the z-axis, oriented counterclockwise as viewed from a point on the z-axis above the circle.

Solution   We have $\operatorname{curl} \vec{F} = 4\vec{k}$. Figure 20.13 shows sketches of $\vec{F}$ and $\operatorname{curl} \vec{F}$.

(a) Let S be the disk enclosed by C. Since S lies in a vertical plane and $\operatorname{curl} \vec{F}$ points vertically everywhere, the flux of $\operatorname{curl} \vec{F}$ through S is zero. Hence, by Stokes' Theorem,

$$\int_C \vec{F} \cdot d\vec{r} = \int_S \operatorname{curl} \vec{F} \cdot d\vec{A} = 0.$$

It makes sense that the line integral is zero. If C is parallel to the yz-plane (even if it is not lying in the plane), the symmetry of the vector field means that the line integral of $\vec{F}$ over the top half of the circle cancels the line integral over the bottom half.

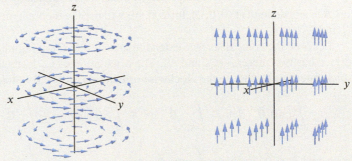

Figure 20.13: The vector fields $\vec{F}$ and $\operatorname{curl} \vec{F}$ (in the planes $z = -1$, $z = 0$, $z = 1$)

---

[1] A proof of Stokes' Theorem using the coordinate definition of curl can be found in the online supplement at www.WileyPLUS.com.
[2] The region containing S and C is open.

(b) Let S be the horizontal disk enclosed by C. Since curl $\vec{F}$ is a constant vector field pointing in the direction of $\vec{k}$, we have, by Stokes' Theorem,

$$\int_C \vec{F} \cdot d\vec{r} = \int_S \text{curl}\,\vec{F} \cdot d\vec{A} = \| \text{curl}\,\vec{F} \| \cdot \text{Area of } S = 4\pi a^2.$$

Since $\vec{F}$ is circling around the z-axis in the same direction as C, we expect the line integral to be positive. In fact, in Example 1 on page 1057, we computed this line integral directly.

Curl-Free Vector Fields

Stokes' Theorem applies to any oriented surface S and its boundary C, even in cases where the boundary consists of two or more curves. This is useful in studying curl-free vector fields.

Example 2   A current I flows along the z-axis in the $\vec{k}$ direction. The induced magnetic field $\vec{B}\,(x, y, z)$ is

$$\vec{B}\,(x, y, z) = \frac{2I}{c}\left(\frac{-y\vec{i} + x\vec{j}}{x^2 + y^2}\right),$$

where c is the speed of light. Example 5 on page 1060 shows that curl $\vec{B} = \vec{0}$.

(a) Compute the circulation of $\vec{B}$ around the circle C_1 in the xy-plane of radius a, centered at the origin, and oriented counterclockwise when viewed from above.

(b) Use part (a) and Stokes' Theorem to compute $\int_{C_2} \vec{B} \cdot d\vec{r}$, where C_2 is the ellipse $x^2 + 9y^2 = 9$ in the plane $z = 2$, oriented counterclockwise when viewed from above.

Solution   (a) On the circle C_1, we have $\| \vec{B} \| = 2I/(ca)$. Since $\vec{B}$ is tangent to C_1 everywhere and points in the forward direction around C_1,

$$\int_{C_1} \vec{B} \cdot d\vec{r} = \| \vec{B} \| \cdot \text{Length of } C_1 = \frac{2I}{ca} \cdot 2\pi a = \frac{4\pi I}{c}.$$

(b) We cannot use Stokes' Theorem on the elliptical disk bounded by C_2 in the plane $z = 2$ because curl $\vec{B}$ is not defined at $(0, 0, 2)$. Instead, we will use the theorem on a conical surface connecting C_1 and C_2.

Let S be the conical surface extending from C_1 to C_2 in Figure 20.14. The boundary of this surface has two pieces, $-C_2$ and C_1. The orientation of C_1 leads to the outward normal on S, which forces us to choose the clockwise orientation on C_2. By Stokes' Theorem,

$$\int_S \text{curl}\,\vec{B} \cdot d\vec{A} = \int_{-C_2} \vec{B} \cdot d\vec{r} + \int_{C_1} \vec{B} \cdot d\vec{r} = -\int_{C_2} \vec{B} \cdot d\vec{r} + \int_{C_1} \vec{B} \cdot d\vec{r}.$$

Since curl $\vec{B} = \vec{0}$, we have $\int_S \text{curl}\,\vec{B} \cdot d\vec{A} = 0$, so the two line integrals must be equal:

$$\int_{C_2} \vec{B} \cdot d\vec{r} = \int_{C_1} \vec{B} \cdot d\vec{r} = \frac{4\pi I}{c}.$$

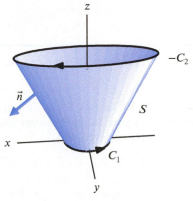

Figure 20.14: Surface joining C_1 to C_2, oriented to satisfy the conditions of Stokes' Theorem

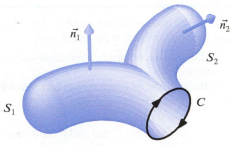

Figure 20.15: The flux of a curl is the same through the two surfaces S_1 and S_2 if they determine the same orientation on the boundary, C

Curl Fields

A vector field $\vec{F}$ is called a *curl field* if $\vec{F} = \operatorname{curl}\vec{G}$ for some vector field $\vec{G}$. Recall that if $\vec{F} = \operatorname{grad} f$, then f is called a potential function. By analogy, if a vector field $\vec{F} = \operatorname{curl}\vec{G}$, then $\vec{G}$ is called a *vector potential* for $\vec{F}$. The following example shows that the flux of a curl field through a surface depends only on the boundary of the surface. This is analogous to the fact that the line integral of a gradient field depends only on the endpoints of the path.

Example 3   Suppose $\vec{F} = \operatorname{curl}\vec{G}$, and that S_1 and S_2 are two oriented surfaces with the same boundary C. Show that, if S_1 and S_2 determine the same orientation on C (as in Figure 20.15), then

$$\int_{S_1} \vec{F} \cdot d\vec{A} = \int_{S_2} \vec{F} \cdot d\vec{A}.$$

If S_1 and S_2 determine opposite orientations on C, then

$$\int_{S_1} \vec{F} \cdot d\vec{A} = -\int_{S_2} \vec{F} \cdot d\vec{A}.$$

Solution   If S_1 and S_2 determine the same orientation on C, then since $\vec{F} = \operatorname{curl}\vec{G}$, by Stokes' Theorem we have

$$\int_{S_1} \vec{F} \cdot d\vec{A} = \int_{S_1} \operatorname{curl}\vec{G} \cdot d\vec{A} = \int_C \vec{G} \cdot d\vec{r}$$

and

$$\int_{S_2} \vec{F} \cdot d\vec{A} = \int_{S_2} \operatorname{curl}\vec{G} \cdot d\vec{A} = \int_C \vec{G} \cdot d\vec{r}.$$

In each case the line integral on the right must be computed using the orientation determined by the surface. Thus, the two flux integrals of $\vec{F}$ are the same if the orientations are the same and they are opposite if the orientations are opposite.

Summary for Section 20.2

- **Stokes' Theorem:** If S is a smooth oriented surface with boundary C, and if $\vec{F}$ is a smooth vector field on a solid region containing S and C, then

$$\int_C \vec{F} \cdot d\vec{r} = \int_S \operatorname{curl}\vec{F} \cdot d\vec{A}.$$

  The orientation of C is determined from the orientation of S according to the right-hand rule.

- A vector field $\vec{F}$ is a **curl vector field** if $\vec{F} = \operatorname{curl}\vec{G}$ for some vector field $\vec{G}$.

Exercises and Problems for Section 20.2

EXERCISES

1. If $\text{curl}\,\vec{F} = \vec{k}$, find the circulation of $\vec{F}$ around C, a circle of radius 1, centered at the origin, with

   (a) C in the xy-plane, oriented counterclockwise when viewed from above.

   (b) C in the yz-plane, oriented counterclockwise when viewed from the positive x-axis.

■ In Exercises 2–3, for the circle C and $\text{curl}\,\vec{F}$ as described, is the circulation of $\vec{F}$ around C positive, negative, or zero?

2. C is in the xy-plane oriented counterclockwise when viewed from above, and $\text{curl}\,\vec{F}$ points parallel to $\vec{j}$ everywhere.

3. C is in the yz-plane oriented clockwise when viewed from the positive x-axis, and $\text{curl}\,\vec{F}$ points parallel to and in the direction of $-\vec{i}$.

■ In Exercises 4–8, calculate the circulation, $\int_C \vec{F} \cdot d\vec{r}$, in two ways, directly and using Stokes' Theorem.

4. $\vec{F} = (x + z)\vec{i} + x\vec{j} + y\vec{k}$ and C is the upper half of the circle $x^2 + z^2 = 9$ in the plane $y = 0$, together with the x-axis from $(3, 0, 0)$ to $(-3, 0, 0)$, traversed counterclockwise when viewed from the positive y-axis.

5. $\vec{F} = y\vec{i} - x\vec{j}$ and C is the boundary of S, the part of the surface $z = 4 - x^2 - y^2$ above the xy-plane, oriented upward.

6. $\vec{F} = (x - y + z)(\vec{i} + \vec{j})$ and C is the triangle with vertices $(0, 0, 0)$, $(5, 0, 0)$, $(5, 5, 0)$, traversed in that order.

7. $\vec{F} = xy\vec{i} + yz\vec{j} + xz\vec{k}$ and C is the boundary of S, the surface $z = 1 - x^2$ for $0 \le x \le 1$ and $-2 \le y \le 2$, oriented upward. Sketch S and C.

8. $\vec{F} = y\vec{i} + z\vec{j} + x\vec{k}$ and C is the boundary of S, the paraboloid $z = 1 - (x^2 + y^2)$, $z \ge 0$ oriented upward. [Hint: Use polar coordinates.]

■ In Exercises 9–12, use Stokes' Theorem to calculate the integral.

9. $\int_C \vec{F} \cdot d\vec{r}$ where $\vec{F} = x^2\vec{i} + y^2\vec{j} + z^2\vec{k}$ and C is the unit circle in the xz-plane, oriented counterclockwise when viewed from the positive y-axis.

10. $\int_C \vec{F} \cdot d\vec{r}$ where $\vec{F} = (y - x)\vec{i} + (z - y)\vec{j} + (x - z)\vec{k}$ and C is the circle $x^2 + y^2 = 5$ in the xy-plane, oriented counterclockwise when viewed from above.

11. $\int_S \text{curl}\,\vec{F} \cdot d\vec{A}$ where $\vec{F} = -y\vec{i} + x\vec{j} + (xy + \cos z)\vec{k}$ and S is the disk $x^2 + y^2 \le 9$, oriented upward in the xy-plane.

12. $\int_S \text{curl}\,\vec{F} \cdot d\vec{A}$ where $\vec{F} = (x + 7)\vec{j} + e^{x+y+z}\vec{k}$ and S is the rectangle $0 \le x \le 3$, $0 \le y \le 2$, $z = 0$, oriented upward.

13. Let $\vec{F} = y\vec{i} - x\vec{j}$ and let C be the unit circle in the xy-plane centered at the origin and oriented counterclockwise when viewed from above.

    (a) Calculate $\int_C \vec{F} \cdot d\vec{r}$ by parameterizing the circle.

    (b) Calculate $\text{curl}\,\vec{F}$.

    (c) Calculate $\int_C \vec{F} \cdot d\vec{r}$ using your result from part (b).

    (d) What theorem did you use in part (c)?

14. **(a)** If $\vec{F} = (\cos x)\vec{i} + e^y\vec{j} + (x - y - z)\vec{k}$, find $\text{curl}\,\vec{F}$.

    (b) Find $\int_C \vec{F} \cdot d\vec{r}$ where C is the circle of radius 3 in the plane $x + y + z = 1$, centered at $(1, 0, 0)$ oriented counterclockwise when viewed from above.

15. Can you use Stokes' Theorem to compute the line integral $\int_C (2x\vec{i} + 2y\vec{j} + 2z\vec{k}) \cdot d\vec{r}$ where C is the straight line from the point $(1, 2, 3)$ to the point $(4, 5, 6)$? Why or why not?

PROBLEMS

16. At all points in 3-space $\text{curl}\,\vec{F}$ points in the direction of $\vec{i} - \vec{j} - \vec{k}$. Let C be a circle in the yz-plane, oriented clockwise when viewed from the positive x-axis. Is the circulation of $\vec{F}$ around C positive, zero, or negative?

17. If $\text{curl}\,\vec{F} = (x^2 + z^2)\vec{j} + 5\vec{k}$, find $\int_C \vec{F} \cdot d\vec{r}$, where C is a circle of radius 3, centered at the origin, with

    (a) C in the xy-plane, oriented counterclockwise when viewed from above.

    (b) C in the xz-plane, oriented counterclockwise when viewed from the positive y-axis.

18. **(a)** Find $\text{curl}(y\vec{i} + z\vec{j} + x\vec{k})$.

    (b) Find $\int_C (y\vec{i} + z\vec{j} + x\vec{k}) \cdot d\vec{r}$ where C is the boundary of the triangle with vertices $(2, 0, 0)$, $(0, 3, 0)$, $(-2, 0, 0)$, traversed in that order.

19. **(a)** Let $\vec{F} = y\vec{i} + z\vec{j} + x\vec{k}$. Find $\text{curl}\,\vec{F}$.

    (b) Calculate $\int_C \vec{F} \cdot d\vec{r}$ where C is

    (i) A circle of radius 2 centered at $(1, 1, 3)$ in the plane $z = 3$, oriented counterclockwise when viewed from above.

    (ii) The triangle obtained by tracing out the path $(2, 0, 0)$ to $(2, 0, 5)$ to $(2, 3, 5)$ to $(2, 0, 0)$.

20. **(a)** Find $\text{curl}(z\vec{i} + x\vec{j} + y\vec{k})$.

    (b) Find $\int_C (z\vec{i} + x\vec{j} + y\vec{k}) \cdot d\vec{r}$ where C is a square of side 2 lying in the plane $x + y + z = 5$, oriented counterclockwise when viewed from the origin.

In Problems **21–26**, find $\int_C \vec{F} \cdot d\vec{r}$ where C is a circle of radius 2 in the plane $x + y + z = 3$, centered at $(1, 1, 1)$ and oriented clockwise when viewed from the origin.

21. $\vec{F} = \vec{i} + \vec{j} + 3\vec{k}$

22. $\vec{F} = -y\vec{i} + x\vec{j} + z\vec{k}$

23. $\vec{F} = y\vec{i} - x\vec{j} + (y - x)\vec{k}$

24. $\vec{F} = (2y + e^x)\vec{i} + ((\sin y) - x)\vec{j} + (2y - x + \cos z^2)\vec{k}$

25. $\vec{F} = -z\vec{j} + y\vec{k}$

26. $\vec{F} = (z - y)\vec{i} + (x - z)\vec{j} + (y - x)\vec{k}$

27. For positive constants a, b, and c, let

$$f(x, y, z) = \ln(1 + ax^2 + by^2 + cz^2).$$

    (a) What is the domain of f?
    (b) Find $\text{grad} f$.
    (c) Find $\text{curl}(\text{grad} f)$.
    (d) Find $\int_C \vec{F} \cdot d\vec{r}$ where C is the helix $x = \cos t$, $y = \sin t$, $z = t$ for $0 \leq t \leq 13\pi/2$ and

$$\vec{F} = \frac{2x\vec{i} + 4y\vec{j} + 6z\vec{k}}{1 + x^2 + 2y^2 + 3z^2}.$$

28. Figure 20.16 shows an open cylindrical can, S, standing on the xy-plane. (S has a bottom and sides, but no top.)

    (a) Give equation(s) for the rim, C.
    (b) If S is oriented outward and downward, find $\int_S \text{curl}(-y\vec{i} + x\vec{j} + z\vec{k}) \cdot d\vec{A}$.

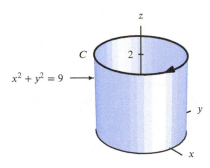

Figure 20.16

29. Evaluate $\int_C (-z\vec{i} + y\vec{j} + x\vec{k}) \cdot d\vec{r}$, where C is a circle of radius 2, parallel to the xz-plane and around the y-axis with the orientation shown in Figure 20.17.

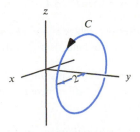

Figure 20.17

30. Evaluate the circulation of $\vec{G} = xy\vec{i} + z\vec{j} + 3y\vec{k}$ around a square of side 6, centered at the origin, lying in the yz-plane, and oriented counterclockwise viewed from the positive x-axis.

31. Find the flux of $\vec{F} = \text{curl}((x^3 + \cos(z^2))\vec{i} + (x + \sin(y^2))\vec{j} + (y^2 \sin(x^2))\vec{k})$ through the upper half of the sphere of radius 2, with center at the origin and oriented upward.

32. Suppose that C is a closed curve in the xy-plane, oriented counterclockwise when viewed from above. Show that $\frac{1}{2} \int_C (-y\vec{i} + x\vec{j}) \cdot d\vec{r}$ equals the area of the region R in the xy-plane enclosed by C.

33. In the region between the circles $C_1 : x^2 + y^2 = 4$ and $C_2 : x^2 + y^2 = 25$ in the xy-plane, the vector field $\vec{F}$ has $\text{curl} \vec{F} = 3\vec{k}$. If C_1 and C_2 are both oriented counterclockwise when viewed from above, find the value of

$$\int_{C_2} \vec{F} \cdot d\vec{r} - \int_{C_1} \vec{F} \cdot d\vec{r}.$$

34. Let $\text{curl} \vec{F} = 3x\vec{i} + 3y\vec{j} - 6z\vec{k}$ and let C_1 and C_2 be the closed curves in the planes $z = 0$ and $z = 5$ in Figure 20.18. Find

$$\int_{C_1} \vec{F} \cdot d\vec{r} + \int_{C_2} \vec{F} \cdot d\vec{r}.$$

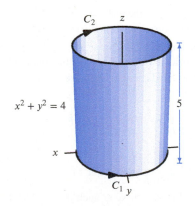

Figure 20.18

35. **(a)** Find $\text{curl}(x^3\vec{i} + \sin(y^3)\vec{j} + e^{z^3}\vec{k})$.
    (b) What does your answer to part (a) tell you about $\int_C (x^3\vec{i} + \sin(y^3)\vec{j} + e^{z^3}\vec{k}) \cdot d\vec{r}$ where C is the circle $(x - 10)^2 + (y - 20)^2 = 1$ in the xy-plane, oriented clockwise?
    (c) If C is any closed curve, what can you say about $\int_C (x^3\vec{i} + \sin(y^3)\vec{j} + e^{z^3}\vec{k}) \cdot d\vec{r}$?

36. For C, the intersection of the cylinder $x^2 + y^2 = 9$ and the plane $z = -2 - x + 2y$ oriented counterclockwise when viewed from above, use Stokes' Theorem to find

$$\int_C \left((x^2 - 3y^2)\vec{i} + (\frac{z^2}{2} + y)\vec{j} + (x + 2z^2)\vec{k} \right) \cdot d\vec{r}.$$

37. Let $\vec{F}(x, y, z) = F_1(x, y)\vec{i} + F_2(x, y)\vec{j}$, where F_1 and F_2 are continuously differentiable for all x, y.

(a) Describe in words how $\vec{F}$ varies through space.

(b) Find an expression for curl $\vec{F}$ in terms of F_1 and F_2.

(c) Let C be a closed curve in the xy-plane, oriented counterclockwise when viewed from above, and let S be the region enclosed by C. Use your answer to part (b) to simplify the statement of Stokes' Theorem for this $\vec{F}$ and C.

(d) The result in part (c) is usually known by another name. What is it?

38. Water in a bathtub has velocity vector field near the drain given, for x, y, z in cm, by

$$\vec{F} = -\frac{y + xz}{(z^2 + 1)^2}\vec{i} - \frac{yz - x}{(z^2 + 1)^2}\vec{j} - \frac{1}{z^2 + 1}\vec{k} \text{ cm/sec.}$$

(a) Rewriting $\vec{F}$ as follows, describe in words how the water is moving:

$$\vec{F} = \frac{-y\vec{i} + x\vec{j}}{(z^2 + 1)^2} + \frac{-z(x\vec{i} + y\vec{j})}{(z^2 + 1)^2} - \frac{\vec{k}}{z^2 + 1}.$$

(b) The drain in the bathtub is a disk in the xy-plane with center at the origin and radius 1 cm. Find the rate at which the water is leaving the bathtub. (That is, find the rate at which water is flowing through the disk.) Give units for your answer.

(c) Find the divergence of $\vec{F}$.

(d) Find the flux of the water through the hemisphere of radius 1, centered at the origin, lying below the xy-plane and oriented downward.

(e) Find $\int_C \vec{G} \cdot d\vec{r}$ where C is the edge of the drain, oriented clockwise when viewed from above, and where

$$\vec{G} = \frac{1}{2} \left(\frac{y}{z^2 + 1}\vec{i} - \frac{x}{z^2 + 1}\vec{j} - \frac{x^2 + y^2}{(z^2 + 1)^2}\vec{k} \right).$$

(f) Calculate curl $\vec{G}$.

(g) Explain why your answers to parts (d) and (e) are equal.

Strengthen Your Understanding

In Problems **39–40**, explain what is wrong with the statement.

39. The line integral $\int_C \vec{F} \cdot d\vec{r}$ can be evaluated using Stokes' Theorem, where $\vec{F} = 2x\vec{i} - 3\vec{j} + \vec{k}$ and C is an oriented curve from $(0, 0, 0)$ to $(3, 4, 5)$.

40. If S is the unit circular disc $x^2 + y^2 \le 1$, $z = 0$, in the xy-plane, oriented downward, C is the unit circle in the xy-plane oriented counterclockwise, and $\vec{F}$ is a vector field, then

$$\int_C \vec{F} \cdot d\vec{r} = \int_S \text{curl} \vec{F} \cdot d\vec{A}.$$

In Problems **41–42**, give an example of:

41. An oriented closed curve C such that $\int_C \vec{F} \cdot d\vec{r} = 0$, where $\vec{F}(x, y, z) = x\vec{i} + y^2\vec{j} + z^3\vec{k}$.

42. A surface S, oriented appropriately to use Stokes' Theorem, which has as its boundary the circle C of radius 1 centered at the origin, lying in the xy-plane, and oriented counterclockwise when viewed from above.

In Problems **43–51**, is the statement true or false? Give a reason for your answer.

43. If curl $\vec{F}$ is everywhere perpendicular to the z-axis, and C is a circle in the xy-plane, then the circulation of $\vec{F}$ around C is zero.

44. If S is the upper unit hemisphere $x^2 + y^2 + z^2 = 1$, $z \ge 0$, oriented upward, then the boundary of S used in Stokes' Theorem is the circle $x^2 + y^2 = 1$, $z = 0$, with orientation counterclockwise when viewed from the positive z-axis.

45. Let S be the cylinder $x^2 + z^2 = 1$, $0 \le y \le 2$, oriented with inward-pointing normal. Then the boundary of S consists of two circles C_1 ($x^2 + z^2 = 1$, $y = 0$) and C_2 ($x^2 + z^2 = 1$, $y = 2$), both oriented clockwise when viewed from the positive y-axis.

46. If C is the boundary of an oriented surface S, oriented by the right-hand rule, then $\int_C \text{curl} \vec{F} \cdot d\vec{r} = \int_S \vec{F} \cdot d\vec{A}$.

47. Let S_1 be the disk $x^2 + y^2 \le 1$, $z = 0$ and let S_2 be the upper unit hemisphere $x^2 + y^2 + z^2 = 1$, $z \ge 0$, both oriented upward. If $\vec{F}$ is a vector field then $\int_{S_1} \text{curl} \vec{F} \cdot d\vec{A} = \int_{S_2} \text{curl} \vec{F} \cdot d\vec{A}$.

48. Let S be the closed unit sphere $x^2 + y^2 + z^2 = 1$, oriented outward. If $\vec{F}$ is a vector field, then $\int_S \text{curl} \vec{F} \cdot d\vec{A} = 0$.

49. If $\vec{F}$ and $\vec{G}$ are vector fields satisfying curl $\vec{F} = $ curl $\vec{G}$, then $\int_C \vec{F} \cdot d\vec{r} = \int_C \vec{G} \cdot d\vec{r}$, where C is any oriented circle in 3-space.

50. If $\vec{F}$ is a vector field satisfying curl $\vec{F} = \vec{0}$, then $\int_C \vec{F} \cdot d\vec{r} = 0$, where C is any oriented path around a rectangle in 3-space.

51. Let S be an oriented surface, with oriented boundary C, and suppose that $\vec{F}$ is a vector field such that $\int_C \vec{F} \cdot d\vec{r} = 0$. Then curl $\vec{F} = \vec{0}$ everywhere on S.

52. The circle C has radius 3 and lies in a plane through the origin. Let $\vec{F} = (2z + 3y)\vec{i} + (x - z)\vec{j} + (6y - 7x)\vec{k}$.

What is the equation of the plane and what is the orientation of the circle that make the circulation, $\int_C \vec{F} \cdot d\vec{r}$, a maximum? [Note: You should specify the orientation of the circle by saying that it is clockwise or counterclockwise when viewed from the positive or negative x- or y- or z-axis.]

20.3 THE THREE FUNDAMENTAL THEOREMS

We have now seen three multivariable versions of the Fundamental Theorem of Calculus. In this section we will examine some consequences of these theorems.

Fundamental Theorem of Calculus for Line Integrals

$$\int_C \operatorname{grad} f \cdot d\vec{r} = f(Q) - f(P).$$

Stokes' Theorem

$$\int_S \operatorname{curl} \vec{F} \cdot d\vec{A} = \int_C \vec{F} \cdot d\vec{r}.$$

Divergence Theorem

$$\int_W \operatorname{div} \vec{F} \, dV = \int_S \vec{F} \cdot d\vec{A}.$$

Notice that, in each case, the region of integration on the right is the boundary of the region on the left (except that for the first theorem we simply evaluate f at the boundary points); the integrand on the left is a sort of derivative of the integrand on the right; see Figure 20.19.

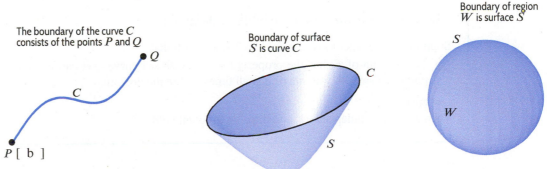

Figure 20.19: Regions and their boundaries for the three fundamental theorems

The Gradient and the Curl

Suppose that $\vec{F}$ is a smooth gradient field, so $\vec{F} = \operatorname{grad} f$ for some function f. Using the Fundamental Theorem for Line Integrals, we saw in Chapter 18 that

$$\int_C \vec{F} \cdot d\vec{r} = 0$$

for any closed curve C. Thus, for any unit vector $\vec{n}$

$$\operatorname{circ}_{\vec{n}} \vec{F} = \lim_{\text{Area} \to 0} \frac{\int_C \vec{F} \cdot d\vec{r}}{\text{Area of } C} = \lim_{\text{Area} \to 0} \frac{0}{\text{Area}} = 0,$$

where the limit is taken over circles C in a plane perpendicular to $\vec{n}$, and oriented by the right-hand rule. Thus, the circulation density of $\vec{F}$ is zero in every direction, so curl $\vec{F} = \vec{0}$, that is,

$$\boxed{\text{curl grad } f = \vec{0}.}$$

(This formula can also be verified using the coordinate definition of curl. See Problem 31 on page 1063.)

Is the converse true? Is any vector field whose curl is zero a gradient field? Suppose that curl $\vec{F} = \vec{0}$ and let us consider the line integral $\int_C \vec{F} \cdot d\vec{r}$ for a closed curve C contained in the domain of $\vec{F}$. If C is the boundary curve of an oriented surface S that lies wholly in the domain of curl $\vec{F}$, then Stokes' Theorem asserts that

$$\int_C \vec{F} \cdot d\vec{r} = \int_S \text{curl } \vec{F} \cdot d\vec{A} = \int_S \vec{0} \cdot d\vec{A} = 0.$$

If we knew that $\int_C \vec{F} \cdot d\vec{r} = 0$ for every closed curve C, then $\vec{F}$ would be path-independent, and hence a gradient field. Thus, we need to know whether every closed curve in the domain of $\vec{F}$ is the boundary of an oriented surface contained in the domain. It can be quite difficult to determine if a given curve is the boundary of a surface (suppose, for example, that the curve is knotted in a complicated way). However, if the curve can be contracted smoothly to a point, remaining all the time in the domain of $\vec{F}$, then it is the boundary of a surface, namely, the surface it sweeps through as it contracts. Thus, we have proved the test for a gradient field that we stated in Chapter 18.

The Curl Test for Vector Fields in 3-Space

Suppose $\vec{F}$ is a smooth vector field on 3-space such that
- The domain of $\vec{F}$ has the property that every closed curve in it can be contracted to a point in a smooth way, staying at all times within the domain.
- curl $\vec{F} = \vec{0}$.

Then $\vec{F}$ is path-independent, and thus is a gradient field.

Example 7 on page 1010 shows how the curl test is applied.

The Curl and the Divergence

In this section we will use the second two fundamental theorems to get a test for a vector field to be a curl field, that is, a field of the form $\vec{F} = \text{curl } \vec{G}$ for some $\vec{G}$.

Example 1   Suppose that $\vec{F}$ is a smooth curl field. Use Stokes' Theorem to show that for any closed surface, S, contained in the domain of $\vec{F}$,

$$\int_S \vec{F} \cdot d\vec{A} = 0.$$

Solution   Suppose $\vec{F} = \text{curl}\,\vec{G}$. Draw a closed curve C on the surface S, thus dividing S into two surfaces S_1 and S_2 as shown in Figure 20.20. Pick the orientation for C corresponding to S_1; then the orientation of C corresponding to S_2 is the opposite. Thus, using Stokes' Theorem,

$$\int_{S_1} \vec{F} \cdot d\vec{A} = \int_{S_1} \text{curl}\,\vec{G} \cdot d\vec{A} = \int_C \vec{G} \cdot d\vec{r} = -\int_{S_2} \text{curl}\,\vec{G} \cdot d\vec{A} = -\int_{S_2} \vec{F} \cdot d\vec{A}.$$

Thus, for any closed surface S, we have

$$\int_S \vec{F} \cdot d\vec{A} = \int_{S_1} \vec{F} \cdot d\vec{A} + \int_{S_2} \vec{F} \cdot d\vec{A} = 0.$$

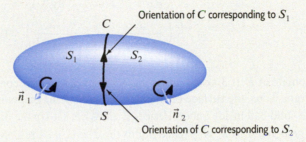

Figure 20.20: The closed surface S divided into two surfaces S_1 and S_2

Thus, if $\vec{F} = \text{curl}\,\vec{G}$, we use the result of Example 1 to see that

$$\text{div}\,\vec{F} = \lim_{\text{Volume}\to 0} \frac{\displaystyle\int_S \vec{F} \cdot d\vec{A}}{\text{Volume enclosed by } S} = \lim_{\text{Volume}\to 0} \frac{0}{\text{Volume}} = 0,$$

where the limit is taken over spheres S contracting down to a point. So we conclude that:

$$\boxed{\text{div}\,\text{curl}\,\vec{G} = 0.}$$

(This formula can also be verified using coordinates. See Problem 30 on page 1063.)

Is every vector field whose divergence is zero a curl field? It turns out that we have the following analogue of the curl test, though we will not prove it.

The Divergence Test for Vector Fields in 3-Space

Suppose $\vec{F}$ is a smooth vector field on 3-space such that
- The domain of $\vec{F}$ has the property that every closed surface in it is the boundary of a solid region completely contained in the domain.
- $\text{div}\,\vec{F} = 0$.

Then $\vec{F}$ is a curl field.

Example 2   Consider the vector fields $\vec{E} = q\dfrac{\vec{r}}{\|\vec{r}\|^3}$ and $\vec{B} = \dfrac{2I}{c}\left(\dfrac{-y\vec{i} + x\vec{j}}{x^2 + y^2}\right)$.

(a) Calculate div $\vec{E}$ and div $\vec{B}$.
(b) Do $\vec{E}$ and $\vec{B}$ satisfy the divergence test?
(c) Is either $\vec{E}$ or $\vec{B}$ a curl field?

Solution   (a) Example 3 on page 1042 shows that div $\vec{E} = 0$. The following calculation shows div $\vec{B} = 0$ also:

$$\text{div } \vec{B} = \frac{2I}{c}\left(\frac{\partial}{\partial x}\left(\frac{-y}{x^2 + y^2}\right) + \frac{\partial}{\partial y}\left(\frac{x}{x^2 + y^2}\right) + \frac{\partial}{\partial z}(0)\right)$$

$$= \frac{2I}{c}\left(\frac{2xy}{(x^2 + y^2)^2} + \frac{-2yx}{(x^2 + y^2)^2}\right) = 0.$$

(b) The domain of $\vec{E}$ is 3-space minus the origin, so a region is contained in the domain if it misses the origin. Thus, the surface of a sphere centered at the origin is contained in the domain of E, but the solid ball inside is not. Hence, $\vec{E}$ does not satisfy the divergence test.

The domain of $\vec{B}$ is 3-space minus the z-axis, so a region is contained in the domain if it avoids the z-axis. If S is a surface bounding a solid region W, then the z-axis cannot pierce W without piercing S as well. Hence, if S avoids the z-axis, so does W. Thus, $\vec{B}$ satisfies the divergence test.

(c) In Example 3 on page 1050 we computed the flux of $\vec{r}/\|\vec{r}\|^3$ through a sphere centered at the origin, and found it was 4π, so the flux of $\vec{E}$ through this sphere is $4\pi q$. Thus, $\vec{E}$ cannot be a curl field, because by Example 1, the flux of a curl field through a closed surface is zero.

On the other hand, $\vec{B}$ satisfies the divergence test, so it must be a curl field. In fact, Problem 26 shows that

$$\vec{B} = \text{curl}\left(\frac{-I}{c}\ln(x^2 + y^2)\vec{k}\right).$$

Summary for Section 20.3

- **Multivariable versions of the Fundamental Theorem of Calculus:**
  - **Fundamental Theorem of Calculus for Line Integrals:** $\int_C \text{grad} f \cdot d\vec{r} = f(Q) - f(P)$
  - **Stokes' Theorem:** $\int_S \text{curl} \vec{F} \cdot d\vec{A} = \int_C \vec{F} \cdot d\vec{r}$
  - **Divergence Theorem:** $\int_W \text{div} \vec{F} \, dV = \int_S \vec{F} \cdot d\vec{A}$

- **The curl test in 3-space:** Suppose $\vec{F}$ is a smooth vector field on 3-space such that
  - The domain of $\vec{F}$ has the property that every closed curve in it can be contracted to a point in a smooth way, staying at all times within the domain.
  - curl $\vec{F} = \vec{0}$.

  Then $\vec{F}$ is **path-independent**, and thus is a **gradient vector field**.

- **The divergence test in 3-space:** Suppose $\vec{F}$ is a smooth vector field on 3-space such that
  - The domain of $\vec{F}$ has the property that every closed surface in it is the boundary of a solid region completely contained in the domain.
  - div $\vec{F} = 0$.

  Then $\vec{F}$ is a **curl vector field**.

Exercises and Problems for Section 20.3 Online Resource: Additional Problems for Section 20.3

EXERCISES

■In Exercises 1–6, is the vector field a gradient field?

1. $\vec{F} = 2x\vec{i} + z\vec{j} + y\vec{k}$
2. $\vec{F} = y\vec{i} + z\vec{j} + x\vec{k}$
3. $\vec{F} = (y + 2z)\vec{i} + (x + z)\vec{j} + (2x + y)\vec{k}$
4. $\vec{F} = (y - 2z)\vec{i} + (x - z)\vec{j} + (2x - y)\vec{k}$
5. $\vec{G} = -y\vec{i} + x\vec{j}$
6. $\vec{F} = yz\vec{i} + (xz + z^2)\vec{j} + (xy + 2yz)\vec{k}$

■In Exercises 7–12, is the vector field a curl field?

7. $\vec{F} = z\vec{i} + x\vec{j} + y\vec{k}$
8. $\vec{F} = z\vec{i} + y\vec{j} + x\vec{k}$
9. $\vec{F} = 2x\vec{i} - y\vec{j} - z\vec{k}$
10. $\vec{F} = (x + y)\vec{i} + (y + z)\vec{j} + (x + z)\vec{k}$
11. $\vec{F} = (-xy)\vec{i} + (2yz)\vec{j} + (yz - z^2))\vec{k}$
12. $\vec{F} = (xy)\vec{i} + (xy)\vec{j} + (xy)\vec{k}$

13. Let $\vec{F}$ be a vector field defined everywhere except the z-axis and with curl $\vec{F} = 0$ at all points of its domain. Determine whether Stokes' Theorem implies that $\int_C \vec{F} \cdot d\vec{r} = 0$ for a circle C of radius 1, where

    (a) C is parallel to the xy-plane with center $(0, 0, 1)$.

    (b) C is parallel to the yz-plane with center $(1, 0, 0)$.

14. Let $\vec{F}$ be a vector field defined everywhere except the origin and with div $\vec{F} = 0$ at all points of its domain. Determine whether the Divergence Theorem implies that $\int_S \vec{F} \cdot d\vec{A} = 0$ for a sphere S of radius 1, where

    (a) S is centered at $(0, 1, 1)$.
    (b) S is centered at $(0.5, 0, 0)$.

■In Exercises 15–18, can the curl test and the divergence test be applied to a vector field whose domain is the given region?

15. All points (x, y, z) such that $z > 0$.

16. All points (x, y, z) not on the y-axis.

17. All points (x, y, z) not on the positive z-axis.

18. All points (x, y, z) except the x-axis with $0 \le x \le 1$.

PROBLEMS

19. Let $\vec{B} = b\vec{k}$, for some constant b. Show that the following are all possible vector potentials for $\vec{B}$:

    (a) $\vec{A} = -by\vec{i}$ (b) $\vec{A} = bx\vec{j}$
    (c) $\vec{A} = \frac{1}{2}\vec{B} \times \vec{r}$.

20. Find a vector field $\vec{F}$ such that curl $\vec{F} = 2\vec{i} - 3\vec{j} + 4\vec{k}$. [Hint: Try $\vec{F} = \vec{v} \times \vec{r}$ for some vector $\vec{v}$.]

21. Find a vector potential for the constant vector field $\vec{B}$ whose value at every point is $\vec{b}$.

22. Express $(3x + 2y)\vec{i} + (4x + 9y)\vec{j}$ as the sum of a curl-free vector field and a divergence-free vector field.

■In Problems 23–24, does a vector potential exist for the vector field given? If so, find one.

23. $\vec{G} = x^2\vec{i} + y^2\vec{j} + z^2\vec{k}$
24. $\vec{F} = 2x\vec{i} + (3y - z^2)\vec{j} + (x - 5z)\vec{k}$

25. An electric charge q at the origin produces an electric field $\vec{E} = q\vec{r} / \|\vec{r}\|^3$.

    (a) Does curl $\vec{E} = \vec{0}$?
    (b) Does $\vec{E}$ satisfy the curl test?
    (c) Is $\vec{E}$ a gradient field?

26. Show that $\vec{A} = \frac{-I}{c}\ln(x^2 + y^2)\vec{k}$ is a vector potential for

$$\vec{B} = \frac{2I}{c}\left(\frac{-y\vec{i} + x\vec{j}}{x^2 + y^2}\right).$$

27. Suppose c is the speed of light. A thin wire along the z-axis carrying a current I produces a magnetic field

$$\vec{B} = \frac{2I}{c}\left(\frac{-y\vec{i} + x\vec{j}}{x^2 + y^2}\right).$$

    (a) Does curl $\vec{B} = \vec{0}$?
    (b) Does $\vec{B}$ satisfy the curl test?
    (c) Is $\vec{B}$ a gradient field?

28. Use Stokes' Theorem to show that if $u(x, y)$ and $v(x, y)$ are two functions of x and y and C is a closed curve in the xy-plane oriented counterclockwise, then

$$\int_C (u\vec{i} + v\vec{j}) \cdot d\vec{r} = \int_R \left(\frac{\partial v}{\partial x} - \frac{\partial u}{\partial y}\right) dx\,dy$$

where R is the region in the xy-plane enclosed by C. This is Green's Theorem.

29. Suppose that $\vec{A}$ is a vector potential for $\vec{B}$.

    (a) Show that $\vec{A} + \text{grad}\,\psi$ is also a vector potential for $\vec{B}$, for any function ψ with continuous second-order partial derivatives. (The vector potentials $\vec{A}$ and $\vec{A} + \text{grad}\,\psi$ are called *gauge equivalent* and the transformation, for any ψ, from $\vec{A}$ to $\vec{A} + \text{grad}\,\psi$

    is called a *gauge transformation*.)

    (b) What is the divergence of $\vec{A} + \text{grad}\,\psi$? How should ψ be chosen such that $\vec{A} + \text{grad}\,\psi$ has zero divergence? (If $\text{div}\,\vec{A} = 0$, the magnetic vector potential $\vec{A}$ is said to be in *Coulomb gauge*.)

Strengthen Your Understanding

■ In Problems 30–31, explain what is wrong with the statement.

30. The curl of a vector field $\vec{F}$ is given by $\text{curl}\,\vec{F} = x\vec{i}$.

31. For a certain vector field $\vec{F}$, we have $\text{curl}\,\text{div}\,\vec{F} = y\vec{i}$.

■ In Problems 32–33, give an example of:

32. A vector field $\vec{F}$ that is not the curl of another vector field.

33. A function f such that $\text{div}\,\text{grad}\,f \neq 0$.

■ In Problems 34–37, is the statement true or false? Give a reason for your answer.

34. There exists a vector field $\vec{F}$ with $\text{curl}\,\vec{F} = \vec{i}$.

35. There exists a vector field $\vec{F}$ (whose components have continuous second partial derivatives) satisfying

$$\text{curl}\,\vec{F} = x\vec{i}.$$

36. Let S be an oriented surface, with oriented boundary C, and suppose that $\vec{F}$ is a vector field such that $\int_S \text{curl}\,\vec{F} \cdot d\vec{A} = 0$. Then $\vec{F}$ is a gradient field.

37. If $\vec{F}$ is a gradient field, then $\int_S \text{curl}\,\vec{F} \cdot d\vec{A} = 0$, for any smooth oriented surface, S, in 3-space.

38. Let $f(x, y, z)$ be a scalar function with continuous second partial derivatives. Let $\vec{F}(x, y, z)$ be a vector field with continuous second partial derivatives. Which of the following quantities are identically zero?

    (a) $\text{curl}\,\text{grad}\,f$

    (b) $\vec{F} \times \text{curl}\,\vec{F}$

    (c) $\text{grad}\,\text{div}\,\vec{F}$

    (d) $\text{div}\,\text{curl}\,\vec{F}$

    (e) $\text{div}\,\text{grad}\,f$

Online Resource: Review Problems and Projects

Chapter 21

PARAMETERS, COORDINATES, AND INTEGRALS

CONTENTS

21.1   COORDINATES AND PARAMETERIZED SURFACES

In Chapter 17 we parameterized curves in 2- and 3-space, and in Chapter 16 we used polar, cylindrical, and spherical coordinates to simplify iterated integrals. We now take a second look at parameterizations and coordinate systems, and see that they are the same thing in different disguises: functions from one space to another.

We have already seen this with parameterized curves, which we view as a function from an interval $a \leq t \leq b$ to a curve in xyz-space. See Figure 21.1.

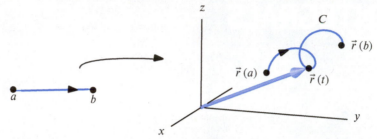

Figure 21.1: The parameterization is a function from the interval, $a \leq t \leq b$, to 3-space, whose image is the curve, C

Polar, Cylindrical, and Spherical Coordinates Revisited

The equations for polar coordinates,

$$x = r\cos\theta$$
$$y = r\sin\theta,$$

can also be viewed as defining a function from the $r\theta$-plane into the xy-plane. This function transforms the rectangle on the left of Figure 21.2 into the quarter disk on the right. We need two parameters to describe this disk because it is a two-dimensional object.

Polar Coordinates as Families of Parameterized Curves

Polar coordinates give two families of parameterized curves, which form the polar coordinate grid. The lines $r = $ Constant in the $r\theta$-plane correspond to circles in the xy-plane, each circle parameterized by θ; the lines $\theta = $ Constant correspond to rays in the xy-plane, each ray parameterized by r.

Cylindrical and Spherical Coordinates

Similarly, cylindrical and spherical coordinates may be viewed as functions from 3-space to 3-space. Cylindrical coordinates take rectangular boxes in $r\theta z$-space and map them to cylindrical regions in xyz-space; spherical coordinates take rectangular boxes in $\rho\phi\theta$-space and map them to spherical regions in xyz-space.

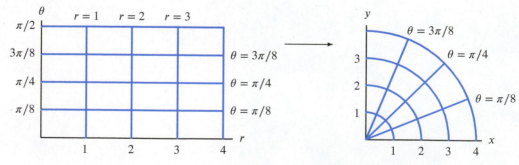

Figure 21.2: A grid in the $r\theta$-plane and the corresponding curved grid in the xy-plane

General Parameterizations

In general, a parameterization or coordinate system provides a way of representing a curved object by means of a simple region in the *parameter space* (an interval, rectangle, or rectangular box), along with a function mapping that region into the curved object. In the next section, we use this idea to parameterize curved surfaces in 3-space.

How Do We Parameterize a Surface?

In Section 17.1 we parameterized a circle in 2-space using the equations

$$x = \cos t, \quad y = \sin t.$$

In 3-space, the same circle in the xy-plane has parametric equations

$$x = \cos t, \quad y = \sin t, \quad z = 0.$$

We add the equation $z = 0$ to specify that the circle is in the xy-plane. If we wanted a circle in the plane $z = 3$, we would use the equations

$$x = \cos t, \quad y = \sin t, \quad z = 3.$$

Suppose now we let z vary freely, as well as t. We get circles in every horizontal plane, forming a cylinder as in the left of Figure 21.3. Thus, we need two parameters, t and z, to parameterize the cylinder.

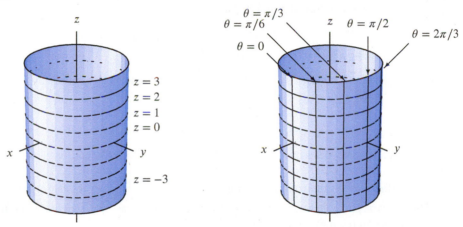

Figure 21.3: The cylinder $x = \cos t$, $y = \sin t$, $z = z$

We can contrast curves and surfaces. A curve, though it may live in two or three dimensions, is itself one-dimensional; if we move along it we can only move backward and forward in one direction. Thus, it only requires one parameter to trace out a curve.

A surface is 2-dimensional; at any given point there are two independent directions we can move. For example, on the cylinder we can move vertically, or we can circle around the z-axis horizontally. So we need *two* parameters to describe it. We can think of the parameters as map coordinates, like longitude and latitude on the surface of the earth. Just as polar coordinates give a polar grid on a circular region, so the parameters for a surface give a grid on the surface. See Figure 21.3 on the right.

In the case of the cylinder our parameters are t and z, so

$$x = \cos t, \quad y = \sin t, \quad z = z, \quad 0 \le t < 2\pi, \quad -\infty < z < \infty.$$

The last equation, $z = z$, looks strange, but it reminds us that we are in three dimensions, not two, and that the z-coordinate on our surface is allowed to vary freely.

In general, we express the coordinates, (x, y, z), of a point on a surface S in terms of two parameters, s and t:

$$x = f_1(s, t), \quad y = f_2(s, t), \quad z = f_3(s, t).$$

As the values of s and t vary, the corresponding point (x, y, z) sweeps out the surface, S. (See Figure 21.4.) The function which sends the point (s, t) to the point (x, y, z) is called the *parameterization of the surface*.

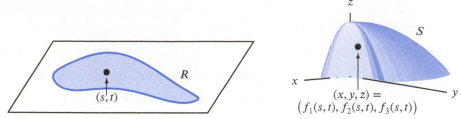

Figure 21.4: The parameterization sends each point (s, t) in the parameter region, R, to a point $(x, y, z) = (f_1(s, t), f_2(s, t), f_3(s, t))$ in the surface, S

Using Position Vectors

We can use the position vector $\vec{r} = x\vec{i} + y\vec{j} + z\vec{k}$ to combine the three parametric equations for a surface into a single vector equation. For example, the parameterization of the cylinder $x = \cos t$, $y = \sin t$, $z = z$ can be written as

$$\vec{r}(t, z) = \cos t\, \vec{i} + \sin t\, \vec{j} + z\vec{k} \qquad 0 \le t < 2\pi, \quad -\infty < z < \infty.$$

For a general parameterized surface S, we write

$$\vec{r}(s, t) = f_1(s, t)\vec{i} + f_2(s, t)\vec{j} + f_3(s, t)\vec{k}.$$

Parameterizing a Surface of the Form $z = f(x, y)$

The graph of a function $z = f(x, y)$ can be given parametrically simply by letting the parameters s and t be x and y:

$$x = s, \quad y = t, \quad z = f(s, t).$$

Example 1

Give a parametric description of the lower hemisphere of the sphere $x^2 + y^2 + z^2 = 1$.

Solution

The surface is the graph of the function $z = -\sqrt{1 - x^2 - y^2}$ over the region $x^2 + y^2 \le 1$ in the plane. Then parametric equations are $x = s$, $y = t$, $z = -\sqrt{1 - s^2 - t^2}$, where the parameters s and t vary inside and on the unit circle.

In practice we often think of x and y as parameters rather than introduce new parameters s and t. Thus, we may write $x = x$, $y = y$, $z = f(x, y)$.

Parameterizing Planes

Consider a plane containing two nonparallel vectors $\vec{v}_1$ and $\vec{v}_2$ and a point P_0 with position vector $\vec{r}_0$. We can get to any point on the plane by starting at P_0 and moving parallel to $\vec{v}_1$ or $\vec{v}_2$, adding multiples of them to $\vec{r}_0$. (See Figure 21.5.)

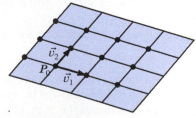

Figure 21.5: The plane $\vec{r}(s, t) = \vec{r}_0 + s\vec{v}_1 + t\vec{v}_2$ and some points corresponding to various choices of s and t

Since $s\vec{v}_1$ is parallel to $\vec{v}_1$ and $t\vec{v}_2$ is parallel to $\vec{v}_2$, we have the following result:

Parameterizing a Plane

The plane through the point with position vector $\vec{r}_0$ and containing the two nonparallel vectors $\vec{v}_1$ and $\vec{v}_2$ has parameterization

$$\vec{r}(s,t) = \vec{r}_0 + s\vec{v}_1 + t\vec{v}_2.$$

If $\vec{r}_0 = x_0\vec{i} + y_0\vec{j} + z_0\vec{k}$, and $\vec{v}_1 = a_1\vec{i} + a_2\vec{j} + a_3\vec{k}$, and $\vec{v}_2 = b_1\vec{i} + b_2\vec{j} + b_3\vec{k}$, then the parameterization of the plane can be expressed with the parametric equations

$$x = x_0 + sa_1 + tb_1, \quad y = y_0 + sa_2 + tb_2, \quad z = z_0 + sa_3 + tb_3.$$

Notice that the parameterization of the plane expresses the coordinates x, y, and z as linear functions of the parameters s and t.

Example 2 Write a parameterization for the plane through the point $(2, -1, 3)$ and containing the vectors $\vec{v}_1 = 2\vec{i} + 3\vec{j} - \vec{k}$ and $\vec{v}_2 = \vec{i} - 4\vec{j} + 5\vec{k}$.

Solution A possible parameterization is

$$\vec{r}(s,t) = \vec{r}_0 + s\vec{v}_1 + t\vec{v}_2 = 2\vec{i} - \vec{j} + 3\vec{k} + s(2\vec{i} + 3\vec{j} - \vec{k}) + t(\vec{i} - 4\vec{j} + 5\vec{k})$$

$$= (2 + 2s + t)\vec{i} + (-1 + 3s - 4t)\vec{j} + (3 - s + 5t)\vec{k},$$

or equivalently,

$$x = 2 + 2s + t, \quad y = -1 + 3s - 4t, \quad z = 3 - s + 5t.$$

Parameterizations Using Spherical Coordinates

Recall the spherical coordinates ρ, ϕ, and θ introduced on page 924 of Chapter 16. On a sphere of radius $\rho = a$ we can use ϕ and θ as coordinates, similar to latitude and longitude on the surface of the earth. (See Figure 21.6.) The latitude, however, is measured from the equator, whereas ϕ is measured from the north pole. If the positive x-axis passes through the Greenwich meridian, the longitude and θ are equal for $0 \leq \theta \leq \pi$.

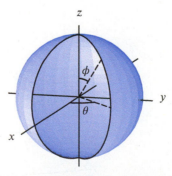

Figure 21.6: Parameterizing the sphere by ϕ and θ

Example 3   You are at a point on a sphere with $\phi = 3\pi/4$. Are you in the northern or southern hemisphere? If ϕ decreases, do you move closer to or farther from the equator?

Solution   The equator has $\phi = \pi/2$. Since $3\pi/4 > \pi/2$, you are in the southern hemisphere. If ϕ decreases, you move closer to the equator.

Example 4   On a sphere, you are standing at a point with coordinates θ_0 and ϕ_0. Your *antipodal* point is the point on the other side of the sphere on a line through you and the center. What are the θ, ϕ coordinates of your antipodal point?

Solution   Figure 21.7 shows that the coordinates are $\theta = \theta_0 + \pi$ if $\theta_0 < \pi$ or $\theta = \theta_0 - \pi$ if $\pi \le \theta_0 \le 2\pi$, and $\phi = \pi - \phi_0$. Notice that if you are on the equator, then so is your antipodal point.

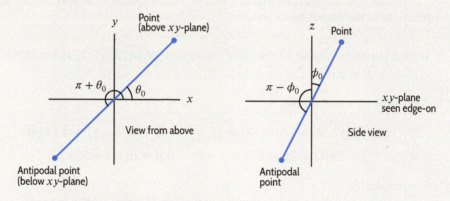

Figure 21.7: Two views of the xyz-coordinate system showing coordinates of antipodal points

Parameterizing a Sphere Using Spherical Coordinates

The sphere with radius 1 centered at the origin is parameterized by

$$x = \sin\phi\cos\theta, \qquad y = \sin\phi\sin\theta, \qquad z = \cos\phi,$$

where $0 \le \theta \le 2\pi$ and $0 \le \phi \le \pi$. (See Figure 21.8.)

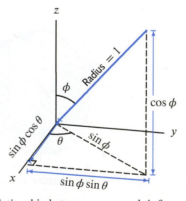

Figure 21.8: The relationship between x, y, z and ϕ, θ on a sphere of radius 1

We can also write these equations in vector form:

$$\vec{r}(\theta, \phi) = \sin\phi\cos\theta\,\vec{i} + \sin\phi\sin\theta\,\vec{j} + \cos\phi\,\vec{k}.$$

Since $x^2 + y^2 + z^2 = \sin^2\phi(\cos^2\theta + \sin^2\theta) + \cos^2\phi = \sin^2\phi + \cos^2\phi = 1$, this verifies that the point with position vector $\vec{r}(\theta, \phi)$ does lie on the sphere of radius 1. Notice that the z-coordinate depends only on the parameter ϕ. Geometrically, this means that all points on the same latitude have the same z-coordinate.

Example 5   Find parametric equations for the following spheres:

(a) Center at the origin and radius 2.
(b) Center at the point with Cartesian coordinates $(2, -1, 3)$ and radius 2.

Solution   (a) We must scale the distance from the origin by 2. Thus, we have

$$x = 2\sin\phi\cos\theta, \qquad y = 2\sin\phi\sin\theta, \qquad z = 2\cos\phi,$$

where $0 \le \theta \le 2\pi$ and $0 \le \phi \le \pi$. In vector form, this is written

$$\vec{r}(\theta, \phi) = 2\sin\phi\cos\theta\,\vec{i} + 2\sin\phi\sin\theta\,\vec{j} + 2\cos\phi\,\vec{k}.$$

(b) To shift the center of the sphere from the origin to the point $(2, -1, 3)$, we add the vector parameterization we found in part (a) to the position vector of $(2, -1, 3)$. (See Figure 21.9.) This gives

$$\vec{r}(\theta, \phi) = 2\vec{i} - \vec{j} + 3\vec{k} + (2\sin\phi\cos\theta\,\vec{i} + 2\sin\phi\sin\theta\,\vec{j} + 2\cos\phi\,\vec{k})$$
$$= (2 + 2\sin\phi\cos\theta)\vec{i} + (-1 + 2\sin\phi\sin\theta)\vec{j} + (3 + 2\cos\phi)\vec{k},$$

where $0 \le \theta \le 2\pi$ and $0 \le \phi \le \pi$. Alternatively,

$$x = 2 + 2\sin\phi\cos\theta, \qquad y = -1 + 2\sin\phi\sin\theta, \qquad z = 3 + 2\cos\phi.$$

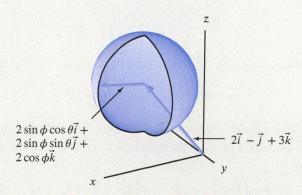

$2\sin\phi\cos\theta\,\vec{i} +$
$2\sin\phi\sin\theta\,\vec{j} +$
$2\cos\phi\,\vec{k}$

$2\vec{i} - \vec{j} + 3\vec{k}$

Figure 21.9: Sphere with center at the point $(2, -1, 3)$ and radius 2

Note that the same point can have more than one value for θ or ϕ. For example, points with $\theta = 0$ also have $\theta = 2\pi$, unless we restrict θ to the range $0 \le \theta < 2\pi$. Also, the north pole, at $\phi = 0$, and the south pole, at $\phi = \pi$, can have any value of θ.

Parameterizing Surfaces of Revolution

Many surfaces have an axis of rotational symmetry and circular cross sections perpendicular to that axis. These surfaces are referred to as *surfaces of revolution*.

Example 6   Find a parameterization of the cone whose base is the circle $x^2 + y^2 = a^2$ in the xy-plane and whose vertex is at height h above the xy-plane. (See Figure 21.10.)

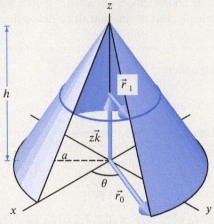
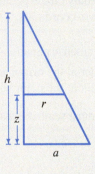

Figure 21.10: The cone whose base is the circle $x^2 + y^2 = a^2$ in the xy-plane and whose vertex is at the point $(0, 0, h)$ and the vertical cross section through the cone

Solution   We use cylindrical coordinates, r, θ, z. (See Figure 21.10.) In the xy-plane, the radius vector, $\vec{r}_0$, from the z-axis to a point on the cone in the xy-plane is

$$\vec{r}_0 = a \cos \theta \vec{i} + a \sin \theta \vec{j}.$$

Above the xy-plane, the radius of the circular cross section, r, decreases linearly from $r = a$ when $z = 0$ to $r = 0$ when $z = h$. From the similar triangles in Figure 21.10,

$$\frac{a}{h} = \frac{r}{h - z}.$$

Solving for r, we have

$$r = \left(1 - \frac{z}{h}\right) a.$$

The horizontal radius vector, $\vec{r}_1$, at height z has components similar to $\vec{r}_0$, but with a replaced by r:

$$\vec{r}_1 = r \cos \theta \vec{i} + r \sin \theta \vec{j} = \left(1 - \frac{z}{h}\right) a \cos \theta \vec{i} + \left(1 - \frac{z}{h}\right) a \sin \theta \vec{j}.$$

As θ goes from 0 to 2π, the vector $\vec{r}_1$ traces out the horizontal circle in Figure 21.10. We get the position vector, $\vec{r}$, of a point on the cone by adding the vector $z\vec{k}$, so

$$\vec{r} = \vec{r}_1 + z\vec{k} = a\left(1 - \frac{z}{h}\right) \cos \theta \vec{i} + a\left(1 - \frac{z}{h}\right) \sin \theta \vec{j} + z\vec{k}, \quad \text{for } 0 \le z \le h \text{ and } 0 \le \theta \le 2\pi.$$

These equations can be written as

$$x = \left(1 - \frac{z}{h}\right) a \cos \theta, \quad y = \left(1 - \frac{z}{h}\right) a \sin \theta, \quad z = z.$$

The parameters are θ and z.

Example 7   Consider the bell of a trumpet. A model for the radius $z = f(x)$ of the horn (in cm) at a distance x cm from the large open end is given by the function

$$f(x) = \frac{6}{(x+1)^{0.7}}.$$

The bell is obtained by rotating the graph of f about the x-axis. Find a parameterization for the first 24 cm of the bell. (See Figure 21.11.)

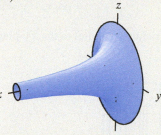

Figure 21.11: The bell of a trumpet obtained by rotating the graph of $z = f(x)$ about the x-axis

Solution   At distance x from the large open end of the horn, the cross section parallel to the yz-plane is a circle of radius $f(x)$, with center on the x-axis. Such a circle can be parameterized by $y = f(x)\cos\theta$, $z = f(x)\sin\theta$. Thus, we have the parameterization

$$x = x, \quad y = \left(\frac{6}{(x+1)^{0.7}}\right)\cos\theta, \quad z = \left(\frac{6}{(x+1)^{0.7}}\right)\sin\theta, \quad 0 \le x \le 24, \quad 0 \le \theta \le 2\pi.$$

The parameters are x and θ.

Parameter Curves

On a parameterized surface, the curve obtained by setting one of the parameters equal to a constant and letting the other vary is called a *parameter curve*. If the surface is parameterized by

$$\vec{r}(s,t) = f_1(s,t)\vec{i} + f_2(s,t)\vec{j} + f_3(s,t)\vec{k},$$

there are two families of parameter curves on the surface, one family with t constant and the other with s constant.

Example 8   Consider the vertical cylinder

$$x = \cos t, \quad y = \sin t, \quad z = z.$$

(a) Describe the two parameter curves through the point $(0, 1, 1)$.
(b) Describe the family of parameter curves with t constant and the family with z constant.

Solution   (a) Since the point $(0, 1, 1)$ corresponds to the parameter values $t = \pi/2$ and $z = 1$, there are two parameter curves, one with $t = \pi/2$ and the other with $z = 1$. The parameter curve with $t = \pi/2$ has the parametric equations

$$x = \cos\left(\frac{\pi}{2}\right) = 0, \quad y = \sin\left(\frac{\pi}{2}\right) = 1, \quad z = z,$$

with parameter z. This is a line through the point $(0, 1, 1)$ parallel to the z-axis.
    The parameter curve with $z = 1$ has the parametric equations

$$x = \cos t, \quad y = \sin t, \quad z = 1,$$

with parameter t. This is a unit circle parallel to and one unit above the xy-plane centered on the z-axis.

(b) First, fix $t = t_0$ for t and let z vary. The curves parameterized by z have equations

$$x = \cos t_0, \quad y = \sin t_0, \quad z = z.$$

These are vertical lines on the cylinder parallel to the z-axis. (See Figure 21.12.)

The other family is obtained by fixing $z = z_0$ and varying t. Curves in this family are parameterized by t and have equations

$$x = \cos t, \quad y = \sin t, \quad z = z_0.$$

They are circles of radius 1 parallel to the xy-plane centered on the z-axis. (See Figure 21.13.)

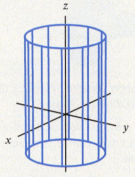
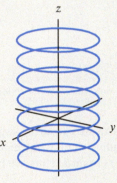

Figure 21.12: The family of parameter curves with $t = t_0$ for the cylinder $x = \cos t, y = \sin t, z = z$

Figure 21.13: The family of parameter curves with $z = z_0$ for the cylinder $x = \cos t, y = \sin t, z = z$

Example 9   Describe the families of parameter curves with $\theta = \theta_0$ and $\phi = \phi_0$ for the sphere

$$x = \sin \phi \cos \theta, \quad y = \sin \phi \sin \theta, \quad z = \cos \phi,$$

where $0 \le \theta \le 2\pi, 0 \le \phi \le \pi$.

Solution   Since ϕ measures latitude, the family with ϕ constant consists of the circles of constant latitude. (See Figure 21.14.) Similarly, the family with θ constant consists of the meridians (semicircles) running between the north and south poles. (See Figure 21.15.)

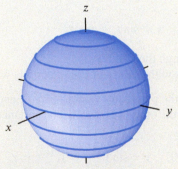
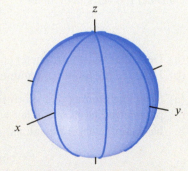

Figure 21.14: The family of parameter curves with $\phi = \phi_0$ for the sphere parameterized by (θ, ϕ)

Figure 21.15: The family of parameter curves with $\theta = \theta_0$ for the sphere parameterized by (θ, ϕ)

We have seen parameter curves before, on pages 703–705 of Section 12.2: The cross sections with $x = a$ or $y = b$ on a surface $z = f(x, y)$ are examples of parameter curves. So are the grid lines on a computer sketch of a surface. The small regions shaped like parallelograms surrounded by nearby pairs of parameter curves are called *parameter rectangles*. See Figure 21.16.

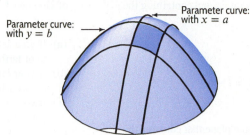

Parameter curve: with $y = b$

Parameter curve: with $x = a$

Figure 21.16: Parameter curves $x = a$ or $y = b$ on a surface $z = f(x, y)$; the darker region is a parameter rectangle

Summary for Section 21.1

- **Parameterized surfaces:** In general, we express the coordinates, (x, y, z), of a point on a surface S in terms of two parameters, s and t:

$$x = f_1(s, t), \quad y = f_2(s, t), \quad z = f_3(s, t).$$

- Using the **position vector**, we can express a parameterization for a general surface S as

$$\vec{r}(s, t) = f_1(s, t)\vec{i} + f_2(s, t)\vec{j} + f_3(s, t)\vec{k}.$$

- **Examples of parameterized surfaces:**
  - A **surface of the form** $z = f(x, y)$ can be parameterized by $x = s, y = t, z = f(s, t)$.
  - A **plane** through the point with position vector $\vec{r}_0$ and containing the two nonparallel vectors $\vec{v}_1$ and $\vec{v}_2$ can be parameterized by $\vec{r}(s, t) = \vec{r}_0 + s\vec{v}_1 + t\vec{v}_2$.
  - A **cylinder** with radius 1 centered around the z-axis can be parameterized by $x = \cos t, y = \sin t, z = z$, where $0 \leq t < 2\pi$ and $-\infty < z < \infty$.
  - A **sphere** with radius 1 centered at the origin can be parameterized by $x = \sin\phi\cos\theta, y = \sin\phi\sin\theta, z = \cos\phi$, where $0 \leq \theta \leq 2\pi$ and $0 \leq \phi \leq \pi$.

- **Parameter curves:** A curve obtained by setting one of the parameters equal to a constant and letting the other vary is called a parameter curve.

Exercises and Problems for Section 21.1

EXERCISES

In Exercises 1–4 decide if the parameterization describes a curve or a surface.

1. $\vec{r}(s) = s\vec{i} + (3 - s)\vec{j} + s^2\vec{k}$
2. $\vec{r}(s, t) = (s + t)\vec{i} + (3 - s)\vec{j}$
3. $\vec{r}(s, t) = \cos s\,\vec{i} + \sin s\,\vec{j} + t^2\vec{k}$
4. $\vec{r}(s) = \cos s\,\vec{i} + \sin s\,\vec{j} + s^2\vec{k}$

Describe in words the objects parameterized by the equations in Exercises 5–8. (Note: r and θ are cylindrical coordinates.)

5. $x = r\cos\theta \qquad y = r\sin\theta \qquad z = 7$
   $0 \leq r \leq 5 \qquad 0 \leq \theta \leq 2\pi$

6. $x = 5\cos\theta \qquad y = 5\sin\theta \qquad z = z$
   $0 \leq \theta \leq 2\pi \qquad 0 \leq z \leq 7$

7. $x = 5\cos\theta \qquad y = 5\sin\theta \qquad z = 5\theta$
   $0 \leq \theta \leq 2\pi$

8. $x = r\cos\theta \qquad y = r\sin\theta \qquad z = r$
   $0 \leq r \leq 5 \qquad 0 \leq \theta \leq 2\pi$

In Exercises 9–12, for a sphere parameterized using the spherical coordinates θ and ϕ, describe in words the part of the sphere given by the restrictions.

9. $0 \leq \theta < 2\pi, \quad 0 \leq \phi \leq \pi/2$

10. $\pi \leq \theta < 2\pi, \quad 0 \leq \phi \leq \pi$

11. $\pi/4 \leq \theta < \pi/3, \quad 0 \leq \phi \leq \pi$

12. $0 \leq \theta \leq \pi, \quad \pi/4 \leq \phi < \pi/3$

PROBLEMS

In Problems 13–16, give parametric equations for the plane through the point with position vector $\vec{r}_0$ and containing the vectors $\vec{v}_1$ and $\vec{v}_2$.

13. $\vec{r}_0 = \vec{i}, \vec{v}_1 = \vec{j}, \vec{v}_2 = \vec{k}$
14. $\vec{r}_0 = \vec{j}, \vec{v}_1 = \vec{k}, \vec{v}_2 = \vec{i}$
15. $\vec{r}_0 = \vec{i} + \vec{j}, \vec{v}_1 = \vec{j} + \vec{k}, \vec{v}_2 = \vec{i} + \vec{k}$
16. $\vec{r}_0 = \vec{i} + \vec{j} + \vec{k}, \vec{v}_1 = \vec{i} - \vec{k}, \vec{v}_2 = -\vec{j} + \vec{k}$

In Problems 17–18, parameterize the plane that contains the three points.

17. $(0, 0, 0), (1, 2, 3), (2, 1, 0)$
18. $(1, 2, 3), (2, 5, 8), (5, 2, 0)$

In Problems 19–20, give two nonparallel vectors and the coordinates of a point in the plane with given parametric equations

19. $x = 2s + 3t, \quad y = s - 5t, \quad z = -s + 2t$
20. $x = 2 + s + t, \quad y = s - t, \quad z = -1 + s + t$

In Problems 21–22, parameterize the plane through the point with the given normal vector.

21. $(3, 5, 7), \vec{i} + \vec{j} + \vec{k}$
22. $(5, 1, 4), \vec{i} + 2\vec{j} + 3\vec{k}$
23. Does the plane $\vec{r}(s, t) = (2 + s)\vec{i} + (3 + s + t)\vec{j} + 4t\vec{k}$ contain the following points?
    (a) $(4, 8, 12)$ (b) $(1, 2, 3)$
24. Are the following two planes parallel?

    $$x = 2 + s + t, \quad y = 4 + s - t, \quad z = 1 + 2s, \quad \text{and}$$

    $$x = 2 + s + 2t, \quad y = t, \quad z = s - t.$$

In Problems 25–28, describe the families of parameter curves with $s = s_0$ and $t = t_0$ for the parameterized surface.

25. $x = s, \quad y = t, \quad z = 1$ for $-\infty < s < \infty, -\infty < t < \infty$
26. $x = s, \quad y = \cos t, \quad z = \sin t$ for $-\infty < s < \infty, 0 \le t \le 2\pi$
27. $x = s \quad y = t, \quad z = s^2 + t^2$ for $-\infty < s < \infty, -\infty < t < \infty$
28. $x = \cos s \sin t, \quad y = \sin s \sin t, \quad z = \cos t$ for $0 \le s \le 2\pi, 0 \le t \le \pi$
29. A city is described parametrically by the equation

    $$\vec{r} = (x_0\vec{i} + y_0\vec{j} + z_0\vec{k}) + s\vec{v}_1 + t\vec{v}_2$$

    where $\vec{v}_1 = 2\vec{i} - 3\vec{j} + 2\vec{k}$ and $\vec{v}_2 = \vec{i} + 4\vec{j} + 5\vec{k}$. A city block is a rectangle determined by $\vec{v}_1$ and $\vec{v}_2$. East is in the direction of $\vec{v}_1$ and north is in the direction of $\vec{v}_2$. Starting at the point (x_0, y_0, z_0), you walk 5 blocks east, 4 blocks north, 1 block west and 2 blocks south. What are the parameters of the point where you end up? What are your x, y and z coordinates at that point?

30. You are at a point on the earth with longitude 80° West of Greenwich, England, and latitude 40° North of the equator.

    (a) If your latitude decreases, have you moved nearer to or farther from the equator?
    (b) If your latitude decreases, have you moved nearer to or farther from the north pole?
    (c) If your longitude increases (say, to 90° West), have you moved nearer to or farther from Greenwich?

31. Describe in words the curve $\phi = \pi/4$ on the surface of the globe.

32. Describe in words the curve $\theta = \pi/4$ on the surface of the globe.

33. A decorative oak post is 48″ long and is turned on a lathe so that its profile is sinusoidal, as shown in Figure 21.17.

    (a) Describe the surface of the post parametrically using cylindrical coordinates.
    (b) Find the volume of the post.

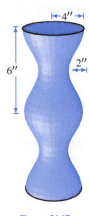

Figure 21.17

34. Find parametric equations for the sphere $(x - a)^2 + (y - b)^2 + (z - c)^2 = d^2$.

35. You are standing at a point on the equator of a sphere parameterized by spherical coordinates θ_0 and ϕ_0. If you go halfway around the equator and halfway up toward the north pole along a longitude, what are your new θ and ϕ coordinates?

36. Find parametric equations for the cone $x^2 + y^2 = z^2$.

37. Parameterize the cone in Example 6 on page 1084 in terms of r and θ.

38. Give a parameterization of the circle of radius a centered at the point (x_0, y_0, z_0) and in the plane parallel to two given unit vectors $\vec{u}$ and $\vec{v}$ such that $\vec{u} \cdot \vec{v} = 0$.

■ For Problems 39–41,
  (a) Write an equation in x, y, z and identify the parametric surface.
  (b) Draw a picture of the surface.

39. $x = 2s \qquad y = s + t \qquad z = 1 + s - t$
$0 \leq s \leq 1 \qquad 0 \leq t \leq 1$

40. $x = s \qquad y = t \qquad z = \sqrt{1 - s^2 - t^2}$
$s^2 + t^2 \leq 1 \qquad s, t \geq 0$

41. $x = s + t \qquad y = s - t \qquad z = s^2 + t^2$
$0 \leq s \leq 1 \qquad 0 \leq t \leq 1$

Strengthen Your Understanding

■ In Problems 42–43, explain what is wrong with the statement.

42. The parameter curves of a parameterized surface intersect at right angles.

43. The parameter curves for constant ϕ on the sphere $\vec{r}(\theta, \phi) = R \sin \phi \cos \theta \vec{i} + R \sin \phi \sin \theta \vec{j} + R \cos \phi \vec{k}$ are circles of radius R.

■ In Problems 44–46, give an example of:

44. A parameterization $\vec{r}(s, t)$ of the plane tangent to the unit sphere at the point where $\theta = \pi/4$ and $\phi = \pi/4$.

45. An equation of the form $f(x, y, z) = 0$ for the plane
$$\vec{r}(s, t) = (s + 1)\vec{i} + (t + 2)\vec{j} + (s + t)\vec{k}.$$

46. A parameterized curve on the sphere $\vec{r}(\theta, \phi) = \sin \phi \cos \theta \vec{i} + \sin \phi \sin \theta \vec{j} + \cos \phi \vec{k}$ that is not a parameter curve.

■ Are the statements in Problems 47–53 true or false? Give reasons for your answer.

47. The equations $x = s + 1$, $y = t - 2$, $z = 3$ parameterize a plane.

48. The equations $x = 2s - 1$, $y = -s + 3$, $z = 4 + s$ parameterize a plane.

49. If $\vec{r} = \vec{r}(s, t)$ parameterizes the upper hemisphere $x^2 + y^2 + z^2 = 1$, $z \geq 0$, then $\vec{r} = -\vec{r}(s, t)$ parameterizes the lower hemisphere $x^2 + y^2 + z^2 = 1$, $z \leq 0$.

50. If $\vec{r} = \vec{r}(s, t)$ parameterizes the upper hemisphere $x^2 + y^2 + z^2 = 1$, $z \geq 0$, then $\vec{r} = \vec{r}(-s, -t)$ parameterizes the lower hemisphere $x^2 + y^2 + z^2 = 1$, $z \leq 0$.

51. If $\vec{r}_1(s, t)$ parameterizes a plane then $\vec{r}_2(s, t) = \vec{r}_1(s, t) + 2\vec{i} - 3\vec{j} + \vec{k}$ parameterizes a parallel plane.

52. Every point on a parameterized surface has a parameter curve passing through it.

53. If $s_0 \neq s_1$, then the parameter curves $\vec{r}(s_0, t)$ and $\vec{r}(s_1, t)$ do not intersect.

54. Match the parameterizations (I)–(IV) with the surfaces (a)–(d). In all cases $0 \leq s \leq \pi/2$, $0 \leq t \leq \pi/2$. Note that only part of the surface may be described by the given parameterization.

  (a) Cylinder
  (b) Plane
  (c) Sphere
  (d) Cone

  I. $x = \cos s$, $y = \sin t$, $z = \cos s + \sin t$
  II. $x = \cos s$, $y = \sin s$, $z = \cos t$
  III. $x = \sin s \cos t$, $y = \sin s \sin t$, $z = \cos s$
  IV. $x = \cos s$, $y = \sin t$, $z = \sqrt{\cos^2 s + \sin^2 t}$

21.2 CHANGE OF COORDINATES IN A MULTIPLE INTEGRAL

In Chapter 16 we used polar, cylindrical, and spherical coordinates to simplify iterated integrals. In this section, we discuss more general changes of coordinate. In the process, we see where the factors r and $\rho^2 \sin \phi$ come from when we convert to polar, cylindrical, or spherical coordinates (see pages 917, 923, and 925).

Polar Change of Coordinates Revisited

Consider the integral $\int_R (x + y) \, dA$ where R is the region in the first quadrant bounded by the circle $x^2 + y^2 = 16$ and the x and y-axes. Writing the integral in Cartesian and polar coordinates, we have

$$\int_R (x + y) \, dA = \int_0^4 \int_0^{\sqrt{16 - x^2}} (x + y) \, dy \, dx = \int_0^{\pi/2} \int_0^4 (r \cos \theta + r \sin \theta) r \, dr \, d\theta.$$

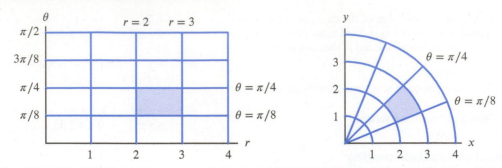

Figure 21.18: A grid in the $r\theta$-plane and the corresponding curved grid in the xy-plane

The integral on the right is over the rectangle in the $r\theta$-plane given by $0 \leq r \leq 4$, $0 \leq \theta \leq \pi/2$. The conversion from polar to Cartesian coordinates changes this rectangle into a quarter-disk. Figure 21.18 shows how a typical rectangle (shaded) in the $r\theta$-plane with sides of length Δr and $\Delta \theta$ corresponds to a curved rectangle in the xy-plane with sides of length Δr and $r\Delta\theta$. The extra r is needed because the correspondence between r, θ and x, y not only curves the lines $r = 1, 2, 3 \ldots$ into circles, it also stretches those lines around larger and larger circles.

General Change of Coordinates

We now consider a general change of coordinates, where x, y coordinates are related to s, t coordinates by the differentiable functions

$$x = x(s, t) \quad y = y(s, t).$$

Just as a rectangular region in the $r\theta$-plane corresponds to a region in the xy-plane, a rectangular region, T, in the st-plane corresponds to a region, R, in the xy-plane. We assume that the change of coordinates is one-to-one, that is, that each point in R corresponds to only one point in T.

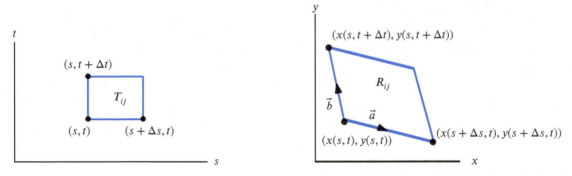

Figure 21.19: A small rectangle T_{ij} in the st-plane and the corresponding region R_{ij} of the xy-plane

We divide T into small rectangles T_{ij} with sides of length Δs and Δt. (See Figure 21.19.) The corresponding piece R_{ij} of the xy-plane is a quadrilateral with curved sides. If we choose Δs and Δt small, then by local linearity of $x(s, t)$ and $y(s, t)$, we know R_{ij} is approximately a parallelogram.

Recall from Chapter 13 that the area of the parallelogram with sides $\vec{a}$ and $\vec{b}$ is $\|\vec{a} \times \vec{b}\|$. Thus, we need to find the sides of R_{ij} as vectors. The side of R_{ij} corresponding to the bottom side of T_{ij} has endpoints $(x(s, t), y(s, t))$ and $(x(s + \Delta s, t), y(s + \Delta s, t))$, so in vector form that side is

$$\vec{a} = (x(s + \Delta s, t) - x(s, t))\vec{i} + (y(s + \Delta s, t) - y(s, t))\vec{j} \approx \left(\frac{\partial x}{\partial s}\Delta s\right)\vec{i} + \left(\frac{\partial y}{\partial s}\Delta s\right)\vec{j}.$$

Similarly, the side of R_{ij} corresponding to the left edge of T_{ij} is given by

$$\vec{b} \approx \left(\frac{\partial x}{\partial t}\Delta t\right)\vec{i} + \left(\frac{\partial y}{\partial t}\Delta t\right)\vec{j}.$$

Computing the cross product, we get

$$\text{Area } R_{ij} \approx \|\vec{a} \times \vec{b}\| \approx \left| \left(\frac{\partial x}{\partial s} \Delta s \right) \left(\frac{\partial y}{\partial t} \Delta t \right) - \left(\frac{\partial x}{\partial t} \Delta t \right) \left(\frac{\partial y}{\partial s} \Delta s \right) \right|$$

$$= \left| \frac{\partial x}{\partial s} \cdot \frac{\partial y}{\partial t} - \frac{\partial x}{\partial t} \cdot \frac{\partial y}{\partial s} \right| \Delta s \Delta t.$$

Using determinant notation,[1] we define the *Jacobian*, $\dfrac{\partial(x, y)}{\partial(s, t)}$, as follows:

$$\frac{\partial(x, y)}{\partial(s, t)} = \frac{\partial x}{\partial s} \cdot \frac{\partial y}{\partial t} - \frac{\partial x}{\partial t} \cdot \frac{\partial y}{\partial s} = \begin{vmatrix} \dfrac{\partial x}{\partial s} & \dfrac{\partial x}{\partial t} \\[2mm] \dfrac{\partial y}{\partial s} & \dfrac{\partial y}{\partial t} \end{vmatrix}.$$

Thus, we can write

$$\text{Area } R_{ij} \approx \left| \frac{\partial(x, y)}{\partial(s, t)} \right| \Delta s \, \Delta t.$$

To compute $\int_R f(x, y) \, dA$, where f is a continuous function, we look at the Riemann sum obtained by dividing the region R into the small curved regions R_{ij}, giving

$$\int_R f(x, y) \, dA \approx \sum_{i,j} f(u_{ij}, v_{ij}) \cdot \text{Area of } R_{ij} \approx \sum_{i,j} f(u_{ij}, v_{ij}) \left| \frac{\partial(x, y)}{\partial(s, t)} \right| \Delta s \, \Delta t.$$

Each point (u_{ij}, v_{ij}) in R_{ij} corresponds to a point (s_{ij}, t_{ij}) in T_{ij}, so the sum can be written in terms of s and t:

$$\sum_{i,j} f(x(s_{ij}, t_{ij}), y(s_{ij}, t_{ij})) \left| \frac{\partial(x, y)}{\partial(s, t)} \right| \Delta s \, \Delta t.$$

This is a Riemann sum in terms of s and t, so as Δs and Δt approach 0, we get

$$\int_R f(x, y) \, dA = \int_T f(x(s, t), y(s, t)) \left| \frac{\partial(x, y)}{\partial(s, t)} \right| ds \, dt.$$

> To convert an integral from x, y to s, t coordinates we make three changes:
> 1. Substitute for x and y in the integrand in terms of s and t.
> 2. Change the xy region R into an st region T.
> 3. Use the absolute value of the Jacobian to change the area element by making the substitution $dx \, dy = \left| \dfrac{\partial(x, y)}{\partial(s, t)} \right| ds \, dt.$

Example 1   Check that the Jacobian $\dfrac{\partial(x, y)}{\partial(r, \theta)} = r$ for polar coordinates $x = r \cos \theta, \ y = r \sin \theta.$

Solution   We have $\dfrac{\partial(x, y)}{\partial(r, \theta)} = \begin{vmatrix} \dfrac{\partial x}{\partial r} & \dfrac{\partial x}{\partial \theta} \\[2mm] \dfrac{\partial y}{\partial r} & \dfrac{\partial y}{\partial \theta} \end{vmatrix} = \begin{vmatrix} \cos \theta & -r \sin \theta \\ \sin \theta & r \cos \theta \end{vmatrix} = r \cos^2 \theta + r \sin^2 \theta = r.$

Example 2   Find the area of the ellipse $\dfrac{x^2}{a^2} + \dfrac{y^2}{b^2} = 1.$

---

[1] See Appendix E. Carl Gustav Jacob Jacobi (1804–1851) was a German mathematician.

Solution   Let $x = as$, $y = bt$. Then the ellipse $x^2/a^2 + y^2/b^2 = 1$ in the xy-plane corresponds to the circle $s^2 + t^2 = 1$ in the st-plane. The Jacobian is $\begin{vmatrix} a & 0 \\ 0 & b \end{vmatrix} = ab$. Thus, if R is the ellipse in the xy-plane and T is the unit circle in the st-plane, we get

$$\text{Area of } xy\text{-ellipse} = \int_R 1 \, dA = \int_T 1 \, ab \, ds \, dt = ab \int_T ds \, dt = ab \cdot \text{Area of } st\text{-circle} = \pi ab.$$

Change of Coordinates in Triple Integrals

For triple integrals, there is a similar formula. Suppose the differentiable functions

$$x = x(s,t,u), \quad y = y(s,t,u), \quad z = z(s,t,u)$$

define a one-to-one change of coordinates from a region S in stu-space to a region W in xyz-space. Then, the Jacobian of this change of coordinates is given by the determinant[2]

$$\frac{\partial(x,y,z)}{\partial(s,t,u)} = \begin{vmatrix} \frac{\partial x}{\partial s} & \frac{\partial x}{\partial t} & \frac{\partial x}{\partial u} \\ \frac{\partial y}{\partial s} & \frac{\partial y}{\partial t} & \frac{\partial y}{\partial u} \\ \frac{\partial z}{\partial s} & \frac{\partial z}{\partial t} & \frac{\partial z}{\partial u} \end{vmatrix}.$$

Just as the Jacobian in two dimensions gives us the change in the area element, the Jacobian in three dimensions represents the change in the volume element. Thus, we have

$$\int_W f(x,y,z) \, dx \, dy \, dz = \int_S f(x(s,t,u), y(s,t,u), z(s,t,u)) \left| \frac{\partial(x,y,z)}{\partial(s,t,u)} \right| ds \, dt \, du.$$

Problem 11 at the end of this section asks you to check that the Jacobian for the change of coordinates to spherical coordinates is $\rho^2 \sin \phi$. The next example generalizes Example 2 to ellipsoids.

Example 3   Find the volume of the ellipsoid $\dfrac{x^2}{a^2} + \dfrac{y^2}{b^2} + \dfrac{z^2}{c^2} = 1$.

Solution   Let $x = as$, $y = bt$, $z = cu$. The Jacobian is computed to be abc. The xyz-ellipsoid corresponds to the stu-sphere $s^2 + t^2 + u^2 = 1$. Thus, as in Example 2,

$$\text{Volume of } xyz\text{-ellipsoid} = abc \cdot \text{Volume of } stu\text{-sphere} = abc \frac{4}{3}\pi = \frac{4}{3}\pi abc.$$

Summary for Section 21.2

- **The Jacobian** is defined using determinant notation as $\dfrac{\partial(x,y)}{\partial(s,t)} = \begin{vmatrix} \frac{\partial x}{\partial s} & \frac{\partial x}{\partial t} \\ \frac{\partial y}{\partial s} & \frac{\partial y}{\partial t} \end{vmatrix}.$

- **Change of coordinates:** To convert an integral from x, y to s, t coordinates we make three changes:

    1. Substitute for x and y in the integrand in terms of s and t.
    2. Change the xy region R into an st region T.
    3. Change the area element by making the substitution $dx \, dy = \left| \dfrac{\partial(x,y)}{\partial(s,t)} \right| ds \, dt$.

---

[2] See Appendix E.

Exercises and Problems for Section 21.2

EXERCISES

■ In Exercises 1–4, find the absolute value of the Jacobian, $\left|\frac{\partial(x,y)}{\partial(s,t)}\right|$, for the given change of coordinates.

1. $x = 5s + 2t, y = 3s + t$
2. $x = s^2 - t^2, y = 2st$
3. $x = e^s \cos t, y = e^s \sin t$
4. $x = s^3 - 3st^2, y = 3s^2 t - t^3$

■ In Exercises 5–7, find positive numbers a and b so that the change of coordinates $s = ax, t = by$ transforms the integral $\int \int_R dx\, dy$ into

$$\int \int_T \left|\frac{\partial(x,y)}{\partial(s,t)}\right| ds\, dt$$

for the given regions R and T.

5. R is the rectangle $0 \le x \le 10, 0 \le y \le 1$ and T is the square $0 \le s, t \le 1$.

6. R is the rectangle $0 \le x \le 1, 0 \le y \le 1/4$ and T is the square $0 \le s, t \le 1$.

7. R is the rectangle $0 \le x \le 50, 0 \le y \le 10$ and T is the square $0 \le s, t \le 1$.

■ In Exercises 8–9, find a number a so that the change of coordinates $s = x + ay, t = y$ transforms the integral $\int \int_R dx\, dy$ over the parallelogram R in the xy-plane into an integral

$$\int \int_T \left|\frac{\partial(x,y)}{\partial(s,t)}\right| ds\, dt$$

over a rectangle T in the st-plane.

8. R has vertices $(0,0), (10,0), (12,3), (22,3)$
9. R has vertices $(0,0), (10,0), (-15,5), (-5,5)$

PROBLEMS

10. Find the region R in the xy-plane corresponding to the region T consisting of points (s,t) with $0 \le s \le 3$, $0 \le t \le 2$ for the change of coordinates $x = 2s-3t$, $y = s - 2t$. Check that

$$\int_R dx\, dy = \int_T \left|\frac{\partial(x,y)}{\partial(s,t)}\right| ds\, dt.$$

11. Compute the Jacobian for the change of coordinates into spherical coordinates:

$$x = \rho \sin \phi \cos \theta, \quad y = \rho \sin \phi \sin \theta, \quad z = \rho \cos \phi.$$

12. For the change of coordinates $x = 3s - 4t$, $y = 5s + 2t$, show that

$$\frac{\partial(x,y)}{\partial(s,t)} \cdot \frac{\partial(s,t)}{\partial(x,y)} = 1$$

13. Use the change of coordinates $x = 2s + t, y = s - t$ to compute the integral $\int_R (x+y)\, dA$, where R is the parallelogram formed by $(0,0), (3,-3), (5,-2)$, and $(2,1)$.

14. Use the change of coordinates $s = x + y, t = y$ to find the area of the ellipse $x^2 + 2xy + 2y^2 = 1$.

15. Use the change of coordinates $s = y, t = y - x^2$ to evaluate $\int \int_R x\, dx\, dy$ over the region R in the first quadrant bounded by $y = 0, y = 16, y = x^2$, and $y = x^2 - 9$.

16. If R is the triangle bounded by $x + y = 1, x = 0$, and $y = 0$, evaluate the integral $\int_R \cos\left(\frac{x-y}{x+y}\right) dx\, dy$.

17. Two independent random numbers x and y from a normal distribution with mean 0 and standard deviation σ have joint density function $p(x,y) = (1/(2\pi\sigma^2))e^{-(x^2+y^2)/(2\sigma^2)}$. The average $z = (x+y)/2$ has a one-variable probability density function of its own.

    (a) Give a double integral expression for $F(t)$, the probability that $z \le t$.

    (b) Give a single integral expression for $F(t)$. To do this, make the change of coordinates: $u = (x+y)/2$, $v = (x - y)/2$ and then do the integral on dv. Use the fact that $\int_{-\infty}^{\infty} e^{-x^2/a^2} dx = a\sqrt{\pi}$.

    (c) Find the probability density function $F'(t)$ of z.

    (d) What is the name of the distribution of z?

18. A river follows the path $y = f(x)$ where x, y are in kilometers. Near the sea, it widens into a lagoon, then narrows again at its mouth. See Figure 21.20. At the point (x,y), the depth, $d(x,y)$, of the lagoon is given by

$$d(x,y) = 40 - 160(y - f(x))^2 - 40x^2 \text{ meters.}$$

The lagoon itself is described by $d(x,y) \ge 0$. What is the volume of the lagoon in cubic meters? [Hint: Use new coordinates $u = x/2, v = y - f(x)$ and Jacobians.]

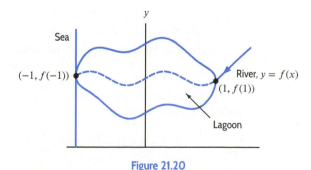

Figure 21.20

Strengthen Your Understanding

In Problems **19–20**, explain what is wrong with the statement.

19. If R is the region $0 \leq x \leq 1$, $0 \leq y \leq 4$ and T is the region $0 \leq s \leq 1$, $-2 \leq t \leq 2$, using the formulas $x = s$, $y = t^2$, we have

$$\int_R f(x, y) \, dx \, dy = \int_T f(s, t^2) \left| \frac{\partial(x, y)}{\partial(s, t)} \right| \, ds \, dt.$$

20. If R and T are corresponding regions of the xy- and st-planes, the change of coordinates $x = t^3$, $y = s$ leads to the formula

$$\int_R (x + 2y) \, dx \, dy = \int_T \left(t^3 + 2s \right) \left(-3t^2 \right) \, ds \, dt.$$

In Problems **21–22**, give an example of:

21. A change of coordinates $x = x(s, t)$, $y = y(s, t)$ where the rectangle $0 \leq s \leq 1$, $0 \leq t \leq 1$ in the st-plane corresponds to a different rectangle in the xy-plane.

22. A change of coordinates $x = x(s, t)$, $y = y(s, t)$ where every region in the st-plane corresponds to a region in the xy-plane with twice the area.

In Problems **23–24**, consider a change of variable in the integral $\int_R f(x, y) \, dA$ from x, y to s, t. Are the following statements true or false?

23. If the Jacobian $\left| \frac{\partial(x, y)}{\partial(s, t)} \right| > 1$, the value of the s, t-integral is greater than the original x, y-integral.

24. The Jacobian cannot be negative.

21.3 FLUX INTEGRALS OVER PARAMETERIZED SURFACES

Most of the flux integrals we are likely to encounter can be computed using the methods of Sections 19.1 and 19.2. In this section, we briefly consider the general case: how to compute the flux of a smooth vector field $\vec{F}$ through a smooth oriented surface, S, parameterized by

$$\vec{r} = \vec{r}(s, t),$$

for (s, t) in some region R of the parameter space. The method is similar to the one used for graphs in Section 19.2. We consider a parameter rectangle on the surface S corresponding to a rectangular region with sides Δs and Δt in the parameter space. (See Figure 21.21.)

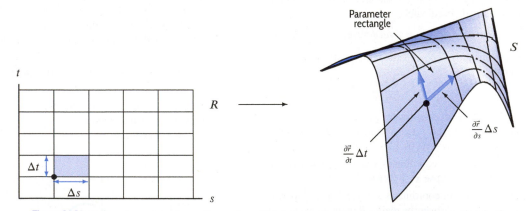

Figure 21.21: Parameter rectangle on the surface S corresponding to a small rectangular region in the parameter space, R

If Δs and Δt are small, the area vector, $\Delta \vec{A}$, of the patch is approximately the area vector of the parallelogram defined by the vectors

$$\vec{r}(s + \Delta s, t) - \vec{r}(s, t) \approx \frac{\partial \vec{r}}{\partial s} \Delta s, \quad \text{and} \quad \vec{r}(s, t + \Delta t) - \vec{r}(s, t) \approx \frac{\partial \vec{r}}{\partial t} \Delta t.$$

Thus,

$$\Delta \vec{A} \approx \frac{\partial \vec{r}}{\partial s} \times \frac{\partial \vec{r}}{\partial t} \Delta s \, \Delta t.$$

We assume that the vector $\partial\vec{r}/\partial s \times \partial\vec{r}/\partial t$ is never zero and points in the direction of the unit normal orientation vector $\vec{n}$. If the vector $\partial\vec{r}/\partial s \times \partial\vec{r}/\partial t$ points in the opposite direction to $\vec{n}$, we reverse the order of the cross product. Replacing $\Delta\vec{A}$, Δs, and Δt by $d\vec{A}$, ds, and dt, we write

$$d\vec{A} = \left(\frac{\partial\vec{r}}{\partial s} \times \frac{\partial\vec{r}}{\partial t}\right) ds\, dt.$$

The Flux of a Vector Field Through a Parameterized Surface

The flux of a smooth vector field $\vec{F}$ through a smooth oriented surface S parameterized by $\vec{r} = \vec{r}(s,t)$, where (s,t) varies in a parameter region R, is given by

$$\int_S \vec{F} \cdot d\vec{A} = \int_R \vec{F}(\vec{r}(s,t)) \cdot \left(\frac{\partial\vec{r}}{\partial s} \times \frac{\partial\vec{r}}{\partial t}\right) ds\, dt.$$

We choose the parameterization so that $\partial\vec{r}/\partial s \times \partial\vec{r}/\partial t$ is never zero and points in the direction of $\vec{n}$ everywhere.

Example 1   Find the flux of the vector field $\vec{F} = x\vec{i} + y\vec{j}$ through the surface S, oriented downward and given by

$$x = 2s, \quad y = s + t, \quad z = 1 + s - t, \qquad \text{where } 0 \le s \le 1, \quad 0 \le t \le 1.$$

Solution   Since S is parameterized by

$$\vec{r}(s,t) = 2s\vec{i} + (s+t)\vec{j} + (1+s-t)\vec{k},$$

we have

$$\frac{\partial\vec{r}}{\partial s} = 2\vec{i} + \vec{j} + \vec{k} \quad \text{and} \quad \frac{\partial r}{\partial t} = \vec{j} - \vec{k},$$

so

$$\frac{\partial\vec{r}}{\partial s} \times \frac{\partial\vec{r}}{\partial t} = \begin{vmatrix} \vec{i} & \vec{j} & \vec{k} \\ 2 & 1 & 1 \\ 0 & 1 & -1 \end{vmatrix} = -2\vec{i} + 2\vec{j} + 2\vec{k}.$$

Since the vector $-2\vec{i} + 2\vec{j} + 2\vec{k}$ points upward, we use $2\vec{i} - 2\vec{j} - 2\vec{k}$ for downward orientation. Thus, the flux integral is given by

$$\int_S \vec{F} \cdot d\vec{A} = \int_0^1 \int_0^1 (2s\vec{i} + (s+t)\vec{j}) \cdot (2\vec{i} - 2\vec{j} - 2\vec{k})\, ds\, dt$$

$$= \int_0^1 \int_0^1 (4s - 2s - 2t)\, ds\, dt = \int_0^1 \int_0^1 (2s - 2t)\, ds\, dt$$

$$= \int_0^1 \left(s^2 - 2st \Big|_{s=0}^{s=1}\right) dt = \int_0^1 (1 - 2t)\, dt = t - t^2 \Big|_0^1 = 0.$$

Area of a Parameterized Surface

The area ΔA of a small parameter rectangle is the magnitude of its area vector $\Delta \vec{A}$. Therefore,

$$\text{Area of } S = \sum \Delta A = \sum \|\Delta \vec{A}\| \approx \sum \left\| \frac{\partial \vec{r}}{\partial s} \times \frac{\partial \vec{r}}{\partial t} \right\| \Delta s \, \Delta t.$$

Taking the limit as the area of the parameter rectangles tends to zero, we are led to the following expression for the area of S.

> ### The Area of a Parameterized Surface
>
> The area of a surface S which is parameterized by $\vec{r} = \vec{r}(s, t)$, where (s, t) varies in a parameter region R, is given by
>
> $$\int_S dA = \int_R \left\| \frac{\partial \vec{r}}{\partial s} \times \frac{\partial \vec{r}}{\partial t} \right\| ds \, dt.$$

Example 2   Compute the surface area of a sphere of radius a.

Solution   We take the sphere S of radius a centered at the origin and parameterize it with the spherical coordinates ϕ and θ. The parameterization is

$$x = a \sin\phi \cos\theta, \quad y = a \sin\phi \sin\theta, \quad z = a \cos\phi, \qquad \text{for} \quad 0 \le \theta \le 2\pi, \quad 0 \le \phi \le \pi.$$

We compute

$$\frac{\partial \vec{r}}{\partial \phi} \times \frac{\partial \vec{r}}{\partial \theta} = (a \cos\phi \cos\theta \vec{i} + a \cos\phi \sin\theta \vec{j} - a \sin\phi \vec{k}) \times (-a \sin\phi \sin\theta \vec{i} + a \sin\phi \cos\theta \vec{j})$$

$$= a^2 (\sin^2\phi \cos\theta \vec{i} + \sin^2\phi \sin\theta \vec{j} + \sin\phi \cos\phi \vec{k})$$

and so

$$\left\| \frac{\partial \vec{r}}{\partial \phi} \times \frac{\partial \vec{r}}{\partial \theta} \right\| = a^2 \sin\phi.$$

Thus, we see that the surface area of the sphere S is given by

$$\text{Surface area} = \int_S dA = \int_R \left\| \frac{\partial \vec{r}}{\partial \phi} \times \frac{\partial \vec{r}}{\partial \theta} \right\| d\phi d\theta = \int_{\phi=0}^{\pi} \int_{\theta=0}^{2\pi} a^2 \sin\phi \, d\theta \, d\phi = 4\pi a^2.$$

Summary for Section 21.3

- **Flux integrals over parameterized surfaces**:

$$\int_S \vec{F} \cdot d\vec{A} = \int_R \vec{F}(\vec{r}(s, t)) \cdot \left(\frac{\partial \vec{r}}{\partial s} \times \frac{\partial \vec{r}}{\partial t} \right) ds \, dt,$$

where we choose the parameterization $\vec{r} = \vec{r}(s, t)$ of the surface S so that $\partial \vec{r}/\partial s \times \partial \vec{r}/\partial t$ is never zero and points in the direction of $\vec{n}$ everywhere.

- **Area of a parameterized surface:** The area of a surface S which is parameterized by $\vec{r} = \vec{r}(s, t)$, where (s, t) varies in a parameter region R, is given by

$$\int_S dA = \int_R \left\| \frac{\partial \vec{r}}{\partial s} \times \frac{\partial \vec{r}}{\partial t} \right\| ds \, dt.$$

Exercises and Problems for Section 21.3

EXERCISES

■ In Exercises 1–4 compute $d\vec{A}$ for the given parameterization for one of the two orientations.

1. $x = s + t, \quad y = s - t, \quad z = st$

2. $x = \sin t, \quad y = \cos t, \quad z = s + t$

3. $x = e^s, \quad y = \cos t, \quad z = \sin t$

4. $x = 0, \quad y = u + v, \quad z = u - v$

■ In Exercises 5–9, compute the flux of the vector field $\vec{F}$ through the parameterized surface S.

5. $\vec{F} = z\vec{k}$ and S is oriented upward and given, for $0 \le s \le 1, \ 0 \le t \le 1$, by
$$x = s + t, \quad y = s - t, \quad z = s^2 + t^2.$$

6. $\vec{F} = x\vec{i} + y\vec{j}$ and S is oriented downward and given, for $0 \le s \le 1, 0 \le t \le 1$, by
$$x = 2s, \quad y = s + t, \quad z = 1 + s - t.$$

7. $\vec{F} = x\vec{i}$ through the surface S oriented downward and parameterized, for $0 \le s \le 4, 0 \le t \le \pi/6$, by
$$x = e^s, \quad y = \cos(3t), \quad z = 6s.$$

8. $\vec{F} = y\vec{i} + x\vec{j}$ and S is oriented away from the z-axis and given, for $0 \le s \le \pi, \ 0 \le t \le 1$, by
$$x = 3\sin s, \quad y = 3\cos s, \quad z = t + 1.$$

9. $\vec{F} = x^2 y^2 z\vec{k}$ and S is the cone $\sqrt{x^2 + y^2} = z$, with $0 \le z \le R$, oriented downward. Parameterize the cone using cylindrical coordinates. (See Figure 21.22.)

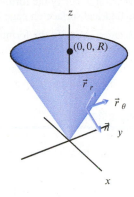

Figure 21.22

■ In Exercises 10–11, find the surface area.

10. A cylinder of radius a and length L.

11. The region S in the plane $z = 3x + 2y$ such that $0 \le x \le 10$ and $0 \le y \le 20$.

PROBLEMS

12. Compute the flux of the vector field $\vec{F} = (x + z)\vec{i} + \vec{j} + z\vec{k}$ through the surface S given by $y = x^2 + z^2$, $1/4 \le x^2 + z^2 \le 1$ oriented away from the y-axis.

13. Find the area of the ellipse S on the plane $2x + y + z = 2$ cut out by the circular cylinder $x^2 + y^2 = 2x$. (See Figure 21.23.)

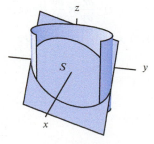

Figure 21.23

14. Consider the surface S formed by rotating the graph of $y = f(x)$ around the x-axis between $x = a$ and $x = b$.

Assume that $f(x) \ge 0$ for $a \le x \le b$. Show that the surface area of S is $2\pi \int_a^b f(x)\sqrt{1 + f'(x)^2}\, dx$.

15. A rectangular channel of width w and depth h meters lies in the $\vec{j}$ direction. At a point d_1 meters from one side and d_2 meters from the other side, the velocity vector of fluid in the channel is $\vec{v} = kd_1 d_2\vec{j}$ meters/sec. Find the flux through a rectangle stretching the full width and depth of the channel, and perpendicular to the flow.

16. The base of a cone is the unit circle centered at the origin in the xy-plane and vertex $P = (a, b, c)$, where $c > 0$.

   (a) Parameterize the cone.
   (b) Express the surface area of the cone as an integral.
   (c) Use a numerical method to find the surface area of the cone with vertex $P = (2, 0, 1)$.

In Problems 17–20, we see how the formula for a flux integral over a parameterized surface can be used to define flux. In Section 19.1, we saw that the limit defining a flux integral might not exist if the surface were subdivided in the wrong way. Using a parameterization can avoid this issue.

17. Use a parameterization to verify the formula for a flux integral over a surface graph on page 1030.

18. Use a parameterization to verify the formula for a flux integral over a cylindrical surface on page 1032.

19. Use a parameterization to verify the formula for a flux integral over a spherical surface on page 1034.

20. One problem with defining the flux integral using a parameterization is that the integral appears to depend on the choice of parameterization. However, the flux through a surface ought not to depend on how the surface is parameterized. Suppose that the surface S has two parameterizations, $\vec{r} = \vec{r}(s, t)$ for (s, t) in the region R of st-space, and also $\vec{r} = r(u, v)$ for (u, v) in the region T in uv-space, and suppose that the two parameterizations are related by the change of coordinates

$$u = u(s, t) \quad v = v(s, t).$$

Suppose that the Jacobian determinant $\partial(u, v)/\partial(s, t)$ is positive at every point (s, t) in R. Use the change of coordinates formula for double integrals on page 1091 to show that computing the flux integral using either parameterization gives the same result.

Strengthen Your Understanding

In Problems 21–22, explain what is wrong with the statement.

21. The area of the surface parameterized by $x = s, y = t, z = f(s, t)$ above the square $0 \leq x \leq 1, 0 \leq y \leq 1$ is given by the integral

$$\text{Area} = \int_0^1 \int_0^1 f(s, t) \, ds \, dt.$$

22. The surface S parameterized by $x = f(s, t), y = g(s, t), z = h(s, t)$, where $0 \leq s \leq 2, 0 \leq t \leq 3$, has area 6.

In Problems 23–24, give an example of:

23. A parameterization $\vec{r} = \vec{r}(s, t)$ of the xy-plane such that $dA = 2 \, ds \, dt$.

24. A vector field $\vec{F}$ such that $\int_S \vec{F} \cdot d\vec{A} > 0$, where S is the surface $\vec{r} = (s - t)\vec{i} + t^2\vec{j} + (s + t)\vec{k}$, $0 \leq s \leq 1$, $0 \leq t \leq 1$, oriented in the direction of $\dfrac{\partial \vec{r}}{\partial s} \times \dfrac{\partial \vec{r}}{\partial t}$.

Are the statements in Problems 25–27 true or false? Give reasons for your answer.

25. If $\vec{r}(s, t), 0 \leq s \leq 1, 0 \leq t \leq 1$ is an oriented parameterized surface S, and $\vec{F}$ is a vector field that is everywhere tangent to S, then the flux of $\vec{F}$ through S is zero.

26. For any parameterization of the surface $x^2 - y^2 + z^2 = 6$, $d\vec{A}$ at $(1, 2, 3)$ is a multiple of $(2\vec{i} - 4\vec{j} + 6\vec{k}) \, dx \, dy$.

27. If you parameterize the plane $3x + 4y + 5z = 7$, then there is a constant c such that, at any point (x, y, z), $d\vec{A} = c(3\vec{i} + 4\vec{j} + 5\vec{k}) \, dx \, dy$.

28. Let S be the hemisphere $x^2 + y^2 + z^2 = 1$ with $x \leq 0$, oriented away from the origin. Which of the following integrals represents the flux of $\vec{F}(x, y, z)$ through S?

(a) $\displaystyle \int_R \vec{F}(x, y, z(x, y)) \cdot \frac{\partial \vec{r}}{\partial x} \times \frac{\partial \vec{r}}{\partial y} \, dx \, dy$

(b) $\displaystyle \int_R \vec{F}(x, y, z(x, y)) \cdot \frac{\partial \vec{r}}{\partial y} \times \frac{\partial \vec{r}}{\partial x} \, dy \, dx$

(c) $\displaystyle \int_R \vec{F}(x, y(x, z), z) \cdot \frac{\partial \vec{r}}{\partial x} \times \frac{\partial \vec{r}}{\partial z} \, dx \, dz$

(d) $\displaystyle \int_R \vec{F}(x, y(x, z), z) \cdot \frac{\partial \vec{r}}{\partial z} \times \frac{\partial \vec{r}}{\partial x} \, dz \, dx$

(e) $\displaystyle \int_R \vec{F}(x(y, z), y, z) \cdot \frac{\partial \vec{r}}{\partial y} \times \frac{\partial \vec{r}}{\partial z} \, dy \, dz$

(f) $\displaystyle \int_R \vec{F}(x(y, z), y, z) \cdot \frac{\partial \vec{r}}{\partial z} \times \frac{\partial \vec{r}}{\partial y} \, dz \, dy$

Online Resource: Review Problems and Projects

READY REFERENCE

READY REFERENCE

Multivariable Functions

Points in 3-space are represented by a system of **Cartesian coordinates** (p. 696). The **distance** between (x, y, z) and (a, b, c) is $\sqrt{(x - a)^2 + (y - b)^2 + (z - c)^2}$ (p. 698).

Functions of two variables can be represented by **graphs** (p. 703), **contour diagrams** (p. 711), **cross-sections** (p. 705), and **tables** (p. 695).

Functions of three variables can be represented by the family of **level surfaces** $f(x, y, z) = c$ for various values of the constant c (p. 732).

A **linear function** $f(x, y)$ has equation

$$f(x, y) = z_0 + m(x - x_0) + n(y - y_0) \text{ (p. 726)}$$
$$= c + mx + ny, \text{ where } c = z_0 - mx_0 - ny_0.$$

Its **graph** is a plane with slope m in the x-direction, slope n in the y-direction, through (x_0, y_0, z_0) (p. 726). Its **table of values** has linear rows (of same slope) and linear columns (of same slope) (p. 727). Its **contour diagram** is equally spaced parallel straight lines (p. 727).

The **limit** of f at the point (a, b), written $\lim_{(x,y) \to (a,b)} f(x, y)$, is the number L, if one exists, such that $f(x, y)$ is as close to L as we please whenever the distance from the point (x, y) to the point (a, b) is sufficiently small, but not zero. (p. 741).

A function f is **continuous at the point** (a, b) if $\lim_{(x,y) \to (a,b)} f(x, y) = f(a, b)$. A function is **continuous on a region** R if it is continuous at each point of R (p. 741).

Vectors

A **vector** $\vec{v}$ has **magnitude** (denoted $\|\vec{v}\|$) and **direction**. Examples are **displacement vectors** (p. 746), **velocity and acceleration vectors** (pp. 755, 757). and **force** (p. 757). We can add vectors, and multiply a vector by a scalar (p. 747). Two non-zero vectors, $\vec{v}$ and $\vec{w}$, are **parallel** if one is a scalar multiple of the other (p. 748).

A **unit vector** has magnitude 1. The vectors $\vec{i}, \vec{j}$, and $\vec{k}$ are unit vectors in the directions of the coordinate axes. A unit vector in the direction of any nonzero vector $\vec{v}$ is $\vec{u} = \vec{v}/\|\vec{v}\|$ (p. 752). We **resolve** $\vec{v}$ **into components** by writing $\vec{v} = v_1\vec{i} + v_2\vec{j} + v_3\vec{k}$ (p. 749).

If $\vec{v} = v_1\vec{i} + v_2\vec{j} + v_3\vec{k}$ and $\vec{w} = w_1\vec{i} + w_2\vec{j} + w_3\vec{k}$ then

$$\|\vec{v}\| = \sqrt{v_1^2 + v_2^2 + v_3^2} \text{ (p. 750)}$$

$$\vec{v} + \vec{w} = (v_1 + w_1)\vec{i} + (v_2 + w_2)\vec{j} + (v_3 + w_3)\vec{k} \text{ (p. 751)},$$

$$\lambda\vec{v} = \lambda v_1\vec{i} + \lambda v_2\vec{j} + \lambda v_3\vec{k} \text{ (p. 751)}.$$

The **displacement vector** from $P_1 = (x_1, y_1, z_1)$ to $P_2 = (x_2, y_2, z_2)$ is

$$\overrightarrow{P_1 P_2} = (x_2 - x_1)\vec{i} + (y_2 - y_1)\vec{j} + (z_2 - z_1)\vec{k} \quad \text{(p. 750)}.$$

The **position vector** of $P = (x, y, z)$ is $\overrightarrow{OP}$ (p. 750). A **vector in n dimensions** is a string of numbers $\vec{v} = (v_1, v_2, \ldots, v_n)$ (p. 759).

Dot Product (Scalar Product) (p. 763).
Geometric definition: $\vec{v} \cdot \vec{w} = \|\vec{v}\|\|\vec{w}\| \cos\theta$ where θ is the angle between $\vec{v}$ and $\vec{w}$ and $0 \le \theta \le \pi$.
Algebraic definition: $\vec{v} \cdot \vec{w} = v_1 w_1 + v_2 w_2 + v_3 w_3$.

Two nonzero vectors $\vec{v}$ and $\vec{w}$ are **perpendicular** if and only if $\vec{v} \cdot \vec{w} = 0$ (p. 765). Magnitude and dot product are related by $\vec{v} \cdot \vec{v} = \|\vec{v}\|^2$ (p. 765). If $\vec{u} = (u_1, \ldots, u_n)$ and $\vec{v} = (v_1, \ldots, v_n)$ then the dot product of $\vec{u}$ and $\vec{v}$ is $\vec{u} \cdot \vec{v} = u_1 v_1 + \ldots + u_n v_n$ (p. 767).

The **equation of the plane** with **normal vector** $\vec{n} = a\vec{i} + b\vec{j} + c\vec{k}$ and containing the point $P_0 = (x_0, y_0, z_0)$ is $\vec{n} \cdot (\vec{r} - \vec{r}_0) = a(x - x_0) + b(y - y_0) + c(z - z_0) = 0$ or $ax + by + cz = d$, where $d = ax_0 + by_0 + cz_0$ (p. 766).

If $\vec{v}_{\text{parallel}}$ and $\vec{v}_{\text{perp}}$ are components of $\vec{v}$ which are parallel and perpendicular, respectively, to a unit vector $\vec{u}$, then $\vec{v}_{\text{parallel}} = (\vec{v} \cdot \vec{u})\vec{u}$ and $\vec{v}_{\text{perp}} = \vec{v} - \vec{v}_{\text{parallel}}$ (p. 768).

The **work**, W, done by a force $\vec{F}$ acting on an object through a displacement $\vec{d}$ is $W = \vec{F} \cdot \vec{d}$ (p. 769).

Cross Product (Vector Product) (p. 775, 775)
Geometric definition

$$\vec{v} \times \vec{w} = \begin{pmatrix} \text{Area of parallelogram} \\ \text{with edges } \vec{v} \text{ and } \vec{w} \end{pmatrix} \vec{n}$$

$$= (\|\vec{v}\|\|\vec{w}\| \sin\theta)\vec{n},$$

where $0 \le \theta \le \pi$ is the angle between $\vec{v}$ and $\vec{w}$ and $\vec{n}$ is the unit vector perpendicular to $\vec{v}$ and $\vec{w}$ pointing in the direction given by the **right-hand rule**.
Algebraic definition

$$\vec{v} \times \vec{w} = (v_2 w_3 - v_3 w_2)\vec{i} + (v_3 w_1 - v_1 w_3)\vec{j} + (v_1 w_2 - v_2 w_1)\vec{k}$$

$$\vec{v} = v_1\vec{i} + v_2\vec{j} + v_3\vec{k}, \vec{w} = w_1\vec{i} + w_2\vec{j} + w_3\vec{k}.$$

To find the **equation of a plane through three points** that do not lie on a line, determine two vectors in the plane and then find a normal vector using the cross product (p. 777). The **area of a parallelogram** with edges $\vec{v}$ and $\vec{w}$ is $\|\vec{v} \times \vec{w}\|$. The **volume of a parallelepiped** with edges $\vec{a}, \vec{b}, \vec{c}$ is $\left| (\vec{b} \times \vec{c}) \cdot \vec{a} \right|$ (p. 779).

The **angular velocity** (p. 779) of a flywheel can be represented by a vector $\vec{\omega}$ whose direction is parallel to the axis of rotation and magnitude is the angular speed of rotation. The velocity vector $\vec{v}$ of a point P on the flywheel is $\vec{v} = \vec{\omega} \times \vec{r}$ where $\vec{r}$ is a vector from the axis of rotation to P.

Differentiation of Multivariable Functions

Partial derivatives of f **(p. 787)**.

$$f_x(a,b) = \begin{array}{c} \text{Rate of change of } f \text{ with respect to } x \\ \text{at the point } (a,b) \end{array}$$

$$= \lim_{h \to 0} \frac{f(a+h,b) - f(a,b)}{h},$$

$$f_y(a,b) = \begin{array}{c} \text{Rate of change of } f \text{ with respect to } y \\ \text{at the point } (a,b) \end{array}$$

$$= \lim_{h \to 0} \frac{f(a,b+h) - f(a,b)}{h}.$$

On the graph of f, the partial derivatives $f_x(a,b)$ and $f_y(a,b)$ give the slope in the x and y directions, respectively (p. 788). The **tangent plane** to $z = f(x,y)$ at (a,b) is

$$z = f(a,b) + f_x(a,b)(x-a) + f_y(a,b)(y-b) \quad \text{(p. 801)}.$$

Partial derivatives can be estimated from a contour diagram or table of values using difference quotients (p. 788), and can be computed algebraically using the same rules of differentiation as for one-variable calculus (p. 795). Partial derivatives for functions of three or more variables are defined and computed in the same way (p 796).

The gradient vector grad f of f is grad $f(a,b) = f_x(a,b)\vec{i} + f_y(a,b)\vec{j}$ (2 variables) (p. 811) or grad $f(a,b,c) = f_x(a,b,c)\vec{i} + f_y(a,b,c)\vec{j} + f_z(a,b,c)\vec{k}$ (3 variables) (p. 819). The gradient vector at P: Points in the direction of increasing f; is perpendicular to the level curve or level surface of f through P; and has magnitude $\|\text{grad } f\|$ equal to the maximum rate of change of f at P (pp. 813, 819). The magnitude is large when the level curves or surfaces are close together and small when they are far apart.

The **directional derivative** of f at P in the direction of a unit vector $\vec{u}$ is (pp. 809, 811)

$$f_{\vec{u}}(P) = \begin{array}{c} \text{Rate of change} \\ \text{of } f \text{ in direction} \\ \text{of } \vec{u} \text{ at } P \end{array} = \text{grad } f(P) \cdot \vec{u}$$

The **tangent plane approximation** to $f(x,y)$ for (x,y) near the point (a,b) is

$$f(x,y) \approx f(a,b) + f_x(a,b)(x-a) + f_y(a,b)(y-b).$$

The right-hand side is the **local linearization** (p. 802). The **differential of** $z = f(x,y)$ at (a,b) is the linear function of dx and dy

$$df = f_x(a,b)\,dx + f_y(a,b)\,dy \quad \text{(p. 804)}.$$

Local linearity with three or more variables follows the same pattern as for functions of two variables (p. 803).

The **tangent plane to a level surface** of a function of three-variables f at (a,b,c) is (p. 822)

$$f_x(a,b,c)(x-a) + f_y(a,b,c)(y-b) + f_z(a,b,c)(z-c) = 0.$$

The Chain Rule for the partial derivative of one variable with respect to another in a chain of composed functions (p. 829):

- Draw a diagram expressing the relationship between the variables, and label each link in the diagram with the derivative relating the variables at its ends.
- For each path between the two variables, multiply together the derivatives from each step along the path.
- Add the contributions from each path.

If $z = f(x,y)$, and $x = g(t)$, and $y = h(t)$, then

$$\frac{dz}{dt} = \frac{\partial z}{\partial x}\frac{dx}{dt} + \frac{\partial z}{\partial y}\frac{dy}{dt} \quad \text{(p. 828)}.$$

If $z = f(x,y)$, with $x = g(u,v)$ and $y = h(u,v)$, then

$$\frac{\partial z}{\partial u} = \frac{\partial z}{\partial x}\frac{\partial x}{\partial u} + \frac{\partial z}{\partial y}\frac{\partial y}{\partial u},$$

$$\frac{\partial z}{\partial v} = \frac{\partial z}{\partial x}\frac{\partial x}{\partial v} + \frac{\partial z}{\partial y}\frac{\partial y}{\partial v} \quad \text{(p. 830)}.$$

Second-order partial derivatives (p. 838)

$$\frac{\partial^2 z}{\partial x^2} = f_{xx} = (f_x)_x, \qquad \frac{\partial^2 z}{\partial x \partial y} = f_{yx} = (f_y)_x,$$

$$\frac{\partial^2 z}{\partial y \partial x} = f_{xy} = (f_x)_y, \qquad \frac{\partial^2 z}{\partial y^2} = f_{yy} = (f_y)_y.$$

Theorem: Equality of Mixed Partial Derivatives. If f_{xy} and f_{yx} are continuous at (a,b), an interior point of their domain, then $f_{xy}(a,b) = f_{yx}(a,b)$ (p. 839).

Taylor Polynomial of Degree 1 Approximating $f(x,y)$ **for** (x,y) **near** (a,b) **(p. 842)**

$$f(x,y) \approx L(x,y) = f(a,b) + f_x(a,b)(x-a) + f_y(a,b)(y-b).$$

Taylor Polynomial of Degree 2 (p. 843)

$$f(x,y) \approx Q(x,y)$$

$$= f(a,b) + f_x(a,b)(x-a) + f_y(a,b)(y-b)$$

$$+ \frac{f_{xx}(a,b)}{2}(x-a)^2 + f_{xy}(a,b)(x-a)(y-b)$$

$$+ \frac{f_{yy}(a,b)}{2}(y-b)^2.$$

Definition of Differentiability (p. 848). A function $f(x,y)$ is **differentiable at the point** (a,b) if there is a linear function $L(x,y) = f(a,b) + m(x-a) + n(y-b)$ such that if the **error** $E(x,y)$ is defined by

$$f(x,y) = L(x,y) + E(x,y),$$

and if $h = x - a, k = y - b$, then the **relative error** $E(a+h, b+k)/\sqrt{h^2 + k^2}$ satisfies

$$\lim_{\substack{h \to 0 \\ k \to 0}} \frac{E(a+h, b+k)}{\sqrt{h^2 + k^2}} = 0.$$

Theorem: Continuity of Partial Derivatives Implies Differentiability (p. 851). If the partial derivatives, f_x and f_y, of a function f exist and are continuous on a small disk centered at the point (a,b), then f is differentiable at (a,b).

Optimization

A function f has a **local maximum** at the point P_0 if $f(P_0) \geq f(P)$ for all points P near P_0, and a **local minimum** at the point P_0 if $f(P_0) \leq f(P)$ for all points P near P_0 (p. 856). A **critical point** of a function f is a point where grad f is either $\vec{0}$ or undefined. If f has a local maximum or minimum at a point P_0, not on the boundary of its domain, then P_0 is a critical point (p. 856). A **quadratic function** $f(x, y) = ax^2 + bxy + cz^2$ generally has one critical point, which can be a local maximum, a local minimum, or a **saddle point** (p. 859).

Second derivative test for functions of two variables (p. 861). Suppose grad $f(x_0, y_0) = \vec{0}$. Let $D = f_{xx}(x_0, y_0)f_{yy}(x_0, y_0) - (f_{xy}(x_0, y_0))^2$.

- If $D > 0$ and $f_{xx}(x_0, y_0) > 0$, then f has a local minimum at (x_0, y_0).
- If $D > 0$ and $f_{xx}(x_0, y_0) < 0$, then f has a local maximum at (x_0, y_0).
- If $D < 0$, then f has a saddle point at (x_0, y_0).
- If $D = 0$, anything can happen.

Unconstrained optimization

A function f defined on a region R has a **global maximum on** R at the point P_0 if $f(P_0) \geq f(P)$ for all points P in R, and a **global minimum on** R at the point P_0 if $f(P_0) \leq f(P)$ for all points P in R (p. 866). For an **unconstrained optimization problem**, find the critical points and investigate whether the critical points give global maxima or minima (p. 866).

A **closed** region is one which contains its boundary; a **bounded** region is one which does not stretch to infinity in any direction (p. 871).

Extreme Value Theorem for Multivariable Functions. If f is a continuous function on a closed and bounded region R, then f has a global maximum at some point (x_0, y_0) in R and a global minimum at some point (x_1, y_1) in R (p. 872).

Constrained optimization

Suppose P_0 is a point satisfying the constraint $g(x, y) = c$. A function f has a **local maximum** at P_0 **subject to the constraint** if $f(P_0) \geq f(P)$ for all points P near P_0 satisfying the constraint (p. 877). It has a **global maximum** at P_0 **subject to the constraint** if $f(P_0) \geq f(P)$ for all points P satisfying the constraint (p. 877). Local and global minima are defined similarly (p. 877). A local maximum or minimum of $f(x, y)$ subject to a constraint $g(x, y) = c$ occurs at a point where the constraint is tangent to a level curve of f, and thus where grad g is parallel to grad f (p. 878).

To optimize f **subject to the constraint** $g = c$ **(p. 878),** find the points satisfying the equations

$$\text{grad } f = \lambda \text{ grad } g \quad \text{and} \quad g = c.$$

Then compare values of f at these points, at points on the constraint where grad $g = \vec{0}$, and at the endpoints of the constraint. The number λ is called the **Lagrange multiplier**.

To optimize f **subject to the constraint** $g \leq c$ **(p. 879),** find all points in the interior $g(x, y) < c$ where grad f is zero or undefined; then use Lagrange multipliers to find the local extrema of f on the boundary $g(x, y) = c$. Evaluate f at the points found and compare the values.

The value of λ is the rate of change of the optimum value of f as c increases (where $g(x, y) = c$) (p. 881). The **Lagrangian function** $\mathcal{L}(x, y, \lambda) = f(x, y) - \lambda(g(x, y) - c)$ can be used to convert a constrained optimization problem for f subject the constraint $g = c$ into an unconstrained problem for $\mathcal{L}$ (p. 881).

Multivariable Integration

The **definite integral** of f, a continuous function of two variables, over R, the rectangle $a \leq x \leq b, c \leq y \leq d$, is called a **double integral**, and is a limit of **Riemann sums**

$$\int_R f \, dA = \lim_{\Delta x, \Delta y \to 0} \sum_{i,j} f(u_{ij}, v_{ij}) \Delta x \Delta y \quad \text{(p. 892)}.$$

The Riemann sum is constructed by subdividing R into subrectangles of width Δx and height Δy, and choosing a point (u_{ij}, v_{ij}) in the ij-th rectangle.

A **triple integral** of f, a continuous function of three variables, over W, the box $a \leq x \leq b, c \leq y \leq d, p \leq z \leq q$ in 3-space, is defined in a similar way using three-variable Riemann sums (p. 909).

Interpretations

If $f(x, y)$ is positive, $\int_R f \, dA$ is the **volume** under graph of f above the region R (p. 892). If $f(x, y) = 1$ for all x and y, then the area of R is $\int_R 1 \, dA = \int_R dA$ (p. 894). If $f(x, y)$ is a **density**, then $\int_R f \, dA$ is the **total quantity** in the region R (p. 890). The **average value** of $f(x, y)$ on the region R is $\frac{1}{\text{Area of } R} \int_R f \, dA$ (p. 894). In probability, if $p(x, y)$ is a **joint density function** then $\int_a^b \int_c^d p(x, y) \, dy \, dx$ is the fraction of population with $a \leq x \leq b$ and $c \leq y \leq d$ (p. 932).

Iterated integrals

Double and triple integrals can be written as **iterated integrals**

$$\int_R f \, dA = \int_c^d \int_a^b f(x, y) \, dx \, dy \quad \text{(p. 899)}$$

$$\int_W f \, dV = \int_p^q \int_c^d \int_a^b f(x, y, z) \, dx \, dy \, dz \quad \text{(p. 909)}$$

Other orders of integration are possible. For iterated integrals over **non-rectangular regions** (p. 900), limits on outer integral are constants and limits on inner integrals involve only the variables in the integrals further out (pp. 902, 911).

Integrals in other coordinate systems

When computing double integrals in polar coordinates, put $dA = r \, dr \, d\theta$ or $dA = r \, d\theta \, dr$ (p. 916). Cylindrical coordinates are given by $x = r \cos \theta, y = r \sin \theta, z = z$, for $0 \leq r < \infty, 0 \leq \theta \leq 2\pi, -\infty < z < \infty$ (p. 921). Spherical coordinates are given by $x = \rho \sin \phi \cos \theta, y = \rho \sin \phi \sin \theta, z = \rho \cos \phi$, for $0 \leq \rho < \infty, 0 \leq \phi \leq \pi, 0 \leq \theta \leq 2\pi$ (p. 924). When computing triple integrals in cylindrical or spherical coordinates, put $dV = r\,dr\,d\theta\,dz$ for cylindrical coordinates (p. 923), $dV = \rho^2 \sin\phi\,d\rho\,d\phi\,d\theta$ for spherical coordinates (p. 925). Other orders of integration are also possible.

For a **change of variables** $x = x(s,t)$, $y = y(s,t)$, the **Jacobian** is

$$\frac{\partial(x,y)}{\partial(s,t)} = \frac{\partial x}{\partial s} \cdot \frac{\partial y}{\partial t} - \frac{\partial x}{\partial t} \cdot \frac{\partial y}{\partial s} = \begin{vmatrix} \frac{\partial x}{\partial s} & \frac{\partial x}{\partial t} \\[6pt] \frac{\partial y}{\partial s} & \frac{\partial y}{\partial t} \end{vmatrix} \quad \text{(p. 1091)}.$$

To convert an integral from x, y to s, t coordinates (p. 1091): Substitute for x and y in terms of s and t, change the xy region R into an st region T, and change the area element by making the substitution $dx\,dy = \left|\frac{\partial(x,y)}{\partial(s,t)}\right| ds\,dt$. For triple integrals, there is a similar formula (p. 1092).

Parameterizations and Vector Fields

Parameterized curves

The motion of a particle is described by **parametric equations** $x = f(t), y = g(t)$ (2-space) or $x = f(t), y = g(t), z = h(t)$ (3-space). The path of the particle is a **parameterized curve** (p. 938). Parameterizations are also written in **vector form** $\vec{r}(t) = f(t)\vec{i} + g(t)\vec{j} + h(t)\vec{k}$ (p. 940). For a **curve segment** we restrict the parameter to to a closed interval $a \le t \le b$ (p. 941). **Parametric equations for the graph** of $y = f(x)$ are $x = t, y = f(t)$.

Parametric equations for a line through (x_0, y_0) in the direction of $\vec{v} = a\vec{i} + b\vec{j}$ are $x = x_0 + at, y = y_0 + bt$. In 3-space, the line through (x_0, y_0, z_0) in the direction of $\vec{v} = a\vec{i} + b\vec{j} + c\vec{k}$ is $x = x_0 + at, y = y_0 + bt, z = z_0 + ct$ (p. 939). In vector form, the equation for a line is $\vec{r}(t) = \vec{r}_0 + t\vec{v}$, where $\vec{r}_0 = x_0\vec{i} + y_0\vec{j} + z_0\vec{k}$ (p. 941).

Parametric equations for a circle of radius R in the plane, centered at the origin are $x = R\cos t$, $y = R\sin t$ (counterclockwise), $x = R\cos t, y = -R\sin t$ (clockwise).

To find the **intersection points** of a curve $\vec{r}(t) = f(t)\vec{i} + g(t)\vec{j} + h(t)\vec{k}$ with a surface $F(x,y,z) = c$, solve $F(f(t), g(t), h(t)) = c$ for t (p. 942). To find the intersection points of two curves $\vec{r}_1(t)$ and $\vec{r}_2(t)$, solve $\vec{r}_1(t_1) = \vec{r}_2(t_2)$ for t_1 and t_2 (p. 942).

The **length of a curve segment** C given parametrically for $a \le t \le b$ with velocity vector $\vec{v}$ is $\int_a^b \|\vec{v}\|\,dt$ if $\vec{v} \ne \vec{0}$ for $a < t < b$ (p. 954).

The **velocity** and **acceleration** of a moving object with position vector $\vec{r}(t)$ at time t are

$$\vec{v}(t) = \lim_{\Delta t \to 0} \frac{\Delta\vec{r}}{\Delta t} \quad \text{(p. 949)}$$

$$\vec{a}(t) = \lim_{\Delta t \to 0} \frac{\Delta\vec{v}}{\Delta t} \quad \text{(p. 952)}$$

We write $\vec{v} = \dfrac{d\vec{r}}{dt} = \vec{r}\,'(t)$ and $\vec{a} = \dfrac{d\vec{v}}{dt} = \dfrac{d^2\vec{r}}{dt^2} = \vec{r}\,''(t)$.

The **components of the velocity and acceleration vectors** are

$$\vec{v}(t) = \frac{dx}{dt}\vec{i} + \frac{dy}{dt}\vec{j} + \frac{dz}{dt}\vec{k} \quad \text{(p. 950)}$$

$$\vec{a}(t) = \frac{d^2x}{dt^2}\vec{i} + \frac{d^2y}{dt^2}\vec{j} + \frac{d^2z}{dt^2}\vec{k} \quad \text{(p. 952)}$$

The **speed** is $\|\vec{v}\| = \sqrt{(dx/dt)^2 + (dy/dt)^2 + (dz/dt)^2}$ (p. 953). Analogous formulas for velocity, speed, and acceleration hold in 2-space.

Uniform Circular Motion (p. 953) For a particle $\vec{r}(t) = R\cos(\omega t)\vec{i} + R\sin(\omega t)\vec{j}$: motion is in a circle of radius R with period $2\pi/\omega$; velocity, $\vec{v}$, is tangent to the circle and speed is constant $\|\vec{v}\| = \omega R$; acceleration, $\vec{a}$, points toward the center of the circle with $\|\vec{a}\| = \|\vec{v}\|^2/R$.

Motion in a Straight Line (p. 953) For a particle $\vec{r}(t) = \vec{r}_0 + f(t)\vec{v}_0$: Motion is along a straight line through the point with position vector $\vec{r}_0$ parallel to $\vec{v}_0$; velocity, $\vec{v}$, and acceleration, $\vec{a}$, are parallel to the line.

Vector fields

A **vector field** in 2-space is a function $\vec{F}(x,y)$ whose value at a point (x,y) is a 2-dimensional vector (p. 958). Similarly, a vector field in 3-space is a function $\vec{F}(x,y,z)$ whose values are 3-dimensional vectors (p. 958). Examples are the **gradient** of a differentiable function f, the **velocity field** of a fluid flow, and **force fields** (p. 958). A **flow line** of a vector field $\vec{v} = \vec{F}(\vec{r})$ is a path $\vec{r}(t)$ whose velocity vector equals $\vec{v}$, thus $\vec{r}'(t) = \vec{v} = \vec{F}(\vec{r}(t))$ (p. 967). The **flow** of a vector field is the family of all of its flow line (p. 967). Flow lines can be approximated numerically using Euler's method (p. 969).

Parameterized surfaces

We **parameterize a surface** with two parameters, $x = f_1(s,t)$, $y = f_2(s,t)$, $z = f_3(s,t)$ (p. 1080). We also use the vector form $\vec{r}(s,t) = f_1(s,t)\vec{i} + f_2(s,t)\vec{j} + f_3(s,t)\vec{k}$ (p. 1080). **Parametric equations for the graph** of $z = f(x,y)$ are $x = s, y = t$, and $z = f(s,t)$ (p. 1080). **Parametric equation for a plane** through the point with position vector $\vec{r}_0$ and containing the two nonparallel vectors $\vec{v}_1$ and $\vec{v}_2$ is $\vec{r}(s,t) = \vec{r}_0 + s\vec{v}_1 + t\vec{v}_2$ (p. 1081). **Parametric equation for a sphere** of radius R centered at the origin is $\vec{r}(\theta, \phi) = R\sin\phi\cos\theta\,\vec{i} + R\sin\phi\sin\theta\,\vec{j} + \cos\phi\,\vec{k}$, $0 \le \theta \le 2\pi, 0 \le \phi \le \pi$ (p. 1081). **Parametric equation for a cylinder** of radius R along the z-axis is $\vec{r}(\theta, z) = R\cos\theta\,\vec{i} + R\sin\theta\,\vec{j} + z\vec{k}$, $0 \le \theta \le 2\pi, -\infty < z < \infty$ (p. 1079). A **parameter curve** is the curve obtained by holding one of the parameters constant and letting the other vary (p. 1085).

Line Integrals

The **line integral** of a vector field $\vec{F}$ along an **oriented curve** C (p. 974) is

$$\int_C \vec{F} \cdot d\vec{r} = \lim_{\|\Delta\vec{r}_i\| \to 0} \sum_{i=0}^{n-1} \vec{F}(\vec{r}_i) \cdot \Delta\vec{r}_i,$$

where the direction of $\Delta\vec{r}_i$ is the direction of the orientation (p. 975).

The line integral measures the extent to which C is going with $\vec{F}$ or against it (p. 976). For oriented curves C, C_1, and C_2, $\int_{-C} \vec{F} \cdot d\vec{r} = -\int_C \vec{F} \cdot d\vec{r}$, where $-C$ is the curve C parameterized in the opposite direction, and $\int_{C_1+C_2} \vec{F} \cdot d\vec{r} = \int_{C_1} \vec{F} \cdot d\vec{r} + \int_{C_2} \vec{F} \cdot d\vec{r}$, where C_1+C_2 is the curve obtained by joining the endpoint of C_1 to the starting point of C_2 (p. 980).

The **work done by a force** $\vec{F}$ along a curve C is $\int_C \vec{F} \cdot d\vec{r}$ (p. 977). The **circulation** of $\vec{F}$ around an oriented closed curve is $\int_C \vec{F} \cdot d\vec{r}$ (p. 979).

Given a **parameterization of** C, $\vec{r}(t)$, for $a \leq t \leq b$, the line integral can be calculated as

$$\int_C \vec{F} \cdot d\vec{r} = \int_a^b \vec{F}(\vec{r}(t)) \cdot \vec{r}'(t)\, dt \quad \text{(p. 985)}.$$

Fundamental Theorem for Line Integrals (p. 992):
Suppose C is a piecewise smooth oriented path with starting point P and endpoint Q. If f is a function whose gradient is continuous on the path C, then

$$\int_C \text{grad } f \cdot d\vec{r} = f(Q) - f(P).$$

Path-independent fields and gradient fields
A vector field $\vec{F}$ is said to be **path-independent**, or **conservative**, if for any two points P and Q, the line integral $\int_C \vec{F} \cdot d\vec{r}$ has the same value along any piecewise smooth path C from P to Q lying in the domain of $\vec{F}$ (p. 994). A **gradient field** is a vector field of the form $\vec{F} = \text{grad } f$ for some scalar function f, and f is called a **potential function** for the vector field $\vec{F}$ (p. 995). A vector field $\vec{F}$ is path-independent if and only if $\vec{F}$ is a gradient vector field (p. 995). A vector field $\vec{F}$ is path-independent if and only if $\int_C \vec{F} \cdot d\vec{r} = 0$ for every closed curve C (p. 1004). If $\vec{F}$ is a gradient field, then $\frac{\partial F_2}{\partial x} - \frac{\partial F_1}{\partial y} = 0$ (p. 1005). The quantity $\frac{\partial F_2}{\partial x} - \frac{\partial F_1}{\partial y}$ is called the 2-dimensional or scalar **curl** of $\vec{F}$.

Green's Theorem (p. 1005):
Suppose C is a piecewise smooth simple closed curve that is the boundary of an open region R in the plane and oriented so that the region is on the left as we move around the curve. Suppose $\vec{F} = F_1\vec{i} + F_2\vec{j}$ is a smooth vector field defined at every point of the region R and boundary C. Then

$$\int_C \vec{F} \cdot d\vec{r} = \int_R \left(\frac{\partial F_2}{\partial x} - \frac{\partial F_1}{\partial y} \right) dx\, dy.$$

Curl test for vector fields in 2-space: If $\frac{\partial F_2}{\partial x} - \frac{\partial F_1}{\partial y} = 0$ and the domain of $\vec{F}$ has no holes, then $\vec{F}$ is path-independent, and hence a gradient field (p. 1007). The condition that the domain have no holes is important. It is not always true that if the scalar curl of $\vec{F}$ is zero then $\vec{F}$ is a gradient field (p. 1008).

Surface Integrals

A surface is **oriented** if a unit normal vector $\vec{n}$ has been chosen at every point on it in a continuous way (p. 1018). For a closed surface, we usually choose the outward orientation (p. 1018). The **area vector** of a flat, oriented surface is a vector $\vec{A}$ whose magnitude is the area of the surface, and whose direction is the direction of the orientation vector $\vec{n}$ (p. 1019). If $\vec{v}$ is the velocity vector of a constant fluid flow and $\vec{A}$ is the area vector of a flat surface, then the total flow through the surface in units of volume per unit time is called the **flux** of $\vec{v}$ through the surface and is given by $\vec{v} \cdot \vec{A}$ (p. 1019).

The **surface integral** or **flux integral** of the vector field $\vec{F}$ through the oriented surface S is

$$\int_S \vec{F} \cdot d\vec{A} = \lim_{\|\Delta\vec{A}\| \to 0} \sum \vec{F} \cdot \Delta\vec{A},$$

where the direction of $\Delta\vec{A}$ is the direction of the orientation (p. 1020). If $\vec{v}$ is a variable vector field and then $\int_S \vec{v} \cdot d\vec{A}$ is the flux through the surface S (p. 1021).

Simple flux integrals can be calculated by putting $d\vec{A} = \vec{n}\, dA$ and using geometry or converting to a double integral (p. 1023).

The **flux through a graph** of $z = f(x, y)$ above a region R in the xy-plane, oriented upward, is

$$\int_R \vec{F}(x, y, f(x, y)) \cdot \left(-f_x\vec{i} - f_y\vec{j} + \vec{k} \right) dx\, dy \quad \text{(p. 1030)}.$$

The **area of the part of the graph** of $z = f(x, y)$ above a region R in the xy-plane is

$$\text{Area of } S = \int_R \sqrt{(f_x)^2 + (f_y)^2 + 1}\, dx\, dy \quad \text{(p. 1031)}.$$

The **flux through a cylindrical surface** S of radius R and oriented away from the z-axis is

$$\int_T \vec{F}(R, \theta, z) \cdot \left(\cos\theta\vec{i} + \sin\theta\vec{j} \right) R\, dz\, d\theta \quad \text{(p. 1032)},$$

where T is the θz-region corresponding to S.

The **flux through a spherical surface** S of radius R and oriented away from the origin is

$$\int_T \vec{F}(R, \theta, \phi) \cdot \left(\sin\phi\cos\theta\vec{i} + \sin\phi\sin\theta\vec{j} + \cos\phi\vec{k} \right)$$
$$R^2\sin\phi\, d\phi\, d\theta, \quad \text{(p. 1034)}$$

where T is the $\theta\phi$-region corresponding to S.

The **flux through a parameterized surface** S, parameterized by $\vec{r} = \vec{r}(s, t)$, where (s, t) varies in a parameter region R, is

$$\int_R \vec{F}(\vec{r}(s, t)) \cdot \left(\frac{\partial\vec{r}}{\partial s} \times \frac{\partial\vec{r}}{\partial t} \right) ds\, dt \quad \text{(p. 1095)}.$$

We choose the parameterization so that $\partial\vec{r}/\partial s \times \partial\vec{r}/\partial t$ is never zero and points in the direction of $\vec{n}$ everywhere.

The **area of a parameterized surface** S, parameterized by $\vec{r} = \vec{r}(s,t)$, where (s,t) varies in a parameter region R, is

$$\int_S dA = \int_R \left\| \frac{\partial \vec{r}}{\partial s} \times \frac{\partial \vec{r}}{\partial t} \right\| ds\, dt \quad \text{(p. 1096)}.$$

Divergence and Curl

Divergence

Definition of Divergence (p. 1039).
Geometric definition: The **divergence** of $\vec{F}$ is

$$\text{div}\, \vec{F}(x,y,z) = \lim_{\text{Volume} \to 0} \frac{\int_S \vec{F} \cdot d\vec{A}}{\text{Volume of } S}.$$

Here S is a sphere centered at (x,y,z), oriented outwards, that contracts down to (x,y,z) in the limit.
Cartesian coordinate definition: If $\vec{F} = F_1\vec{i} + F_2\vec{j} + F_3\vec{k}$, then

$$\text{div}\, \vec{F} = \frac{\partial F_1}{\partial x} + \frac{\partial F_2}{\partial y} + \frac{\partial F_3}{\partial z}.$$

The divergence can be thought of as the outflow per unit volume of the vector field. A vector field $\vec{F}$ is said to be **divergence free** or **solenoidal** if $\text{div}\,\vec{F} = 0$ everywhere that $\vec{F}$ is defined. Magnetic fields are divergence free (p. 1042).

The Divergence Theorem (p. 1049). If W is a solid region whose boundary S is a piecewise smooth surface, and if $\vec{F}$ is a smooth vector field which is defined everywhere in W and on S, then

$$\int_S \vec{F} \cdot d\vec{A} = \int_W \text{div}\, \vec{F}\, dV,$$

where S is given the outward orientation. In words, the Divergence Theorem says that the total flux out of a closed surface is the integral of the flux density over the volume it encloses.

Curl
The **circulation density** of a smooth vector field $\vec{F}$ at (x,y,z) around the direction of the unit vector $\vec{n}$ is defined to be

$$\text{circ}_{\vec{n}}\, \vec{F}(x,y,z) = \lim_{\text{Area} \to 0} \frac{\text{Circulation around } C}{\text{Area inside } C}$$

$$= \lim_{\text{Area} \to 0} \frac{\int_C \vec{F} \cdot d\vec{r}}{\text{Area inside } C} \quad \text{(p. 1056)}.$$

Circulation density is calculated using the **right-hand rule** (p. 1056).

Definition of curl (p. 1057).
Geometric definition The curl of $\vec{F}$, written $\text{curl}\, \vec{F}$, is the vector field with the following properties

- The direction of $\text{curl}\, \vec{F}(x,y,z)$ is the direction $\vec{n}$ for which $\text{circ}_{\vec{n}}(x,y,z)$ is greatest.

- The magnitude of $\text{curl}\, \vec{F}(x,y,z)$ is the circulation density of $\vec{F}$ around that direction.

Cartesian coordinate definition If $\vec{F} = F_1\vec{i} + F_2\vec{j} + F_3\vec{k}$, then

$$\text{curl}\, \vec{F} = \left(\frac{\partial F_3}{\partial y} - \frac{\partial F_2}{\partial z} \right)\vec{i} + \left(\frac{\partial F_1}{\partial z} - \frac{\partial F_3}{\partial x} \right)\vec{j} + \left(\frac{\partial F_2}{\partial x} - \frac{\partial F_1}{\partial y} \right)\vec{k}.$$

Curl and circulation density are related by $\text{circ}_{\vec{n}}\, \vec{F} = \text{curl}\, \vec{F} \cdot \vec{n}$ (p. 1059). A vector field is said to be **curl free** or **irrotational** if $\text{curl}\, \vec{F} = \vec{0}$ everywhere that $\vec{F}$ is defined (p. 1060).

Given an oriented surface S with a boundary curve C we use the right-hand rule to determine the orientation of C (p. 1064).

Stokes' Theorem (p. 1065). If S is a smooth oriented surface with piecewise smooth, oriented boundary C, and if $\vec{F}$ is a smooth vector field which is defined on S and C, then

$$\int_C \vec{F} \cdot d\vec{r} = \int_S \text{curl}\, \vec{F} \cdot d\vec{A}.$$

Stokes' Theorem says that the total circulation around C is the integral over S of the circulation density. A **curl field** is a vector field $\vec{F}$ that can be written as $\vec{F} = \text{curl}\, \vec{G}$ for some vector field $\vec{G}$, called a **vector potential** for $\vec{F}$ (p. 1067).

Relation between divergence, gradient, and curl
The curl and gradient are related by $\text{curl}\,\text{grad}\, f = 0$ (p. 1072). Divergence and curl are related by $\text{div}\,\text{curl}\, \vec{F} = 0$ (p. 1073).

The curl test for vector fields in 3-space (p. 1072) Suppose that $\text{curl}\, \vec{F} = \vec{0}$, and that the domain of $\vec{F}$ has the property that every closed curve in it can be contracted to a point in a smooth way, staying at all times within the domain. Then $\vec{F}$ is path-independent, so $\vec{F}$ is a gradient field and has a potential function.

The divergence test for vector fields in 3-space (p. 1073) Suppose that $\text{div}\, \vec{F} = 0$, and that the domain of $\vec{F}$ has the property that every closed surface in it is the boundary of a solid region completely contained in the domain. Then $\vec{F}$ is a curl field.

Section 12.1

1 Q

3 $(-4, 2, 7)$

5 $(1, -1, 1)$; Front, left, above

7 $(2, 4.5, 3)$

9 $(-1, 1, 0), (-2, -2, 4)$

11 $(2, 2, 4), (-2, -2, 4)$

13

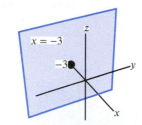

15

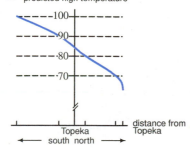

17 $x^2 + y^2 + z^2 = 25$

19 $y = 3$

21 predicted high temperature

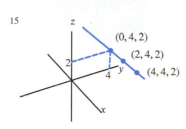

23 $f(w, 60)$: 23.4, 27.3, 31.2, 35.2, 39.1

25 25

29 $f(20, p)$: 2.65, 2.59, 2.51, 2.43
$f(100, p)$: 5.79, 5.77, 5.60, 5.53
$f(I, 3.00)$: 2.65, 4.14, 5.11, 5.35, 5.79
$f(I, 4.00)$: 2.51, 3.94, 4.97, 5.19, 5.60

31 Increasing function

33 57.9 kg

37 (a) $R = 100s + 5m$
   (b) 125,000 dollars

39 (b) Increasing
   (c) Decreasing

41 (b) Increasing
   (c) Increasing

43 $(1.5, 0.5, -0.5)$

45 Cone, tip at origin, along x-axis with slope of 1

47 Yes; $(2, 5, 4)$

49 (a) $z = 7, z = -1$
   (b) $x = 6, x = -2$
   (c) $y = 7, y = -1$

51 (a) $(12, 7, 2); (5, 7, 2); (12, 1, 2)$
   (b) $(5, 1, 4); (5, 7, 4); (12, 1, 4)$

53 (a) $(3, 9, 13)$
   (b) $(2, 7, 10)$
   (c) $(4, 11, 16)$

55 xy-plane is $z = 0$
   $xy = 0$ is yz-plane and xz-plane

57 $f(x, y) = x - y$

59 True

61 False

63 True

65 False

67 False

69 False

71 True

Section 12.1 (online problems)

73 (a) yz-plane: circle $(y + 3)^2 + (z - 2)^2 = 3$
       xz-plane: none
       xy-plane: point $(1, -3, 0)$

   (b) Does not intersect

Section 12.2

1 (III)

3 (I), (IV)

5 (a) (IV)
   (b) (II)
   (c) (I)
   (d) (V)
   (e) (III)

7 (a) Decreases

   (b) Increases

9 Sphere, radius 3

11 Upside-down bowl, vertex $(0, 0, 5)$

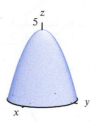

13 Plane, x-intercept 6, y-intercept 3, z-intercept 4

15 Circular cylinder extended in the y-direction

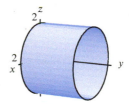

17 $x^2 + (y - \sqrt{7})^2 + z^2 = 9$

19 (a)

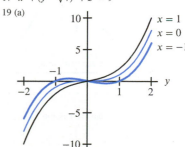

   (b)

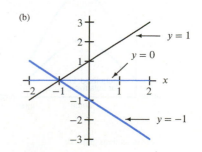

1108

21 (a)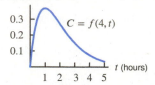

(b)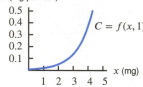

23 $f(2, 10) = 100$ joules

25 $f(2, 12) = 731.9$ millibars.

27 (I) cross-sections with x fixed, (II) cross-sections with y fixed

29 (a) Bowl
(b) Neither
(c) Plate
(d) Bowl
(e) Plate

31 (a) (II)
(b) (I)
(c) (III)
(d) (VI)
(e) (V)
(f) (IV)

33 Cross-sections graph I:

(a)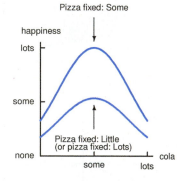

(b)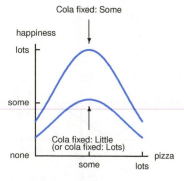

Cross-sections graph II:

(a)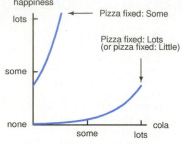

(b)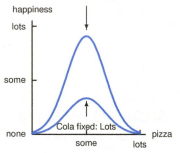

Cross-sections graph III:

(a)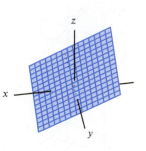

(b)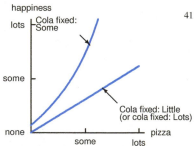

Cross-sections graph IV:

(a)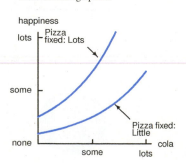

35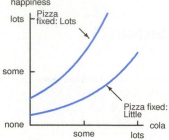

37

39 (a) $y = 0$
(b) $x = 0$
(c) $z = 1$

41 (a)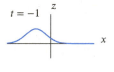
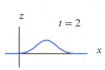

(b) Increasing x
(c)

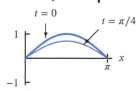

43 Graph is surface in 3-space

45 $f(x, y) = x^2 + y^2 + 2$

47 $f(x, y) = 1 - x^2 - y^2$

49 True

51 False

53 False

55 True

57 True

59 False

61 True

Section 12.2 (online problems)

63 (a)

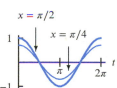

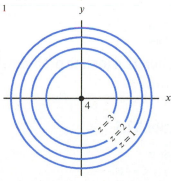

(b) $f = 0$; ends of string don't move

Section 12.3

1

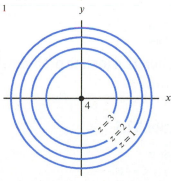

3

---

5 -1

7 $(0, 0)$, other answers possible.

9 $3x^2 y + 7x + 20 = 805$

11 (a) (III)
   (b) (I)
   (c) (V)
   (d) (II)
   (e) (IV)

13 Table 12.6 matches (II)
   Table 12.7 matches (III)
   Table 12.8 matches (IV)
   Table 12.9 matches (I)

15 Contours evenly spaced parallel lines

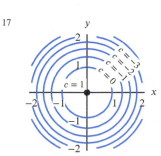

17

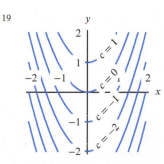

19

21

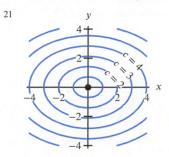

23

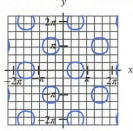

25 (a) is (II)
   (b) is (I)
   (c) is (III)

27 (a) -2 Grapes/Cherry
   (b) No change in happiness when replacing 2 grapes with one cherry

29 Underweight: below 18.5, Normal: 18.5-25

31 (a) is III
   (b) is VI
   (c) is I
   (d) is IV
   (e) is II
   (f) is V

33 Answers in °C:

   (a)

   (b)

   (c)

   (d)

35

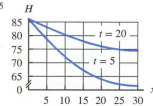

37 (a) 2
   (b) -2
   (c) -1

39 (a) $\sqrt{5}$
   (b) 0
   (c) 0

41 (a) About \$137
   (b) About \$250

   (a) About \$122

(b) About $350

43 Other answers possible

T°F

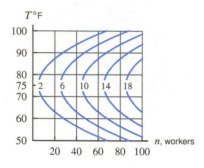

45 (a) II
   (b) IV
   (c) III
   (d) I

47 (a) $\pi = 3q_1 + 12q_2 - 4$ (thousands)
   (b) q_2

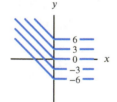

49 (a) (II) (E)
    (b) (I) (D)
    (c) (III) (G)

51 (a) (I) g
         (II) f
    (b) $0.2 < \alpha < 0.8$

53 (a)

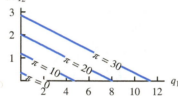

(b)

(c)

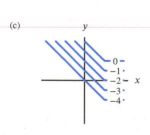

(d)

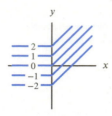

55 $y = 2x$, $y = (1/3)x$

59 (a) $P(d, v) = kd^2 v^3$
   (b) $1/\sqrt[3]{4}$
   (c)

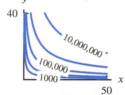

61 Spacing of the contours of f and g are different

63 $f(x, y) = y - x^2$

65 Might be true

67 Not true

69 True

71 True

73 False

75 False

Section 12.4

1 -1.0

3 Not a linear function

5 Linear function

7 $z = \frac{4}{3}x - \frac{1}{2}y$

9 $z = -2y + 2$

11 $\Delta z = 0.4$; $z = 2.4$

13 Linear

15 Basic subscription cost $8
    Premium subscription cost $12

17 (a) Linear
   (b) Linear
   (c) Not linear

19 180 lb person at 8 mph
    120 lb person at 10 mph

21 $g(x, y) = 3x + y$

23 Not linear

25 $f(x, y) = 2x - 0.5y + 1$

27 Could be linear; $z = -4 + x + 4y$

29 Could be linear; $z = 5 + (3/2)x$

31 Could be linear; $z = -5 + (3/2)x - 2y$

33 Could be linear; $z = 2x + y + 3$

35

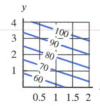

Other answers possible

37

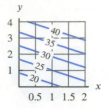

Other answers are possible

39

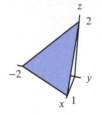

41

43 8

45 (a) Impossible
   (b) Impossible
   (c) 20

47 $f(x, y) = xy$ has linear cross-sections

49 $z = -2x + y$

51 False

53 True

55 True

57 False

59 False

61 False

Section 12.4 (online problems)

65 (a) $7/\sqrt{29}$
   (b) $-5/\sqrt{104}$

Section 12.5

1 (a) I
  (b) II

3 $f(x, y) = \frac{1}{3}(5 - x - 2y)$

5 $f(x, y) = (1 - x^2 - y)^2$

7 Elliptical and hyperbolic paraboloid, plane

9 Hyperboloid of two sheets

11 Ellipsoid

13 Yes, $f(x, y) = (2x + 3y - 10)/5$

15 No

17 $f(x, y) = 2x - (y/2) - 3$
    $g(x, y, z) = 4x - y - 2z = 6$

19 $f(x, y) = -\sqrt{2(1 - x^2 - y^2)}$
    $g(x, y, z) = x^2 + y^2 + z^2/2 = 1$

21 (a) 1596 kcal/day
   (b) 1284 kcal/day
   (c) Plane; weight, height, age combinations of woman whose BMR is 2000 kcal/day
   (d) Lose weight

23 (a) $P(1 + 0.01r)^t = 2653.3$
   (b) $(P, r, t) = (1628.9, 5, 10)$; other answers possible

25 (a) $r^2 h\rho = 120$
   (b) $(r, h, \rho) = (2, 5, 6)$; other answers possible

27 $f(x, y) = 3\sqrt{1 - x^2 - y^2/4}$;
   $g(x, y) = -3\sqrt{1 - x^2 - y^2/4}$

29 $f(x, y, z), r(x, y, z), m(x, y, z)$

31 (a) Graph of f is the graph of
   $y^2 + z^2 = 1, z \geq 0$
   (b) $\sqrt{1 - y^2} - z = 0$

33 Elliptical cylinder along y-axis

35 Parallel planes

37 Surface of rotation

39 Spheres

43
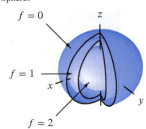

45 Vertical shifts

47 Graph of $f(x, y, z)$ needs 4 dimensions

49 Level surfaces cylinders

51 $f(x, y, z) = x^2 + z^2$

53 $f(x, y, z) = x^2 + y^2 - z$

55 False

57 False

59 False

61 True

63 True

Section 12.6

1 Not continuous

3 Continuous

5 Not continuous

7 1

9 0

11 1

13 2

15 Does not exist

23 No

25 $c = 1$

27 (c) No

29 For quotient, need $g(a, b) \neq 0$

31 $f(x, y) = (x^2 + 2y^2)/(x^2 + y^2)$

33 $f(x, y) = 1/((x - 2)^2 + y^2)$

35 False

37 True

39 True

Section 13.1

1 $\vec{a} = \vec{i} + 3\vec{j}$
   $\vec{b} = 3\vec{i} + 2\vec{j}$
   $\vec{v} = -2\vec{i} - 2\vec{j}$
   $\vec{w} = -\vec{i} + 2\vec{j}$

3 $-3\vec{i} - 4\vec{j}$

5 $\vec{a} = \vec{b} = \vec{c} = 3\vec{k}$
   $\vec{d} = 2\vec{i} + 3\vec{k}$
   $\vec{e} = \vec{j}$
   $\vec{f} = -2\vec{i}$

7 $\vec{i} + 3\vec{j}$

9 $-4.5\vec{i} + 8\vec{j} + 0.5\vec{k}$

11 $-3\vec{i} - 12\vec{j} + 3\vec{k}$

13 $0.9\vec{i} + 0.2\vec{j} - 0.7\vec{k}$

15 $\sqrt{6}$

17 $\sqrt{11}$

19 5.6

21 $-6\vec{i} + 20\vec{j} + 13\vec{k}$

23 $21\vec{j}$

25 $2\sqrt{73}$

27 $0.6\vec{i} - 0.8\vec{k}$

29 $-\vec{i}/2 + \vec{j}/4 + \sqrt{11}\vec{k}/4$

31 $0, -10$

33 (a) $(3/5)\vec{i} + (4/5)\vec{j}$
   (b) $6\vec{i} + 8\vec{j}$

35 (a) $\sqrt{2}\vec{i} + \sqrt{2}\vec{j}$
   (b) $(\sqrt{3}/2)\vec{i} + \vec{k}/2$

37 $\vec{p} = -\frac{4\sqrt{5}}{5}\vec{i} - \frac{2\sqrt{5}}{5}\vec{j}$

39 (a) $\vec{a} = \vec{i}$ and $\vec{b} = -\vec{i}$; other answers possible
   (b) $\vec{a} = (1/\sqrt{2})\vec{i} - (1/\sqrt{2})\vec{j}$ and $\vec{b} = (1/\sqrt{2})\vec{j} + (1/\sqrt{2})\vec{k}$; other answers possible
   (c) $\vec{a} = \vec{i}$ and $\vec{b} = \vec{i}$; other answers possible
   (d) Not possible

41 (a) $t = 1$
   (b) No t values
   (c) Any t values

43 $(\vec{i} + \vec{j})/\sqrt{2}, (\vec{i} - \vec{j})/\sqrt{2}, (-\vec{i} + \vec{j})/\sqrt{2}, (-\vec{i} - \vec{j})/\sqrt{2}$

45 (a) $(a, 0, 0)$
   (b) $\left(b/\sqrt{b^2 + c^2}\right)\vec{j} + \left(c/\sqrt{b^2 + c^2}\right)\vec{k}$

47 $\|\vec{u} + \vec{v}\|$ could be less than 1

49 Longer diagonal if angle between $\vec{u}$ and $\vec{v}$ more than 90°

51 $\vec{v} = \vec{j} + \sqrt{3}\vec{k}$

53 $\vec{u} = \vec{i}, \vec{v} = 3\vec{i} + 3\vec{j}$

55 False

57 False

59 False

61 False

63 False

Section 13.2

1 Scalar

3 Scalar

5 Vector

7 $-37.59\vec{i}, -13.68\vec{j}$

9 $21\vec{i} + 35\vec{j}$

11 (a) 8.64 km/hr
   (b) 0.093 radian or about 5° off course

13 (a) $17.93\vec{i} - 7.07\vec{j}$
   (b) 19.27 km/hr
   (c) 21.52° south of east

15 48.3° east of north
   744 km/hr

17 4.87° north of east
   540.63 km/hr

19 38.7° south of east

21 $-98.76\vec{i} + 18.94\vec{j} + 2998.31\vec{k}$
   2998.31 newtons directly up

23 $0.4v\vec{i} + 0.7v\vec{j}$

25 $0.1\vec{i} + 0.08\vec{j} + 0.1625\vec{k}$ or $(0.1, 0.08, 0.1625)$

37 $42.265\vec{i} + 42.265\vec{j} - 5.229\vec{k}$ mph

39 Not if $|\vec{j}$-component$| \geq 2$

41 Let $\vec{v} = c\vec{v}$ for any $c > 0$

43 No

45 No

47 No

Section 13.2 (online problems)

49 3.4° north of east

Section 13.3

1 14

3 29

5 $7.5\sqrt{2}$

7 -14

9 14

11 -2

13 $28\vec{j} + 14\vec{k}$

15 185

17 $\vec{i} + 3\vec{j} + 2\vec{k}$ (multiples of)

19 $3\vec{i} + 4\vec{j} - \vec{k}$ (multiples of)

21 $3x - y + 4z = 6$

23 $x - y + z = 3$

25 $2x + 4y - 3z = 5$

27 $2x - 3y + 5z = -17$

29 $2\pi/3$ radians (120°)

31 $\pi/6$ radians (30°)

33 (a) - (II)
   (b) - (I)
   (c) - (III)

35 (a) $(2/\sqrt{13})\vec{i} + (3/\sqrt{13})\vec{j}$
   (b) Multiples of $3\vec{i} - 2\vec{j}$

37 $-1/5$

39 (a) $\lambda = -2.5$
   (b) $a = -6.5$

41 (a) $(21/5, 0, 0)$
   (b) $(0, -21, 0); (0, 0, 3)$ (for example)
   (c) $\vec{n} = 5\vec{i} - \vec{j} + 7\vec{k}$ (for example)
   (d) $21\vec{j} + 3\vec{k}$ (for example)

43 Possible answers:
   (a) $2\vec{i} + 3\vec{j} - \vec{k}$
   (b) $3\vec{i} - 2\vec{j}$

45 (a) is (I); (b) is (III), (IV); (c) is (II), (III); (d) is (II)

47 $\vec{v}_1, \vec{v}_4, \vec{v}_8$ all parallel
   $\vec{v}_3, \vec{v}_5, \vec{v}_7$ all parallel
   $\vec{v}_1, \vec{v}_4, \vec{v}_8$ perpendicular to $\vec{v}_3, \vec{v}_5, \vec{v}_7$
   $\vec{v}_2$ and $\vec{v}_9$ perpendicular

49 $\vec{u} \perp \vec{v}$ for $t = 2$ or -1.
No values of t make $\vec{u}$ parallel to $\vec{v}$

51 2

53 $\sqrt{20}$

55 $\vec{a} = -\frac{8}{21}\vec{d} + (\frac{79}{21}\vec{i} + \frac{10}{21}\vec{j} - \frac{118}{21}\vec{k})$

57 Lengths: $\sqrt{34}, \sqrt{29}, \sqrt{13}$
Angles: $37.235°, 64.654°, 78.111°$

59 39 joules; 28.765 foot-pounds

61 120 foot-pounds; 162.698 joules

63 (a) $\vec{F}_{\text{parallel}} = -0.168\vec{i} - 0.224\vec{j}$
(b) $\vec{F}_{\text{perp}} = 0.368\vec{i} - 0.276\vec{j}$
(c) $W = -1.4$

65 (a) $\vec{F}_{\text{parallel}} = \vec{0}$
(b) $\vec{F}_{\text{perp}} = \vec{F}$
(c) $W = 0$

67 (a) $\vec{F}_{\text{parallel}} = \vec{F}$
(b) $\vec{F}_{\text{perp}} = \vec{0}$
(c) $W = -50$

69 (a) $\vec{F}_{\text{parallel}} = 3.846\vec{i} - 0.769\vec{j}$
(b) $\vec{F}_{\text{perp}} = -3.846\vec{i} - 19.231\vec{j}$
(c) $W = 20$

71 (a) $\vec{F}_{\text{parallel}} = \vec{0}$
(b) $\vec{F}_{\text{perp}} = \vec{F}$
(c) $W = 0$

73 $70.529°$

75 $\vec{w}_4$ increases most
$\vec{w}_3$ decreases most

77 (a) $(20, 20, -10)$
(b) 108.167

79 $710 revenue

87 Can't take dot product of a scalar and a vector

89 Normal vector is $2\vec{i} + 3\vec{j} - \vec{k}$

91 $f(x, y) = (-1/3)x + (-2/3)y$

93 True

95 False

97 True

99 False

101 True

103 True

Section 13.4

1 $-\vec{i}$

3 $-\vec{i} + \vec{j} + \vec{k}$

5 $\vec{i} + 3\vec{j} + 7\vec{k}$

7 $7\vec{i} + \vec{j} + 4\vec{k}$

9 $-2\vec{k}$

11 $\vec{i} - \vec{j}$

13 $\vec{v} \times \vec{w} = -6\vec{i} + 7\vec{j} + 8\vec{k}$
$\vec{w} \times \vec{v} = 6\vec{i} - 7\vec{j} - 8\vec{k}$
$\vec{v} \times \vec{w} = -(\vec{w} \times \vec{v})$

15 $x - y - z = -3$

17 4

19 0

21 $-\vec{i} - \vec{j} - \vec{k}$

23 $\vec{0}$

25 $x + 2y + 2z = 0$

27 $3x - y - 2z = 0$

29 $4\vec{i} + 26\vec{j} + 14\vec{k}$

---

31 $4(x - 4) + 26(y - 5) + 14(z - 6) = 0$

33 (a) $\vec{u}$ and $-\vec{u}$ where
$\vec{u} = \frac{12}{13}\vec{i} - \frac{4}{13}\vec{j} - \frac{3}{13}\vec{k}$
(b) $\theta \approx 49.76°$
(c) $13/2$
(d) $13/\sqrt{29}$

35 (a) $(4, 0, 0)$
(b) $(0, 2, 0)$
(c) $(0, 0, 4)$
(d) 9.798

37 (a) 0.6
(b) 0.540

39 (a) 1.625
(b) 1.019

41 (a) Increases force
(b) Ball moves down and to the left

43 (b) $(-y, x)$

45 $\vec{i} - 3\vec{j} - 5\vec{k}$

47 (a) $4\vec{k}$
(b) $3\vec{j}$
(c) $2\vec{i}$

49 $\theta = \pi/4$ or $3\pi/4$

51 $0 \leq \theta < \pi/4$ or $3\pi/4 < \theta \leq \pi$

55 $4\pi\vec{i}$

57 (a) $((u_2 v_3 - u_3 v_2)^2 + (u_3 v_1 - u_1 v_3)^2 + (u_1 v_2 - u_2 v_1)^2)^{1/2}$
(b) $|u_1 v_2 - u_2 v_1|$
(c) $m = (u_2 v_3 - u_3 v_2)/(u_2 v_1 - u_1 v_2)$,
$n = (u_3 v_1 - u_1 v_3)/(u_2 v_1 - u_1 v_2)$

59 Parallel, not perpendicular

61 $\vec{v} = (8\vec{i} - 6\vec{j})/5$

63 False

65 True

67 True

69 True

71 False

Section 14.1

1 $f_x(3, 2) \approx -2/5$; $f_y(3, 2) \approx 3/5$

3 $-0.0493, -0.3660$
$-0.0501, -0.3629$

5 $\partial P/\partial t$:
dollars/month
Rate of change in payments with time
negative
$\partial P/\partial r$:
dollars/percentage point
Rate of change in payments with interest rate
positive

7 (a) Payment $376.59/mo at 1% for 24 mos
(b) 4.7¢ extra/mo for $1 increase
(c) Approx $44.83 increase for 1% interest increase

9 (a) Negative
(b) Positive

11 (a) $f(A) = 15$
(b) Zero
(c) Negative

13 (a) $f(A) = 88$
(b) Negative
(c) Negative

15 (a) $f(A) = 40$
(b) Negative
(c) Positive

17 $f_x > 0, f_y < 0$

---

19 $f_x < 0, f_y > 0$

21 Positive, Negative, 10, 2, -4

23 (i)(c); (ii)(a)

25 (a) Both negative
(b) Both negative

27 $f_T(5, 20) \approx 1.2°\text{F}/°\text{F}$

29 -1.5 and -1.22

31 (a)

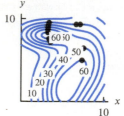

(b)

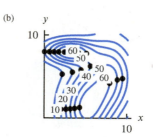

33 (a) Negative
(b) Positive

35 (a) $2.5, 0.02$
(b) $3.33, 0.02$
(c) $3.33, 0.02$

37 -2.5

39

	w (gm/m³)		
	0.1	0.2	0.3
T (°C) 10	1300	900	1200
20	800	800	900
30	800	700	800

41 (a)

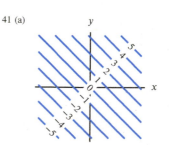

(b)

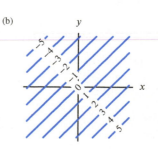

(c)

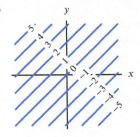

(d)

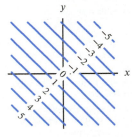

43 There are many possibilities.

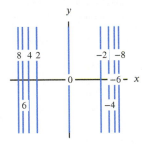

45 There are many possibilities.

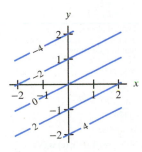

47 $f_y = 0$

49 $f(x, y) = 4x - y$

51 False

53 False

55 True

57 True

59 True

Section 14.2

1 (a) 7.01
   (b) 7

3 $f_x(1, 2) = 15$,
   $f_y(1, 2) = -5$

---

5 $\partial z/\partial x = \frac{14x+7}{(x^2+x-y)^{-6}}$
   $\partial z/\partial y = -7(x^2 + x - y)^6$

7 $f_x = 0.6/x$
   $f_y = 0.4/y$

9 $2xy + 10x^4 y$

11 $V_r = \frac{2}{3}\pi rh$

13 $e^{\sqrt{xy}}(1 + \sqrt{xy}/2)$

15 g

17 $(a + b)/2$

19 $2B/u_0$

21 $2mv/r$

23 $(15x^2 y - 3y^2) \cos(5x^3 y - 3xy^2)$

25 y

27 $z_x = 7x^6 + yx^{y-1}$
   $z_y = 2^y \ln 2 + x^y \ln x$

29 Gm_1/r^2

31 $-e^{-x^2/a^2}(a^2 - 2x^2)/a^4$

33 $\cos(\pi\theta\phi) \cdot \pi\phi + \frac{1}{\theta^2 + \phi} \cdot 2\theta$

35 $f_a = e^a \sin(a + b) + e^a \cos(a + b)$

37 $\partial V/\partial r = \frac{8}{3}\pi rh$,
   $\partial V/\partial h = \frac{4}{3}\pi r^2$

39 $-(x - \mu)e^{-(x-\mu)^2/(2\sigma^2)}/(\sqrt{2\pi}\sigma^3)$

41 (a) $f_x(1, 1) = 2; f_y(1, 1) = 2$
   (b) (II)

43 (a) $f_x(1, 1) = 2; f_y(1, 1) = 2e$
   (b) (I)

45 (a) $f_w(2, 2) \approx 2.78$
       $f_z(2, 2) \approx 4.01$
   (b) $f_w(2, 2) \approx 2.773$
       $f_z(2, 2) = 4$

47 (a) Pre^{rt}
   (b) e^{rt}

49 1.277 m^2, 0.005 m^2/kg, 0.006 m^2/cm

51 $h_x(2, 5) \approx -0.38$ cm/meter
   $h_t(2, 5) \approx 0.76$ cm/second

55 No such points exist

57 $(0, 1), (0, -1), (-2, 1)$ and $(-2, -1)$

59 $\partial f/\partial x$ or $\partial f/\partial y$?

61 $f(x, y) = 2x + 3y + x^2$

63 $f(x, y) = y^2 + 1$

65 True

67 False

69 False

71 False

Section 14.3

1 $z = ex$

3 $z = 6y - 9$

5 $z = -4 + 2x + 4y$

7 $z = -36x - 24y + 148$

9 $df = y\cos(xy)\,dx + x\cos(xy)\,dy$

11 $dz = -e^{-x}\cos(y)dx - e^{-x}\sin(y)dy$

13 $dg = 4\,dx$

15 $dP \approx 2.395\,dK + 0.008\,dL$

17 -5

19 0.99

21 95

23 12.005

---

25 (a) Dollars/Square foot
   (b) Larger plots at same distance $3/ft^2 more
   (c) Dollars/Foot
   (d) Farther from beach but same area $2/ft
       less
   (e) 998 ft^2

27 (b) $f(x, y) \approx$
       $0.3345 - 0.33(x - 1) - 0.15(y - 2)$
   (c) $f(x, y) \approx 0.3345 -$
       $0.3345(x - 1) - 0.1531(y - 2)$

29 (a) $f_x(1, 2) = 3; f_y(1, 2) = 2$
   (b) 2
   (c) 2.1

31 $df = -3dx + 2dy$ at $(2, -4)$

33 376

35 $df = \frac{1}{3}dx + 2dy$
   $f(1.04, 1.98) \approx 2.973$

37 136.09°C

39 $P(r, L) \approx$
       $80 + 2.5(r - 8) + 0.02(L - 4000)$
   $P(r, L) \approx$
       $120 + 3.33(r - 8) + 0.02(L - 6000)$
   $P(r, L) \approx$
       $160 + 3.33(r - 13) + 0.02(L - 7000)$

43 (a) $nRT/(V - nb) - n^2 a/V^2$
   (b) $\Delta P \approx (nR/(V_0 - nb))\Delta T + (2n^2 a/V_0^3 -$
       $nRT_0/((V_0 - nb)^2))\Delta V$

45 (a) $d\rho = -\beta\rho\,dT$
   (b) 0.00015, $\beta \approx 0.0005$

47 $-43200\Delta t$
   Slow if $\Delta t > 0$; fast if $\Delta t < 0$

49 (a) $4x\,dx = 2y\,dy + 6z\,dz$
   (b) $dz = \frac{2}{3}dx - \frac{1}{2}dy$
   (c) $z = 2 + \frac{2}{3}(x - 2) - \frac{1}{2}(y - 3)$

51 (a) $e^y\,dx + xe^y\,dy + 2z\,dz = -\sin(x-1)\,dx + \frac{z}{\sqrt{z^2+3}}\,dz$
   (b) $dz = -\frac{2}{3}dx - \frac{2}{3}dy$
   (c) $z = 1 - \frac{2}{3}(x - 1) - \frac{2}{3}(y - 0)$

53 $z = f(3, 4) + f_x(3, 4)(x - 3) + f_y(3, 4)(y - 4)$

55 Equation not linear

57 sphere of radius 3 centered at the origin

59 True

61 False

63 True

65 False

Section 14.4

1 $(\frac{15}{2}x^4)\vec{i} - (\frac{24}{7}y^5)\vec{j}$

3 $2m\vec{i} + 2n\vec{j}$

5 $\left(\frac{5\alpha}{\sqrt{5\alpha^2+\beta}}\right)\vec{i} + \left(\frac{1}{2\sqrt{5\alpha^2+\beta}}\right)\vec{j}$

7 $\nabla z = e^y\vec{i} + e^y(1 + x + y)\vec{j}$

9 $\sin\theta\vec{i} + r\cos\theta\vec{j}$

11 $\nabla z = \frac{1}{y}\cos\left(\frac{x}{y}\right)\vec{i} - \frac{x}{y^2}\cos\left(\frac{x}{y}\right)\vec{j}$

13 $\left(\frac{-12\beta}{(2\alpha - 3\beta)^2}\right)\vec{i} + \left(\frac{12\alpha}{(2\alpha - 3\beta)^2}\right)\vec{j}$

15 $60\vec{i} + 85\vec{j}$

17 $10\pi\vec{i} + 4\pi\vec{j}$

19 $(\pi/2)^{1/2}\vec{i}$

21 $\frac{1}{100}(2\vec{i} - 6\vec{j})$

23 $\vec{i}$

25 $\vec{i} + \vec{j}$

27 $\vec{i} - \vec{j}$

29 Negative

31 Negative

33 Approximately zero

35 $-\vec{i}$

37 $\vec{i}$

39 $-\vec{i} + \vec{j}$

41 $\vec{i} + \vec{j}$

43 6.325

45 $-46/5$

47 $22/5$

49 $84/5$

51 $(2x + 3e^y)dx + 3xe^y dy$

53 $(x+1)ye^x\vec{i} + xe^x\vec{j}$

55 50.2

57 (a) Should be number
   (b) 11/5

59 -2

61 $\vec{i} + 2\vec{j}$ or any multiple

63 0.316

65 1

67 100

69 (a) $-\sqrt{2}/2$
   (b) $\sqrt{3} + 1/2$

71 (a) $2/\sqrt{13}$
   (b) $1/\sqrt{17}$
   (c) $\vec{i} + \frac{1}{2}\vec{j}$

73 (a) $5/\sqrt{2}$
   (b) 510

75 (a) $-16\vec{i} + 12\vec{j}$
   (b) $16\vec{i} - 12\vec{j}$
   (c) $12\vec{i} + 16\vec{j}$; answers may vary

77 1.7; closer estimate is 1.35

79 2.5; better estimate is 1.8

81 -0.9; better estimate is -1.8

83 Fourth quadrant

85 (a) Negative
   (b) Positive
   (c) Positive
   (d) Negative

87 $f_{\vec{u}}(P) < f_{\vec{w}}(P) < f_{\vec{v}}(P)$

89 $f(P) \approx 6$, $f(Q) \approx -24$

91 $3\vec{i} + 2\vec{j}$; $3(x - 2) + 2(y - 3) = 0$

93 $-5\vec{i}$; $x = 2$

95 $5/\sqrt{2}$

97 (a) ellipses centered at $(0, 0)$
   (b) decreasing at 49.9°C per meter
   (c) $-\vec{i} - 2\vec{j}$. Other answers possible

99 (a) $\sqrt{13}$ meters ascended/horizontal meter
   (b) 3.54 meters ascended/horizontal meter
   (c) $\vec{u} = 3\vec{i} + 2\vec{j}$; $\vec{u} = -3\vec{i} - 2\vec{j}$

101 grad $f(0, 0)$ is vector, not scalar

103 $-\vec{i}$

105 False

107 False

109 True

111 False

113 True

115 True

Section 14.4 (online problems)

117 (a) Perpendicular to contour of f at P
   (b) Maximum directional derivative of f at P
   (c) Directional derivative $f_{\vec{u}}(P)$

119 19.612

121 356.5

123 (a) -3.268
   (b) -4.919

125 Yes

127 Yes

129 (a)

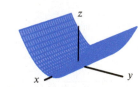

(b)

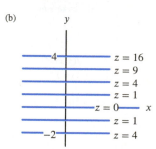

(c) $\vec{j}$

131 (a) Circles centered at P
   (b) away from P
   (c) 1

133 $(3\sqrt{5} - 2\sqrt{2})\vec{i} + (4\sqrt{2} - 3\sqrt{5})\vec{j}$

135 $4\sqrt{2}$,
   $6\vec{i} + 2\vec{j}$

139 (a) $\sqrt{m^2 + n^2}$
   (b) $(C_2 - C_1)/\sqrt{m^2 + n^2}$

141 (a) $a\vec{i} + 2b\vec{j}$
   (b) $\sqrt{(2+a)/(2-a)}$
   (c) $\sqrt{(2-a)/(2+a)}$

143 True

145 True

147 False

Section 14.5

1 $2x\vec{i}$

3 $e^x e^y e^z(\vec{i} + \vec{j} + \vec{k})$

5 $\dfrac{-2xyz^2}{(1+x^2)^2}\vec{i} + \dfrac{z^2}{1+x^2}\vec{j} + \dfrac{2yz}{1+x^2}\vec{k}$

7 $(x\vec{i} + y\vec{j} + z\vec{k})/\sqrt{x^2 + y^2 + z^2}$

9 $y\vec{i} + x\vec{j} + e^z\cos(e^z)\vec{k}$

11 $e^p\vec{i} + (1/q)\vec{j} + 2re^{r^2}\vec{k}$

13 $\vec{0}$

15 $6\vec{i} + 4\vec{j} - 4\vec{k}$

17 $-\pi\vec{i} - \pi\vec{k}$

19 $9/\sqrt{3}$

21 $-1/\sqrt{2}$

23 $-\sqrt{77/2}$

25 $-2\vec{i} - 2\vec{j} + 4\vec{k}$;
   $-2(x+1) - 2(y-1) + 4(z-2) = 0$

27 $2\vec{j} - 4\vec{k}$; $2(y-1) - 4(z-2) = 0$

29 $-2\vec{i} + \vec{k}$; $-2(x+1) + (z-2) = 0$

31 $6(x-1) + 3(y-2) + 2(z-1) = 0$

33 $2x + 3y + 2z = 17$

35 $z = 2x + y + 3$

37 $x + 4y + 10z = 18$

39 $10/3$

41 $2z + 3x + 2y = 17$

43 $x + 3y + 7z = -9$
   $\vec{i} + 3\vec{j} + 7\vec{k}$

45 grad $g(-1, -1)$ lies directly under path of steepest descent

47 (a) $(x - 2) + 4(y - 3) - 6(z - 1) = 0$
   (b) $z = 1 + (1/6)(x-2) + (2/3)(y-3)$

49 $3x + 10y - 5z + 19 = 0$

51 $16/\sqrt{14}$

53 (a) Spheres centered at the origin
   (b) $2x\sin(x^2 + y^2 + z^2)\vec{i} + 2y\sin(x^2 + y^2 + z^2)\vec{j} + 2z\sin(x^2 + y^2 + z^2)\vec{k}$
   (c) $0, 180°$

55 (a) Circle: $(y+1)^2 + (z-3)^2 = 10$
   (b) Yes
   (c) Multiples of $-10\vec{i} + 4\vec{j} - 12\vec{k}$

57 (a) $(-3\vec{i} + 6\vec{j} + 12\vec{k})/\sqrt{21}$
   (b) $(8.345, 2.309, 4.619)$

59 Any multiple of $2\vec{i} + 2\vec{j} + \vec{k}$

61 $(-1/6, 1/3, -1/12)$

63 (a) $6.33\vec{i} + 0.76\vec{j}$
   (b) -34.69

65 (a) 23
   (b) -9.2
   (c) $-16\vec{i} + 6\vec{j}$
   (d) $16x - 6y - z = 23$

67 (a) Parallel planes: $2x - 3y + z = T - 10$
   (b) $f_z(0,0,0) = 1$, temp increases 1°C per unit in z-direction
   (c) $2\vec{i} - 3\vec{j} + \vec{k}$
   (d) Yes; 27°C

69 (a) $-25/\sqrt{21}$
   (b) $-8\vec{i} + 7\vec{j} + 4\vec{k}$
   (c) $\sqrt{129}$

71 1.131 atm/sec

73 (a) is (V); (b) is (IV); (c) is (V)

75 $f_x(0,0,0)x + f_y(0,0,0)y + f_z(0,0,0)z = 0$

77 $f(x, y, z) = 2x + 3y + 4z + 100$

79 False

81 False

Section 14.6

1 $\dfrac{dz}{dt} = e^{-t}\sin(t)(2\cos t - \sin t)$

3 $2\cos\left(\dfrac{2t}{1-t^2}\right)\dfrac{1+t^2}{(1-t^2)^2}$

5 $2e^{1-t^2}(1 - 2t^2)$

7 $\dfrac{\partial z}{\partial u} = \dfrac{1}{vu}\cos\left(\dfrac{\ln u}{v}\right)$

$$\frac{\partial z}{\partial v} = -\frac{\ln u}{v^2}\cos\left(\frac{\ln u}{v}\right)$$

9   $\dfrac{\partial z}{\partial u} = \dfrac{e^v}{u}$

    $\dfrac{\partial z}{\partial v} = e^v \ln u$

11   $\dfrac{\partial z}{\partial u} = 2ue^{(u^2-v^2)}(1+u^2+v^2)$

    $\dfrac{\partial z}{\partial v} = 2ve^{(u^2-v^2)}(1-u^2-v^2)$

13   $\dfrac{\partial z}{\partial u} =$

    $(e^{-v\cos u} - v(\cos u)e^{-u\sin v})\sin v$
    $- (-u(\sin v)e^{-v\cos u} + e^{-u\sin v})v\sin u$

    $\dfrac{\partial z}{\partial v} =$

    $(e^{-v\cos u} - v(\cos u)e^{-u\sin v})u\cos v$
    $+ (-u(\sin v)e^{-v\cos u} + e^{-u\sin v})\cos u$

15   $\dfrac{\partial z}{\partial u} = \dfrac{-2uv^2}{u^4 + v^4}$

    $\dfrac{\partial z}{\partial v} = \dfrac{2vu^2}{u^4 + v^4}$

17   $-2\rho\cos 2\phi,\ 0$

19   (a) $\partial f/\partial t$
    (b) $(\partial f/\partial x)(dx/dt)$
    (c) $(\partial f/\partial y)(dy/dt)$

21   -5 pascal/hour

23   -0.6

25   (a) $1/\sqrt{10} = 0.316\,°\text{F/mile}$
    (b) $2.5/\sqrt{10} = 0.791\,°\text{F/hr}$
    (c) $2.5\,°\text{F/hr}$

27   Three

29   $\dfrac{dw}{dt} = \dfrac{\partial w}{\partial x}\dfrac{dx}{dt} + \dfrac{\partial w}{\partial y}\dfrac{dy}{dt} + \dfrac{\partial w}{\partial z}\dfrac{dz}{dt}$

31   (a) $F_u(x,3)$
    (b) $F_v(3,x)$
    (c) $F_u(x,x) + F_v(x,x)$
    (d) $F_u(5x,x^2)(5) + F_v(5x,x^2)(2x)$

33   $b \cdot e + d \cdot p$

35   $b \cdot e + d \cdot p$

37   (a) $\dfrac{\partial z}{\partial r} = \cos\theta\dfrac{\partial z}{\partial x} + \sin\theta\dfrac{\partial z}{\partial y}$

       $\dfrac{\partial z}{\partial \theta} = r(\cos\theta\dfrac{\partial z}{\partial y} - \sin\theta\dfrac{\partial z}{\partial x})$

    (b) $\dfrac{\partial z}{\partial y} = \sin\theta\dfrac{\partial z}{\partial r} + \dfrac{\cos\theta}{r}\dfrac{\partial z}{\partial \theta}$

       $\dfrac{\partial z}{\partial x} = \cos\theta\dfrac{\partial z}{\partial r} - \dfrac{\sin\theta}{r}\dfrac{\partial z}{\partial \theta}$

39   $\left(\dfrac{\partial U_3}{\partial P}\right)_V$

41   $\left(\dfrac{\partial U}{\partial T}\right)_V = 7/2$

    $\left(\dfrac{\partial U}{\partial V}\right)_T = 11/4$

45   $dz/dt = f_x(g(t),h(t))g'(t) +$
    $f_y(g(t),h(t))h'(t)$

47   $dz/dt|_{t=0} = f_x(2,3)g'(0) + f_y(2,3)h'(0)$

49   $f(x,y) = 4x + 2y$

51   $w = uv$, $u = 2s^2 + t$ and $v = e^{st}$, many other
    answers are possible

Section 14.6 (online problems)
    (c)

57   $\int_0^b F_u(x,y)\,dy$

Section 14.7

1   $f_{xx} = 2$
   $f_{yy} = 2$
   $f_{yx} = 2$
   $f_{xy} = 2$

3   $f_{xx} = 6y$
   $f_{xy} = 6x + 15y^2$

   $f_{yx} = 6x + 15y^2$
   $f_{yy} = 30xy$

5   $f_{xx} = 0$
   $f_{yx} = e^y = f_{xy}$
   $f_{yy} = e^y(x + 2 + y)$

7   $f_{xx} = -\left(\sin\left(\frac{x}{y}\right)\right)\left(\frac{1}{y^2}\right)$

   $f_{xy} = -\left(\sin\left(\frac{x}{y}\right)\right)\left(\frac{-x}{y^2}\right)\left(\frac{1}{y}\right)$

     $+ \left(\cos\left(\frac{x}{y}\right)\right)\left(\frac{-1}{y^2}\right) = f_{yx}$

   $f_{yy} = -\left(\sin\left(\frac{x}{y}\right)\right)\left(\frac{-x}{y^2}\right)^2$

     $+ \left(\cos\left(\frac{x}{y}\right)\right)\left(\frac{2x}{y^3}\right)$

9   $f_{xx} = 30xy^2 + 18$
   $f_{xy} = 30x^2y - 21y^2$
   $f_{yx} = 30x^2y - 21y^2$
   $f_{yy} = 10x^3 - 42xy$

11   $f_{xx} = -12\sin 2x\cos 5y$
   $f_{xy} = -30\cos 2x\sin 5y$
   $f_{yx} = -30\cos 2x\sin 5y$
   $f_{yy} = -75\sin 2x\cos 5y$

13   $Q(x,y) = 1 + 2x - 2y + x^2 - 2xy + y^2$

15   $Q(x,y) = 1 + x + x^2/2 - y^2/2$

17   $Q(x,y) = 1 - x^2/2 - 3xy - (9/2)y^2$

19   $Q(x,y) = -y + x^2 - y^2/2$

21   $1 + x - y/2 - x^2/2 + xy/2 - y^2/8$

23   (a)   Negative
    (b)   Zero
    (c)   Negative
    (d)   Zero
    (e)   Zero

25   (a)   Positive
    (b)   Zero
    (c)   Positive
    (d)   Zero
    (e)   Zero

27   (a)   Zero
    (b)   Negative
    (c)   Zero
    (d)   Negative
    (e)   Zero

29   (a)   Positive
    (b)   Positive
    (c)   Zero
    (d)   Zero
    (e)   Zero

31   (a)   Positive
    (b)   Negative
    (c)   Negative
    (d)   Negative
    (e)   Positive

33   (a)   Positive
    (b)   Negative
    (c)   Positive
    (d)   Negative
    (e)   Negative

35   -8

37   3

39   Not possible

41   6

43   $L(x,y) = y$
    $Q(x,y) = y + 2(x-1)y$
    $L(0.9,0.2) = 0.2$
    $Q(0.9,0.2) = 0.16$
    $f(0.9,0.2) = 0.162$

49   $a = -b^2$

51   Positive or zero, negative or zero

53   (a) $z_{yx} = 4y$

    (b) $z_{xyx} = 0$
    (c) $z_{xyy} = 4$

55   $d = e = f = 0$

57   $d = 0,\ e > 0,\ f < 0$

59   (a)

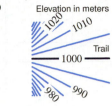

Elevation in meters

    (b) $\partial h/\partial x = 0$, $\partial h/\partial y > 0$, $(\partial^2 h)/(\partial x\partial y) < 0$
    (c) $(\partial^2 h)/(\partial x\partial y)$

61   (a) A
    (b) B

63   (a)   (II)
    (b)   (I)
    (c)   (III)

65   (a) xy
    $1 - \frac{1}{2}(x - \frac{\pi}{2})^2 - \frac{1}{2}(y - \frac{\pi}{2})^2$
    (b)

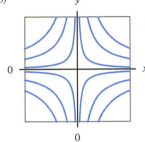

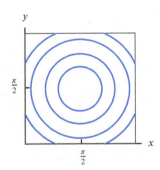

67   $f(x,y)$:

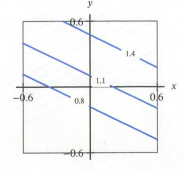

$L(x, y)$:

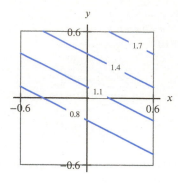

$Q(x, y)$:

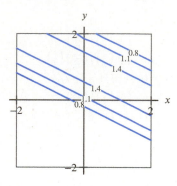

$Q(x, y)$:

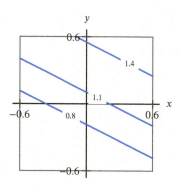

$f(x, y)$:

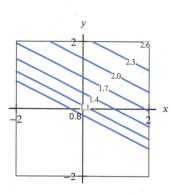

$L(x, y)$:

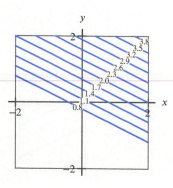

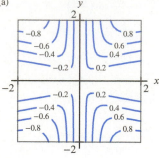

69 None since $f_{xy} \neq f_{yx}$

71 $f(x, y) = 2x + y^2$, $g(x, y) = 2x + y^2 + x^3$

Section 14.8

1 $(0, 0)$

3 x-axis and y-axis

5 None

7 None

9 $(1, 2)$

11 (a)

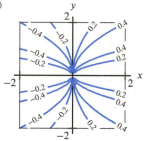

(b) No
(c) No
(d) No
(e) Exist, not continuous

13 (a)

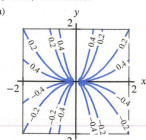

(b) Yes
(c) Yes
(d) No
(e) Exist, not continuous

15 (a)

(b) Yes
(d) No
(f) No

17 (a)

(c) No, no

19 (a) $f_x(x, y) =$
        $(x^4 y + 4x^2 y^3 - y^5)/(x^2 + y^2)^2$
        $f_y(x, y) =$
        $(x^5 - 4x^3 y^2 - xy^4)/(x^2 + y^2)^2$
   (c) Yes
   (d) Yes

21 Counterexample: $\sqrt{x^2 + y^2}$

23 $f(x, y) = \sqrt{x^2 + y^2}$

25 (a) Differentiable
   (b) Not differentiable
   (c) Not differentiable
   (d) Differentiable

Section 15.1

1 (I) and (V) Local maximum, (II) and (VI) Local minimum, (III) and (IV) Saddle point

3 (a) None
  (b) E, G
  (c) D, F

5 (a)

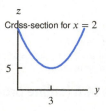

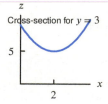

(b)

z
Cross-section for $x = 2$

5

3

y

z
Cross-section for $y = 3$

5

2

x

(c)

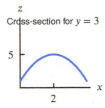

z
Cross-section for $x = 2$

5

3

y

z
Cross-section for $y = 3$

5

2

x

7 Saddle point

9 Local minimum

11 Local maximum

13 Local minimum

15 Local max: $(4, 2)$

17 Local max: $(1, 5)$

19 Saddle point: $(0, 0)$
    Saddle point: $(2, 0)$
    Local min: $(1, 0.25)$

21 Saddle pts: $(1, -1), (-1, 1)$
    Local max: $(-1, -1)$
    Local min: $(1, 1)$

23 Local max: $(-1, 0)$
    Saddle pts: $(1, 0), (-1, 4)$
    Local min: $(1, 4)$

25 Saddle point: $(0, 0)$
    Local max: $(1, 1), (-1, -1)$

27 Local min: $(0, 0)$

29 (a) All values of k
    (b) None
    (c) None

31 $a = -9, b = -12, c = 50$

33 (a) $k < 4$
    (b) None
    (c) $k \geq 4$

35

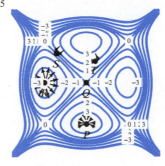

37 Saddle point: $(0, 0)$.

39 Critical points: $(0, 0), (\pm\pi, 0)$,
    $(\pm 2\pi, 0), (\pm 3\pi, 0), \cdots$
    Local minima: $(0, 0)$,
    $(\pm 2\pi, 0), \pm 4\pi, 0), \cdots$
    Saddle points: $(\pm\pi, 0)$,
    $(\pm 3\pi, 0), (\pm 5\pi, 0), \cdots$

41 (a) $(1, 3)$ is a local minimum
    (b)

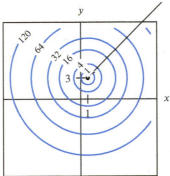

47

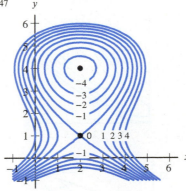

49 (a) $(0, 0)$
    (b) $D = -24x^2$
    (c) Saddle point

51

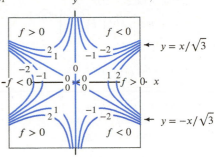

53 $(1, 3)$ could be saddle point

55 Can be saddle if f_{xy} large

57 $f(x, y) = 4 - (x - 2)^2 - (y + 3)^2$

59 False

61 True

63 True

65 True

67 False

69 False

Section 15.2

1 Mississippi:
    $87 - 88$ (max), $83 - 87$ (min)
    Alabama:
    $88 - 89$ (max), $83 - 87$ (min)
    Pennsylvania:
    $89 - 90$ (max), 80 (min)
    New York:
    $81 - 84$ (max), $74 - 76$ (min)
    California:
    $100 - 101$ (max), $65 - 68$ (min)
    Arizona:
    $102 - 107$ (max), $85 - 87$ (min)
    Massachusetts:
    $81 - 84$ (max), 70 (min)

3 Max: 30.5 at $(0, 0)$
   Min: 20.5 at $(2.5, 5)$

5 High: $(0, 0, 8)$
   Low: $(0, 0, -6)$

7 High: None
   Low: $(5, \pi, 2\pi)$

9 None

11 Min $= 0$ at $(0, 0)$
    (not on boundary)
    Max $= 2$ at $(1, 1), (1, -1)$,
    $(-1, -1)$ and $(-1, 1)$
    (on boundary)

13 max$= 1$ at $(1, 0)$ and $(-1, 0)$
    (on boundary)
    min$= -1$ at $(0, 1), (0, -1)$
    (on boundary)

15 Global min; no global max

17 Global max; no global min

19 Global max; no global min

21 Global min at $(0, 2\pi n)$, all n
    No global max

23 Saddle at $(0, 1/2)$; local min at $(-2/3, 7/6)$;
    no global max or min

25 Saddle at $(0, 0)$; local max at $(2/9, 4/27)$; no
    global max or min

27 All edges $(32)^{1/3}$ cm

29 $l = w = h = 45$ cm

31 $(3/14, 1/7, 1/14)$

35 $q_1 = 300, q_2 = 225$.

37 (a) $L = \left[pA \left(\frac{a}{k} \right)^a \left(\frac{l}{b} \right)^{(a-1)} \right]^{1/(1-a-b)}$

$K = \frac{la}{kb} L$

(b) No

39 $y = 24x^2/49 - 2/7$

41 $y = \frac{25}{6} - \frac{3}{2}x$

43 (b) $f(\sqrt{1/2}, \sqrt{3/5}) = 4\sqrt{2} + 2\sqrt{15} \approx 13.403$

45 (a) Decrease; increase

(d) Both zero

47 Some do, like $f(x, y) = x^2 + y^2$; some don't

49 $f(x, y) = x + y$

51 True

53 True

55 True

57 False

59 True

Section 15.3

1 Min $= -\sqrt{2}$, max $= \sqrt{2}$

3 Max: 20 at $(-1, 2)$;
Min: 0 at $(1, -2)$

5 Min $= -22$, max $= 22$

7 Maximum $f(10, 12.5) = 250$;
No minimum

9 Min $= \frac{3}{4}$, no max

11 Max $= 0$, no min

13 Max: $f(0, 2) = f(0, -2) = 8$
Min: $f(0, 0) = 0$

15 Max $= \frac{\sqrt{2}}{4}$, min $= -\frac{\sqrt{2}}{4}$

17 Max: 32 at $(1, -1)$;
Min: 8 at $(-1, 1)$

19 (a) P minimum, Q minimum, R neither, S
maximum

(b) P minimum, Q neither, R neither, S maximum

21 $K = 4, L = 5, \lambda = 0.072$

23 $K = 100, L = 400, \lambda = 4$

25 Global max $(12/5, 8/5)$; global min $(1, 3)$

27 0.5

29 (a)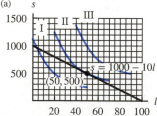

(b) $s = 1000 - 10l$

31 (a) Min; max at endpt of constraint
λ neg

(b) Max; min at endpt of constraint
λ pos

33 $\Delta c/4$; $-\Delta c/4$

37 (a) $C = \$4349$
(b) \$182

39 (a) $W = 225$
$K = 37.5$

(c) $W = 225$
$K = 37.5$
$\lambda = 0.29$

41 (a) No
(b) Yes
(c) $a + b = 1$

43 $x_1 = ((v_1)^{1/2} + (v_2)^{1/2})/(m(v_1)^{1/2})$
$x_2 = ((v_1)^{1/2} + (v_2)^{1/2})/(m(v_2)^{1/2})$

45 (a) $f_1 = \frac{k_1}{k_1+k_2} mg, f_2 = \frac{k_2}{k_1+k_2} mg$
(b) Distance the mass stretches the top spring
and compresses the lower spring

49 (a) Cost of producing quantity u when prices
are p, q
(b) $2\sqrt{pqu}$

51 (a) $-5\lambda^2 + 15\lambda$
(b) 1.5, 11.25
(c) 11.25, 1.5
(d) same

53 (a) $S = \ln(a^a(1-a)^{(1-a)}) + \ln b - a \ln p_1 - (1-a) \ln p_2$
(b) $b = e^c p_1^a p_2^{(1-a)}/(a^a(1-a)^{(1-a)})$

55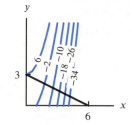

57 $f(x, y) = x^2 + y^2$

59 $f(x, y) = 10 - x^2 - y^2$

63 True

65 True

67 False

69 False

71 True

73 False

75 False

Section 15.3 (online problems)

77 (a)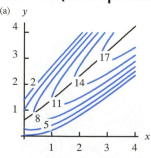

(b)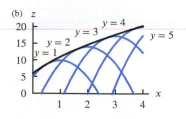

Section 16.1

1 24; 43.5

3 Over: Approx 137
Under: Approx 60

5 about 2300

7 Average height of a tent in meters

9 Positive

11 Zero

13 Zero

15 Positive

17 About 4.888 km^3

19 120

21 Need f nonnegative everywhere

23 $f(x, y) = 5 - x - y$; R is square with vertices
$(\pm 1, \pm 1)$

25 False

27 False

29 True

31 True

33 False

Section 16.1 (online problems)

35 25.2°C

Section 16.2

1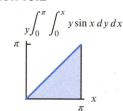
$\int_0^\pi \int_0^x y \sin x \, dy \, dx$

3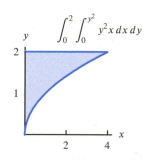
$\int_0^2 \int_0^{y^2} y^2 x \, dx \, dy$

5 150

7 54

9 $e - 2$

11 $3 - \sin 3$

13 $(e^4 - 1)(e^2 - 1)e$

15 -2.678

17 15

19 $\frac{2}{3}(e^8 - 1)$

21 $\int_1^4 \int_1^2 f \, dy \, dx$ or $\int_1^2 \int_1^4 f \, dx \, dy$

23 $\int_{-1}^3 \int_{-2}^{(1-3x)/4} f \, dy \, dx$
or $\int_{-2}^1 \int_{-1}^{(1-4y)/3} f \, dx \, dy$

25 $\int_1^3 \int_{\frac{1}{2}(y-1)}^{-\frac{1}{2}(y-5)} f \, dx \, dy$ or

$\int_0^1 \int_1^{2x+1} f \, dy \, dx +$
$\int_1^2 \int_1^{-2x+5} f \, dy \, dx$

27 $\int_1^2 \int_0^x f \, dy \, dx$ or

$\int_0^1 \int_1^2 f \, dx \, dy +$
$\int_1^2 \int_y^2 f \, dx \, dy$

29 $\frac{4}{15}(9\sqrt{3} - 4\sqrt{2} - 1) = 2.38176$

31 $32/9$

33 $13/6$

35 0

37 $2/3$

39 $\int_0^6 \int_0^{x/2} f(x,y) \, dy \, dx$

41 $\int_0^9 \int_{-\sqrt{9-y}}^{\sqrt{9-y}} f(x,y) \, dx \, dy$

43 $(e-1)/2$

45 $\frac{2}{9}(3\sqrt{3} - 2\sqrt{2})$

47 $\frac{1}{2}(e^2 - 1)$

49 $\ln(17)/4$

51 $\{(I),(IV),(V)\}, \{(II),(III),(VI)\}$

53 (a) $8/3$
   (b) $16/3$

55 (a) $\int_0^{1/2} \int_y^{1-y} f(x,y) \, dx \, dy$

$\int_0^{1/2} \int_0^x f(x,y) \, dy \, dx$ +
$\int_{1/2}^1 \int_0^{1-x} f(x,y) \, dy \, dx$
   (b) $1/8$

57 15

59 (a) Plate 1
   (b) Plate 1: 5 coulombs; Plate 2: 4 coulombs

61 18 gm

63 $\int_{-3}^3 \int_{-\sqrt{9-y^2}}^{\sqrt{9-y^2}} (9 - x^2 - y^2) \, dx \, dy$

65 4

67 117.45

69 Volume $= 1/(6abc)$

71 (a) Circles centered at $(1,0)$
   (b) $\int_{-\sqrt{3}}^{\sqrt{3}} e^{-y^2} \, dy$
   (c) $\int_{-2}^2 \int_{-\sqrt{4-x^2}}^{\sqrt{4-x^2}} e^{-(x-1)^2 - y^2} \, dy \, dx$

73 (a)

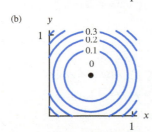

   (b)

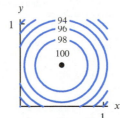

75 (a) $(4/3)a + b + (4/3)c = 20$

(b) $f(x,y) = x^2 + \frac{44}{3}xy + 3y^2$:

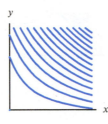

$f(x,y) = -3x^2 + 24xy$:

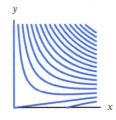

77 Outside limits on right should be constants

79 $f(x,y) = 12x$

81 False

83 True

85 True

87 True

Section 16.2 (online problems)

89 Volume $= 6$

91 $k(a^3 b + ab^3)/3$

Section 16.3

1 2

3 -8

5

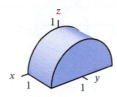

7

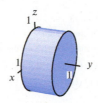

9

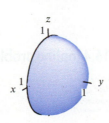

11

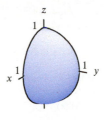

13

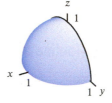

15 Mass of E in kg

17 Positive

19 Positive

21 Zero

23 Positive

25 Zero

27 Positive

29 Positive

31 Positive

33 Zero

35 1

37 $4/3$

39 $500/3$

41 $V = \int_{-1}^1 \int_{-\sqrt{1-x^2}}^{\sqrt{1-x^2}} \int_{x^2+y^2}^{\sqrt{4-x^2-y^2}} 1 \, dz \, dy \, dx$
   Can reverse order x, y

43 $V = \int_0^2 \int_y^{(y+2)/2} \int_0^{\sqrt{9-x^2-y^2}} 1 \, dz \, dx \, dy$

45 $V = \int_0^1 \int_0^{\sqrt{4-x^2}} \int_0^{\sqrt{4-x^2-y^2}} 1 \, dz \, dy \, dx$

47 $\int_0^1 \int_{-2}^2 \int_0^{\sqrt{4-z^2}} f(x,y,z) \, dy \, dz \, dx$

49 $\int_0^r \int_{-\sqrt{r^2-x^2}}^{\sqrt{r^2-x^2}} \int_0^{\sqrt{r^2-x^2-y^2}} f(x,y,z) \, dz \, dy \, dx$

51 $125/3$

53 $2/3$

55 $15/2$

57 (a) Mass of pyramid in grams
   (b) Four
   (c) 27 grams

59 $\int_0^{3/4} \int_{\frac{2y}{3}}^{2-2y} \int_0^{4-2x-4y} f(x,y,z) \, dz \, dx \, dy$

61 $\int_0^{\frac{1}{2}} \int_0^{4-8x} \int_{\frac{3x}{2}}^{1-\frac{x}{2}-\frac{z}{4}} f(x,y,z) \, dy \, dz \, dx$

63 (a) $z = \sqrt{1-x^2}, 0 \le y \le 10$
   (b) $\int_0^{10} \int_{-1}^1 \int_0^{\sqrt{1-x^2}} f(x,y,z) \, dz \, dx \, dy$

65 $\int_0^2 \int_0^{\sqrt{12-3y^2}} \int_0^{6y^2} f(x,y,z) \, dz \, dx \, dy$

67 $\int_0^{\sqrt{12}} \int_0^{24-2x^2} \int_{\sqrt{\frac{z}{6}}}^{\sqrt{\frac{12-x^2}{3}}} f(x,y,z) \, dy \, dz \, dx$

69 $\int_0^2 \int_0^{(3/2)\sqrt{4-y^2}} \int_{(15-5x)/3}^5 f(x,y,z)\,dz\,dx\,dy$

71 $\int_0^5 \int_0^{(15-3z)/5} \int_0^{(2/3)\sqrt{9-x^2}} f(x,y,z)\,dy\,dx\,dz$

73 $\int_0^2 \int_0^{4-x^2} \int_0^{2-x} f(x,y,z)\,dy\,dz\,dx$

75 $m = 2$;
   $(\bar{x}, \bar{y}, \bar{z}) = (13/24, 13/24, 25/24)$

77 Not true for $f(x,y,z) = z$

79 $f(x,y,z) = 7/(12\pi)$

81 False

83 False

85 True

87 False

89 False

Section 16.3 (online problems)

91 4

93 1

95 (a) $\int_0^2 \int_{\sqrt{\frac{x}{2}}}^{\frac{4-x}{2}} \int_0^{4-x-2y} f(x,y,z)\,dz\,dy\,dx$

   (b) $\int_0^2 \int_0^{4-x-\sqrt{2x}} \int_{\sqrt{\frac{x}{2}}}^{\frac{4-x-z}{2}} f(x,y,z)\,dy\,dz\,dx$

97 $m(b^2 + c^2)/3$

Section 16.4

1 $\int_0^{\pi/2} \int_0^{1/2} f\,r\,dr\,d\theta$

3 $\int_{\pi/4}^{3\pi/4} \int_0^2 f\,r\,dr\,d\theta$

5 $\int_1^5 \int_2^4 f(x,y)\,dy\,dx$

7 $\int_\pi^{2\pi} \int_2^4 f(r\cos\theta, r\sin\theta)\,r\,dr\,d\theta$

9
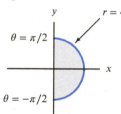

11
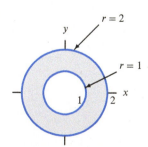

13
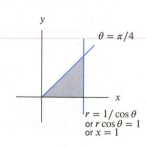

15
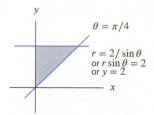

17 $\pi(1 - \cos 4)$

19 $-2/3$

21 1.

23 $\int_{-1}^0 \int_0^{-\sqrt{3}x} 2y\,dy\,dx$; $\int_0^{\sqrt{3}} \int_{-1}^{-y/\sqrt{3}} 2y\,dx\,dy$

25 16

27 (a) y

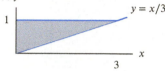

   (b) $\int_0^1 \int_0^{3y} f(x,y)\,dx\,dy$
   (c) $\int_{\tan^{-1}(1/3)}^{\pi/2} \int_0^{1/\sin\theta} f(r\cos\theta, r\sin\theta)\,r\,dr\,d\theta$

29 $2/\sqrt{3}$

31 $625\pi/2$

33 (a) $\pi(1 - e^{-a^2})$
   (b) Volume tends to π

35 (a) $\int_{\pi/2}^{3\pi/2} \int_1^4 \delta(r,\theta)\,r\,dr\,d\theta$
   (b) (i)
   (c) About 39,000

37 Total charge $= 2k\pi R$

39 (a)

   (b) $\sqrt{3}/2 + \pi/3$

41 Integrand r^3 instead of r^2

43 Regions of integration are not the same

45 Quarter disk $0 \le x \le 1, 0 \le y \le \sqrt{1-x^2}$

47 $f(x,y) = 1/\sqrt{x^2 + y^2}$

49 (a), (c), (e)

51 True

53 False

55 False

Section 16.4 (online problems)

57 (a) $\int_{-\sqrt{3}/2}^{\sqrt{3}/2} \int_{1-\sqrt{1-y^2}}^{\sqrt{1-y^2}} dx\,dy$

   (b) $\int_0^1 \int_{-\arccos(r/2)}^{\arccos(r/2)} r\,d\theta\,dr$

Section 16.5

1 (a) is (IV); (b) is (II); (c) is (VII); (d) is (VI);
   (e) is (III); (f) is (V)

3 $z = \sqrt{1 - r^2}$

5 $\phi = \pi/4$

7 $\rho = 4/\cos\phi$

9 (a) Cones opening vertically.
       One vertex at origin, opening upward.
       One vertex at $(0, 0, 6)$ opening downward
   (b) $z = r$; $z = 6 - r$
   (c) Circle, horizontal
   (d) $z = 3$; $r = 3$
   (e) $x^2 + y^2 = 9$; $z = 3$

11 $200\pi/3$

13 25π

15 $\int_0^1 \int_0^{2\pi} \int_0^4 f \cdot r\,dr\,d\theta\,dz$

17 $\int_0^\pi \int_0^\pi \int_2^3 f \cdot \rho^2 \sin\phi\,d\rho\,d\phi\,d\theta$

19 $\int_0^5 \int_0^2 \int_0^{x/5} f\,dz\,dy\,dx$

21 $\int_0^{2\pi} \int_0^2 \int_{2r}^4 f(r,\theta,z)\,r\,dz\,dr\,d\theta$

23 $\int_{-2}^2 \int_{-\sqrt{4-x^2}}^{\sqrt{4-x^2}} \int_{2}^{4}_{\sqrt{x^2+y^2}} h(x,y,z)\,dz\,dy\,dx$

25 $\int_0^\pi \int_0^K \int_0^{2\pi} \rho^2 \sin\phi\,d\theta\,d\rho\,d\phi$

27 (a) $\int_{-1/\sqrt{2}}^{1/\sqrt{2}} \int_{-\sqrt{(1/2)-x^2}}^{\sqrt{(1/2)-x^2}}$
       $\int_{\sqrt{x^2+y^2}}^{\sqrt{1-x^2-y^2}} dz\,dy\,dx$
   (b) $\int_0^{2\pi} \int_0^{1/\sqrt{2}} \int_r^{\sqrt{1-r^2}} r\,dz\,dr\,d\theta$
   (c) $\int_0^{2\pi} \int_0^{\pi/4} \int_0^1 \rho^2 \sin\phi\,d\rho\,d\phi\,d\theta$

29 (a) $\int_0^{2\pi} \int_0^{\sqrt{2}} \int_r^{\sqrt{4-r^2}} r\,dz\,dr\,d\theta$
   (b) $\int_0^{2\pi} \int_0^{\pi/4} \int_0^2 \rho^2 \sin\phi\,d\rho\,d\phi\,d\theta$

31 $V = \int_0^{2\pi} \int_0^{\pi/3} \int_0^3 \rho^2 \sin\phi\,d\rho\,d\phi\,d\theta$
   Order of integration can be altered;
   other coordinates can be used

33 $V = \int_0^\pi \int_{\sqrt{2}}^{\sqrt{3}} \int_5^{10} r\,dz\,dr\,d\theta$;
   Order of integration can be altered;
   other coordinates can be used

35 $V = \int_0^{2\pi} \int_1^3 \int_1^{\sqrt{10-r^2}} r\,dz\,dr\,d\theta$
   or
   $V = \int_0^{2\pi} \int_1^{\sqrt{10}} \int_0^{\sqrt{10-z^2}} r\,dr\,dz\,d\theta$ Order of
   integration can be altered;
   other coordinates can be used

37 (a) $\int_0^{2\pi} \int_0^{1/\sqrt{3}} \int_{\sqrt{3}r}^1 r\,dz\,dr\,d\theta$
   (b) $\pi/9$

39 $16\pi(\sqrt{2} - 1)/(3\sqrt{2})$

41 $28\pi/15$

43 $\int_0^{2\pi} \int_0^{5/\sqrt{2}} \int_r^{5/\sqrt{2}} r\,dz\,dr\,d\theta = $
   $125\pi/(6\sqrt{2}) = 46.28$ cm^3

45 (a) Positive
   (b) Zero

47 $\int_0^{2\pi} \int_0^l \int_a^{a+h} r\,dr\,dz\,d\theta = $
   $\pi l((a+h)^2 - a^2)$

49 $\int_0^{2\pi} \int_0^a \int_{hr/a}^h r\,dz\,dr\,d\theta = \pi ha^2/3$

51 (a) $\int_0^{2\pi} \int_1^5 \int_{-\sqrt{25-r^2}}^{\sqrt{25-r^2}} r\,dz\,dr\,d\theta$
   (b) $64\sqrt{6}\pi = 492.5$ mm^3

53 $81\pi(-\sqrt{2} + 2)/4$

55 $324\pi/5$ gm

57 1702π gm

59 Mass $= \int_{-2}^{2} \int_{-\sqrt{4-x^2}}^{\sqrt{4-x^2}}$
$\int_{0}^{4-x^2-y^2} e^{-x-y} \, dz \, dy \, dx$ gm

61 $1/27$

63 Total charge $= 2\pi k R^2$

65 (a) $\pi/5$
(b) $5/6$

67 $3a/8b$ above center of base

69 Limits of outer integral not constant

71 $\int_{0}^{2\pi} \int_{0}^{\pi/2} \int_{0}^{5} \rho^2 \sin\phi \, d\rho \, d\phi \, d\theta$

73 (c)

Section 16.5 (online problems)

75 $W \;=\; \int_{0}^{1} \int_{0}^{2\pi} \int_{\sqrt{1-r^2}}^{(\sqrt{9-r^2})-1} r \, dz \, d\theta \, dr \;+$
$\int_{1}^{2\sqrt{2}} \int_{0}^{2\pi} \int_{0}^{(\sqrt{9-r^2})-1} r \, dz \, d\theta \, dr$

77 $3I = \frac{6}{5}a^2; \; I = \frac{2}{5}a^2$

79 $(q^2/8\pi\epsilon)((1/a) - (1/b))$

81 $r^2 \sin\theta \, dr \, d\theta \, d\phi$

Section 16.6

1 Is a joint density function

3 Not a joint density function

5 Is joint density function

7 $1/16$

9 $3/16$

11 0.28

13 0.19

15 0

17 1

19 $7/8$

21 $1/16$

23 (a) $20/27$
(b) $199/243$

25 (a) $k = 3/8$
(b) $15/32$
(c) $1/16$

27 (a) 0.60
(b) 0.70
(c) 0.32

29 (a) $\lambda/(\lambda + \mu)$

31 (a) 0 if $t \leq 0$, $2t^2$ if $0 < t \leq 1/2$,
$1 - 2(1 - t)^2$ if $1/2 < t \leq 1$,
1 if $1 < t$
(b) 0 if $t \leq 0$, $4t$ if $0 < t \leq 1/2$,
$4 - 4t$ if $1/2 < t \leq 1$,
0 if $1 < t$

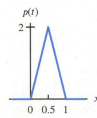

(c) x, y: All equally likely
z: Near $1/2$

33 $p(60, 170)$ not a probability

35 $g(y) = y$

37 False

39 True

Section 17.1

1 $x = 0, y = t, -2 \leq t \leq 1$

3 $x = 1 + 2t, \quad y = 1 + t, \quad 0 \leq t \leq 1$

5 $x = t, \quad y = 3 - 3t, \quad 0 \leq t \leq 1$

7 $x = t, y = 1, z = -t$

9 $x = 1, \quad y = 0, \quad z = t$

11 $x = 1 + 3t, \quad y = 2 - 3t, \quad z = 3 + t$

13 $x = -3 + 2t, y = -2 - t, z = 1 - 2t$

15 $x = 2 + 3t, \quad y = 3 - t, \quad z = -1 + t$

17 $x = 3 - 3t, y = 0, z = -5t$

19 $x = 3\cos t, y = 3\sin t, z = 5, 0 \leq t < 2\pi$

21 $x = 2\cos t, y = -2\sin t, z = 0$

23 $x = 2\cos t, y = 0, z = 2\sin t$

25 $x = 0, y = 3\cos t, z = 2 + 3\sin t$

27 $x = t^2, y = t, z = 0$

29 $x = -3t^2, y = 0, z = t$

31 $x = t, y = 4 - 5t^4, z = 4$

33 $x = 3\cos t, y = 2\sin t, z = 0$

35 $x = -1 + 3t, y = 2, z = -3 + 5t$

37 $\vec{r}(t) = \vec{i} - 3\vec{j} + 2\vec{k} + t(3\vec{i} + 4\vec{j} - 5\vec{k})$,
$0 \leq t \leq 1$, $x = 1 + 3t, y = -3 + 4t, z = 2 - 5t$,
$0 \leq t \leq 1$

39 $x = \cos t, y = \sin t, z = 0, 0 \leq t \leq \pi$

41 Two arcs:
$\vec{r}(t) = 5\vec{i} + 5(-\cos t\vec{i} + \sin t\vec{j})$,
$0 \leq t \leq \pi$ or
$\vec{r}(t) = 5\vec{i} + 5(\cos t\vec{i} + \sin t\vec{j})$,
$\pi \leq t \leq 2\pi$

43 $x = 10\cos t, y = 10\sin t, z = t$

45 $x = 2\cos t, y = t, z = 2\sin t$

47 $\vec{r}(t) = (2 + 10t)\vec{i} + (5 + 4t)\vec{j}$

49 $\vec{r}(t) = (2 + ((t - 20)/10)10)\vec{i}$
$+(5 + ((t - 20)/10)4)\vec{j}$

51 $\vec{r}(t) = (2 - 10t)\vec{i} + (5 - 4t)\vec{j}$

53 No

55 (b) $-\vec{i} - 10\vec{j} - 7\vec{k}$
(c) $\vec{r} = (1 - t)\vec{i} + (3 - 10t)\vec{j} - 7t\vec{k}$

57 (a) $\vec{r} = (\vec{i} + 3\vec{j} + 7\vec{k}) + t(2\vec{i} - 3\vec{j} - \vec{k})$
(b) $(3, 0, 6)$
(c) $\sqrt{14}$

59 (b) Reverse of part (a)

61 $(-9, -2, 1)$

63 Not the same

65 Not the same

67 Same lines

69 $c = 2$

71 $x = \frac{8}{3}, \quad y = 3t - \frac{1}{3}, \quad z = 3t$

73 $x = 1 + 2t, \quad y = 2 + 4t, \quad z = 5 - t$

75 Yes

77 (a) (II)
(b) (III)
(c) (I)
(d) (IV)

79 $25/\pi$

81 10

83 (a) Repeats every year
(b) Mid-August
(c) Mid-April
(d) 2°C per month

85 (a) Center: (1, 2), Radius: 3
(b) $x = 1 + 3\cos t$
$y = 2 + 3\sin t$
$0 \leq t \leq 2\pi$
(c) $x = 1 + 3\cos t$
$y = 2 + 3\sin t$
$z = 14 + 6\cos t + 12\sin t$
$0 \leq t \leq 2\pi$

87 (a) II, $y = x$
(b) IV, $x + y = a$
(c) V, $x^2 - y^2 = a^2$
(d) I, $x^2 + y^2 = a^2$
(e) III, $x^2 + y^2 = a^2$

89 Many possible answers
(a) $a = -2, b = 7, c = 4, d = 0$
(b) $a = -2, b = 7, c = 4, d = 11$
(c) $a = 7, b = 2, c = 0, d = 41$

91 Line Equation:
$x = 1 + 2t$
$y = 2 + 3t$
$z = 3 + 4t$
Shortest distance: $\sqrt{6/29}$

93 (a) (vii)
(b) (ii)
(c) (iv)
95 (a) (i) is (C); (ii) is (A); (iii) is (D); (iv)
is (G)
(b) (iii)

97 (a) Parallel
(b) (i) Perpendicular
(ii) Parallel

99 Distance $|R|$ from z-axis
Distance $\sqrt{R^2 + t^2}$ from origin

101 $\vec{i} + 2\vec{j} + 3\vec{k} + t\left(\vec{i} + 2\vec{j}\right)$
$\vec{i} + 2\vec{j} + 3\vec{k} + t\left(\vec{i} - \vec{k}\right)$

103 False

105 False

107 False

109 True

111 True

113 True

Section 17.1 (online problems)

115 (a) Center: $(a/2, b/2)$, Radius:
$\sqrt{c + (a^2 + b^2)/4}$.
(b) $x = a/2 + \sqrt{c + (a^2 + b^2)/4}\cos t$
$y = b/2 + \sqrt{c + (a^2 + b^2)/4}\sin t$
$0 \leq t \leq 2\pi$
(c) $x = a/2 + \sqrt{c + (a^2 + b^2)/4}\cos t$
$y = b/2 + \sqrt{c + (a^2 + b^2)/4}\sin t$
$z = (a^2 + b^2)/2 + c + a\sqrt{c + (a^2 + b^2)/4}\cos t + b\sqrt{c + (a^2 + b^2)/4}\sin t$
$0 \leq t \leq 2\pi$

117 (a) $-2e^{-1}/3 \; \mu g/m^3/m$
(b) $t = \pm\sqrt{3}/2$ sec

Section 17.2

1 $\vec{v} = 3\vec{i} + \vec{j} - \vec{k}, \; \vec{a} = \vec{0}$

3 $\vec{v} = \vec{i} + 2t\vec{j} + 3t^2\vec{k}, \; \vec{a} = 2\vec{j} + 6t\vec{k}$

5 $\vec{v} = -3\sin t\vec{i} + 4\cos t\vec{j}$,
$\vec{a} = -3\cos t\vec{i} - 4\sin t\vec{j}$

7 $\vec{v} = \vec{i} + 2t\vec{j} + 3t^2\vec{k}$,
Speed $= \sqrt{1 + 4t^2 + 9t^4}$,
Particle never stops

9 $\vec{v} = 6t\vec{i} + 3t^2\vec{j}$,
$\|\vec{v}\| = 3|t| \cdot \sqrt{4 + t^2}$,
Stops when $t = 0$

11 $\vec{v} = 6t\cos(t^2)\vec{i} - 6t\sin(t^2)\vec{j}$,
$\|\vec{v}\| = 6|t|$,
Stops when $t = 0$

13 Length $= \sqrt{42}$

15 Length $= e - 1$

17 $\vec{v} = -6\pi\sin(2\pi t)\vec{i} + 6\pi\cos(2\pi t)\vec{j}$,
$\vec{a} = -12\pi^2\cos(2\pi t)\vec{i} - 12\pi^2\sin(2\pi t)\vec{j}$,
$\vec{v} \cdot \vec{a} = 0$, $\|\vec{v}\| = 6\pi$, $\|\vec{a}\| = 12\pi^2$

19 Line through $(2, 3, 5)$ in direction of
$\vec{i} - 2\vec{j} - \vec{k}$,
$\vec{v} = 2t(\vec{i} - 2\vec{j} - \vec{k})$, $\vec{a} = 2(\vec{i} - 2\vec{j} - \vec{k})$

21 $x = 1 + 2(t - 2), y = 2, z = 4 + 12(t - 2)$

23 Vertical: $t = 3$
Horizontal: $t = \pm 1$
As $t \to \infty$, $x \to \infty$, $y \to \infty$
As $t \to -\infty$, $x \to \infty$, $y \to -\infty$

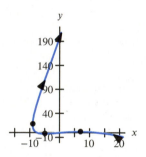

25 (a) $\vec{v}(2) \approx -4\vec{i} + 5\vec{j}$,
Speed $\approx \sqrt{41}$
    (b) About $t = 1.5$
    (c) About $t = 3$

27 (a) $x = 2 + 0.6t, y = -1 + 0.8t, z = 5 - 1.2t$,
$0 \le t \le 5$
    (b) $x = 2 + 1.92t, y = -1 + 2.56t, z = 5 - 3.84t, 0 \le t \le 1.56$

29 (a) 6.4 meters
    (b) 1.14 sec
    (c) 15.81 m/sec
    (d) $(11.4, -5.7, 0)$
    (e) -9.8 m/sec^2

31 (a) 5 secs; $(10, 15, 100)$
    (b) $t = 0, 10$ secs, $\sqrt{113}$ cm/sec
    (c) 5 secs, $\sqrt{13}$ cm/sec

33 (a) $t = 5.181$ sec
    (b) $x = 103.616$ meters
    (c) 2 meters
    (d) 9.8 meters/sec^2
    (e) $\theta = 0.896$; $v = 32.016$ meters/sec

35 (a) (IV); 4.5 sec; $(0, 8.9 \text{ m}, 0)$
    (b) (II); 3.2 sec; base of tower
    (c) (V); 10 sec; halfway up

37 (a) $-2\vec{i}$
    (b) $(0, 3)$
    (c) π

39 (a) π m/sec
    (b) 2.45 m
    (c) 3.01 m

41 (a) $(x, y) = (t, 1)$

    (b) $(x, y) = (t + \cos t, 1 - \sin t)$

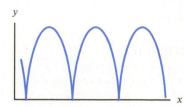

45 (a) R, counterclockwise, $2\pi/\omega$
    (b) $\vec{v} = -\omega R\sin(\omega t)\vec{i} + \omega R\cos(\omega t)\vec{j}$
    (c) $\vec{a} = -\omega^2\vec{r}$

47 Same path, B moves twice as fast

49 Counterclockwise

51 Orthogonal only if speed is constant

53 Length $= \int_A^B \|\vec{v}(t)\| \, dt$

55 $0 \le t \le 10/\sqrt{2}$

57 True

59 False

61 True

63 False

65 False

67 False

Section 17.2 (online problems)

69 (a) $x - \sqrt{6}y + z = 3 - 7\sqrt{6}i$
    (b) $\pi/3$
    (c) 4 ppm/sec

Section 17.3

1 $\vec{V} = x\vec{i}$

3 $\vec{V} = x\vec{i} + y\vec{j} = \vec{r}$

5 $\vec{V} = -x\vec{i} - y\vec{j} = -\vec{r}$

7 (a) y-axis
    (b) Increasing
    (c) Neither

9 (a) x-axis
    (b) Increases
    (c) Decreases

11

13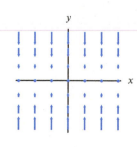

15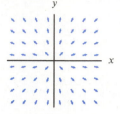

17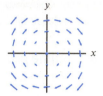

19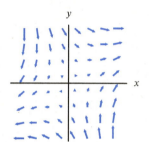

21 (a) III
    (b) II
    (c) IV
    (d) VI

23 $3\vec{i} - 4\vec{j}$, other answers possible

25 $(1/\sqrt{1 + x^2})(\vec{i} - x\vec{j})$, other answers possible

27 $\vec{F}(x, y) = (y + \cos x)((1 + y^2)\vec{i} - (x + y)\vec{j})$,
other answers possible

29 I, II, III

31 (a) (III)
    (b) (II)
    (c) (VI)
    (d) (V)
    (e) (IV)
    (f) (I)

33 $\vec{F}(x, y) = \dfrac{-x\vec{i} - y\vec{j}}{\sqrt{x^2+y^2}}$ (for example)

35 (a) $(1, -3, -7)$; other answers possible
    (b) $(0, 0, 0)$; other answers possible
    (c) $-4x + y - 3z = 0$; plane through origin

37 (a) Radiates out from origin

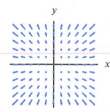

    (b) Spirals outward counterclockwise around origin

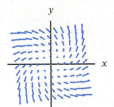

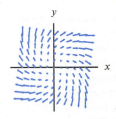

(c) Spirals outward clockwise around origin

z = g(x, y):

39 (a) $z = f(x, y)$:

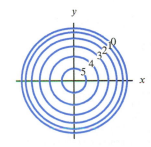

$z = g(x, y)$:

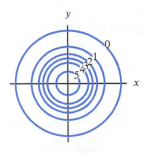

(b)

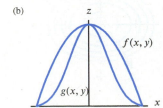

41 To plot $\vec{G}(x, y, z)$ move arrows of $\vec{F}(2x, 2y, 2z)$ halfway to origin

43 $(x^2 + 1)\left(\vec{i} + \vec{j} + \vec{k}\right)$

Section 17.4

1 Field:

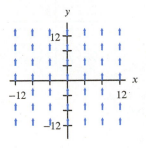

Flow, $x =$ constant:

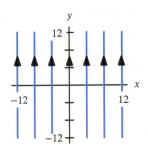

3 Field:

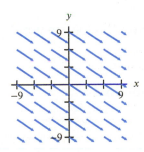

Flow, $y = -(2/3)x + c$:

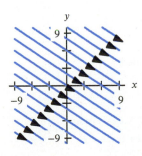

5 Field:

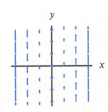

Flow:

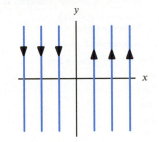

7 Field:

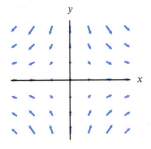

Flow:

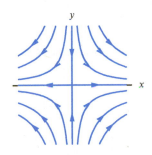

9 Field:

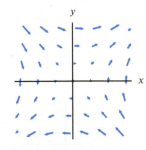

Flow:

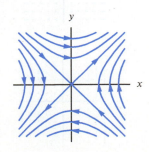

15 (a) Same directions, different magnitudes
   (b) Same curves, different parameterizations

23 (a) $\vec{v} = \pi(-y\vec{i} + x\vec{j})/12$
   (b) Horizontal circles

25 Counterexample: $\vec{F} = -y\vec{i} + x\vec{j}$

27 $\vec{F}(x, y, z) = \vec{i} + 2x\vec{j} + 3y\vec{k}$

29 True

31 False

33 True

35 True

37 True

Section 18.1

1 Negative

3 Zero

5 Zero

7 0

9 0

11 28

13 16

15 −48

17 19/3

19 20

21 28

23 −10

25 −9

27 0

29 C_1 is zero; C_2 is pos; C_3 is neg

31 C_1 is 0; C_2 is neg; C_3 is pos

33 −8

35 C_1, C_2

37 $a < 0$

39 $c > 1$

41 (a) (i)

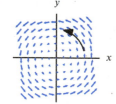

(ii)

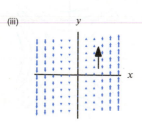

(iii)

(iv)

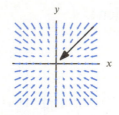

(b) (i), (iii)

43 Positive

45 0

47 11

49 6

51 13

53 $-1.2 \cdot 10^7 \text{ meter}^2/\text{sec}$

55 14π; tangent to C
   Same direction, $||\vec{F}|| = 7$
   -14π; tangent to C
   Opposite direction, $||\vec{F}|| = 7$

59 $-2.5 \cdot 10^{-5} GMm$

61 If $\int_C \vec{F} \cdot d\vec{r} < 0$, then $\int_{-C} \vec{F} \cdot d\vec{r} > 0$

63 $\vec{F} = \vec{i} - \vec{j}$

65 False

67 True

69 False

71 True

73 False

Section 18.1 (online problems)

75 $-\int_C \vec{E} \cdot d\vec{r}$

77 Spheres centered at origin

Section 18.2

1 $\int_0^\pi (\cos^2 t - \sin^2 t)\, dt$
   Other answers are possible

3 $\int_0^{2\pi}(-\sin t \cos(\cos t) + \cos t \cos(\sin t)) dt$

5 24

7 −4

9 −6

11 9

13 82/3

15 12

17 116.28

19 12

21 21

23 0

25 $\int_C 3x\,dx - y\sin x\,dy$

27 $(x + 2y)\vec{i} + x^2 y\vec{j}$

29 3124

31 144

33 77,000/3

35 (a) 11/6
   (b) 7/6

37 (a) 3/2
   (b) 3/2

39 200π

43 (a) −5
   (b) 5
   (c) 0

45 F could point with C at some points and against C at others

47 $y = \pi/2$, $x = t$, $0 \le t \le 3$, $\int_C \vec{F} \cdot d\vec{r} = 3$

49 True

51 True

53 False

55 False

57 (a)

Section 18.3

1 12

3 Negative, not path-independent

5 Negative, not path-independent

7 Path-independent

9 Path-independent

11 Path-independent

13 $f(x, y) = x^2 y + K$

15 $f(x, y, z) = e^{xyz} + \sin(xz^2) + C$
   $C = $ constant

17 −2

19 2

21 0

23 $e^3 - 1$

25 0

27 PQ

31 Yes

33 Yes.

35 $5xy + y^2/2$
   (a) 50
   (b) 50

37 (a) 50π
   (b) No, integral over closed path not zero

39 Use $f_{xy} = f_{yx}$

41 (a) e
   (b) e

43 9/2

45 $\frac{3}{\sqrt{2}} \ln(\frac{3}{\sqrt{2}} + 1)$

47 $e^{(1.25\pi)^2/2} - 1$

49 (a) $7e^3 - 2e$
   (b) $7e^3 - 2e$

51 (a) 9
   (b) 0

53 (a)

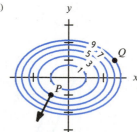

(b) Shorter
(c) 6

55 (a) Positive
   (b) Not gradient
   (c) $\vec{F}_2$

57 (a) $(8, 9)$

(b) 50

59 (a) $2\pi mg$
   (b) Yes

61 $f(Q) - f(P)$ where $\vec{F} = \text{grad } f$

63 Methods other than Theorem 18.1 can be used

65 Gradient of any function

71 True

73 True

75 False

77 True

79 False

Section 18.3 (online problems)

83 If $A'(x) = a(x)$, then $f(x, y) = A(x)$ is potential function
   $x + x^2/2 + x^3/3 + C$, any C

85 (a) $\vec{F} - \text{grad } \phi = -y \, \text{grad } h$
   (b) 30

87 (a) $\vec{F} - \text{grad } \phi = -(x + 2y) \, \text{grad } h$
   (b) −50

89 (a) Increases

Section 18.4

1 No

3 No

5 $f(x, y) = x^3/3 + xy^2 + C$

7 Yes, $f = \ln A|xyz|$ where $A > 0$

9 No

11 -2π

13 -6

15 -12

17 $-3\pi m^2$

19 (a)

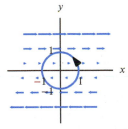

   (b) $-\pi$

21 $e - \cos 1$

23 $1/4$

25 $1/24$

27 $-9\pi/8$

29 (a) 0
   (b) 0
   (c) 0
   (d) -6π
   (e) -6π
   (f) 0
   (g) -6π

31 (a) 0
   (b) 10
   (c) -8π
   (d) 7

33 (b) 0
   (c) $\vec{G} = \nabla(xyz + zy + z)$
   (d) $\vec{H}_1 = \nabla(yx^2)$, $\vec{H}_2 = \nabla(y(x + z))$

35 πab

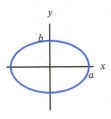

37 $3/2$

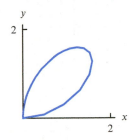

39

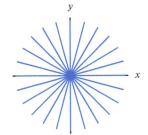

41 (a) Possible answers are:
   $\vec{F} = \text{grad}(xy)$
   $\vec{G} = \text{grad}(\arctan(x/y))$, $y \neq 0$
   $\vec{H} = \text{grad}\left((x^2 + y^2)^{1/2}\right)$, $(x, y) \neq (0,0)$
   (b) $0, -2\pi, 0$
   (c) Does not apply to $\vec{G}$, $\vec{H}$; holes in domain

43 $L_1 < L_2 < L_3$

45 (a) $21\pi/2$
   (b) 2

47 (a) $\vec{0}$
   (b) $q/\|\vec{r}\|$

49 Green's Theorem does not apply;
   Line integral depends on $\vec{F}$

51

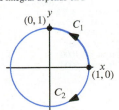

53 True

55 True

57 True

59 True

Section 19.1

1 $-3\vec{i}$

3 $15\vec{j}$

5 Rectangle in xz plane with area 150, oriented pos y direction

7 (a) Positive
   (b) Negative
   (c) Zero
   (d) Zero
   (e) Zero

9 (a) Zero
   (b) Zero
   (c) Zero
   (d) Negative
   (e) Zero

11 (a) 45
   (b) −45

13 −12

15 4

17 12π

19 4

21 0

23 $10\pi/\sqrt{3}$

25 6

27 -75π

29 $-\pi^3$

31 2000π

33 32000π

35 0

37 12.8

39 Zero

41 24

43 -160π

45 $130/\sqrt{2}$

47 42

49 -96π

51 $4\sqrt{3}$

53 $\pi \sin 9$

55 0

57 −27

61 (a) Zero
   (b) Zero

63 4π

65 (a) 0
   (b) 32π

67 (a) Zero
   (b) Zero

71 (a)

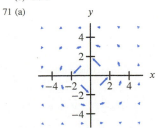

   (b) 0
   (c) $Ih \ln |b/a|/2\pi$

73 Sign of $\int_S \vec{F} \cdot d\vec{A}$ depends on both $\vec{F}$ and S

75 $\vec{F} = z\vec{k}$
   $S: 0 \leq x \leq 1, 0 \leq y \leq 1, z = 1$, oriented upwards

77 True

79 False

81 True

83 True

85 True

Section 19.1 (online problems)

87 (a) 0

(b) 0

Section 19.2

1 $\left(-3\vec{i} + 5\vec{j} + \vec{k}\right) dx\, dy$

3 $\left(-4x\vec{i} + 6y\vec{j} + \vec{k}\right) dx\, dy$

5 $\int_{-2}^{3}\int_{0}^{5} 70\, dy\, dx$

7 $\int_{0}^{5}\int_{0}^{5-x}(yz\sin x - 2xy\cos 2y + xy)\, dy\, dx$

9 -500

11 $-5/3 - \sin 1 = -2.508$

13 $\int_{0}^{\pi/2}\int_{0}^{5} 10\,(\cos\theta + 2\sin\theta)\, dz\, d\theta$

15 $\int_{0}^{2\pi}\int_{-8}^{8}\left(6z^2\cos\theta + 6\sin\theta e^{6\cos\theta}\right) dz\, d\theta$

17 2000

19 $100\sqrt{2}/3$

21 $\int_{0}^{2\pi}\int_{0}^{\pi/2} 100\,(\sin\phi\cos\theta + 2\sin\phi\sin\theta + 3\cos\phi)\sin\phi\, d\phi\, d\theta$

23 $\int_{-\pi/2}^{\pi/2}\int_{0}^{\pi} 16\cos^2\phi\sin^2\phi\cos\theta\, d\phi\, d\theta$

25 8000/3

27 $(8 - 5\sqrt{2})\pi/6 = 0.486$

29 6

31 6

33 18

35 36π

37 7/3

39 $\pi\sin 25$

41 $\pi/2$

43 1296π

45 π

47 $100\sqrt{27}$

49 2.228

51 (a) $\int_{R} a/\sqrt{a^2 - x^2 - y^2}\, dx\, dy$

(b) $\int_{0}^{2\pi}\int_{0}^{a} ar/\sqrt{a^2 - r^2}\, dr\, d\theta.$

(c) $2\pi a^2$

53 36π

55 $2\pi/3$

57 $4\pi a^3$

59 -1

61 $11\pi/2$

65 (a) Constant inside cylinder radius a

(b) $\vec{E} = \begin{cases} \frac{1}{2}k\delta_0 r\vec{e}_r & \text{if } r \leq a \\ \frac{1}{2}k\delta_0\frac{a^2}{r}\vec{e}_r & \text{if } r > a \end{cases}$

67 $\vec{n} = \left(-f_x\vec{i} - f_y\vec{j} + \vec{k}\right)/\sqrt{f_x^2 + f_y^2 + 1}$

$dA = \sqrt{f_x^2 + f_y^2 + 1}\, dx\, dy$

69 $r = 10, 0 \leq \theta \leq 2, 0 \leq z \leq 3$, oriented outwards

71 False

Section 19.3

1 Scalar; $2x + xe^z$

3 (I)

5 0

7 $4x$

9 $2x/(x^2 + 1) - \sin y + xye^z$

11 0

13 (a) Positive

(b) Zero

(c) Negative

15

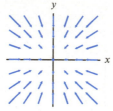

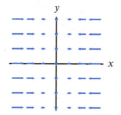

17 -0.030

19 (a) (i) $0.016\pi/3$

(ii) -0.08

(b) Flux positive at $(2,0,0)$ and negative at $(0,0,10)$

21 (a) $4w^3$

(b) 4

(c) 4

23 div $\vec{v} = -6$

25 (a) $-1/3, 1$

(b) 1/3

27 Undefined

29 (b)

31 (a) 0

(b) Undefined

33 (a) $\rho(0) < \rho(1000) < \rho(5000)$

(b) cars/hour

(d) $\rho(x) = 4125/(55 - x/50)$
  if $0 \leq x < 2000$
  $\rho(x) = 4125/15 = 275$
  if $2000 \leq x < 7000$
  $\rho(x) = \dfrac{4125}{(15 + (x - 7000)/25)}$
  if $7000 \leq x < 8000$
  $\rho(x) = 4125/55 = 75$
  if $x \geq 8000$

(e) 139 ft. at $x = 0$
  89 ft. at $x = 1000$
  38 ft. at $x = 5000$

35 (a) 0

(b) 0

39 0

41 $\vec{b} \cdot (\vec{a} \times \vec{r})$

43 (d)

Section 19.4

1 24

3 8

5 Zero

7 24

9 72

11 288

13 36π

15 620π

17 5π

19 420

21 8

23 20/3

25 $10\pi a^3$

27 Yes; -3.22

29 $\int_{S}\vec{F}\cdot d\vec{A} = \int_{W} \text{div}\,\vec{F}\, dV = 0$

31 (a) $cb(12a - a^2)$

(b) 6, 10, 10; 3600

33 (a) 4π

(b) 0

(c) 4π

35 4π

37 (a) 2

(b) 0.016

(c) $0.016053\cdots$

39 (a) 30 watts/km^3

(b) $\alpha = 10$ watts/km^3

(d) $6847°$C

41 (a) 0

(c) No

43 S not the boundary of a solid region

45 Any sphere

47 False

49 False.

51 True.

53 True

55 True

57 True

59 True

45 div $\vec{F} = 2x + 2 - 2z$

47 $\vec{F}(x, y, z) = 2x\vec{i} + 3y\vec{j} + 4z\vec{k}$

49 $\vec{F}(x, y) = 2x\vec{i}$

51 False

53 False

55 False

57 False

59 False

61 False

Section 20.1

1 Vector; $\vec{i} + \vec{j} - \vec{k}$

3 Vector; $(x + 1)\vec{i} - (y + 2)\vec{j}$

5 Vector; $\vec{0}$

7 $4y\vec{k}$

9 $4x\vec{i} - 5y\vec{j} + z\vec{k}$

11 $\vec{0}$

13 $\vec{0}$

15 Zero curl

17 Nonzero curl

19 0

21 $50\vec{i} + 300\vec{j} + 2\vec{k}$

23 (a) $(f-c)\vec{i} + (be^z - e\cos x)\vec{j} + (2dx - 3ay^2)\vec{k}$
   (b) $f = c$
   (c) $f = c, b = e = 0$

25 (a) Horizontal
   (b) Vertical
   (c) Parallel to the yz-plane, making angle t with horizontal

27 (a) $w = 1$

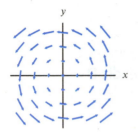

$w = -1$

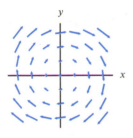

   (b) $|\omega| \cdot \sqrt{x^2 + y^2}$
   (c) $\operatorname{div} \vec{v} = 0$
        $\operatorname{curl} \vec{v} = 2\omega\vec{k}$
   (d) $2\pi\omega R^2$

35 Counterexample: $\vec{F} = y\vec{i}$

37 $\vec{F} = z\vec{i}$

39 True.

41 True

43 False

45 False

47 (a)-z, (c)-y, (d)-x

Section 20.2

1 (a) π
  (b) 0

3 Positive

5 -8π

7 -2

9 0

11 18π

13 (a) -2π
   (b) $-2\vec{k}$
   (c) -2π
   (d) Stokes' Theorem

15 No

17 (a) 45π
   (b) $81\pi/2$

19 (a) $-\vec{i} - \vec{j} - \vec{k}$
   (b) (i) -4π
        (ii) $15/2$

21 0

23 0

25 $8\pi/\sqrt{3}$

27 (a) All 3-space
   (b) $\frac{2ax\vec{i} + 2by\vec{j} + 2cz\vec{k}}{1 + ax^2 + by^2 + cz^2}$
   (c) 0
   (d) $\ln(3 + 507\pi^2/4) - \ln(2)$

29 -8π

31 4π

33 63π

35 (a) $\vec{0}$
   (b) 0
   (c) 0

37 (a) Parallel to xy-plane; same in all horizontal planes
   (b) $(\partial F_2/\partial x - \partial F_1/\partial y)\vec{k}$
   (d) Green's Theorem

39 C not the boundary of a surface

41 Any oriented circle

43 True

45 False

47 True

49 True

51 False

Section 20.3

1 Yes

3 Yes

5 No

7 Yes

9 Yes

11 Yes

13 (a) No
   (b) Yes

15 Curl yes; Divergence yes

17 Curl yes; Divergence yes

21 $(1/2)\vec{b} \times \vec{r}$

23 No

25 (a) Yes
   (b) Yes
   (c) Yes

27 (a) Yes
   (b) No
   (c) No

29 (b) $\nabla^2\psi = -\operatorname{div}\vec{A}$

31 Curl of scalar function not defined

33 $f(x, y, z) = x^2$

35 False

37 True

Section 20.3 (online problems)

39 (a) $\operatorname{curl}\vec{E} = \vec{0}$
   (b) 3-space minus a point if $p > 0$
        3-space if $p \le 0$.
   (c) Satisfies test for all p.
        $\phi(r) = r^{2-p}$ if $p \ne 2$.
        $\phi(r) = \ln r$ if $p = 2$.

Section 21.1

1 Curve

3 Surface

5 Horizontal disk of radius 5 in plane $z = 7$

7 Helix radius 5 about z-axis

9 Top hemisphere

11 Vertical segment

13 $x = 1$
    $y = s$
    $z = t$

15 $x = 1 + t$
    $y = 1 + s$
    $z = s + t$

17 $\vec{r}(s, t) = (s + 2t)\vec{i} + (2s + t)\vec{j} + 3s\vec{k}$, other answers possible

19 $(0, 0, 0), 2\vec{i} + \vec{j} - \vec{k}, 3\vec{i} - 5\vec{j} + 2\vec{k}$

21 $\vec{r}(s, t) = (3 + s + t)\vec{i} + (5 - s)\vec{j} + (7 - t)\vec{k}$, other answers possible

23 (a) Yes
   (b) No

25 $s = s_0$: lines parallel to y-axis with $z = 1$
    $t = t_0$: lines parallel to x-axis with $z = 1$

27 $s = s_0$: parabolas in planes parallel to yz-plane
    $t = t_0$: parabolas in planes parallel to xz-plane

29 $s = 4, t = 2$
    $(x, y, z) = (x_0 + 10, y_0 - 4, z_0 + 18)$

31 Horizontal circle

33 (a) $x = \left(\cos\left(\frac{\pi}{3}t\right) + 3\right)\cos\theta$
        $y = \left(\cos\left(\frac{\pi}{3}t\right) + 3\right)\sin\theta$
        $z = t$ $0 \le \theta \le 2\pi, 0 \le t \le 48$
   (b) 456π in^3

35 If $\theta < \pi$, then $(\theta + \pi, \pi/4)$
    If $\theta \ge \pi$, then $(\theta - \pi, \pi/4)$

37 $x = r\cos\theta,$ $0 \le r \le a$
    $y = r\sin\theta,$ $0 \le \theta \le 2\pi$
    $z = (1 - r/a)h$

39 (a) $-x + y + z = 1,$
        $0 \le x \le 2,$
        $-1 \le y - z \le 1$
   (b)

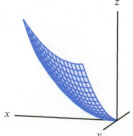

41 (a) $z = (x^2/2) + (y^2/2)$
        $0 \le x + y \le 2$
        $0 \le x - y \le 2$
   (b)

43 Radius: $R \sin \phi$

45 $x + y - z - 3 = 0$

47 True

49 True

51 True

53 False

Section 21.2

1 1

3 e^{2s}

5 $a = 1/10, b = 1$

7 $a = 1/50, b = 1/10$

9 3

11 $\rho^2 \sin \phi$

13 13.5

15 72

17 (a) $(1/(2\pi\sigma^2)) \int_{-\infty}^{\infty} \int_{-\infty}^{2t-x} e^{-(x^2+y^2)/(2\sigma^2)} dy\,dx$

(b) $(1/(\sqrt{\pi}\sigma)) \int_{-\infty}^{t} e^{-u^2/\sigma^2} du$

(c) $(1/(\sqrt{\pi}\sigma)) e^{-t^2/\sigma^2}$

(d) Normal, mean 0, standard deviation $\sigma/\sqrt{2}$

19 R does not correspond to T

21 $x = 2s, y = 3t$

23 False

Section 21.3

1 $((s+t)\vec{i} - (s-t)\vec{j} - 2\vec{k})\,ds\,dt$

3 $-e^s(\cos t\,\vec{j} + \sin t\,\vec{k})\,ds\,dt$

5 4/3

7 $6(e^4 - 1)$

9 $-\pi R^7/28$

11 $200\sqrt{14}$

13 $\sqrt{6}\pi$

15 $khw^3/6$ meter^3/sec.

21 Integral gives volume

23 $\vec{r}(s,t) = 2s\vec{i} + t\vec{j}$

25 True

27 False

Appendix A

1 (a) $y \le 30$

(b) two zeros

3 -1.05

5 2.5

7 $x = -1.1$

9 0.45

11 1.3

13 (a) $x = -1.15$

(b) $x = 1, x = 1.41$, and $x = -1.41$

15 (a) $x \approx 0.7$

(b) $x \approx 0.4$

17 (a) 4 zeros

(b) $[0.65, 0.66], [0.72, 0.73],$ $[1.43, 1.44], [1.7, 1.71]$

19 (b) $x \approx 5.573$

21 Bounded $-5 \le f(x) \le 4$

23 Not bounded

Appendix B

1 $2e^{i\pi/2}$

3 $\sqrt{2}e^{i\pi/4}$

5 $0e^{i\theta}$, for any θ.

7 $\sqrt{10}e^{i(\arctan(-3)+\pi)}$

9 $-3 - 4i$

11 $-5 + 12i$

13 $1/4 - 9i/8$

15 $-1/2 + i\sqrt{3}/2$

17 $-125i$

19 $\sqrt{2}/2 + i\sqrt{2}/2$

21 $\sqrt{3}/2 + i/2$

23 -2^{50}

25 $2i\sqrt[3]{4}$

27 $(1/\sqrt{2})\cos(-\pi/12) + (i/\sqrt{2})\sin(-\pi/12)$

29 $-i, -1, i, 1$ $i^{-36} = 1, i^{-41} = -i$

31 $A_1 = 1 + i$ $A_2 = 1 - i$

37 True

39 False

41 True

Appendix C

1 (a) $f'(x) = 3x^2 + 6x + 3$

(b) At most one

(c) $[0, 1]$

(d) $x \approx 0.913$

3 $\sqrt[4]{100} \approx 3.162$

5 $x \approx 0.511$

7 $x \approx 1.310$

9 $x \approx 1.763$

11 $x \approx 0.682328$

Appendix D

1 3, 0 radians

3 2, $3\pi/4$ radians

5 $7\vec{j}$

7 $\|3\vec{i} + 4\vec{j}\| = \| - 5\vec{i}\| = \|5\vec{j}\|, \|\vec{i} + \vec{j}\| = \|\sqrt{2}\vec{j}\|$

9 $5\vec{j}$ and $-6\vec{j}$; $\sqrt{2}\vec{j}$ and $-6\vec{j}$

11 (a) $(-3/5)\vec{i} + (4/5)\vec{j}$

(b) $(3/5)\vec{i} + (-4/5)\vec{j}$

13 $8\vec{i} - 6\vec{j}$

15 $\vec{i} + 2\vec{j}$

17 Equal

19 Equal

21 $\vec{i} + \vec{j}$, $\sqrt{2}, \vec{i} - \vec{j}$

23 Pos: $(1/\sqrt{2})\vec{i} + (1/\sqrt{2})\vec{j}$ Vel: $(-1/\sqrt{2})\vec{i} + (1/\sqrt{2})\vec{j}$ Speed: 1

Differentiation Formulas

1. $(f(x) \pm g(x))' = f'(x) \pm g'(x)$

2. $(kf(x))' = kf'(x)$

3. $(f(x)g(x))' = f'(x)g(x) + f(x)g'(x)$

4. $\left(\dfrac{f(x)}{g(x)}\right)' = \dfrac{f'(x)g(x) - f(x)g'(x)}{(g(x))^2}$

5. $(f(g(x)))' = f'(g(x)) \cdot g'(x)$

6. $\dfrac{d}{dx}(x^n) = nx^{n-1}$

7. $\dfrac{d}{dx}(e^x) = e^x$

8. $\dfrac{d}{dx}(a^x) = a^x \ln a \quad (a > 0)$

9. $\dfrac{d}{dx}(\ln x) = \dfrac{1}{x}$

10. $\dfrac{d}{dx}(\sin x) = \cos x$

11. $\dfrac{d}{dx}(\cos x) = -\sin x$

12. $\dfrac{d}{dx}(\tan x) = \dfrac{1}{\cos^2 x}$

13. $\dfrac{d}{dx}(\arcsin x) = \dfrac{1}{\sqrt{1 - x^2}}$

14. $\dfrac{d}{dx}(\arctan x) = \dfrac{1}{1 + x^2}$

A Short Table of Indefinite Integrals

■ I. Basic Functions

1. $\displaystyle\int x^n \, dx = \dfrac{1}{n+1}x^{n+1} + C, \quad n \neq -1$

2. $\displaystyle\int \dfrac{1}{x} \, dx = \ln|x| + C$

3. $\displaystyle\int a^x \, dx = \dfrac{1}{\ln a}a^x + C, \quad a > 0$

4. $\displaystyle\int \ln x \, dx = x \ln x - x + C$

5. $\displaystyle\int \sin x \, dx = -\cos x + C$

6. $\displaystyle\int \cos x \, dx = \sin x + C$

7. $\displaystyle\int \tan x \, dx = -\ln|\cos x| + C$

■ II. Products of e^x, $\cos x$, and $\sin x$

8. $\displaystyle\int e^{ax} \sin(bx) \, dx = \dfrac{1}{a^2 + b^2} e^{ax}[a\sin(bx) - b\cos(bx)] + C$

9. $\displaystyle\int e^{ax} \cos(bx) \, dx = \dfrac{1}{a^2 + b^2} e^{ax}[a\cos(bx) + b\sin(bx)] + C$

10. $\displaystyle\int \sin(ax) \sin(bx) \, dx = \dfrac{1}{b^2 - a^2}[a\cos(ax)\sin(bx) - b\sin(ax)\cos(bx)] + C, \quad a \neq b$

11. $\displaystyle\int \cos(ax) \cos(bx) \, dx = \dfrac{1}{b^2 - a^2}[b\cos(ax)\sin(bx) - a\sin(ax)\cos(bx)] + C, \quad a \neq b$

12. $\displaystyle\int \sin(ax) \cos(bx) \, dx = \dfrac{1}{b^2 - a^2}[b\sin(ax)\sin(bx) + a\cos(ax)\cos(bx)] + C, \quad a \neq b$

■ III. Product of Polynomial $p(x)$ with $\ln x$, e^x, $\cos x$, $\sin x$

13. $\displaystyle\int x^n \ln x \, dx = \dfrac{1}{n+1}x^{n+1}\ln x - \dfrac{1}{(n+1)^2}x^{n+1} + C, \quad n \neq -1$

14. $\displaystyle\int p(x)e^{ax} \, dx = \dfrac{1}{a}p(x)e^{ax} - \dfrac{1}{a}\int p'(x)e^{ax} \, dx$

$$= \dfrac{1}{a}p(x)e^{ax} - \dfrac{1}{a^2}p'(x)e^{ax} + \dfrac{1}{a^3}p''(x)e^{ax} - \cdots$$

$$(+ - + - \ldots)$$

(signs alternate)

15. $\displaystyle\int p(x)\sin ax\,dx = -\frac{1}{a}p(x)\cos ax + \frac{1}{a}\int p'(x)\cos ax\,dx$

$$= -\frac{1}{a}p(x)\cos ax + \frac{1}{a^2}p'(x)\sin ax + \frac{1}{a^3}p''(x)\cos ax - \cdots$$
$$(- + + - - + + \ldots)$$
(signs alternate in pairs after first term)

16. $\displaystyle\int p(x)\cos ax\,dx = \frac{1}{a}p(x)\sin ax - \frac{1}{a}\int p'(x)\sin ax\,dx$

$$= \frac{1}{a}p(x)\sin ax + \frac{1}{a^2}p'(x)\cos ax - \frac{1}{a^3}p''(x)\sin ax - \cdots$$
$$(+ + - - + + - - \ldots) \quad \text{(signs alternate in pairs)}$$

■ IV. Integer Powers of sin x and cos x

17. $\displaystyle\int \sin^n x\,dx = -\frac{1}{n}\sin^{n-1} x \cos x + \frac{n-1}{n}\int \sin^{n-2} x\,dx, \quad n \text{ positive}$

18. $\displaystyle\int \cos^n x\,dx = \frac{1}{n}\cos^{n-1} x \sin x + \frac{n-1}{n}\int \cos^{n-2} x\,dx, \quad n \text{ positive}$

19. $\displaystyle\int \frac{1}{\sin^m x}\,dx = \frac{-1}{m-1}\frac{\cos x}{\sin^{m-1} x} + \frac{m-2}{m-1}\int \frac{1}{\sin^{m-2} x}\,dx, \quad m \neq 1,\ m \text{ positive}$

20. $\displaystyle\int \frac{1}{\sin x}\,dx = \frac{1}{2}\ln\left|\frac{(\cos x)-1}{(\cos x)+1}\right| + C$

21. $\displaystyle\int \frac{1}{\cos^m x}\,dx = \frac{1}{m-1}\frac{\sin x}{\cos^{m-1} x} + \frac{m-2}{m-1}\int \frac{1}{\cos^{m-2} x}\,dx, \quad m \neq 1,\ m \text{ positive}$

22. $\displaystyle\int \frac{1}{\cos x}\,dx = \frac{1}{2}\ln\left|\frac{(\sin x)+1}{(\sin x)-1}\right| + C$

23. $\displaystyle\int \sin^m x \cos^n x\,dx$: If m is odd, let $w = \cos x$. If n is odd, let $w = \sin x$. If both m and n are even and positive,

convert all to $\sin x$ or all to $\cos x$ (using $\sin^2 x + \cos^2 x = 1$), and use IV-17 or IV-18. If m and n are even and one of them is negative, convert to whichever function is in the denominator and use IV-19 or IV-21. If both m and n are even and negative, substitute $w = \tan x$, which converts the integrand to a rational function that can be integrated by the method of partial fractions.

■ V. Quadratic in the Denominator

24. $\displaystyle\int \frac{1}{x^2 + a^2}\,dx = \frac{1}{a}\arctan\frac{x}{a} + C, \quad a \neq 0$

25. $\displaystyle\int \frac{bx + c}{x^2 + a^2}\,dx = \frac{b}{2}\ln|x^2 + a^2| + \frac{c}{a}\arctan\frac{x}{a} + C, \quad a \neq 0$

26. $\displaystyle\int \frac{1}{(x-a)(x-b)}\,dx = \frac{1}{a-b}(\ln|x-a| - \ln|x-b|) + C, \quad a \neq b$

27. $\displaystyle\int \frac{cx + d}{(x-a)(x-b)}\,dx = \frac{1}{a-b}[(ac+d)\ln|x-a| - (bc+d)\ln|x-b|] + C, \quad a \neq b$

■ VI. Integrands Involving $\sqrt{a^2 + x^2}$, $\sqrt{a^2 - x^2}$, $\sqrt{x^2 - a^2}$, $\quad a > 0$

28. $\displaystyle\int \frac{1}{\sqrt{a^2 - x^2}}\,dx = \arcsin\frac{x}{a} + C$

29. $\displaystyle\int \frac{1}{\sqrt{x^2 \pm a^2}}\,dx = \ln\left|x + \sqrt{x^2 \pm a^2}\right| + C$

30. $\displaystyle\int \sqrt{a^2 \pm x^2}\,dx = \frac{1}{2}\left(x\sqrt{a^2 \pm x^2} + a^2\int \frac{1}{\sqrt{a^2 \pm x^2}}\,dx\right) + C$

31. $\displaystyle\int \sqrt{x^2 - a^2}\,dx = \frac{1}{2}\left(x\sqrt{x^2 - a^2} - a^2\int \frac{1}{\sqrt{x^2 - a^2}}\,dx\right) + C$